PHYSICAL GEOGRAPHY

SCIENCE AND SYSTEMS OF THE HUMAN ENVIRONMENT

PHYSICAL GEOGRAPHY
SCIENCE AND SYSTEMS OF THE HUMAN ENVIRONMENT

ALAN STRAHLER
Boston University

ARTHUR STRAHLER
Columbia University (emeritus)

John Wiley & Sons, Inc.
New York • Chichester • Brisbane • Toronto • Singapore • Weinheim

Acquisitions Editor	Nanette Kauffman
Developmental Editor	Kathleen Dolan-Davies
Marketing Managers	Catherine Faduska
Production Editor	Sandra Russell
Designer	Kevin Murphy
Manufacturing Manager	Dorothy Sinclair
Senior Photo Editor	Mary Ann Price
Photo Editor	Alexandra Truit
Photo Researcher	Jerry Marshall
Photo Assistant	Kim Khatchatourian
Illustration Editor	Edward Starr
Illustrations/Maps	J/B Woolsey/Maryland Cartographics
Cover Photos	From top to bottom: Clouds: © Gnass/West Stock Volcano: © Gregory G. Dimijian/Photo Researchers Sand Dunes: © David Muench Cape Cod: Ric Ergenbright Photography

This book was set in New Baskerville by Ruttle, Shaw, & Wetherill, Inc. and
printed and bound by Von Hoffmann Press, Inc. The cover was printed by Lehigh Press Inc.
The color separations were prepared by Lehigh Press Colortronics, Inc.

Recognizing the importance of preserving what has been written, it is a policy of John Wiley & Sons, Inc.
to have books of enduring value published in the United States printed on acid-free paper, and we exert our
best efforts to that end.

The paper in this book was manufactured by a mill whose forest management programs include sustained
yield harvesting of its timberlands. Sustained yield harvesting principles ensure that the numbers of trees
cut each year does not exceed the amount of new growth.

Library of Congress Cataloging in Publication Data:

Strahler, Alan H.
    Physical geography : Science and systems of the Human Environment / Alan
Strahler, Arthur Strahler.
        p.   cm.
    Includes index.
    ISBN 0-471-11299-2 (cloth : alk. paper)
    1. Physical geography.   I. Strahler, Arthur Newell. 1918–   .
    II. Title.
    GB54.5.S784   1996   96-18934
    910'.02–dc20   CIP

Printed in the United States of America

10 9 8 7 6 5 4 3 2 1

PREFACE

In writing *Physical Geography: Science and Systems of the Human Environment,* our objective was to provide a view of physical geography as a key discipline in understanding the earth's diverse environments and how they are modified by global change. This goal required a new and fresh approach to the subject matter, one that emphasized not only the basic science of physical geography, but also the interrelations among the diverse processes that act at the earth's surface and thus condition the human environment. In response, we chose to rely on the concept of systems—more particularly, flow systems of energy and matter—as a unifying theme. In this way, we provide a paradigm for understanding the underlying scientific principles and common elements of the many processes that constitute the science of physical geography.

As we developed the book chapter by chapter, it became clear that with systems as an ongoing theme, it would be easy to integrate some simple quantitative materials into the text. Accordingly, we have added a sampling of quantitative concepts as a new feature. These concepts are selected to explore quantitative relationships and the reasons why they are useful, rather than to cover all, or even the most important, quantitative relationships in physical geography. Accordingly, we hope to open the door for students in a population that is traditionally less oriented toward mathematics to realize how quantitative methods can contribute to the understanding of scientific principles.

Global change is also a recurring theme in our text. Geographers, perhaps more than scientists from other fields, are particularly aware of human impact on the earth and its environment. Global change is, of course, a two-way street. Humans change the earth in many ways—by clearing and cultivating its lands, harvesting its wildland resources, damming its rivers, and polluting its air. But in return the earth changes the situations of its human inhabitants when volcanic eruptions bury human settlements, tornadoes and hurricanes level towns and villages, and floods erode agricultural lands. An understanding of global change requires not just a familiarity with the latest climatic change projections, but an understanding of the many processes that shape the human physical environment, both natural and human-induced. Our text focuses on these processes directly, providing both the breadth and depth necessary to appreciate how humans are changing, and are changed by, the earth.

COURSE GOALS

Physical Geography: Science and Systems of the Human Environment is designed for a one- or two-semester course in introductory physical geography. Its overall objective is to introduce the science of physical geography and its related themes of systems and environment. A secondary goal is to expose the student to quantitative tools used by physical geographers to explore and model the phenomena they observe. It is targeted at the student with some science preparation at the secondary school level, and it can be used for courses suited to both majors and nonmajors.

In structure and organization, our text follows the standard sequence of topics that has evolved over the years, beginning with atmosphere-surface interactions and concluding with ecosystem geography. This pattern is eminently teachable and should fit existing lecture structures quite easily. An Introduction, stressing systems concepts, provides an opening orientation to the major themes of the text. The sequence of 21 chapters is broken into four major parts: Weather and Climate Systems; Systems and Cycles of the Solid Earth; Systems of Landform Evolution; and Systems and Cycles of Soils and the Biosphere. An Epilogue on physical geography, environment, and global change completes the book.

Although the sequence of topics may be traditional to the instructor, students need continuing orientation and explanation of the structure of the material they are learning. Throughout the text, particular emphasis is placed on establishing this context at all levels. Within chapters, major topics are introduced by statements that provide the requisite context. Each chapter begins and ends with paragraphs that place the chapter in the perspective of the book as a whole. Each part opener includes text that explains the structure of the chapter sequence to come and relates it to the preceding parts. The Introduction and Epilogue open and close the entire book with unifying overviews and perspectives. Our experience as classroom instructors has shown the importance of providing this context to help students master the course as a whole, not just as a sum of individual parts.

THE LEARNING ENVIRONMENT

The learning environment in *Physical Geography: Science and Systems of the Human Environment* includes a number of features expressly designed to further its goals.

Systems Treatment

Throughout the text, systems concepts are treated as unifying concepts that aid in understanding key principles and ideas, not as separate topics that merely add another level of complexity to the book. Systems treatment begins with the Introduction, which provides basic ideas about flow systems and cycles of energy and matter. It continues through the main text, in contextual and transitional paragraphs, where it is applied to relevant topics as they occur. Systems concepts are highlighted in *Focus on Systems* boxes. These features provide specific applications and examples of systems appropriate to the material at hand. And, systems are called out specifically in the text of part openers, which provides the orientation and perspective on the next set of chapters to come.

Environment Features

Although physical geography is concerned largely with systems and processes that shape the human environment, specific topics can be targeted as primarily environmental. These are identified as *Eye on the Environment* sections within the text. They are distinguished by a special design treatment—a color bar to the left of the text and a special icon marking the closing of the section—and are highlighted in each chapter's list of section headings.

Quantitative Features

Since physical geography is a quantitative science, it is appropriate to introduce students to simple quantitative concepts and techniques as a way of enhancing the understanding of particular science principles. However, the tradition in introductory physical geography has been to provide a largely descriptive treatment at the beginning level. We have broken with that tradition but have been careful to separate quantitative material from the main text in a series of *Working It Out* boxes that are independent of the text. In this way, the instructor has the freedom to select any or all of these, depending on the readiness of the students and the quantitative topics to be emphasized.

The majority of the *Working It Out* boxes are accessible with simple algebraic skills—being able to substitute values in a computational formula, for example. A few require natural and common logarithms and exponentiation. All have associated problems that are placed in the closing matter of the chapter in which they appear. Problems are solved in a Problem Answers section in the back of the book. In developing the boxed features, we tried to focus on the application of the quantitative concept in some useful way, rather than simply presenting a formula to be evaluated. The problems that accompany each feature are designed to extend this focus on the quantitative approach as an aid to learning rather than as a test of arithmetic skills.

Canadian Features

Our previous physical geography texts have been quite popular in Canada, and in this book we have paid special attention to Canadian needs. In particular, we have greatly expanded our coverage of periglacial processes and landforms in Chapter 13. We have also revised and expanded our treatment of permafrost. Wherever possible, we have extended U.S. maps into southern Canada, providing coverage of most of Canada's populated areas, or have used North American maps showing the full extent of both countries. We have also increased our use of Canadian photos and examples in the text. In addition, we provide a six-page supplement on the Canadian system of soil classification as Appendix 3.

Terms

Vocabulary is an essential part of any science. To help organize the terms that we have included, we have set off the Key Terms—those dozen or more terms that are most important in each chapter—in **boldface**, and listed them at the close of each chapter. Terms of lesser importance are set in *italics*. All terms, whether bold or italic, are defined in the Glossary at the close of the book.

End of Chapter Student Aids

Each chapter concludes with a number of aids to facilitate student learning. The Chapter Summary is worded like a scientific abstract, succinctly covering all the major concepts of the chapter. The list of Key Terms indicates the most important concepts to study to understand the chapter material. The Review Questions are designed as oral or written exercises that require description or explanation of important ideas and concepts. Some questions utilize sketching or graphing as a way of motivating students to visualize key concepts. Questions are provided for both text and *Focus on Systems* boxes. Also provided are Essay Questions that require more synthesis or the reorganization of knowledge in a new context. Last of all are problems for *Working It Out* boxes, which invite the student to apply quantitative skills in a manner that emphasizes learning from quantitative principles rather than computational skills.

Text and Illustrations

Today, science must be taught in an open, accessible manner to reach students, and written material needs to communicate directly and simply. Moreover, today's students are accustomed to visual learning and to an interactive methodology. The text and illustrations of *Physical Geography: Science and Systems of the Human Environment* draw on the heritage of *Introducing Physical Geography* (Wiley, 1994), which devised entirely new treatments of topics in physical geography that are accessible and inviting, yet provide clear descriptions of scientific principles. In our text, each sentence is worded to ensure that it communicates its meaning clearly and simply. Every map and line drawing is carefully designed and styled to make sure that its message is clear, obvious, and direct. Every photo is selected not only to provide a fine illustration of the relevant science concept, but also to demonstrate it effectively and strikingly. In summary, we have worked very hard to produce an illustrated text of which we are very proud.

SUPPLEMENTS

Exercise Manual. By Arthur N. Strahler and Alan H. Strahler. Designed to build on students' knowledge of physical geography developed through class lectures and textbook study, the manual puts this knowledge to use in exercises based on course topics. The Exercise Manual to accompany *Physical Geography: Science and Systems of the Human Environment* provides all data, figures, and maps needed for the exercises that are not otherwise in the parent textbook.

Student's Companion. By Lawrence W. Martz of the University of Saskatchewan. Provides insights to text chapters and emphasizes topics discussed in class through chapter overviews and discussion of difficult terminology and important terms. The **Student's Companion** to accompany **Physical Geographical: Science and Systems of the Human Environment** also includes approximately 500 study and practice test questions to help students prepare for exams.

Instructor's Manual and Test Bank. By James Duvall of Contra Costa College. This combination instructor's manual/test bank provides instructors with chapter objectives, lecture outlines, lists of supplementary readings and videos, and an extensive chapter-by-chapter directory of world wide websites listing subjects of interest to both instructors and physical geography students. The test bank portion of this supplement provides between 30 and 40 multiple choice, true/false, short answer, and essay questions corresponding to each textbook chapter.

Test Bank ASCII Files for Mac and Windows. For instructors' convenience, test bank questions are available in both Mac and Windows files which can be downloaded into individual word programs. These allow more flexibility than traditional computerized test banks because they can be annotated, revised, and do not require instructors to learn any new software programs.

Full-Color Overhead Transparencies. One hundred full-color transparencies, including both figures and photos, drawn from both **Physical Geography: Science and Systems of the Human Environment** and **Introducing Physical Geography,** are available to instructors.

Full-Color Slides. The same figures available as transparencies are also available as full-color slides.

The Wiley Geography Website. Be sure to visit Wiley's website at ftp://ftp.wiley.com/public/college/geography to access additional materials to supplement your physical geography course, chat with other instructors in the discipline, and find out the latest news in the field.

Check with your Wiley representative regarding other materials which may be available as supplements for **Physical Geography: Science and Systems of the Human Environment.**

ACKNOWLEDGMENTS

The preparation of *Physical Geography: Science and Systems of the Human Environment* was greatly aided by many reviewers who read and evaluated the various parts of the manuscript. They include

Joseph Ashley, Montana State University
Bryan Baker, Sonoma State University
Ian Campbell, University of Alberta
Robert Crane, Pennsylvania State University
Dirk de Boer, University of Saskatchewan
Nicholas Dunning, University of Cincinnati
Douglas Goodin, Kansas State University
Donald Green, Lock Haven University
John Harrington, Jr., Kansas State University
Steven Jennings, Texas A&M University
Guy King, California State University—Chico
Jeffrey Lee, Texas Tech University
Lawrence Martz, University of Saskatchewan
Wayne Pollard, McGill University

Catherine Source, Indiana University—Purdue
Jeffrey Torguson, University of Wisconsin—Oshkosh

STUDENT REVIEWERS
Wendy Grove, Kansas State University
Robert Densmore, Lock Haven University

It is with particular pleasure that we thank the staff at Wiley for their careful work, encouragement, and sense of humor in the preparation and production of *Physical Geography: Science and Systems of the Human Environment.* They include our editor Nanette Kauffman, developmental editor Kathleen Davies, photo editors Mary Ann Price and Alexandra Truitt, photo research assistant Kim Khatchatourian, designer Kevin Murphy, illustration editor Edward Starr, marketing managers Cathy Faduska and Rebecca Hope, supplements editor Francine Banner, manufacturing manager Dorothy Sinclair, and our production editor Sandra Russell. We give special thanks to John Woolsey and his studio, J/B Woolsey Associates, and to Maryland Cartographics, who produced the final illustrations with skill, care and artistry. We also thank John Hodges of Boston University, who ably checked the quantitative sections and problems. We are deeply indebted to Kristi Strahler for her help in so many phases of the preparation of the book, and for her support and endurance through the many long months of work this book required from start to finish.

Alan Strahler
Cambridge, Massachusetts

Arthur Strahler
Santa Barbara, California

July 1, 1996

ABOUT THE AUTHORS

Alan Strahler (b. 1943) received his B.A. degree in 1964 and his Ph.D. degree in 1969 from The Johns Hopkins University, Department of Geography and Environmental Engineering. He has held academic positions at the University of Virginia, the University of California at Santa Barbara, and Hunter College of the City University of New York, and is now Professor of Geography at Boston University. With Arthur Strahler, he is a coauthor of seven textbook titles with six revised editions on physical geography and environmental science. He has published over one hundred articles in the refereed scientific literature, largely on the theory of remote sensing of vegetation, and has also contributed to the fields of plant geography, forest ecology, and quantitative methods. His work has been supported by over $3 million in grant and contract funds, primarily from NASA. In 1993, he was awarded the Association of American Geographers/Remote Sensing Specialty Group Medal for Outstanding Contributions to Remote Sensing.

Arthur Strahler (b. 1918) received his B.A. degree in 1938 from the College of Wooster, Ohio, and his Ph.D. degree in geology from Columbia University in 1944. He was appointed to the Columbia University faculty in 1941, serving as Professor of Geomorphology from 1958 to 1967 and as Chair of the Department of Geology from 1959 to 1962. He was elected as a Fellow of

both the Geological Society of America and the Association of American Geographers for his pioneering contributions to quantitative and dynamic geomorphology, contained in over 30 major papers in leading scientific journals. He is the author or coauthor with Alan Strahler of 16 textbook titles with 12 revised editions in physical geography, environmental science, the earth sciences, and geology. His most recent new title, *Understanding Science: An Introduction to Concepts and Issues,* published by Prometheus Books in 1992, has been widely reviewed as an introductory college text on the philosophy of science.

STUDENT TO STUDENT

Hey There,

If you have been kind enough to direct your attention to this page, I thank you for sharing in my five or so minutes of "fame." Wow, to be published! Actually, I had hoped to be some day—but never imagined it would happen during my undergrad years or have anything to do with geography. But, after taking an environmental geography course at Kansas State University, I was told that the publishers were interested in getting some feedback from a typical student (a nongeography major) on the first few chapters of a new textbook. So, if you are any kind of the science-skeptic/phobic that I am, let me assure you that the authors have your best interest in mind. They are *for* you and not *against* you—and do consider the "two-cents" of the common student.

I am a junior psychology major, planning to attend graduate school for counseling. In other words, I am not a geography whiz, but I did find quite a bit of the information to be interesting and actually applicable to everyday life. In reviewing the new chapters, I commented on what I liked most, and not so much and what seemed confusing. I was even given the figures, pictures, and questions to constructively criticize. So, without rambling any further, I'll share with you what I found and what you will hopefully find helpful as you embark on earning a few more credit hours with a decent grade.

This book offers itself, if you're willing to use it, as a complete study guide. Not only are there questions and mathematical problems to allow you to check your attained knowledge after you read, but the material is often presented in special sections to help you focus your reading. This organization also gives you a break in the monotony of reading straight text, page after page. So, take note:

- The *Eye on the Environment* sections, marked with the vertical line to the side, were my favorite as they specifically report on how all this information affects our daily lives—or even how we are affecting this information (for example, the greenhouse effect).
- The *Working It Out* boxes demonstrate how they're coming up with these calculations, and afterward, you've got a reference for figuring out later problems.
- The *Focus on Systems* boxes are separated and highlighted so you can note while you're reading how all these separate pieces of information connect in the flow of different systems. In the Introduction they even use the analogy of the Interstate highway system to help explain the idea of geography systems. The authors use many other effective analogies throughout the book.
- The figures and pictures were also carefully considered. The ones included were designed to especially help you understand the reading material. Do take time to look these over. Again, they will offer you a break from reading page after page and should allow some of the information to "sink in." I have found them to be good imaging tools at test time when pondering the correct answers.
- The key terms at the end of each chapter list the dozen or so most important terms, which were boldface in the chapter. After looking them over, you may be prompted to go back and reread them if their definitions don't come to mind. There's also a glossary at the end of the book with definitions of all terms—boldface and italics. And, of course, the essay and review questions are good checks for what you've learned and also reinforce the information.

Well, I hope I've shown the useful aspects of this book as a study guide, and maybe have given you some new strategies or ways of looking at the layout that will help you. I hope you come away from this textbook and course with as much as I did. Everyday "geographical" occurrences like thunderstorms and the seasons, which I used to take for granted, have more meaning for me now that I understand better the causes and processes behind them. I am much more aware and appreciative of the role that the systems of physical geography play in my life, and vice versa. I wish you motivation and good luck in your pursuits.

Your fellow collegian,

Wendy L. K. Grove

Wendy L. K. Grove
Kansas State University

BRIEF CONTENTS

CONTENTS

To Amy
and the new generation
to whom we will entrust the earth

Spheres, Scales, Systems, and Cycles

Physical geography is the science of life environments, from the local to the global scale. It is devoted to the study of the systems and cycles that interact within the earth's **life layer**—the shallow surface layer that contains most forms of life and is located where lands and oceans meet the atmosphere. The life layer includes the solid land surface itself, which is largely covered by soil and vegetation and is sculpted into landforms by running water, wind, waves, and glaciers. Also included in the life layer is the lower part of the atmosphere, in which winds, clouds, precipitation, and storms occur. Another important part of the life layer is the uppermost water layer of the oceans. Physical geography investigates the processes that take place as part of the systems and cycles of the life layer. These processes modify and diversify the life layer through time and from place to place, thus creating the varied environments of the earth.

Because natural processes are constantly active, the earth's environments are constantly changing. Sometimes the changes are slow and subtle, as in the movement of crustal plates over geologic time to create continents and ocean basins. At other times, the changes are rapid, as when hurricane winds flatten vast areas of forests or even human habitations.

Environmental change is now produced not only by the natural processes that have acted on our planet for millions of years but also by human activity. The human race has populated our planet so thoroughly that few places remain free of some form of human impact. Global change, then, involves not only natural processes, but also human processes that interact with them. Physical geography is the key to understanding this interaction.

The Four Great Realms

The natural systems that we will encounter in the study of physical geography operate within the four great realms, or spheres, of the earth. These are the atmosphere, the lithosphere, the hydrosphere, and the biosphere (Figure I.1). The four realms interact within the life layer.

The **atmosphere** is a gaseous layer that surrounds the earth. It receives heat and moisture from the surface and redistributes them, returning some heat and all the moisture to the surface. The atmosphere also supplies vital elements—carbon, hydrogen, oxygen, and nitrogen—that are needed to sustain life-forms.

The solid earth, or **lithosphere**, forms the platform for the life layer. The solid rock of the lithosphere bears a shallow layer of soil in which nutrient elements

The Weminuche Wilderness in the Rocky Mountains near Durango, Colorado.

Figure I.1 This dramatic view of the Napali Coast, Kauai, Hawaii, shows the four realms—lithosphere, atmosphere, hydrosphere, and biosphere—of our planet, Earth.

become available to organisms. The lithosphere is sculpted into landforms. These features—such as mountains, hills, and plains—provide varied habitats for plants, animals, and humans.

Water in all its forms constitutes the **hydrosphere**. The main mass of the hydrosphere lies in the world's oceans, but water also occurs in the atmosphere as gaseous vapor, liquid droplets, and solid ice crystals. Water is found in or atop the uppermost layers of the lithosphere, forming groundwater reservoirs, lakes, streams, and rivers. It is essential to life.

The **biosphere** encompasses all living organisms of the earth. Life-forms on earth utilize the gases of the atmosphere, the water of the hydrosphere, and the nutrients of the lithosphere, and so the biosphere is dependent on all three of the other great realms. Figure I.2 diagrams this relationship.

As we noted earlier, most of the biosphere is contained in the shallow surface zone called the life layer. It includes the surface of the lands and the upper 100 meters or so (about 300 ft) of the ocean (Figure I.3). On land, the life layer is the zone of interactions among the biosphere, lithosphere, and atmosphere. The hydrosphere is represented on land by rain, snow, still water in ponds and lakes, and running water in

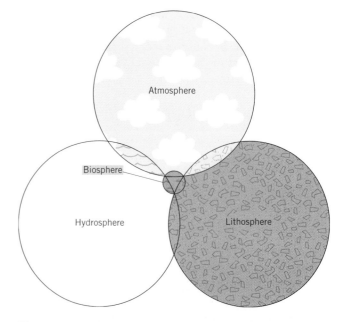

Figure I.2 The earth realms, shown as intersecting circles.

Figure I.3 As this sketch shows, the life layer is the layer of the earth's surface that supports nearly all of the earth's life. This includes the land and ocean surfaces and the atmosphere in contact with them.

rivers. In the ocean, the life layer is the zone of interactions among the hydrosphere, biosphere, and atmosphere. The lithosphere is represented by nutrients dissolved in the upper layer of sea water.

Throughout our exploration of physical geography, we will often refer to the life layer and the four realms that interact within it. We will find that the themes and subjects of physical geography concern systems and cycles that involve the atmosphere, hydrosphere, lithosphere, or biosphere—and usually two or three of these realms simultaneously.

Scales

The processes of the life layer and the four great realms operate on various scales. On a *global scale*, our planet is a nearly spherical body that turns endlessly within the glow of our sun. Because the sun is the energy source that powers most of the phenomena that occur within the life layer, earth–sun relationships are very important. In studying these relationships, we need to consider the planet and its global energy balance system as a whole and view the sun and earth from a vantage point far from the earth itself.

The sun's energy is not absorbed evenly by the earth's land and water surface. Unequal solar heating produces currents of air and water, just as a candle flame creates a current of hot, rising air. These currents constitute the global atmospheric and oceanic circulation system. To study this system, we need to move to the *continental scale*, where we can distinguish continents and oceans and track winds and ocean currents.

Moving still closer to the earth to the *regional scale*, we see the cloud patterns of weather systems and their regular movements over time. These movements, along with solar control of surface temperature, form the basis for the climates of the world.

Regional climate strongly influences the nature of the earth's cover of plants and the soil layers beneath them. But factors at the *local scale* are important in de-

termining the exact patterns of vegetation and soils. Is the location on a mountain slope or in a flat valley? Is it being farmed or used as grazing land? This information is too specific to use at the global, continental, or regional scales, but it is important when we consider the nature of the vegetation cover and soils within a specific region.

At the very finest scale, we see features such as a grassy sand dune on a beach or a moss-covered bank on the side of a river. These *individual landforms* are produced by unique activities of wind or water, and develop distinctive biological communities and soil properties.

Time scales vary widely, too. It may take millions of years for earth forces to produce a chain of mountains like the Himalayas. But a fracturing of the earth's crust and the resulting earthquake may last only a minute or two.

The point here is that the systems and cycles seen in nature act over a wide range of scales in time and space. Some systems are best understood by considering the earth as a whole. The global circulation of the atmosphere is a good example of this. Still others are concerned with very local phenomena, such as sand grains being blown along a beach.

If you look at the table of contents for this book, you will note that our study of physical geography will generally take us from the coarse scale to the fine—from global phenomena to local phenomena. For example, in Part I, Weather and Climate Systems, we begin with earth–sun relationships and global atmospheric circulation, then move to precipitation and weather systems, and finally consider individual climates within specific environments. Similarly, in Part II, we consider the structure of the earth as a whole, discuss the formation of continents, ocean basins, and other crustal features, and then describe specific landforms of volcanic and tectonic activity. There are exceptions to this overall plan, but understanding the focus of this organization is helpful in learning about the systems and processes of physical geography.

Physical Geography and the Human Habitat

The lands of the earth are the habitat of the human species as well as other forms of terrestrial life. Since physical geography is the study of the earth's surface and its environments, physical geography is also the study of the **human habitat**—your habitat. This personal involvement is one aspect of physical geography that makes it especially interesting and rewarding to learn. Everyday concerns like weather systems are a good example. The weather is always changing, and physical geography will help you understand how and why it changes.

Still other aspects of physical geography help explain the landscape of your region. The shapes of hills and valleys normally change too slowly for the changes to be noticed on a daily basis. But if you watch soil being washed down a slope during a sudden rain shower, you can extend this phenomenon to the land around you and come to understand how years of rainfall slowly carve the landscape into unique landforms.

If you have the opportunity to travel, studying physical geography will help you understand more about the landscapes you visit. While considering a Peace Corps assignment to the country of Mali, on the African continent, you might expect to find a tropical jungle. But your study of physical geography will tell you that much of Mali looks like the desert of the American Southwest!

EYE ON THE ENVIRONMENT:
PHYSICAL GEOGRAPHY, ENVIRONMENT, AND GLOBAL CHANGE

Physical geography also encompasses a concern for the quality of our environment and how it changes in response to natural and human influences. The survey of physical geography in this book focuses on the natural processes and systems that shape global environments, and this information will help you to understand environmental problems and global environmental change. However, not all aspects of environmental problems and global change are well understood. In some cases, the full impact of human activity is not yet known, and scientific predictions of change may vary significantly. In discussing these cases, we will present the scientific data and the contrasting opinions drawn from current studies without conclusions. When you consider these scientific arguments, and others that will be aired in the future, your knowledge of physical geography will make you better equipped to understand them and so to make intelligent decisions about environmental issues.

Environment and global change are sufficiently important that we have set off these topics from ordinary text in this book by placing a color bar alongside the text. This section is an example. As you read the book, watch for the color bar as your key to the application of physical geography in environmental and global change topics. 🐾

Systems in Physical Geography

The processes that interact within the four realms to shape the life layer and differentiate global environments are varied and complex. A helpful way to understand the relationships among these processes is to study them as **systems**. You've probably heard a lot about "systems," and you may already have some good ideas about what systems are. But since we are going to use the concept of systems throughout this book, we need to start by looking at systems in more detail. The remainder of this chapter will introduce some simple ideas about systems that we will refer to often in the chapters to come.

Let's begin with a familiar example—the U.S. interstate highway system. In this human-engineered system, individual roadways serve as pathways or tracks for the flow of vehicles (Figure I.4). The roadways are connected by interchanges that permit vehicles to move from one roadway to another. The vehicles on the roads and interchanges—cars, trucks, and motorcycles—are part of the system, too. However, unlike the roadways and interchanges, they can move from place to place.

The interstate highway system is actually a special sort of system called a flow system. A **flow system** is a system in which something moves from one place to another. Nearly all the systems we will encounter in physical geography are flow systems. In the case of the interstate highway system, it is the vehicles containing people and goods that "flow."

Vehicles are tangible objects—pieces of matter—that are in motion in the interstate highway system. Thus, we can distinguish this system as a *matter flow system*, a system in which matter is in motion. In some systems of physical geography, it will be energy that is in motion. For example, we may be concerned with a system that describes the flow of solar energy from the sun to the earth and its atmosphere. In that case, we will be studying an *energy flow system*.

Flow systems have a structure in which **pathways** of flow are connected. For the interstate highway system, the individual roadways are the pathways. However, pathways are not always quite so obvious. For example, we may regard the reflection of solar radiation from the top of a bright, white cloud as a pathway by which a flow of solar energy is turned away from the earth and back toward space.

We refer to the pattern of the pathways and their interconnections as the **structure** of the system. It is also convenient to use the term *components* to refer to the parts of a system, such as the pathways, their connections, and the types of matter and energy that flow within the system.

An important characteristic of many flow systems is their **inputs** and **outputs**. In the interstate highway system, the flow of vehicles fed in by on-ramps is the input, and the flow of vehicles leaving the system by exit ramps is the output.

An essential feature of flow systems is the need of each system for some sort of **power source**. For the interstate highway system, it is easy to identify the power

Figure I.4   The interstate highway system is a simple example of a flow system that has many of the properties of the flow systems of energy and matter that we will study in this text.

source—the motor fuels that are burned by the cars, trucks, and motorcycles on the highways.

The flow systems we will study in this book are *natural flow systems* rather than artificial, human-engineered systems such as the interstate highway system. Natural systems are powered largely or completely by natural sources. These sources include the flow of energy from the sun to the earth, the power stored in the inertia of the earth's rotation, and the outward flow of heat from the earth's interior that ultimately produces movements of the earth's crust.

What types of natural flow systems will we encounter in physical geography? An example is a river system, a matter flow system of water in a set of connected stream channels. In this system, the stream channels—pathways—are connected in a structure—the channel network—that organizes the flow of water from high lands to low lakes or oceans. Another is the food chain of an ecosystem, in which energy in the form of food flows among plant and animal components. Yet a third example is the global energy balance system. In this energy flow system, the heat of the sun is distributed around the earth by currents of warm water and moist air. Thus, the flow in this system is one of heat energy.

To summarize, a flow system consists of a structure of connected pathways within which matter and/or en-

ergy flows—light, heat, air, water, ozone, pollutants, or sand grains, for example. Flow systems can have input and output streams in which flows enter and exit from the system. All flow systems require power sources to run them.

Open and Closed Flow Systems

The interstate highway system is an example of an **open flow system**, one in which there are inputs and outputs of matter. In this system, cars enter and leave the system freely. Among natural systems, a river system is also an open flow system. Precipitation provides an input, and water is output to the ocean (or lake) where the river terminates. However, there is also a type of matter flow system in which there are no input or output flows of matter. Instead, the flowing materials in the system move endlessly in a series of interconnected paths or loops. Because there is no input or output, we refer to this system as a **closed flow system**. This type of system is also referred to as a **cycle**, or more fully, a **material cycle**.

The closed flow system is rather like an automobile race around a track—the Indy 500, for example. The vehicles move around the track, making loop after loop, powered by the flow of motor fuel in their tanks

(Figure I.5). During the race, the cars remain on the track (except for brief stops in the pit), so there is no input or output flow of cars as there is in the interstate highway system. Most natural closed flow systems are not so simple as this example, of course. They have more complicated structures consisting of many looping pathways that are interconnected in many places.

What are some closed flow systems that we will encounter in physical geography? One is the hydrologic cycle, which describes global flows of water. The loops in this system are flow paths of water in solid, liquid, and gaseous forms—for example, water moving as solid ice in glaciers, water as a liquid in rivers, streams, or ocean currents, and as a gas as water vapor in flows of moist air. Another example is the global carbon cycle. Here, the loops describe how the element carbon travels in different chemical forms between carbon-bearing rocks, ocean waters, the atmosphere, and the bodies of plants and animals.

Whether a material flow system is open or closed depends partly on where we draw the boundary around the system. As an example, consider a single river network as a simple system, represented in part (*a*) of Figure I.6. Water enters the network when it falls on the land as precipitation and runs off into the river system. We regard this water as an input that crosses the system boundary and enters the system. Water exits from the network at the river mouth, where it enters the ocean. This is a system output, crossing another system boundary located here. Thus, the river system is an open material flow system in which water enters and leaves the system as input and output flows.

Let's now redraw the system boundary to include the whole earth and its atmosphere. This situation is shown symbolically in part (*b*) of the figure. For this case, a new pathway must be added—the return flow of water from the oceans to the atmosphere by evaporation. There is no input or output, since water does not leave the earth or enter from space. Thus, the system is closed. The river system becomes the global hydrologic cycle, a closed matter flow system, which has been described earlier.

Thinking a bit further, we find that any global matter flow system must be closed, since only a minuscule amount of matter flows from earth to space (wandering gas molecules at the edge of the atmosphere) or from space to the earth (meteors and meteorites). Thus, the global carbon, nitrogen, and oxygen cycles (described further in Chapter 21) are all closed material flow systems.

What about energy flow systems? They are always open. Why? Because, as we shall see in Chapter 2, all objects that are warmer than the depths of space emit radiant energy, and some fraction of that energy ultimately leaves the earth. Thus, there is always an output energy flow, even when the system boundary is drawn around the whole earth and atmosphere. All objects also absorb some portion of the radiant energy they receive (such as solar energy), so there is always an energy input.

In short, matter flow systems may be open (have inputs and outputs) or closed (lacking inputs and outputs). Closed systems are referred to as cycles. Global matter flow systems, such as the carbon cycle, are always closed, while energy flow systems are always open.

Feedback and Equilibrium in Flow Systems

There are two other important concepts regarding flow systems that we will touch on here. The first, **feedback**

Figure I.5 A road race on a closed circuit, such as the Monte Carlo Grand Prix shown here, demonstrates the concept of a closed system.

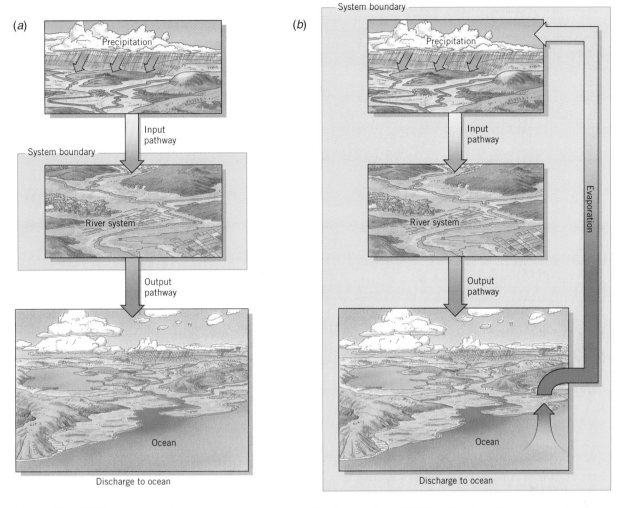

(*a*)

(*b*)

Figure I.6 (*a*) Diagram of a river system as an open matter flow system. (*b*) When the boundary is moved to enclose the earth and atmosphere, the system becomes a closed, global flow system—the hydrologic cycle.

in a flow system, occurs when the flow in one pathway acts either to reduce or to increase the flow in another pathway. Feedback is probably most familiar as the squealing noise made by a public address system when the volume is turned up too high. In this case, a voice is picked up by a microphone, and the sound is amplified and transmitted by loudspeakers. If the sound from the speakers is too great, it reaches the microphone and becomes reamplified. The process instantly repeats itself, making the sound even louder and producing an unpleasant squeal. This is a *positive feedback*, since it reinforces the flow of sound energy in the pathway between the microphone and the speakers.

An example of *negative feedback* is provided by a thermostat that controls a home heating system. When the heating system is on, the room warms. Eventually the warming trips a temperature-sensitive switch that turns the heating system off. Thus, the heat energy of the furnace provides a negative feedback that reduces its amount.

The second concept is **equilibrium**. By this term, we mean a steady state in which the flow rates in the vari-

ous pathways of a system stay about the same. A lake within a closed basin in an arid climate—like the Great Salt Lake, Utah—is a simple example of an equilibrium system (Figure I.7, part *a*). This type of lake has no stream outlets and would dry up completely if not fed by streams that arise in nearby high mountains. Water enters the lake from rivers and streams that feed it. Water leaves the lake by evaporation. Note that the amount of evaporation depends on the surface area of the lake—the larger the surface area, the more evaporation occurs.

Suppose that the climate becomes a bit wetter, and the input of rivers that feed the lake increases (part *b*). The water level then rises, and the area of the lake expands. Because of the greater area, evaporation is greater. Eventually, the level rises to the point where the increased evaporation rate equals the increased inflow rate. That is, it reaches an equilibrium. If the climate changes again, and the input is reduced, the lake level will fall, surface area will decrease, and evaporation will decrease. Eventually, the lake will move to a new, and lower, equilibrium level. You will also recog-

Figure I.7 (*a*) A lake in a closed basin as a matter flow system in equilibrium. In (*b*), precipitation increases, the lake level rises, and because the surface area of the lake increases, evaporation also increases to balance the greater input.

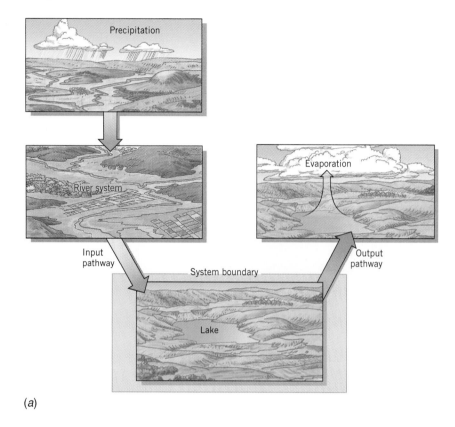

(*a*)

nize the coupling between input, surface area, and evaporation as a negative feedback. Systems that come to an equilibrium are normally stabilized by negative feedback loops or pathways.

An example of feedback and equilibrium occurs in the global climate system. This system has numerous pathways of energy flow that include several important negative feedbacks. Thus, the global climate system tends to an equilibrium in which global surface temperature fluctuates around a mean. Climate researchers have recently agreed that human activity has modified the global climate system. They predict that global temperatures will soon be rising in response.

The effect of clouds on the global climate system demonstrates the role of negative and positive feedback on an equilibrium system. What is the effect of clouds on surface temperature? Low clouds are large white bodies that reflect sunlight back to space much more efficiently than dark land or ocean surfaces below (see Figure I.8). They provide an energy flow pathway in which a portion of the solar energy flow is turned backward and redirected toward space. This pathway tends to cool the surface and so acts as a negative feedback. High clouds are different, however. They tend to absorb the outgoing flow of heat from the earth to space and redirect it earthward. Thus, high clouds provide a positive feedback that warms the surface.

Now consider a small increase in the global surface temperature. If the earth is warmer, then more water will evaporate from the oceans and moist land surfaces, and so more clouds will form. Will this increase in cloud cover cause surface cooling or warming? The best calculations now suggest that if surface temperatures increase, more high clouds than low clouds will form. Thus, the effect will be a positive feedback that will tend to make the surface even warmer. The effect of clouds on climate is complex, and we will return to this topic in Chapter 6.

Time Cycles

Any system, whether open or closed, can undergo a change in the rates of energy or matter flow within its pathways. Flow rates may grow faster and faster, or may slow down. These changes in activity can be reversed at intervals of time—that is, a rate can alternately speed up and slow down—in what we call a **time cycle**.

In many natural systems, there is a rhythm of increasing and decreasing flow. The annual revolution of the earth around the sun generates a time cycle of energy flow in many natural systems, and we speak of this cycle as the rhythm of the seasons. The rotation of the earth on its axis sets up the night-and-day cycle of darkness and light. The moon, in its monthly orbit around the earth, sets up its own time cycle. We see the lunar cycle in the range of tides, with higher high tides and lower low tides ("spring tides") occurring both at full moon and at new moon.

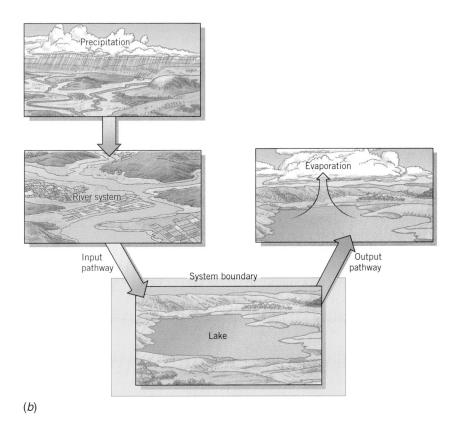

(*b*)

The astronomical time cycles of earth rotation and solar revolution will appear at several places in our early chapters. Other time cycles with durations of tens to hundreds of thousands of years describe the alternate growth and shrinkage of the great ice sheets. Still others, with durations of millions of years, describe cycles of the solid earth in which supercontinents form, break apart, and reform anew.

Systems Thinking

Systems and cycles are ideas that help us to understand and organize earth processes and phenomena. Put another way, a system is actually a way of thinking about how things move, change, and interact. By identifying the components, pathways, connections, structure, and power source of a system, we can better understand how it works. This helps us to predict the behavior of the system. The global climate system, discussed above, is an example. By placing the role of clouds in systems terms, we can easily see how clouds can work as negative or positive feedback mechanisms to modify surface temperatures.

As a further aid to mastering systems concepts, we have included a series of text features—*Focus on Systems* boxes—that are placed within the chapter text to help

Figure I.8 Low white clouds, such as these over the Hawaiian islands of Molokai, Lanai, and Maui, reflect much more of the sun's energy back to space than the darker surfaces of land and water below them.

Focus on Systems I.1 • The Value of Systems Thinking

Scientists spend a lot of time on systems thinking. They observe systems as a whole, as well as the pathways, interconnections, power sources, and other components of systems. They formulate theories about how systems are organized, and they design experiments to test their understanding of systems. They use math and computers to model the behavior of systems. In fact, there is a body of knowledge called *systems theory* that attempts to explain how systems work generally and how system characteristics can be predicted from components and interconnections.

Systems thinking is very important for the scientists who study natural processes—geographers, geologists, ecologists, oceanographers, atmospheric scientists, and others. Because they study earth and life phenomena, their ability to conduct experiments is limited. For example, it is not possible to reduce the sun's energy output by, say, 10 percent in order to see how climate would be affected. The ecosystem of even a small pond or bog is too complex to be reproduced in the laboratory for any meaningful experiments on the ecosystem as a whole. Thus, scientists who study natural systems must do their work largely by treating the earth as their laboratory—that is, by observing the earth and its processes at different times and places and on different scales. By using a systems approach to guide these observations, natural scientists can understand the components and connections within the systems they study and come to understand and predict the behavior of these systems as wholes.

Whatever the application, thinking critically about how things work and how they are related to one another is a skill that everyone relies on. One objective of this book is to help you build those skills and apply them to the principles and processes of natural systems.

you develop systems thinking and understanding as you read this book. These features elaborate on systems concepts, providing explanations and examples that help clarify how systems are structured and how they work. *Focus on Systems I.1 • The Value of Systems Thinking* provides more perspectives on the value of systems thinking. The text also includes *Working It Out* boxes— short features that are designed to help you develop quantitative skills in using the laws and principles of science that apply to systems understanding and physical geography in general.

The past few pages have presented a rather heavy dose of some fairly abstract ideas about systems. But as you will see in further chapters of this book, examining the phenomena of physical geography as natural flow systems of energy and matter will enhance your understanding of the processes that shape the world around us, from the daily changes in weather to the generation of earthquakes in the earth's crust.

Humans are now the dominant species on the planet. Nearly every part of the earth has felt human impact in some way. As our population continues to grow, our impact on natural systems will continue to increase. Each of us is charged with the responsibility to treat the earth well and respect its finite nature. Understanding the processes that shape the human habitat as they are described by physical geography will help you carry out this mission.

CHAPTER SUMMARY

Physical geography is a branch of science that investigates the earth's surface and how and why it changes. This discipline focuses on the life layer, the shallow surface layer where lands and oceans meet the atmosphere and where most forms of life are found.

Physical geography deals with the four great realms—atmosphere, hydrosphere, lithosphere, and biosphere. The systems of interaction between the realms can be examined at different scales, including global, continental, regional, and local. Since physical geography focuses on the land surface, it includes the study of the human habitat—the environment in which we live and interact with natural processes.

A helpful way to understand the relationships among the processes that occur in the life layer is to study them as systems. Most natural systems are flow systems. We can distinguish matter flow systems and energy flow systems, in which we are concerned with flows of matter and energy, respectively. Flow systems are composed of pathways of energy or matter flow that are interconnected in a structure. All flow systems have a power source.

Open flow systems have inputs and outputs, while closed flow systems do not. Closed matter flow systems are also called cycles. In a cycle, materials move in an endless series of interconnected pathways or loops. The hydrologic cycle and the carbon cycle are examples of cycles encountered in physical geography. Although matter flow systems may be open or closed, energy flow systems are always open.

Feedback in a flow system occurs when the flow in one pathway acts either to reduce or to increase the flow in another pathway. Positive feedback enhances or increases the flow within a pathway, while negative feedback reduces it. A system with negative feedback loops or pathways tends to be self-stabilizing and move to an equilibrium—a steady state in which flow rates in system pathways remain about the same. The role of clouds in the global climate system provides an example of feedback and equilibrium.

Systems may undergo periodic, repeating changes in flow rates that constitute time cycles. Important time cycles in physical geography range in length from hours to millions of years.

Systems thinking helps us to understand and predict natural processes and events. For natural scientists, systems thinking is an important tool for studying earth and life phenomena.

KEY TERMS

physical geography	human habitat	inputs	cycle
life layer	systems	outputs	material cycle
atmosphere	flow system	power source	feedback
lithosphere	pathways	open flow system	equilibrium
hydrosphere	structure	closed flow system	time cycle
biosphere			

REVIEW QUESTIONS

1. What is physical geography, and what is the life layer?

2. Name and describe each of the four great realms of earth.

3. Provide two examples of processes or systems that operate at each of the following scales: global, continental, regional, and local.

4. What relation does physical geography have to the human habitat?

5. What is a flow system? Provide a simple example.

6. Identify the key components of a flow system.

7. What distinguishes an open flow system from a closed flow system?

8. What is a cycle (material cycle)? Identify one or more cycles that are studied in physical geography.

9. Describe the concepts of feedback and equilibrium as applied to systems. Provide an example drawn from a natural system.

10. What is a time cycle as applied to a system? Give an example of a time cycle evident in natural systems.

11. Why is systems thinking important and useful for natural scientists?

Focus on Systems I.1 • The Value of Systems Thinking

1. What is meant by systems thinking? Why is it especially important for natural scientists?

ESSAY QUESTION

1. Select a flow system with which you are familiar. Identify its components and describe any negative or positive feedback pathways or loops it may contain. Does the system tend to an equilibrium? Possible examples: the system of electrical wiring or plumbing in a house or apartment; a bodily system, such as the respiratory system; a small ecosystem, such as an aquarium. Imagine some changes to the structure of the system, and describe how they might affect it.

Part I

Weather and Climate Systems

The flow of energy from the sun to the earth powers a vast and complex system of energy and matter flows within the atmosphere, oceans, and at the land surface. In this part, we explore how these flows are linked to weather and climate. We begin in Chapter 1 with an examination of how the earth's rotation and revolution induce the daily and seasonal rhythms in energy flow that mark the passage of time on the human scale. Although the pathways of energy flow are complex, their main features can be easily understood using simple physical principles—conservation of energy and matter, for example. We present these principles and the broad picture of global energy flows in Chapter 2. Flows of energy into and out of the layer of air close to the surface control a characteristic of the weather that is very important in our daily lives—air temperature—which we treat in Chapter 3.

An important pathway for solar energy flow lies in the evaporation of water from the ocean or from moist land surfaces. The latent heat absorbed in evaporation is later released in the process of condensation to form precipitation, the subject of Chapter 4. Another feature of earth–sun relationships is that some parts of the earth are heated more intensely than others—such as the equatorial regions. This unequal heating, coupled

with the earth's eastward rotation, produces a pattern of global wind circulation that is discussed in Chapter 5.

The global pattern of oceans and continents, coupled with the global pattern of atmospheric circulation, produces organized weather systems in which warm and cool air masses are in contact, often causing condensation and latent heat release. These phenomena, which we know as fronts and storm systems, are the subject of Chapter 6.

The last three chapters of Part I are devoted to climate—the average cycle of weather experienced at a location through the year. When average weather is measured by annual cycles of monthly temperature and precipitation, it is easy to group these cycles into a set of distinctive climates that cover the earth's land surface. Chapter 7 provides the basic principles for climate classification, while Chapters 8 and 9 cover specific climates from the equator to the poles.

As you study the chapters of Part I, keep in mind that nearly all the phenomena we describe are connected in some way to the grand, global flow system of energy and matter at the earth's surface that is powered by the sun.

Chapter 1

The Earth As a Rotating Planet

Our planet is a rotating body that revolves around the sun. Its shape is very close to that of a sphere, as we all learn early in school. Pictures taken from space by astronauts and by orbiting satellites also show us that the earth is a round body. We learn this fact so early in life that it seems quite unremarkable.

Many of our ancestors, however, were not aware of our planet's spherical shape. To sailors of the Mediterranean Sea in ancient times, the shape and breadth of the earth's oceans and lands were unknown. On their ships and out of sight of land, the sea surface looked perfectly flat and bounded by a circular horizon. Given

this view, many sailors concluded that the earth had the form of a flat disk and that their ships would fall off if they traveled to its edge.

High in a jet airplane at cruising altitude, you could come to the same conclusion. Peering out the window, you see below a vast expanse of earth or ocean fading into a distant and level horizon. However, at sunset, you might notice an interesting phenomenon (Figure 1.1). While the sun is still visible to you, illuminating the clouds at your altitude with an orange glow, the surface below you is in shadow. This means that the sun is below the horizon to an observer at the surface but above the horizon to an air traveler. This only happens when the curvature of the earth blocks the sun for the surface observer but not for you, the air traveler.

Actually, the earth is not perfectly spherical. Its true shape is described as an *oblate ellipsoid*. The outward force of the earth's rotation causes the earth to bulge slightly at the equator. As a result, the earth's circum-

---

Midnight in June, Lake Clark National Park, Alaska. The sun is still above the horizon to the right, while the moon is perched atop lofty peaks.

Figure 1.1 This magnificent sunset, seen from the passenger cabin of a high-flying jet aircraft over the southern Pacific Ocean, demonstrates the earth's curvature. Since the sun is still visible above the horizon at the level of the aircraft, it lights the clouds and sky. At the same moment, the ocean below is in shadow. The sun has already left the sky for an observer on the surface.

ference measured at the equator is about 0.3 percent greater than the polar circumference.

Scientists still study the earth's shape and attempt to measure it as precisely as possible. This is important since the information is needed by satellite navigation systems for aircraft, ocean vessels, and ground vehicles seeking to determine their exact location. The more precisely the earth's shape is known, the more accurately location can be determined.

EARTH ROTATION

Another fact about our planet that we learn early in life is that it spins slowly, making a full turn with respect to the sun every day. We use the term **rotation** to describe this motion. One complete rotation with respect to the sun defines the *solar day*. By convention, the solar day is divided into exactly 24 hours.

The earth rotates on its *axis,* an imaginary straight line through its center. The intersections of the axis of rotation and the earth's surface are defined as the poles. To distinguish between the two poles, one is called the *north pole* and the other, the *south pole.*

Direction of Rotation

The direction of earth rotation can be determined by using one of the following guidelines (Figure 1.2):

- Imagine yourself looking down on the north pole of the earth. From this position, the earth is turning in a counterclockwise direction (Figure 1.2*a*).
- Imagine yourself off in space, viewing the planet much as you would view a globe in a library, with the north pole on top. The earth is rotating from left to right, or in an eastward direction (Figure 1.2*b*).

The earth's rotation is important for three reasons. First, the axis of rotation serves as a reference in setting up the geographic grid of latitude and longitude, which we will discuss later in the chapter. Second, it provides the day as a convenient measure of the passage of time. Third, it influences physical and life processes on earth, as we will now describe.

Environmental Effects of Earth Rotation

The effects of the earth's rotation are of great importance to us and our environment. The first—and perhaps most obvious—effect of rotation is that it imposes a daily, or *diurnal*, rhythm in daylight, air temperature, air humidity, and air motion. Plants and animals respond to this diurnal rhythm. Green plants store energy during the day and consume some of it at night. Among animals, some prefer the day, others the night, for food gathering. The daily cycle of incoming solar energy and the corresponding cycle of fluctuating air temperature will be topics for analysis in Chapters 2 and 3.

Second, flow paths of both air and water are turned consistently in one direction because of the earth's rotation. When viewed from the starting point, flows are turned toward the right in the northern hemisphere and toward the left in the southern hemisphere. This phenomenon is called the *Coriolis effect*. It is of great importance in studying the earth's systems of winds and ocean currents. We will investigate both the Coriolis effect and its influence on winds and currents in Chapter 5.

A third physical effect of the earth's rotation is the movement of the tides. The moon exerts a gravitational attraction on the earth, while at the same time the earth is turning with respect to the moon. These forces induce a rhythmic rise and fall of the ocean surface known as the *tide*. The tide in turn causes water currents of alternating direction to flow in the shallow salty waters of the coastal zone. The ocean tide may have no importance for a grain farmer in Kansas, but for the clam digger or charter boat captain on Cape Cod, the tidal cycle is a clock regulating daily activities. The ebb and flow of tidal currents is a life-giving pulse for many plants and animals that live in coastal saltwater environments. The tide and its currents are discussed further in Chapter 17. 🦐

THE GEOGRAPHIC GRID

The geographic grid provides a system for locating places on the earth's surface. Without such a system, human society would literally be lost. Because the earth's surface is curved, and not flat, we cannot divide it into a rectangular grid, like a sheet of graph paper. Instead, we divide it using imaginary circles set on the surface that are perpendicular to the axis of rotation in one direction and parallel to the axis of rotation in the other direction.

Parallels and Meridians

Imagine a point on the earth's surface. As the earth rotates, the point traces out a path in space, following an *arc*—that is, a curved line that forms a portion of a circle. With the completion of one rotation, the line forms a full circle. This is known as a parallel of latitude, or a **parallel** (Figure 1.3*a*). Parallels cut the globe much as you might slice an onion to produce onion rings—that is, perpendicular to the onion's main axis. The largest parallel of latitude lies midway between the two poles and is designated the **equator**. The equator is a fundamental reference line for measuring the position of points on the globe.

Imagine now slicing the earth with a plane that passes through the axis of rotation, instead of across it. This is the way you might cut up a lemon to produce

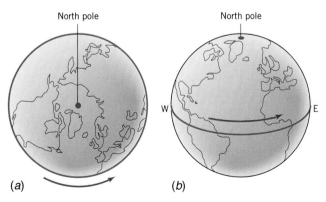

Figure 1.2 The direction of rotation of the earth can be thought of as (*a*) counterclockwise at the north pole, or (*b*) from left to right (eastward) at the equator.

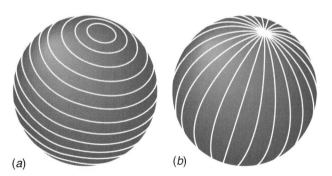

Figure 1.3 (*a*) Parallels of latitude divide the globe crosswise into rings. (*b*) Meridians of longitude divide the globe from pole to pole.

Focus on Systems 1.1 • The Word *System*

System is a common English word that we use in everyday speech. It typically means a set or collection of things that are somehow related or organized. An example is the *solar system*—a collection of planets that revolve around the sun. Another is the geographic grid, which we refer to in this chapter as a system of meridians and parallels. In the text, we will use *system* in this way quite often.

Sometimes the word *system* refers to a scheme for naming things. For example, we will introduce a climate system in Chapter 7 and a soil classification system in Chapter 19.

These general uses of *system* are somewhat different from our use of the word in *flow system*. Recall from the Introduction that a flow system is a system in which something moves from one place to another. Open flow systems have inputs and outputs, while closed flow systems, or *cycles*, do not. A flow system re-quires a power source to keep it running.

Understanding flow systems in physical geography is important in part because it also helps you to understand how things are connected—that is, how the all the processes that affect climate, for example, are related and influence each other. So remember as you read the text that when we use the term *flow system*, we have this special meaning of *system* in mind.

---

wedges. The cut outlines a circle on the globe passing through both poles. Half of this circular outline, connecting one pole to the other, is known as a meridian of longitude, or, more simply, a **meridian** (Figure 1.3*b*).

Meridians and parallels define compass directions. Meridians are north-south lines, so you are following a meridian if you walk north or south. Parallels are east-west lines, and so you are following a parallel if you walk east or west. There can be any number of parallels and meridians.

Every point on the globe is associated with a unique combination of one parallel and one meridian. The position of the point is defined by their intersection. The total system of parallels and meridians forms a network of intersecting circles called the **geographic grid**.

Note here that we have used the word *system* in describing the geographic grid. This use of the term is somewhat different from its use in *flow system*, which we defined in the Introduction. *Focus on Systems 1.1 • The Word System* discusses the ways in which the word system is used.

Latitude and Longitude

We use a special system to label parallels and meridians—latitude and longitude. Parallels are identified by latitude and meridians by longitude. Since parallels are circles and meridians are half-circles, we use degrees of arc as latitude and longitude measures to mark our place along particular parallels and meridians. What do we mean by *degrees of arc?* Figure 1.4 illustrates this concept. A full circle consists of a 360° angle. A lesser angle is associated with an arc, or portion of a circle. We can measure the arc by measuring the degrees of that angle.

Latitude measures the position of a given point in terms of its angular distance from the equator (Figure 1.5). That is, latitude is an indicator of how far north or

south of the equator a given point is situated. Latitude is measured in degrees of arc from the equator (0°) toward either pole, where the value reaches 90°. The equator divides the globe into two equal portions, or hemispheres. All points north of the equator—that is, in the *northern hemisphere*—are designated as north latitude, and all points south of the equator—in the *southern hemisphere*—are designated as south latitude.

Figure 1.5 shows how latitude is measured. The point *P* lies at the latitude of 50° north, which we can abbreviate as lat. 50° N. Notice that latitude arc angle is actually measured along the meridian that passes through the point.

Longitude is a measure of the position of a point eastward or westward from a reference meridian, called the *prime meridian*. As Figure 1.5 shows, longitude is the arc angle, measured in degrees, between the meridian of a given point and the prime meridian. This arc is measured east to west along the parallel that passes through the point. In this example, the point *P* lies at longitude 60° west (long. 60° W).

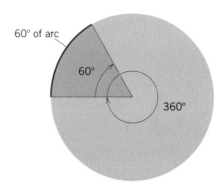

Figure 1.4 A circle consists of 360°, measured as an angle from its center. An arc is a part of a circle. An angle is associated with an arc that measures the arc in degrees. Thus, we state that the arc angle is 60°.

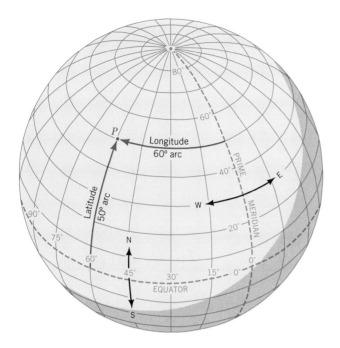

Figure 1.5 The point *P* has a latitude of 50°. This means that the arc between *P*'s parallel and the equator has an angle of 50°. We say that *P* is located on the 50th north parallel. *P* also has a longitude of 60°. Therefore, the arc between *P*'s meridian and the prime meridian, in Greenwich, England is 60°. We speak of *P* as being on the 60th west meridian.

The prime meridian passes through the location of the old Royal Observatory at Greenwich, near London, England (Figure 1.6). For this reason it is also referred to as the Greenwich meridian. It has the value long. 0°. The longitude of any given point on the globe is measured eastward or westward from the prime meridian, depending on which direction gives the shorter arc. Longitude thus ranges from 0° to 180°, east or west. When both the latitude and longitude of a place are known, it can be accurately and precisely located on the geographic grid.

When arcs of latitude or longitude are measured other than in full-degree increments, *minutes* and *seconds* can be used. A minute is 1/60 of a degree, and a second is 1/60 of a minute, or 1/3600 of a degree. Thus, the latitude 41°, 27 minutes ('), and 41 seconds (") north (lat. 41° 27' 41" N) means 41° north plus 27/60 of a degree plus 41/3600 of a degree. This cumbersome system has now largely been replaced by decimal notation. In this example, the latitude 41° 27' 41"N translates to 41 + 27/60 + 41/3600 = 41 + 0.4500 + 0.0114 = 41.4614° N.

Degrees of latitude and longitude can also be used as distance measures. A degree of latitude, which measures distance in a north-south direction, is equal to about 111 km (69 mi). The distance associated with a degree of longitude, however, will vary over the globe

because meridians converge toward the poles. *Working It Out 1.2 • Distances from Latitude and Longitude* provides more information on this topic and shows how to convert change in latitude and longitude into distance.

Map Projections

With a working understanding of the geographic grid, we can consider how to display the locations of continents, rivers, cities, islands, and other geographic features on maps. This will take us briefly into the realm of *cartography*, the art and science of making maps. Our discussion will focus on a few simple types of maps used in this text.

As we observed earlier, the earth's surface is nearly spherical. However, maps are flat. It is impossible to copy a curved surface onto a flat surface without cutting, stretching, or otherwise distorting the curved surface in some way. So, making a map means devising an orderly way of changing the globe's geographic grid of curved parallels and meridians into a grid that lies flat. We refer to a system for changing the geographic grid to a flat grid as a **map projection**.

Associated with every map is a *scale fraction*—a ratio that relates distance on the map to distance on the

Figure 1.6 This photograph, taken at dusk at the old Royal Observatory at Greenwich, England, shows the prime meridian, which has been marked as a stripe on the forecourt paving. The dome covering the observatory is partly opened, and the telescope, located exactly on the meridian, is pointed toward the sky as if making an observation.

Working It Out 1.2 • Distances from Latitude and Longitude

Statements of latitude and longitude do not describe distances in kilometers or miles directly. However, for latitude, you can estimate conversions from degrees into kilometers quite easily. One degree of latitude is approximately equivalent to 111 km (69 mi) of surface distance in the north-south direction. This value can be rounded off to 110 km (or 70 mi) for multiplying in your head. For example, if you live at lat. 40° N (on the 40th parallel north), you are located about 40 × 110 = 4400 km (40 × 70 = 2800 mi) north of the equator.

East-west distances cannot be converted so easily from degrees of longitude into kilometers or miles because the meridians converge toward the poles (Figure 1.3b). Only at the equator is a degree of longitude equivalent to 111 km (69 mi). At lat. 60° N or S, meridians are twice as close as at the equator, so a degree of longitude is reduced to half its equatorial length, or about 56 km (35 mi). As the pole is approached, the length goes to zero.

We can use a simple formula to determine the length of a degree of longitude:

$$L_{LONG} = \cos(lat) \times L_{LAT}$$

where L_{LONG} is the length of a degree of longitude; cos is the trigonometric cosine function, evaluated for degrees; *lat* is the latitude of the location at which the length is to be calculated, in degrees; and L_{LAT} is the length of a degree of latitude, that is, 111 km or 69 mi.

For example, a degree of longitude at 30° lat. has length

$$L_{LONG} = \cos(lat) \times L_{LAT}$$
$$= \cos(30°) \times 111 \text{ km}$$
$$= (0.866) \times 111 \text{ km} = 96.1 \text{ km}$$

Here's another: suppose we want to find the distance between two cities located on the 35th parallel that are separated by 6° of longitude. First, we determine the length of a degree of longitude at 35° lat. That is,

$$L_{LONG} = \cos(lat) \times L_{LAT}$$
$$= \cos(35°) \times 111 \text{ km}$$
$$= (0.819) \times 111 \text{ km} = 90.9 \text{ km}$$

Then,

$$6° \text{ long} \times \frac{90.0 \text{ km}}{1° \text{ long}} = 546 \text{ km}$$

These simple conversions help demonstrate the nature of the geographic grid and the convergence of meridians toward the poles.

---

earth's surface. For example, a scale fraction of 1:50,000 means that one unit of map distance equals 50,000 units of distance on the earth.

Because a curved surface cannot be projected onto a flat surface without some distortion, the scale fraction of a map holds only for one point or a single line on the map. At a location on the map away from that point or line, the scale fraction will be different. However, the variation in scale is only important for maps showing large regions, such as continents or hemispheres.

We will concentrate on the three most useful map projections. The first is the polar projection, which is essential today for scientific uses like weather maps of the polar regions. Second is the Mercator projection, a navigator's map invented in 1569 by Gerhardus Mercator. It is a classic that has never gone out of style. Third is the Goode projection, named for its designer, Dr. J. Paul Goode. It has special qualities not found in the other two projections.

Polar Projection

The *polar projection* (Figure 1.7) can be centered on either the north or the south pole. Meridians are straight lines radiating outward from the pole, and parallels are nested circles centered on the pole. Spacing of the parallels increases outward from the center. The map is usually cut off to show only one hemisphere, so that the

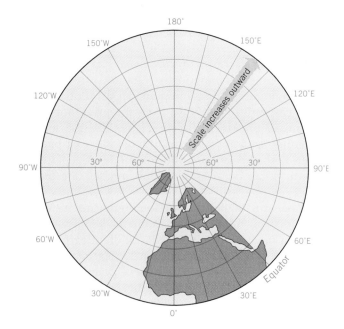

Figure 1.7 A polar projection. The map is centered on the north pole. All meridians are straight lines radiating from the center point, and all parallels are concentric circles. The scale fraction increases in an outward direction, making shapes toward the edges of the map appear larger.

equator forms the outer edge of the map. Because the intersections of the parallels with the meridians always form true right angles, this projection shows the true shapes of all small areas. If you drew a small circle anywhere on the globe, it would reproduce as a perfectly shaped circle on this map. However, because the scale fraction increases in an outward direction, the circle would be larger toward the edge of the map than near the center.

Mercator Projection

The *Mercator projection* (Figure 1.8) is a rectangular grid of meridians as straight vertical lines and parallels as straight horizontal lines. Meridians are evenly spaced, but the spacing between parallels increases at higher latitude so that the spacing at 60° is double that at the equator. Closer to the poles, the spacing increases even more, and the map must be cut off at some arbitrary parallel, such as 80° N. This change of scale enlarges features when they near the pole, as can easily be seen in Figure 1.8. There Greenland appears larger than Australia and is nearly the size of Africa! In fact, it is very much smaller, as you can see on a globe.

The Mercator projection has several special properties. One is that a straight line drawn anywhere on the map is a line of constant compass direction. A navigator can therefore simply draw a line between any two points on the map and measure the *bearing*, or direction angle of the line, using a protractor. Once aimed in that compass direction, a ship or an airplane can be held to the same compass bearing to reach the final point or destination.

This line will not necessarily follow the shortest actual distance between two points. The shortest path on the globe follows the arc of a *great circle*. (A great circle connects two points on the globe and has the earth's center as the circle's center.) On a Mercator projection, a great circle line usually curves and can (falsely) seem to represent a much longer distance than a compass line.

Because the Mercator projection shows the true compass direction of any straight line on the map, it is used to show many types of straight-line features. Among these are flow lines of winds and ocean currents, directions of crustal features (such as chains of volcanoes), and lines of equal values, such as lines of equal air temperature or equal air pressure. This explains why the Mercator projection is chosen for maps of temperatures, winds, and pressures.

Goode Projection

The *Goode projection* (Figure 1.9) uses two sets of mathematical curves (sine curves and ellipses) to form its meridians. Between the 40th parallels, sine curves are

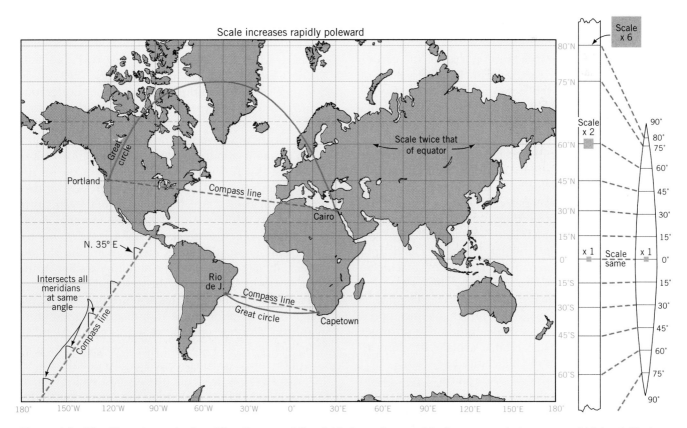

Figure 1.8 The Mercator projection. The diagram at the right shows how rapidly the map scale increases at higher latitudes. At lat. 60°, the scale is double the equatorial scale. At lat. 80°, the scale is six times

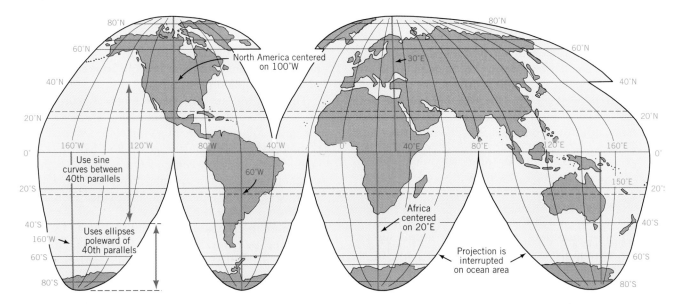

Figure 1.9 The Goode projection. The meridians in this projection follow sine curves between lat. 40° N and lat. 40° S and ellipses between lat. 40° and the poles. Although the shapes of continents are distorted, their areas are properly shown. (Copyright © by the University of Chicago. Used by permission of the Committee on Geographical Studies, University of Chicago.)

used, and beyond the 40th parallel, toward the poles, ellipses are used. Since the ellipses converge to meet at the pole, the entire globe can be shown. The straight, horizontal parallels make it easy to scan across the map at any given level to compare regions most likely to be similar in climate.

The Goode projection has one very important property—it indicates the true sizes of areas of the earth's surface. That is, if we drew a small circle on a sheet of clear plastic and moved it over all parts of the Goode world map, all the areas enclosed by the circle would have the same value in square kilometers or square miles. Because of this property, we use the Goode map to show geographical features that occupy surface areas. Examples of useful Goode projections include maps of the world's climates, soils, and vegetation.

The Goode map suffers from a serious defect, however. The Goode projection distorts the shapes of areas, particularly in high latitudes and at the far right and left edges. To minimize this defect, Dr. Goode split his map apart into separate, smaller sectors, each centered on a different vertical meridian. These were then assembled at the equator. This type of split map is called an *interrupted projection*. Although the interrupted projection greatly reduces shape distortion, it separates parts of the earth's surface that actually lie close together, particularly in the high latitudes.

Maps and map projections are a practical application of the earth's geographic grid. Another practical application, which involves both the grid and the earth's rotation, is global time, a subject we turn to next.

GLOBAL TIME

Our planet requires 24 hours for a full rotation with respect to the sun. Put another way, humans long ago decided to divide the solar day into 24 units, called hours, and devised clocks to keep track of hours in groups of 12. Yet different regions set their clocks differently. For example, when it is 10:03 A.M. in New York, it is 9:03 A.M. in Chicago, 8:03 A.M. in Denver, and 7:03 A.M. in Los Angeles. Note that these times differ by exactly one hour. How did this system come about? How does it work?

Our global time system is oriented to the sun. Think for a moment about how the sun appears to move across the sky. In the morning, the sun is low on the eastern horizon, and as the day progresses, it rises higher until at *solar noon* it reaches its highest point in the sky. If you check your watch at that moment, it will read a time somewhere near 12 o'clock (12:00 noon). After solar noon, the sun's elevation in the sky decreases. By late afternoon, the sun appears low in the sky, and at sunset it rests on the western horizon.

Imagine for a moment that you are in Chicago, the time is noon, and the sun is at or near its highest point in the sky. Further imagine that you call a friend in New York and ask about the time there and the position of the sun. You will receive a report that the time is 1:00 P.M. and that the sun is already past solar noon, its highest point. Calling a friend in Los Angeles, you hear that it is 10:00 A.M. there and that the sun is still working its way up to its highest point. However, a friend in

Mobile, Alabama, will tell you that the time in Mobile is the same as in Chicago and that the sun is at about solar noon. How can we explain these different observations?

The difference in time between Chicago, New York, and Los Angeles makes sense because solar noon can occur simultaneously only at locations with the same longitude. In other words, only one meridian can be directly under the sun and experience solar noon at a given moment. Locations on meridians to the east of Chicago, like New York, already will have passed solar noon, and locations to the west of Chicago, like Los Angeles, will not yet have reached solar noon. Since Mobile and Chicago have nearly the same longitude, they experience solar noon at approximately the same time.

Figure 1.10 indicates how time varies with longitude. In this figure, the inner disk shows a polar projection of the world, centered on the north pole. Meridians are straight lines (radii) ranging out from the pole. The outer ring indicates the time in hours. The figure shows the moment in time when the prime meridian is directly under the sun—that is, the 0° meridian is directly on the 12:00 noon mark. This means that, at this instant, the sun is at the highest point of its path in the sky in Greenwich, England. The alignment of meridians with hour numbers tells us the time in other loca-

tions around the globe. For example, the time in New York, which lies roughly on the 75° W meridian, is about 7:00 A.M. In Los Angeles, which lies roughly on the 120° W meridian, the time is about 4:00 A.M.

Notice that 15° of longitude equates to an hour of time. Since the earth turns 360° in a 24-hour day, the rotation rate is 360 + 24 = 15° per hour.

Standard Time

We've just seen that locations with different longitudes experience solar noon at different times. But consider what would happen if each town or city set its clocks to read 12:00 at its own local solar noon. Because no two towns or cities lie exactly on the same meridian, all would have different local time systems. In these days of instantaneous global communication, chaos would soon result.

The use of standard time simplifies the global time-keeping problem. In the **standard time system**, the globe is divided into 24 **time zones**. All inhabitants within a zone keep time according to a *standard meridian* that passes through their zone. Since the standard meridians are usually 15 degrees apart, the difference in time between adjacent zones is normally one hour. In some states and nations, however, the difference is only one-half hour.

Eight time zones cover the United States and its Caribbean possessions. Five zones cover Canada. Their names and standard meridians of longitude are as follows:

U.S. Zones	Meridian	Canadian Zones
Atlantic	60°	
	67½°	Newfoundland
Eastern	75°	Eastern
Central	90°	Central
Mountain	105°	Mountain
Pacific	120°	Pacific-Yukon
Alaska	135°	
Hawaii, Bering	150°	

If carried out strictly, the standard time system would consist of belts exactly 15°, extending to meridians 7½° east and west of each standard meridian. However, this system could be inconvenient since the boundary meridians could divide a state, county, or city into two different time zones. As a result, time zone boundaries are often routed to follow agreed-upon natural or political boundaries.

Figure 1.11 presents a map of time zones for the contiguous United States and southern Canada. From this map, you can see that most time zone boundaries are conveniently located along an already existing and widely recognized line. For example, the eastern time–central time boundary line follows Lake Michigan

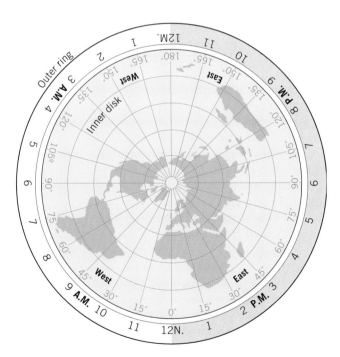

Figure 1.10 The relation of longitude to time. The outer ring shows the time at locations identified by longitude meridians on the central map. The diagram is set to show noon conditions in Greenwich, England—that is, on the prime meridian. Clock time is earlier to the west of Greenwich and later to the east.

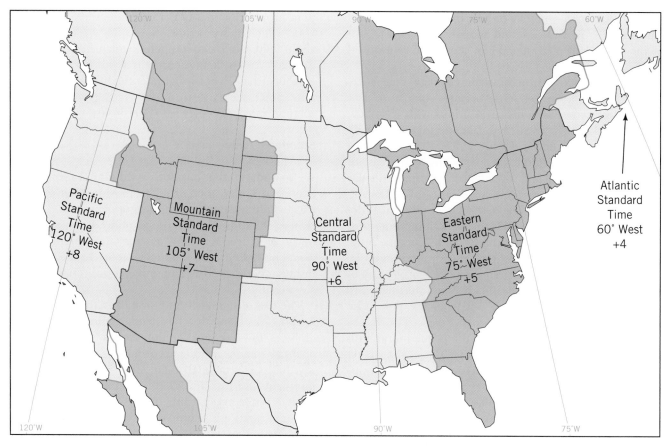

Figure 1.11 Time zones of the contiguous United States and southern Canada. The name, standard meridian, and number code are shown for each time zone. Note that time zone boundaries often follow preexisting natural or political boundaries.

down its center, and the mountain time–Pacific time boundary follows a ridge-crest line also used by the Idaho–Montana state boundary.

World Time Zones

Figure 1.12 shows the 24 principal standard time zones of the world. In the figure, 15° meridians are dashed lines, while the 7½° meridians, which form many of the boundaries between zones, are bold lines. The figure also shows the time of day in each zone when it is noon at the Greenwich meridian. The country spanning the greatest number of time zones from east to west is Russia, with 11 zones, but these are grouped into 8 standard time zones. China spans five time zones but runs on a single national time using the standard meridian of Beijing.

A few countries keep time by a meridian that is midway between standard meridians, so that their clocks depart from those of their neighbors by 30 or 90 minutes. India and Iran are examples. The Canadian province of Newfoundland and the interior Australian states of South Australia and Northern Territory are ex-

amples of regions within countries that keep time by 7½° meridians.

World time zones are numbered, and given the time zone numbers of two different time zones, it is easy to convert the time in one of them to the time in the other. *Working It Out 1.3 • Global Timekeeping* shows how to do this, and provides some practical examples for the world traveler.

International Date Line

When we take a world map or globe with 15° meridians and count them in an eastward direction, starting with the Greenwich meridian as 0, we find that the 180th meridian is number 12 and that the time at this meridian is therefore 12 hours later than Greenwich time. Counting in a similar manner westward from the Greenwich meridian, we find that the 180th meridian is again number 12 but that the time is 12 hours earlier than Greenwich time. How can the same meridian be both 12 hours ahead of Greenwich time and 12 hours behind? This paradox is explained by the fact that on the 180th meridian at the exact instant of midnight,

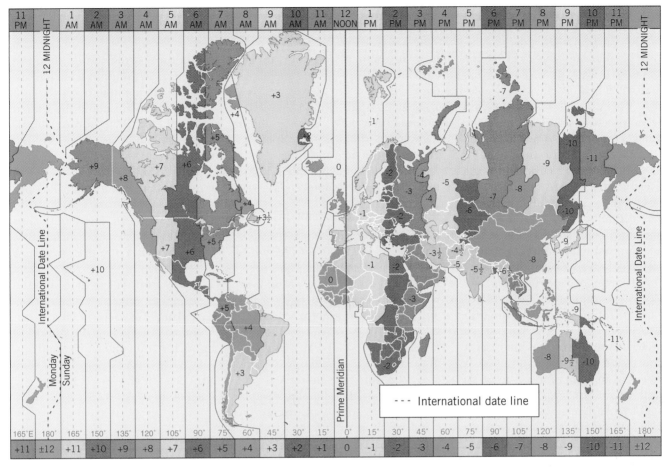

Figure 1.12 Time zones of the world. Dashed lines represent 15° meridians, and bold lines represent 7½° meridians. Alternate zones appear in color. (U.S. Navy Oceanographic Office.)

the same calendar day covers the entire globe. At any other time, the calendar day observed just west of the 180th meridian is one day ahead of that just to the east. For example, if the day on the Asian side of the date-line is Tuesday, June 8, then the day on the American side will be Monday, June 7.

Over a century ago, an international conference voted that the 180th meridian should be designated the *International Date Line*. To make this dividing line practical, it had to be modified by a zig-zag offset between Asia and North America, as well as an eastward offset in the South Pacific to keep clear of New Zealand and several island groups. (See Figure 1.12.)

Crossing the date line in either direction normally requires adjusting your calendar. On a cruise ship, you would either skip a whole day (going westward) or repeat a day (going eastward). Flying from North America to Asia may also require skipping a day. For example, flying from Los Angeles to Sydney, Australia, you may depart on a Tuesday evening and arrive on a Thursday morning after a flight that lasts only 14 hours.

On a flight from Tokyo to San Francisco, you may actually arrive the day before you take off, taking the date change into account!

Daylight Saving Time

Especially in urban areas, many human activities begin well after sunrise and continue long after sunset. Therefore, we adjust our clocks during the part of the year that has a longer daylight period to correspond more closely with the modern pace of society. This adjusted time system, called *daylight saving time*, is obtained by setting ahead all clocks by one hour. The effect of the time change is to transfer the early morning daylight period, theoretically wasted while schools, offices, and factories are closed, to the early evening, when most people are awake and busy. Daylight saving time also yields a considerable savings in power used for electric lights. In the United States, daylight saving time comes into effect on the first Sunday in April and is discontinued on the last Sunday of October.

Working It Out 1.3 • Global Timekeeping

Your eight-hour flight to Rome leaves New York at 8:30 P.M. What time will it be in Rome when you arrive? To help you answer this and other practical problems involving timekeeping, time zones are numbered away from the time zone of the Greenwich meridian—negatively in an eastward direction and positively in a westward direction. These numbers are shown on the map in Figure 1.12. The numbers tell you how many hours to change your clock to get Greenwich time.

For example, New York is in zone +5, so you must add five hours to New York time to get the time in Greenwich. Rome is in zone -1, so in Rome you subtract an hour to get Greenwich time. Thus, when it is 8:30 P.M. in New York, it must be 1:30 A.M. in Greenwich. And if it is 1:30 A.M. in Greenwich, it must be 2:30 A.M. in Rome, since Romans

have to subtract an hour to get Greenwich time. With the knowledge that you are departing New York at 2:30 A.M. Rome time, you can see that after an eight-hour flight, you will arrive in Rome at 10:30 A.M. the next morning.

It is easy to come up with a simple formula for global timekeeping:

$$D = Z_{HOME} - Z_{AWAY}$$

Here D is the difference—positive for ahead, negative for behind—that you add to change the time; Z_{HOME} is the time zone whose time you already know; and Z_{AWAY} is the time zone whose time you want to know. So, for New York and Rome, $D = +5 - (-1) = +6$, meaning you must add six hours to New York time to get the time in Rome.

Of course, this simple formula works only if both locations are on standard time or are on daylight

saving time. If they differ, then you need to take the difference into account. For example, if it's daylight time in Rome but not in the United States, then they've "sprung ahead" and there will be seven hours difference.

This formula works when you cross the date line as well. Consider a 10-hour flight from Tokyo (-9) to Los Angeles (+8). The flight leaves at 4:00 P.M. Tokyo time on Tuesday, December 12. From the formula, $D = (-9) - (+8) = -17$, so you subtract 17 hours from Tokyo time, and the time in Los Angeles when you depart from Tokyo is 11:00 P.M. on Monday, December 11. Add 10 hours for your flight time, and you will arrive in Los Angeles at 9:00 A.M. on Tuesday—well before you left Tokyo!

THE EARTH'S REVOLUTION AROUND THE SUN

So far, we have discussed the importance of the earth's rotation on its axis. Another important motion of the earth is its **revolution**, or its movement in orbit around the sun.

The earth completes a revolution around the sun in about 365¼ days—one-fourth day more than the calendar year of 365 days. Every four years, the extra one-fourth days add up to about one whole day. By inserting a 29th day in February in leap years, we largely correct the calendar for this effect. Further minor corrections are necessary to perfect the system.

The earth's orbit around the sun is shaped like an ellipse, or oval. This means that the distance between the earth and sun varies somewhat through the year. At *perihelion*, which occurs on or about January 3, the earth is nearest to the sun. At *aphelion*, on or about July 4, the earth is farthest away from the sun. However, the distance between sun and earth varies only by about 3 percent during one revolution, because the elliptical orbit is shaped very much like a circle. For most purposes we can regard the orbit as circular.

In which direction does the earth revolve? Imagine yourself in space, looking down on the north pole of

the earth. From this viewpoint, the earth travels counterclockwise around the sun (Figure 1.13). This is the same direction as the earth's rotation. It is also the same direction as the moon's rotation and revolution around the earth. In fact, nearly all the planets and major satellites in our solar system have the same direction of rotation and revolution.

Tilt of the Earth's Axis

The seasons we experience on earth are related to the orientation of the earth's axis of rotation and the position of the sun. We usually describe this situation by

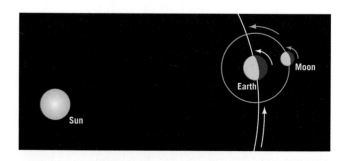

Figure 1.13   The earth from above the north pole.

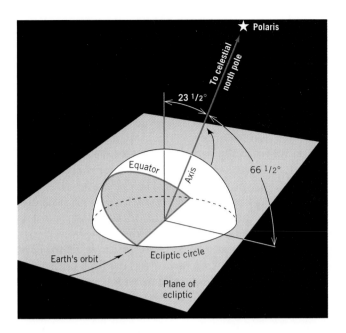

Figure 1.14 The tilt of the earth's axis of rotation with respect to its orbital plane. As the earth moves in its orbit on the plane of the ecliptic around the sun, its rotational axis remains pointed toward Polaris, the north star, and makes an angle of 66½° with the ecliptic plane.

stating that the earth's axis is tilted. Figure 1.14 shows the earth within a portion of the *plane of the ecliptic*—the plane containing the earth's orbit around the sun. Notice that the earth's orbit is shown lying within the plane.

Now, consider the axis of the earth's rotation. Rather than being at a right angle to the plane of the ecliptic, the axis is tilted at an angle of 23½° away from a right angle. That is, the angle between the axis and the plane of the ecliptic is 66½°, not 90°. In addition, the direction in which the axis points is fixed in space. The north polar end of the axis is aimed toward Polaris, the north star. The direction of the axis does not change as the earth revolves. As a result, the north pole is tilted away from the sun during one part of the year (the situation shown in the diagram) and is tilted toward the sun during the remainder.

Solstice and Equinox

Figure 1.15 diagrams the earth as it revolves in its orbit through the four seasons. Consider first the event on December 22, which is pictured on the far right. On this day, the earth is positioned so that the north polar end of its axis leans at the maximum angle away from the sun, 23½°. This event is called the **winter solstice**. (While it is winter in the northern hemisphere, it is

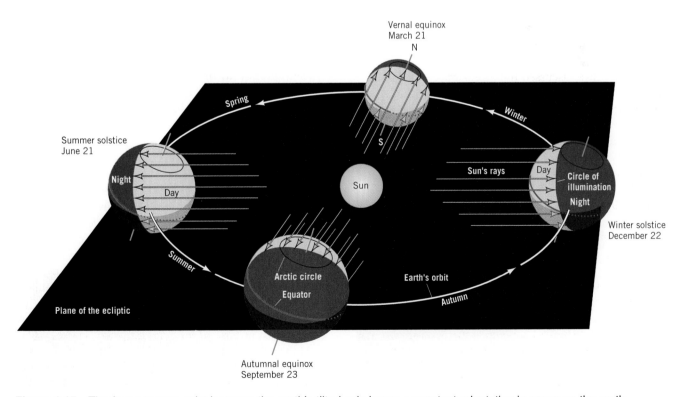

Figure 1.15 The four seasons exist because the earth's tilted axis keeps a constant orientation in space as the earth revolves about the sun. This tips the northern hemisphere toward the sun for the summer solstice, and away from the sun for the winter solstice. Both hemispheres are illuminated equally at the spring equinox and the fall equinox.

summer in the southern hemisphere, so you can use the term *December solstice* to avoid any confusion.) At this time, the southern hemisphere is tilted toward the sun and enjoys strong solar heating.

Six months later, on June 21, the earth is on the opposite side of its orbit in an equivalent position. At this event, known as the **summer solstice** (*June solstice*), the north polar end of the axis is tilted at its maximum angle of $23\frac{1}{2}°$ toward the sun. Thus, the north pole and northern hemisphere are tilted toward the sun, while the south pole and southern hemisphere are tilted away.

Midway between the solstice dates, the equinoxes occur. At an **equinox**, the earth's axis makes a right angle with a line drawn to the sun, and neither the north nor south pole is tilted toward the sun. The *vernal equinox* occurs on March 21, and the *autumnal equinox* occurs on September 23. Conditions are identical as far as earth–sun relationships are concerned on the two equinoxes. We should also note that the date of any solstice or equinox in a particular year may vary by a day or so, since the revolution period is not exactly 365 days.

Equinox Conditions

Let's look at equinoxes and solstices in more detail. The conditions at an equinox, shown in Figure 1.16, form the simplest case. The figure illustrates two important concepts of global illumination that we use for describing equinoxes and solstices. The first concept is the *circle of illumination*. Note that the earth is always divided into two hemispheres with respect to the sun's

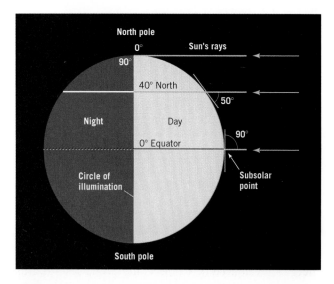

Figure 1.16 Equinox conditions. At this time, the earth's axis of rotation is exactly at right angles to the direction of solar illumination. The subsolar point lies on the equator. At both poles, the sun is seen at the horizon.

rays. One hemisphere (day) is lit by the sun, and the other (night) lies in the darkness of the earth's shadow. The circle of illumination is the circle that separates the day hemisphere from the night hemisphere. The second concept is the *subsolar point*. The sun is directly overhead at this single point on the earth's surface.

At equinox, the circle of illumination passes through the north and south poles, as we see in Figure 1.16. The sun's rays graze the surface at either pole, and the surface there receives little or no solar energy. The subsolar point falls on the equator. Here, the angle between the sun's rays and the earth's surface is 90°, and the solar illumination is received in full force. At an intermediate latitude, such as 40° N, the rays of the sun at noon strike the surface at a lesser angle. This *noon angle*—the elevation of the sun above the horizon at noon—can be easily determined. Some simple geometry shows that the noon angle is equal to 90° minus the latitude, or 50° in this example, for equinox conditions.

Imagine yourself at a point on the earth, say, at a latitude of 40° N. Visualize the earth rotating from left to right, so that you turn with the globe, completing a full circuit in 24 hours. At the equinox, you spend 12 hours in darkness and 12 hours in sunlight. This is because the circle of illumination passes through the poles, dividing every parallel exactly in two. Thus, one important feature of the equinox is that day and night are of equal length everywhere on the globe.

Solstice Conditions

Now examine the solstice conditions shown in Figure 1.17. Summer solstice is on the left. Consider yourself back at a point on the lat. 40° N parallel. The circle of illumination does not divide your parallel in equal halves because of the tilt of the northern hemisphere toward the sun. Instead, the larger part is in daylight. For you, the day is now considerably longer (about 15 hours) than the night (about 9 hours).

The farther north you go, the more the effect increases. In fact, the entire area of the globe north of lat. $66\frac{1}{2}°$ is on the daylight side of the circle of illumination. This parallel is known as the **arctic circle**. Even though the earth rotates through a full cycle during a 24-hour period, the area north of the arctic circle remains in continuous daylight. We also see that the subsolar point is at a latitude of $23\frac{1}{2}°$ N. This parallel is known as the **tropic of cancer**. Because the sun is directly over the tropic of cancer at this solstice, solar energy is most intense here.

At the winter solstice, conditions are exactly reversed from those of the summer solstice. If you imagine yourself back at lat. 40° N, you find that the night is now about 15 hours long while daylight lasts about 9 hours. All the area south of lat. $66\frac{1}{2}°$ S lies under the sun's rays, inundated with 24 hours of daylight. This parallel

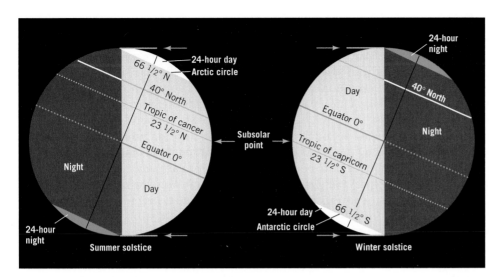

Figure 1.17 Solstice conditions. At the solstice, the north end of the earth's axis of rotation is fully tilted either toward or away from the sun. Because of the tilt, polar regions experience either 24-hour day or 24-hour night. The subsolar point lies on one of the tropics, at lat. 23½° N or S.

is known as the **antarctic circle**. The subsolar point has shifted to a point on the parallel at lat. 23½° S, known as the **tropic of capricorn**.

The solstices and equinoxes represent conditions at only four times of the year. Between these times, the subsolar point travels northward and southward in an annual cycle. We refer to the latitude of the subsolar point as it varies from -23½° (23½° S lat.) to +23½° (23½° N lat.) as the sun's *declination*. Figure 1.18 plots the sun's *declination* through the year.

As the year progresses, the subsolar point and the latitude of maximum solar intensity move across the belt between the two tropics. Areas of 24-hour daylight or 24-hour night shrink and then grow as the seasonal cycle progresses. Except at the equator, the length of daylight varies a little from one day to the next. In this way, the earth experiences the rhythm of the seasons as it continues its revolution around the sun.

THE SUN–EARTH ENERGY FLOW SYSTEM

The biggest and most important flow system for us, the land dwellers of the earth, is the sun–earth energy flow system. As we will see in many later chapters, the sun's flow of radiant energy to the earth is the power source for many other systems and processes that occur at or near the earth's surface. The sun–earth system is so important that we will spend much of the next chapter discussing it.

What have we learned about the sun–earth energy flow system thus far? Here are two simple, important facts:

- *Half of the earth is always receiving solar energy.* So, there is a constant flow of light and heat from sun to earth.
- *However, the solar energy flow is not received uniformly over the surface.* This is because of the earth's spherical shape, its constant rotation on its axis, and its constant revolution around the sun.

We will take up the second point in more detail in the next chapter.

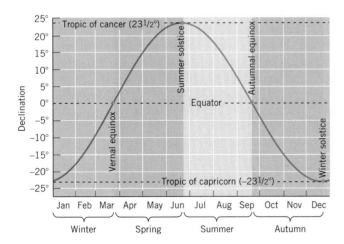

Figure 1.18 The sun's declination through the year. The latitude of the subsolar point marks the sun's declination, which changes slowly through the year from -23½° to +23½° to -23½°. Solstices and seasons are labeled for the northern hemisphere.

An understanding of the causes for the seasons is essential to the topic we treat next—the global energy balance. In Chapter 2 we will examine in detail solar energy—the source that powers life processes—and its interaction with the earth's atmosphere and surface. This discussion will include an explanation of why the solar energy available varies through the seasons and with latitude.

CHAPTER SUMMARY

The rotation of the earth on its axis and the revolution of the earth in its orbit about the sun are very fundamental topics in physical geography. The rotation provides the first great rhythm of our planet—the daily alternation of sunlight and darkness. The tides, and a sideward turning of ocean and air currents, are further effects of the earth's rotation.

The earth's axis of rotation provides a reference for the system of location on the earth's surface—the geographic grid, which consists of meridians and parallels. This system is indexed by our system of latitude and longitude, which uses the equator and the prime meridian as references. We require a map projection to display the earth's curved surface on a flat map. The polar projection is centered on either pole and pictures the globe as we might view it from the top or bottom. The Mercator projection converts the geographic grid into a flat, rectangular one and best displays directional features. The Goode projection distorts the shapes of continents and coastlines but preserves the areas of landmasses in their correct proportion.

We monitor the earth's rotation by daily timekeeping. Each hour the earth rotates by 15°. In the standard time system, we keep time according to a nearby standard meridian. Since standard meridians are normally 15° apart, clocks around the globe usually differ by whole hours.

The seasons are the second great earthly rhythm. They arise from the revolution of the earth in its orbit around the sun, combined with the fact that the earth's rotational axis is tilted with respect to its orbital plane. The solstices and equinoxes mark the cycle of this revolution. At the summer (June) solstice, the northern hemisphere is tilted toward the sun. At the winter (December) solstice, it is the southern hemisphere that is tilted toward the sun. At the equinoxes, day and night are of equal length.

KEY TERMS

rotation	longitude	summer solstice
parallel	map projection	equinox
equator	standard time system	arctic circle
meridian	time zone	tropic of cancer
geographic grid	revolution	antarctic circle
latitude	winter solstice	tropic of capricorn

REVIEW QUESTIONS

1. What is the shape of the earth? How do you know?

2. What is meant by earth rotation? Describe three physical effects of the earth's rotation.

3. Describe the geographic grid, including parallels and meridians.

4. How do latitude and longitude determine position on the globe? In what units are they measured?

5. Name three types of map projections and describe each briefly. Give reasons why you might choose different map projections to display different types of geographical information.

6. Explain the global timekeeping system. Define and use the terms *standard time*, *standard meridian*, and *time zone* in your answer.

7. What is meant by the "tilt of the earth's axis"? How is the tilt responsible for the seasons?

8. Sketch a diagram of the earth at an equinox. Show the north and south poles, the equator, and the circle of illumination. Indicate the direction of the sun's incoming rays and shade the night portion of the globe.

9. Sketch a diagram of the earth at the summer (June) solstice, showing the same features. Also include the tropics of cancer and capricorn, and the arctic and antarctic circles.

Focus on Systems 1.1 • The Word *System*
1. Provide three or four examples of different uses of the term *system*, contrasting the various meanings.

ESSAY QUESTION

1. Suppose that the earth's axis were tilted at 40° to the plane of the ecliptic instead of $23\frac{1}{2}°$. What would be the global effects of this change? How would the seasons change at your location?

PROBLEMS

Working It Out 1.2 • Distances from Latitude and Longitude
1. Chicago and Mobile are located on about the same meridian, but they are about 11° lat. apart. What is the approximate distance between Chicago and Mobile?

2. Ottawa and Portland, Oregon, are both located very close to the 45th parallel, but their longitudes are 76° W and 124° W, respectively. What is the approximate distance between the two cities, measured along the parallel? (In using the formula, be sure that your calculator is set to use degrees when finding the cosine.)

3. A map of a region close to the equator shows an area of 1° of latitude by 1° of longitude. About how many square kilometers are covered by the map? How many square miles? How do these areas compare with those of a 1° by 1° map near Winnipeg (50° N lat.)?

Working It Out 1.3 • Global Timekeeping
1. A flight from Chicago (+6) to Beijing (−8) takes 16 hours and 30 minutes (since it has a three-hour layover in Tokyo). You leave Chicago at 3:30 P.M. on Saturday, June 14. When are you scheduled to arrive, Beijing time?

2. A one-stop flight from Anchorage to Washington leaves at 1:59 A.M. and arrives at 3:51 P.M.. The flight includes a 1-hour-and-25-minute layover in Denver. How long will you be in the air (or taxiing), if the flight holds to the schedule? (Use Figure 1.12 to obtain time zone numbers.)

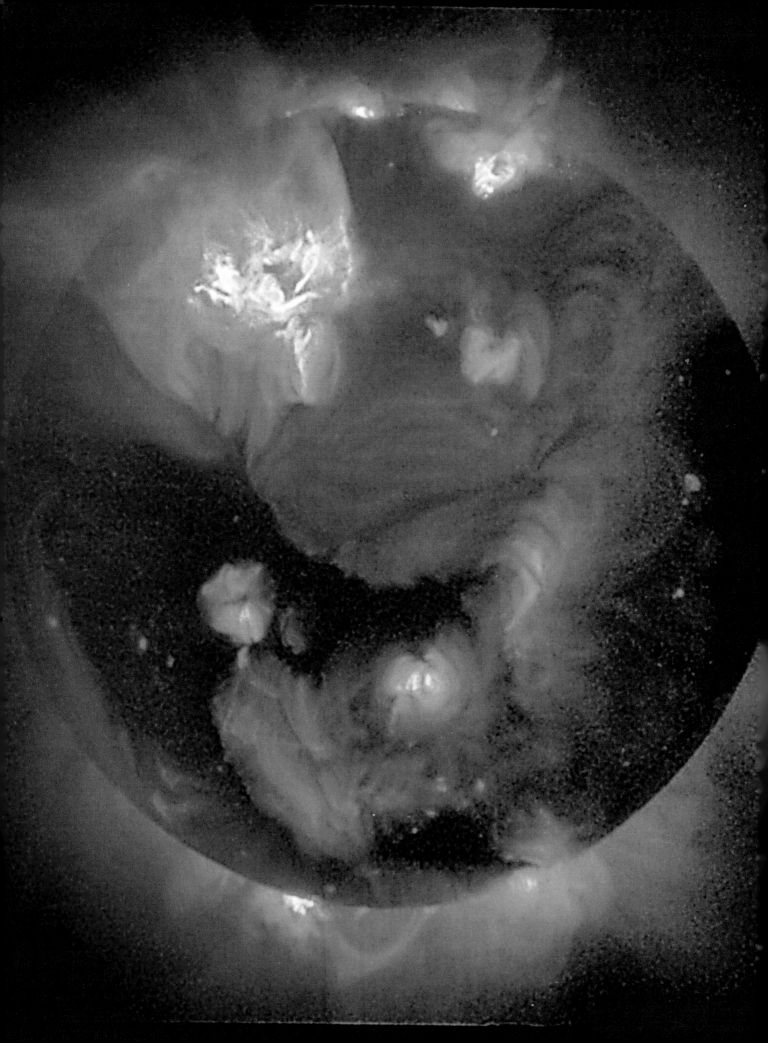

Chapter 2

The Global Energy System

This chapter explains how solar radiation is intercepted by our planet, flows through the earth's atmosphere, and interacts with the earth's land and ocean surfaces. Solar radiation, as we will see, is the driving power source for wind, waves, weather, rivers, and ocean currents. Most natural phenomena that take place at the earth's surface are directly or indirectly solar-powered—from the downhill flow of a river to the movement of a sand dune to the growth of a forest.

A primary topic of this chapter is the *energy balance* of the earth, which includes land and ocean surfaces and the atmosphere. The energy balance of a surface describes the balance between the flows of energy reaching that surface and flows of energy leaving it. The earth's energy balance controls the seasonal and daily changes in the earth's surface temperature. Differences in energy flow rates from place to place, as measured by the surface energy balance, also drive currents of air and ocean water. These, in turn, produce the changing weather and rich diversity of climates we experience on the earth's surface.

Human activities now dominate many regions of the earth, and we have irreversibly modified our planet by farming much of its surface and adding carbon dioxide

An X ray image of the sun showing intense sunspot activity. This image was acquired by the Japanese Yohkoh ("Sun-ray") astronomical satellite.

to its atmosphere. Have we shifted the balance of energy flows? Is our earth absorbing more solar energy and becoming warmer? Is it absorbing less and becoming cooler? We must examine the global energy balance in detail before we can understand human impact on the earth–atmosphere system.

ELECTROMAGNETIC RADIATION

Our study of the global energy system begins with the subject of radiation—that is, **electromagnetic radiation.** This form of radiation is emitted by all objects. Think of electromagnetic radiation as a collection, or spectrum, of waves of a wide range of wavelengths traveling quickly away from the surface of the object. *Wavelength* describes the distance separating one wave crest from the next wave crest (Figure 2.1). The unit used to measure wavelength is the *micrometer* (formerly called the micron). A micrometer is one millionth of a meter. This is such a small unit that the tip of your little finger is about 15,000 micrometers wide. In this text, we use the abbreviation μm for the micrometer. The first letter is the Greek letter μ, or mu. It is used in the metric system to denote micro, meaning one-millionth. Radiant energy can exist at any wavelength. Heat and light, two familiar examples of electromagnetic radiation, are identical except for their wavelengths.

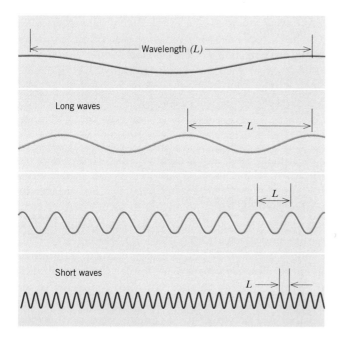

Figure 2.1 Electromagnetic radiation can be described as a collection of energy waves with different wavelengths. Wavelength *L* is the crest-to-crest distance between successive wave crests.

Electromagnetic radiation is one of several fundamental forms of energy. Some other forms of energy that are important in the study of physical geography are discussed in *Focus on Systems 2.1 • Forms of Energy.*

Two important physical principles concern the emission of electromagnetic radiation. The first is that there is an inverse relationship between the wavelength of the radiation that an object emits and the temperature of the object. For example, the sun, a very hot object, emits radiation with short wavelengths. In contrast, the earth, a much cooler object, emits radiation with longer wavelengths.

The second principle is that hot objects radiate more energy than cooler objects—much more. In fact, the flow of radiant energy from the surface of an object is directly related to absolute temperature of the surface raised to the fourth power. (Absolute temperature is temperature measured on the Kelvin scale, with zero being the absence of all heat.) Because of this relationship, a small increase in temperature can mean a large increase in the rate at which radiation is given off by an object. For example, if the absolute temperature of an object is doubled, the flow of radiant energy from its surface will be 16 times larger. These two principles are described in more depth in *Working It Out 2.2 • Radiation Laws.* Understanding them will help you to master global energy balance.

Solar Radiation

Our sun is a ball of constantly churning gases that are heated by continuous nuclear reactions. It is about average in size compared to other stars, and it has a surface temperature of about 6000°C (about 11,000°F). Like all objects, it emits energy in the form of electromagnetic radiation. The energy travels outward in straight lines, or rays, from the sun at a speed of about 300,000 km (186,000 mi) per second. At that rate, it takes the energy about 8⅓ minutes to travel the 150 million km (93 million mi) from the sun to the earth.

As solar radiation travels through space, none of it is lost. However, the rays spread apart as they move away from the sun. This means that a planet farther from the sun, like Mars, receives less radiation than one located nearer to the sun, like Venus. The earth intercepts only about one-half of one-billionth of the sun's total energy output.

The portion of the electromagnetic spectrum that contains radiation emitted by the sun and earth can be divided into four major portions, based on wavelength. These spectral regions are shown at the top of Figure 2.2. At the left lies the *ultraviolet* radiation portion of the spectrum. Its wavelengths range from about 0.2 to 0.4 mm. These short waves are high in energy. Solar ultraviolet radiation is nearly all absorbed by the gas molecules of the earth's atmosphere. However, when this

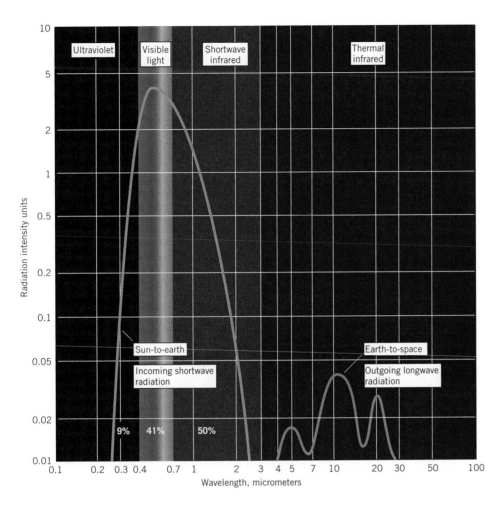

Figure 2.2 There are four important divisions within the electromagnetic spectrum—ultraviolet, visible light, shortwave infrared, and thermal infrared. The curve on the left shows the intensity of incoming solar shortwave radiation as it reaches the surface. The curve on the right shows the outgoing longwave radiation that travels from earth and atmosphere to outer space.

radiation penetrates to the surface of the earth, it can damage living tissues, such as human skin.

Toward the center of the solar spectrum is the **visible light** portion of the spectrum, with wavelengths ranging from 0.4 to 0.7 µm. Our eyes sense this radiation as light energy. The color of the light is determined by its wavelength. The shorter wavelengths are perceived as violet, while the longer ones are perceived as red. In between is the rest of our familiar color spectrum, including blue, green, yellow, and orange. Only a small amount of visible light is absorbed by atmospheric gases or particles in passing through the atmosphere. Unless intercepted by clouds, nearly all of this radiation reaches the earth's surface.

Next is the *shortwave infrared* radiation region, ranging from 0.7 to 3 µm. This radiation acts much like visible light, although our eyes are not sensitive to it. Shortwave infrared radiation also penetrates the atmosphere easily. Taken together, we refer to ultraviolet, visible light and shortwave infrared radiation as **shortwave radiation**.

At the right are the *thermal infrared* wavelengths, longer than 3 µm, which are emitted by cooler objects. We perceive this type of radiation as heat, such as you might feel sitting beside the glowing coals of a campfire. Thermal infrared radiation is also referred to as **longwave radiation**.

The sun does not emit all wavelengths of radiation equally. The tall curve at the left on the graph of Figure 2.2 shows the relative intensity of solar energy entering the outer atmosphere, plotted by wavelength. Energy intensity is shown on the graph on the vertical scale. Note that it is a logarithmic scale—that is, each whole unit marks an intensity 10 times greater than the one below.

The ultraviolet portion of the spectrum accounts for about 9 percent of the total incoming solar energy. Also included in the ultraviolet division are X rays and gamma rays, forms of radiation that have shorter wavelengths than the ultraviolet. They are not shown on the graph. The solar energy curve peaks in the visible light range, which accounts for about 41 percent of the total energy received by the earth. The shortwave infrared portion of the spectrum accounts for 50 percent of the total energy. Very little energy arrives in wavelengths longer than 2 µm.

The sun's interior is the source generating solar energy. Here, hydrogen is converted to helium at very

Focus on Systems 2.1 • Forms of Energy

Cosmologists—those scientists who study the universe—tell us that everything that exists is composed of either matter or energy. Energy is what causes matter to change through time within the universe. That goes, of course, for our planet and its inhabitants. On earth, energy drives the life cycle of every individual organism as it comes into existence, grows, reproduces, and dies. Every action of an organism through that life span requires the intake and expenditure of energy. Whatever you learn here about energy and its relationship to matter will apply not only to our physical geography topics but also to every personal life experience and every event you observe going on in the environment around you.

The term *energy* is a comparative newcomer to physics. It was first used in 1807, but the idea is much older, going back to Sir Isaac Newton's laws of motion, published in 1687. Leibnitz, a great philosopher and mathematician, proposed in 1695 the existence of a "vital force." This force, acting over distance (force times distance), is equal to one-half of mass times velocity squared ($\frac{1}{2}\,mv^2$). Today we call the vital force *kinetic energy*—the energy of mass (or "matter") in motion.

Physicists define *energy* as the ability to do work. So what is work? Work, again according to physicists, is equal to the change in kinetic energy in a body. This obviously circular maneuver is of no real help to us in understanding the nature of energy. We already know that energy is somehow tied in with matter in motion through space and time, but what *is* energy? Keep looking for clues as we go from one chapter to the next.

Whatever energy is, it does seem to "move" or "travel" with (or within) forms of matter. For example, we understand that the gasoline engine in our car transfers the energy of fuel combustion into forward motion of the entire structure of the car, including the engine, by means of the transmission and wheels. This entire moving object is said to possess kinetic energy. Kinetic energy is one of the energy forms coming under the general heading of *mechanical energy*. Should this speeding automobile crash head-on into a massive power pole, it will demonstrate its ability to do work upon its own body, upon its passengers, and upon the pole as well. The energy released in the collision will increase in direct proportion to the mass of the car, and it will also increase with the square of the auto's speed, since, as Liebnitz

suggested, "vital force," or kinetic energy, is equal to $\frac{1}{2}\,mv^2$.

Now the crushed car lies silent and still. What happened to all that energy of motion? Energy cannot just be destroyed and disappear. If you had been on the spot immediately after the crash, you would have found the car body to be extremely hot, even though it didn't catch fire. Kinetic energy of motion has turned into *heat energy*. In this form of energy, the atoms and molecules within a solid (or liquid, or gas) are in extremely fast motion invisible to us. In the metal of the car, the individual molecules are "vibrating" in place. The hotter the metal, the faster they vibrate. So here we find another form of kinetic energy. The proper name for it is "kinetic energy of atoms in motion," but it is also called *sensible heat*, because it can be sensed by touch as well as being measured by a thermometer. This form of energy is extremely important in physical geography, and we will encounter it over and over in studying temperature changes in air, water, and soil.

If you had stood close to the collapsed auto just after the impact, you might have felt a glow of heat on the exposed skin of your face and hands. This is yet another form of energy: *radiant energy*. It is de-

high temperatures and pressures. In this process of nuclear fusion, a vast quantity of energy is generated and finds its way to the sun's surface. Because the rate of production of energy is nearly constant, the output of solar radiation is also nearly constant. So, given the average distance of earth from the sun, the amount of solar energy received on a small, fixed area of surface held at right angles to the sun's rays is almost constant. This rate of incoming shortwave energy, known as the *solar constant,* is measured beyond the outer limits of the earth's atmosphere, before energy has been lost in passing through the earth's atmosphere. The solar con-

stant has a value of 1400 watts per square meter (W/m^2).

Let's take a moment to examine the units of energy flow for radiation, as used, for example, in the definition of the solar constant. The *watt* (W), which describes a rate of energy flow, is a familiar measure of power consumption. You've probably seen it applied to light bulbs or to home appliances, such as stereo amplifiers or microwave ovens. When we use the watt to measure the intensity of a flow of radiant energy, we must specify the unit of surface area that receives the energy. This cross section is assigned the area of 1

scribed and explained in detail in this chapter, because it is the form of energy we receive from the sun and it "runs the whole show" on our planet. We know radiant energy best in the form of sunlight, which travels in straight lines through space and easily penetrates transparent substances such as air, clear water, and even clear solids such as glass.

Let's move to another form of energy, one that depends on the attractive force of gravity. Imagine dragging or rolling a heavy object—a boulder, for example—to the top of a hill. Feeling the spirit of Galileo in the Leaning Tower of Pisa, as he dropped objects to the ground, you nudge the boulder off a cliff at the top of the hill and watch it drop to the ground below. It lands with a satisfying thud.

Now, analyze the energy changes so far. In sweating the boulder to the top of the hill (an expenditure of energy released from your body) you overcame the ever-present downward pull of gravity. In so doing, you endowed the boulder with a certain amount of *stored energy*. This quantity of stored energy belongs to the energy class known as *potential energy*. With each meter of vertical distance the boulder falls, it converts a certain fixed

quantity of that potential energy into kinetic energy. (We can disregard the small energy loss by friction with the surrounding air.) Upon impact, the kinetic energy is converted to heat energy, which quickly dissipates.

An important point is that potential energy, present because of gravity, must always be evaluated in terms of a given reference level, or *base level*. The standard base level is the average level of the ocean, called "mean sea level." In nature, any kind of matter on the land or in the atmosphere can be said to hold potential energy with respect to sea level. Examples are raindrops and hailstones formed in a storm cloud and the mineral particles of ash or cinder spewed from an erupting volcano. As these substances fall toward the earth, their potential energy is converted to kinetic energy. Potential energy can also be converted to kinetic energy as substances descend along sloping and winding pathways. Flow of water in a river and of ice in a glacier are two examples.

Finally, in this introduction to energy found in the natural realms of our life environment, we must recognize *chemical energy*, which is absorbed or released by matter when chemical reactions take place.

These reactions involve the coming together of atoms to form simple molecules, the recombining of simple molecules into new and more complex compounds, and the reverse changes back into simpler forms. Most of the time, chemical energy is in the stored form, held within the molecule. An important example is the way in which green plants absorb radiant energy of the sun's rays to produce those complex organic molecules we refer to as carbohydrates.

Chemical energy is also stored for long periods of geologic time in the hydrocarbon compounds we use as fuels. The gasoline that powered our automobile to its tragic end represents stored chemical energy. As the gasoline burns, its molecules combine with oxygen of the air, releasing heat energy as well as many simpler molecules, such as carbon dioxide and carbon monoxide.

At this point, the realm of energy in all its forms may seem complicated and hard to grasp. However, we will return to the topic of energy again many times in this book, and as you work with concepts of energy, they will become clearer.

square meter (1 m^2). Thus, the measure of intensity of received (or emitted) radiation is given as watts per square meter (W/m^2). Because there are no common equivalents for this energy flow rate in the English system, we will use only metric units.

Longwave Radiation from the Earth

Recall that both the wavelength and intensity of radiation emitted by an object depend on the object's temperature. Since the earth's surface and atmosphere are

much colder than the sun's surface, we can deduce that our planet radiates less energy than the sun and that the energy emitted has longer wavelengths.

Figure 2.2 confirms these deductions. The curve on the right shows outgoing radiation flow from the earth as measured at the top of the atmosphere. Notice that the overall height is much lower than for the solar curve on the left, showing that the planet emits radiation at a lower rate. What about wavelength? The graph shows that the energy is emitted at wavelengths of about 3 to 30 μm, which is longer than the wavelengths of solar radiation and is consistent with the principle

Working It Out 2.2 • Radiation Laws

Physicists use two simple laws, expressed as formulas, to describe the principles that hotter objects radiate more energy and that the energy is of shorter wavelengths. First is the Stefan-Boltzmann Law. It describes the relation of flow of energy emitted by a surface to its temperature:

$$M = \sigma T^4$$

Here, M is the energy flow from the surface, in W/m^2; σ (the Greek letter sigma) is the Stefan-Boltzmann constant, 5.67×10^{-8} $W/m^2/K^4$; and T is the temperature of the surface, in K.

The watt, W, is a measure of energy flow that is explained within the chapter. The symbol K denotes absolute temperature, which is measured in Kelvins. On this scale, 0 indicates the absence of heat energy. Absolute temperature is related to temperature in °C by the formula $K = C + 273$, where K is absolute temperature and C is Celsius temperature.

The second law applied to radiant energy is Wein's Law, which identifies the wavelength at which radiation of a surface will be greatest, given the temperature:

$$\lambda_{MAX} = b/T$$

where λ_{MAX} is the wavelength at which radiation is at a maximum, in μm; b is a constant equal to 2898 μm K; and T is the temperature of the surface in K.

Both of these laws apply only to a perfectly radiating surface, known as a *blackbody*. However, most natural surfaces are good radiators and do not differ too much from a blackbody.

As an example of the use of these formulas, let's calculate the flow of energy emitted by a surface and then determine the wavelength at which radiation from the surface is greatest. We will use a temperature of 450 K (177°C, 351°F) for the surface, which is about that of a home

oven while baking. Using the Stefan-Boltzman equation, we find that substituting 450 K for T yields

$$\begin{aligned} M = \sigma T^4 &= (5.67 \times 10^{-8} \text{ W/m}^2\text{K}^4) \\ &\quad \times (450 \text{ K})^4 \\ &= 5.67 \times 10^{-8} \text{ W/m}^2\text{K}^4 \times 4.10 \times \\ &\quad 10^{10} \text{ K}^4 \\ &= 2.32 \times 10^3 \text{ W/m}^2 \\ &= 2320 \text{ W/m}^2 \end{aligned}$$

Applying Wein's Law, we have

$$\lambda_{max} = \frac{b}{T} = \frac{2898 \text{ μm K}}{450 \text{ K}} = 6.44 \text{ μm}$$

In comparison, a similar calculation for a surface at room temperature (293 K, 20°C, 68°F) shows an emission rate of 418 W/m^2 and a wavelength of maximum emission of 9.89 μm. Thus, the oven surface emits more than five times as much energy as the room temperature surface, and the wavelength of maximum emission is somewhat shorter.

that a cooler surface emits radiation at longer wavelengths.

Note that the curve of outgoing longwave radiation shows three distinct peaks at wavelengths of about 5, 10, and 20 μm. Between these wavelengths, atmospheric gases, primarily water vapor and carbon dioxide, absorb much of the radiation leaving the surface.

Thermal infrared radiation is no different in principle from visible light radiation. Although photographic film is not sensitive to thermal radiation, it is possible to acquire an image of thermal radiation using a special sensor. Figure 2.3 shows a thermal infrared image of a suburban scene obtained at night. Windows and pavement appear bright because they are warm and radiate

Figure 2.3 A thermal infrared image of a suburban scene at night. Black and blue tones show lower temperatures, while yellow and red tones show higher temperatures. Ground and sky are coldest, while the heated windows of the homes are warmest.

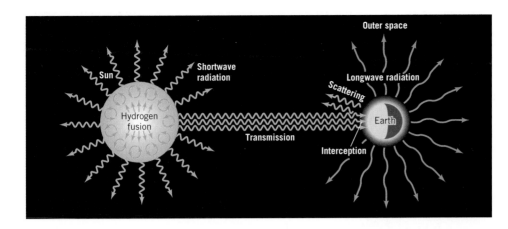

Figure 2.4 A diagram of the global radiation balance. Shortwave radiation from the sun is transmitted through space, where it is intercepted by the earth. The absorbed radiation is then ultimately emitted as longwave radiation to outer space. (From A. N. Strahler, *Journal of Geography*, vol. 69. Copyright 1970 © by *The Journal of Geography*. Used by permission.)

more intensely. In contrast, the moist surfaces of lawns are cooler and appear darker.

The Global Radiation Balance

As we have seen, the earth constantly absorbs solar shortwave radiation and emits longwave radiation. Figure 2.4 presents a diagram of this energy flow process, which we refer to as the earth's **radiation balance**.

The sun provides a nearly constant flow of shortwave radiation toward earth. Part of this radiation, as we noted earlier, is scattered back into space without absorption. Clouds and dust particles in the atmosphere contribute to this scattering, as do land and ocean surfaces. The remaining shortwave energy from the sun is absorbed by either atmosphere, land, or ocean. Once absorbed, solar energy raises the temperature of the atmosphere, as well as the surfaces of the oceans and lands.

The atmosphere, land, and ocean also emit energy in the form of longwave radiation. This radiation ultimately leaves the planet, headed for outer space. The longwave radiation outflow tends to lower the temperature of the atmosphere, ocean, and land, and thus cool the planet. In the long run, these flows balance—incoming energy absorbed and outgoing radiation emitted are equal. Since the temperature of a surface is determined by the amount of energy it absorbs and emits, the earth's overall temperature tends to remain constant.

Using the radiation laws presented in *Working It Out 2.2*, we find it easy to calculate the magnitude of the energy flows involved in the global radiation balance. *Working It Out 2.3 • Calculating the Global Radiation Balance* shows how this can be done.

INSOLATION OVER THE GLOBE

The flow of solar radiation to the earth as a whole remains constant, but it varies from place to place and

time to time. We turn now to this variation and its causes. In this discussion, we will refer to **insolation** (**in**coming **sol**ar radi**ation**)—the flow of solar energy intercepted by an exposed surface for the case of a uniformly spherical earth with no atmosphere. Insolation is a flow rate and has units of watts per square meter (W/m^2). Daily insolation refers to the average of this flow rate over a 24-hour day. Annual insolation, discussed in a later section, is the average flow rate over the entire year.

Insolation depends on the angle of the sun above the horizon. Figure 2.5 shows that insolation is greatest where the sun's rays strike vertically. When the sun's horizon angle diminishes, the same amount of solar energy spreads over a greater area of ground surface. So when the sun is low in the sky overhead, solar radiation is less intense. When the sun is high in the sky, solar radiation is greater.

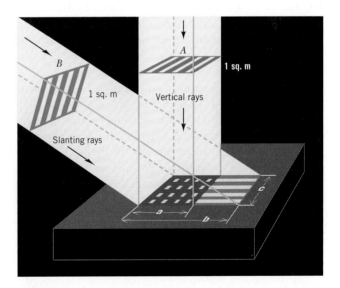

Figure 2.5 The angle of the sun's rays determines the intensity of insolation. The energy of vertical rays *A* is concentrated in square *a* by *c*, but the same energy in the slanting rays *B* is spread over a larger rectangle, *b* by *c*.

Working It Out 2.3 • Calculating the Global Radiation Balance

The earth's global radiation balance, as diagrammed in Figure 2.4, is an example of an energy flow system. With a few simple facts and formulas, it is easy to estimate the magnitude of flow in the two principal pathways of this system—incoming shortwave radiation and outgoing longwave radiation.

First, what is the input flow of solar radiation to the earth? To determine that, we need to know the rate of flow and the area that receives the flow. Recall that the rate is described by the solar constant, which has a value of about 1400 W/m^2 or 1.40 kW/m^2, where kW is the kilowatt, equal to 10^3 W. Now, the earth's radius is about 6400 km. That means that the earth presents to the parallel rays of the sun a disk of radius 6400 km (see figure at left). The disk has an area $\pi r^2 = 3.24 \times (6.40 \times 10^3 \, km)^2 = 1.29 \times 10^8 \, km^2$. Since 1 $km^2 = 10^6 \, m^2$, this is equal to $1.29 \times 10^{14} \, m^2$. Each square meter of the disk intercepts solar energy flow at the rate of the solar constant, so the energy flow from the sun to the earth is about 1.40 $kW/m^2 \times 1.29 \times 10^{14} \, m^2 = 1.81 \times 10^{14}$ kW. In other words,

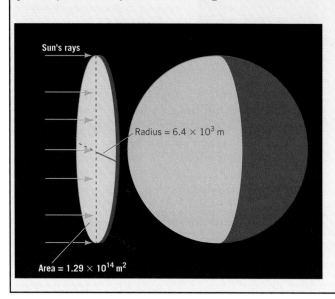

Sun's rays

Radius = 6.4×10^3 m

Area = $1.29 \times 10^{14} \, m^2$

The earth intercepts the parallel rays of the sun in an area equal to a disk of earth radius.

$$\frac{\text{Sun–Earth}}{\text{Energy Flow}} = \frac{\text{Flow per}}{\text{Unit Area}}$$
$$(1.81 \times 10^{14} \, kW) = 1.40 \, kW/m^2$$
$$\times \frac{\text{Area of Earth}}{\text{Presented to Sun}}$$
$$(1.29 \times 10^{14} \, m^2)$$

Daily insolation at a location depends on two factors: (1) the angle at which the sun's rays strike the earth, and the length of time of exposure to the rays (2). Recall from Chapter 1 that both of these factors are controlled by the latitude of the location and the time of year. For example, in midlatitude locations in summer, days are long and the sun rises to a position high in the sky, thus heating the surface more intensely. Let's investigate these variations in more detail.

Refer back to Chapter 1 and Figure 1.16, showing equinox conditions. Only at the subsolar point does the earth's spherical surface present itself at a right angle to the sun's rays. To a viewer on the earth at the subsolar point, the sun is directly overhead. But as we move away from the subsolar point toward either pole, the earth's curved surface becomes turned at an angle with respect to the sun's rays. To the earthbound viewer at a new latitude, the sun appears to be nearer the horizon. When the circle of illumination is reached, the sun's rays are parallel with the surface. That is, the sun is viewed just at the horizon. Thus, the angle of the sun in the sky at a particular location depends on the latitude of the location, the time of day, and the time of year.

The Path of the Sun in the Sky

How does the angle of the sun vary during the day? The angle depends on the sun's path in the sky. When the sun is high above the horizon—near noon, for example—the sun's angle is greater and so insolation will be greater. Figure 2.6 shows the sun's daily path in the sky and how this path changes from season to season. The diagram is drawn for latitude 40° N and is typical of conditions found in midlatitudes in the northern hemisphere—for example, at New York or Denver. The diagram shows a small area of the earth's surface bounded by a circular horizon. This is the way things appear to an observer standing on a wide plain. The earth's surface appears flat, and the sun seems to travel inside a vast dome in the sky.

At equinox, the sun rises directly to the east and sets directly to the west. The noon sun is positioned at an angle of 50° above the horizon in the southern sky. The sun is above the horizon for exactly 12 hours, as shown by the hour numbers on the sun's path. At summer solstice, the sun's path rises much higher in the sky—at noon it will be $73\frac{1}{2}°$ above the horizon. In addition, the sun is above the horizon for about 15 hours. At win-

What about the outflows? These take the form of (1) reflected shortwave radiation and (2) emitted longwave radiation. The earth as a planet reflects about one-third of the solar radiation that it receives, or about 1.81 $\times 10^{14}$ kW ÷ 3 = 0.60 $\times 10^{14}$ kW. The remaining two-thirds, 1.21 $\times 10^{14}$ kW, is absorbed by the earth and atmosphere and ultimately emitted as longwave radiation.

A simple diagram of this radiant energy flow system is shown in the figure at right. Energy flows are shown by arrows. The earth and its atmosphere receive an inflow of solar shortwave radiation. Within the earth and atmosphere, some of the energy flow is transformed from shortwave radiation to longwave radiation. Both shortwave and longwave radiation flow outward to space.

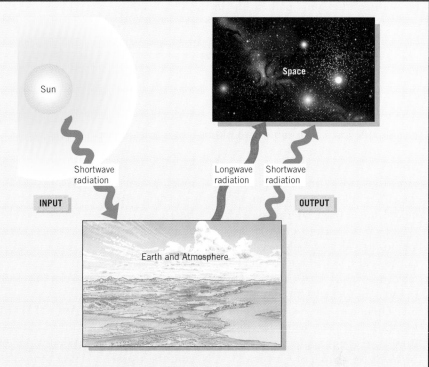

Global radiant energy flow system for the earth and its atmosphere. The input is the flow of shortwave radiation from the sun. Outputs are flows of reflected shortwave radiation and emitted longwave radiation.

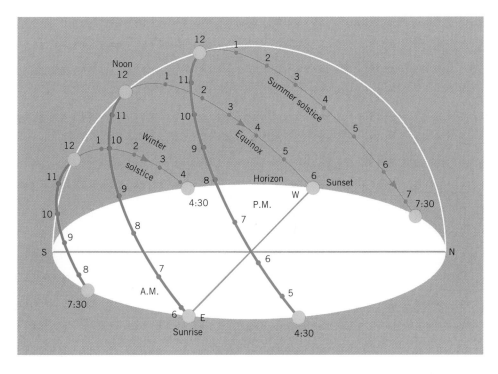

Figure 2.6 The sun's path in the sky changes greatly in position and height above the horizon from summer to winter. This diagram is for a location at lat. 40° N (New York or Denver). At the winter solstice, the sun is low in the sky, and the period of daylight is short. At the summer solstice, the sun is high, and the daylight period is long.

ter solstice, the sun's path is low in the sky, reaching only $26\frac{1}{2}°$ above the horizon, and the sun is visible for only about 9 hours. Clearly, the solar radiation reaching the surface is much greater during the day at summer solstice than at winter solstice.

Insolation Through the Year

When daily insolation values are plotted for a full year, they form a wavelike curve on a graph, as shown in Figure 2.7. The curve for latitude 40° N shows the features just discussed—a much greater daily insolation rate at the summer solstice (June) and much lower daily insolation rate at the winter solstice (December).

Not every latitude shows a simple progression from low to high to low daily insolation through the course of the year. Notice that the equator has two periods of maximum daily insolation. These periods correspond to the equinoxes, when the sun is overhead at the equator. There are also two minimum periods corresponding to the solstices, when the subsolar point moves farthest north and south from the equator. All latitudes between the tropic of cancer ($23\frac{1}{2}°$ N) and the tropic of capricorn ($23\frac{1}{2}°$ S) have two maximum and minimum values. However, as either tropic is approached in a poleward direction, the two maximum periods get

closer and closer in time, and then merge into a single maximum.

Poleward of the two tropics is a single daily insolation cycle with a maximum at one solstice and a minimum at the other. At the arctic circle ($66\frac{1}{2}°$ N), daily insolation is reduced to zero on the day of the winter solstice. At latitudes farther poleward, this seasonal period of no insolation increases from 1 day to 182 days.

From this discussion, we can see that the seasonal pattern of daily insolation is directly related to latitude. This pattern is important because it is the driving force for the annual cycle of climate. Nowhere on earth is insolation constant from day to day. Even at the equator, Figure 2.7 shows that there are significant seasonal differences in the solar energy available to warm the earth and atmosphere. Poleward of the tropics, the large differences produce the large variations in temperature from winter to summer.

Annual Insolation by Latitude

What is the effect of latitude on annual insolation—the rate of insolation averaged over an entire year? Figure 2.8 shows two curves of annual insolation by latitude— one for the actual case of the earth's axis tilted at $23\frac{1}{2}°$, the other for an earth with an untilted axis. Let's look first at the real case of a tilted axis. Annual insolation varies smoothly from the equator to the pole. As we might expect, annual insolation is greater at lower latitudes. But the high latitudes still receive a considerable amount of solar radiation—the annual insolation value

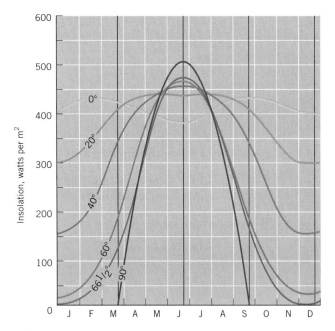

Figure 2.7 Insolation curves at various latitudes in the northern hemisphere at the top of the atmosphere. Black lines mark the equinoxes and solstices. Latitudes between the equator (0°) and tropic of cancer ($23\frac{1}{2}°$ N) show two maximum values; others show only one. Poleward of the arctic circle ($66\frac{1}{2}°$ N), insolation is zero for at least some period of the year. (Copyright © A. N. Strahler.)

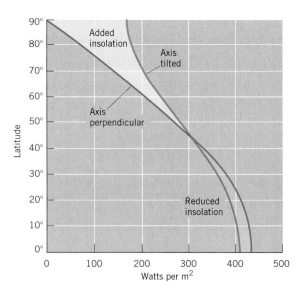

Figure 2.8 Total annual insolation from equator to pole for the earth at the top of the atmosphere. The red line shows the actual curve. The blue line shows what the annual insolation would be if the axis were perpendicular to the plane of the ecliptic.

at the pole is about 40 percent of the value at the equator.

The case of the untilted axis—that is, for which the earth's axis is perpendicular to the plane of the ecliptic—is quite different. In this case, there are no seasons. The sun traces the same path in the sky every day of the year at every location. Under these circumstances, annual insolation changes from a high value at the equator, where the sun is directly overhead at noon every day, to zero at the poles, where the sun is always just on the horizon. The curve of annual insolation thus shows a progression from a value near 450 W/m^2 at the equator to zero at 90° latitude.

The comparison of the two cases shows that the tilting of the earth's axis redistributes a very significant portion of the earth's insolation from the equatorial regions toward the poles. Although the pole is in the dark for six months of the year, it still receives close to half the amount of solar radiation received at the equator. Without a tilted axis, our planet would be quite a different place!

WORLD LATITUDE ZONES

The seasonal pattern of insolation can be used as a basis for dividing the globe into broad latitude zones (Figure 2.9). The zone limits shown in the figure should not be taken as absolute and binding, however. Rather, this system of names is a convenient way to identify general world geographic belts throughout this book.

The *equatorial zone* encompasses the equator and covers the latitude belt roughly 10° north to 10° south. The sun provides intense insolation throughout the year, and days and nights are of roughly equal length within this zone. Spanning the tropics of cancer and capricorn are the *tropical zones*, ranging from latitudes 10° to 25° north and south. A marked seasonal cycle exists in these zones, combined with a large total annual insolation.

Moving toward the poles from each of the tropical zones are transitional regions called the *subtropical zones*. For convenience, we assign these zones the latitude belts 25° to 35° north and south. At times we may extend the label "subtropical" a few degrees farther poleward or equatorward of these parallels.

The *midlatitude zones* are next, lying between 35° and 55° north and south latitude. In these belts, the sun's height in the sky shifts through a wide range annually. Differences in day length from winter to summer are also large. Thus, seasonal contrasts in insolation are quite strong. In turn, these regions can experience a large range in annual surface temperature.

Bordering the midlatitude zones on the poleward side are the *subarctic zone* and *subantarctic zone*, 55° to

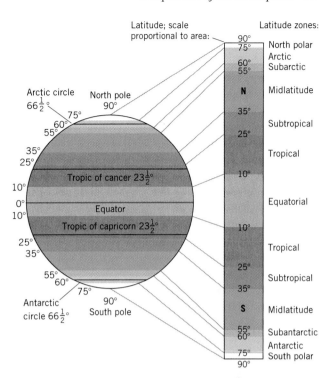

Figure 2.9 A geographer's system of latitude zones. These are based on the seasonal patterns of insolation observed over the globe.

60° north and south latitudes. Astride the arctic and antarctic circles, 66½° north and south latitudes, lie the *arctic zone* and *antarctic zone*. These zones have an extremely large yearly variation in day lengths, yielding enormous contrasts in insolation from solstice to solstice.

The *polar zones*, north and south, are circular areas between about 75° latitude and the poles. Here the polar regime of a six-month day and six-month night is predominant. These zones experience the greatest seasonal contrasts of insolation of any area on earth.

COMPOSITION OF THE ATMOSPHERE

A main function of this chapter is to explain the flows of energy within the atmosphere and between the atmosphere and the earth's surface. However, an understanding of this topic first requires some knowledge of basic facts about the atmosphere.

The earth's atmosphere consists of *air*—a mixture of various gases surrounding the earth to a height of many kilometers. This envelope of air is held to the earth by gravitational attraction. Almost all the atmosphere (97 percent) lies within 30 km (19 mi) of the earth's surface. The upper limit of the atmosphere is at a height of approximately 10,000 km (about 6000 mi) above the earth's surface, a distance approaching the diameter of the earth itself.

From the earth's surface upward to an altitude of about 80 km (50 mi), the chemical composition of air is highly uniform in terms of the proportions of its gases. Pure, dry air consists largely of nitrogen, about 78 percent by volume, and oxygen, about 21 percent (Figure 2.10). Other gases account for the remaining 1 percent.

Nitrogen in the atmosphere exists as a molecule consisting of two nitrogen atoms (N_2). Nitrogen gas does not enter easily into chemical union with other substances and can be thought of mainly as a neutral substance. Very small amounts of nitrogen are extracted by soil bacteria and made available for use by plants. Otherwise, it is largely a "filler," adding inert bulk to the atmosphere.

In contrast to nitrogen, *oxygen* gas (O_2) is highly active chemically. It combines readily with other elements in the process of oxidation. Combustion of fuels represents a rapid form of oxidation, whereas certain forms of rock decay (weathering) represent very slow forms of oxidation. Living tissues require oxygen to convert foods into energy.

These two main component gases of the lower atmosphere are perfectly diffused so as to give the pure, dry air a definite set of physical properties, as though it were a single gas.

The remaining 1 percent of dry air is mostly argon, an inactive gas of little importance in natural processes. In addition, there is a very small amount of *carbon dioxide* (CO_2), amounting to about 0.033 percent. This gas is of great importance in atmospheric processes because of its ability to absorb radiant heat passing through the atmosphere from the earth. Carbon dioxide thus adds to the warming of the lower atmosphere. Carbon dioxide is also used by green plants in the photosynthesis process. During photosynthesis, CO_2 is converted into chemical compounds that build up the plant's tissues, organs, and supporting structures.

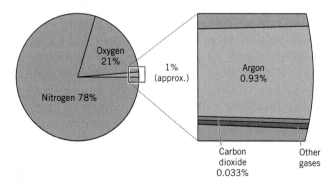

Figure 2.10 Components of the lower atmosphere. Values show percentage by volume for dry air. Nitrogen and oxygen form 99 percent of our air, with other gases, principally argon and carbon dioxide, accounting for the final 1 percent.

Another important component of the atmosphere is *water vapor*, the gaseous form of water. Individual water molecules in the form of vapor are mixed freely throughout the atmosphere, just like the other atmospheric component gases. Unlike the other component gases, however, water vapor can vary highly in concentration. Usually, water vapor makes up less than 1 percent of the atmosphere. Under very warm, moist conditions, as much as 2 percent of the air can be water vapor. Since water vapor, like carbon dioxide, is a good absorber of heat radiation, it also plays a major role in warming the lower atmosphere.

The atmosphere also contains dust and tiny floating particles that absorb and scatter radiation. These are discussed more fully in Chapter 4.

Ozone in the Upper Atmosphere

Another small, but important, constituent of the atmosphere is ozone. **Ozone** (O_3) is found mostly in the upper part of the atmosphere, in a layer termed the *stratosphere.* This layer of the atmosphere lies about 14–50 km (9–31 mi) above the surface, and is described in more detail in the next chapter. Ozone in the stratosphere absorbs ultraviolet radiation from the sun as this radiation passes through the atmosphere, thus shielding the life layer from its harmful effects.

Ozone is most concentrated in a layer that begins at an altitude of about 15 km (9 mi) and extends to about 55 km (34 mi) above the earth. It is produced by gaseous chemical reactions in the stratosphere. The details of these chemical reactions are complicated, but in summary they are quite simple. An oxygen molecule (O_2) absorbs ultraviolet light energy and splits into two oxygen atoms (O + O). A free oxygen atom (O) then combines with an O_2 molecule to form ozone, O_3. Once formed, ozone can also be destroyed. Like O_2, ozone absorbs ultraviolet radiation, and it splits to form O_2 + O. Furthermore, two oxygen atoms (O + O) can join to form O_2. The net effect is that ozone (O_3), molecular oxygen (O_2), and atomic oxygen (O) are constantly formed, destroyed, and reformed in the ozone layer, absorbing ultraviolet radiation with each transformation.

The absorption of ultraviolet radiation by the ozone layer protects the earth's surface from this damaging form of radiation. If the concentration of ozone is reduced, fewer transformations among O, O_2, and O_3 occur, and so ultraviolet absorption is reduced. If solar ultraviolet radiation were to reach the earth's surface at full intensity, all bacteria exposed on the earth's surface would be destroyed, and animal tissues would be severely damaged. The presence of the ozone layer is thus an essential protection in maintaining a viable environment for life on this planet.

Threats to the Ozone Layer

A serious threat to the ozone layer is posed by the release of *chlorofluorocarbons*, or *CFCs*, from the earth's surface into the atmosphere. CFCs are synthetic industrial chemical compounds containing chlorine, fluorine, and carbon atoms. Before 1976, when the Environmental Protection Agency in the United States issued a ban on aerosol spray cans charged with CFCs, many households used such aerosols. CFCs are still in wide use as cooling fluids in refrigeration systems. When these appliances are disposed of, their CFCs are released to the air.

Molecules of CFCs in the atmosphere are very stable close to the earth. They move upward by diffusion without chemical change until they eventually reach the ozone layer. As these compounds absorb ultraviolet radiation, they are decomposed, and chlorine oxide molecules are formed. The chlorine oxide molecules in turn attack molecules of ozone, converting them in large numbers into ordinary oxygen molecules by a chain reaction. In this way, the ozone concentration within the stratospheric ozone layer is reduced, and so there are fewer molecules to absorb ultraviolet energy in the cycle of transformations among O, O_2, and O_3. The intensity of ultraviolet radiation reaching the earth's surface can be increased.

Other gaseous molecules can act to reduce the concentrations of ozone in the stratosphere as well. Nitrogen oxides, bromine oxides, and hydrogen oxides are examples. As with chlorine oxides, the concentrations of these molecules in the stratosphere are influenced by human activity.

Not all threats to the ozone layer can be attributed to human activity. Volcanic dust, inserted into the stratosphere, also can act to reduce ozone concentrations. The June 1991 eruption of Mount Pinatubo, in the Philippines, reduced global ozone in the stratosphere by 4 percent during the following year, with reductions over midlatitudes of up to 9 percent.

How might declining ozone concentrations affect the earth? A marked increase in the incidence of skin cancer in humans is one of the predicted effects. Other possible effects include reduction of crop yields and the death of some forms of aquatic life.

Declining Ozone and the Ozone Hole

As studies of the ozone layer based on satellite data accumulated during the 1980s, a substantial decline in the total global ozone was noted. By the late years of the decade, scientists had agreed that the decrease in the ozone layer was escalating far faster than had been predicted.

Compounding the problem of global ozone decrease was the discovery in the mid-1980s of a "hole" in the ozone layer over the continent of Antarctica. Here, seasonal thinning of the ozone layer occurs during the early spring of the southern hemisphere, and ozone reaches a minimum during the month of October.

The thinning occurs after the formation of a polar vortex in the stratosphere during the winter period. This vast whirlpool of wind traps the air it contains and keeps this air out of the sun during the months of the long polar night. The air in the vortex eventually becomes very cold, and clouds of crystalline water (ice) or other water-containing compounds form within it. The crystals are important because they provide a surface on which chemical reactions can take place. One of these reactions allows chlorine to be converted from a stable form to chlorine oxide, ClO, which is highly reactive in the presence of sunlight. As the southern hemisphere spring approaches, the polar vortex slowly becomes illuminated by the sun, and the chlorine oxide reacts with ozone. Ozone concentrations are reduced, and the ozone hole is formed.

Figure 2.11 presents a satellite-derived map of ozone concentration over the south pole on October 2, 1994. The ozone hole is clearly visible in the image. A polar vortex forms in the northern hemisphere as well, but is much weaker than in the southern hemisphere and is less stable. As a result, no early-spring ozone hole is normally observed in the Arctic.

Although the most dramatic reductions in stratospheric ozone levels have occurred during the antarctic spring, reductions have been noted elsewhere as well—including the northern hemisphere. In 1992, researchers reanalyzed 12 years of weather satellite data and compared the results with ground measurements. They concluded that at various points in the northern hemisphere the ozone shield had thinned by as much as 6 percent. Damage to the ozone layer was noted as far south as Florida, and some loss persisted from the early spring into the early summer. In another study, ozone levels over North America dropped by about a half-percent per year between 1979 and 1990. By 1993, the total loss had reached 7.5 percent. Figure 2.12 is a satellite-derived map of the change in northern hemisphere ozone levels between March 1979 and March 1995. It shows a region of strong ozone depletion near the pole and a band of lesser depletion in the midlatitudes, including the United States.

We have seen that as the global ozone layer thins, the rate of incoming ultraviolet solar radiation should increase. In fact, scientists estimate that for each 1 percent decrease in global ozone, ultraviolet radiation should increase by 2 percent. Measurements of ultraviolet radiation reaching the earth's surface, however, have not always shown increases in recent years. Several

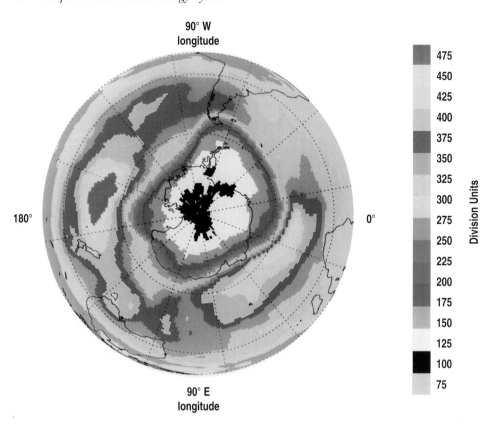

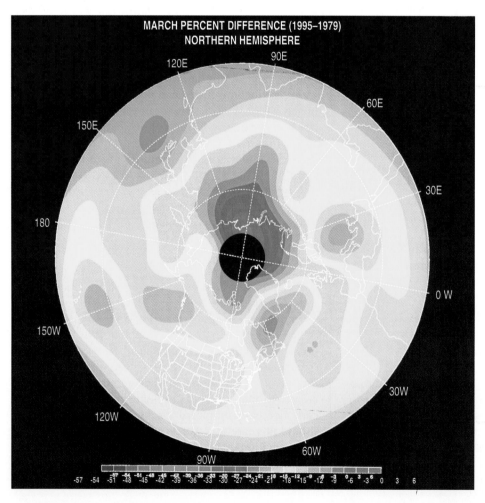

Figure 2.11 Map of ozone concentration over the south pole on October 2, 1994. Colors indicate ozone levels, measured on a special scale (Dobson units). The average global concentration of ozone on this scale is about 300. High ozone concentrations are shown in red and orange. Purple, pink, and white values are lowest. Outlines of the continents are shown in black. The ozone hole over Antarctica shows very clearly in the center of the image. Produced from data of the Nimbus-7 satellite by NASA and NOAA researchers.

Figure 2.12 Map of total ozone difference between March 1995 and March 1979. Areas where values for March 1995 are at least 20 percent below values for March 1979 are shown in greens and blues. The black region at the pole is an area where no data are available. (This figure was provided by the NOAA National Environmental Satellite, Data, and Information Service and the National Weather Service/National Center for Environmental Prediction.)

other factors affect year-to-year changes in recorded ultraviolet radiation, including variations in average cloud cover and air pollution levels. Because of such factors, we need to use caution before linking any increase in ultraviolet radiation to ozone depletion.

Responding to the global threat of ozone depletion and its anticipated impact on the biosphere, 23 nations endorsed a United Nations plan in 1987 for cutting global CFC consumption by 50 percent by the year 1999. Since then, international agreements on CFC control have been strengthened. Although the manufacture and consumption of CFCs have now been halted or greatly reduced, the ozone layer will continue to thin for some time as CFCs already in the lower atmosphere continue to work their way upward to the stratosphere. Still, the CFC treaty is something of an environmental success story. Because of overwhelming public support for these measures, implementation has been quick. Global cooperation is resulting in a global solution to a global problem. 🐾

SENSIBLE HEAT AND LATENT HEAT TRANSFER

Thus far, we have discussed energy in the form of shortwave and longwave radiation. However, energy can be transported in other ways. Two of them—sensible heat and latent heat transfer—are extremely important in discussing the global energy balance.

The most familiar form of heat storage and transport is known as **sensible heat**—the quantity of heat held by an object that can be sensed by touching or feeling. This kind of heat is measured by a thermometer. When the temperature of an object or a gas increases, its store of sensible heat increases. When two objects of unlike temperature contact each other, heat energy moves by conduction from the warmer to the cooler. This type of heat flow is referred to as **sensible heat transfer**.

Unlike sensible heat, **latent heat**—or hidden heat—cannot be measured by a thermometer. It is heat that is taken up and stored when a substance changes state from a liquid to a gas or from a solid directly to a gas. Water provides an important example. For liquid water to make the transition to water vapor, energy is required. As the water evaporates from liquid to gas, it draws up heat from the surroundings. This is the latent heat. It is carried in the form of the fast, random motion of the free water vapor molecules as they enter the gaseous state during the process of evaporation. (*Working It Out 4.1 • Energy and Latent Heat*, in Chapter 4, provides more information about the amount of heat required or released in changes of state of water.)

Although the latent heat of water vapor in air cannot be measured by a thermometer, energy is stored there just the same. When water vapor turns back to a liquid, the latent heat is released. In the earth–atmosphere system, **latent heat transfer** occurs when water evaporates from a moist land surface or open water surface. This process transfers latent heat from the surface to the atmosphere. On a global scale, latent heat transfer by movement of moist air provides a very important mechanism for transporting large amounts of energy from one region of the earth to another.

THE GLOBAL ENERGY SYSTEM

The flow of energy from the sun to the earth's atmosphere and surface and then back out into space is a complex flow system with a number of important pathways. The system involves not only radiation flows, but also the storage and transport of energy as sensible and latent heat. As we will see, the atmosphere both absorbs and scatters radiation flows passing through it, no matter whether they are moving downward or upward. This property has the important effect of trapping longwave radiation at the earth's surface, making the earth warmer than we might otherwise expect. Because all incoming radiation must be accounted for as it flows from one location to another and changes form, we can develop a budget of energy flows of energy among the sun, atmosphere, surface, and space. We will investigate this complex system in more detail to provide the basis for the study of weather and climate in later chapters.

Insolation Losses in the Atmosphere

Consider the flow of solar radiation through the atmosphere on its way toward the surface. As the shortwave radiation penetrates the atmosphere, its energy is absorbed or diverted in various ways. This is shown in Figure 2.13, which gives typical values for losses of incoming solar energy to the nearest 5 percent.

At an altitude of 150 km (93 mi), the solar radiation spectrum possesses almost 100 percent of its original energy. However, by the time rays penetrate to an altitude of 88 km (55 mi), absorption of X rays is nearly complete, and some of the ultraviolet radiation has been absorbed as well.

As radiation moves through deeper and denser atmospheric layers, gas molecules cause some visible light rays to be turned aside in all possible directions, a process known as **scattering**. Dust and other particles in the air cause further scattering. Scattered radiation is unchanged, except for the direction of its path. Scattered radiation moving in all directions through the atmosphere is known as *diffuse radiation*. Because scattering occurs in all directions, some energy is sent back into outer space, and some flows down to the earth's surface. When the scattering turns radiation back to

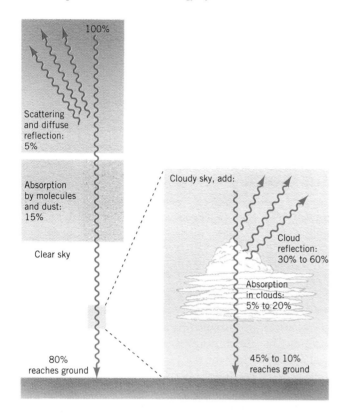

Figure 2.13 Losses of incoming solar energy are much lower with clear skies (left) than with cloud cover (right). (Copyright © A. N. Strahler.)

space, we call this process *diffuse reflection*. Under clear sky conditions, this diffuse reflection sends about 5 percent of the incoming solar radiation back to space.

Another form of energy loss in the atmosphere is *absorption*. In this process, solar rays strike gas molecules and dust, and their energy is absorbed. This accounts for about 15 percent of the incoming solar radiation. Thus, when skies are clear, reflection and absorption combined total about 20 percent, leaving as much as 80 percent of the solar radiation to reach the ground.

Both carbon dioxide and water vapor are capable of directly absorbing some wavelengths of solar radiation. Carbon dioxide is present fairly uniformly in the air, but, as we noted earlier, the water vapor content of the air can vary greatly. Thus, absorption can vary from one global environment to another. Absorption results in a rise in air temperature. In this way, incoming solar radiation causes some direct heating of the lower atmosphere.

The presence of clouds can greatly increase the amount of incoming solar radiation reflected back to space. The bright white surfaces of thick, low clouds are extremely good reflectors of shortwave radiation. Cloud reflection can account for a direct turning back to space of 30 to 60 percent of incoming radiation (Figure 2.13). Clouds also absorb radiation, perhaps as

much as 5 to 20 percent. Under conditions of a heavy cloud layer, as little as 10 percent of incoming solar radiation can actually reach the ground.

Albedo

The percentage of shortwave radiant energy scattered upward by a surface is termed its **albedo**. For example, a surface that reflects 40 percent of the shortwave radiation it receives has an albedo of 40 percent. Albedo is an important property of a surface because it determines how fast the surface heats up when exposed to insolation. A surface with a high albedo, such as snow or ice (45 to 85 percent), reflects much or most of the solar radiation and so heats up slowly. A surface with a low albedo, such as black pavement (3 percent), absorbs nearly all the incoming insolation and so heats up quickly. The albedo of a water surface varies with the angle of incoming radiation. It is very low (2 percent) for nearly vertical rays. However, when the sun shines on a calm water surface at a low angle, much of the radiation is directly reflected as sun glint, producing a higher albedo. For fields, forests, and bare ground, albedos are of intermediate value, ranging from as low as 3 percent to as high as 25 percent.

Certain orbiting satellites are equipped with instruments to measure the energy levels of shortwave and longwave radiation above the top of the atmosphere. Both incoming radiation from the sun and outgoing radiation from the atmosphere and earth's surfaces below are measured. Data from these satellites have been used to estimate the earth's average albedo. This albedo includes reflection by the earth's atmosphere as well as its surfaces, so it includes upward scattering by clouds, dust, and atmospheric molecules. The albedo values obtained in this way vary between 29 and 34 percent. This means that the earth–atmosphere system directly returns back to space slightly less than one-third of the solar radiation it receives.

Counterradiation and the Greenhouse Effect

The ground can only radiate longwave energy upward, but the atmosphere radiates longwave energy in all directions—both upward to space, where it is lost, and downward to the ground. We have already seen that water vapor and carbon dioxide can directly absorb some wavelengths of shortwave radiation. In fact, water vapor and carbon dioxide in the atmosphere also absorb much of the longwave radiation emitted upward from the earth's surface. When they emit this radiation, a part is radiated back down toward the earth surface. This return of upwelling radiation is called **counterradiation**. Thus, the lower atmosphere, with its longwave-absorbing gases, acts like a blanket that returns heat to the earth. This mechanism helps to keep surface tem-

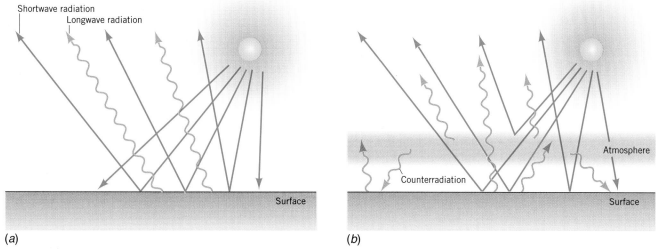

Figure 2.14 Effect of the atmosphere on the surface energy budget. Without an atmosphere (*a*), some solar radiation is reflected by the surface back to space, while the remainder is absorbed by the surface. The surface also emits longwave radiation to space. In (*b*), the atmosphere absorbs much upwelling longwave energy and emits some back to the surface as counterradiation. The atmosphere also scatters some shortwave radiation back to space.

peratures from dropping excessively during the night or in winter at middle and high latitudes. Cloud layers are even more important than carbon dioxide and water vapor in producing a blanketing effect to retain heat in the lower atmosphere, since clouds are excellent absorbers and emitters of longwave radiation.

The principle of counterradiation is shown in Figure 2.14. In part (*a*), no atmosphere is present. Solar energy is either absorbed or reflected upward by scattering, and longwave energy is emitted by the surface directly to space. In part (*b*), an atmospheric layer absorbs upwelling longwave radiation as well as a fraction of incoming shortwave radiation. The portion that is absorbed and reemitted downward is the counterradiation.

A similar principle is used in greenhouses and in homes with large windows to entrap solar heat. Here, the glass windows permit entry of shortwave energy but block the exit of longwave energy. This phenomenon is even better demonstrated by the intense heating of air in a closed automobile left parked in the sun. We use the expression **greenhouse effect** to describe this atmospheric heating principle.

Global Energy Budgets of the Atmosphere and Surface

Thus far we have described a number of important pathways by which incoming solar radiation is scattered, absorbed, and reradiated. It is an important principle of physics that, except for nuclear reactions, energy is neither created nor destroyed. It is therefore possible to follow the initial stream of solar energy and account for its diversion into different system pathways and its conversion into different energy forms. A full

accounting of all the important energy flows among the sun, atmosphere, surface, and space forms the global energy budget of the atmosphere and surface. Like a household budget, it must balance, at least over the long term. *Focus on Systems 2.4 • Energy and Matter Budgets* discusses energy and matter budgets more fully.

The global energy budget for the earth's surface and atmosphere is diagrammed in Figure 2.15. We begin with a discussion of shortwave radiation, shown in part (*a*) of the figure. Since shortwave radiation comes from the sun, we are concerned here only with the downward flow of solar radiation and its fate.

As shown on the left, reflection by molecules and dust, clouds, and the surface (including the oceans) totals 31 percentage units. This is the albedo of the earth–atmosphere system. The right side of part (*a*) shows values for absorption in the atmosphere. The combined losses through absorption by molecules, dust, and clouds average 21 percentage units. With 31 percentage units of the incoming solar energy flow reflected and 21 percentage units absorbed in the atmosphere, 48 units are left. This flow is absorbed by the earth's land and water surfaces. We have now accounted for all of the incoming flow of insolation.

Part (*b*) of Figure 2.15 shows the components of outgoing longwave radiation for the earth's surface, the atmosphere, and the planet as a whole. The large arrow on the left shows total longwave radiation leaving the earth's land and ocean surface. The long, thin arrow indicates that 6 percentage units are lost to space, while 107 units are absorbed by the atmosphere. The loss is therefore equivalent to 113 percentage units.

This means that the surface emits 113 percent of the total incoming solar radiation! How can this be cor-

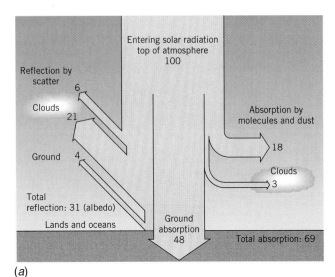

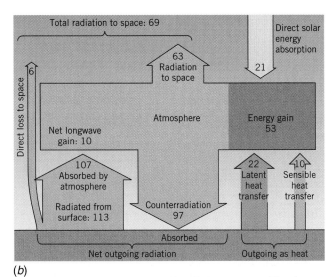

(a)   (b)

Figure 2.15 Diagram of the global energy balance. Values are percentage units based on total insolation as 100. The fate of incoming solar radiation is shown in (*a*), while (*b*) shows longwave energy flows occurring among the surface, atmosphere, and space. The transfers of latent heat, sensible heat, and direct solar absorption that balance the budget for earth and atmosphere are shown on the right side of (*b*).

rect? To answer this question, note that the surface receives a flow of 97 units of longwave counterradiation from the atmosphere. This will be in addition to the 48 units of shortwave radiation that are absorbed (upper figure). Thus, the surface receives 97 (longwave) + 48 (shortwave) = 145 units in all. Thus, a loss of 113 units by longwave radiation out of a gain of 145 units is certainly possible.

On the far right of Figure 2.15*b* are two smaller arrows. These show the flow of energy away from the surface as latent heat and sensible heat. Evaporation of water from moist soil or ocean transfers latent heat from the surface to the atmosphere, as explained earlier. Sensible heat transfer occurs when heat is directly conducted from the surface to the adjacent air layer. When air is warmed by direct contact with a surface and rises, it takes its heat along with it. In this process, heat flows from the surface to the atmosphere.

These last two heat flows are not part of the radiation balance, since they are not in the form of radiation. However, they are a very important part of the total energy budget of the surface, which includes all forms of energy. Taken together, these two flows account for 32 units leaving the surface. With their contribution, the surface energy balance is complete. Total gains are 97 (longwave) + 48 (shortwave) = 145. Total losses are 113 (longwave) + 22 (latent heat) + 10 (sensible heat) = 145.

Like the earth's surface, the atmosphere also emits longwave radiation (see Figure 2.14*b*). This radiation is divided into two parts. One part is directed into space (63 units), while the other part is counterradiation directed back to the earth's surface (97 units). The total of this emitted radiation is 63 + 97 = 160 percentage

units. This figure seems absurdly large until we note that the atmosphere receives not only 107 units of longwave radiation but also 22 units of latent heat, 10 units of sensible heat from the surface, and 21 units of shortwave radiation by direct absorption. These units total to 160. So the energy budget balances for the atmosphere as well as for the surface.

This analysis helps us to understand the vital role of the atmosphere in trapping heat though the greenhouse effect. Because of counterradiation, the surface receives 145 units and radiates away 113 of them. In order to do so, the surface must be warm indeed. Without the counterradiation of the atmosphere, the surface would have only 48 units of absorbed shortwave radiation to lose. The temperature needed to radiate so few units is well below freezing. So without the greenhouse effect, our planet would be a very cold, forbidding place.

NET RADIATION, LATITUDE, AND THE ENERGY BALANCE

Solar energy is intercepted by our planet, and because some of it is absorbed, the heat level of our planet tends to rise. At the same time, our planet radiates energy into outer space, a process that tends to reduce its level of heat energy. Over time, as we have seen, these flows must balance for the earth as a whole. However, incoming and outgoing radiation flows for any given surface do not have to be in balance. At night, for example, there is no incoming radiation, yet the earth's surface and atmosphere still emit outgoing radiation.

Focus on Systems 2.4 • Energy and Matter Budgets

An important tool in the understanding of systems is a *budget*—an accounting of the energy and matter flows that enter, move within, and leave a system. It relies on the principle that matter and energy are conserved. So when energy or matter flow into a system at a certain rate, we should be able to account for the entire flow in pathways within the system and in flows leaving the system.

As a familiar example of a flow system, consider a checking account. Money flows into the account as you make deposits and flows out as you make withdrawals or write checks. Given a monthly salary of $2000, a budget might record $500 spent for rent, $300 for food, and so forth.

Suppose that energy or matter flows into a system at a fixed rate but flows out of a system at a lesser rate. This can happen if energy or matter is flowing into storage inside the system. For a checking account, if you deposit more than you withdraw, the account balance grows. Your budget will show this as inputs

exceed outflows. Conversely, if outflows exceed inflows, energy or matter must flow out of storage to make up the difference. If you withdraw more than you deposit, the account balance will shrink. The budget will show that your stock of money is decreasing.

Recall from the Introduction that a flow system in equilibrium is one in which the flows in its various pathways tend toward a steady value. Unless you are accumulating money in your account, or spending it down over a period of time, your checking account is probably like this. Each month, inflow and output are about the same, and so the amount of money in the account stays about the same. At equilibrium, a budget will show all the flows entering a system to be balanced by flows leaving the system.

Energy and matter budgets are complicated by the fact that energy and matter can change forms inside a system. For example, radiant energy becomes sensible heat when it is absorbed by matter, as when the sun warms the earth's surface.

When the surface cools, energy is transformed from sensible heat to radiant energy. Matter changes form when it changes state (that is, changes among solid, liquid, or gas) and when it enters into chemical reactions. As you may know from studying chemistry, the products of a chemical reaction may be very different—in state, color, or density, for example—from the reactants. However, the atoms of the reactants are fully conserved in the products, and no matter is lost. Because energy and matter change form, we must account in our budget for flow in all forms.

One of the most important reasons for constructing a budget for a flow system is that it directs us toward finding out what happens to all the matter and energy entering the system. In this way, new pathways of flow and new conversions of matter and energy to other forms are uncovered, leading to a better understanding of the workings of the entire flow system.

In any one place and time, then, more radiant energy is being gained than lost, or vice versa.

Net radiation is the difference between all incoming radiation and all outgoing radiation. In places where radiant energy is flowing in faster than it is flowing out, net radiation is a positive quantity, providing a surplus. In other places, where radiant energy is flowing out faster than it is flowing in, net radiation is a negative quantity, yielding a radiant energy deficit. Our analysis of the earth's radiation balance has already shown that for the entire earth and atmosphere as a unit, the net radiation is zero on an annual basis.

In Figure 2.7, we saw that solar energy input varies strongly with latitude. What is the effect of this difference on net radiation? A study of the net radiation profile spanning the entire latitude range 90° N to 90° S will answer this question. In this analysis, we will use yearly averages for each latitude, so that the effect of seasons is concealed.

The lower part of Figure 2.16 presents a global profile of net radiation from pole to pole. Between about lat. 40° N and lat. 40° S there is a net radiant energy gain labeled "energy surplus." In other words, incoming solar radiation exceeds outgoing longwave radiation throughout the year as long as there are other means to carry away the excess heat energy. Similarly, poleward of 40° N and 40° S, the net radiation is negative and is labeled "deficit"—meaning that outgoing longwave radiation exceeds incoming shortwave radiation. This can happen only if energy is brought into these middle and high latitudes from the equatorial and tropical zones.

What mechanism brings energy from the lower latitudes poleward? Two processes are responsible. First, the circulation pattern of the oceans brings warm water poleward, moving heat energy across latitude boundaries in a sensible heat transfer. Second, the movement of warm, moist air poleward carries latent heat energy

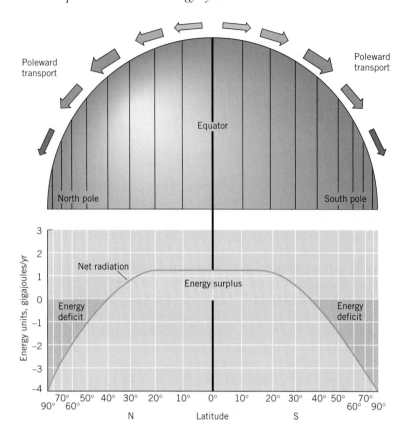

Figure 2.16 Annual surface net radiation from pole to pole. Where net radiation is positive, incoming solar radiation exceeds outgoing longwave radiation. There is an energy surplus, and so energy moves poleward as latent heat and sensible heat.

to higher latitudes in a latent heat transfer. These two energy transport processes taken together are called **poleward heat transport**, since the transport moves heat energy from lower to higher latitudes. The upper part of Figure 2.16 diagrams this flow.

If you carefully examine the lower part of Figure 2.16, you will find that the area on the graph labeled "surplus" is equal in size to the combined areas labeled "deficit." The matching of these areas confirms the fact that net radiation for the whole earth is zero and that global incoming shortwave radiation exactly balances global outgoing longwave radiation.

EYE ON THE ENVIRONMENT: SOLAR POWER

Our earth intercepts solar energy at the rate of $1\frac{1}{2}$ quadrillion megawatt-hours per year. (A megawatt is one million—10^6—watts.) This quantity of energy is about 28,000 times as much as all the energy presently being consumed each year by human society. Thus, an enormous source of energy is near at hand waiting to be used.

The simplest form of solar heating involves the use of large glass panes to admit sunlight into a room—the greenhouse principle. Practical solar heating of interior building space and hot-water systems makes use of solar collectors placed on the roof of the building. Solar power can also be used to produce electricity—ei-

ther directly, using solar cells, or indirectly, by focusing sunlight on a steam boiler using movable mirrors called heliostats.

Note that the sun also powers the natural motion of wind and water and that these moving fluids can be harnessed for power generation as well. We will touch on wind, water, and wave power in later chapters.

Although our analysis of the earth's energy balance is not yet complete, it is clear that the balance is a sensitive one involving a number of factors that determine how energy is transmitted and absorbed. Has industrial activity already altered the components of the planetary radiation balance? An increase in carbon dioxide will increase the absorption of longwave radiation by the atmosphere, enhancing the greenhouse effect. Will this change cause a steady and permanent rise in average atmospheric temperature?

An increase in atmospheric particles at upper levels of the atmosphere will increase the scattering of incoming shortwave radiation and thus reduce the shortwave energy available to warm the surface. On the other hand, increased dust content at low levels will act to absorb more longwave radiation and so raise the surface temperature. Has either change occurred?

Human habitation, through cultivation and urbanization, has profoundly altered the earth's land surfaces. Have these changes, which affect surface albedo and the capacity of the ground to absorb and to emit longwave radiation, modified the global energy balance? Answers to such questions require further study of the processes of heating and cooling of the earth's atmosphere, lands, and oceans.

CHAPTER SUMMARY

Electromagnetic radiation is a form of energy emitted by all objects. The wavelength of the radiation determines its characteristics. The hotter an object, the shorter the wavelength of the radiation and the greater the amount of radiation that it emits. Shortwave radiation, emitted by the sun, includes ultraviolet, visible, and shortwave infrared radiation. Longwave radiation, emitted by earth surfaces, is familiar as heat. Radiation flows are measured in watts per square meter. The earth continuously absorbs solar (shortwave) radiation and emits longwave radiation. In the long run, the gain and loss of radiant energy remain in balance, and the earth's average temperature remains constant.

Insolation, the rate of solar radiation flow available at a location at a given time, is greater when the sun is higher in the sky. Daily insolation is also greater when the period of daylight is longer. Annual insolation is greatest at the equator and least at the poles. However, the poles still receive 40 percent of the radiation received at the equator. The pattern of annual insolation with latitude leads to a natural naming convention for latitude zones: equatorial, tropical, subtropical, midlatitude, subarctic (subantarctic), arctic (antarctic), and polar.

The earth's atmosphere is dominated by nitrogen and oxygen gases. Carbon dioxide and water vapor, though minor constituents by volume, are very important because they absorb longwave radiation and enhance the greenhouse effect.

Ozone (O_3) is a tiny but important constituent of the atmosphere that is concentrated in a layer of the upper atmosphere. It is formed from oxygen (O_2) by chemical reactions that absorb ultraviolet radiation, thus sheltering the organisms of the earth's surface from the damaging effects of ultraviolet rays. The ozone layer is threatened by chlorofluorocarbons (CFCs), which react chemically in the upper atmosphere to destroy ozone. The ozone hole, which forms over the south pole in the southern hemisphere spring, documents the loss of ozone from CFCs. Although CFC manufacture has now all but ceased, damage to the ozone layer from CFCs already in the atmosphere will continue for many years. Global levels of ozone appear to have decreased in recent years, but increases in ultraviolet radiation reaching the surface have not been consistently observed.

Latent heat and sensible heat are additional forms of energy. Latent heat is taken up or released when a change of state occurs. Sensible heat is contained within a substance and can flow to a cooler substance that it contacts.

The global energy system includes a number of important pathways by which insolation is transferred and transformed. Part of the solar radiation passing through the atmosphere is absorbed or scattered by molecules, dust, and larger particles. Some of the scattered radiation returns to space as diffuse reflection. The land surfaces, ocean surfaces, and clouds also reflect some solar radiation back to space. The proportion of radiation that a surface absorbs is termed its albedo. The albedo of the earth and atmosphere as a whole planet is about 30 percent.

The atmosphere absorbs longwave energy emitted by the earth's surface, causing the atmosphere to counterradiate some of that longwave radiation back to earth, thereby creating the greenhouse effect. Because of this heat trapping, the earth's surface temperature is considerably warmer than we might expect for an earth without an atmosphere.

Although absorbed shortwave radiation balances longwave radiation emitted by the earth's surface and atmosphere for the earth as a whole, this net radiation does not have to balance at every surface location on the globe. At latitudes lower than 40 degrees, annual net radiation is positive, while it is negative at higher latitudes.

This imbalance can be sustained only if heat is exported from low to high latitudes by poleward heat transport. This transport is achieved through sensible heat transfer in currents of warm ocean water, and as latent heat transfer in flows of warm, moist air.

The sun's radiation can serve as a source of power to help meet the needs of human civilization for heating and electricity.

KEY TERMS

electromagnetic radiation	insolation	scattering
visible light	ozone	albedo
shortwave radiation	sensible heat	counterradiation
longwave radiation	sensible heat transfer	greenhouse effect
radiation balance	latent heat	net radiation
	latent heat transfer	poleward heat transport

REVIEW QUESTIONS

1. What is electromagnetic radiation? How does the temperature of the hood of a car influence the nature of the electromagnetic radiation it emits?

2. Describe the types of radiation emitted by the sun and their characteristics.

3. Compare the terms *shortwave radiation* and *longwave radiation*. What are their sources?

4. What is the solar constant? What is its value? What are the units with which it is measured?

5. How does the sun's path in the sky influence the average daily insolation at a location?

6. What is the influence of latitude on the annual cycle of daily insolation? on annual insolation?

7. Sketch the world latitude zones on a circle representing the globe and give their approximate latitude ranges.

8. Why are carbon dioxide and water vapor important atmospheric constituents?

9. How does the ozone layer protect the life layer?

10. What are CFCs and how do they affect the ozone layer?

11. When and where have ozone reductions been reported? Have corresponding reductions in ultraviolet radiation been noted?

12. Describe latent heat transfer and sensible heat transfer.

13. What is the fate of incoming insolation? Define *scattering* and *absorption*, including the role of clouds.

14. Define albedo and give two examples.

15. Describe the counterradiation process and how it relates to the greenhouse effect.

16. Discuss the energy balance of the earth's surface. Identify the types and sources of energy that the surface receives, and do the same for energy that it loses.

17. Discuss the energy balance of the atmosphere. Identify the types and sources of energy that the atmosphere receives and do the same for energy that it loses.

18. What is net radiation? How does it vary with latitude?

19. Identify four different ways of harnessing solar power.

Focus on Systems 2.1 • Forms of Energy

1. What is energy?

2. Distinguish among these forms of energy: kinetic, potential, mechanical, radiant, chemical.

Focus on Systems 2.4 • Energy and Matter Budgets

1. What important physical principle permits the construction of budgets of energy and matter flow?

2. Using the example of a checking account, explain how monitoring the flows of money with a budget can show whether or not a flow system is in equilibrium.

3. Why is budgeting an important tool for studying flow systems?

ESSAY QUESTIONS

1. Suppose the earth's axis of rotation were perpendicular to the orbital plane instead of tilted at $23\frac{1}{2}°$. How would global insolation be affected? How would insolation vary with latitude? How would the path of the sun in the sky change with the seasons?

2. Imagine that you are following either (a) a beam of shortwave solar radiation entering the earth's atmosphere heading toward the surface or (b) a beam of longwave radiation emitted from the surface heading toward space. How will the atmosphere influence the beam?

PROBLEMS

Working It Out 2.2 • Radiation Laws

1. Assuming that the sun behaves as a blackbody, what is the flow rate of energy leaving its surface if the surface is at a temperature of 5950°K (5677°C, 10,251°F)? At what wavelength will the emitted radiation be greatest?

2. In 1995, the average surface temperature of the earth was reported as 15.4°C (59.7°F). Assuming that the surface radiates perfectly, what is the flow rate of energy emitted by the surface? At what wavelength will the radiation be greatest?

3. Using the answers to problems 1 and 2, determine the ratio of the flow rate of energy emitted by the sun's surface to the flow rate of energy emitted by the earth's surface.

Working It Out 2.3 • Calculating the Global Radiation Balance

1. Venus has a radius of about 6050 km. Because it is closer to the sun than the earth, the solar radiation flow it receives is 1.92 times stronger than that received by the earth. The atmosphere of Venus consists of dense clouds of carbon dioxide vapor, which reflect about 65 percent of the incoming solar radiation. Diagram the energy flow system for Venus and calculate the rates of inflow and outflow of shortwave and longwave radiation.

Chapter 3

Air Temperature and Air Temperature Cycles

Temperature is a familiar concept. It is a measure of the level of sensible heat of matter, whether it is gaseous (air), liquid (water), or solid (rock or dry soil). When a substance absorbs a flow of heat energy, its temperature rises. Similarly, when heat flows away from a substance, its temperature falls.

Heat may flow to and from the substance by several means. One of these is *radiant energy transfer* by absorp-

tion and emission (net radiation), as we discussed in Chapter 2. Two other means, also discussed in Chapter 2, are *sensible heat transfer* and *latent heat transfer*. Sensible heat transfer occurs by **conduction,** in which heat flows from a warmer substance to a colder one when the two are touching. Latent heat transfer requires the presence of water and occurs when water changes state. For example, liquid water can evaporate from the surface of a moist substance, cooling the substance by removing the latent heat required for evaporation. Heat may also be transferred within a substance if the substance is a fluid—liquid or gas—and mixing occurs. This form of heat transfer, in which heat is distributed by mixing, is **convection**. (We'll return to convection in Chapters 4 and 5.)

The formation and break-up of sea ice with the seasons is one of the many environmental rhythms produced by the annual cycle of air temperature. Here a Russian icebreaker breaks through early summer pack ice in the Weddell Sea, Antarctica.

The subject of this chapter is air temperature. The temperature of the layer of air close to the ground changes when heat flows into or out of the layer by the heat transfer mechanisms described above. Because flows of heat, particularly net radiation, change in natural cycles, air temperatures measured close to the ground surface follow natural cycles. These include both daily and seasonal rhythms of temperature rise and fall, which are produced by the daily and seasonal variations in insolation described in Chapter 2.

Another important influence on air temperature cycles that we will see in this chapter is location—maritime or continental. Because oceans heat and cool more slowly than continents, coastal regions generally experience smaller temperature cycles than inland regions. However, keep in mind as you study this chapter that the main cause of air temperature change is still the change in the amount of the sun's energy that strikes the earth's surface from day to day, season to season, and latitude to latitude.

MEASUREMENT OF AIR TEMPERATURE

Air temperature is a piece of weather information that we encounter daily. In the United States, temperature is still widely measured and reported using the Fahrenheit scale. The freezing point of water on the Fahrenheit scale is 32°F, and the boiling point is 212°F. This represents a range of 180°F. In this book, we use the Celsius temperature scale, which is the international standard. On the Celsius scale, the freezing point of water is 0°C and the boiling point is 100°C. Thus, 100 Celsius degrees are the equivalent of 180 Fahrenheit degrees (1°C = 1.8°F; 1°F = 0.56°C). Conversion formulas between these two scales are given in Figure 3.1 and in *Working It Out 3.1 • Temperature Conversion.*

You are probably familiar with the *thermometer* as an instrument for measuring temperature. In this device, a liquid inside a glass tube expands and contracts with temperature, and the position of the liquid surface indicates the temperature. Since air temperature can vary with height, it is measured at a standard level—1.2 m (4 ft) above the ground. A *thermometer shelter* (Figure 3.2a) is a louvered box that holds thermometers or other weather instruments at proper height while sheltering them from the direct rays of the sun. Air circulates freely through the louvers, ensuring that temperatures inside the shelter are the same as the outside air.

Liquid-filled thermometers are being replaced by newer instruments for the routine measuring of temperatures. These instruments make use of the *thermistor*, a device that changes its electrical resistance with temperature. By measuring this resistance, the temperature may be obtained automatically. Many weather stations are now equipped with temperature measurement systems that use thermistors (Figure 3.2b).

Although some weather stations report temperatures hourly, most stations only report the highest and lowest temperatures recorded during a 24-hour period. These are most important values in observing long-term trends in temperature. This network comprises more than 5000 stations in the United States.

Daily, monthly, and yearly temperature statistics for a station are produced using the daily maximum and minimum temperature. The mean daily temperature is defined as the average of the maximum and minimum daily values. Records of mean daily temperatures averaged for monthly and yearly intervals of many years' duration are compiled for each observing station. These averages, along with other weather statistics, are used to describe the climate of the station and its surrounding area.

THE DAILY CYCLE OF AIR TEMPERATURE

Because the earth rotates on its axis, incoming solar energy at a location can vary widely throughout the 24-hour period. Insolation is greatest in the middle of the

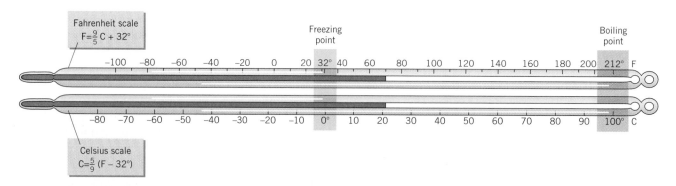

Figure 3.1 A comparison of the Celsius and Fahrenheit temperature scales. At sea level, the freezing point of water is at Celsius temperature (C) 0°, while it is 32° on the Fahrenheit (F) scale. Boiling occurs at 100°C, or 212°F.

Working It Out 3.1 • Temperature Conversion

Converting from Fahrenheit to Celsius and the reverse can be done simply and quickly using the formulas below:

$$C = \tfrac{5}{9}(F - 32)$$
$$F = \tfrac{9}{5} C + 32$$

Here C is the temperature in degrees Celsius, and F is the temperature in degrees Fahrenheit. The fractions $\tfrac{5}{9}$ and $\tfrac{9}{5}$ arise from the fact that 100 Celsius degrees covers the span of 180 degrees Fahrenheit. These values are in the ratio of 5 to 9 and vice versa.

As an example, let's convert 50°F to Celsius and 50°C to Fahrenheit. For the first conversion, we have

$$C = \tfrac{5}{9}(F - 32) = \tfrac{5}{9}(50 - 32)$$
$$= \tfrac{5}{9}(18) = 10°C$$

and for the second,

$$F = \tfrac{9}{5} C + 32 = \tfrac{9}{5}(50) + 32$$
$$= 90 + 32 = 122°F$$

daylight period, when the sun is at its highest position in the sky, and falls to zero at night. Recall from Chapter 2 that net radiation is the balance between incoming energy absorbed by a surface and outgoing radiation that is emitted from that surface. Since received solar radiation varies greatly, so does net radiation. During the day, net radiation is positive, and the surface gains heat. At night, net radiation is negative, and the surface loses heat by radiating it to the sky and space. Since the air next to the surface is warmed or cooled as well, air temperatures follow the same cycle. This results in the daily cycle of rising and falling air temperatures.

Daily Insolation and Net Radiation

Let's look in more detail at how insolation, net radiation, and air temperature are linked in this daily cycle. The three graphs in Figure 3.3 show average curves of daily insolation, net radiation, and air temperature that we might expect for a typical observing station at lat. 40° to 45° N in the interior of North America. The time scale is set so that 12:00 noon occurs when the sun is at its highest elevation in the sky.

Graph (*a*) shows daily insolation. At the equinox (middle curve), insolation begins at about sunrise (6 A.M.), rises to a peak value at noon, and declines to zero

(a)

(b)

Figure 3.2 (*a*) A thermometer shelter. This white wooden louvered box houses maximum-minimum thermometers and other instruments. (*b*) A maximum-minimum temperature instrument system appears on the left side of this photo. To the right is a large recording rain gauge.

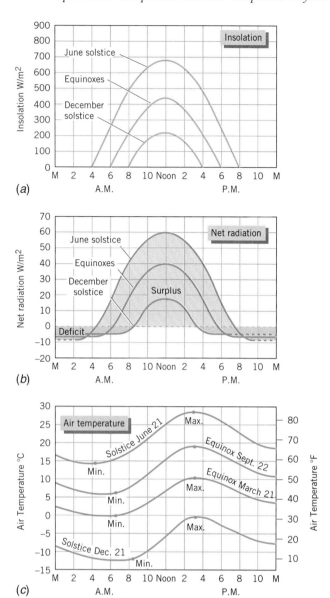

(a)

(b)

(c)

Figure 3.3 Idealized cycles of insolation, net radiation, and air temperature for a midlatitude station in the interior United States. Insolation (*a*) is a strong determiner of net radiation (*b*). Air temperatures (*c*) respond by generally increasing while net radiation is positive and decreasing while net radiation is negative.

at sunset (6 P.M.). At the June solstice, insolation begins about two hours earlier (4 A.M.) and ends about two hours later (8 P.M.). The June peak is much greater than at equinox, and the total insolation for the day is also much greater. At the December solstice, insolation begins about two hours later than the equinox curve (8 A.M.) and ends about two hours earlier (4 P.M.). Both the peak intensity and daily total insolation are greatly reduced in the winter solstice season.

Graph (*b*) shows net radiation for the surface. Recall that when net radiation is positive, the surface gains heat, and when negative, it loses heat. The curves for

the solstices and equinox generally resemble those of insolation. Unlike insolation, net radiation begins the 24-hour day as a negative value—a deficit—at midnight. The deficit continues into the early morning hours. Net radiation shows a positive value—a surplus—shortly after sunrise and rises sharply to a peak at noon. In the afternoon, net radiation decreases as insolation decreases. A value of zero is reached shortly before sunset. With no incoming insolation, net radiation then becomes negative, showing a deficit, where it remains until midnight.

Although the three net radiation curves show the same general daily pattern—negative to positive to negative—they differ greatly in magnitude. For the June solstice, the positive values are quite large, and careful inspection of the figure will show that the area of surplus is much larger than the area of deficit. This means that for the day as a whole, net radiation is positive. At the December solstice, the surplus period is short and the surplus is small. The total deficit, which extends nearly 18 hours, outweighs the surplus. This means that the net radiation for the entire day is negative. As we will see, this pattern of positive daily net radiation in the summer and negative daily net radiation in the winter drives the annual cycle of temperatures.

Daily Temperature

Graph (*c*) shows the typical, or average, daily air temperature cycle. The minimum daily temperature usually occurs about a half hour after sunrise. Since net radiation has been negative during the night, heat has flowed from the ground surface, and the ground has cooled the surface air layer to its lowest temperature. As net radiation becomes positive, the surface warms quickly and transfers heat to the air above. Air temperature rises sharply in the morning hours and continues to rise long after the noon peak of net radiation.

We should expect the air temperature to rise as long as net radiation is positive. However, another process begins in the early afternoon. Mixing of the lower air by vertical currents distributes heat upward, offsetting the temperature rise. Therefore, the temperature peak usually occurs in the midafternoon. It is shown in the figure at about 3 P.M., but it usually occurs between 2 and 4 P.M., depending on local climatic conditions. By sunset, air temperature is falling rapidly. It continues to fall, but at a decreasing rate, throughout the night.

Note also that the general level of the temperature curves varies with the seasons. In the summer, the daily curve is high, showing warm temperatures. In winter, the curve is low, showing cold temperatures. In between are the equinoxes, with their intermediate temperatures. Since the temperatures lag behind the seasonal changes in net radiation, the September equinox shows temperatures considerably warmer than the March equinox. Even though net radiation is the same for the two, each curve lags behind, reflecting earlier seasonal conditions.

Temperatures Close to the Ground

The air temperatures we have discussed thus far are measured under standard conditions, at a short distance from the ground surface. As we get closer to the ground surface, the temperature trends get more extreme. Under the direct rays of the sun, a pavement surface heats to much higher temperatures than the air at standard height. At night, the pavement surface cools to temperatures that are lower than the air temperature at standard height. These effects are weaker under a forest cover where the ground is shaded and moist. On a dry desert floor or on the pavements of city streets and parking lots, however, the surface temperature extremes can be large.

Figure 3.4 shows a series of average temperature profiles of soil and the air directly above it as observed at various times of the day in summer. Before sunrise (5 A.M.), the soil has cooled significantly under the influence of negative net radiation. Soil temperature increases from the surface downward, and air temperature increases from the surface upward, showing an inversion. By 8 A.M., positive net radiation has warmed the soil surface, and heat has raised the temperature of the upper few centimeters of the soil. The inversion of air temperature has disappeared, and the air temperature is nearly isothermal—that is, of uniform temperature. At 10 A.M., the soil surface is much warmer, as is air temperature directly above the soil. Surface warming continues through noon and on to 3 P.M., strongly heating the air in a layer about 15 cm above the surface. The heat flow into the soil continues the warming of the soil layer. By 8 P.M., with net radiation near zero, the soil surface has cooled, along with the air directly above it. However, a warm layer is still present within the soil. By sunrise, this warm layer has disappeared. The daily cycle shows its greatest range near the soil surface and gradually weakens with depth.

The temperature of a surface depends on the balance between the energy flow that it receives and the energy flow that it loses. Energy can flow between a surface and the atmosphere as net radiation, latent heat, or sensible heat, and between the surface and the soil as sensible heat. The *surface energy balance equation* describes these flows. It is discussed more fully in *Focus on Systems 3.2 • The Surface Energy Balance Equation.*

Urban and Rural Temperature Contrasts

Human activity has altered much of the earth's land surface. Cities are an obvious example. To build a city, vegetation is removed, and soils are covered with pavement or structures. Figure 3.5 illustrates some of the differences between urban and rural surfaces. In rural areas, the land surface is normally covered with a layer of vegetation. A natural activity of plants is **transpiration**. In this process, water is taken up by plant roots and moved to the leaves, where it evaporates. Since evaporation cools a surface by removing heat, we would expect the rural surface to be cooler.

The cooling effect of vegetation is even greater in a forest. Not only are transpiring leaves abundant, but also solar radiation is intercepted by a layer of leaves stretching from the tops of the trees to the forest floor. Thus, solar warming is not concentrated intensely at the ground surface, but rather serves to warm the whole forest layer.

Another feature of the rural environment is that the soil surfaces tend to be moist. During rainstorms, precipitation seeps into the soil. When sunlight reaches the soil surface, the water evaporates, thus removing heat and keeping the soil cool.

In contrast are the typical surfaces of the city. Rain runs easily off the impervious surfaces of roofs, sidewalks, and streets. The rainwater is channeled into sewer systems, where it flows directly into rivers, lakes, or oceans, instead of soaking into the soil beneath the city surfaces. Because the impervious surfaces are dry, the full energy of insolation warms the surfaces, with evaporation taking place only where street trees, lawns, and bare soil patches occur. Thus, air temperatures measured above urban surfaces during the daytime are usually higher than those of rural surfaces.

Another factor contributing to warmer urban temperature is the ability of concrete, stone, and asphalt to

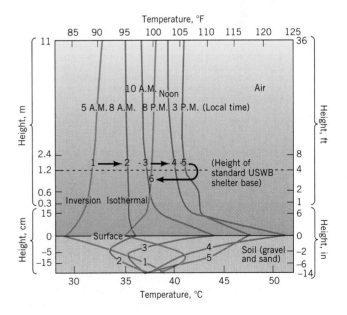

Figure 3.4 Air and soil temperatures at various levels above and below the ground surface throughout the day and night. Data are averages for July and August during a single summer. Height follows a square-root scale. (Copyright © A. N. Strahler.)

Focus on Systems 3.2 • The Surface Energy Balance Equation

Using the principles of budgeting, it is easy to devise a simple equation for the energy balance of a surface as an energy flow system. For this equation, we consider the surface as a thin layer positioned between the atmosphere and the soil, as shown in the figure on the right. It acts as an energy transfer layer that changes one type of energy flow to another. The surface is too thin to hold any heat itself. There are five flows of energy to and from the surface. They are related by the surface energy balance equation:

$$0 = R_{SHORT} + R_{LONG} + H_{LATENT} + H_{SENSIBLE} + H_{SOIL}$$

Here, R_{SHORT} is the flow of short-wave radiation absorbed by the surface; R_{LONG} is the net flow of long-wave radiation between the surface and the atmosphere; H_{LATENT} is the latent heat flow between the surface and the atmosphere; $H_{SENSIBLE}$ is the sensible heat flow between the surface and the atmosphere; and H_{SOIL} is the sensible heat flow between the surface and the soil. In using the equation, we consider

outgoing flows to be positive and incoming flows to be negative. From the principle of conservation of energy, incoming energy flow must balance outgoing energy flow, so the terms of the equation must add up to zero.

The equation includes two radiation terms. R_{SHORT} is the flow of shortwave radiation to the surface from the sun. Note that this flow includes only radiation that is absorbed by the surface. Shortwave radiation that is reflected directly back toward the atmosphere is not included. R_{SHORT} will be negative during the day, since by our convention flows coming into the surface are negative. At night, this flow will drop to zero. R_{LONG} is the net longwave radiation. Because it is a net term, it represents the balance between outgoing longwave radiation emitted by the surface and incoming longwave radiation from the atmosphere above. Because the surface is usually warmer than the atmosphere, even at night, the flow will normally be away from the surface, which is taken as positive. As

we saw in Chapter 2, the sum of R_{SHORT} and R_{LONG} is the net radiation flow.

Three energy flows remain. H_{LATENT} is the latent heat flow, which arises when water changes state from a solid or liquid at the surface and diffuses into the atmosphere as water vapor. During the day, it will be positive as soil water evaporates. At night, condensation or deposition may occur, yielding dew or frost. If so, latent heat will be released at the surface, providing a heat flow to the surface (negative).

$H_{SENSIBLE}$ is the sensible heat flux to the atmosphere, which arises when the surface conducts heat to the air at the surface-atmosphere boundary, and that heat is carried upward by convection. It will be positive when the air is warmed by the surface, which is the normal condition during the day. At night, the surface can become colder than the air it contacts, so the heat flow may be from the air to the surface (negative).

The last flow, H_{SOIL}, is the flow of heat by conduction from the sur-

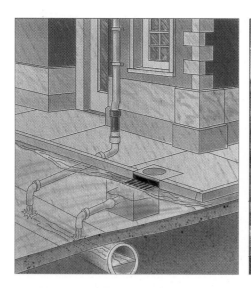

Figure 3.5 Urban surfaces (left) are composed of asphalt, concrete, building stone, and similar materials. Sewers drain away rainwater, keeping urban surfaces dry. Rural surfaces (right) are composed of moist soil, covered largely by vegetation.

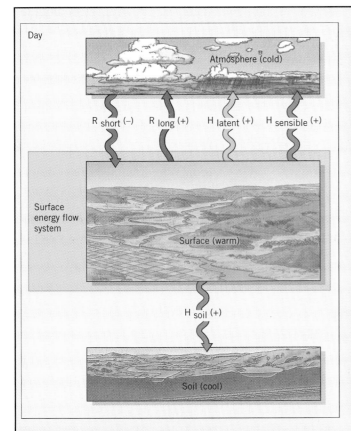
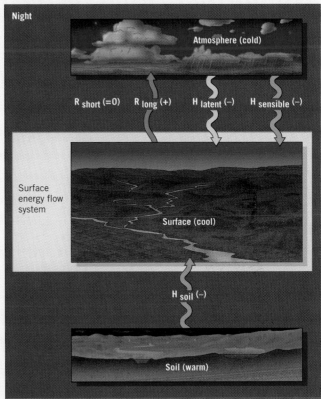

Diagrams of the surface energy balance equation for typical day and night conditions.

face to the soil below. This flow will normally be away from the surface (positive) during the day, as heat is conducted from the warm surface into the soil. At night, however, the surface will normally cool enough that heat will be conducted upward. This will then be a negative flow.

conduct and hold heat better than soil, even when the soil is dry. Therefore, during the day, more heat can be stored by the city's building materials. At night, the heat is conducted back to the surface, keeping nighttime temperatures warmer. Another factor is that heat absorption is enhanced by the many vertical surfaces in cities, which reflect radiation from one surface to another. Since some radiation is absorbed with each reflection, the network of vertical surfaces tends to trap heat more effectively than a single horizontal surface.

EYE ON THE ENVIRONMENT:
The Urban Heat Island

As a result of these effects, air temperatures in the central region of a city are typically several degrees warmer than those of the surrounding suburbs and countryside. Figure 3.6 shows city air temperatures on a typical summer evening. The lines of equal air temperature delineate a **heat island**. The heat island persists through the night because of the availability of a large quantity of heat stored in the ground during the daytime hours.

Another important factor in warming the city is fuel consumption. In summer, city temperatures are raised through the use of air conditioning. The machinery pumps heat out of buildings, releasing the heat to the air. The power used to run the air conditioning systems is also released as heat. Although Figure 3.6 shows the urban heat island in summer, the heat island effect is usually more pronounced in winter. In that season, furnaces warm buildings. Interior heat is conducted to outside walls and roofs, which radiate heat into the urban environment.

Note that the heat island effect does not necessarily apply to cities in desert climates. In the desert, the

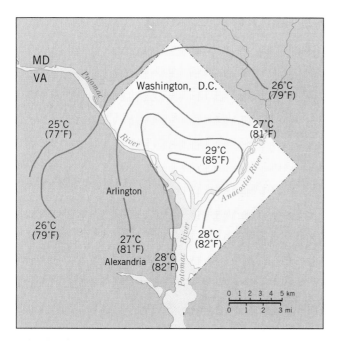

Figure 3.6 A heat island over Washington, D.C., and surrounding areas. Air temperatures were taken at 10:00 P.M. on an evening in early August. (Data of H. E. Landsberg.)

evapotranspiration of the irrigated vegetation of the city may actually keep the city cooler than the surrounding barren region. 🐾

TEMPERATURE STRUCTURE OF THE ATMOSPHERE

Thus far, we have discussed temperatures in the range of the atmosphere that surrounds us—up to about 2 m (6 ft) above the ground surface. How different are air temperatures at increasing heights through the atmosphere? In general, temperatures are lower. The decrease in measured temperature is described by the **lapse rate**, which indicates the drop in temperature in degrees C per 1000 m (or degrees F per 1000 ft).

Why is air cooler at higher altitudes? Recall from Chapter 2 that most of the incoming solar radiation penetrates the atmosphere and is absorbed at the surface. Thus, the atmosphere is largely warmed from below. In general, the farther the air is away from the earth's surface, the cooler the air will be.

Figure 3.7 shows how temperature varies with altitude for a typical summer day in the midlatitudes. Altitude is plotted on the vertical axis and temperature on the horizontal axis. The curve is an average one—if we sent up a balloon that radioed the temperature of the air back to us every minute or two, and sent up many balloons over a long period of time, we would obtain an average profile of temperature very much like that

shown in the figure. Temperature drops with altitude at an average rate of 6.4°C/1000 m (3.5°F/1000 ft). This average value is known as the **environmental temperature lapse rate**. For example, when the air temperature near the surface is a pleasant 21°C (70°F), the air at an altitude of 12 km (40,000 ft) will be a bone-chilling −55°C (−67°F). Keep in mind that the environmental temperature lapse rate is an average value and that on any given day the observed lapse rate might be quite different.

Figure 3.7 shows another important feature of the atmosphere. For the first 12 km (7 mi) or so, temperature falls with increasing elevation. However, at 12 to 15 km (7 to 9 mi) in height, the temperature stops decreasing. In fact, above that height, temperature slowly increases with elevation. This feature has led atmospheric scientists to distinguish two different layers in the lower atmosphere—the troposphere and the stratosphere.

Troposphere

The **troposphere** is the lowest atmospheric layer, in which temperature decreases with increasing elevation. It extends from the ground up to 12–15 km (7–9 mi). Since almost all human activity occurs in this layer, it is of primary importance to us. Everyday weather phenomena, such as clouds or storms, occur mainly in this layer.

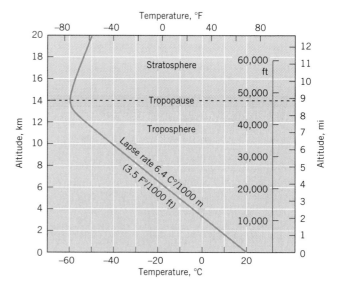

Figure 3.7 A typical environmental temperature lapse-rate curve for a summer day in the midlatitudes. The rate of temperature decrease with elevation, or lapse rate, is shown at the average value of 6.4°C/1000 m (3.5°F/1000 ft). The tropopause separates the troposphere, where temperature decreases with increasing elevation, from the stratosphere, where temperature is constant or slightly increases with elevation.

One important feature of the troposphere is that it contains significant amounts of water vapor. When the water vapor content is high, vapor can condense into water droplets, forming low clouds and fog, or can be deposited as ice crystals, forming high clouds. When condensation or deposition is rapid, rain, snow, hail, or sleet—collectively termed *precipitation*—may be produced and fall to earth. Regions where water vapor content is high throughout the year will therefore have moist climates. In desert regions, where water vapor is present only in small amounts, precipitation is infrequent. In Chapter 2, we also described the important role of water vapor in absorbing and reradiating heat emitted by the earth's surface. In this way, water vapor helps to create the greenhouse effect, which warms the earth to habitable temperatures.

The troposphere contains countless tiny dust particles, so small and light that the slightest movements of the air keep them aloft. These are called *aerosols.* They are swept into the air from dry desert plains, lakebeds, and beaches, or they are released by exploding volcanoes. Oceans are also a source of aerosols. Strong winds blowing over the ocean lift droplets of spray into the air. These droplets of spray lose most of their moisture by evaporation, leaving tiny particles of salt as residues that are carried high into the air. Forest fires and brushfires are another important source of aerosols, contributing particles of soot as smoke. Meteors contribute dust particles as they vaporize from the heat of friction upon entering the upper layers of air. Industrial processes that burn coal or fuel oil incompletely release aerosols to the air as well.

The most important function of aerosols is that certain types serve as nuclei, or centers, around which water vapor condenses to form tiny droplets. When these droplets grow large and occur in high concentration, they are visible to the eye as clouds or fog.

Aerosols in the troposphere also scatter sunlight, thus brightening the whole sky while reducing slightly the intensity of the solar beam. This scattering behavior is strongest for the longer wavelengths in the visible part of the spectrum, and thus the red colors of sunrise and sunset are accentuated by dust.

The height at which the troposphere gives way to the stratosphere above is known as the *tropopause.* Here, temperatures stop decreasing with altitude and start to increase. The altitude of the tropopause varies somewhat with latitude and season. This means that the troposphere is not uniformly thick.

Stratosphere and Upper Layers

Above the troposphere lies the **stratosphere** in which the air becomes slightly warmer as altitude increases. The stratosphere extends to a height of roughly 50 km (about 30 mi) above the earth's surface. It is the home of strong, persistent winds that blow from west to east.

There is little mixing of air between the troposphere and stratosphere, and so the stratosphere normally holds very little water vapor or dust.

One important feature of the stratosphere is that it contains the ozone layer. As we saw in Chapter 2, the ozone layer absorbs solar ultraviolet radiation and thus shields earthly life from this intense, harmful form of energy. In fact, the warming of the stratosphere with altitude is caused largely by the absorption of solar energy by ozone molecules.

Above the stratosphere are two other layers, identified by the curve of temperature with increasing altitude—the mesosphere and thermosphere. They are shown in Figure 3.8. In the *mesosphere,* temperature falls with elevation. This layer begins at the *stratopause,* the altitude at which the stratospheric temperature ceases to increase with altitude. It ends at the *mesopause,* the level at which temperature ceases to fall with altitude. Above the mesosphere is the *thermosphere,* a region of increasing temperature. However, at the altitude of the thermosphere, the density of air is very thin and the air holds little heat.

From the viewpoint of gas composition, the atmosphere is uniform for about the first 100 km of altitude, which includes the troposphere, stratosphere, mesosphere, and lower portion of the thermosphere. This region is referred to as the *homosphere.* Above 100 km, gas molecules tend to become increasingly sorted into layers by molecular weight and electric charge. This region is referred to as the *heterosphere.*

High-Mountain Environments

As we've noted, atmospheric temperatures generally drop with increasing altitude. This happens because upper layers are farther away from the surface, which is

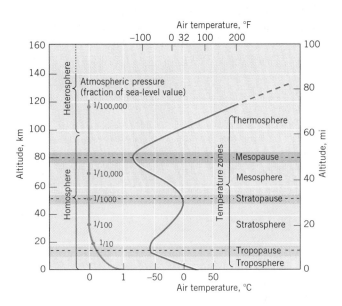

Figure 3.8 Temperature structure of the atmosphere.

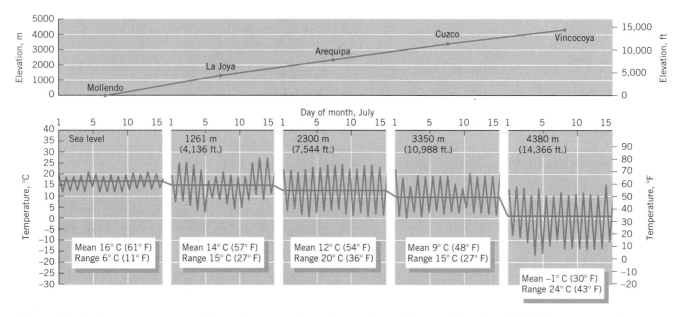

Figure 3.9 Daily maximum and minimum air temperatures for mountain stations in Peru, lat. 15° S. All data cover the same 15-day observation period in July. As elevation increases, the mean daily temperature decreases and the temperature range increases. (Data from Mark Jefferson.)

the primary source of heating for the atmosphere. At high elevations—on mountains or high plateaus—we can expect temperatures to be colder than at sea level, since these elevations are exposed to cooler air that has been circulating far above the surface. Figure 3.9 shows temperature graphs for five stations ascending the Andes Mountain range in Peru. The July monthly mean for each station is shown as a horizontal line. Mean temperatures clearly decrease with elevation, from 16°C (61°F) at sea level to –1°C (30°F) at 4380 m (14,360 ft).

The density of the air also decreases with elevation. At sea level, the air is compressed by the weight of the air above it and so is relatively dense. At high eleva-

Figure 3.10 Mount Whitney (elevation 4418 m, 14,495 ft) and surrounding peaks of the Southern Sierra Nevada. The excellent visibility and dark blue sky are products of the thin air at this elevation.

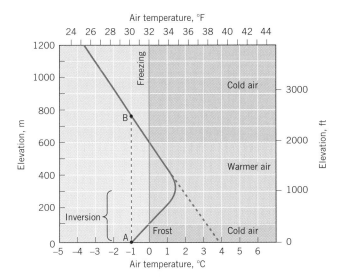

Figure 3.11 A low-level temperature inversion with frost. Instead of temperature decreasing with elevation (dashed line) from, say, 4°C (39°F), the surface temperature is at –1°C (30°F) and temperature increases with elevation (solid line) for several hundred meters (1000 ft or so) above the ground. Since the surface is below the freezing point, frost can form.

tions, the air is less compressed, and less dense, because there is a lesser weight of air above. This means that above a mountain or high plateau, there will be fewer air molecules and dust particles to scatter and absorb the sun's light. The air appears clearer, and the sky darkens (Figure 3.10).

Because the air is thinner at high elevations, incoming shortwave radiation will be more intense, and temperatures will rise more quickly during the day. And there will be less carbon dioxide and water vapor in the air above, so the greenhouse effect will be reduced. With the reduced warming effect, temperatures will tend to drop lower at night. These effects are shown in

the daily temperature ranges of the Andean locations in Figure 3.9. The daily range clearly increases with elevation, except for Cuzco. This large city does not experience nighttime temperatures as low as you might expect because of its urban heat island.

Temperature Inversion and Frost

There are exceptions to the rule that temperatures decrease with height above the surface. Think about what happens on a clear, calm night. Under clear conditions, the ground surface radiates longwave energy to the sky, net radiation becomes negative, and the surface cools. This means that the air near the surface will also be cooled, so as you move away from the cool surface, the air becomes warmer.

This situation is illustrated in Figure 3.11. The straight, slanting line of the normal environmental lapse rate bends to the left in a "J" hook. In this example, the air temperature at the surface, point A, has dropped to –1°C (30°F). This value is the same as at point B, some 750 m (about 2500 ft) aloft. As we move up from ground level, temperatures become warmer up to about 300 m (about 1000 ft). Here the curve reverses itself, and normal lapse rate takes over. The lower portion of the lapse rate curve is called a low-level **temperature inversion**. Here, the normal cooling trend is reversed, and temperatures increase with height.

In the case shown, the temperature of the lowermost air has fallen below the freezing point, 0°C (32°F). For sensitive plants during the growing season, this temperature condition is called a killing frost (even though actual frost may not form). Perhaps you have seen news reports about killing frosts damaging fruit trees or crops in Florida or California. Growers commonly use several methods to reduce the inversion. Oil-burning heaters are used to warm the surface air layer and create air circulation (Figure 3.12). Or large fans are used

Figure 3.12 Kerosene burners are used to warm the cold air of an inversion, protecting delicate pear blossoms from frost. Hood River, Oregon.

to mix the cool air at the surface with the warmer air above.

Low-level temperature inversions often occur over snow-covered surfaces in winter. Inversions of this type are very intense and can extend thousands of meters into the air. They build up over many long nights in arctic and polar regions, where the solar heat of the short winter day cannot completely compensate for nighttime cooling. Inversions can also result when a warm air layer overlies a colder one. This type of inversion is often found along the west coasts of major continents, and we will discuss it in more detail in Chapter 4.

THE ANNUAL CYCLE OF AIR TEMPERATURE

As the earth revolves around the sun, the tilt of the earth's axis causes an annual cycle of variation in insolation. This cycle produces an annual cycle of net radiation, which, in turn, causes an annual cycle to occur in mean monthly air temperatures. Although the annual cycle of net radiation is the most important factor in determining the annual temperature cycle, another important consideration is location—maritime or continental. That is, places located well inland and far from oceans generally experience a stronger temperature contrast from winter to summer. We will begin our study of the annual temperature cycle, however, with the relationship between net radiation and temperature for four examples, ranging from the equator almost to the arctic circle.

Net Radiation and Temperature

Graph (*a*) of Figure 3.13 shows the yearly cycle of the net radiation rate for four stations. The average value of net radiation flow rate for the month is plotted, in units of W/m^2. Graph (*b*) shows mean monthly air temperatures for these same stations. We will compare the net radiation graph with the air temperature graph for each station, beginning with Manaus, a city on the Amazon River in Brazil.

At Manaus, located nearly on the equator, the average net radiation rate is strongly positive in every month. However, there are two minor peaks. These coincide approximately with the equinoxes, when the sun is nearly straight overhead. A look at the temperature graph of Manaus shows uniform air temperatures, aver-

aging about 27°C (81°F) for the year. The annual temperature range, or difference between the highest and lowest mean monthly temperature, is only 1.7°C (3°F). In other words, near the equator the temperature is similar each month. There are no temperature seasons.

We go next to Aswan, Egypt, a very dry desert location on the Nile River at lat. 24° N. The positive net radiation rate curve shows that a large radiation surplus exists for every month. Furthermore, the net radiation rate curve has a much stronger annual cycle, with values for June and July that are almost double those of December and January. The temperature graph shows a corresponding annual cycle, with an annual range of about 17°C (31°F). June, July, and August are terribly hot, averaging over 32°C (90°F).

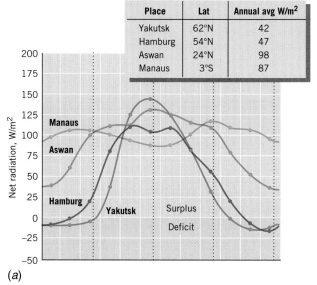

Place	Lat	Annual avg W/m^2
Yakutsk	62°N	42
Hamburg	54°N	47
Aswan	24°N	98
Manaus	3°S	87

(*a*)

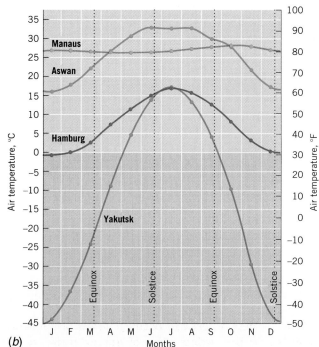

(*b*)

Figure 3.13 Net radiation (*a*) and temperature (*b*) cycles for four stations ranging from the equatorial zone to the arctic zone. Aswan (Egypt), Hamburg (Germany), and Yakutsk (Siberia) all show strong cycles with summer maxima and winter minima. In contrast is Manaus (Brazil), very near the equator, which shows two net radiation peaks and a nearly uniform temperature. (Data courtesy of David H. Miller.)

Moving farther north, we come to Hamburg, Germany, lat. 54° N. The net radiation rate cycle here is also strongly developed. The rate is positive for nine months, providing a radiation surplus. When the rate becomes negative for three winter months, a deficit is produced. The temperature cycle reflects the reduced total insolation at this latitude. Summer months reach a maximum of just over 16°C (61°F), while winter months reach a minimum of just about freezing (0°C or 32°F). The annual range is about 17°C (31°F), the same as at Aswan.

Finally, we travel to Yakutsk, Siberia, lat. 62° N. During the long, dark winters, the net radiation rate is negative, and there is a radiation deficit that lasts about six months. During this time, air temperatures drop to extremely low levels. For three of the winter months, monthly mean temperatures are between –35 and –45°C (about –30 and –50°F). This is actually one of the coldest places on earth. In summer, when daylight lasts most of a 24-hour day, the net radiation rate rises to a strong peak. In fact, this peak value is higher than those of the other three stations. As a result, air temperatures show a phenomenal spring rise to summer-month values of over 13°C (55°F). In July, the temperature is about the same as for Hamburg. Because of Yakutsk's high latitude and continental interior location, its annual temperature range is enormous—over 60°C (108°F).

Land and Water Contrasts

Have you ever visited San Francisco? If so, you probably noticed that this magnificent city has quite a unique climate. Fog is frequent, and cool, damp weather prevails for most of the year (see Figure 4.13). The cool climate is due to its location—on the tip of a peninsula, with the Pacific Ocean on one side and San Francisco Bay on the other. Ocean and bay water temperatures are quite cool, since a southward-flowing current sweeps cold water from Alaska down along the northern California coast. Winds from the west move cool, moist ocean air, as well as clouds and fog, across the peninsula, keeping summer air temperatures low and winter temperatures above freezing. Figure 3.14 shows a typical record of temperatures for San Francisco for a week in the summer. Temperatures hover around 13°C (55°F) and change only a little from day to night.

In contrast is a location far from the water, like Yuma, Arizona, also shown in Figure 3.14. Located in the Sonoran Desert, air temperatures here are much warmer on the average—about 28°C (82°F). Clearly, no ocean cooling is felt in Yuma! The daily range is also much greater, nearly 20°C (36°F). Hot desert days become cool desert nights, with the clear, dry air allowing the ground to lose heat rapidly.

Why do these differences occur? The important principle is this: the surface layer of any extensive, deep

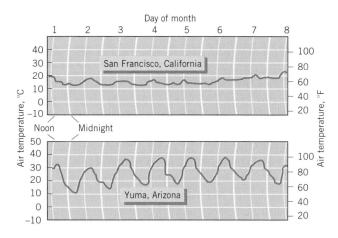

Figure 3.14 A recording thermometer made these continuous records of the rise and fall of air temperature over a period of a week in summer at San Francisco, California, and at Yuma, Arizona. The daily cycle is strongly developed at Yuma, a station in the desert. In contrast, the graph for San Francisco, on the Pacific Ocean, shows a very weak daily cycle.

body of water heats more slowly and cools more slowly than the surface layer of a large body of land when both are subjected to the same intensity of insolation. Because of this principle, daily and annual air temperature cycles will be quite different at coastal locations than at interior locations.

Four important thermal differences between land and water surfaces account for the land–water contrast (see Figure 3.15). The most important difference is that much of the downwelling solar radiation penetrates water, distributing the absorbed heat throughout a substantial water layer. In contrast, solar radiation does not penetrate soil or rock, so its heating effect is concentrated at the surface. The radiation therefore

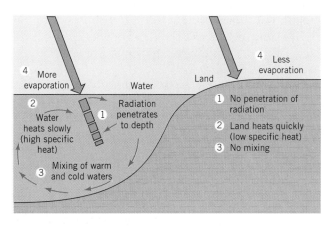

Figure 3.15 Four differences that illustrate why a land surface heats more rapidly and more intensely than the surface of a deep water body. Locations near the ocean have more uniform air temperatures—cooler in summer and warmer in winter—because of these differences.

warms a thick water surface layer only slightly, while a thin land surface layer is warmed more intensely.

A second thermal factor is that water is slower to heat than dry soil or rock. The *specific heat* of a substance describes how the temperature of a substance changes with a given input of heat. Consider an experiment using a small volume of water and the same volume of rock. Let's suppose we warm each volume by the same amount—perhaps 1°C (1.8°F). If we measure the amount of heat required, we will find that it takes about five times as much heat to raise the temperature of the water one degree as it does to raise the temperature of the rock one degree. That is, the specific heat of water is about five times greater than that of rock. The same will be true for cooling—after losing the same amount of heat, the temperature of water falls less than the temperature of rock.

A third difference between land and water surfaces is related to mixing. That is, a warm water surface layer can mix with cooler water below, producing a more uniform temperature throughout. For the open ocean, the mixing is produced by wind-generated waves. Clearly, no such mixing occurs on land surfaces.

A fourth thermal difference is that an open water surface can be cooled easily by evaporation. Land surfaces can also be cooled by evaporation, but only if water is present at or near the soil surface. When the surface dries, evaporation stops. In contrast, a free water surface can always provide evaporation.

Daily Temperature Cycle

What is the effect of the contrast in thermal characteristics of land and water on temperatures? Let's examine the daily cycle of air temperature first. Two sets of daily air temperature curves are shown in Figure 3.16—El Paso, Texas, and North Head, Washington—and four months are presented—January, April, July, and October.

The El Paso curves show the temperature environment of an interior desert in midlatitudes. Because soil moisture content is low and vegetation is sparse, evaporation and transpiration are not important cooling effects. Cloud cover is generally sparse. Under these circumstances, the ground surface heats intensely during the day and cools rapidly at night. Air temperatures show an average daily range of 11 to 14°C (20 to 25°F). This type of variation represents an interior temperature environment—typical of a station located in the interior of a continent, far from the ocean's influence.

North Head is located on the Washington coast. Here, prevailing westerly winds sweep cool, moist air off the adjacent Pacific Ocean. The average daily range at North Head is a mere 3°C (5°F) or less. Persistent fogs and cloud cover also contribute to the minimal daily range. Note further that the annual range is much restricted, especially when compared to El Paso. North Head exemplifies a coastal temperature environment, typical of a station located in the path of oceanic air.

Annual Temperature Cycle

Let's now turn to the effects of land–water surface contrasts on the temperature cycle for the year. We have already noted for El Paso and North Head that the temperature cycle for the four months plotted shows a greater range for the interior station than for the coastal one. Let's look in more detail at the annual cycle for another pair of stations—Winnipeg, Manitoba, located in the interior of the North American continent, and the Scilly Islands, off the southwestern tip of England, which are surrounded by the waters of the Atlantic Ocean. This time, the two stations chosen are at the same latitude, 50° N. As a result, they have the same insolation cycle and receive the same potential amount of solar energy for surface warming. Figure 3.17 shows their annual cycles of temperature as well as the insolation curve common to both.

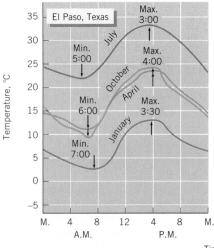
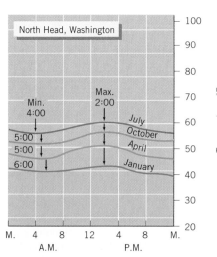

Figure 3.16 The average daily cycle of air temperature for four different months shows the effect of continental and maritime location. Daily and seasonal ranges are great at El Paso, a station in the continental interior, but only weakly developed at North Head, Washington, which is on the Pacific coast. The seasonal effect on overall temperatures is stronger at El Paso.

Time of day

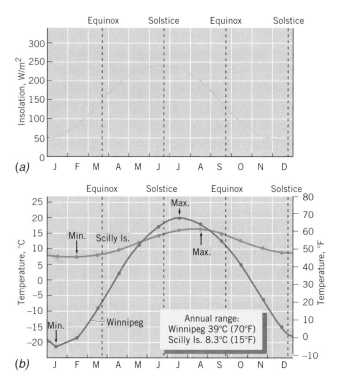

(a)

(b)

Figure 3.17 Annual cycles of insolation (*a*) and monthly mean air temperature (*b*) for two stations at lat. 50° N: Winnipeg, Canada, and Scilly Islands, England. Insolation is identical for the two stations. Winnipeg temperatures clearly show the large annual range and earlier maximum and minimum that are characteristic of its continental location. Scilly Islands temperatures show its maritime location in the small annual range and delayed maximum and minimum.

The temperature graphs for the Scilly Islands and Winnipeg confirm the effects we have already noted for North Head and El Paso—that the annual range in temperature is much larger for the interior station (39°C, 70°F) than for the coastal station (8°C, 14°F). Note that the nearby ocean waters, warmed by the North Atlantic drift current (Chapter 5), keep the air temperature at the Scilly Islands well above freezing in the winter. In January, temperatures at Winnipeg fall to near −20°C (−4°F)!

Another important effect concerns the timing of maximum and minimum temperatures. Insolation reaches a maximum at summer solstice, but it is still strong for a long period afterward. This means that heat energy continues to flow into the ground well after the solstice. Therefore, the hottest month of the year for interior regions is July, the month following the solstice. Similarly, the coldest month of the year for large land areas is January, the month after the winter solstice. This is because the ground continues to lose heat even after insolation begins to increase.

Over the oceans and at coastal locations, maximum and minimum air temperatures are reached a month later than on land—in August and February, respec-

tively. Because water bodies heat or cool more slowly than land areas, the air temperature changes more slowly. This effect is clearly shown in the Scilly Islands graph, where February is slightly colder than January.

WORLD PATTERNS OF AIR TEMPERATURE

Thus far, we have presented some important principles about air temperatures in this chapter. Some of these principles are local—such as the effect of urban and rural surfaces on temperatures—and others are global—such as the effect of latitude or elevation on temperature, or coastal-interior location. We now turn to world temperature patterns, tying our discussion closely to these principles. First, however, we will need a quick explanation of air temperature maps and their meaning.

The distribution of air temperatures is often shown on a map by **isotherms**—lines drawn to connect locations having the same temperature. Figure 3.18 shows a map on which the observed air temperatures have been recorded and placed at their proper location on the map. These may be single readings, such as a daily maximum or minimum, or they may be averages of many years of records for a particular day or month of a year, depending on the purposes of the map. The isotherms are constructed by drawing smooth lines through and among the points in a way that best indicates a uniform

Figure 3.18 Isotherms are used to make temperature maps. Each line connects points having the same temperature. Where temperature changes along one direction, a temperature gradient exists. Where isotherms close in a tight circle, a center exists. This example shows a center of low temperature.

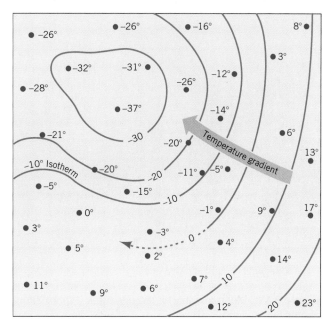

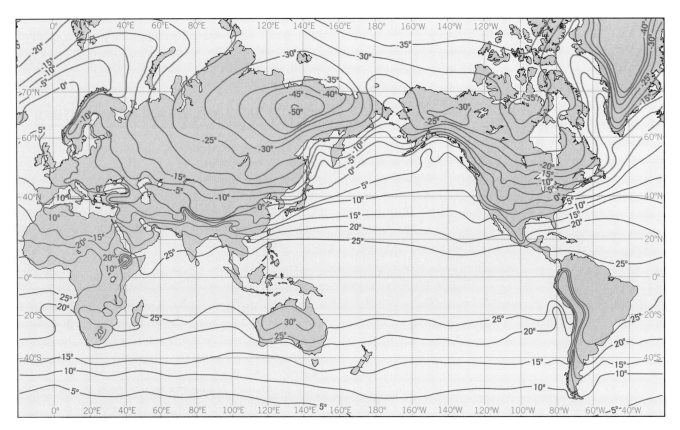

JANUARY

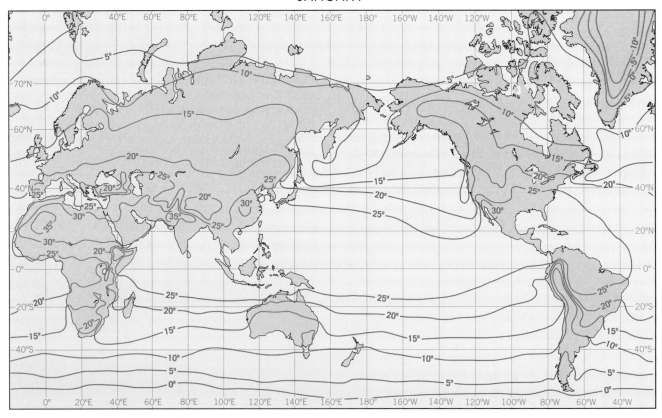

JULY

Figure 3.19 Mean monthly air temperatures (°C) for January and July, Mercator and polar projections. (Compiled by John E. Oliver.)

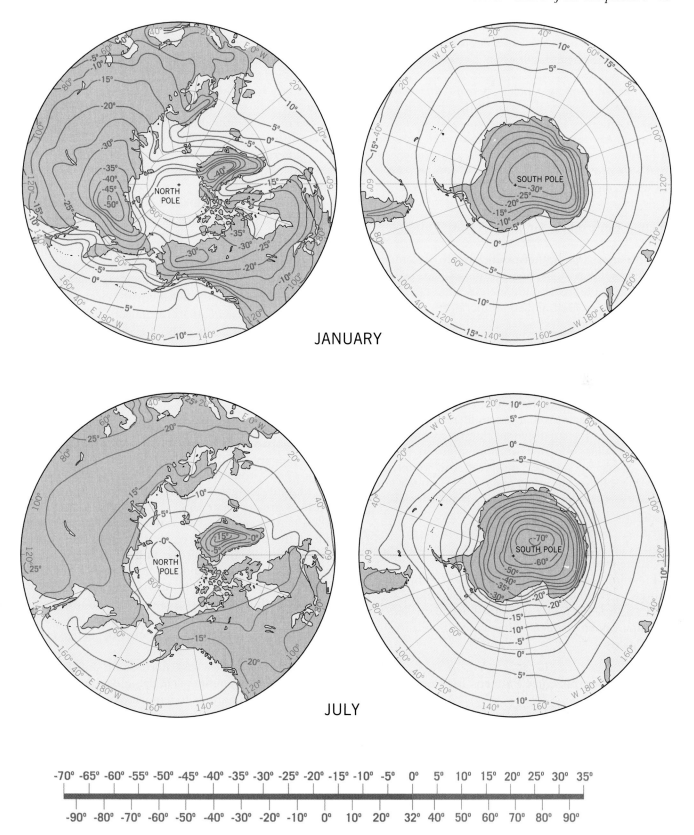

JANUARY

JULY

-70° -65° -60° -55° -50° -45° -40° -35° -30° -25° -20° -15° -10° -5° 0° 5° 10° 15° 20° 25° 30° 35°

-90° -80° -70° -60° -50° -40° -30° -20° -10° 0° 10° 20° 32° 40° 50° 60° 70° 80° 90°

temperature, given the observations at hand. Usually, isotherms representing 5- or 10-degree differences are chosen, but they can be drawn at any convenient temperature interval.

Isothermal maps are valuable because they make the important features of the temperature pattern clearly visible. Centers of high or low temperatures are clearly outlined. Also visible are *temperature gradients*—directions along which temperature changes. Centers and gradients give the broad pattern of temperature, as shown in our discussion of world patterns below.

Factors Controlling Air Temperature Patterns

World patterns of isotherms are largely explained by three factors that we have already discussed. The first of these is latitude. As latitude increases, average annual insolation decreases, and so temperatures decrease as well, making the poles colder than the equator. The effect of latitude on seasonal variation is also important. For example, more solar energy is received at the poles at the summer solstice than at the equator. We must therefore consider the time of year as well as the latitude in understanding world temperature patterns.

The second factor is that of coastal-interior contrasts. As we've noted, coastal stations that receive marine air from prevailing winds have more uniform temperatures—cooler in summer and warmer in winter. Interior stations, on the other hand, show a much larger annual variation in temperature. Ocean currents, discussed further in Chapter 5, can also have an effect. By keeping coastal waters somewhat warmer or cooler than expected, temperatures at maritime stations will be influenced in a similar manner.

Elevation is the third important factor. At higher elevations, temperatures will be cooler. Therefore, we expect world temperature maps to show the presence of mountain ranges, which will be cooler than surrounding regions.

World Air Temperature Patterns for January and July

With these factors in mind, let's look at some world temperature maps in more detail. Figure 3.19 consists of maps of world temperatures for two months—January and July. The Mercator projections show temperature trends from the equator to the midlatitude zones, and the polar projections give the best picture for high latitudes. From the maps, we can make six important points about the temperature patterns and the factors that produce them.

1. **Temperatures decrease from the equator to the poles.** Annual insolation decreases from the equator to the poles, thus causing temperatures to decrease.

This temperature gradient is most clearly seen in the polar maps for the southern hemisphere in January and July. On these maps, the isotherms are nearly circular, decreasing to a center of low temperature on Antarctica near the south pole. The center is much colder in July, when most of the polar region is in perpetual night. We can also see this same general trend in the north polar maps, but the continents complicate the pattern. The general temperature gradient from the equator poleward is also evident on the Mercator maps.

2. **Large landmasses located in the subarctic and arctic zones develop centers of extremely low temperatures in winter.** The two large landmasses we have in mind are North America and Eurasia. The January north polar map shows these low temperature centers very well. The cold center in Siberia, reaching –50°C (–58°F), is strong and well defined. The cold center over northern Canada is also quite cold (–35°C, –31°F) but is not as well defined. Both features are visible on the January Mercator map. Greenland shows a low temperature center as well, but it has a high dome of glacial ice, as discussed in point 6. An important factor in keeping winter temperatures low in these regions is the high albedo of snow cover, which reflects much of the winter insolation back to space.

3. **Temperatures in equatorial regions change little from January to July.** Note the broad space between 25°C (77°F) isotherms, which is evident on both January and July Mercator maps. In this region, the temperature is greater than 25°C (77°F) but less than 30°C (86°F). Although the two isotherms move a bit from winter to summer, the equator always falls between them. This demonstrates the uniformity of equatorial temperatures. (An exception is the northern end of the Andes Mountains in South America, where high elevations, and thus cooler temperatures, exist at the equator.) Equatorial temperatures are uniform primarily because insolation at the equator does not change greatly with the seasons.

4. **Isotherms make a large north-south shift from January to July over continents in the midlatitude and subarctic zones.** Figure 3.20 demonstrates this principle. In the winter, isotherms dip equatorward, while in the summer, they arch poleward. This effect is shown in North America and Eurasia in the January and July Mercator maps. In January the isotherms drop down over these continents, and in June they curve upward. For example, the 15°C (59°F) isotherm lies over central Florida in January. But by July this same isotherm has moved far north, cutting the southern shore of Hudson Bay and then looping far up into northwestern Canada. In contrast are the isotherms over oceans, which shift much less. This striking difference is due to the contrast between oceanic and continental surface properties, which cause continents to heat and cool more rapidly than oceans.

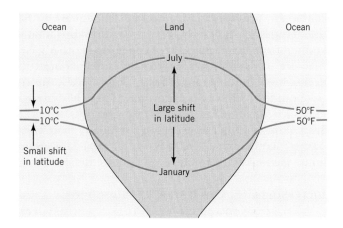

Figure 3.20 The seasonal migration of isotherms is much greater over continents than over oceans. This difference occurs because oceans heat and cool much more slowly than continents.

5. **Highlands are always colder than surrounding lowlands.** You can see this by looking at the pattern of isotherms around the Rocky Mountain chain, in western North America, on the Mercator maps. In winter, the −5°C (23°F) and −10°C (14°F) isotherms dip down around the mountains, indicating that the center of the range is colder. In summer, the 20°C (68°F) and 25°C (77°F) isotherms also dip down, showing the

same effect even though temperatures are much warmer. The Andes Mountains in South America show the effect even more strongly. The principle at work here is that temperatures decrease with an increase in elevation.

6. **Areas of perpetual ice and snow are always intensely cold.** Greenland and Antarctica contain our planet's two great ice sheets. Notice how they stand out on the polar maps as cold centers in both January and July. They are cold for two reasons. First, their surfaces are high in elevation, rising to over 3000 m (about 10,000 ft) in their centers. Second, the white snow surfaces reflect much of the insolation. Since little solar energy is absorbed, little is available to warm the snow surface and the air above it. The Arctic Ocean, bearing a cover of floating ice, also maintains its cold temperatures throughout the year. However, the cold is much less intense in January than on the Greenland Ice Sheet, since ocean water underneath the ice acts as a heat reservoir to keep the ice above from getting extremely cold.

The Annual Range of Air Temperatures

Figure 3.21 is a world Mercator map showing the annual range of air temperatures. The lines, resembling isotherms, show the difference between the January

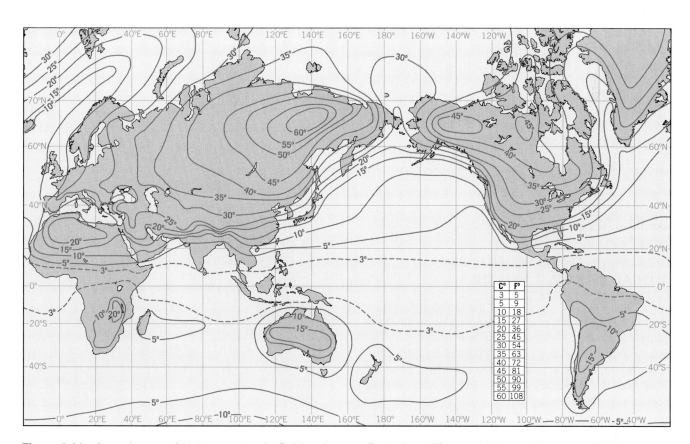

Figure 3.21 Annual range of air temperature in Celsius degrees. Data show differences between January and July means. The inset box shows the conversion of Celsius degrees to Fahrenheit for each isotherm value. (Compiled by John E. Oliver.)

and July monthly means. We can explain the features of this map by using the same effects of latitude, interior-maritime location, and elevation.

1. **The annual range increases with latitude, especially over northern hemisphere continents.** This trend is most clearly shown for North America and Asia. This is due to the contrast between summer and winter insolation, which increases with latitude.

2. **The greatest ranges occur in the subarctic and arctic zones of Asia and North America.** The map clearly shows two very strong centers of large annual range—one in northeast Siberia and the other in northwest Canada–eastern Alaska. In these regions, summer insolation is nearly the same as at the equator, while winter insolation is very low.

3. **The annual range is moderately large on land areas in the tropical zone, near the tropics of cancer and capricorn.** These are regions of large deserts—North Africa (Sahara), southern Africa (Kalahari), and Australia (interior desert) are examples. Dry air and the absence of clouds and moisture allow these continental locations to cool strongly in winter and warm strongly in summer, even though insolation contrasts with the season are not as great as at higher latitudes.

4. **The annual range over oceans is less than that over land at the same latitude.** This can be clearly seen by following a parallel of latitude—40° N, for example. Starting from the right, we see that the range is between 5 and 10°C (9 and 18°F) over the Atlantic but increases to about 30°C (54°F) in the interior of North America. In the Pacific, the range falls to 5°C (9°F) just off the California coast and increases to 15°C (27°F) near Japan. In Central Asia, the range is near 35°C (63°F). Again, these major differences are due to the contrast between land and water surfaces. Since water heats and cools much more slowly than land, a narrower range of temperatures is experienced.

5. **The annual range is very small over oceans in the tropical zone.** As shown on the map, the range is less than 3°C (5°F) since insolation varies little with the seasons near the equator and water heats and cools slowly.

EYE ON THE ENVIRONMENT: GLOBAL WARMING AND THE GREENHOUSE EFFECT

One of the most important topics of global change is climate warming. Many scientists have concluded that the temperature of the planet is warming significantly because of human activities. Other scientists, however, argue that the warming of the last few decades could be part of a natural climate cycle.

Carbon dioxide, produced by human activities, is a major cause of concern in climate warming. This gas is released to the atmosphere in large quantities by fossil fuel burning. Recall that the greenhouse effect is caused by atmospheric absorption of longwave radiation, largely by carbon dioxide and water vapor, that is emitted from the earth. Also of concern are other gases that are normally present in very small concentrations—methane (CH_4), nitrous oxide (NO), ozone (O_3), and the chlorofluorocarbons. These also absorb longwave radiation and enhance the greenhouse effect, even though they are even less abundant than CO_2. Taken together with CO_2, they are referred to as **greenhouse gases**. Let's examine CO_2 in more detail.

Increasing CO_2 Levels in the Atmosphere

In the centuries before global industrialization, carbon dioxide concentration in the atmosphere was at a level slightly less than 300 parts per million (ppm), or about 3/100ths of a percent by volume. During the last hundred years or so, that amount has been substantially increased by fossil fuel burning. When fuels like coal, oil, or natural gas are burned, they yield water vapor and carbon dioxide. The release of water vapor does not present a problem, because a large amount of water vapor is normally present in the global atmosphere. But because the normal amount of CO_2 is so small, fossil fuel burning has raised the level to about 350 ppm. This is a 22 percent increase.

Figure 3.22 shows how CO_2 has increased with time since 1860. Until 1940 or so, the level remained nearly stable. But after 1940 CO_2 began a rapid rise—so rapid that at the present rate of increase (about 4 percent per year), the amount of CO_2 in the air will double by 2030! Even if worldwide fossil fuel combustion is cut in half, the prediction is that doubling will occur by 2050, only about a half-century away.

The increase of CO_2 since 1940 follows a pattern known as *exponential growth*. This is the growth principle behind the compound interest that causes a savings ac-

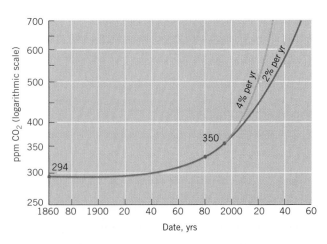

Figure 3.22 Increase in atmospheric carbon dioxide, observed until 1990 and predicted into the twenty-first century.

count to increase in value. Many scientific phenomena show exponential growth. *Working It Out 3.3 • Exponential Growth* provides a closer look.

Not all the carbon dioxide emitted into the air by fossil fuel burning remains there. Instead, a complex cycle moves CO_2 throughout the life layer. *Photosynthesis* is one part of this carbon cycle. Plants use CO_2 in photosynthesis, the process by which they use light energy to build their tissues. So, plants take in carbon dioxide, removing it from the atmosphere. When plants die, their remains are digested by decomposing organisms, which release CO_2. This process returns CO_2 to the atmosphere. Normally, these processes are in balance.

Humans have tilted this balance by clearing land and burning the vegetation cover as new areas of forest are opened for development. This practice increases the amount of CO_2 in the air. When agricultural land is allowed to return to its natural forest state, CO_2 is removed from the air by growing trees. At present, scientists calculate that forests are growing more rapidly than they are being destroyed in the midlatitude regions of the northern hemisphere. This plant growth helps to counteract the buildup of CO_2 produced by fossil fuel burning. But it may be outweighed by the tropical deforestation that is taking place in South America, Africa, and Asia.

Laboratory experiments have shown that plants exposed to increased concentrations of CO_2 will grow faster and better. The faster they grow, the more CO_2 they can take in, which helps to reduce the amount in the atmosphere. However, scientists are unsure whether increased CO_2 will stimulate plant growth under natural conditions.

Another part of the cycle involves the oceans. The ocean's surface layer contains microscopic plant life that takes in carbon dioxide. The CO_2 in the ocean water initially comes from the atmosphere and is mixed into the ocean by surface waves. When these microscopic floating plants die, their bodies sink to the ocean bottom. There they decompose and release CO_2, enriching waters near the ocean floor.

This CO_2 eventually returns to the surface through a system of global ocean current flows involving both bottom and surface currents (Figure 3.23). In this system, cold CO_2-rich bottom waters rise to the surface in the northernmost Pacific, where CO_2 is released. Meanwhile, warmer, CO_2-poor waters sink in the northernmost Atlantic. In fact, the ocean acts like a slow con-

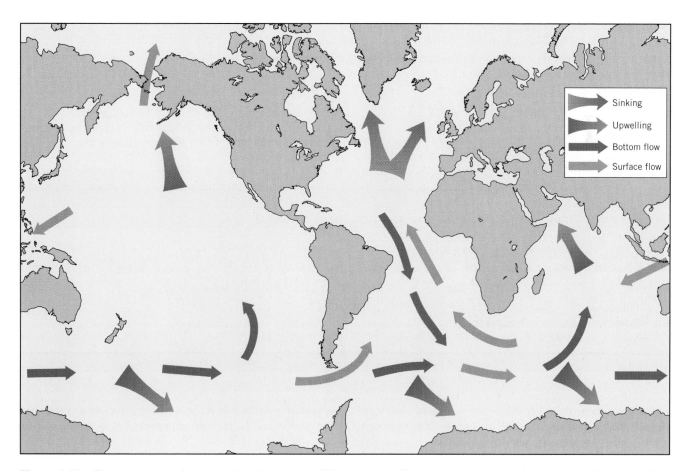

Figure 3.23 The movement of deep and surface ocean CO_2 currents is like a giant conveyor belt, carrying cold, salty, CO_2-rich waters eastward along ocean floors (blue arrows). The returning warm surface current, which loses CO_2 and nutrients as they are used by tiny floating plants, flows westward (red arrows). (After NASA.)

Working it Out 3.3 • Exponential Growth

Exponential growth is a type of growth in which something grows by a constant percentage during each growth period. For example, the money in a bank account will grow according to the rate of interest paid, such as 5 percent per year. The important thing about this type of growth is that it is compounded. That is, each dollar will provide $1.05 at the end of the first year, but the interest on the second year is paid on the $1.05 accumulated from the first year, not on the original amount of $1.

The figure below shows an example of exponential growth at two rates—3 percent and 6 percent per time period. The vertical axis shows the multiplier that is applied to the original amount. For example, after 20 time periods, the multiplier for the 6 percent growth rate is slightly more than 3, meaning the original quantity will have more than tripled in that time period at that growth rate. In contrast, the

multiplier for the 3 percent growth rate is less than 2.

Here is an approximate formula for exponential growth:

$$M = e^{(R \times T)}$$

where M is the multiplier; R is the rate, expressed as a decimal fraction of increase per time period (i.e., $R = 0.04$ for 4 percent per year); and T is the number of time periods (i.e., years) to elapse. The symbol e stands for the base of natural logarithms, which has a value of 2.718. To evaluate the expression, first find $(R \times T)$, then raise e to that power. Most scientific calculators will have a key labeled "exp" or "e^x" to evaluate this function.

Here's an example. Suppose the CO_2 concentration in the atmosphere is 350 ppm at present and is increasing at a rate of 4 percent per year. What concentration can we expect in 20 years if the present rate of growth continues?

$$M = e^{(R \times T)} = e^{(0.04 \times 20)}$$
$$= 2.718^{0.80} = 2.26$$

Thus, the concentration of 350 ppm will grow to about $350 \times 2.26 = 779$ ppm.

Sometimes it is convenient to think in terms of a doubling time—that is, the time it will take for a quantity to double given that it is growing exponentially at a fixed percentage rate. The doubling time is shown graphically for growth rates of 3 and 6 percent in the figure at right. It will be the time associated with the multiplier 2 (for doubling). For 3 and 6 percent, the doubling time is 11.6 and 23.1, respectively. There is a handy rule to figure the doubling time from the growth rate—divide the interest rate in percent into 70, and the result will be the doubling time. So, a 4 percent annual growth rate would lead to a doubling time of $70/4 = 17.5$ years.

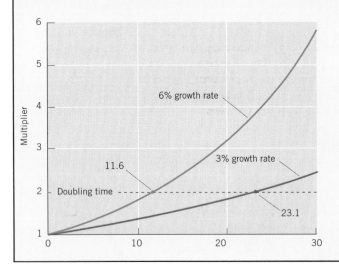

Exponential growth. The multiplier for a quantity growing at a constant percentage rate increases with time. The higher the growth rate, the more quickly the multiplier increases.

veyor belt, moving CO_2 from the surface to ocean depths and releasing it again in a cycle lasting about 1500 years. At present, scientists estimate that ocean surface waters absorb more CO_2 than they release, owing to increased atmospheric levels of CO_2. Therefore, carbon dioxide may be accumulating in ocean depths. However, current studies of global climate

computer simulations indicate that the oceans may not be as effective in removing excess CO_2 as scientists previously thought.

Although there is a great deal of uncertainty about the movements and buildup rate of excess CO_2 released to the atmosphere by fossil fuel burning, one thing is certain. Without conversion to alternative energy sources, fuel consumption will continue to release carbon dioxide, and it is only a matter of time before the earth will feel its impact.

The Temperature Record

Has the buildup of CO_2 caused global temperatures to rise? Before addressing this important question, we need to review the record of the earth's surface temperature over the last few centuries. Figure 3.24 shows the earth's mean annual surface temperature from 1880 to 1994. The temperature is expressed as a difference from the average annual temperature for the period 1951–1980. Two curves are shown—yearly data and smoothed data, taken from a five-year moving average. Although temperature has increased since the last century, and especially toward the end, there have been wide swings in the mean annual surface temperature.

Several theories have been proposed to explain the variations. For example, some measures of solar activity seem to suggest that varying solar output induces similarly varying temperatures. If this theory is correct, then the present warming may not be the result of greenhouse gas accumulation. Instead, solar cycles may be responsible.

Another factor to consider is volcanic activity. Dust and gases—especially sulfur dioxide, SO_2—are emitted in major eruptions and are typically carried upward into the stratosphere. SO_2 molecules rapidly take up water and form tiny sulfate particles. Strong stratospheric winds spread the volcanic particles quickly throughout the entire atmospheric layer. These parti-

cles have a cooling effect on the troposphere because they block incoming solar radiation.

For example, the eruption of Mount Pinatubo in the Philippines in the spring of 1991 lofted 15 to 20 million tons of sulfuric acid particles into the stratosphere (Figure 3.25). The layer of particles produced by the eruption reduced solar radiation reaching the earth's surface between 2 and 3 percent for the year or so following the blast. In response, global temperatures fell 0.5 to 0.7°C (0.9 to 1.3°F). Thus, volcanic activity may offset warming of the earth's surface from other causes.

Although the period of direct air temperature measurement does not extend past the middle of the last century, the record can be extended further back by using tree-ring analysis. The principle is simple: Each year, trees grow in diameter. In climates where the seasons are distinct, this growth can appear as an annual ring. If growing conditions are good, the annual ring is wide. If poor, the annual ring is narrow. For trees along the timberline in North America, the width is related to temperature—the trees grow better when temperatures are warmer. Since only one ring is formed each year, the date of each ring is easy to determine by counting backward from the present. Because the trees are quite old, the temperature record may be extended backward several centuries.

Figure 3.26 shows a reconstruction of northern hemisphere temperatures from 1700 to about 1980 using tree-ring analysis. Like Figure 3.24, it is expressed as a difference from a recent average (1950–1965). From 1880 to 1970, the temperatures reconstructed from tree-ring analysis seem to fit the observed temperatures (Figure 3.24) quite well. Analysis of an earlier period shows us another cycle of temperature increases and decline. The low point, around 1840, marks a cold event during which European alpine glaciers became more active and advanced. Other evidence indicates that the two cycles in Figure 3.26 are part of a natural global cycle of temperature warming and cooling lasting about 150 to 200 years. These cycles have occurred regularly over the last thousand years. What causes these cycles? Again, a number of theories have been offered, but no consensus has yet been reached as to the actual cause or causes.

Still longer cycles of temperature change are evident in the fossil record, inducing the advance and retreat of continental and mountain glaciers during the Ice Age. These cycles are discussed in more detail in Chapter 18.

Thus, the temperature record shows that the earth's climate is naturally quite variable, responding to many different influences on many different time scales. This fact makes it difficult to conclude that human activity has a role in changing global climate. However, convincing new evidence has now demonstrated human influence on global climate.

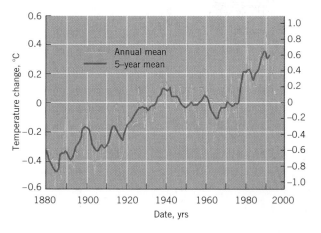

Figure 3.24 Mean annual surface temperature of the earth, 1880–1994. The vertical scale shows departures in degrees from a zero line of reference representing the average for the years 1951–1980. The yellow line shows the mean for each year. The red line shows a running five-year average. Note the effect of Mount Pinatubo in 1992–1993. (James Hansen/NASA Goddard Institute for Space Studies.)

Figure 3.25 Eruption of Mount Pinatubo, Philippine Islands, April 1991. Volcanic eruptions like this can inject particles and gases into the stratosphere, influencing climate for several years afterward.

Future Scenarios

The year 1995 was a record year in two respects. First, it was the warmest year on record since the middle of the nineteenth century, with an average temperature of 14.8°C (58.6°F) measured by one dataset and 15.4°C (59.7°F) by another. Second, it was the year in which the Intergovernmental Panel on Climate Change, a United Nations-sponsored group of more than 2000 scientists, concluded for the first time that human activity has caused climatic warming. This judgment was based largely on computer simulations of global climate that account for the release of CO_2 and SO_2 from fossil fuel burning occurring since the turn of the century. The simulations agreed well with the patterns of warming observed over that period, leading to the con-

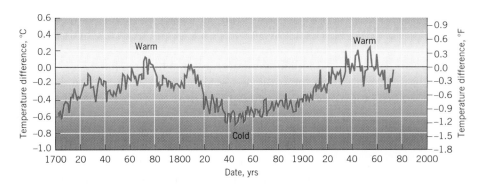

Figure 3.26 A reconstruction of the departures of northern hemisphere temperatures from the 1950–1965 mean, based on analyses of tree rings sampled along the northern tree limit of North America. (Courtesy of Gordon C. Jacoby of the Tree-Ring Laboratory of the Lamont-Doherty Geological Observatory of Columbia University.)

clusion that, in spite of natural variability, human influence has been felt in the climate record of the twentieth century.

Given that CO_2 and other greenhouse gases are warming the planet, what will be the effect? Using various projections of the continuing release of greenhouse gases coupled with computer climate models, this same group of scientists projected that global temperatures will warm between 1°C (1.8°F) and 3.5°C (6.3°F) by the year 2100, with the most likely value of 2°C (3.6°F).

Why does this temperature rise cause concern? The problem is that many other changes may accompany a rise in temperature. One of these is a rise in sea level, as glaciers and sea ice melt in response to the warming. Current predictions call for a rise of 19 to 90 cm (7.5 to 35.4 in.) in sea level by the year 2100, with a best estimate of 48 cm (18.9 in.). This would place as many as 92 million people within the risk of annual flooding. Climate change could also promote the spread of insect-borne diseases such as malaria. Furthermore, climate boundaries may shift their positions, making some regions wetter while others become drier. Thus, agricultural patterns could shift, displacing large human populations as well as natural ecosystems.

A recently discovered effect of climatic warming is the enhancement of variability in climate. Events such as very high 24-hour precipitation—extreme snowstorms, rainstorms, sleet and ice storms, for example—appear to be occurring more frequently since 1980. More frequent and more intense spells of hot and cold weather may also be related to climatic warming.

The world has become widely aware of the problem of the buildup in CO_2 and other greenhouse gases. At the Rio de Janeiro Earth Summit in 1992, nearly 150 nations signed a treaty limiting emissions of greenhouse gases. Under the terms of the treaty, emissions of the large industrial nations were to be reduced and held to 1990 levels by the year 2000. These nations planned to accomplish this objective largely by making more efficient use of fuels and by curbing the production of chlorofluorocarbons. As of early 1995, however, only three of the top ten emitters—Russia, Germany and Britain—were expected to achieve their targets. Worldwide emissions of CO_2 continued to rise at around 20 percent per decade. In a followup climate convention, held in Berlin in April 1995, 120 nations agreed to begin negotiations on greenhouse gas reductions beyond the year 2000. A key issue is the role of the emissions of developing nations, which have thus far remained unrestricted. Although these efforts will slow the buildup of greenhouse gases, the ultimate solution is probably greater reliance on solar and geothermal energy sources, which produce power without releasing CO_2.

In this chapter we have developed an understanding of both air temperatures and temperature cycles, along with the factors that influence them. Air temperatures are a very important part of climate, and looking ahead, we will study the climates of the earth in Chapters 7–9. The other key ingredient of climate is precipitation—the subject covered in the next chapter.

CHAPTER SUMMARY

The two major cycles of air temperature—daily and annual—are controlled by the cycles of insolation produced by the rotation and revolution of the earth. These cycles induce cycles of net radiation at the surface. When net radiation in the daily cycle is positive, air temperatures increase. When negative, air temperatures decrease. This principle applies for both daily and annual temperature cycles. Days are warm and nights are cool as net radiation goes from positive to negative. Summers are warm and winters are cold because average net radiation is high in the summer and low in the winter. Surface characteristics affect temperatures, too. Rural surfaces are generally moist and slow to heat, while urban surfaces are dry and absorb heat readily. This difference creates an urban heat island effect.

Air temperatures normally fall with altitude in the troposphere. At the tropopause, this decrease stops. In the stratosphere above, temperatures increase slightly with altitude. Air temperatures observed at mountain locations are lower with higher elevation, and day–night temperature differences increase with elevation.

Daily and annual air temperature cycles are influenced by maritime or continental location. Ocean temperatures vary less than land temperatures because water heats more slowly and can both mix and evaporate freely. Maritime locations that receive oceanic air therefore show smaller ranges of daily and annual temperature.

Global temperature patterns for January and July show the effects of latitude and maritime–continental location. Equatorial temperatures vary little from season to season. Poleward, temperatures decrease with latitude, and continental surfaces at high latitudes can become very cold in winter. Isotherms over continents swing widely north and south with the seasons, while isotherms over oceans move through a much smaller range of latitude.

Our planet's global temperature changes from year to year. Within the last few decades, global temperatures have been increasing. Researchers now attribute some portion of that increase to the human-induced buildup of carbon dioxide and other gases that enhance the greenhouse effect. Global temperatures are projected to rise significantly if we continue to release large quantities of greenhouse gases. Climatic warming could lead to a rise in sea level, threatening coastal populations, and also cause agricultural patterns and natural ecosystems to shift in response to changing climate boundaries. An increase in the number of extreme climatic events has been linked to the change in climate and could increase in the future. The reduction in release of greenhouse gases by world nations is the subject of ongoing international diplomatic activity.

KEY TERMS

temperature	heat island	stratosphere
conduction	lapse rate	temperature inversion
convection	environmental tempera-	isotherm
air temperature	ture lapse rate	greenhouse gases
transpiration	troposphere	

REVIEW QUESTIONS

1. How are mean daily temperature and mean monthly temperature determined?

2. Sketch graphs showing how insolation, net radiation, and temperature might vary from midnight to midnight during a 24-hour cycle at a midlatitude station such as Chicago.

3. How does the daily temperature cycle measured within a few centimeters or inches of the surface differ from the cycle at normal air temperature measurement height?

4. Compare the characteristics of urban and rural surfaces and describe how the differences affect urban and rural air temperatures. Include a discussion of the urban heat island.

5. What are the two layers of the lower atmosphere? How are they distinguished?

6. How and why are the temperature cycles of high mountain stations different from those of lower elevations?

7. Sketch a graph of air temperature with height showing a low-level temperature inversion. Where and when is such an inversion likely to occur?

8. Why do large water bodies heat and cool more slowly than land masses? What effect does this have on daily and annual temperature cycles for coastal and interior stations?

9. What three factors are most important in explaining the world pattern of isotherms? Explain how and why each factor is important, and what effect it has.

10. Turn to the January and July world temperature maps shown in Figure 3.19. Make six important observations about the patterns and explain why each occurs.

11. Turn to the world map of annual temperature range in Figure 3.21. What five important observations can you make about the annual temperature range patterns? Explain each.

12. Why has the atmospheric concentration of CO_2 increased in recent years?

13. How does plant life affect the level of atmospheric CO_2?

14. What is the role of the ocean in influencing atmospheric levels of CO_2?

15. Describe how global air temperatures have changed in the recent past. Identify some factors or processes that influence global air temperatures on this time scale.

16. Has human activity thus far influenced global climate? On what evidence is this conclusion based? How could human-induced climatic warming affect our environment?

Focus on Systems 3.2 • The Surface Energy Balance Equation

1. Consider the surface energy balance during the day. What happens if the surface falls under the shadow of a cloud? How will energy flows to and from the surface be affected? What do you expect will happen to the temperature of the surface?

2. Suppose that on a hot sunny day, a surface layer of soil dries out so that water is no longer available for evaporation. How will energy flows to and from the surface be affected? What will happen to the surface temperature?

3. Suppose that the air temperature is measured at standard height (1.2 m, 4 ft) above the surface. Would you expect this air temperature to be warmer or colder than the surface during the day? Why? What about at night? Why?

ESSAY QUESTIONS

1. Portland, Oregon, on the north Pacific coast, and Minneapolis, Minnesota, in the interior of the North American continent, are at about the same latitude. Sketch the annual temperature cycle you would expect for each location. How do they differ and why? Select one season, summer or winter, and sketch a daily temperature cycle for each location. Again, describe how they differ and why.

2. Many scientists have concluded that human activities are acting to raise global temperatures. What human processes are involved? How do they relate to natural processes? Are global temperatures increasing now? What other effects could be influencing global temperatures? What are the consequences of global warming?

PROBLEMS

Working It Out 3.1 • Temperature Conversion

1. On a summer day, a Toronto radio station broadcasts a weather report forecasting a high of 38°C. What is this temperature in °F? On a winter day, a radio station in Buffalo broadcasts a weather report forecasting the same high (38°), but in degrees Fahrenheit. What is this temperature in °C?

2. A maximum-minimum thermometer records a high of 46°F and a low of 28°F. What is the temperature range in °C?

3. At what temperature will both Celsius and Fahrenheit thermometers display the same value?

(Hint: Check your work using Figure 3.1.)

Working It Out 3.3 • Exponential Growth

1. If the present concentration of CO_2 is 350 ppm and it increases at a rate of 2 percent per year, what will the concentration be in 50 years? What will the concentration be if CO_2 is increasing at a rate of 3 percent?

2. The population of Singapore is about 2.8 million, and it is increasing at an annual rate of 1.3 percent. The Republic of Congo has a population of about 2.4 million, now increasing at an annual rate of 3.0 percent. What is the doubling time of each population? How large will the populations of these two countries be in 25 years if growth continues at the present rates?

Chapter 4

Atmospheric Moisture and Precipitation

Water exists in the air in the form of water vapor, clouds, fog, and precipitation. Water vapor is unseen but felt as an uncomfortable dampness in the air on hot, humid days. Clouds are visible as fleecy white or ominous gray masses in the sky overhead. Fog is a low cloud that surrounds you like a cold, damp blanket. And precipitation forms as raindrops and snowflakes that fall from the sky, providing our source of fresh water.

How are water vapor, clouds, fog, and precipitation related? How does water vapor, an invisible gas, be-

come transformed into a torrent of rain that falls to earth? This subject is the main focus of this chapter. But before we begin our study of atmospheric moisture and precipitation, we will briefly review the three states of water and the conversion of one state to another.

As shown in Figure 4.1, water can exist in three states—solid (ice), liquid (water), and gas (water vapor). A change of state from solid to liquid, liquid to gas, or solid to gas requires the input of heat energy. As we noted in Chapter 2, this energy is *latent heat*, which is drawn in from the surroundings. When the change goes the other way, from liquid to solid, gas to liquid, or gas to solid, this latent heat is released to the surroundings.

A rainy day in Tokyo.

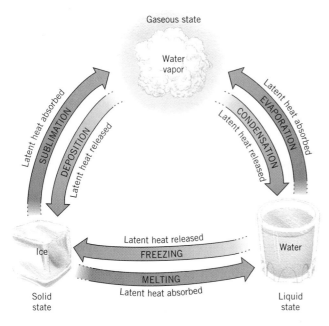

Figure 4.1 A schematic diagram of the three states of water. Arrows show the ways that any one state can change into either of the other two states. Heat energy is absorbed or released, depending on the direction of change.

For each type of transition, there is a specific name, shown in the figure. Melting, freezing, evaporation, and condensation are all familiar terms. **Sublimation** describes the direct transition from solid to vapor. Perhaps you have noticed that old ice cubes in your refrigerator's freezer seem to shrink away from the sides of the ice cube tray and even get smaller over time. In this case, they shrink through sublimation—which is induced by the constant circulation of cold, dry air through the freezer. The ice cubes never melt, yet they lose bulk directly as vapor. **Deposition** is the reverse process, when water vapor crystallizes directly as ice. Frost formed on a cold winter night is an example. The amounts of latent heat taken up and released with changes of state of water are presented in *Working It Out 4.1 • Energy and Latent Heat.*

WATER—THE GLOBAL PERSPECTIVE

Water plays several key roles on our planet. First, the oceans cover more than two-thirds of the planet's surface. They act as a huge heat storage reservoir and redistribute heat from low to high latitudes by ocean currents. The oceans also act as a storage reservoir for dissolved compounds, ranging from salts to nutrients. Second, water falls on land in the form of rain or snow. As it runs off to the sea, water erodes rocks and soils and creates landscapes and landforms. This flow moves nutrients from one location to another, which also influences the distribution of plant and animal life. Third, water in the air moves huge quantities of heat from one place to another by absorbing surface heat in evaporation over warm oceans and releasing that latent heat in condensation or deposition over cooler regions. Like the flow of heat in ocean currents, this movement is generally poleward.

Let's now turn to the nature of the hydrosphere and the flows of water among ocean, land, and atmosphere.

The Hydrosphere and the Hydrologic Cycle

Recall from Chapter 1 that the hydrosphere includes water on the earth in all its forms. About 97.2 percent of the hydrosphere consists of ocean saltwater, as shown in Figure 4.2. The remaining 2.8 percent is fresh water. The next largest reservoir of fresh water is stored as ice in the world's ice sheets and mountain glaciers. This water accounts for 2.15 percent of total global water.

Fresh liquid water is found both on top of and beneath the earth's land surfaces. Water occupying openings in soil and rock is called *subsurface water*. Most of it is held in deep storage as *groundwater*, at a level where plant roots cannot access it. Groundwater makes up 0.63 percent of the hydrosphere, leaving 0.02 percent of the water remaining.

The right-hand portion of Figure 4.2 shows how the small remaining proportion of the earth's water is distributed. This proportion is important to us because it includes the water available for plants, animals, and human use. *Soil water*, which is held in the soil within

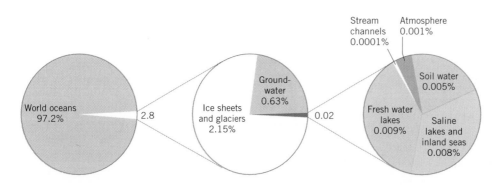

Figure 4.2 Volumes of global water in each reservoir of the hydrosphere. Nearly all the earth's water is contained in the world ocean. Fresh surface and soil water make up only a small fraction of the total volume of global water.

Working It Out 4.1 • Energy and Latent Heat

In the metric system, energy is measured in joules. The *joule* (J) is defined as one newton-meter, where a newton is the force produced by an acceleration of 1 m/s² applied to a 1-kg mass. The joule thus has units of force times distance, which is a measure of work or energy. The joule is related to the watt, since one watt is defined as the flow of one joule of energy per second. From this definition, the energy emitted by a 100-watt bulb in one second is 100 joules. In the English system, energy is measured by the British thermal unit (Btu). However, this unit is little-used, and thus we do not indicate conversions of joules to Btu in the text.

The amount of latent heat taken up or released by water when a change of state occurs depends on

the change of state and the temperature of the water. For melting and freezing, the energy amounts to about 335 kilojoules per kilogram (kJ/kg) when water changes between a liquid at 0°C and a solid at the same temperature. For evaporation and condensation, the energy required or released is about 2260 kJ/kg at a temperature of 100°C. Note also that energy is required to heat the water from 0°C to 100°C. The amount required is 4.19 kJ/kg for each Celsius degree, or 419 kJ/kg for 100°. (The values given above depend partly on atmospheric pressure and are taken for sea-level conditions.)

To determine the amount of energy required for sublimation from solid to gas, we can add the amounts of heat required first for

melting, then for warming to the boiling point, and finally, for evaporating. That is, 335 + 419 + 2260 = 3014 kJ/kg. This will also be the amount of energy released when deposition occurs.

What quantity of energy is required to evaporate a kilogram of liquid water at 25°C? This quantity will be the sum of the energy required first to bring the water to the evaporation point and then the latent heat required to convert it to vapor. For the first quantity, each degree requires 4.19 kJ/kg, and there are 100 − 25 = 75°C, so 4.19 × 75 = 314 kJ/kg are needed. Added to the 2260 kg necessary for the change of state, we have 314 + 2260 = 2574 kJ/kg.

reach of plant roots, comprises 0.005 percent of the global total. Water held in streams, lakes, marshes, and swamps is called *surface water*. Most of this surface water is about evenly divided between freshwater lakes and salty (saline) lakes. An extremely small proportion is held in streams and rivers as they flow toward the sea or inland lakes.

Note that the quantity of water held as vapor and cloud water droplets in the atmosphere is also very small—0.001 percent of the hydrosphere. Though small, this reservoir of water is of enormous importance. It provides the supply of precipitation that replenishes all freshwater stocks on land. And, as we will see in the next chapter, the flow of water vapor from warm tropical oceans to cooler regions provides a global flow of heat, in latent form, from low to high latitudes.

The movements of water among the great global reservoirs constitute the **hydrologic cycle**. In this cycle, water moves from land and ocean to the atmosphere as water vapor and returns as precipitation. Because precipitation over land exceeds evaporation, water also runs off the land to the oceans. We will return to the hydrologic cycle in more detail in Chapter 15.

For now, we can summarize the main features of the hydrologic cycle in the global water balance, shown in Figure 4.3. The cycle begins with evaporation from water or land surfaces, in which water changes state

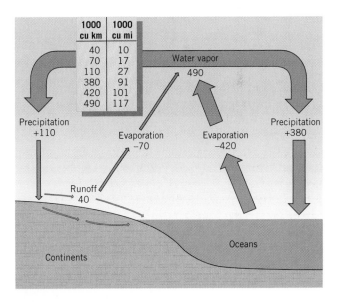

Figure 4.3 The global water balance. Figures give average annual water flows in and out of world land areas and world oceans. Values are given in thousands of cubic kilometers (cubic miles). Global precipitation equals global evaporation. (Based on data of John R. Mather.)

Focus on Systems 4.2 • The Global Water Balance as a Matter Flow System

Since our planet contains only a fixed amount of water, a global balance must be maintained among flows of water to and from the lands, oceans, and atmosphere. Let's examine this idea in more detail. For our analysis, we will assume that the volume of ocean waters and the overall volume of fresh water in surface and subsurface water remains constant from year to year. This is probably quite reasonable, unless climate is changing rapidly.

The global water balance can be seen as a flow system diagrammed below, which is derived from Figure 4.3. In this flow system, the material that is flowing is water, primarily as liquid and solid water in precipitation and as water vapor in evaporation. For our analysis, we will consider the surface as consisting of two components—land and oceans. Water on the land portion of the earth's surface is largely in the form of liquid water, but significant amounts are also present in solid form as snow and ice. The ocean-covered portion of the surface is largely liquid water, but the Arctic Ocean and some parts of the ocean near Antarctica have a cover of sea ice that is often topped with snow.

Flows of water link the atmosphere and the two surface components. Following Figure 4.3, these are marked with their values in units of thousands of cubic kilometers per year. For budgeting purposes, we adopt the convention that flows to the surface (land or oceans) are positive and flows leaving the surface are negative.

Let's first look at the flows into and out of the atmosphere. Since the cycle is in balance, they sum to zero. That is,

$$P_{LAND} + E_{LAND} + P_{OCEAN} + E_{OCEAN} = 0$$
$$(+110) + (-70) + (+380) + (-420) = 0$$

Here, P_{LAND} and P_{OCEAN} are precipitation flows from the atmosphere to the land and oceans, and E_{LAND} and E_{OCEAN} are evaporation flows from the land and oceans to the atmosphere. (Note that in the term *precipitation,* we include here flows of water in both liquid (rain) and solid (snow) forms as well as direct deposition of water vapor as frost. Similarly, we take evaporation here to include sublimation of snow and ice to water vapor.)

In looking at these values, we also note that although the global balance for the atmosphere is zero, there is a positive balance for land

(+40), indicating more precipitation than evaporation, and a negative balance for oceans (-40), indicating more evaporation than precipitation. Thus, each year land gains 40,000 km^3 of water while oceans lose 40,000 km^3 of water. Without a link between land and water, this would mean that the oceans would eventually empty and all the water would accumulate over land! Of course, this doesn't happen because a flow pathway connects land and oceans—runoff of water from rivers, glaciers and ice sheets into the ocean. For the oceans to stay at the same level, and for the same amount of water to be stored on the land year after year, the flow in runoff must equal this difference—40 units, or 40,000 km^3.

Recall from earlier discussions of flow systems that all matter flow systems need a power source. What is the power source for this vast flow system of water in all its forms? The answer, of course, is the sun. Solar energy evaporates water and moves it into the atmosphere, where it can later fall to earth as precipitation. Gravity also plays a role in moving streams of water and ice off the lands and into the oceans as runoff.

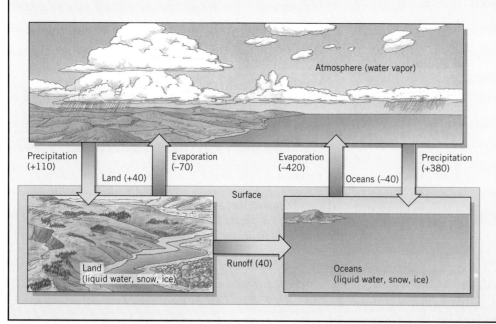

The global water balance as a flow system.

from liquid to vapor and enters the atmosphere. Total evaporation is about six times greater over oceans than land, however. This is because the oceans cover most of the planet and because land surfaces are not always wet enough to yield much evaporated water. Once in the atmosphere, water vapor can condense or deposit to form precipitation, which falls to earth as rain, snow, or hail. Precipitation over the oceans is nearly four times greater than precipitation over land.

Upon reaching the land surface, precipitation has three fates. First, it can evaporate and return to the atmosphere as water vapor. Second, it can sink into the soil and then into the surface rock layers below. As we will see in later chapters, this subsurface water emerges from below to feed rivers, lakes, and even ocean margins. Third, precipitation can run off the land, concentrating in streams and rivers that eventually carry it to the ocean or to a lake in a closed inland basin. This flow of water is known as *runoff*.

Focus on Systems 4.2 • The Global Water Balance as a Matter Flow System provides a systems view of the hydrologic cycle. Evaporation, precipitation, and runoff are flow pathways for water in its various forms that are interconnected in this closed matter flow system.

For most of this chapter, we will be concerned with one aspect of the hydrologic cycle—the flow of water from the atmosphere to the surface in the form of precipitation. The examination of this aspect of the hydrologic cycle will involve descriptions of how the water vapor content of air is measured and how clouds and precipitation form.

HUMIDITY

The amount of water vapor present in the air varies widely from place to place and time to time. It ranges from almost nothing in the cold, dry air of arctic regions in winter to as much as 4 or 5 percent of a given volume of air in the warm wet regions near the equator. The general term **humidity** refers to the amount of water vapor present in the air.

Understanding humidity and how the moisture content of air is measured involves an important principle—namely, that the maximum quantity of moisture that can be held at any time in the air is dependent on temperature. Warm air can hold more water vapor than cold air—a lot more. Air at room temperature (20°C, 68°F) can hold about three times as much water vapor as air at freezing (0°C, 32°F).

The measure of humidity that we encounter every day is *relative humidity*. This measure compares the amount of water vapor present to the maximum amount that the air can hold at that temperature, expressed as a percentage. For example, if the air currently holds half the moisture possible at the present temperature, then the relative humidity is 50 percent.

When the humidity is 100 percent, the air holds the maximum amount possible and is *saturated*.

A change in relative humidity of the atmosphere can happen in one of two ways. The first is by evaporation. If an exposed water surface or wet soil is present, additional water vapor can enter the air. This process is slow because the water vapor molecules must diffuse upward from the surface into the air layer above.

The second way is through a change of temperature. Even though no water vapor is added, a lowering of temperature results in a rise of relative humidity. Recall that the capacity of air to hold water vapor is dependent on temperature. When the air is cooled, this capacity is reduced. The existing amount of water vapor then represents a higher percentage of the total capacity.

Relative Humidity Through the Day

An example may help to illustrate these principles (Figure 4.4). Beginning in the early morning hours, at 4 A.M. the temperature is 5°C (41°F), and the relative humidity of the air is 100 percent. That is, the air is saturated and cannot hold any additional water vapor. By 10 A.M., the temperature has risen to 16°C (61°F). The relative humidity has dropped to 50 percent, even though the amount of water vapor in the air remains the same. By 3 P.M., the air has been warmed by the sun to 32°C (90°F). The relative humidity has dropped to 20 percent, which is very dry air. The same amount of water vapor is present in the air, but the capacity of the air to hold water vapor has greatly increased. As air temperature falls in the evening, relative humidity again rises, and the cycle repeats itself.

Another way of describing the water vapor content of air is by its **dew-point temperature**. If air is slowly chilled, it eventually will reach saturation, with a relative humidity of 100 percent. At this temperature, the

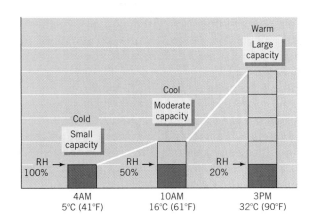

Figure 4.4 Relative humidity changes with temperature because the capacity of warm air to hold water vapor is greater than that of cold air. In this example, the amount of water vapor stays the same, and only the capacity changes.

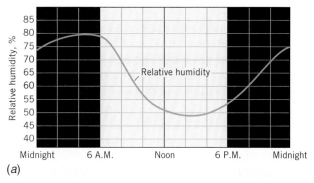

(a)

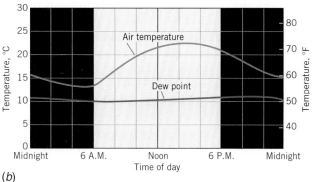

(b)

Figure 4.5 (*a*), Relative humidity, and (*b*), air temperature and dew-point temperature throughout the day at Washington, D.C. The curves show average values for the month of May. (Data from National Weather Service.)

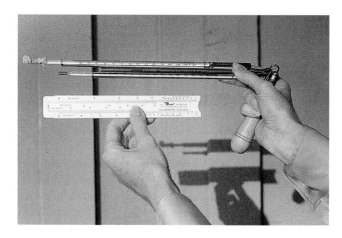

Figure 4.6 A standard sling psychrometer. Pictured below the psychrometer is a sliding scale that enables rapid determination of relative humidity from wet and dry bulb readings.

air holds the maximum amount of water vapor possible. If further cooling continues, condensation will begin and dew will form. The temperature at which saturation occurs is therefore known as the dew-point temperature. Moist air will have a higher dew-point temperature than drier air.

The principle of relative humidity change caused by temperature change is further illustrated by a graph of these two properties throughout the day (Figure 4.5). Values shown are averages for May at Washington, D.C. As air temperature rises during the day, relative humidity falls, and vice versa. The dew-point temperature, which indicates the true moisture content, remains nearly constant.

How is humidity measured? A simple method uses two thermometers mounted together side by side in an instrument called a *sling psychrometer* (Figure 4.6). The wet-bulb thermometer bulb is covered with a sleeve of cotton fibers and is wetted. The dry-bulb thermometer remains dry. To operate the sling psychrometer, water is applied to the sleeve covering the wet bulb, and the two thermometers are then whirled in the open air. Evaporation cools the wet bulb. If the air is saturated, then evaporation cannot occur. In this case, no cooling results, and the temperature of the two thermometers is the same. If the air is not saturated, evaporation occurs and the wet-bulb thermometer is cooled. The drier the air, the greater is the cooling. To determine

the relative humidity, a special sliding scale is used (Figure 4.6). Wet-bulb and dry-bulb temperatures are set on the scale, and relative humidity is read off directly. There are also instruments that read relative humidity directly. One such instrument uses a thin layer of a special material bonded to a metal film. The material absorbs water vapor in an amount depending on the relative humidity. The water vapor affects the ability of the metal film to hold an electric charge. This ability is sensed by an electronic circuit and converted to a direct reading of relative humidity.

Specific Humidity

Relative humidity measures only the amount of water vapor present relative to the amount held at saturation. The actual quantity of water vapor held by the air is its *specific humidity*. This quantity is stated as the mass of water vapor contained in a given mass of air and is expressed as grams of water vapor per kilogram of air (g/kg).

We stated earlier that the amount of water vapor that air can hold depends on its temperature. Figure 4.7 shows this relationship. We see, for example, that at 20°C (68°F), the maximum amount of water vapor that the air can hold—that is, the maximum specific humidity—is about 15 g/kg. At 30°C (86°F), it is almost doubled—about 26 g/kg. For cold air, the values are quite small. At –10°C (14°F), the maximum is only about 2 g/kg.

Specific humidity is often used to describe the moisture characteristics of a large mass of air. For example, extremely cold, dry air over arctic regions in winter may have a specific humidity as low as 0.2 g/kg. In comparison, the extremely warm, moist air of equatorial regions often holds as much as 18 g/kg. The total natural range on a worldwide basis is very wide. In fact, the largest values of specific humidity observed are from 100 to 200 times as great as the smallest.

Specific humidity is a geographer's yardstick for a basic natural resource—water—that can be applied from equatorial to polar regions. It is a measure of the quantity of water that can be extracted from the atmosphere as precipitation. Cold, moist air can supply only a small quantity of rain or snow, but warm, moist air is capable of supplying large quantities.

Figure 4.8 is a set of global profiles showing how specific humidity varies with latitude and how it relates to mean surface air temperature. Both humidity and temperature are measured at the same locations in standard thermometer shelters the world over, and also on ships at sea. Note that the horizontal axis of the graphs has an uneven scale of units—they get closer together toward both left and right edges, which represent the poles. In this case, the units are adjusted so that they reflect the amount of the earth's surface present at that latitude. That is, since there is much more surface area between 0° and 10° latitude than between 60° and 70°, the region between 0° and 10° is allocated more width on the graph.

Let's look first at specific humidity (*a*). The curve clearly shows the largest values for the equatorial zones, with values falling off rapidly toward both poles. This curve follows the pattern of insolation quite nicely (see Figure 2.7). More insolation is available at lower latitudes, on average, to evaporate water in oceans or on moist land surfaces. Therefore, specific humidity values are higher at low latitudes than at high latitudes.

The global profile of mean (average) surface air temperature, graph (*b*), shows a similar shape to the specific humidity profile. We would expect this to be the case, since air temperature and maximum specific humidity vary together, as shown in Figure 4.7.

THE ADIABATIC PROCESS

Given that ample water vapor is present in a mass of air, how is that related to precipitation? In other words, how is the water vapor turned into liquid or solid particles that can fall to earth? The answer is by natural cooling of the air. Since the ability of air to hold water vapor is dependent on temperature, the air must give up water vapor if it is cooled to the dew point and below. Think about a moist sponge. To extract the water, you have to squeeze the sponge—that is, reduce its ability to hold water. Chilling the air is like squeezing the sponge.

How is air chilled sufficiently to produce precipitation? One mechanism for chilling air is nighttime cooling. As we have seen, the ground surface can become quite cold on a clear night through loss of longwave radiation. Thus, still air near the surface can be cooled below the condensation point, producing dew or frost. However, this mechanism is not sufficient to form pre-

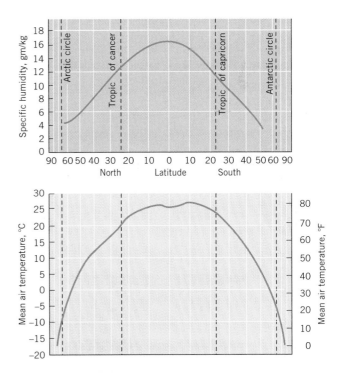

Figure 4.8 Pole-to-pole profiles of specific humidity (*above*) and temperature (*below*). The two are similar because the ability of air to hold water vapor (measured by specific humidity) is limited by temperature. (Data of J. von Hann, R. Süring, and J. Szava-Kovats as shown in Haurwitz and Austin, *Climatology*.)

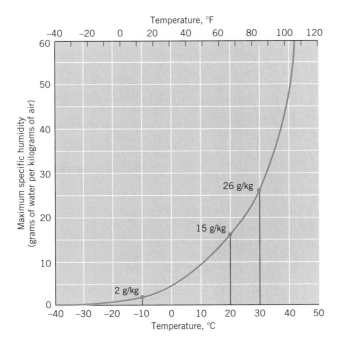

Figure 4.7 The maximum specific humidity of a mass of air increases sharply with rising temperature.

cipitation. Precipitation is only formed when a substantial mass of air experiences a steady drop in temperature below the dew point. This happens when the air mass is uplifted to a higher level in the atmosphere, as we will see in the following section.

Dry Adiabatic Rate

An important principle of physics is that when a gas is allowed to expand, its temperature drops. Conversely, when a gas is compressed, its temperature increases. You have probably observed this latter effect yourself, if you have ever pumped up a bicycle tire using a hand pump. In this process, the pump gets hot because air inside the pump is being compressed and therefore heated. In the same way, a small jet of air escaping from a high-pressure hose feels cool. Perhaps you have flown on an airplane and noticed a small nozzle overhead that directs a stream of cool air toward you. In moving from higher pressure in the air hose that feeds the nozzle to lower pressure in the cabin, the air expands and cools. Physicists use the term **adiabatic process** to refer to a heating or cooling process that occurs solely as a result of pressure change.

Given that air cools or heats when the pressure on it changes, how does that relate to uplift and precipitation? The missing link is simply that atmospheric pressure decreases with an increase in altitude. So, if a parcel of air is uplifted, atmospheric pressure on the parcel will be lower, and it will expand and cool. And if a parcel of air descends, atmospheric pressure will be higher, and it will be compressed and warmed (Figure 4.9). The **dry adiabatic lapse rate** describes this behavior. This rate has a value of about 10°C per 1000 m

(5.5°F per 1000 ft) of vertical rise. That is, if a mass of air is raised 1 km, its temperature will drop by 10°C. Or, in English units, if raised 1000 ft, its temperature will drop by 5.5°F. This is the "dry" rate because condensation does not occur.

Note that in the previous chapter we encountered the environmental temperature lapse rate. This is quite different from the dry adiabatic lapse rate. The dry adiabatic rate is always constant and is determined by physical laws. It applies to a mass of air in vertical motion. The temperature lapse rate is simply an expression of how the temperature of still air varies with altitude. This rate will vary from time to time and from one place to another, depending on the state of the atmosphere. No motion of air is implied for the temperature lapse rate.

Wet Adiabatic Rate

With the adiabatic process firmly in mind, let's examine the fate of a parcel of moist air that is moved upward in the atmosphere (Figure 4.10). We will assume that the parcel starts with a temperature of 20°C (68°F). As the parcel is moved upward, its temperature drops at the dry adiabatic rate, 10°C/1000 m (5.5°F/1000 ft). At 500 m (1600 ft), the temperature has dropped by 5°C (9°F) to 15°C (59°F). At 1000 m (3300 ft), the temperature has fallen to 10°C (50°F).

If the rising process continues, the air will be cooled to its dew point and thus to saturation. Condensation can then occur. This is shown on the figure as the *lifting condensation level*. A complication arises, however, because the dew-point temperature of the parcel depends partly on the atmospheric pressure and there-

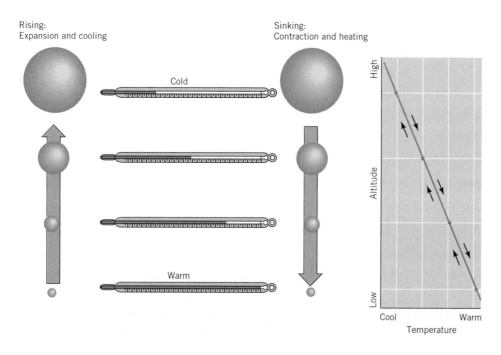

Rising:
Expansion and cooling

Sinking:
Contraction and heating

Cold

Warm

Altitude
High
Low

Cool   Warm
Temperature

Figure 4.9 A schematic diagram of adiabatic cooling and heating that accompanies the rising and sinking of a mass of air. When air is forced to rise, it expands and its temperature decreases. When air is forced to descend, its temperature increases. (A. N. Strahler.)

Working It Out 4.3 • The Lifting Condensation Level

When a parcel of air moves upward, it cools at the dry adiabatic rate of about 10°C/1000 m. This cooling occurs because the parcel is exposed to a lesser atmospheric pressure as it rises, and under the adiabatic principle, it expands and cools. The change in pressure also affects the dew-point temperature, which falls at a rate of about 1°C/1000 m. When the temperature of the cooling air parcel reaches the dew-point temperature, condensation will begin to occur. The elevation at which condensation begins to occur is referred to as the *lifting condensation level*.

Suppose an air parcel is at 20°C and is raised 500 m. What will its temperature be? Since the dry adiabatic rate is 10°C/1000 m, its temperature will drop by 500/1000 x 10 = 5°C. So, the temperature will be 20 − 5 = 15°C. Putting this in equation form,

$$T = T_0 - \frac{H}{1000} \times R_{DRY}$$

where T is the temperature of the parcel, T_0 is the starting temperature at height 0 m, H is the height in meters, and R_{DRY} is the dry adiabatic rate, 10°C/1000 m.

Let's assume that the dew point of that same parcel of air is 11°C. What will the dew-point temperature be at 500 m? Since the dew-point lapse rate is 1.8°C/1000 m, the dew-point temperature will fall by 500/1000 × 1.8 = 0.9°C. So, the dew-point temperature will be 11 − 0.9 = 10.1°C. Similarly, we can write

$$T_D = T_{DEW} - \frac{H}{1000} \times R_{DEW}$$

where T_D is the dew-point temperature of the parcel at height H, T_{DEW} is the starting dew-point temperature, and R_{DEW} is the dew-point lapse rate, 1.8°C/1000 m.

At what level will condensation occur? Condensation will occur when the parcel's temperature reaches the dewpoint—that is, at the height at which $T = T_D$. For this level, then, we can write

$$T_0 - \frac{H}{1000} \times R_{DRY} = T_{DEW} - \frac{H}{1000} \times R_{DEW} \quad (1)$$

After some algebraic rearrangement and substitution of the values for the dry adiabatic rate and the dew-point lapse rate, we have the formula

$$H = 1000 \times \frac{T_0 - T_{DEW}}{8.2} \quad (2)$$

which gives the lifting condensation level directly in meters.

What will be the lifting condensation level for the air parcel described above? Substituting,

$$H = 1000 \times \frac{20 - 11}{8.2} = 1000 \times \frac{9}{8.2} = 1098 \text{ m}$$

What will the temperature of the air parcel be at the lifting condensation level? By substituting in the first expression,

$$T = T_0 - \frac{H}{1000} \times R_{DRY} = 20 - \frac{1098}{1000} \times 10 = 9.0°C$$

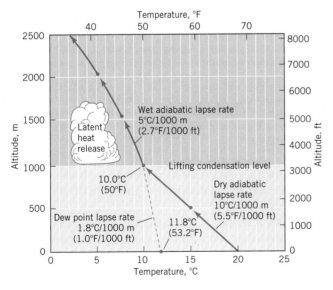

Figure 4.10 Adiabatic decrease of temperature in a rising parcel of air leads to condensation of water vapor into water droplets and the formation of a cloud. (Copyright © A. N. Strahler.)

fore the elevation of the parcel. Instead of remaining constant with elevation, the dewpoint temperature will fall at the *dew-point lapse rate* of about 1.8°C/1000 m (1.0°F/1000 ft). Suppose the dew-point temperature of the air mass is 11.8°C (53.2°F). Then at 1000 m (3300 ft), the dew-point temperature will be 11.8 − 1.8 = 10°C (50°F). That will also be the temperature of the parcel for the preceding example, so condensation will begin at that level. The lifting condensation level is thus determined by the initial temperature of the air and its dew point. *Working It Out 4.3 • The Lifting Condensation Level* shows how to calculate the lifting condensation level, given these two quantities.

If cooling continues, water droplets will form, producing a cloud. If the parcel of saturated air continues to rise, however, a new principle comes into effect—latent heat release. That is, when condensation occurs, latent heat is released and warms the uplifted air. In other words, two effects are occurring at once. First, the uplifted air is being cooled by the reduction in atmospheric pressure. Second, it is being warmed by the release of latent heat from condensation.

Which effect is stronger? As it turns out, the cooling effect is stronger, so the air will continue to cool as it is uplifted. But because of the release of latent heat, the cooling will occur at a lesser rate. This cooling rate is called the **wet adiabatic lapse rate**, and ranges between 4 and 9°C per 1000 m (2.2 and 4.9°F per 1000 ft). It is variable because it depends on the temperature and pressure of the air and its moisture content. For most situations, however, we can use a value of 5°C/1000 m (2.7°F/1000 ft). The higher rates apply only to cold, relatively dry air that contains little moisture and therefore little latent heat. In Figure 4.10, the wet adiabatic rate is shown as a slightly curving line to indicate that its value increases with altitude.

CLOUDS

A **cloud** is made up of water droplets or ice particles suspended in air. These particles have a diameter in the range of 20 to 50 mm (0.0008 to 0.002 in.). Recall from Chapter 2 that mm denotes the *micrometer*, or one-millionth of a meter. Each cloud particle is formed on a tiny center of solid matter, called a *condensation nucleus*. This nucleus has a diameter in the range 0.1 to 1 mm (0.000004 to 0.00004 in.).

An important source of condensation nuclei is the surface of the sea. When winds create waves, droplets of spray from the crests of the waves are carried rapidly upward in turbulent air. Evaporation of sea water droplets leaves a tiny residue of crystalline salt suspended in the air. This aerosol strongly attracts water molecules. Even very clean and clear air contains enough condensation nuclei for the formation of clouds. However, the heavy load of dust carried by polluted air over cities substantially aids condensation and the formation of clouds and fog.

In our everyday life at the earth's surface, liquid water turns to ice when the surrounding temperature falls to the freezing point, 0°C (32°F), or below. However, the water in tiny cloud particles can remain in the liquid state at temperatures far below freezing. Such water is described as *supercooled*. Clouds consist entirely of water droplets at temperatures down to about –12°C (10°F). As cloud temperatures grow colder, a mix of water droplets and ice crystals occur. The coldest clouds, with temperatures below –40°C (–40°F), occur at altitudes of 6 to 12 km (20,000 to 40,000 ft) and are formed entirely of ice particles.

Cloud Forms

Clouds come in many shapes and sizes—from the small, white puffy clouds often seen in summer to the dark layers that produce a good, old-fashioned rainy day. Meteorologists classify clouds into four families, arranged by height: high, middle, and low clouds, and clouds with vertical development. These are shown in Figures 4.11 and 4.12. Some individual types with their names are also shown.

Clouds are grouped into two major classes on the basis of form—stratiform, or layered clouds, and cumuliform, or globular clouds. *Stratiform* clouds are blanket-like and cover large areas. A common type is *stratus*, which covers the entire sky. Stratus clouds are formed when large air layers are forced to rise gradually. This can happen when one air layer overrides another. As the rising layer is cooled, condensation occurs over a large area, and a blanket-like cloud forms. If the layer is quite moist and rising continues, dense, thick stratiform clouds result that can produce abundant rain or snow.

Cumuliform clouds are globular masses of cloud that are associated with small to large parcels of rising air. The air parcels rise because they are warmer than the surrounding air. Like bubbles in a fluid, they move upward. Of course, as they are buoyed upward, they are cooled by the adiabatic process. When condensation occurs, a cloud is formed. The most common cloud of this type is the *cumulus* cloud (Figure 4.12). Sometimes the upward movement yields dense, tall clouds that produce thunderstorms. This form of cloud is the *cumulonimbus*. ("Nimbus" is the latin word for rain cloud, or storm.)

Fog

Fog is simply a cloud layer at or very close to the surface. In our industrialized world, fog is a major environmental hazard. Dense fog on high-speed highways can cause chain-reaction accidents, sometimes involving dozens of vehicles and causing injury or death to their occupants. At airports, landing delays and shutdowns because of fog bring economic losses to airlines and can inconvenience thousands of travelers in a single day. For centuries, fog at sea has been a navigational hazard, increasing the danger of ship collisions and groundings. In addition, polluted fogs, like those of London in the early part of this century, can injure urban dwellers' lungs and take a heavy toll in lives.

One type of fog, known as *radiation fog*, is formed at night when temperature of the air layer at the ground level falls below the dew point. This kind of fog is associated with a low-level temperature inversion (Figure 3.11). Another fog type, *advection fog*, results when a warm, moist air layer moves over a cold or surface. As the air layer loses heat to the surface, its temperature drops below the dew point, and condensation sets in. Advection fog commonly occurs over oceans where warm and cold currents occur side by side. When warm, moist air above the warm current moves over the cold current, condensation occurs. Fogs of the Grand

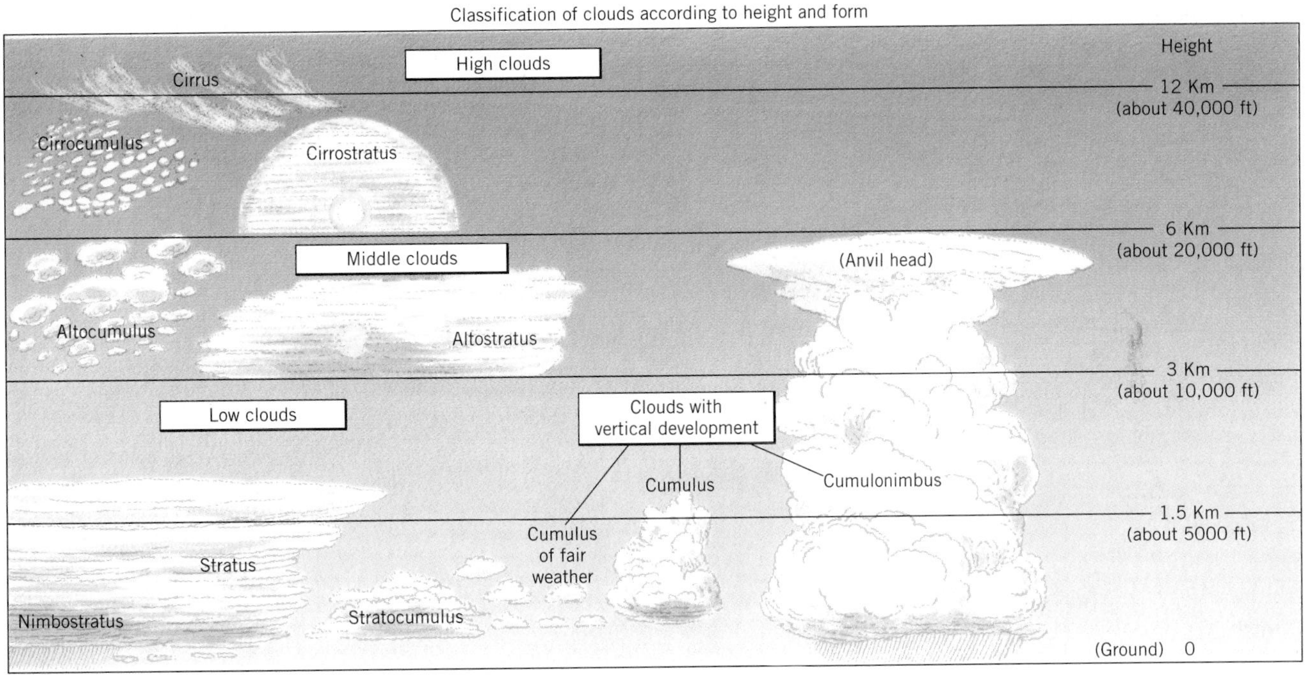

Classification of clouds according to height and form

High clouds

Cirrus

Cirrocumulus

Cirrostratus

Height

— 12 Km —
(about 40,000 ft)

— 6 Km —
(about 20,000 ft)

Middle clouds

(Anvil head)

Altocumulus

Altostratus

— 3 Km —
(about 10,000 ft)

Low clouds

Clouds with
vertical development

Cumulus

Cumulonimbus

Cumulus
of fair
weather

— 1.5 Km —
(about 5000 ft)

Stratus

Nimbostratus

Stratocumulus

(Ground) 0

Figure 4.11 Clouds are grouped into families on the basis of height. Individual cloud types are named according to their form.

Figure 4.12 (*left*) Rolls of altocumulus clouds, also known as a "mackerel sky." (*right*) High cirrus clouds lie above puffy cumulus clouds in this scene from Holland.

Figure 4.13 Coastal fog under the Golden Gate Bridge, San Francisco.

Banks off Newfoundland are formed in this way, because here the cold Labrador current comes in contact with the warm waters of the Gulf Stream.

Sea fog is frequently found along the California coast. It forms within a cool marine air layer in direct contact with cold water of the California current (Figure 4.13). Similar fogs are found on continental west coasts in the tropical latitude zones where cool, equatorward currents parallel the shoreline.

PRECIPITATION

With a firm understanding of the adiabatic process and how it produces cooling, condensation or deposition, and clouds, we can now turn to the final topic of our chapter—precipitation. Precipitation can form in two ways. In the first, cloud droplets collide and coalesce into larger and larger water droplets that fall as rain. In the second, ice crystals form and grow in a cloud that contains a mixture of both ice crystals and water droplets.

In the first process, saturated air rises rapidly, and cooling forces additional condensation. Cloud particles grow by added condensation and attain a diameter of 50 to 100 mm (0.002 to 0.004 in.). In collisions with one another, the particles grow into droplets of about 500 mm in diameter (about 0.02 in.). This is the size of water droplets in drizzle. Further collisions increase drop size and yield *rain*. Average raindrops have diameters of about 1000 to 2000 mm (about 0.04 to 0.1 in.), but they can reach a maximum diameter of about 7000 mm (about 0.25 in.). Above this value they become unstable and break into smaller drops while falling. This type of precipitation formation occurs in warm clouds typical of the equatorial and tropical zones.

Snow is produced by the second process in clouds that are a mixture of ice crystals and supercooled water droplets. When an ice crystal collides with a droplet of supercooled water, it induces freezing of the droplet. The ice crystals further coalesce to form snow particles, which can become heavy enough to fall from the cloud. Some ice crystals grow directly by deposition. Snowflakes formed entirely by deposition can have intricate crystal structures (Figure 4.14), but most particles of snow are simply tiny lumps of ice.

When the underlying air layer is below freezing, snow reaches the ground as a solid form of precipitation. Otherwise, it melts and arrives as rain. A reverse process, the fall of raindrops through a cold air layer, results in the freezing of rain and produces pellets or grains of ice. These are commonly referred to in North

Figure 4.14 These individual snow crystals, greatly magnified, were selected for their beauty. Most snowflakes are lumps of ice without such a delicate structure.

America as *sleet*. (Among the British, sleet refers to a mixture of snow and rain.)

Perhaps you have experienced an *ice storm*. This occurs when the ground is frozen and the lowest air layer is also below freezing. Rain falling through the layer is chilled and freezes onto ground surfaces as a clear glaze. Ice storms cause great damage, especially to telephone and power lines and to tree limbs pulled down by the weight of the ice. In addition, roads and sidewalks are made extremely hazardous by the slippery glaze. Actually, the ice storm is more accurately named an "icing" storm, since it is not ice that is falling but supercooled rain.

Hail, another form of precipitation, consists of lumps of ice ranging from pea- to grapefruit-sized—that is, with a diameter of 5 mm (0.2 in.) or larger. Most hail particles are roughly spherical in shape. The formation of hail will be explained in our discussion of the thunderstorm.

Precipitation is measured in units of depth of fall per unit of time—for example, centimeters or inches per hour or per day. A centimeter (inch) of rainfall would cover the ground to a depth of 1 cm (1 in.), if the water did not run off or sink into the soil. Rainfall is measured with a *rain gauge*. The simplest rain gauge is a straight-sided, flat-bottomed pan, which is set outside before a rainfall event. (An empty coffee can does nicely.) After the rain, the depth of water in the pan is measured.

A very small amount of rainfall, such as 2 mm (0.1 in.), makes too thin a layer to be accurately measured in a flat pan. To avoid this difficulty, a typical rain gauge is constructed from a narrow cylinder with a funnel at the top (Figure 4.15). The funnel gathers rain from a wider area than the mouth of the tube, so the cylinder fills more quickly. The water level gives the amount of precipitation, which is read on a graduated scale.

Snowfall is measured by melting a sample column of snow and reducing it to an equivalent in rainfall. In this way rainfall and snowfall records may be combined in a single record of precipitation. Ordinarily, a 10 cm (or 10 in.) layer of snow is assumed to be equivalent to 1 cm (or 1 in.) of rainfall, but this ratio may range from 30 to 1 in very loose snow to 2 to 1 in old, partly melted snow.

Precipitation Processes

So far, we have seen how air that is moving upward will be chilled by the adiabatic process to the saturation point and then to condensation, and how eventually

Figure 4.15 This rain gauge consists of two clear plastic cylinders, here partly filled with red-tinted water. The inner cylinder receives rainwater from the funnel top. When it fills, the water overflows into the larger, outer cylinder.

precipitation will form. However, one key piece of the precipitation puzzle has been missing—what causes air to move upward? Air can move upward in three ways. First, it can be forced upward as a through-flowing wind. Consider the case of a mass of air moving up and over a mountain range, for example. If the mountain range is high enough and the air is moist enough, then precipitation will occur. This is called **orographic precipitation**, and we'll describe it more fully in the following section.

A second way for air to be forced upward is through *convection*. In this process, a parcel of air is heated, perhaps by a patch of warm ground, so that it is warmer and therefore less dense than the air around it. Like a bubble, it rises. If the air is quite moist and condensation sets in, then the release of latent heat of condensation will ensure that the bubble of air remains warmer than the surrounding air as it rises. This type of precipitation is **convectional precipitation**, and it will be described after orographic precipitation.

A third way for air to be forced upward is through the movement of air masses. As we will see in the next chapter, air masses are large bodies of air, each with a set of relatively uniform temperature and moisture properties. Air masses move normally from west to east in midlatitudes, and one can overtake another. When this happens, one of them will be forced aloft. Since this overtaking usually occurs in a common type of

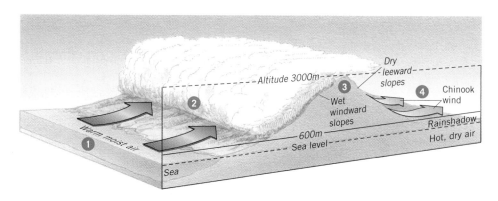

Figure 4.16 The forced ascent of a warm, moist oceanic air mass over a mountain barrier produces precipitation and a rainshadow desert. As the air moves up the mountain barrier, it loses moisture through precipitation. As it descends the far slope, it is warmed.

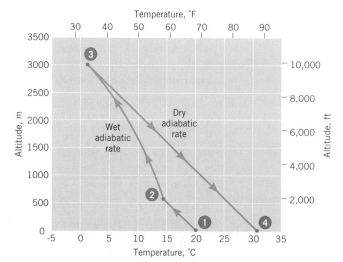

storm called a cyclone, this type of precipitation is known as *cyclonic precipitation*. We will return to this subject in Chapter 6.

Orographic Precipitation

In orographic precipitation, through-flowing winds move moist air up and over a mountain barrier. (The term *orographic* means "related to mountains.") Figure 4.16 shows how this process works. Moist air arrives at the coast after passing over a large ocean surface (1). As the air rises on the windward side of the range, it is cooled at the dry adiabatic rate. When cooling is sufficient, the lifting condensation level is reached, condensation sets in, and clouds begin to form (2). Cooling now proceeds at the wet adiabatic rate. Eventually, precipitation begins. As the air continues up the slope, precipitation continues to fall.

After passing over the mountain summit, the air begins to descend down the leeward slopes of the range (3). However, as it descends, it is compressed. This means that it is warmed, according to the adiabatic principle. Since cooling is no longer occurring, cloud droplets and ice crystals evaporate or sublimate. The air clears rapidly. As it continues to descend, it warms further. At the base of the mountain on the far side (4), the air is now hot—and dry, since its moisture has been removed.

The net result of the trip over the mountain is that the warm, moist air has become hot and dry. It was warmed by the latent heat released to the surrounding air as water vapor formed water droplets and ice crystals in the precipitation process. Its water was shed to the mountain slopes below, on the uphill journey. This effect creates a *rainshadow* on the far side of the mountain—a belt of dry climate that extends on down the leeward slope and beyond. Several of the earth's great deserts are of this type.

California's rainfall patterns provide an excellent example of orographic precipitation and the rainshadow effect. A map of California (Figure 4.17) shows mean annual precipitation. It uses lines of equal precipitation, called *isohyets*. Focus now on central California and follow the arrow on the map. Prevailing westerly winds bring moist air in from the Pacific Ocean, first over the Coast Ranges of central and northern California, and then, after rain is deposited on the Coast Ranges, the air descends into the broad Central Valley. Here, precipitation is low, averaging less than 25 cm (10 in.) per year.

Next the air continues up and over the great Sierra Nevada, whose summits rise to 4200 m (about 14,000 ft) above sea level. Heavy precipitation, largely in the form of winter snow, falls on the western slopes of these ranges and nourishes rich forests. Passing down the steep eastern face of the Sierra Nevada, the air descends quickly into the Owens Valley, at about 1000 m

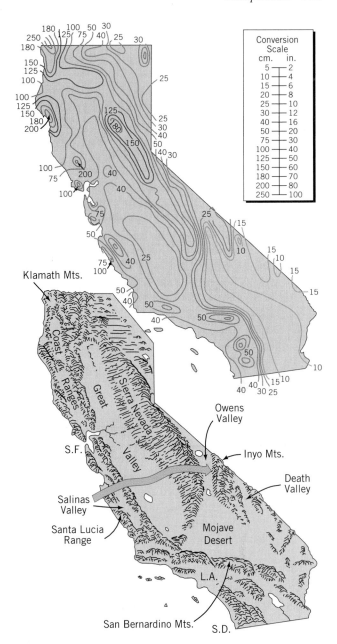

Figure 4.17 The effect of mountain ranges on precipitation is strong in the state of California, because of the prevailing flow of moist oceanic air from west to east. Centers of high precipitation coincide with the western slopes of mountain ranges, including the coast ranges and Sierra Nevada. To the east, in their rainshadows, lie desert regions.

(3300 ft) elevation. The adiabatic heating warms and dries the air, creating a rainshadow desert. In the Owens Valley, annual precipitation is less than 10 cm (4 in.). In this way, the orographic effect on air moving across the mountains of California produces a part of America's great interior desert zone, extending from eastern California and across Nevada.

Convectional Precipitation

The second process of inducing uplift in a parcel of air is *convection*, the upward motion of a parcel of heated air. In this process, strong updrafts occur within *convection cells*. Air rises in a convection cell because it is warmer, and therefore less dense, than the surrounding air.

The convection process starts when a surface is heated unequally. For example, consider an agricultural field surrounded by a forest. Since the field surface consists largely of bare soil and only a low layer of vegetation, it will be warmer under steady sunshine than the adjacent forest. This means that as the day progresses the air above the field will grow warmer than the air above the forest.

Now, the density of air depends on its temperature—warm air is less dense than cooler air. The hot-air balloon operates on this principle. The balloon is open at the bottom, and in the basket below a large gas burner forces heated air into the balloon. Because the heated air is less dense than the surrounding air, the balloon rises. The same principle will cause a bubble of air to form over the field, rise, and break free from the surface. Figure 4.18 diagrams this process.

As the bubble of air rises, it is cooled adiabatically, and its temperature will decrease as it rises. However, we know that the temperature of the surrounding air will normally decrease with altitude as well. Nonetheless, as long as the bubble is still warmer than the surrounding air, it will be less dense and will therefore continue to rise.

If the bubble remains warmer than the surrounding air and uplift continues, adiabatic cooling chills the bubble below the dew point. Condensation occurs, and the rising air column becomes a puffy cumulus cloud. The flat base shows the level at which condensation be-

gins. The bulging "cauliflower" top of the cloud is the top of the rising warm-air column, pushing into higher levels of the atmosphere. Normally, the small cumulus cloud will encounter winds aloft that mix it into the local air. After drifting some distance downwind, the cloud evaporates.

Unstable Air

Sometimes, however, convection continues strongly, and the cloud develops into a dense cumulonimbus mass, or thunderstorm, from which heavy rain will fall. Two conditions encourage the development of thunderstorms: (1) air that is very warm and moist, and (2) an environmental temperature lapse rate for which temperature decreases more rapidly with altitude than it does for either the dry or wet adiabatic lapse rates. Figure 4.19, which diagrams the convection process in unstable air, shows this type of lapse rate as having a flatter slope than the dry or wet adiabatic rates. Air with these characteristics is referred to as **unstable air.**

For the first condition, recall that there are two adiabatic rates—dry and wet (Figure 4.10). The wet rate is about half the dry rate. While condensation is occurring, the lesser wet rates apply. Thus, air in which condensation is occurring cools less rapidly with uplift. Furthermore, the wet rate is smallest for warm, moist air. This means that the temperature decrease experienced by warm, moist, rising, condensing air is quite small. And, since the temperature decrease is small, the rising air is more likely to stay warmer than the surrounding air. Thus, uplift continues.

With regard to the second condition—an environmental temperature lapse rate value that exceeds the adiabatic rates—keep in mind that this rate describes the temperature of the surrounding, still air. Therefore, the temperature of the still air falls quite rapidly

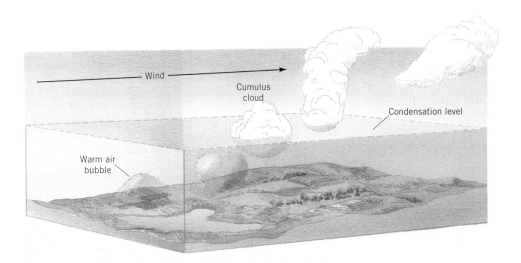

Figure 4.18 Rise of a bubble of heated air to form a cumulus cloud.

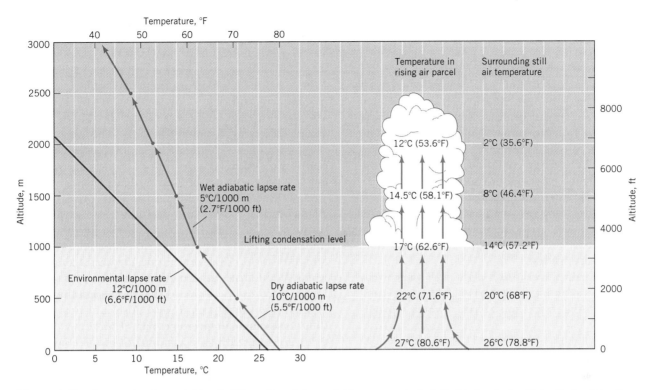

Figure 4.19 Convection in unstable air. When the air is unstable, a parcel of air that is heated sufficiently to rise will continue to rise to great heights.

with altitude. This means that the condensing air in a rising parcel will stay warmer than the surrounding air. Again, uplift continues.

A simple example may make convection in unstable air clearer. Figure 4.19 shows how temperature changes with altitude for a parcel moving upward by convection in unstable air. The surrounding air is at a temperature of 26°C (78.8°F) at ground level and has a lapse rate of 12°C/1000 m (6.6°F/1000 ft). A parcel of air is heated by 1°C (1.8°F) to 27°C (80.6°F), and it begins to rise. At first, it cools at the dry adiabatic rate. At 500 m (1640 ft), the parcel is at 22°C (71.6°F), while the surrounding air is at 20°C (68°F). Since it is still warmer than the surrounding air, uplift continues. At 1000 m, the lifting condensation level is reached. The temperature of the parcel is 17°C (62.6°F).

As the parcel rises above the condensation level, the wet adiabatic rate applies, here using the value 5°C/1000 m (2.7°F/1000 ft). Now the parcel cools more slowly as it rises. At 1500 m (4920 ft), the parcel is 14.5°C (58.1°F), while the surrounding air is 8°C (46.4°F). Since the parcel is still warmer than the surrounding air, it continues to rise. Note that the difference in temperature between the rising parcel and the surrounding air now actually increases with altitude. This means that the parcel will be buoyed upward ever more strongly, forcing even more condensation and precipitation.

The key to the convectional precipitation process is latent heat. When water vapor condenses into cloud droplets or ice particles, it releases latent heat to the rising air parcel. By keeping the parcel warmer than the surrounding air, this latent heat fuels the convection process, driving the parcel ever higher. When the parcel reaches a high altitude, most of its water will have condensed. As adiabatic cooling continues, less latent heat will be released. As a result, the convection cell will weaken. Eventually, uplift will stop, since the energy source, latent heat, is gone. The cell dies and dissipates into the surrounding air.

Unstable air is typical of summer air masses in the central and southeastern United States. As we will see later, summer weather patterns sweep warm, humid air from the Gulf of Mexico over the continent. The intense summer insolation, over a period of days, strongly heats the air layer near the ground, producing a steep lapse rate. Thus, both of the conditions that create unstable air are present. As a result, thunderstorms are very common in these regions during the summer.

Unstable air is also likely to be found in the vast, warm and humid regions of the equatorial and tropical zones. In these regions, convective showers and thundershowers are common. At low latitudes, much of the orographic rainfall is actually in the form of heavy showers and thundershowers produced by convection. In this case, the forced ascent of unstable air up a mountain slope easily produces rapid condensation, which then triggers the convection mechanism.

Thunderstorms

A **thunderstorm** is an intense local storm associated with a tall, dense cumulonimbus cloud in which there are very strong updrafts of air (Figure 4.20). A single thunderstorm consists of several individual convection cells. A single convection cell is diagrammed in Figure 4.21. Air rises within the cell as a succession of bubble-like air parcels. Intense adiabatic cooling within the bubbles produces precipitation. It can be in the form of water at the lower levels, mixed water and snow at intermediate levels, and snow at high levels where cloud temperatures are coldest.

As the rising air parcels reach high levels, which may be 6 to 12 km (about 20,000 to 40,000 ft) or even higher, the rising rate slows. Strong winds typically present at such high altitudes drag the cloud top downwind, giving the thunderstorm cloud its distinctive shape—resembling an old-fashioned blacksmith's anvil.

Ice particles falling from the cloud top act as nuclei for freezing and deposition at lower levels. Large ice crystals form and begin to sink rapidly. Melting, they coalesce into large, falling droplets. The rapid fall of raindrops adjacent to the rising air bubbles pulls the air downward and feeds a downdraft. Emerging from the cloud base, the downdraft of cool air, laden with precipitation, approaches the surface and spreads out in all directions. Part of the downdraft moves forward, causing warm, moist surface air to rise and enter the updraft portion of the storm. This effect helps perpetuate the storm. The downdraft creates strong local winds. Wind gusts can sometimes be violent enough to topple trees and raise the roofs of weak buildings.

In addition to powerful wind gusts and heavy rains, thunderstorms can produce hail. Hailstones (Figure 4.22) are formed by the accumulation of ice layers on an ice pellet suspended in the strong updrafts of the thunderstorm. In this process, they can reach diameters of 3 to 5 cm (1.2 to 2.0 in.). When they become too heavy for the updraft to support, they fall to earth.

Annual losses from crop destruction caused by hailstorms amount to several hundred million dollars.

Figure 4.20 This majestic cumulonimbus cloud marks a thunderstorm, moving from left to right. Lightning strikes lead the storm. A dark plume of falling rain follows, on the left.

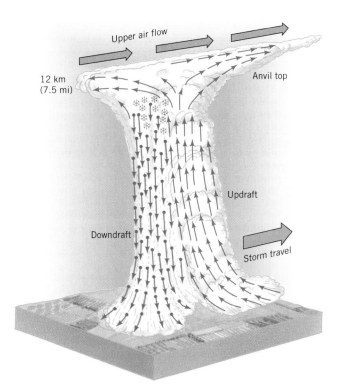

Figure 4.21 Anatomy of a thunderstorm cell. Successive bubbles of moist condensing air push upward in the cell. Their upward movement creates a corresponding downdraft, expelling rain, hail, and cool air from the storm as it moves forward.

Figure 4.22 Close-up view of hailstones.

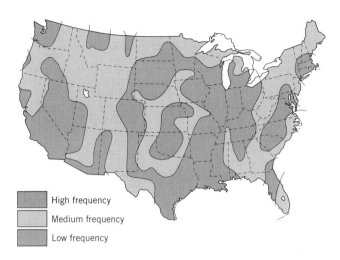

High frequency

Medium frequency

Low frequency

Figure 4.23 Map of frequency of severe hailstorms in the 48 contiguous states of the United States. A severe hailstorm is defined as a local convective storm producing hailstones equal to or greater than 1.9 cm (0.75 in.) in diameter.(From Richard H. Skaggs, *Proc. Assoc. American Geographers*, vol. 6, Figure 2. Used by permission.)

Damage to wheat and corn crops is particularly severe in the Great Plains, running through Nebraska, Kansas, Missouri, Oklahoma, and northern Texas (Figure 4.23). Figure 4.24 shows a hailstorm in progress.

Another effect of convection cell activity is to generate lightning. This occurs when updrafts and downdrafts cause positive and negative static charges to accumulate within different regions of the cloud. Lightning is a great electric arc—a series of gigantic sparks—passing between differently charged parts of the cloud mass or between cloud and ground (see Figure 4.20). During a lightning discharge, a current of as much as 60,000 to 100,000 amperes may develop. This current heats the air intensely, which makes it expand very rapidly—much like an explosion. This expansion sends out sound waves, which we recognize as a thunderclap. Most lightning discharges occur within the cloud, but a significant proportion strike land. In the United States, lightning causes a yearly average of about 150 human deaths and property damage of hundreds of millions of dollars, including loss by structural and forest fires set by lightning.

The thunderstorm is a good example of a flow system of matter and energy—one that is a bit more complicated than our previous examples. *Focus on Systems 4.4 • The Thunderstorm as a Flow System* views the thunderstorm from the systems perspective.

Microbursts

The downdraft that accompanies a thunderstorm can sometimes be very intense—so intense that it is capable of causing low-flying aircraft to crash. This type of intense downdraft is called a *microburst*. A diagram of a microburst is shown in Figure 4.25. The downward-moving air flows outward in all directions. It is often, but not always, accompanied by rain.

An aircraft flying through the microburst first encounters strong headwinds. They may cause a bumpy ride but do not interfere with the ability of the airplane to fly. However, the aircraft encounters a strong tailwind as it passes through the far side of the microburst. The lift of the airplane's wings depends on the speed of the air flowing across them, and the tailwind greatly reduces the air speed. This causes a loss of lift. If the tailwind is strong enough, the airplane cannot hold its altitude and may crash. Microbursts can be detected by special radar instruments that measure horizontal wind speeds. They are quite expensive, however, and are installed only at some major airports.

New training procedures for pilots, instituted in 1987, have reduced the incidence of aviation accidents in the United States attributed to microbursts and associated wind shear. However, in 1994 a jet aircraft attempting to land at Charlotte, North Carolina, crashed during a violent thunderstorm, killing 37 and injuring 20 others. A microburst was detected just as the airplane approached the airport, but the warning was not broadcast to the pilots on the radio channel they were following. Although they attempted to abort the landing, a pilot error occurred and their aircraft was brought down by the severe tailwind of the microburst.

Figure 4.24 A hailstorm in progress, Santa Cruz County, Arizona. Although the hail is falling from a cloud directly overhead, the low afternoon sun still illuminates the scene at an angle. Marble-sized hailstones are accumulating on the ground.

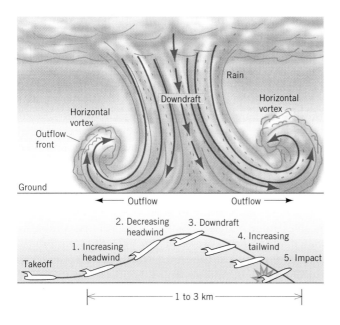

Figure 4.25 Anatomy of a microburst. *(above)* Schematic cross section showing a downdraft reaching the ground and producing a horizontal outflow. *(below)* Schematic profile of an aircraft taking off through a microburst, leading to a loss of lift and eventual crash. (Adapted from diagrams by Research Applications Program, National Center for Atmospheric Research, Boulder, Colorado.)

Focus on Systems 4.4 • The Thunderstorm as a Flow System

A thunderstorm is a dramatic and spectacular example of a system in which matter, in the form of air and water, moves from one location to another under the influence of a coupled flow of energy. Let's look at the energy and matter flows in a thunderstorm in more detail. A diagram of these flows is shown below. Part (*a*) shows the structure of the storm, similar to Figure 4.21. Part (*b*) shows how water flows and changes state within the storm. Part (*c*) diagrams the flow of energy within the storm.

Examining the flow of water in the thunderstorm system (*b*), we see that cells of warm, moist air rise in the forward part of the storm. They serve to bring in water vapor at low altitude, which condenses to liquid and solid forms as it moves upward. At the top of the cloud, some ice particles are carried away in

through-flowing winds at high altitude. The remainder are joined by water droplets and descend on the rearward side of the storm, striking the ground as rain, hail, or even snow. Thus, the water flow forms a loop upward as vapor, then downward as precipitation.

Part (*c*) shows the flow of energy within the storm. As the convection cells within the storm cloud move moist air aloft, condensation converts latent heat to sensible heat, enhancing the convective uplift. At the top of the storm, jet stream winds carry a portion of the sensible heat forward and away from the storm. As liquid and solid water particles are carried downward on the trailing side of the storm, some evaporation and sublimation occur, shifting a portion of the sensible heat released to latent heat. At the surface, the net effect is to lose a

larger quantity of latent heat than is gained, resulting in generally cooler air. Aloft, the air remains warmer than before the storm, owing to the gain in sensible heat.

The power source for the storm is the latent heat in the water vapor that moves aloft. By changing from the vapor to liquid or solid state, the vapor releases the energy that powers the storm. Some of that energy is recovered when solid or liquid water is converted to vapor in the downdraft cell, but a larger portion is dissipated at higher altitudes. Thus, the storm acts like a heat pump, moving heat from the surface to upper altitudes. Of course, the ultimate power source for the storm is really the sun, which heated the surface and powered the evaporation that made the air moist to begin with.

(*a*)

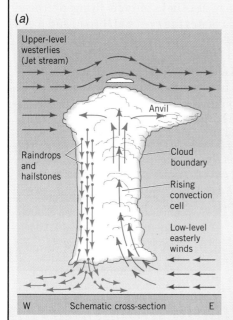

(*b*)

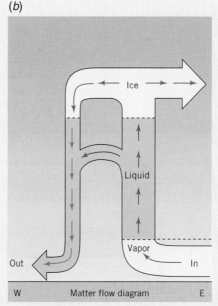

(*c*)

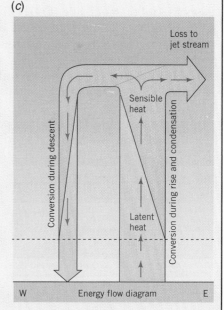

The thunderstorm as a flow system of energy and matter.

EYE ON THE ENVIRONMENT:
AIR POLLUTION

So far, we have discussed a number of atmospheric phenomena, including air temperature and precipitation. Yet another phenomenon of the atmosphere is air pollution. Unlike the others, this phenomenon is largely a result of human activity. Here, we provide a short introduction to the topic.

In preceding chapters, we have discussed two kinds of substances in the air—aerosols and gases. Aerosols are small bits of matter in the air, so small that they float freely with normal air movements. Gases are molecular compounds that are mixed together to form the main body of air. To these two categories, we add a third category, *particulates*—larger, heavier particles that sooner or later fall back to earth.

An *air pollutant* is an unwanted substance injected into the atmosphere from the earth's surface by either natural or human activities. Air pollutants come as aerosols, gases, and particulates. Most pollutants generated by human activity arise in two ways. The first is through the day-to-day activities of large numbers of people, for example, in driving automobiles within urban centers. The second is through industrial activities, such as fossil fuel combustion or the smelting of mineral ores to produce metals.

Figure 4.26 shows the relative proportions of the most common air pollutants and their sources. The gases are carbon monoxide (CO), sulfur oxides, and nitrogen oxides. The nitrogen oxides can be NO, NO_2, or NO_3; this mixture is usually referred to as NO_X. Similarly, the sulfur oxides SO_2 and SO_3 are referred to as SO_X. Hydrocarbon compounds occur both as gases and as aerosols. The remaining pollutants are in the form of particulates.

Combustion of fossil fuel, in transportation or as stationary fuel combustion, is the most important source of all these pollutants. As shown in the right half of the figure, it accounts for 73 percent of the emissions. Exhaust from gasoline and diesel engines contributes most of the carbon monoxide, half the hydrocarbons, and about a third of the nitrogen oxides.

Stationary sources of fuel combustion include power plants that generate electricity and various industrial plants that also burn fossil fuels. They contribute most of the sulfur oxides because they burn coal or lower-grade fuel oils. These fuels are richer in sulfur than gasoline and diesel fuel. Stationary sources also supply most of the particulate matter. Some of it is fly ash—coarse soot particles emitted from smokestacks of power plants. These particles settle out quite quickly within close range of the source. Smaller and finer carbon particles, however, can remain suspended almost indefinitely and become aerosols. Industrial processes, such as the smelting of ore, are also important sources of air pollutants (Figure 4.27). The burning of solid wastes makes a small contribution to air pollution.

Smog and Haze

When aerosols and gaseous pollutants are present in considerable density over an urban area, the resultant mixture is known as **smog** (Figure 4.28). This term was coined by combining the words "smoke" and "fog." Typically, smog allows hazy sunlight to reach the ground, but it may also hide aircraft flying overhead from view. Smog irritates the eyes and throat, and it can corrode structures over long periods of time.

Modern urban smog has three main toxic ingredients: nitrogen oxides, hydrocarbons, and ozone. Nitrogen oxides and hydrocarbons are largely automobile pollutants, although significant amounts can be released by industrial power plants and petroleum processing and storage facilities. Ozone is not normally produced directly by pollution sources in the city but

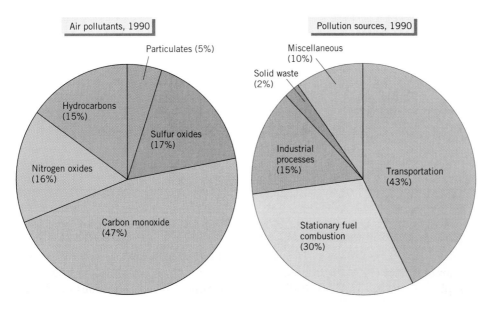

Figure 4.26 Air pollution emissions in the United States in percentage by weight. *Left*: pollutants emitted. *Right*: sources of emission. (Data from National Air Pollution Control Administration, HEW.)

Figure 4.27 This satellite image shows the devastating effects of pollutants emitted by a smelter near Wawa, Ontario, Canada. Healthy vegetation is shown in red to pink tones. The blue streak on the right-hand side of the image is barren of vegetation.

Figure 4.28 Smog in the city. New York's World Trade Center projects skyward through a brownish-pink layer of smog.

forms through a photochemical reaction in the air. In this reaction, nitrogen oxides react with hydrocarbons in the presence of sunlight to form ozone. Ozone in urban smog can harm plant tissues and eventually kill sensitive plants. It is also harmful to human lung tissue, and it aggravates bronchitis, emphysema, and asthma. This is the same compound that absorbs ultraviolet radiation in the ozone layer within the stratosphere. So, ozone in the stratosphere is desirable, but if close to earth, it is undesirable indeed. Photochemical reactions can also produce other toxic compounds in smog.

Haze is a condition of the atmosphere in which aerosols obscure distant objects. Haze builds up naturally in stationary air as a result of human and natural activity. When the air is humid and abundant water vapor is available, water films grow on suspended nuclei. This creates aerosol particles large enough to obscure and scatter light, reducing visibility. Natural haze aerosols include soil dust, salt crystals from the sea surface, hydrocarbon compounds from plants, plant pollen, and smoke from forest and grass fires. Some regions are noted for naturally occurring haze—the Great Smoky Mountains are an example.

Fallout and Washout

Pollutants generated by a combustion process are contained within hot exhaust air emerging from a factory smokestack or an auto tailpipe. Since hot air rises, the pollutants are at first carried aloft by convection. However, the larger particulates soon settle under gravity and return to the surface as *fallout*. Particles too small to settle out are later swept down to earth by precipitation in a process called *washout*. Through a combination of fallout and washout, the atmosphere tends to be cleaned of pollutants. Although a balance between input and output of pollutants is achieved in the long run, the quantities stored in the air at a given time fluctuate widely.

Pollutants are also eliminated from the air over their source areas by wind. Strong, through-flowing winds will disperse pollutants into large volumes of cleaner

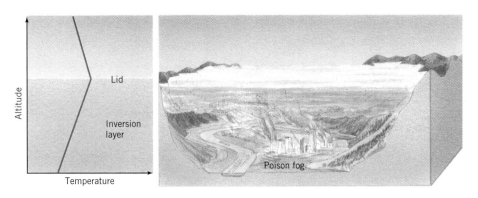

Figure 4.29 A low-level inversion held air pollutants close to the ground, inducing a poison fog accumulation at Donora, Pennsylvania, in October 1948.

air in the downwind direction. Strong winds can quickly sweep away most pollutants from an urban area, but, during periods when winds are light or absent, concentrations of pollutants can rise to high values.

Inversion and Smog

The concentration of pollutants over a source area rises to its highest levels when vertical mixing (convection) of the air is inhibited. This happens in an inversion—a condition in which the temperature of the air increases with altitude.

Why does the inversion inhibit mixing? Recall that a heated air parcel, perhaps emerging from a smokestack or chimney, will rise as long as it is warmer than the surrounding air. Recall, too, that as it rises, it is cooled according to the adiabatic principle. In an inversion, however, the surrounding air gets warmer, not colder, with altitude. So, the parcel will cool as it rises, while the surrounding air becomes warmer. Thus, the parcel will quickly arrive at the temperature of the surrounding air and uplift will stop. Under these conditions, heated air will move only a short distance upward. Pollutants in the air parcel will disperse at low levels, keeping concentrations high near the ground.

Two types of inversions are important in causing high air pollutant concentrations—low-level and high-level inversions. When a *low-level temperature inversion* develops over an urban area with many air pollution sources, pollutants are trapped under the "inversion lid." Heavy smog or highly toxic fog can develop. An

example is the tragedy that occurred in Donora, Pennsylvania, in late October 1948, when a persistent low-level inversion developed. The city occupies a valley floor hemmed in by steeply rising valley walls. The walls prevented the free mixing of the lower air layer with that of the surrounding region (Figure 4.29) and helped keep the inversion intact. Industrial smoke and gases from factories poured into the inversion layer for five days, increasing the pollution level. A poisonous fog formed and began to take its toll. Twenty persons died, and several thousand persons were stricken before a change in weather patterns dispersed the smog layer.

Another type of inversion is responsible for the smog problem experienced in the Los Angeles basin and other California coastal regions, ranging north to San Francisco and south to San Diego. Here special climatic conditions produce prolonged inversions and smog accumulations (Figure 4.30). Off the California coast is a persistent fair-weather system, which is especially strong in the summer. This system produces a layer of hot, dry air at upper elevations. However, a cold current of upwelling ocean bottom water runs along the coast, just offshore. Moist ocean air at the surface moves across this cool current and is chilled, creating a cool, marine air layer.

Now, the Los Angeles basin is a low, sloping plain lying between the Pacific Ocean and a massive mountain barrier on the north and east sides. Weak winds from the south and southwest move the cool, marine air inland over the basin. Further landward movement

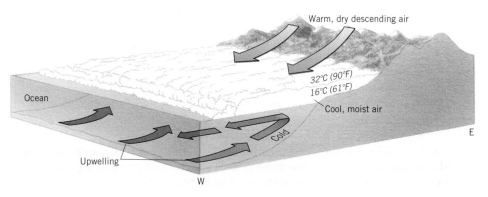

Figure 4.30 A layer of warm, dry descending air from a persistent fair-weather system rides over a cool, moist marine air layer at the surface to create a persistent temperature inversion.

is blocked by the mountain barrier. Since there is a warm layer above this cool marine air layer, the result is an inversion. Pollutants accumulate in the cool air layer and produce smog. The upper limit of the smog stands out sharply in contrast to the clear air above it, filling the basin like a lake and extending into valleys in the bordering mountains. Since this type of inversion persists to a higher level in the atmosphere, we can refer to it as a *high-level temperature inversion.*

An actual temperature inversion, in which temperature close to the surface increases with altitude, is not essential for building a high concentration of pollutants above a city. All that is required is light or calm winds and stable air. *Stable air* has a temperature profile that decreases with altitude, but at a slow rate. Some convectional mixing occurs in stable air, but convectional precipitation is inhibited. At certain times of the year, slow-moving masses of dry, stable air occupy the central and eastern portions of the North American continent. Under these conditions, a broad *pollution dome* can form over a city or region, and air quality will suffer (Figure 4.31). When there is a regional wind, the pollution from a large city will be carried downwind to form a *pollution plume.*

Climatic Effects of Urban Air Pollution

Urban air pollution reduces visibility and illumination. Specifically, a smog layer can cut illumination by 10 percent in summer and 20 percent in winter. Ultraviolet radiation is absorbed by ozone in smog. At times, ultraviolet radiation is completely prevented from reaching the ground. While this reduces the risk of human skin cancer, it may also permit increased viral and bacterial activity at ground level.

Winter fogs are much more frequent over cities than over the surrounding countryside. The fog is enhanced by the abundance of urban aerosols and particulates. Coastal airports, such as those of New York City, Newark, and Boston, suffer from an increased frequency of fog resulting from urban air pollution. Cities also show an increase in cloudiness and precipitation as compared to the surrounding countryside. This increase results from intensified convection generated when the lower air is heated by human activities.

Acid Deposition

Yet another air pollution problem is acid deposition. Perhaps you've heard about acid rain killing fish and poisoning trees. Acid rain is part of the phenomenon of **acid deposition**, which includes not only acid rain, but also dry acidic dust particles. Acid rain simply consists of raindrops that have been acidified by air pollutants. Dry acidic particles are dust particles that are acidic in nature. They fall to earth and coat the surface in a thin dust layer. When wetted by rain or fog, they acidify the water on leaves and soils. In winter, acid particles can mix with snow as it accumulates. In spring, when the snow melts, a surge of acid water is then released to soils and streams.

Acid deposition is produced by the release of sulfur dioxide (SO_2) and nitric oxide (NO_2) into the air. This occurs when fossil fuels are burned, for example, as gasoline and diesel fuels for transportation or as coal and oil for electric and industrial power generation. Once in the air, SO_2 and NO_2 readily combine with oxygen and water in the presence of sunlight and dust particles to form sulfuric and nitric acid aerosols. These aerosols serve as condensation nuclei and acidify the tiny water droplets created. When the droplets coalesce in precipitation, acid raindrops or ice crystals result. Sulfuric and nitric acids can also be formed on dust particles, creating dry acid particles. These can be as damaging to plants, soils, and aquatic life as acid rain.

What are the effects of acid deposition? A primary effect is acidification of lakes and streams, injuring aquatic plants and animals and perhaps eventually sterilizing a lake, pond, or stream. Another problem is soil damage. Through a series of chemical effects, acid deposition can cause soils to lose nutrients, affecting the natural ecosystem within a region.

To measure acidity, scientists use the pH scale. Values on this scale range from 1 to 14. High values indicate alkalinity, and low values indicate acidity. Pure water has a pH of 7. A value of 1 is very acidic—like battery acid. At this pH, acid will eat through clothes and burn the skin. Normally, rainwater is slightly acid, showing pH values of 5 to 6. In the 1960s European water chemists observed that the pH of rain in northwestern Europe had dropped to values as low as 3 in

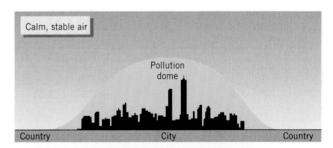

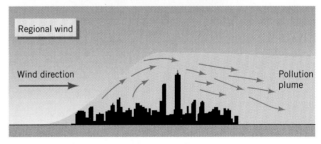

Figure 4.31 If calm, stable air overlies a major city, a pollution dome can form. When a wind is present, pollutants are carried away as a pollution plume.

some samples, similar to household vinegar. Because pH numbers are on a logarithmic scale, these values mean that rain in these samples was 100 to 1000 times more acid than normal.

In North America, scientists studying the chemical quality of rainwater have reported that since 1975 rainwater over a large area of the northeastern United States has had an average pH of about 4. Values less than pH 4 occur at times over many heavily industrialized U.S. cities, including Boston, New York, Philadelphia, Birmingham, Chicago, Los Angeles, and San Francisco. Values between pH 4 and 5 have been observed near smaller cities such as Tucson, Arizona; Helena, Montana; and Duluth, Minnesota—localities we do not usually associate with heavy air pollution.

Figure 4.32 is a map of the United States and Canada showing the average acidity of rainwater. Compare this with Figure 4.33, showing emissions of sulfur dioxide and nitrogen oxides by states. There is a strong relationship, especially between SO_2 and the average acidity of rainwater.

An important factor in the level of impact of acid deposition on the environment is the ability of the soil and surface water to absorb and neutralize acid. This factor ranges widely. In dry climates, acids can be readily neutralized because surface waters are normally somewhat alkaline. Areas where soil water is naturally somewhat acidic are most sensitive to acid deposition. Such areas are associated with moist climates generally and include the eastern United States, high-mountain regions of the western states, and the Pacific Northwest.

Acid deposition in Europe and North America has severely impacted some ecosystems. In Norway, acidification of stream water has virtually eliminated salmon runs by inhibiting salmon egg development. At present, the acidity of Norwegian streams has leveled off as air pollution levels have stabilized. However, salmon kills continue to occur, especially when rainstorms carry large amounts of sulfate into streams.

The increased fish mortality observed in Canadian lakes has also been attributed to acidification. In 1980, Canada's Department of Environment reported that 140 Ontario lakes had no fish and that thousands of other lakes of that province were threatened with a similar fate. In 1990, American scientists estimated that 14 percent of Adirondack lakes were heavily acidic, along with 12 to 14 percent of the streams in the Mid-Atlantic states.

Forests, too, have been damaged by acid deposition. In western Germany, the impact has been especially severe in the Harz Mountains and the Black Forest. In 1983, it was reported that about one-third of West Germany's forests showed visible damage. In the eastern United States, pine and spruce trees have experienced damage in recent years. In the early 1990s, Europe continued to experience heavy losses in timber, resulting from fallout of sulfur dioxide in combination with nitrogen oxide emissions from vehicles, industry, and farm wastes. Fallout of sulfur oxides was most heavily concentrated in Germany, the Czech Republic, Slovakia, Poland, and Hungary.

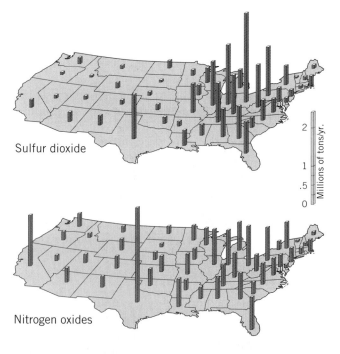

Figure 4.33 Emissions of sulfur dioxide and nitrogen oxides by states for the year 1993. Sulfur dioxide releases are especially heavy in the states that burn Appalachian coal. Releases of nitrogen oxides are high in Texas, due to oil processing, and in California, due to auto emissions. (Data from EPA, 1994.)

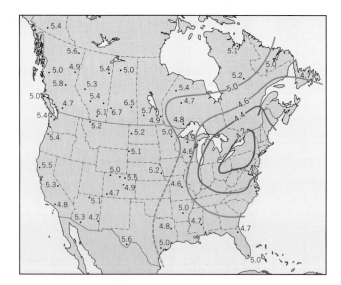

Figure 4.32 Acidity of rainwater for the United States and Canada, averaged over a year. Values are pH units. The northeastern United States and southeastern Canada are most strongly affected.

In 1990, the National Acid Precipitation Assessment Program (NAPAP) issued a report after its ten-year study of the causes and effects of acid precipitation in the United States. Emissions of sulfur dioxide and volatile organic compounds had declined 30 percent since 1970. Furthermore, nitrogen oxides had also declined over the period by 8 to 16 percent. These effects were due primarily to improved industrial emission controls. In 1990, electrical power plants continued to be the major source of sulfur oxides. However, new controls on sulfur dioxide emissions called for by the 1990 Clean Air Act should further reduce SO_2 releases and continue this improvement in air quality through the remainder of the decade.

Although the situation is improving, acid deposition is still a very important problem in many parts of the world—especially Eastern Europe and the states of the former Soviet Union. There, air pollution controls have been virtually nonexistent for decades. Reducing pollution levels and cleaning up polluted areas will be a major task for these nations as they enter the next century.

Air Pollution Control

The United States and Canada have some of the strictest laws in the world limiting air pollution. As a result, many strategies have been developed to reduce emissions. Sometimes these strategies involve trapping and processing pollutants after they are generated. The "smog controls" on automobiles are an example. Sometimes the use of alternative, nonpolluting technology is in order—for example, substituting solar, wind, or geothermal power for coal burning to generate electricity.

In any event, both as individuals and as a society we can do much to preserve the quality of the air we breathe. Reducing fossil fuel consumption is an example. It will be a considerable challenge to our society and others to preserve and increase global air quality in the face of an expanding human population and increasing human resource consumption. 🌿

Precipitation has been the major focus of this chapter. Keep in mind that no matter what the cause of precipitation, latent heat is always released. Because a significant amount of oceanic evaporation results in precipitation over land, there is a very significant flow of latent heat from oceans to land. And, as we will see in the following chapter, global air circulation patterns move this latent heat poleward, helping to warm the continents and creating the distinctive pattern of climates that differentiates our planet.

CHAPTER SUMMARY

Water moves freely between ocean, atmosphere, and land in the hydrologic cycle. The global water balance describes these flows. The fresh water in the atmosphere and on land in lakes, streams, rivers, and groundwater is only a very small portion of the total water in the hydrosphere.

Humidity describes the amount of water vapor present in air. The ability of air to hold water vapor depends on temperature. Warm air can hold much more water vapor than cold air. Relative humidity measures water vapor in the air as the percentage of the maximum amount of water vapor that can be held, at the given air temperature. Specific humidity simply measures grams of water vapor per kilogram of air.

The adiabatic principle states that when a gas is compressed, it warms, and when a gas expands, it cools. When an air parcel moves upward in the atmosphere, it encounters a lower pressure and so expands and cools. The dry adiabatic rate describes the rate of cooling with altitude. If the air is cooled below the dew point, condensation or deposition occurs, and latent heat is released. This heat reduces the rate of cooling with altitude, which is described as the wet adiabatic rate if condensation or deposition is occurring.

Clouds are composed of droplets of water or crystals of ice that form on condensation nuclei. Clouds typically occur in layers, as stratiform clouds, or in globular masses, as cumuliform clouds. Fog occurs when a cloud forms at ground level. Precipitation from clouds occurs as rain, hail, snow, and sleet. When supercold rain falls on a surface below freezing, it produces an ice storm.

There are three types of precipitation—orographic, convectional, and cyclonic. In orographic precipitation, air moves up and over a mountain barrier. As it moves up, it is cooled adiabatically and rain forms. As it descends the far side of the mountain, it is warmed. In convectional precipitation, unequal heating of the surface causes an air parcel to become warmer and less dense than the surrounding air. Because it is less dense, it rises. As it moves upward, it cools, and condensation with precipitation may occur. Under conditions of unstable air, thunderstorms can form, yielding hail and lightning. Cyclonic precipitation is described in Chapter 6.

Air pollution is defined as unwanted gases, aerosols, and particulates injected into the air by human and natural activity. Polluting gases are generated largely by fuel combustion. Aerosols and particulates are also released by fossil fuel burning. Smog, a common form of air pollution, contains nitrogen oxides, hydrocarbons, and ozone. Inversions can trap smog and other pollutant mixtures in a layer close to the ground, creating unhealthy air. Acid deposition, either as acid precipitation or as acidic dust particles, can acidify lakes and streams, killing aquatic life. Soils can lose fertility under acid attack. However, increasingly stringent pollution control has improved North American air quality significantly in the past decade.

KEY TERMS

sublimation	adiabatic process	convectional precipitation
deposition	dry adiabatic lapse rate	unstable air
hydrologic cycle	wet adiabatic lapse rate	thunderstorm
humidity	cloud	smog
dew-point temperature	orographic precipitation	acid deposition

REVIEW QUESTIONS

1. What is the hydrosphere? Where is water found on our planet? In what amounts? How does water move in the hydrologic cycle?

2. Define the terms *relative humidity* and *specific humidity*. How is relative humidity measured? Sketch a graph showing relative humidity and temperature through a 24-hour cycle.

3. Use the terms *saturation*, *dew point*, and *condensation* properly in describing what happens when an air parcel of moist air is chilled.

4. What is the adiabatic process? Why is it important?

5. Distinguish between dry and wet adiabatic lapse rates. In a parcel of air moving upward in the atmosphere, when do they apply? Why is the wet adiabatic lapse rate less than the dry adiabatic rate? Why is the wet adiabatic rate variable in amount?

6. How are clouds classified? Name four cloud families, two broad types of cloud forms, and three specific cloud types.

7. What is fog? Explain how radiation fog and advection fog form.

8. How is precipitation formed? Describe the process for warm and cold clouds.

9. Describe the orographic precipitation process. What is a rainshadow? Provide an example of the rainshadow effect.

10. What is unstable air? What are its characteristics?

11. Describe the convectional precipitation process. What is the energy source that powers this source of precipitation? Explain.

12. Sketch a convection cell within a thunderstorm. Show rising bubbles of air, updraft, downdraft, precipitation, and other features.

13. What are the most abundant air pollutants, and what are their sources?

14. What is smog? What important pollutant forms within smog, and how does this happen?

15. Distinguish between low-level and high-level inversions. How are they formed? What is their effect on air pollution? Give an example of a pollution situation of each type.

16. What type of pollution does the term *acid deposition* refer to? What are its causes? What are its effects?

Focus on Systems 4.2 • The Global Water Balance as a Matter Flow System

1. Sketch a diagram showing five pathways of matter flow for water in the global water balance flow system. Indicate their magnitudes on the diagram.

2. Suppose that global climate warming increases evaporation over land and oceans. Provided that the cycle remains in balance (that is, $P_{LAND} + E_{LAND} + P_{OCEAN} + E_{OCEAN} = 0$), what will be the effect on precipitation?

3. At present, oceans cover 71 percent of the earth's surface. Suppose this value were 50 percent. How do you think that the global water balance might be affected?

Focus on Systems 4.4 • The Thunderstorm as a Flow System

1. Compare and contrast the energy and matter flows in updraft cells and downdraft cells within a convective storm.

2. How would the convective precipitation process be affected by stable surrounding air?

ESSAY QUESTIONS

1. Compose an essay on water in the atmosphere. What part of the global supply of water is atmospheric? Why is it important? What is its global role? How does the capacity of air to hold water vapor vary? How is the moisture content of air measured? Clouds and fog visibly demonstrate the presence of atmospheric water. What are they? How do they form?

2. Compare and contrast orographic and convectional precipitation. Begin with a discussion of the adiabatic process and the generation of precipitation within clouds. Then compare the two processes, paying special attention to the conditions that create uplift. Can convectional precipitation occur in an orographic situation? Under what conditions?

PROBLEMS

Working It Out 4.1 • Energy and Latent Heat

1. Figure 4.3 shows that about 490 km^3 of water are evaporated from land and water surfaces each year. If all this evaporation occurs from a water surface at 15°C, how much energy is required? (1 m^3 of water has a mass of about 10^3 kg.)

2. U.S. energy consumption is about 8.5×10^{13} kj/yr. How does this value compare with the annual energy of evaporation computed in Problem 1?

Working It Out 4.3 • The Lifting Condensation Level

1. What is the lifting condensation level for a parcel of air at an initial temperature of 25°C and a dew-point temperature of 18°? What will be the temperature of the air parcel at that level?

2. Suppose that the same parcel of air is warmed to 30°C before it begins its ascent. What will be the lifting condensation level for the warmed parcel? What will be its temperature at that level?

3. Using algebra, show how expression (2) is derived from expression (1).

Chapter 5

Winds and the Global Circulation System

Wind is the primary subject of this chapter. Winds are caused by unequal heating of the earth's surface. As explained in Chapter 4, heating causes air to expand and become less dense. When this occurs, cooler, denser, surrounding air presses inward to buoy up the lighter, heated air, and a wind along the surface is created. This principle directly explains many types of local winds, such as the sea breeze you might experi-

ence on an afternoon at the beach. It is also the cause of global wind motions, when we consider that the equatorial and tropical regions are heated more intensely by the sun than are the mid- and higher latitudes. In response to this difference in global heating, vast tongues of warm air move poleward, and huge pools of cool air shift equatorward, influencing our day-to-day weather strongly. For global wind motions, however, wind direction is strongly influenced by the earth's rotation—a factor that is also a subject of this chapter.

 Another topic of this chapter is ocean currents, which are driven largely by surface winds and also act to

A racing yacht—powered by the global atmospheric circulation system.

move heat energy across parallels of latitude. We will also describe the phenomenon of El Niño—a recurring pattern of Pacific Ocean circulation that is accompanied by changes in global weather patterns.

Before we examine local and global wind patterns, however, we will need to examine atmospheric pressure and how wind develops in response to pressure differences. These are the subjects for the next few pages.

ATMOSPHERIC PRESSURE

Objects and organisms on the earth's land surface are actually located at the bottom of a vast ocean. However, it's an ocean of air—the earth's atmosphere. Like the water in the ocean, the air in the atmosphere is constantly pressing on the solid or liquid surface beneath it. This pressure is exerted in all directions. Atmospheric pressure exists because air has mass and is constantly being pulled toward the earth by the force of gravity. Pressure increases downward through the atmosphere as the mass of air above each layer increases. Thus, atmospheric pressure is greatest at the earth's surface. Unlike water, air is easily compressed. This means that air near the earth's surface is denser than air aloft.

Measuring Atmospheric Pressure

Pressure is a force per unit area. In the metric system, the unit of pressure is the *pascal* (Pa), which is defined as one newton per square meter. (Recall from Chapter 4 that a newton is the force produced when a 1-kg mass is accelerated by 1 m/s^2.) For the atmosphere, the force is produced by the acceleration of gravity acting on the mass of a column of air (the weight of a column of air). At sea level, the average pressure of air is 1.013×10^5 Pa. Another way to express sea-level pressure is as the weight (due to gravity) of about 1 kg of mass per square centimeter (Figure 5.1). In the English system, sea-level atmospheric pressure is about 15 lb/in.^{2}, which is also diagrammed in the figure.

Although the pascal is the proper metric unit for pressure, many atmospheric pressure measurements are based on the *bar* and the *millibar* (mb) (1 bar = 10^3 mb), two former metric units that are still in wide use. In this text, we will use the *millibar* as the metric unit of atmospheric pressure, following this long tradition. One millibar is equal to 100 pa. Standard sea-level atmospheric pressure is 1013.2 mb.

Atmospheric pressure is measured by a **barometer**. One simple type of barometer is the *mercury barometer* (Figure 5.2). To construct the barometer, a glass tube about 1 m (3 ft) long, sealed at one end, is completely filled with mercury. The open end is temporarily held

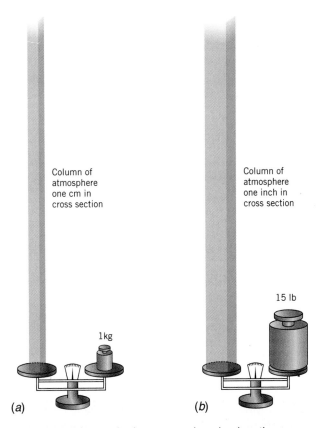

Column of atmosphere one cm in cross section

Column of atmosphere one inch in cross section

15 lb

1 kg

(a)   (b)

Figure 5.1 Atmospheric pressure imagined as the weight of a column of air. (*a*) Metric system. The weight of a column of air 1 cm on a side is balanced by the weight of a mass of about 1 kg. (*b*) English system. The weight of a column of air 1 in. on a side is balanced by a weight of about 15 pounds.

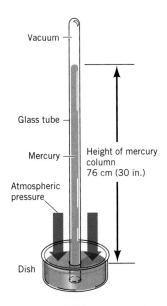

Vacuum

Glass tube

Mercury

Atmospheric pressure

Height of mercury column 76 cm (30 in.)

Dish

Figure 5.2 A simple diagram of a mercury barometer. Atmospheric pressure pushes the mercury upward into the tube, balancing the pressure exerted by the weight of the mercury column. As atmospheric pressure changes, the level of mercury in the tube changes. At sea level, the average height of the mercury level is about 76 cm (30 in.).

closed. Then the tube is inverted, and the end is immersed into a dish of mercury. When the opening of the tube is uncovered, the mercury in the tube falls some distance but then remains fixed at a level about 76 cm (30 in.) above the surface of the mercury in the dish. What is happening here? At first, mercury flows out of the tube, under the force of gravity, and into the dish. Since there is no air in the top of the tube, a vacuum is formed. At the same time, atmospheric pressure pushes on the surface of the mercury in the dish, pressing the mercury upward into the tube. Since there is no air in the tube, there is no air pressure to push back against it. The pull of gravity on the mercury column then balances the pressure of the air, and the mercury level stays at a stable height. If air pressure changes, so will the height of the mercury column. So this device will measure air pressure directly—but as a height in centimeters or inches rather than a pressure. The mercury barometer has become the standard instrument for weather measurements because of its high accuracy.

The widespread use of the mercury barometer has led to the common measurement of air pressure in heights of mercury, as centimeters or inches. On this scale, standard sea-level pressure is 76 cm Hg (29.92 in. Hg). The letters Hg are the chemical symbol for mercury. In this text, we will use in. Hg as the English unit for atmospheric pressure.

A more common type of barometer is the *aneroid barometer* (Figure 5.3). It uses a sealed canister of air, much like a small, flat can you might see in a supermarket. The canister has a slight partial vacuum, tens-

ing the lid, or diaphragm, against a spring. The diaphragm flexes as air pressure changes, and through a mechanical linkage this fine movement drives a hand indicator that moves along a scale.

Atmospheric pressures at a location vary from day to day by only a small proportion. On a cold, clear winter day, the barometric pressure at sea level might be as high as 1030 mb (30.4 in. Hg), while in the center of a storm system, pressure might drop to 980 mb (28.9 in. Hg)—a difference of about 5 percent. Pressure changes are associated with traveling weather systems, which we will discuss later in this chapter.

How Air Pressure Changes with Altitude

Figure 5.4 shows how atmospheric pressure decreases with altitude. Moving upward from the earth's surface, the decrease of pressure with altitude is initially quite rapid. At higher altitudes, the decrease is much slower. Because atmospheric pressure decreases rapidly with altitude near the surface, a small change in elevation will often produce a significant change in air pressure. For

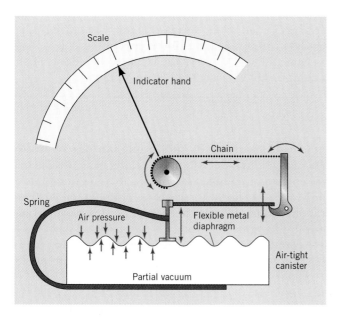

Figure 5.3 Diagram of an aneroid barometer. As air pressure varies, the diaphragm moves up and down. This motion is transmitted mechanically to move an indicator hand along a scale.

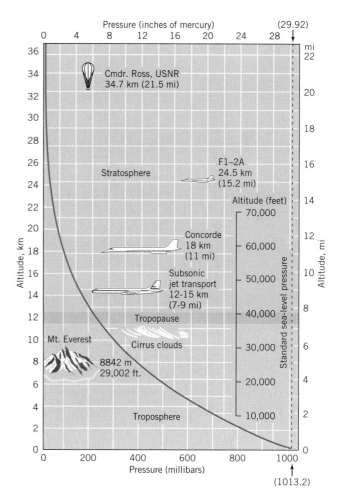

Figure 5.4 Atmospheric pressure decreases with increasing altitude above the earth's surface.

Working It Out 5.1 • Pressure and Density in the Oceans and Atmosphere

Although it is sometimes useful to think of the earth's surface as the bottom of a vast "ocean of air," the atmosphere is quite different from the oceans because the oceans are composed of an incompressible liquid—water with dissolved compounds—and the atmosphere is composed of a mixture of compressible gases. Recall that both liquids and gases are classed as fluids.

The pressure at any level in a fluid is created by the "weight"—gravitational attraction—of the mass of the fluid above that level. In the oceans, it's relatively simple to determine that weight. Let's consider a square meter (1 m × 1 m) of horizontal area at depth D within the ocean. Pressing down on that square meter is the weight of $D \times 1 \times 1$ m^3 of water. Each cubic meter of water weighs about 1000 kg or 1 metric ton (t). So as a quick rule of thumb, we can regard the pressure at any depth in the ocean as about $1 \times D = D$ t/m^2. That is,

$$P_{OCEAN} = D$$

where P is the pressure in t/m^2 and D is the depth in m. For example, the pressure at 500 m depth is about 500 t/m^2; at 5 km = 5000 m, it is about 5000 t/m^2.

These values are based on a constant density of 1 t/m^3. To be pre-cise, there is a slight variation in the density of sea water with pressure, temperature, and salinity. That is, colder, saltier, and deeper water will be more dense. At the surface, the density of ocean water is 1.028 t/m^3, while at 5000 km depth, the density is 1.051 t/m^3, an increase of about 2.2 percent. The left graph in the figure, which plots pressure with depth in the ocean, uses a constant density value of 1.035 t/m^3.

Note that at any depth in the ocean, there will be not only the weight of the water above, but also the weight of the atmosphere above the water. Sea-level pressure is about 1000 millibars, which converts to 1 t/m^2. Thus, the atmosphere adds the pressure of about one more meter of water depth.

Although small, the density differences in water with temperature and salinity can vary the pressure at a given depth enough to create pressure gradients and therefore induce movement of water from lower to higher pressure. An example is the convection loop circulation that occurs in a container of water when it is heated from below. This loop is described in more detail in *Focus on Systems 5.2 • The Convection Loop as an Energy Flow System.*

In contrast to the oceans, the atmosphere is readily compressible.

What is the weight of a cubic meter of air? There is no rule-of-thumb answer for that question, because density varies freely with pressure. When compressed, more molecules of the gases that compose air are present in a given volume, and so the volume will have a greater mass and weigh more. At the earth's surface, the density of a cubic meter of air is about 1.225 kg/m^3, while at 11 km (about 36,000 ft), the height of a cruising jet aircraft, it is about 0.364 kg/m^3, or a little less than one-third of the surface value.

The difference is nearly all due to pressure, but temperature is also important. At a given pressure, warmer air will be less dense, while colder air will be more dense. As in the case of the oceans, the change in density with temperature is enough to create pressure gradients that induce movements of air, and, in fact, sustain the winds that constantly move and mix the troposphere.

Atmospheric pressure is plotted against altitude in the right graph of the figure. Note that the graph curves smoothly, whereas the graph of pressure against depth for the oceans is straight. The curve of the atmospheric pressure graph is a result of the fact that the density of air decreases with altitude.

example, you may have noticed that your ears sometimes "pop" from a pressure change during an elevator ride in a tall building. You may also have noticed your ears popping on an airplane flight. Aircraft cabins are pressurized to about 800 mb (24 in. Hg), which corresponds to an elevation of about 1800 m (5900 ft).

The decrease of atmospheric pressure with height above the surface, as well as the increase of water pressure in the ocean with depth, are explained by some simple physical principles that are given in *Working It Out 5.1 • Pressure and Density in the Oceans and Atmosphere.*

The decrease of pressure with elevation also affects human physiology. At high altitudes, travelers are often tired and find breathing difficult. With the decreased air pressure of higher altitudes, oxygen moves into

lung tissues more slowly, producing shortness of breath and fatigue. These symptoms, along with headache, nosebleed, or nausea, are referred to as mountain sickness and are likely to occur at altitudes of 3000 m (about 10,000 ft) or higher. However, most people adjust to the reduced air pressure after several days.

As atmospheric pressure decreases, the boiling point of water becomes lower. For example, at 3000 m (about 10,000 ft), the boiling point, which is 100°C (212°F) at sea level, is reduced to 90°C (194°F). At 5000 m (16,000 ft), the boiling point is 84°C (183°F). As the boiling point decreases, the time required to cook foods by boiling increases. As many experienced campers know, boiling potatoes is a slow process at altitudes above 2000 m (about 6600 ft)!

Atmospheric pressure at any given altitude can be approximated by the following formula:

$$P_Z = 1014[1 - 0.0226Z]^{5.26}$$

where P_Z is the pressure in millibars at height Z in km. This formula takes into account the fact that both temperature and pressure decrease with altitude. It is derived by atmospheric physicists for certain standard atmospheric conditions and is used in plotting the figure. Note that the value of 1014 is the surface atmospheric pressure in millibars for these standard conditions.

As an example, let's find the pressure at 12 km, which is approximately the altitude of a passenger jet on a transcontinental flight (about 39,000 ft). Substituting for Z, we have

$$P_Z = 1014 \times [1 - (0.0226 \times 12)]^{5.26}$$
$$= 1014 \times (1 - 0.271)^{5.26}$$
$$= 1014 \times (0.729)^{5.26}$$
$$= 1014 \times 0.189 = 192 \text{ mb}$$

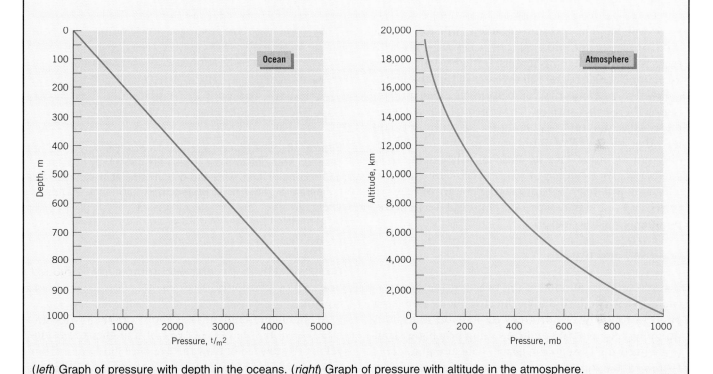

(*left*) Graph of pressure with depth in the oceans. (*right*) Graph of pressure with altitude in the atmosphere.

WINDS AND PRESSURE GRADIENTS

Wind is defined as air motion with respect to the earth's surface. Wind movement is dominantly horizontal. Air motions that are dominantly vertical are not called winds, but are known by other terms, such as updrafts or downdrafts.

Wind is caused by differences in atmospheric pressure from place to place. Air tends to move from high to low pressure until the air pressures are equal. This follows the simple physical principle that any flowing fluid (such as air) subjected to gravity will move until the pressure at every level is uniform.

Figure 5.5 illustrates this principle. It is a simple map showing air pressure near two locations, taken here as

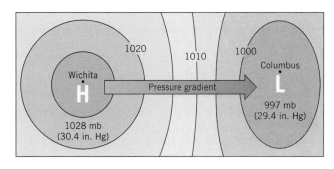

Figure 5.5 Isobars and a pressure gradient. High pressure is centered at Wichita, and low pressure is centered at Columbus.

Wichita and Columbus. At Wichita, the pressure is high (H), and the barometer (adjusted to sea-level pressure) reads 1028 mb (30.4 in. Hg). At Columbus, the barometer is low (L) and reads 997 mb (29.4 in. Hg). Thus, there is a **pressure gradient** between Wichita and Columbus. The lines drawn on the map around Wichita and Columbus are lines of equal pressure, called **isobars**. As you move between Wichita and Columbus, you cross these isobars on the map, encountering changing pressures as a result of the drop from 1025 to 1000 mb (30.3 to 29.5 in. Hg).

Because atmospheric pressure is unequal at Wichita and Columbus, a *pressure gradient force* will push air from Wichita toward Columbus. The greater the pressure difference between the two locations, the greater this force will be. This pressure gradient will produce a wind. However, the wind will not necessarily move directly from Wichita to Columbus. It will be deflected by the Coriolis effect, as we will see later in this chapter.

Pressure gradients develop because of unequal heating of the earth's surface. This unequal heating produces air of different temperatures. Recall from Chapter 4 that air parcels of different temperatures will have different densities. That is, warm air is less dense than cool air. Thus, a column of warmer, less dense air over a location will weigh less, producing low pressure. Conversely, a column of colder, denser air will weigh more, producing high pressure.

Sea and Land Breezes

If you've ever vacationed at the beach in the summer, you may have noticed that in the afternoons a *sea breeze* often sets in. This wind brings cool air off the water, dropping temperatures and refreshing beachgoers and residents living close to the beach. Late at night, a *land breeze* may develop. This wind moves cooler air, which is chilled over land by nighttime radiant cooling, toward the water. Sea and land breezes are simple examples of how heating and cooling produce a pressure gradient that causes a wind. Figure 5.6 shows them in more detail.

On a clear day, the sun heats the land surface rapidly, and a shallow air layer near the ground is strongly warmed (Figure 5.6*a*). The warm air expands. Because the warm air is less dense than the cool air off-shore, it weighs less and exerts less pressure. Thus, low pressure is formed over the coastal belt. Pressure remains higher over the water. This means that there is a pressure gradient from ocean to land, and a wind is set in motion. The sea breeze can become very strong in a shallow air layer close to shore.

Because the air over the land is warmer and less dense, it will also tend to rise. Over the ocean, the air moving landward will be replaced by air moving downward from above, causing a sinking motion. The rising over land and the sinking over ocean will create a re-

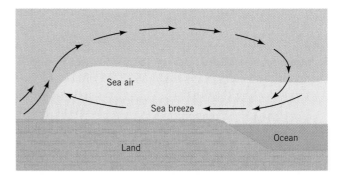

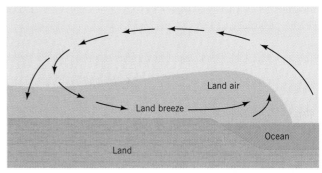

Figure 5.6 Sea breeze and land breeze alternate in direction from day to night. These are caused by heating and cooling of land, while the ocean remains at an even temperature. (After R. G. Barry and R. J. Chorley, *Atmosphere, Weather and Climate*, 6th ed., Routledge. Used by permission.)

turn flow aloft that is opposite in direction to the sea breeze. We refer to this as a *convection loop*, in which unequal heating causes a circuit of moving air.

At night, the land surface cools rapidly if skies are clear. A cooler, denser air layer of higher pressure develops (Figure 5.6*b*). Now the pressure gradient is from land to ocean, setting up a land breeze. Like the sea breeze, the convection loop is completed by a wind aloft moving in the opposite direction.

Measurement of Winds

To describe winds fully, we must measure both wind speed and wind direction (Figure 5.7). Wind direction can be determined by a *wind vane*, the most common of the weather instruments. The design of the wind vane keeps it facing into the wind. Wind direction is given as the direction from which the wind is coming. So, an east wind comes from the east and moves to the west (Figure 5.8).

Wind speed is measured by an *anemometer*. The most common type is the cup anemometer, which consists of three funnel-shaped cups mounted on the ends of spokes of a horizontal wheel. The cups rotate with a speed proportional to that of the wind. One version of the anemometer turns a small electric generator. The more rapidly the wheel rotates, the more current is

Figure 5.7 A combination cup anemometer and wind vane. Wind speed and compass direction are displayed on the meter below. Also shown are a barometer and maximum-minimum thermometer. Normally, the wind vane and anemometer will be mounted outside, with a cable from the instruments leading to the meter, which is located indoors.

generated. The current is measured by a meter, which is calibrated in units of wind speed. Units are either meters per second or miles per hour.

The Coriolis Effect and Winds

The pressure gradient force tends to move air from high pressure to low pressure. For sea and land breezes, which are local in nature, this push produces a wind motion in about the same direction as the pressure gradient. For larger wind systems, however, the direction of air motion will be somewhat different. The difference is due to the earth's rotation, through the **Coriolis effect**. This effect notes that an object in motion on the earth's surface always appears to be deflected away from its course. The object moves as though a force were pulling it sidewards. The apparent deflection is to the right in the northern hemisphere and to the left in the southern hemisphere. The effect is strongest near the poles and decreases to no apparent deflection at the equator. Figure 5.9 diagrams the effect. Note that the apparent force does not depend on direction of motion—the deflection occurs whether the object is moving toward the north, south, east or west.

The Coriolis effect is a result of the earth's rotation. Imagine that a rocket is launched from the north pole toward New York, following the 74° W longitude meridian (Figure 5.10). However, as it travels toward New York, the earth is rotating beneath its trajectory, moving from west to east. The rocket's path will appear to curve to the right with respect to the earth underneath.

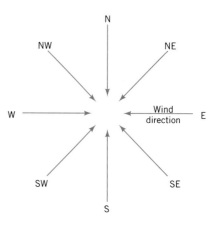

Figure 5.8 Winds are designated according to the compass point from which the wind comes. An east wind comes from the east, but the air is moving westward.

Using reasonable values for calculation, we find that the rocket might well fall to earth near Chicago!

What would happen if the rocket were launched from New York heading along the 74° W meridian toward the pole? Because of the earth's rotation, a point on the earth's surface at the latitude of New York (about 40° N) moves eastward at about 1300 km/hr (800 mi/hr). Although the rocket is aimed properly along the meridian, its motion will have an eastward component of 1300 km/hr. However, as the rocket

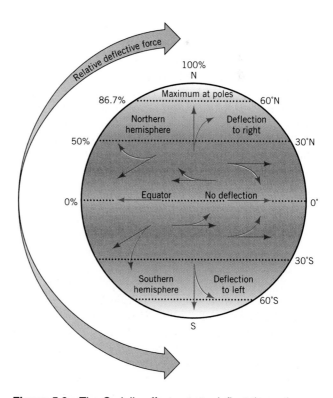

Figure 5.9 The Coriolis effect acts to deflect the paths of winds or ocean currents to the right in the northern hemisphere and to the left in the southern hemisphere as viewed from the starting point.

Figure 5.10 The Coriolis effect. The path of a rocket launched from the north pole toward New York will apparently be deflected to the right by the earth's rotation, and might land near Chicago. If launched from New York toward the north pole, its path will apparently be deflected to the right as well.

moves northward, it passes over land that is moving less swiftly eastward. At 60° N latitude, the eastward velocity of the earth's surface is only about 800 km/hr (500 mi/hr). So, the rocket will be moving more quickly eastward than the land underneath, and its path will appear to be deflected to the right.

Cyclones and Anticyclones

If the Coriolis effect deflects air moving from high to low pressure, how does that affect the wind pattern? A parcel of air in motion near the surface is subjected to three influences: (1) the pressure gradient that propels the parcel toward low pressure; (2) the Coriolis effect that turns the parcel to the side; and (3) friction with the ground surface. As a result of all these influences, air will move toward low pressure but at an angle to the pressure gradient.

Figure 5.11 shows wind patterns around surface high- and low-pressure centers in northern and southern hemispheres. When low pressure is at the center, the pressure gradient is straight inward (left side of figure). When high pressure is at the center, the gradient is straight outward (right side). But when air actually moves in response to these pressure gradients, it will move at an angle across the gradient.

For low-pressure centers (left side), the wind will be deflected into an inward-moving spiral pattern, or *inspiral*. In the northern hemisphere (upper portion), the inspiral will be counterclockwise, since the Coriolis deflection is to the right. In the southern hemisphere

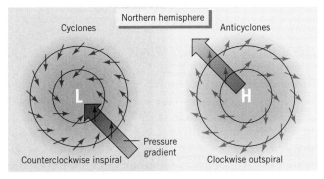

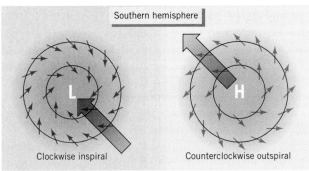

Figure 5.11 Surface winds spiral inward toward the center of a cyclone, but outward and away from the center of an anticyclone. Because the Coriolis effect deflects moving air to the right in the northern hemisphere, and to the left in the southern hemisphere, the direction of inspiraling and outspiraling reverses from one hemisphere to the other.

(lower portion), the inspiral will be clockwise, since the Coriolis deflection is to the left. Meteorologists use the term **cyclone** to refer to a center of low pressure with inspiraling air.

For high-pressure centers, the situation is reversed. Air moving outward forms an *outspiral*, clockwise in the northern hemisphere and counterclockwise in the southern hemisphere. This outspiral is referred to as an **anticyclone** (Figure 5.11).

Recall that in the case of the sea and land breezes, air moves upward at the site of low pressure and downward at the site of high pressure. The same is true of cyclones and anticyclones. In the cyclone, air spiraling inward converges, causing a rising motion at the center. Recall from Chapter 4 that when air is forced aloft, it is cooled according to the adiabatic principle, and condensation and precipitation can occur. This is why cyclones, or low-pressure centers, are associated with cloudy or rainy weather. In the anticyclone, air spiraling outward diverges, causing a sinking motion at the center. When air descends, we know from Chapter 4 that it is warmed according to the adiabatic principle, and thus condensation cannot occur. So, anticyclones, or high-pressure centers, are associated with clear weather.

Cyclones and anticyclones are typically large features of the atmosphere—perhaps a thousand kilometers (about 600 mi) across, or more. For example, a fair

weather system, which is an anticyclone, may stretch from the Rockies to the Appalachians. Cyclones and anticyclones can remain more or less in one location, or they can move, sometimes rapidly, to create weather disturbances.

The vertical motions of air in adjacent cyclones and anticyclones are often connected in a convection loop. This loop can be viewed as an energy flow system, and is described in more detail in *Focus on Systems 5.2 • The Convection Loop as an Energy Flow System.* Because a spiraling motion is involved, it is more complex than the simple convection loops we have discussed in this chapter and in Chapter 2.

Surface Winds on an Ideal Earth

Consider wind and pressure patterns on an ideal earth—one without a complicated pattern of land and water and no seasonal changes. Pressure and winds for such an ideal earth are shown in Figure 5.12. The surface wind patterns are shown on the globe's surface. On the right, a cross section through the atmosphere shows the wind patterns aloft.

The most important features for understanding the pattern at low latitudes are shown on the cross section to the right, labeled "Hadley cells." Recall that when air is heated, low pressure forms and air rises. Since inso-

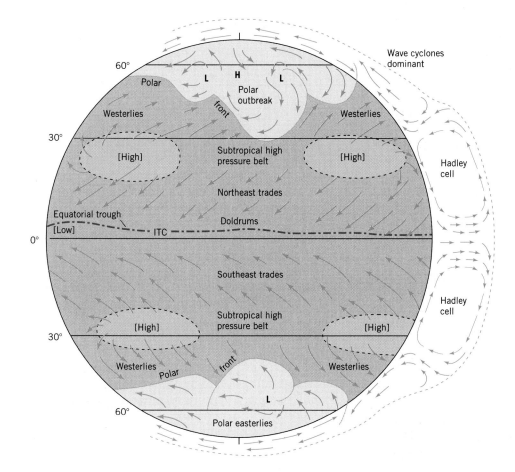

Figure 5.12 This schematic diagram of global surface winds and pressures shows the features of an ideal earth, without the disrupting effect of oceans and continents and the variation of the seasons. Surface winds are shown on the disk of the earth, while the cross section at the right shows winds aloft.

Focus on Systems 5.2 • The Convection Loop as an Energy Flow System

In Chapters 2 and 4, we established the principle that when air is heated, it rises. The rising occurs because the density of air at any given level depends on temperature, and if an air parcel is warmer than the surrounding air, it will be less dense and therefore buoyed upward. This phenomenon powers the *convection loop*, discussed in the text.

The convection loop is actually a flow system in which a closed system of matter flow is linked with an open system of energy flow. To make this clearer, think of a laboratory beaker of water gently heated by a single, low flame, as sketched on the right. The flame heats the water directly above it, which becomes less dense than the surrounding fluid and is buoyed upward in a column. Meanwhile, cooler water moves in from the sides to take its place. Soon a convection loop is formed, with warm water ascending in a cylinder at the center of the beaker, spreading out across the top of the liquid, descending along the sides of the

beaker, and finally converging above the flame to be heated and rise again.

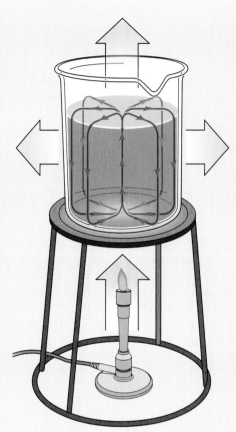

What happens to the sensible heat that was absorbed by the water? Some of the heat will be lost to the air above the beaker when the warm liquid spreads out over the top surface. Another portion will be lost from the sides of the beaker as the water descends and warms the glass and the air that surrounds the glass. Another portion will go to warm the liquid as a whole, since some mixing will occur even though the flow in the loop may be fairly smooth.

If we wait for a while, the system will come to a steady state in which the liquid constantly moves in a three-dimensional convection loop, absorbing heat at the bottom of the

The convectional circulation in a beaker of water heated by a low flame constitutes a flow system in which the fluid cycles in a three-dimensional loop and heat (broad arrows) is transported from the bottom center of the vessel to the water surface and sides of the container, where it warms the surrounding air. (Copyright © A. N. Strahler.)

lation is strongest when the sun is directly overhead, the equator will be heated more strongly than other places on this featureless earth. The result will be two convection loops, the **Hadley cells**, which form in the northern and southern hemispheres. In each Hadley cell, air rises over the equator, flows poleward, and descends at about a 30° latitude. (The Hadley cell is named for George Hadley, who first proposed its existence in 1735.)

Since air is heated and rises at the equator, a zone of low pressure will result. This zone is referred to as the *equatorial trough*. As shown by the wind arrows, air moves toward the equatorial trough, where it converges and moves aloft as part of the Hadley cell circulation. Convergence occurs in a narrow zone, named the **intertropical convergence zone (ITC).** Because the air is generally rising, winds in the ITC are light and variable. In earlier centuries, mariners on sailing vessels referred to this region as the *doldrums*, where they were sometimes becalmed for days at a time.

On the poleward side of the Hadley cell circulation, air descends and surface pressures are high. This produces two **subtropical high-pressure belts**, each centered at about 30° latitude. Within the belts, two, three, or four very large and stable anticyclones are formed. Air is descending, and winds are weak at the center of these anticyclones. Calm prevails as much as one-quarter of the time. Because of the high frequency of calms, mariners named this belt the *horse latitudes*. The name is said to have originated in colonial times from the experiences of New England traders carrying cargoes of horses to the West Indies. When their ships were stranded for long periods of time, the freshwater supplies ran low and the horses were thrown overboard.

Winds around the subtropical high-pressure centers are outspiraling and move toward equatorial as well as middle latitudes. The winds moving equatorward are the strong and dependable *trade winds*. North of the equator, these are from the northeast and are referred to as the *northeast trades*. To the south of the equator,

beaker and releasing heat at the top surface and sides. The net effect is then a continuous cycling of matter while energy flows in at one location and out at others.

When large convection loops are set up in the earth's atmosphere by unequal surface heating, they are not so simple as the example of water heating in the beaker. The rotation of the earth, through the Coriolis effect, induces a spiraling motion to the flows. The figure to the right shows an anticyclone and a cyclone linked together in a convection loop, as they might be in a sequence of eastward-moving weather systems. In the anticylone (high-pressure center), air converges aloft and descends in a spiraling motion. Near the surface, the flow diverges, fanning out and moving away from the center of the anticyclone. In the cyclone, air converges near the surface and spirals upward. Aloft, the air diverges and spirals outward.

The upward and downward flows are linked as a portion of the moving air flows from the anticyclone to the cyclone near the surface, and another portion flows from the cyclone to the anticyclone aloft. The loop is not perfect. Air escapes and enters from the sides of the flow, as shown by the arrows.

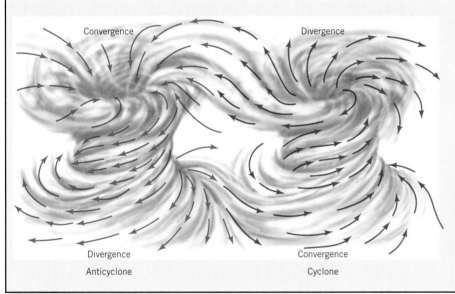

A cyclone and anticyclone linked together in a convection loop. (Copyright © A. N. Strahler.)

they are from the southeast and are the *southeast trades*. Poleward of the subtropical anticyclones, air spiraling outward produces southwesterly winds in the northern hemisphere and northwesterly winds in the southern hemisphere.

Between about 30° and 60° latitude, the pressure and wind pattern becomes more complex. This latitudinal belt is a zone of conflict between air bodies with different characteristics. Masses of cool, dry air move into the region, heading eastward and equatorward. These form *polar outbreaks*. The border of the polar outbreak is known as the *polar front*. (We'll have much more to say about air masses and the polar front later in this chapter.) As a result of this activity, pressures and winds can be quite variable in the midlatitudes from day to day and week to week. On average, however, winds are more often from the west, so the region is said to have *prevailing westerlies*.

At the poles, the air is intensely cold. As a result, high pressure occurs. Outspiraling of winds around a polar anticyclone should create surface winds from a generally easterly direction, known specifically as *polar easterlies*. As we will see in discussing actual pressure and wind patterns, this situation exists only in the south polar region. In the north polar region, winds tend to have an eastward component, but there is too much variation in direction to consider polar easterlies as dominant winds there.

GLOBAL WIND AND PRESSURE PATTERNS

So far, we've discussed the wind pattern for a seasonless, featureless earth. Let's turn now to actual global wind and pressure patterns. Here, we will use the global maps of wind and pressure for January and July, shown in Figures 5.13 and 5.14. The pressures and winds shown are averages over many years for all daily observations in either January or July. They are corrected for the elevation of the recording station, so that pressures

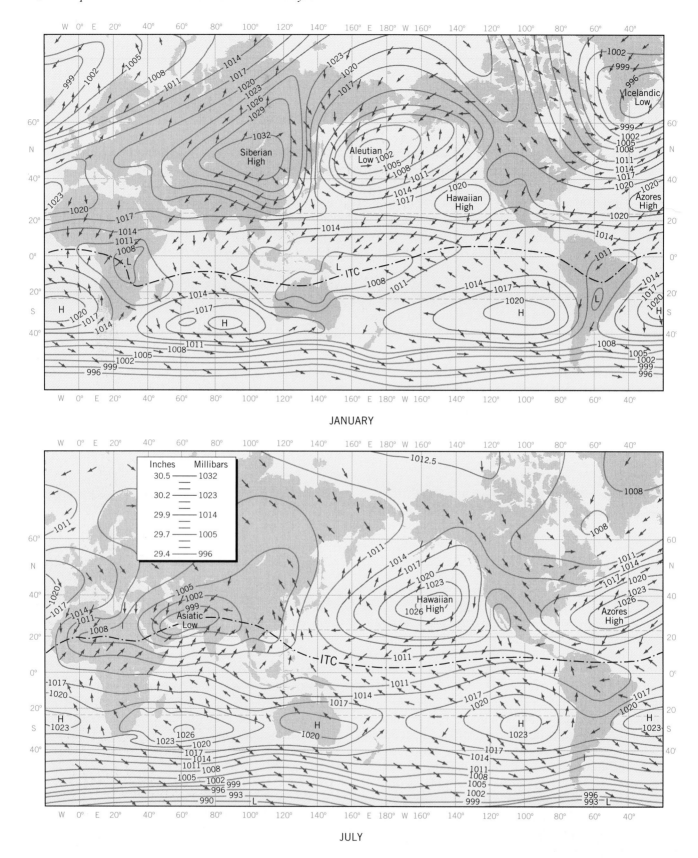

JANUARY

JULY

Figure 5.13 Mean monthly atmospheric pressure and prevailing surface winds for January and July. Pressure units are millibars reduced to sea level. Many of the wind arrows are inferred from isobars. See also Figure 5.14. (Data compiled by John E. Oliver.)

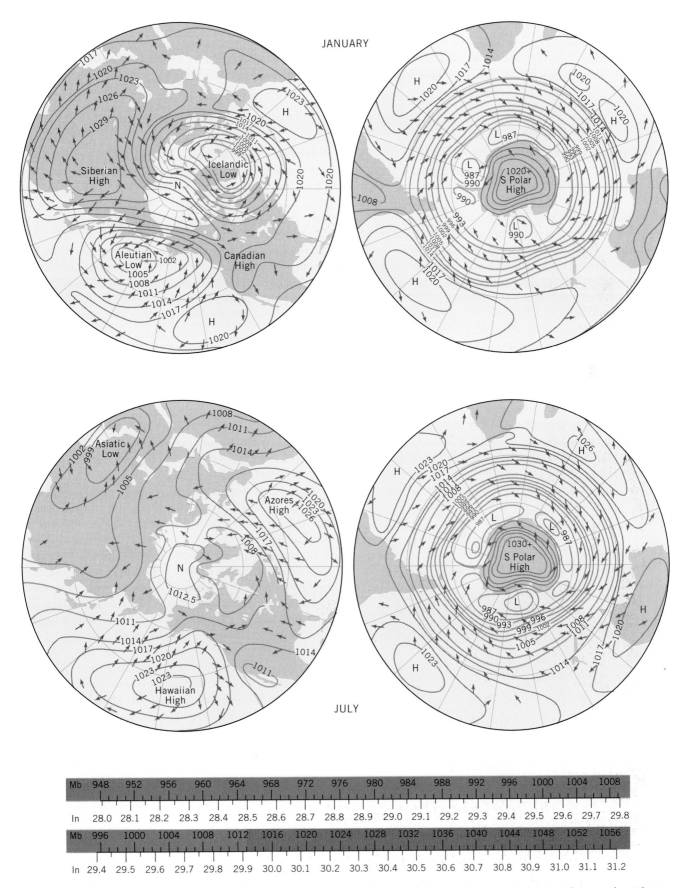

Figure 5.14 Mean monthly atmospheric pressure and prevailing surface winds for January and July, northern and southern hemispheres. Pressure units are millibars reduced to sea level. See also Figure 5.13. (Data compiled by John E. Oliver.)

are shown for sea level. Average barometric pressure is 1013 mb (29.9 in. Hg). Values greater than that are "high" (red lines), while those lower are "low" (green lines). The maps make use of both Mercator and polar projections. As we discuss the various features of global pressure and winds, be sure to examine both sets of maps.

Subtropical High-Pressure Belts

The largest and most prominent features of the maps are the subtropical high-pressure belts, created by the Hadley cell circulation. The high-pressure belt in the southern hemisphere conforms well to the pattern of Figure 5.12. It has three large high-pressure cells, each developed over oceans, that persist year round. A fourth, weaker high-pressure cell forms over Australia in July, as the continent cools during the southern hemisphere winter.

In the northern hemisphere, the situation is somewhat different. The subtropical high-pressure belt shows two large anticyclones centered over oceans—the Hawaiian High in the Pacific and the Azores High in the Atlantic. From January to July, these intensify and move northward. From their positions off the east and west coasts of the United States, they have a dominant influence on summer weather in North America.

Figure 5.15 is a schematic map of two large high-pressure cells bounded by continents, showing the features of circulation around the cells. As we noted earlier, the outspiraling circulation produces the trade winds in the tropical and equatorial zones, and the westerlies in the subtropical zones and poleward. In the high-pressure cells, air on the east side subsides more intensely. This means that winds spiraling outward on the eastern side are drier. On the west side, subsidence is less strong. In addition, these winds travel long distances across warm, tropical ocean surfaces before reaching land, and so are warm and moist.

When the Hawaiian and Azores Highs intensify and move northward in the summer, our east and west coasts feel their effects. On the west coast, dry subsiding air from the Hawaiian High dominates, so fair weather and rainless conditions prevail. On the east coast, warm, moist air from the Azores High flows across the continent from the southeast, producing generally hot, humid weather for the central and eastern United States. In winter, these two anticyclones weaken and move to the south—leaving our weather to the mercy of colder winds and air masses from the north and west.

The ITC and the Monsoon Circulation

Recall from Chapter 2 that insolation is most intense when the sun is directly overhead. Remember, too, that the latitude at which the sun is directly overhead changes with the seasons, migrating between the tropics of cancer and capricorn. Since the Hadley cell circulation is driven by this heating, we can expect that the elements of the Hadley cell circulation—ITC and subtropical high-pressure belts—will also shift with the seasons. We have already noted this shift for the Hawaiian and Azores Highs, which intensify and move northward as July approaches.

In general, the ITC also shifts with the seasons, as we might expect. The shift is moderate in the western hemisphere. The ITC moves a few degrees north from January to July over the oceans. In South America, the ITC lies across the Amazon in January and swings northward by about 20° to the northern part of the continent. But what happens in Africa and Asia? By comparing the two Mercator maps of Figure 5.13, you can see that there is a huge shift indeed! In January, the ITC runs south across eastern Africa and crosses the Indian Ocean to northern Australia at a latitude of about 15° S. In July, it swings north across Africa along the southern side of the Sahara, and then it rises further north to lie along the south rim of the Himalayas, in India, at a latitude of about 25° N. This is a shift of about 40 degrees of latitude!

Why does such a large shift occur in Asia? To answer this question, look at the pressure and wind pattern over Asia. In January, an intense high-pressure system, the Siberian High, is found there. Recall from Chapter 3 that, in winter, temperatures in northern Asia are very cold, so this high-pressure is to be expected. In July, this high-pressure center is absent, replaced instead by a low centered over the Middle Eastern desert

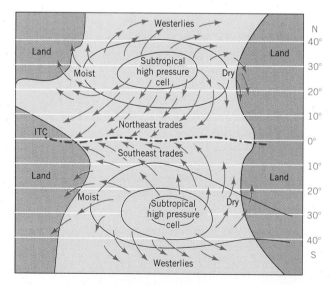

Figure 5.15 Over the oceans, surface winds spiral outward from the subtropical high-pressure cells, feeding the trades and the westerlies. On the eastern side of the cells, air subsides more strongly, producing dry winds. On the western sides, subsidence is not as strong, and a long passage over oceans brings warm, moist air to the continents.

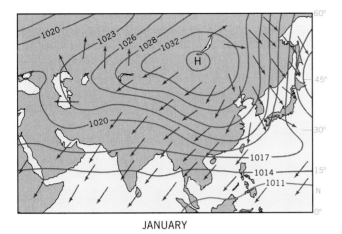

JANUARY

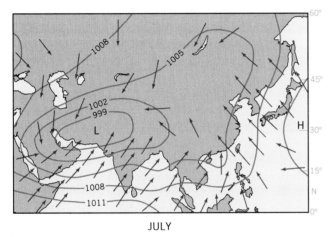

JULY

Figure 5.16 The Asiatic monsoon winds alternate in direction from January to July, responding to reversals of barometric pressure over the large continent.

region. The Asiatic low is produced by the intense summer heating of the landscape.

The movement of the ITC and the change in the pressure pattern with the seasons create a reversing wind pattern in Asia known as the **monsoon** (Figure 5.16). In the winter, there is a strong outflow of dry, continental air from the north across China, Southeast Asia, India, and the Middle East. During this *winter monsoon*, dry conditions prevail. In the summer, warm, humid air from the Indian Ocean and the southwestern Pacific moves northward and northwestward into Asia, passing over India, Indochina, and China. This airflow is known as the *summer monsoon* and is accompanied by heavy rainfall in southeastern Asia.

North America does not have the remarkable extremes of monsoon winds experienced in Asia. Even so, in summer there is a prevailing tendency for warm, moist air originating in the Gulf of Mexico to move northward across the central and eastern part of the United States. This Gulf air sometimes moves westward across the southwestern deserts of New Mexico and Arizona, occasionally reaching the California coast. This

airflow is referred to locally as a "monsoon." In winter, the airflow across North America generally reverses, and dry, continental air from Canada moves south and eastward.

Wind and Pressure Features of Higher Latitudes

The northern and southern hemispheres are quite different in their geography. As you can easily see from Figure 5.17, the northern hemisphere has two large continental masses, separated by oceans. An ocean is also at the pole. In the southern hemisphere, we find a large ocean with a cold, glacier-covered continent at the center. These differing land–water patterns strongly influence the development of high- and low-pressure centers with the seasons.

Let's examine the northern hemisphere first (Figure 5.14). Recall from Chapter 3 that continents will be cold in winter and warm in summer, as compared to oceans at the same latitude. But we know from our analysis earlier in the chapter that cold air will be associated with high pressure and warm air with low pressure. Thus, continents will show high pressure in winter and low pressure in summer.

This pattern is very clear in the northern hemisphere polar maps of Figure 5.14. In winter, the strong Siberian High is found in Asia, and a weaker cousin, the Canadian High, is found in North America. From these high-pressure centers, air spirals outward, bringing cold air to the south. Over the oceans, two large centers of low pressure are striking—the Icelandic Low and the Aleutian Low. These two low-pressure centers are not actually large stable features that we would expect to find on every daily world weather map in January. Rather, they are regions of average low pressure where winter storm systems are spawned.

In summer, the pattern reverses. The continents show generally low pressure, while high pressure builds over the oceans. This pattern is easily seen on the northern hemisphere July polar map. The Asiatic Low

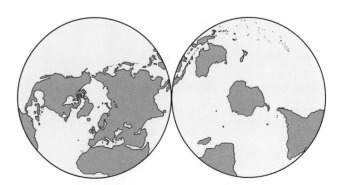

Figure 5.17 Northern and southern hemispheres—a study in contrasts in the distribution of lands and oceans. One pole bears a deep ocean and the other a great continental landmass.

is strong and intense. Inspiraling winds, forming part of the monsoon cycle, bring warm, moist Indian Ocean air over India and Southeast Asia. A lesser low forms over the deserts of the southwestern United States and northwestern Mexico. The two subtropical highs, the Hawaiian Highs and the Azores Highs, strengthen and dominate the Atlantic and Pacific Ocean regions. Outspiraling winds, following the pattern of Figure 5.15, keep the west coasts of North America and Europe warm and dry, and the east coasts of North America and Asia warm and moist (see Figure 5.13 also).

The higher latitudes of the southern hemisphere present a polar continent surrounded by a large ocean. Since Antarctica is covered by a glacial ice sheet and is cold at all times, a permanent anticyclone, the South Polar High, is centered there. Easterly winds spiral outward from the high-pressure center. Surrounding the high is a band of deep low pressure, with strong, inward-spiraling westerly winds. As early mariners sailed southward, they encountered this band, in which wind strength intensifies toward the pole. Because of the strong prevailing westerlies, they named these southern latitudes the "roaring forties," "flying fifties," and "screaming sixties."

LOCAL WINDS

Not every wind is produced by global pressure gradients in large weather systems. In some locations, *local winds* are important. These are often influenced by local terrain. We have already discussed one class of local winds—sea and land breezes—earlier in this chapter.

Mountain winds and *valley winds* are local winds that alternate in direction in a manner similar to the land and sea breezes. During the day, mountain hill slopes are heated intensely by the sun, causing air to rise. To take the place of the rising air, a current moves up valleys from the plains below—upward over rising mountain slopes, toward the summits. At night, the hill slopes are chilled by radiation. The cooler, denser hill slope air then moves valleyward, down the hill slopes, to the plain below. These winds are responding to local pressure gradients, set up by heating or cooling of the local air.

Another group of local winds are known as *drainage winds*, in which cold, dense air flows under the influence of gravity from higher to lower regions. In a typical situation, cold, dense air accumulates in winter over a high plateau or high interior valley. Under favorable conditions, some of this cold air spills over low divides or through passes, flowing out on adjacent lowlands as a strong, cold wind.

Drainage winds occur in many mountainous regions of the world and go by various local names. The *mistral* of the Rhône Valley in southern France is a well-known example—it is a cold, dry local wind. On the ice sheets of Greenland and Antarctica, powerful drainage winds move down the gradient of the ice surface and are fun-

Figure 5.18 Strong downslope winds of hot, dry air fan this brushfire in Oakland, California.

neled through coastal valleys. Picking up loose snow, these winds produce blizzard-like conditions that last for days at a time.

Another type of local wind occurs when the outward flow of dry air from an anticyclone is combined with the local effects of mountainous terrain. An example is the Santa Ana—a hot, dry easterly wind that sometimes blows from the interior desert region of southern California across coastal mountain ranges to reach the Pacific coast. This wind is funneled through local mountain gaps or canyon floors where it gains great force. At times the Santa Ana wind carries large amounts of dust. Because this wind is dry, hot, and strong, it can easily fan wildfires in brush or forest out of control (Figure 5.18).

Still another type of local wind, sometimes bearing the name *chinook*, results when strong regional winds pass over a mountain range and descend on the lee side. (See Figure 4.16.) As we know from studying orographic precipitation in Chapter 4, the descending air is heated and dried. A chinook wind can sublimate snow or dry out soils very rapidly.

WINDS ALOFT

The surface wind systems we have examined describe airflows at or near the surface. How does air move at the higher levels of the troposphere? Just as we saw for air near the surface, air aloft will move in response to pressure gradients and will be influenced by the Coriolis effect.

Pressure gradients aloft are generated by large, slow-moving, high- and low-pressure systems. Figure 5.19 shows a typical pattern of high and low pressure and winds that might be found at an altitude of 5 to 7 km (about 16,000 to 23,000 ft). Notice that the wind arrows are shown as following the isobars, not cutting across them as they do at the surface. The difference is that a parcel of air at high altitude moves freely, without the drag of surface friction. As a result, air at upper levels in the atmosphere flows smoothly around centers of high and low pressure, following a curving path. This type of wind, in which the wind direction is perpendicular to the pressure gradient, is called the *geostrophic wind*.

Global Circulation at Upper Levels

Figure 5.20 sketches the general pattern of airflows at higher levels in the troposphere. The pattern has four major features—equatorial easterlies, tropical high-pressure belts, upper-air westerlies, and a polar low.

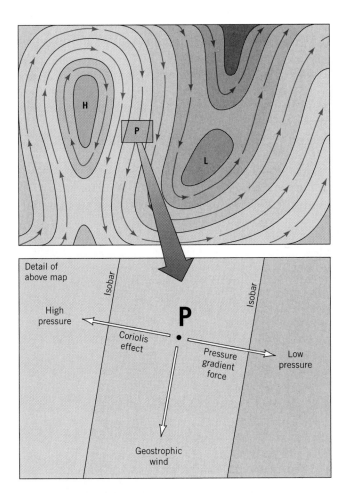

Figure 5.19 At high levels above the earth's surface, the wind blows parallel to the isobars. The northern hemisphere is shown here.

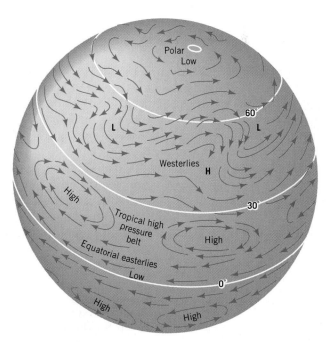

Figure 5.20 A generalized plan of global winds high in the troposphere. Easterly winds flow in the band in between the subtropical high-pressure belts. Poleward of these belts, winds are westerly, but often sweep to the north or south around centers of high and low pressure aloft. (Copyright © A. N. Strahler.)

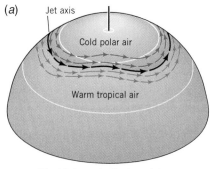

(a) The jet stream begins to undulate.

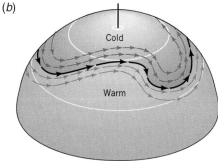

(b) Rossby waves begin to form.

Figure 5.21 Development of upper-air Rossby waves in the westerlies of the northern hemisphere. (Copyright © A. N. Strahler.)

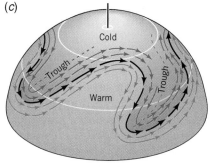

(c) Waves are strongly developed. The cold air occupies troughs of low pressure.

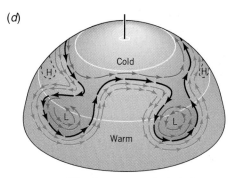

(d) When the waves are pinched off, they form cyclones of cold air.

The *upper-air westerlies* blow in a complete circuit about the earth, from about 25° lat. almost to the poles. They comprise the first major tropospheric circulation system. At high latitudes these westerlies form a huge circumpolar spiral, circling a great polar low-pressure center. Toward lower latitudes, atmospheric pressure rises steadily, forming a *tropical high-pressure belt* at 15°– 20° N and S lat. This is the high-altitude part of the surface subtropical high-pressure belt, but it is shifted somewhat equatorward. Between the high-pressure ridges is a zone of weak low pressure in which the winds are easterly. These winds comprise the second major circulation system of the troposphere. They are called the *equatorial easterlies*.

Thus, the overall picture of upper-air wind patterns is really quite simple—a band of easterly winds in the equatorial zone, belts of high pressure near the tropics of cancer and capricorn, and westerly winds, with some variation in direction, spiraling around polar lows.

Rossby Waves

The smooth westward flow of the upper-air westerlies frequently forms undulations, called **Rossby waves**. Figure 5.21 shows how these waves develop and grow in the northern hemisphere. The waves arise in a zone of contact between cold polar air and warm tropical air.

For a period of several days or weeks, the flow may be fairly smooth. Then, an undulation develops, and warm air pushes poleward while a tongue of cold air is brought to the south. Eventually, the tongue is pinched off, leaving a pool of cold air at a latitude far south of its normal location. This cold pool may persist for some days or weeks, slowly warming with time.

Jet Streams

Jet streams are important features of upper-air circulation. They are narrow zones at a high altitude in which

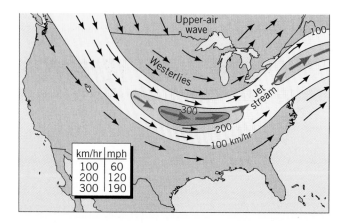

Figure 5.22 The polar jet stream is shown on this map by lines of equal wind speed. (National Weather Service.)

Figure 5.23 A schematic diagram of wind directions and jet streams along an average meridian from pole to pole. The four polar and subtropical jets are westerly in direction, in contrast to the single tropical easterly jet. (Copyright © A. N. Strahler.)

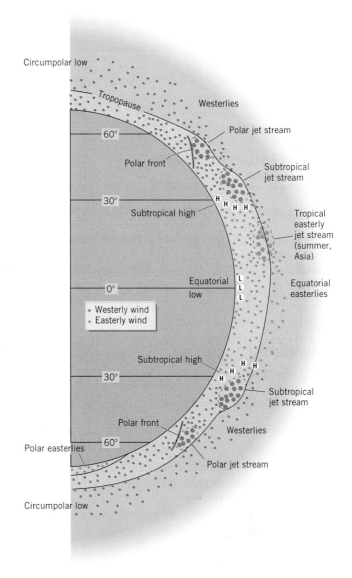

wind streams reach great speeds. Along a jet stream, pulselike movements of air follow broadly curving tracks (Figure 5.22). The greatest wind speeds occur in the center of the stream, with velocities decreasing away from the center. There are three kinds of jet streams. Two are westerly wind streams. The third is a weaker jet with easterly winds that develops in Asia as part of the summer monsoon circulation. These are shown in Figure 5.23.

The most poleward type of jet stream is located along the polar front—the fluctuating boundary between cold polar air and warmer tropical air (Figure 5.12). It is designated the *polar-front jet stream* (or simply, the "polar jet"). Located generally between 35° and 65° latitude, it is present in both hemispheres. The polar jet is typically found at altitudes of 10 to 12 km (about 30,000 to 40,000 ft), and wind speeds in the jet range from 350 to 450 km/hr (about 200 to 250 mi/hr). In the northern hemisphere, traveling aircraft often use the polar jet to increase ground speed when flying eastward. In the westward direction, flight paths are chosen to avoid the strong winds of the polar jet.

A second type of jet stream forms in the subtropical latitude zone—the *subtropical jet stream*. It occupies a position at the tropopause just above the Hadley cell in northern and southern hemispheres (Figure 5.23). Here, westerly wind speeds reach maximum values of 345 to 385 km/hr (215 to 240 mph). Cloud bands of the subtropical jet stream are pictured in Figure 5.24.

Figure 5.24 A strong subtropical jet stream is marked in this space photo by a narrow band of cirrus clouds, generated by turbulence in the air stream. This jet stream is moving from west to east at an altitude of about 12 km (about 40,000 ft). The cloud band lies at about 25° N lat. In this view, astronauts aimed their camera toward the southeast, taking in the Nile River Valley and the Red Sea. At the left is the tip of the Sinai Peninsula.

A third type of jet stream is found at even lower latitudes. Known as the *tropical easterly jet stream*, it runs from east to west—opposite in direction to that of the polar-front and subtropical jet streams. The tropical easterly jet occurs only in the summer season and is limited to a northern hemisphere location over Southeast Asia, India, and Africa.

TEMPERATURE LAYERS OF THE OCEAN

As with the atmosphere, the ocean has a layered structure. Ocean layers are recognized in terms of temperature, with temperatures generally highest at the sea surface and decreasing with depth. This trend is not surprising, since the source of heat that warms the ocean is from the sun's rays and from heat supplied by the overlying atmosphere, both of which act to warm the water at or near the surface.

The ocean's layered temperature structure is shown in Figure 5.25. At low latitudes throughout the year and in middle latitudes in the summer, a warm surface layer develops. Here wave action mixes heated surface water with the water below it to give a warm layer that may be as thick as 500 m (about 1600 ft) with a temperature of 20° to 25°C (68° to 77°F) in oceans of the equatorial belt. Below the warm layer, temperatures drop rapidly in a zone known as the *thermocline*. Below the thermocline is a layer of very cold water extending to the deep ocean floor. Temperatures near the base of the deep layer range from 0°C to 5°C (32° to 41°F). In arctic and antarctic regions, the warm layer and thermocline are absent, as Figure 5.25 shows.

OCEAN SURFACE CURRENTS

Just as there is a circulation pattern to the atmosphere, so there is a circulation pattern to the oceans. An *ocean current* is any persistent, dominantly horizontal flow of ocean water. Surface currents of the oceans are broadly divided into warm currents, which are within the warm layer above the thermocline, and cold currents, which are found in the cold surface layer in arctic and polar latitudes. Because ocean currents move warm water poleward and cold waters toward the equator, they are important regulators of air temperatures. Warm currents keep winter temperatures in the British Isles from falling much below freezing in winter. Cold currents keep weather on the California coast cool, even in the height of summer. Current systems act to exchange heat between low and high latitudes and are essential in sustaining the global energy balance.

Nearly all large-surface currents of the oceans are set in motion by prevailing surface winds. Energy is transferred from wind to water by the friction of the air blowing over the water surface. Because of the Coriolis effect, the actual direction of water drift is deflected about 45° from the direction of the driving wind.

Figure 5.26 shows the general features of the circulation of an idealized ocean between two continental

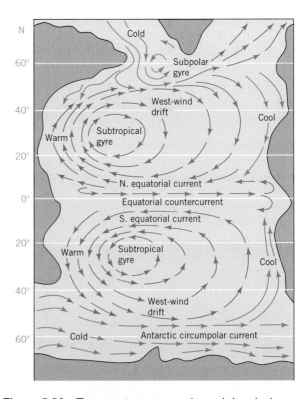

Figure 5.26 Two great gyres, one in each hemisphere, dominate the circulation of shallow ocean waters.

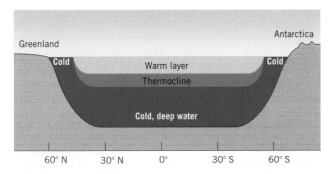

Figure 5.25 A schematic north-south cross section of the world ocean shows that the warm surface water layer disappears in arctic and antarctic latitudes, where very cold water lies at the surface.

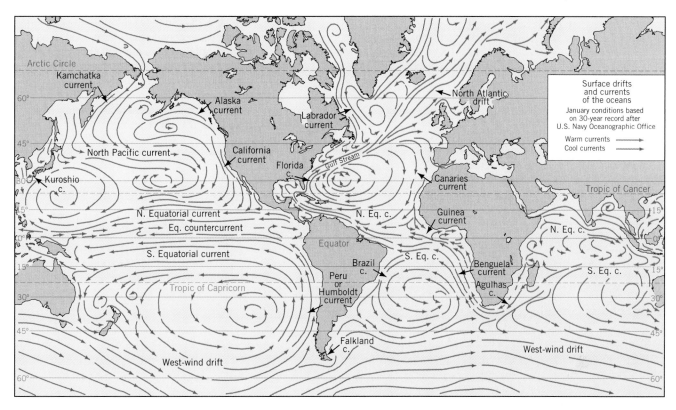

Figure 5.27 Surface drifts and currents of the oceans in January. (Based on data from U.S. Navy Oceanographic Office. Redrawn and revised by A. N. Strahler.)

masses. The circulation includes two large circular movements, called **gyres**, that are centered at latitudes of 20° to 30°. These motions track the movements of air around the subtropical high-pressure cells. An *equatorial current* with westward flow marks the belt of the trade winds. Although the trades blow to the southwest and northwest, at an angle across the parallels of latitude, the water movement follows the parallels. The equatorial currents are separated by an equatorial countercurrent. A slow, eastward movement of water over the zone of the westerlies is named the *west-wind drift*. It covers a broad belt between lat. 35° and 45° in the northern hemisphere and between lat. 30° and 60° in the southern hemisphere.

Figure 5.27 is a world map that shows ocean currents for the month of January. In the equatorial region, westward-flowing equatorial currents approach land and are turned poleward along the west sides of the oceans, forming warm currents paralleling the coast. Examples are the Gulf Stream of eastern North America and the Kuroshio Current of Japan. In the Pacific, a thin equatorial countercurrent flows eastward. It separates the northern and southern hemisphere equatorial westward flows.

Further poleward, west-wind drift waters approach the western sides of the continents and are deflected

equatorward to form cool currents along the coast. These currents are often accompanied by *upwelling* along continental margins. In this process, colder water from greater depths rises to the surface. Examples of cool currents with upwelling are the Humboldt (or Peru) Current, off the coast of Chile and Peru, the Benguela Current, off the coast of southern Africa, and the California Current.

West-wind drift water also moves poleward to join arctic and antarctic circulations. In the northeastern Atlantic Ocean, the west-wind drift forms a relatively warm current. This is the North Atlantic Drift, which spreads around the British Isles, into the North Sea, and along the Norwegian coast. The Russian port of Murmansk, on the arctic circle, remains ice-free year round because of this warm drift current.

In the northern hemisphere, where the polar sea is largely landlocked, cold currents flow equatorward along the east sides of continents. Two examples are the Kamchatka Current, which flows southward along the Asian coast across from Alaska, and the Labrador Current, which flows between Labrador and Greenland to reach the coasts of Newfoundland, Nova Scotia, and New England.

The strong west winds around Antarctica produce an Antarctic circumpolar current of cold water, shown in

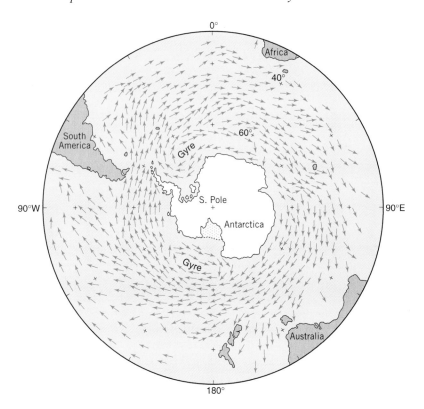

Figure 5.28 Surface water of the Southern Ocean flows continuously from west to east around Antarctica as a broad circumpolar current. The map is greatly simplified from a computer-generated model. (Based on NOAA data as presented by D. Olbers and M. Wenzel in *EOS*, American Geophysical Union, vol. 71, no. 1.)

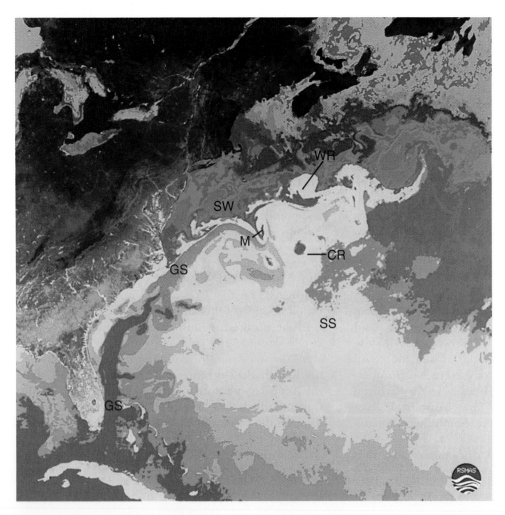

Figure 5.29 A satellite image in false colors showing the Gulf Stream (GS) and its interactions with cold water of the Continental Slope (SW), brought down by the Labrador Current, and the warm water of the Sargasso Sea (SS). Other features include a meander (M), a warm-core ring (WR), and a cold-core ring (CR). The image was made from data collected by the NOAA-7 orbiting satellite. Data of 35 satellite passes during a week in April were processed together to remove the effects of clouds on single images.

Figure 5.28. Some of this flow branches equatorward along the west coast of South America, adding to the Humboldt Current (Figure 5.27). Two small gyres lie close to the Antarctic continent.

Oceanic Streams and Their Eddies

Where warm currents of the great middle-latitude gyres flow poleward in close proximity to cold currents that are moving equatorward from their arctic sources, a sharp line of demarcation usually separates the two. This line can be likened to the polar front of the troposphere, since both air and water are fluids and behave similarly in many respects. Like the polar front, the boundary between cold and warm currents can develop features resembling Rossby waves, with their cutoff lows and highs, as depicted in Figure 5.21.

Figure 5.29 shows a satellite image of ocean temperature along the east coast of North America for a week in April. The Gulf Stream stands out as a tongue of red and yellow (warm) color, moving along the coast and heading off from North Carolina in a northeasterly direction. Cooler water from the Labrador Current, in green and blue tones, hugs the northern Atlantic coast. This current heads south and then turns to follow the Gulf Stream to the northeast. Instead of mixing, the two flows remain quite distinct. Note the warm-core and cold-core "rings" that have formed, similar to cutoff lows and highs in the Rossby waves' upper atmosphere.

EYE ON THE ENVIRONMENT: EL NIÑO

At intervals of about three to eight years, a remarkable disturbance of ocean and atmosphere occurs. It begins in the eastern Pacific Ocean and spreads its effects widely over the globe. This disturbance lasts for more than a year, bringing droughts, heavy rainfalls, severe spells of heat and cold, or a high incidence of cyclonic storms to various parts of the Pacific and its eastern coasts. This phenomenon is called *El Niño*. The expression comes from Peruvian fishermen, who refer to the *Corriente del Niño,* or the "Current of the Christ Child," in describing an invasion of warm surface water that occurs once every few years around Christmas time and greatly depletes their catch of fish. El Niño occurs at irregular intervals and with varying degrees of intensity. Notable El Niño events occurred in 1891, 1925, 1940–1941, 1965, 1972–1973, 1982–1983, 1989–1990, 1991–1992, and 1994–1995.

Normally, the cool Humboldt (Peru) Current flows northward off the South American coast, and then at about the equator it turns westward across the Pacific as the south equatorial current (see Figure 5.27). The Humboldt Current is characterized by upwelling of cold, deep water, bringing with it nutrients that serve as food for plankton. Fish feed on these high concentrations of plankton. The anchoveta, a small fish used commercially to produce fish meal for animal feed, thrives in great numbers and is harvested by the Peruvian fishermen. With the onset of El Niño, upwelling ceases, the cool water is replaced by warm water from the west, and the plankton and their anchoveta predators disappear. Vast numbers of birds that feed on the anchoveta die of starvation.

El Niño Currents and Winds

In an El Niño year, a major change in barometric pressures occurs across the entire stretch of the equatorial zone as far west as southeastern Asia. Normally, low pressure prevails over northern Australia, the East Indies, and New Guinea, where the largest and warmest body of ocean water can be found (Figure 5.30a). This low pressure is associated with a pool of warm water and a deep thermocline. Abundant rainfall normally occurs in this area during December, which is the high-sun period in the southern hemisphere.

During an El Niño event, the low-pressure system is replaced by weak high pressure and drought ensues. In contrast, pressures become lower than normal in the equatorial zone of the eastern Pacific, strengthening the equatorial trough. Rainfall is abundant in this new low-pressure region (Figure 5.30b). The shift in barometric pressure patterns is known as the *Southern Oscillation.*

Surface winds and currents also change with this change in pressure. During normal conditions, the strong, prevailing trade winds blow westward, causing very warm ocean water to move to the western Pacific and to "pile up" near the western equatorial low. This westward motion causes the normal upwelling along the South American coast, as bottom water is carried up to replace the water dragged to the west. During an El Niño event, the easterly trade winds die with the change in atmospheric pressure. A weak westerly wind flow sometimes occurs, completely reversing the normal wind direction. Without the pressure of the trade winds to hold them back, warm waters surge eastward. Sea-surface temperatures and actual sea levels rise off the tropical western coasts of the Americas.

The major change in sea-surface temperatures that accompanies an El Niño can also shift rainfall patterns dramatically in large regions across the globe. During a typical El Niño event, southeastern and interior northwestern United States, southeastern South America, and the southern tip of India receive more rainfall. Intense rainfall can also occur along the southern California coast in winter and in the Andean highlands of Peru, Bolivia, and Colombia. Drought regions include the western Pacific, northern South America, southeast Africa, and northern India.

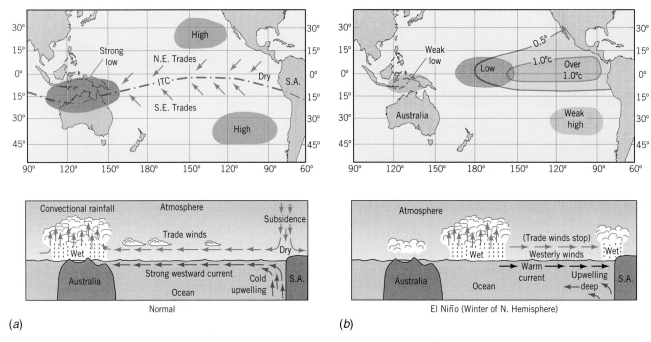

Figure 5.30 Maps of pressures in the tropical Pacific and eastern Indian Ocean in November during normal and El Niño years. In a normal year (*a*), low pressure dominates in Malaysia and northern Australia. In an El Niño year (*b*), low pressure moves eastward to the central part of the western Pacific and sea-surface temperatures become warmer in the eastern Central Pacific. (Copyright © A. N. Strahler.)

As research on the global changes in winds and currents linked to the El Niño has continued, scientists have recently recognized a condition called La Niña (the girl child). This event presents a situation roughly opposite to El Niño. During a La Niña period, sea-surface temperatures in the central and western Pacific Ocean fall to lower than average levels. This happens because the South Pacific subtropical high becomes very strongly developed during the high-sun season. The result is abnormally strong southeast trade winds. The force of these winds drags a more-than-normal amount of warm surface water westward, bringing cooler water to the surface off western continental coasts. La Niña conditions were recognized during 1988 and may have been related to a drought experienced in some parts of North America in the summer of 1988.

Figure 5.31 shows two satellite images of sea-surface temperature observed during El Niño and La Niña years. During La Niña conditions, the cool water of the Humboldt Current, moving northward along the coast of Chile and Peru, is carried westward into the Pacific in a long plume. Much cold, upwelling water (dark green) is brought to the surface along the Peruvian coast. During an El Niño year, the eastward motion of warm water holds the Humboldt Current in check. Some slight upwelling occurs (yellow color), but the amount is greatly reduced compared to normal or La Niña years.

Possible Causes for El Niño

What causes the El Niño/Southern Oscillation (also known as ENSO)? At first, meteorologists tried to explain the changes in winds and pressures by the changes in ocean surface currents, while oceanographers tried to explain the changes in surface currents by changes in winds and pressures. It wasn't until the 1950s that scientists realized that the two phenomena were linked. One view is that the cycle is simply a natural oscillation caused by the way in which the atmosphere and oceans are coupled by energy exchange. Another possibility has been proposed by geologists, who have noticed that volcanic activity along the East Pacific Rise, an underwater zone of sea-floor spreading off the west coast of South America, correlates with El Niño events. They hypothesize that the emergence of upwelling lavas from the sea floor releases vast quantities of heat into the ocean, and that this heat in some way triggers an El Niño event. In any event, scientists now have good computer models that accept sea-surface temperature along with air temperature and pressure data and can predict El Niño events reasonably well some months before they occur.

(a)

(b)

Figure 5.31   Sea-surface temperatures in the eastern tropical Pacific during La Niña and El Niño years as measured by the NOAA-7 satellite. Green tones indicate cooler temperatures, while red tones indicate warmer temperatures. (*a*) La Niña year. (*b*) El Niño year.

El Niño is an example of how dynamic our planet really is. As a grand-scale, global phenomenon, it shows how the circulation patterns of the ocean and atmosphere are linked and interact. What are the human influences on El Niño? Will the global buildup of carbon dioxide change the frequency or nature of El Niño events? At this time, there are no clear answers. In the 1990s, El Niño events of varying intensity occurred in four of the five years between 1990 and 1995. Some scientists speculated that global warming produced by emission of greenhouse gases might be the cause, but the effects of the 1991 eruption of Mount Pinatubo on the stratosphere, as well as natural, decades-long variations in the ocean–atmosphere system, could not be ruled out.

EYE ON THE ENVIRONMENT:
WIND POWER, WAVE POWER, AND CURRENT POWER

Wind power is an indirect form of solar energy that has been used for centuries. In the Low Countries of Europe, the windmill played a major role in pumping water from the polders as they were reclaimed from tidal land. The windmill was also used to grind grain in low, flat areas where streams could not be adapted to waterpower.

The design of new forms of windmills has intrigued inventors for many decades. The total supply of wind energy is enormous. The World Meteorological Organization has estimated that the combined electrical-generating wind power of favorable sites throughout the world comes to about 20 million megawatts, a figure about 100 times greater than the total electrical-generating capacity of the United States. Many problems must be solved, however, in developing wind power as a major resource.

Small wind turbines are a promising source of supplementary electric power for individual farms, ranches, and homes. The Darrieus rotor, with circular blades turning on a vertical axis, is well adapted to small generators—those with less than 50 kilowatts output. Wind turbines presently in operation and being developed are adjusted in scale according to the purposes they serve. Turbines capable of producing power in the range of 50 to 200 kilowatts are already in operation to serve small communities.

Figure 5.32 A windfarm in San Gorgonio Pass, near Palm Springs, California. Wind speeds here average 27 km/hr (17 mi/hr).

Wind turbines with a generating capacity in the range from 50 to 100 kilowatts have been assembled in large numbers at favorable locations to form "windfarms." Arranged in rows along ridge crests, the turbines intercept local winds of exceptional frequency and strength. One such locality is the Tehachapi Pass in California, where winds moving southeastward in the San Joaquin Valley are funneled through a narrow mountain gap before entering the Mohave Desert. The San Gorgonio Pass, near Palm Springs, presents a similar situation favorable to the development of windfarms (Figure 5.32). Another important windfarm locality lies about 80 km (50 mi) east of San Francisco in the Altamont Pass area of Alameda and Contra Costa counties. Here, daytime westerly winds of great persistence develop in response to the buildup of surface low pressure in the Great Valley to the east. A group of windfarms built here with a total of about 3000 turbines

provides the Pacific Gas and Electric Company with a yield of over 500 million kilowatt-hours of electricity per year. Larger wind turbines, each capable of generating from 2 to 4 megawatts, have been constructed at a number of sites.

Wave energy is another indirect form of solar energy. Nearly all ocean waves are produced by the stress of wind blowing over the sea surface. Water waves are characterized by motion of water particles in vertical orbits. Energy is extracted from wave motion by use of a floating object tethered to the sea floor. As the floating mass rises and falls, a mechanism is operated and drives a generator. Pneumatic systems use the principle of the bellows, operated by the changing pressure of the surrounding water as the water level rises and falls. A number of such devices are in the planning and testing stages.

Harnessing the vast power of a great current stream, such as the Gulf Stream or the Kuroshio, is a prospect that has not gone unnoticed. One rather grandiose plan, called the Coriolis Program after the effect that serves to direct the Gulf Stream flow against the continental margins, proposes a large number of current-driven turbines to be tethered to the ocean floor so as to operate below the ocean surface. Each turbine would be 170 m (558 ft) in diameter and capable of generating 83 megawatts of power. An array of 242 such turbines could deliver 10,000 megawatts—a large part of Florida's power requirement and equivalent to the use of about 130 million barrels of crude oil per year.

The global circulation of winds and currents paves the way for our next subject—weather systems. Recall from Chapter 4 that when warm, moist air rises, precipitation can occur. This happens in the centers of cyclones, where air converges. Although cyclones and anticyclones are generally large features, they move from day to day and are steered by the global pattern of winds. Thus, your knowledge of global winds and pressures will help you to understand how weather systems and storms develop and migrate. Your knowledge will also be very useful for the study of climate, which we take up in Chapter 7.

CHAPTER SUMMARY

The term *atmospheric pressure* describes the weight of air pressing on a unit of surface area. Atmospheric pressure decreases rapidly as altitude increases. Wind occurs when air moves with respect to the earth's surface. Air motion is produced by pressure gradients that form between one region and another. A pressure gradient is formed when air in one location is heated to a temperature that is warmer than another. Because warm air is less dense than cold air, low pressure is associated with warm air and high pressure with cold air. The sea breeze is an example of a wind caused by a pressure gradient that is produced by a difference in surface heating. A convection loop is formed as warm air rises over land and sinks over the nearby water surface.

At a global scale, atmospheric circulation is strongly influenced by the earth's rotation. The Coriolis effect deflects winds and ocean currents, producing circular or spiraling flow paths around cyclones (centers of low pressure) and anticyclones (centers of high pressure).

Because the equatorial and tropical regions are heated more intensely than the higher latitudes, two convection loops develop—the Hadley cells. These loops drive the northeast and southeast trade winds, the convergence and lifting of air at the intertropical convergence zone (ITC), and the sinking and divergence of air in the subtropical high-pressure cells. The monsoon circulation of Asia responds to a reversal of atmospheric pressure over the continent with the seasons. A winter monsoon flow of cool, dry air from the northeast alternates with a summer monsoon flow of warm, moist air from the southwest.

In the midlatitudes and poleward, westerly winds prevail. In winter, continents develop high pressure, and intense oceanic low-pressure centers are found off the Aleutian Islands and near Iceland in the northern hemisphere. In the summer, the continents develop low pressure as oceanic subtropical high-pressure cells intensify and move poleward.

Local winds are generated by local pressure gradients. The sea and land breezes, as well as mountain and valley winds, are examples caused by local surface heating. Other local winds include drainage winds, Santa Ana, and chinook winds.

Winds aloft follow a simple basic pattern—easterlies between two belts of tropical high pressure, and westerlies to the north and south of these belts. Rossby waves develop in the westerlies, bringing cold, polar air equatorward and warmer air poleward. The polar and subtropical jet streams are concentrated westerly wind streams with high wind speeds.

Ocean surface currents are dominated by huge gyres that are driven by the global surface wind pattern. Equatorial currents move warm water westward and then poleward along the east coasts of continents. Return flows bring cold water equatorward along the west coasts of continents.

El Niño events occur when atmospheric pressure changes in the western Pacific and East Indies region produce a weakening of the Pacific trade winds. Warm water moves eastward to the coasts of South and Central America, suppressing the normal northward flow of the Humboldt Current. Upwelling along the Peruvian coast is greatly reduced. El Niño events normally occur on a three- to eight-year cycle and affect global patterns of precipitation in many regions. The causes of El Niño are not well understood.

Wind power is an enormous source of energy that remains to be tapped. It can be harnessed easily and cheaply with arrays of windmills, or windfarms. The vertical motion of ocean waves can also be used to generate power.

KEY TERMS

barometer	anticyclone	monsoon
wind	Hadley cells	Rossby waves
pressure gradient	intertropical convergence zone (ITC)	jet stream
isobars		gyres
Coriolis effect	subtropical high-pressure belts	
cyclone		

REVIEW QUESTIONS

1. Explain atmospheric pressure. Why does it occur? How is atmospheric pressure measured, and in what units? What is the normal value of atmospheric pressure at sea level? How does atmospheric pressure change with altitude?

2. Describe land and sea breezes. How do they illustrate the concepts of pressure gradient and convection loop?

3. What is the Coriolis effect, and why is it important? What produces it? How does it influence the motion of wind and ocean currents in the northern hemisphere? In the southern hemisphere?

4. Define cyclone and anticyclone. How does air move within each? What is the direction of circulation of each in the northern and southern hemispheres? What type of weather is associated with each and why?

5. Sketch an ideal earth (without seasons or ocean–continent features) and its global wind system. Label the following on your sketch: doldrums, equatorial trough, Hadley cell, ITC, northeast trades, polar easterlies, polar front, polar outbreak, southeast trades, subtropical high-pressure belts, and westerlies.

6. What is the Asian monsoon? Describe the features of this circulation in summer and winter. How is the ITC involved? How is the monsoon circulation related to the high- and low-pressure centers that develop seasonally in Asia?

7. Compare the winter and summer patterns of high and low pressure that develop in the northern hemisphere with those that develop in the southern hemisphere.

8. List four types of local winds and describe three of them.

9. Describe the basic pattern of global atmospheric circulation at upper levels.

10. What are Rossby waves? How are they formed?

11. Identify five jet streams. Where do they occur? In which direction do they flow?

12. What is the general pattern of ocean surface current circulation? How is it related to global wind patterns? To air patterns?

13. Compare the normal pattern of wind, pressure, and ocean currents in the equatorial Pacific with the pattern during an El Niño event.

14. What are some of the global weather changes reported for El Niño events?

15. What is La Niña, and how does it compare with the normal pattern?

16. Discuss wind power as a source of energy.

Focus on Systems 5.2 • The Convection Loop as an Energy Flow System

1. Diagram a simple convection loop and show how it uses a flow of energy to cycle matter. Show energy inputs and outputs.

2. Describe how the cyclone and anticylone are coupled in atmospheric convection loop. In what respects does this loop differ from the simple convection loop described in 1. above?

ESSAY QUESTIONS

1. An airline pilot is planning a nonstop flight from Los Angeles to Sydney, Australia. What general wind conditions can the pilot expect to find in the upper atmosphere as the airplane travels? What jet streams will be encountered? Will they slow or speed the aircraft on its way?

2. You are planning to take a round-the-world cruise, leaving New York in October. Your vessel's route will take you through the Mediterranean Sea to Cairo, Egypt, in early December. Then you will pass through the Suez Canal and Red Sea to the Indian Ocean, calling at Bombay, India, in January. From Bombay, you will sail to Djakarta, Indonesia, and then go directly to Perth, Australia, arriving in March. Rounding the southern coast of Australia, your next port of call is Auckland, New Zealand, which you will reach in April. From Auckland, you head directly to San Francisco, your final destination, arriving in June. Describe the general wind and weather conditions you will experience on each leg of your journey.

PROBLEMS

Working It Out 5.1 • Pressure and Density in the Oceans and Atmosphere

1. A diving pool is 5 m deep. What will be the pressure on a person swimming at the bottom of the pool, including sea-level atmospheric pressure? What fraction of the pressure is due to the water, and what fraction is due to the atmosphere? Suppose the swimmer is now a deep-sea diver at a depth of 100 m. What are the fractions for this case?

2. Mount Washington, in New Hampshire, has a summit elevation of 1917 m (6288 ft). Mount Whitney, in California, has a summit elevation of 4418 m (14,494 ft). Using either the pressure formula above or reading directly from the figure, what barometric pressures would you expect to find at these two summits? What percentage of sea-level pressure is each value? (Use 1014 mb as surface atmospheric pressure.) (To obtain the fractional power 5.26, use the y^x button found on most calculators.)

3. In the course of the passage of a hurricane, a weather observer notes a change in barometric pressure from 1016 mb to 979 mb, a difference of 37 mb. Using the formula for atmospheric pressure with altitude, would a barometer reading 1014 mb at sea level change by at least that amount if taken up to an elevation of 350 m?

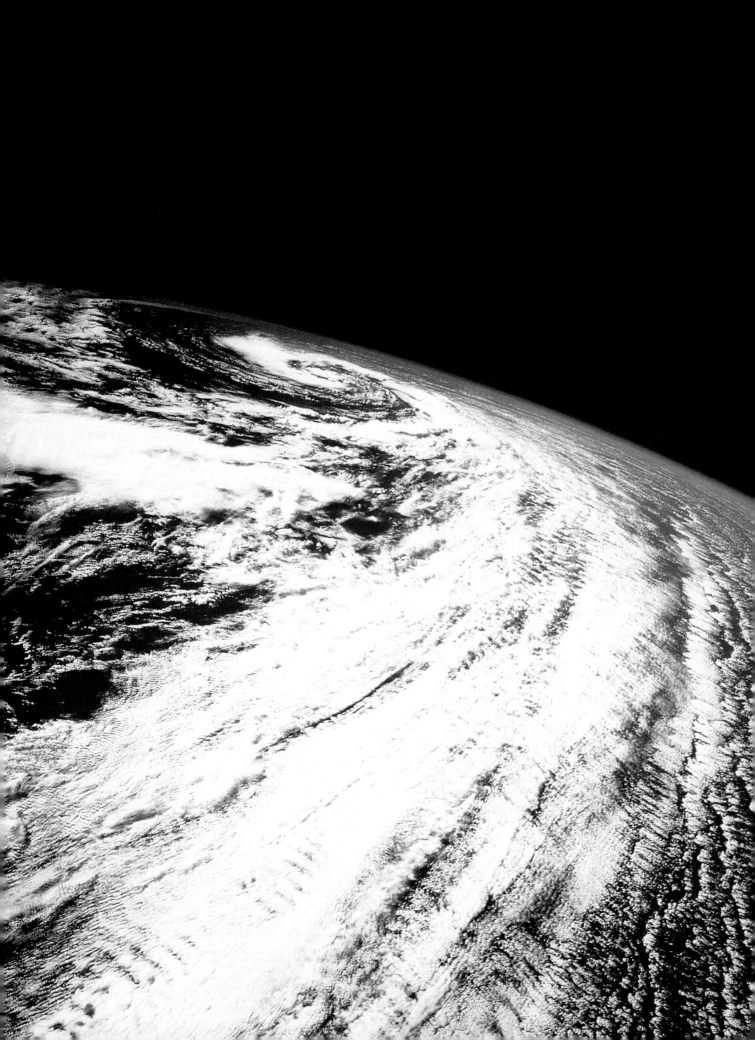

Chapter 6

Weather Systems

Cold fronts, warm fronts, low-pressure centers, cyclonic storms, hurricanes—these are all familiar weather phenomena that are associated with weather systems, the subject of this chapter. A **weather system** is an organized state of the atmosphere associated with a characteristic weather pattern. Some of the weather systems we will encounter in this chapter include wave cyclones, traveling anticyclones, tornadoes, tropical cyclones (including hurricanes), easterly waves, and weak equatorial lows.

Astronaut's view of a tropical cyclone over the Pacific Ocean.

TRAVELING CYCLONES AND ANTICYCLONES

Cyclones and anticyclones are weather systems that involve masses of air moving in a spiraling motion. In a *cyclone*, air spirals inward, whereas in an *anticyclone*, air spirals outward. Most types of cyclones and anticyclones are large features that move slowly across the earth's surface, bringing changes in the weather as they move. These are referred to as *traveling cyclones* and *traveling anticyclones*.

In a cyclone, low pressure at the center draws air inward and carries it aloft. And when moist air rises, condensation or deposition can occur. Thus, cyclones normally have dense, stratiform clouds, often releasing rain or snow. This is **cyclonic precipitation**. A familiar example of the cyclone is the wave cyclone—a low-pressure center in the midlatitudes that is associated with cold and warm fronts and normally moves eastward. (Wave cyclones and fronts are described in more detail later in this chapter.) Many cyclones are weak and pass overhead with little more than a period of cloud cover and light precipitation. However, when pressure gradients are steep and the inspiraling motion is strong, intense winds and heavy rain or snow can accompany the cyclone. In this case, the disturbance is called a **cyclonic storm**.

Traveling cyclones fall into three types. First is the wave cyclone of midlatitude, arctic, and antarctic zones. This type of cyclone ranges in intensity from a weak disturbance to a powerful storm. Second is the tropical cyclone of tropical and subtropical zones. This type of cyclone ranges in intensity from a mild disturbance to the highly destructive hurricane, or typhoon. A third type is the tornado—a small, intense cyclone of enormously powerful winds. The tornado is much, much smaller in size than other cyclones, and it is related to strong convectional activity.

In an anticyclone, high pressure at the center pushes surface air outward, causing air to descend from above. Since descending air warms adiabatically, condensation does not occur and skies are fair, except for occasional puffy cumulus clouds that can develop in a moist surface air layer. Because of these characteristics, anticyclones are sometimes termed fair-weather systems. Toward the center of an anticyclone, the pressure gradient is weak, and winds are light and variable. Traveling anticyclones are found in the midlatitudes. They are typically associated with ridges or domes of clear, dry air that move equatorward and eastward.

Air Masses

Cyclonic and anticyclonic weather systems are often associated with the motion of air masses. An **air mass** is a large body of air with fairly uniform temperature and moisture characteristics. It can be several thousand kilometers or miles across, and extend upward to the top of the troposphere. A given air mass is characterized by a distinctive combination of surface temperature, environmental temperature lapse rate, and surface-specific humidity. Air masses range widely in temperature, from searing hot to icy cold, as well as in moisture content.

Air masses acquire their characteristics in *source regions*. In a source region, air moves slowly or stagnates, which allows the air to acquire temperature and moisture characteristics from the surface. For example, an air mass with warm temperatures and a high water vapor content develops over a warm equatorial ocean. Elsewhere, slowly subsiding air forms a hot air mass with low relative humidities over a large tropical desert. A very cold air mass with a low water vapor content is generated over cold, snow-covered land surfaces in the arctic zone in winter.

Air masses move from one region to another under the influence of barometric pressure gradients and are sometimes pushed or blocked by high-level jet stream winds. When an air mass moves to a new area, its properties will begin to change because it is influenced by the new surface environment. For example, the air mass may lose heat or take up water vapor.

Air masses are classified on the basis of the latitudinal position and the nature of the underlying surface of their source regions. Latitudinal position primarily determines surface temperature and environmental temperature lapse rate of the air mass, while the nature of the underlying surface—continent or ocean—usually determines the moisture content. For latitudinal position, five types of air masses are distinguished as shown in the following table.

Air Mass	Symbol	Source Region
Arctic	A	Arctic ocean and fringing lands
Antarctic	AA	Antarctica
Polar	P	Continents and oceans, lat. 50–60° N and S
Tropical	T	Continents and oceans, lat. 20–35° N and S
Equatorial	E	Oceans close to equator

For the type of underlying surface, two subdivisions are used:

Air Mass	Symbol	Source Region
Maritime	m	Oceans
Continental	c	Continents

Combining these two types of labels produces a list of six important types of air masses, shown in Table 6.1. The table also gives some typical values of surface temperature and specific humidity, although these can vary widely, depending on season. Air mass temperatures can range from –46°C (–51°F) for continental arctic (cA) air masses to 27°C (81°F) for maritime equatorial air masses (mE). Specific humidities show a very high range—from 0.1 gm/kg for the cA air mass to as much as 19 gm/kg for the mE air mass. In other words, maritime equatorial air can hold about 200 times as much moisture as continental arctic air.

Figure 6.1 shows schematically the global distribution of source regions of these air masses. An idealized continent is shown in the center of the figure surrounded by ocean. Note from the figure that the polar air masses (mP, cP) originate in the subarctic latitude zone, not in the polar latitude zone. Recall that in Chapter 2 we defined "polar" as a region containing one of the poles. However, meteorologists use the word "polar" to describe air masses from the subarctic and

Table 6.1   Properties of Typical Air Masses

Air Mass	Symbol	Source Region	Properties	Temperature °C	(°F)	Specific Humidity (gm/kg)
Maritime equatorial	mE	Warm oceans in the equatorial zone	Warm, very moist	27°	(81°)	19
Maritime tropical	mT	Warm oceans in the tropical zone	Warm, moist	24°	(75°)	17
Continental tropical	cT	Subtropical deserts	Warm, dry	24°	(75°)	11
Maritime polar	mP	Midlatitude oceans	Cool, moist (winter)	4°	(39°)	4.4
Continental polar	cP	Northern continental interiors	Cold, dry (winter)	–11°	(12°)	1.4
Continental arctic (and continental antarctic)	cA (cAA)	Regions near north and south poles	Very cold, very dry (winter)	–46°	(–51°)	0.1

subantarctic zones, and we will follow their usage when referring to air masses.

The maritime tropical air mass (mT) and maritime equatorial air mass (mE) originate over warm oceans in the tropical and equatorial zones. They are quite similar in temperature and water vapor content. With their high values of specific humidity, both are capable of very heavy yields of precipitation. The continental tropical air mass (cT) has its source region over subtropical deserts of the continents. Although this air mass may

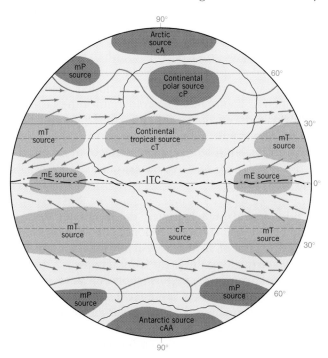

Figure 6.1   A schematic diagram of global air masses and source regions. An idealized continent, producing continental (c) air masses, is shown at the center. It is surrounded by oceans, producing maritime air masses (m). Tropical (T) and equatorial (E) source regions provide warm or hot air masses, while polar (P), arctic (A), and antarctic (AA) source regions provide colder air masses with lower specific humidities.

have a substantial water vapor content, it tends to be stable and has low relative humidity when highly heated during the daytime.

The maritime polar air mass (mP) originates over midlatitude oceans. Since the quantity of water vapor it holds is not as large as maritime tropical air masses, the mP air mass yields only moderate precipitation. Much of this precipitation is orographic and occurs over mountain ranges on the western coasts of continents. The continental polar air mass (cP) originates over North America and Eurasia in the subarctic zone. It has low specific humidity and is very cold in winter. Last is the continental arctic (and continental antarctic) air mass type (cA, cAA), which is extremely cold and holds almost no water vapor.

North American Air Masses

The air masses that form in and near North America have a strong influence on North American weather. The source regions of these air masses are shown in Figure 6.2. The continental polar (cP) air mass of North American originates over north-central Canada. This air mass forms tongues of cold, dry air that periodically extend south and east from the source region to produce anticyclones accompanied by cool or cold temperatures and clear skies. The arctic air mass (cA) that develops over the Arctic Ocean and its bordering lands of the arctic zone in winter is extremely cold and stable. When this air mass moves southward, it produces a severe cold wave.

The maritime polar air mass (mP) originates over the North Pacific and Bering Strait, in the region of the persistent Aleutian low-pressure center. This air mass is characteristically cool and moist, with a tendency in winter to become unstable, giving heavy precipitation over coastal ranges. Another maritime polar air mass of the North American region originates over the North Atlantic Ocean. It, too, is cool and moist, and is felt especially in Canada's maritime provinces in winter.

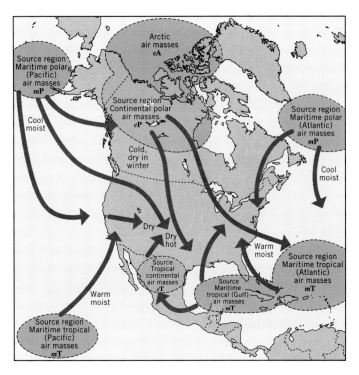

Figure 6.2 North American air-mass source regions and trajectories. (Data from U.S. Department of Commerce.)

Of the tropical air masses, the most common visitor to the central and eastern United States is the maritime tropical air mass (mT) from the Gulf of Mexico. It moves northward, bringing warm, moist, unstable air over the eastern part of the country. In the summer, particularly, this air mass brings hot, sultry weather to the central and eastern United States. It also produces many thunderstorms. Closely related is a maritime tropical air mass from the Atlantic Ocean east of Florida, over the Bahamas.

Over the Pacific Ocean, a source region of another maritime tropical air mass (mT) lies in the cell of high pressure located to the southwest of lower California. Occasionally, in summer, this moist unstable air mass penetrates the southwestern desert region, bringing severe thunderstorms to southern California and southern Arizona. In winter, a tongue of mT air frequently reaches the California coast, bringing heavy rainfall that is intensified when forced to rise over coastal mountain ranges.

A hot, dry continental tropical air mass (cT) originates over northern Mexico, western Texas, New Mexico, and Arizona during the summer. This air mass does not travel widely, but governs weather conditions over the source region.

Cold, Warm, and Occluded Fronts

A given air mass usually has a sharply defined boundary between itself and a neighboring air mass. This bound-

ary is termed a **front**. We saw an example of a front in the contact between polar and tropical air masses, shown in Figures 5.12, 5.21, and 5.23. This feature is the *polar front*, and it is located below the axis of the jet stream in the upper-air waves. If one air mass is advancing on another, the surface of contact between the two will be at a relatively low angle. If the two air masses are stationary with respect to each other, the frontal contact will be nearly vertical.

Figure 6.3 shows the structure of a front along which a cold air mass invades a zone occupied by a warm air mass. A front of this type is called a **cold front**. Because the colder air mass is denser, it remains in contact with the ground. As it moves forward, it forces the warmer air mass to rise above it. If the warm air is unstable, severe thunderstorms may develop. Thunderstorms near a cold front often form a long line of massive clouds stretching for tens of kilometers (Figure 6.4).

Figure 6.5 diagrams a **warm front** in which warm air moves into a region of colder air. Here, again, the cold air mass remains in contact with the ground because it is denser. The warm air mass is forced to rise on a long ramp over the cold air below. The rising motion causes stratus clouds to form, and precipitation often follows. If the warm air is stable, the precipitation will be steady (Figure 6.5). If the warm air is unstable, convection cells can develop, producing cumulonimbus clouds with heavy showers or thunderstorms (not shown in the figure).

Cold fronts normally move along the ground at a faster rate than warm fronts. Thus, when both types are in the same neighborhood, a cold front can overtake a warm front. The result is an **occluded front**, diagrammed in Figure 6.6. The colder air of the fast-moving cold front remains next to the ground, forcing both the warm air and the less cold air to rise over it. The warm air mass is lifted completely free of the ground.

Wave Cyclones

In middle and high latitudes, the dominant form of weather system is the **wave cyclone**. The wave cyclone is a large inspiral of air that repeatedly forms, intensifies, and dissolves along the polar front. Figure 6.7 shows a situation favorable to the formation of a wave cyclone. Two large anticyclones are in contact on the polar front. One contains a cold, dry polar air mass, and the other a warm, moist maritime air mass. Airflow converges from opposite directions on the two sides of the front, setting up an unstable situation. The wave cyclone will begin to form between the two high-pressure cells in a zone of lower pressure referred to as a *low-pressure trough*.

How does a wave cyclone form, grow, and eventually dissolve? Figure 6.8 shows the life history of a wave cyclone. In Block A (early stage), the polar-front region shows a wave beginning to form. Cold air is turned in a

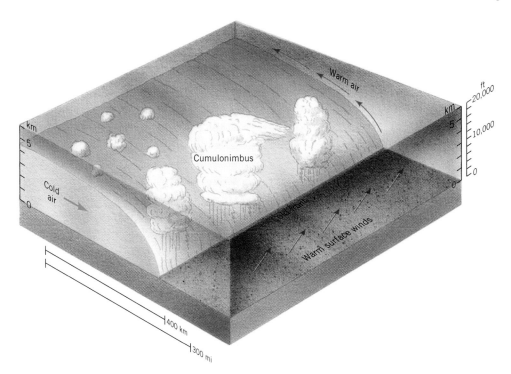

Figure 6.3 A cold front, along which a cold air mass lifts a warm air mass aloft. The upward motion sets off a line of thunderstorms. The frontal boundary is actually much less steep than shown in this schematic drawing. (Drawn by A. N. Strahler.)

Figure 6.4 A line of cumulus clouds marks the advance of a cold front, moving from left to right. The cold air pushes warmer, moister air aloft, triggering cloud formation.

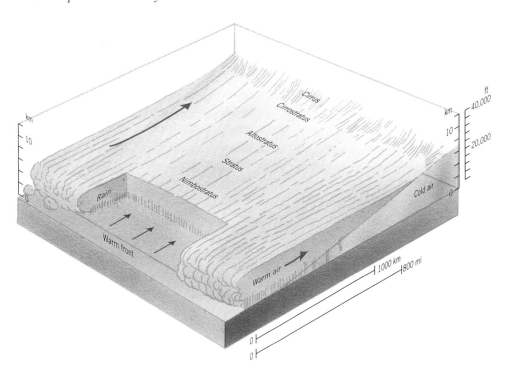

Figure 6.5 A warm front. Warm air advances toward cold air and rides up and over the cold air. A notch of cloud is cut away to show rain falling from the dense stratus cloud layer. (Drawn by A. N. Strahler.)

southerly direction and warm air in a northerly direction, so that each advances on the other. As these frontal motions develop, precipitation will form.

In Block B (open stage), the wave disturbance along the polar front has deepened and intensified. Cold air actively pushes southward along a cold front, and warm air actively moves northeastward along a warm front. The zones of precipitation along the two fronts are now strongly developed but are wider along the warm front than along the cold front.

In Block C (occluded stage), the cold front has overtaken the warm front, producing an occluded front. The warm air mass at the center of the inspiral is forced off the ground, intensifying precipitation. Eventually, the polar front is reestablished (Block D, dissolving stage), but a pool of warm, moist air remains aloft. As the moisture content of the pool is reduced, precipitation dies out, and the clouds gradually dissolve.

Keep in mind that a wave cyclone is quite a large feature—1000 km (about 600 mi) or more across. Also,

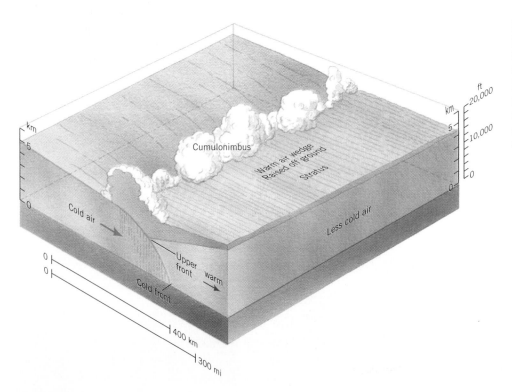

Figure 6.6 An occluded front. A warm front is overtaken by a cold front. The warm air is pushed aloft, and it no longer contacts the ground. Abrupt lifting by the denser cold air produces precipitation. (Drawn by A. N. Strahler.)

the cyclone normally moves eastward as it develops, propelled by prevailing westerlies aloft. Therefore, Blocks A–D are like three-dimensional snapshots taken at intervals along an eastbound track.

You may recall from Chapter 4 that we recognized three types of precipitation—orographic, convectional, and cyclonic. With your knowledge of how precipitation occurs in cold fronts, warm fronts, and occluded fronts, you now understand cyclonic precipitation as the precipitation associated with air mass movements, fronts, and cyclonic storms.

Weather Changes Within a Wave Cyclone

How does weather change as a wave cyclone passes through a region? Figure 6.9 shows two simplified weather maps of the eastern United States depicting conditions on successive days. The structure of the storm is defined by the isobars, labeled in millibars. The three kinds of fronts are shown by special line symbols. Areas of precipitation are shown in gray.

Map *a* shows the cyclone in an open stage, similar to Figure 6.8, block B. The isobars show that the cyclone is a low-pressure center with inspiraling winds. The cold front is pushing south and east, supported by a flow of cold, dry continental polar air from the northwest filling in behind it. Note that the wind direction

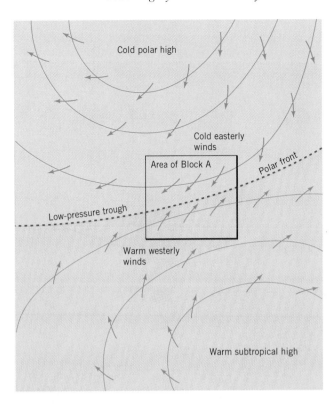

Figure 6.7 Conditions for formation of a wave cyclone. Warm subtropical air and cold polar air are in contact on the polar front. The shaded area is shown in (*a*) of the following figure as the early stage of development.

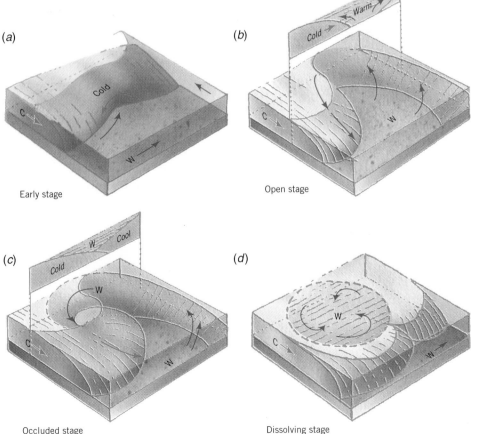

(*a*) Early stage

(*b*) Open stage

(*c*) Occluded stage

(*d*) Dissolving stage

Figure 6.8 Development of a wave cyclone. In (*a*), a cyclonic motion begins at a point along the polar front. Cold and warm fronts intensify in (*b*). In (*c*), the cold front overtakes the warm front, producing an occluded front in the center of the cyclone. Later, the polar front is reestablished with a mass of warm air isolated aloft (*d*). (Drawn by A. N. Strahler.)

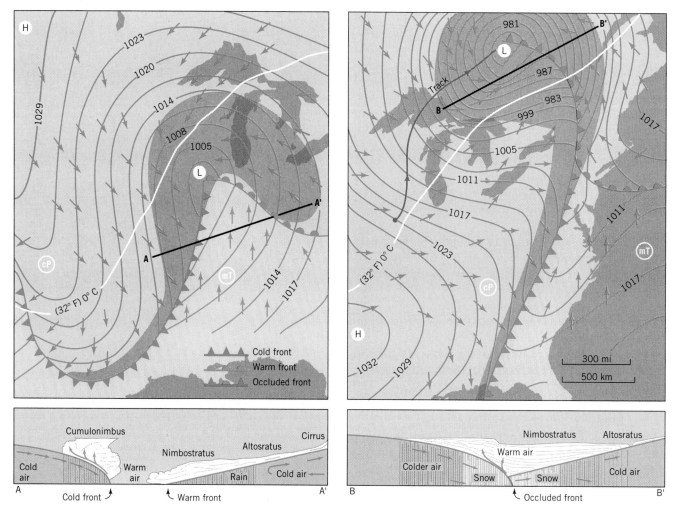

Figure 6.9 Simplified surface weather maps and cross sections through a wave cyclone. In the open stage (*left*), cold and warm fronts pivot around the center of the cyclone. In the occluded stage (*right*), the cold front has overtaken the warm front, and a large pool of warm, moist air has been forced aloft.

changes abruptly as the cold front passes. There is also a sharp drop in temperature behind the cold front as cP air fills in. The warm front is moving north and somewhat east, with warm, moist maritime tropical air following. The precipitation pattern includes a broad zone near the warm front and the central area of the cyclone. A thin band of precipitation extends down the length of the cold front. Cloudiness generally prevails over much of the cyclone.

A cross section through Map *a* along the line A–A' shows how the fronts and clouds are related. Along the warm front is a broad layer of stratus clouds. These take the form of a wedge with a thin leading edge of cirrus. (See Figures 4.11 and 4.12 for cloud types.) Westward, this wedge thickens to altostratus, then to stratus, and finally to nimbostratus with steady rain. Within the sector of warm air, the sky may partially clear with scattered cumulus. Along the cold front are cumulonimbus clouds associated with thunderstorms. These yield heavy rains but only along a narrow belt.

The second weather map, Map *b*, shows conditions 24 hours later. The cyclone has moved rapidly northeastward, its track shown by the red line. The center has tracked about 1600 km (1000 mi) in 24 hours—a speed of just over 65 km (40 mi) per hour. The cold front has overtaken the warm front, forming an occluded front in the central part of the disturbance. A high-pressure area, or tongue of cold polar air, has moved in to the area west and south of the cyclone, and the cold front has pushed far south and east. Within the cold air tongue, the skies are clear. A cross section below the map shows conditions along the line B–B', cutting through the occluded part of the storm. Notice that the warm air mass is lifted well off the ground and yields heavy precipitation.

Cyclone Tracks and Cyclone Families

Wave cyclones tend to form in certain areas and travel common paths until they dissolve. Figure 6.10 is a

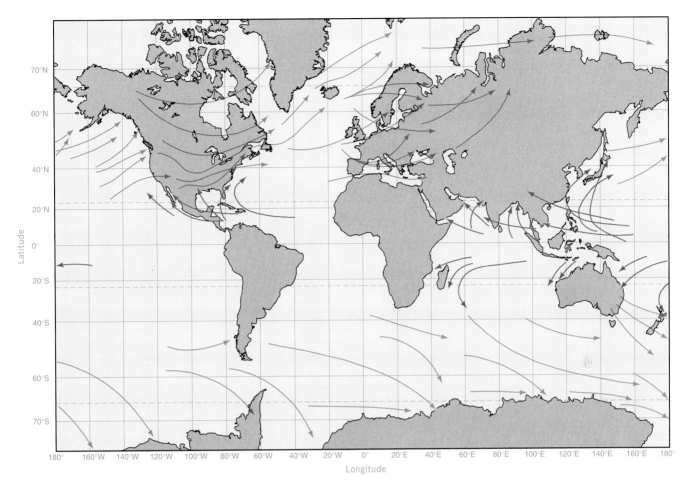

Figure 6.10 Typical paths of tropical cyclones (red lines) and wave cyclones of middle latitudes (blue lines). (Based on data of S. Pettersen, B. Haurwitz and N. M. Austin, J. Namias, M. J. Rubin, and J-H. Chang.)

Figure 6.11 A daily weather map of the world for a given day during July or August might look like this map, which is a composite of typical weather conditions. (After M. A. Garbell.)

world map showing common paths of wave cyclones and tropical cyclones. (We will discuss tropical cyclones in detail shortly.) The western coast of North America commonly receives wave cyclones arising in the North Pacific Ocean. Wave cyclones also originate over land, shown by the tracks starting in Alaska, the Pacific Northwest, the south central United States, and along the Gulf coast. Most of these tracks converge toward the northeast and pass out into the North Atlantic, where they tend to concentrate in the region of the Icelandic Low.

In the northern hemisphere, wave cyclones are heavily concentrated in the neighborhood of the Aleutian and Icelandic Lows. These cyclones commonly form in a succession, traveling as a chain across the North Atlantic and North Pacific oceans. Figure 6.11, a world weather map, shows several such cyclone families. Each cyclone moves northeastward, deepening in low pressure and occluding to form an upper-air low. For this reason, intense cyclones arriving at the western coasts of North America and Europe are usually occluded.

In the southern hemisphere, storm tracks are more nearly along a single lane, following the parallels of latitude. Three such cyclones are shown in Figure 6.11. This track is more uniform because of the uniform pattern of ocean surface circling the globe at these lati-

tudes. Only the southern tip of South America projects southward to break the monotonous expanse of ocean.

The Tornado

A **tornado** is a small but intense cyclonic vortex in which air spirals at tremendous speed. It is associated with thunderstorms spawned by fronts in midlatitudes of North America. Tornadoes also occur inside tropical cyclones (hurricanes).

The tornado appears as a dark funnel cloud hanging from the base of a dense cumulonimbus cloud (Figure 6.12). At its lower end, the funnel may be 100 to 450 m (about 300 to 1500 ft) in diameter. The base of the funnel appears dark because of the density of condensing moisture, dust, and debris swept up by the wind. Wind speeds in a tornado exceed speeds known in any other storm. Estimates of wind speed run as high as 400 km (about 250 mi) per hour. As the tornado moves across the countryside, the funnel writhes and twists. Where it touches the ground, it can cause the complete destruction of almost anything in its path.

Tornadoes occur as parts of cumulonimbus clouds traveling in advance of a cold front. They seem to originate where turbulence is greatest. They are most common in the spring and summer, but can occur in any

Figure 6.12   This tornado touched down near Clearwater, Kansas, on May 16, 1991. Surrounding the funnel is a cloud of dust and debris carried into the air by the violent winds.

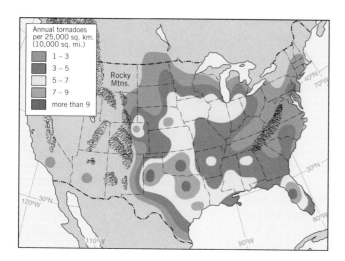

Figure 6.13 Frequency of occurrence of observed tornadoes in the 48 contiguous United States. The data span a 30-year record, 1960–1989. (Courtesy of Edward W. Ferguson, National Severe Storms Forecast Center, National Weather Service.)

month. Where a cold front of maritime polar air lifts warm, moist maritime tropical air, conditions are most favorable for tornadoes. As shown in Figure 6.13, they occur in greatest numbers in the central and southeastern states and are rare over mountainous and forested regions. They are almost unknown west of the Rocky Mountains and are relatively less frequent on the eastern seaboard. Tornadoes are a typically American phenomenon, since they are most frequent and violent in the United States. They also occur in Australia in substantial numbers and are occasionally reported from other midlatitude locations.

Devastation from a tornado is often complete within the narrow limits of its path (Figure 6.14). Only the strongest buildings constructed of concrete and steel can resist major structural damage from the extremely violent winds. The National Weather Service maintains a tornado forecasting and warning system. Whenever weather conditions favor tornado development, the danger area is alerted, and systems for observing and reporting a tornado are set in readiness.

Figure 6.14 A swath of destruction left by a tornado that swept through Plainfield, Illinois, in August of 1990. Undamaged houses delimit the tornado path.

TROPICAL AND EQUATORIAL WEATHER SYSTEMS

So far, the weather systems we have discussed are those that are important in the midlatitudes and poleward. Weather systems of the tropical and equatorial zones show some basic differences from those of the midlatitudes. The Coriolis effect is weak close to the equator, and there is a lack of strong contrast between air masses. So, clearly defined fronts and large, intense wave cyclones are missing. On the other hand, there is intense convectional activity because of the high moisture content of low-latitude maritime air masses. Slight convergence and uplifting can be sufficient to trigger convectional precipitation in these very moist air masses.

Easterly Waves and Weak Equatorial Lows

One of the simplest forms of tropical weather systems is an *easterly wave*, a slowly moving trough of low pressure within the belt of tropical easterlies (trades). These waves occur in latitudes 5° to 30° N and S over oceans, but not over the equator itself. Figure 6.15 is a simplified upper-air map of an easterly wave showing wind patterns, the axis of the wave, isobars, and the zone of showers. At the surface, a zone of weak low pressure underlies the axis of the wave. The wave travels westward at a rate of 300 to 500 km (about 200 to 300 mi) per day. Surface airflow converges on the eastern, or rear, side of the wave axis. This convergence causes the moist air to be lifted, producing scattered showers and

thunderstorms. The rainy period may last a day or two as the wave passes.

Another related weather system is the *weak equatorial low*, a disturbance that forms near the center of the equatorial trough. Moist equatorial air masses converge on the center of the low, causing rainfall from many individual convectional storms. Several such weak lows are shown on the world weather map (Figure 6.11), lying along the ITC. Because the map is for a day in July or August, the ITC is shifted well north of the equator. At this season, the rainy monsoon is in progress in Southeast Asia. It is marked by a weak equatorial low in northern India.

Polar Outbreaks

Another distinctive feature of low-latitude weather is the occasional penetration of powerful tongues of cold polar air from the midlatitudes into very low latitudes. These tongues are known as *polar outbreaks*. The leading edge of a polar outbreak is a cold front with squalls, which is followed by unusually cool, clear weather with strong, steady winds. The polar outbreak is best developed in the Americas. Outbreaks that move southward from the United States into the Caribbean Sea and Central America are called "northers" or "nortes," while those that move north from Patagonia into tropical South American are called "pamperos." One such outbreak is shown over South America on the world

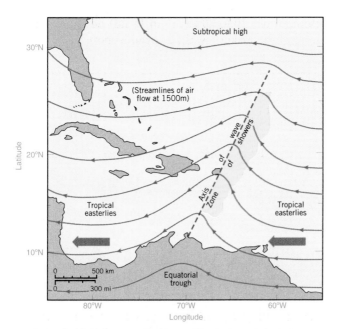

Figure 6.15 An easterly wave passing over the West Indies. (Data from H. Riehl, *Tropical Meteorology*, McGraw-Hill, New York.)

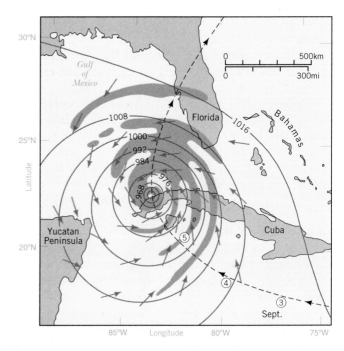

Figure 6.16 A simplified weather map of a hurricane passing over the western tip of Cuba. Daily locations, beginning on September 3, are shown as circled numerals. The recurving path will take the storm over Florida. Shaded areas show dense rain clouds as seen in a satellite image.

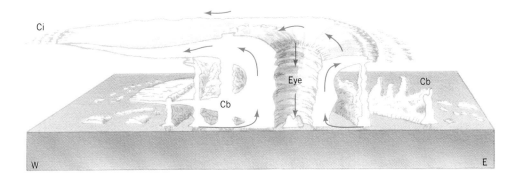

Figure 6.17 Schematic diagram of a hurricane. Cumulonimbus (Cb) clouds in concentric rings rise through dense stratiform clouds. Cirrus clouds (Ci) fringe out ahead of the storm. Width of diagram represents about 1000 km (about 600 mi). (Redrawn from NOAA, National Weather Service.)

weather map (Figure 6.11). A severe polar outbreak may bring subfreezing temperatures to the highlands of South America and severely damage such essential crops as coffee.

Tropical Cyclones

The most powerful and destructive type of cyclonic storm is the **tropical cyclone**, which is known as the *hurricane* in the Atlantic Ocean or the *typhoon* in the Pacific and Indian Oceans. This type of storm develops over oceans in 8° to 15° N and S latitudes, but not closer to the equator. It originates as a weak low, which deepens and intensifies, growing into a deep, circular low. High sea-surface temperatures, over 27°C (81°F), are required for tropical cyclones to form. Once formed, the storm moves westward through the trade-wind belt, often intensifying as it travels. It can then curve northwest, north, and northeast, steered by westerly winds aloft. Tropical cyclones can penetrate well into the midlatitudes, as many residents of the southern and eastern coasts of the United States who have experienced hurricanes can attest.

The tropical cyclone is an almost circular storm center of extremely low pressure. Winds spiral inward at high speed, accompanied by very heavy rainfall (Figure 6.16). The storm gains its energy through the release of latent heat as the intense precipitation forms. The storm's diameter may be 150 to 500 km (about 100 to 300 mi). Wind speeds range from 120 to 200 km (75 to 125 mi) per hour and sometimes much higher. Barometric pressure in the storm center commonly falls to 950 mb (28.1 in. Hg) or lower.

A characteristic feature of the tropical cyclone is its central eye, in which clear skies and calm winds prevail (Figure 6.17). The eye is a cloud-free vortex produced by the intense spiraling of the storm. In the eye, air descends from high altitudes and is adiabatically warmed. As the eye passes over a site, calm prevails, and the sky clears. Passage of the eye may take about half an hour, after which the storm strikes with renewed ferocity, but with winds in the opposite direction.

Figure 6.10, presented earlier, shows typical paths of wave cyclones and tropical cyclones. As you can see, tropical cyclones always form over oceans. In the western hemisphere, hurricanes originate in the Atlantic off the west coast of Africa, in the Caribbean Sea, or off the west coast of Mexico. Curiously, tropical cyclones do not form in the South Atlantic or southeast Pacific regions, so South America is never threatened by these severe storms. In the Indian Ocean, typhoons originate both north and south of the equator, moving north and east to strike India, Pakistan, and Bangladesh, as well as south and west to strike the eastern coasts of Africa and Madagascar. Typhoons of the western Pacific also form both north and south of the equator, moving into northern Australia, Southeast Asia, China, and Japan.

Tracks of tropical cyclones of the North Atlantic are shown in detail in Figure 6.18. Most of the storms originate at 10° to 20° N latitude, travel westward and northwestward through the trades, and then turn northeast at about 30° to 35° N latitude into the zone of the westerlies. Here their intensity lessens, especially if they move over land. In the trade-wind belt, the cyclones travel 10 to 20 km (6 to 12 mi) per hour. In the zone of the westerlies, their speed is more variable.

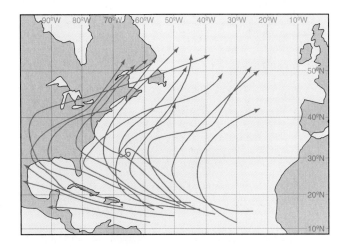

Figure 6.18 Tracks of some typical hurricanes occurring during August. The storms arise in warm tropical waters and move northwest. On entering the region of prevailing westerlies, the storms change direction and move toward the northeast.

Tropical cyclones occur only during certain seasons. For hurricanes of the North Atlantic, the season runs from May through November, with maximum frequency in late summer or early autumn. In the southern hemisphere, the season is roughly the opposite. These periods follow the annual migrations of the ITC to the north and south with the seasons, and correspond to times of the year when ocean temperatures are warmest.

EYE ON THE ENVIRONMENT:
Impacts of Tropical Cyclones

Tropical cyclones can be tremendously destructive storms. Islands and coasts feel the full force of the high winds and flooding as tropical cyclones move onshore. For example, Hurricane Andrew, striking the Florida coast near Miami, was the most devastating storm ever to occur in the United States, claiming as much as $30 billion in property damage and 43 lives. Figure 6.19, a computer-enhanced satellite image, shows the storm as it advanced toward the Louisiana coast after passing across the Florida peninsula.

The most serious effect of tropical cyclones is coastal destruction by storm waves and very high tides. Since the atmospheric pressure at the center of the cyclone is so low, sea level rises toward the center of the storm. High winds create a damaging surf and push water toward the coast, raising sea level even higher. Waves attack the shore at points far inland of the normal tidal range. Low pressure, winds, and the underwater shape of a bay floor can combine to produce a sudden rise of water level, known as a **storm surge**. During surges ships are lifted bodily and carried inland to become stranded.

If high tide accompanies the storm, the limits reached by inundation are even higher. Enormous

Figure 6.19 Hurricane Andrew approaching the Louisiana coast on August 25, 1992, as depicted in a computer-enhanced satellite image.

death tolls can be created by this flooding. At Galveston, Texas, in 1900, a sudden storm surge generated by a severe hurricane flooded the low coastal city and drowned about 6000 persons. Low-lying coral atolls of the western Pacific may be entirely swept over by wind-driven sea water, washing away palm trees and houses and drowning the inhabitants.

Also important is the large amount of rainfall produced by tropical cyclones. For some coastal regions, these storms provide much of the summer rainfall. Although this rainfall is a valuable water resource, it can also produce freshwater flooding, raising rivers and streams out of their banks. On steep slopes, soil saturation and high winds can topple trees and produce disastrous mudslides and landslides. 🪶

POLEWARD TRANSPORT OF HEAT AND MOISTURE

Recall from Chapter 2 that a vast global heat flow extends from the equatorial and tropical regions, where the sun's energy is concentrated and net radiation is positive, to the middle and upper latitudes, where net radiation is negative. This heat flow was referred to as *poleward heat transport*. With our knowledge of the global circulation patterns of the atmosphere and oceans, and our knowledge of precipitation processes, we can now examine this flow of heat in more detail.

Atmospheric Heat and Moisture Transport

Figure 6.20 is a schematic diagram of global heat and moisture flow within the atmosphere. An important feature is the Hadley cell circulation—a global convection loop in which moist air converges and rises in the intertropical convergence zone (ITC) while subsiding and diverging in the subtropical high pressure belts. *Focus on Systems 6.1 • Hadley Cell Circulation as a Convection Loop Flow System* shows the anatomy of a Hadley cell in more detail.

The Hadley cell convection loop acts to pump heat from warm equatorial oceans poleward to the subtropical zone. Near the surface, air flowing toward the ITC picks up water vapor evaporated by sunlight from warm ocean surfaces, increasing the moisture and latent heat content of the air. The latent heat is released as sensible heat in the condensation that occurs with convergence and uplift of the air at the ITC. Air traveling poleward in the return circulation aloft retains much of this heat, although some is lost to space by radiant cooling. When the air descends in the subtropical high pressure belts, the sensible heat becomes available at the surface. The net effect is thus to gather heat from tropical and equatorial zones and release the heat in the subtropical zone, where it can be conveyed further

poleward by the motion of mT and cT air masses into the midlatitudes.

In the mid- and high latitudes, poleward heat transport is also produced by the Rossby wave mechanism (see Figure 5.21). Lobes of cold, dry polar or arctic air (cP, cA, and cAA air masses) plunge toward the equator, while tongues of warmer, moister air (mT and mP air masses) flow toward the poles. At their margins, cyclones develop, releasing latent heat in precipitation and providing a heat flow that warms the mid- and higher latitudes well beyond the capabilities of the sun.

Oceanic Heat Transport

Just as atmospheric circulation plays a role in moving heat from one region of the globe to another, so does oceanic circulation. Although no Hadley cells are present in the ocean, there is an overall circulation pattern involving surface and deep waters (see Figure 3.23). This very large circulation pattern is superimposed on the ocean surface currents shown in Figure 5.28, and it is more like a slow drifting of huge volumes of water that contain more rapidly moving currents on their surfaces.

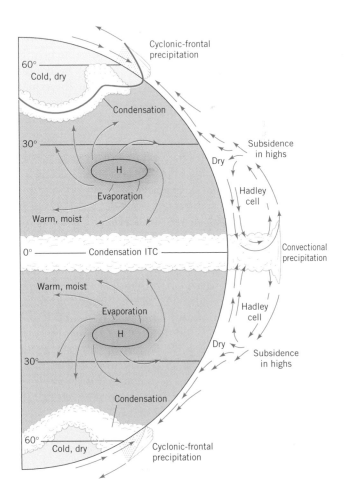

Figure 6.20 Global atmospheric transport of heat and moisture.

Focus on Systems 6.1 • Hadley Cell Circulation as a Convection Loop Flow System

The atmosphere is a largely clear, fluid layer that transmits most of the sun's energy to the surface of the earth. Because solar energy does not warm the earth's surface uniformly, the earth's atmosphere and oceans exhibit a complex circulation pattern of air and water flows that acts to redistribute absorbed solar radiation more evenly across the globe. In this circulation, convection loops play an important role.

The Hadley cell is a good example of a global-scale convectional circulation pattern. The figure on the right shows the Hadley cell circulation on a day in December. The diagram uses lines of equal wind velocity, called isotachs. Two kinds of isotachs are shown. The upper diagram (*a*) shows wind speed in a horizontal (north-south) direction, while the lower diagram (*b*) shows wind speed in a vertical direction.

Looking first at the horizontal component of the wind (*a*), a nest of centered ovals at the bottom of the figure shows a center of strong horizontal motion near the surface. This flow is southward, as shown by the direction of the small arrow. At the top of the figure is another nest of solid isotachs, indicating a strong wind flow in the northern direction. These two flows are the connecting legs of horizontal flow in the convection loop. The flows are strongest near 10° N latitude. The vertical component of the wind, presented in (*b*), also shows two centers of wind strength. Near the equator, motion is upward. At about 15° N latitude, the motion is downward.

Taken together, the four flows shown by the isotachs provide a single convection loop, diagrammed by the broad arrows. In this loop, sensible heat is released from latent heat by condensation in rising air near the equator, and is carried by the air in motion to the upper part of the troposphere. Here, a portion of the heat is lost by direct radiation to space. As the air descends, a portion leaks out of the cycle and heads poleward. This air delivers a flow of heat to higher latitudes. In its place, cooler air enters the loop near the surface.

Although the Hadley cell circulation may fit a simple convection loop fairly well overall, the motions of the atmosphere, induced by solar heating of the earth's surface and influenced by the earth's rotation, are generally quite complex. These motions also power the surface currents of the ocean, and, ultimately, the deep currents as well. We can view all the motions of the earth's oceans and atmosphere as a continuous cycle of interlinked fluid flows—a cycle powered by an energy transport system that moves heat from low to high latitudes.

Figure 6.21 diagrams the features of this flow. Surface water moves northward from the South Atlantic toward the North Atlantic. Exposed to intense insolation in the tropical zones, the water is warmed. Evaporation also concentrates salt in the surface layer, making the water more dense. As it travels northward, the water cools, losing heat and warming the atmosphere above it. Finally, the water is chilled almost to its freezing point. Because of its cold temperature and saltiness, the North Atlantic water is dense enough to sink to the bottom of the Atlantic basin. This cold, Atlantic bottom water then heads slowly southward.

Note that this circulation pattern is also a type of convection loop. In the other convection loops we have discussed, the fluid is heated, becomes less dense, and rises. In this case, however, motion results when the fluid loses heat, becomes more dense, and sinks.

Figure 6.21 Warm surface waters flow into the North Atlantic, cool, and sink to the deep Atlantic basin, creating a circulation that warms westerly winds moving onto the European continent.

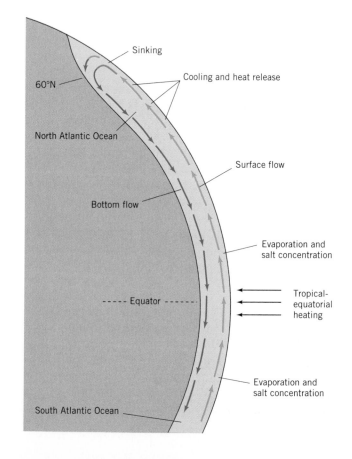

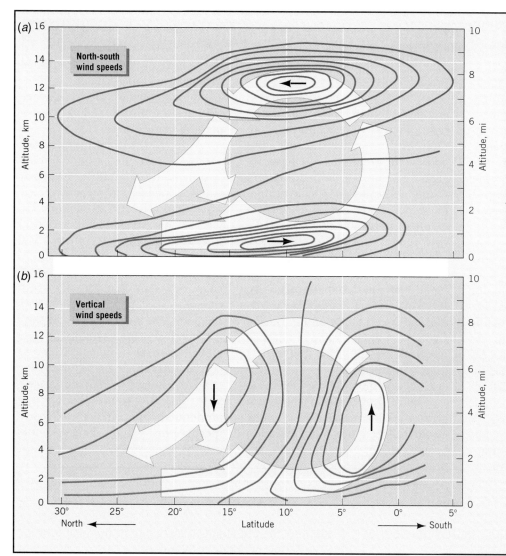

A cross section of a Hadley cell, shown for a day in December. Contour lines are isotachs, or lines of equal velocity. (*a*) North-south component of velocity, ranging up to about 4 m/sec. (*b*) Upward-downward component, ranging up to about 1 cm/sec. Note that the height of the cell is greatly exaggerated with respect to its width. It is really a flat feature that hugs the surface. (Copyright © A. N. Strahler.)

The effect of this sinking convection loop of oceanic waters is to act like a heat pump. Sensible heat is acquired in tropical and equatorial regions and moved northward into the North Atlantic, where it is transferred to the air. Since wind patterns move air eastward at higher latitudes, this heat ultimately warms Europe. The amount of heat released is quite large. In fact, a recent calculation shows that it accounts for about 35 percent of the total solar energy received by the Atlantic Ocean north of 40° latitude! This type of circulation does not occur in the Pacific or Indian oceans.

EYE ON THE ENVIRONMENT:
CLOUD COVER, PRECIPITATION, AND GLOBAL WARMING

Clouds and precipitation, the subjects of Chapters 4–6, are everyday weather phenomena that change rapidly in time and vary widely from region to region. As part of the weather patterns that change from season to sea-

son, these phenomena are an important ingredient of the climate that is experienced at a given location. On a global scale, however, clouds and precipitation are also factors that can influence climate globally. To see this effect, let's explore the possible impact of global warming on cloud cover and precipitation.

Recall that global temperatures have been rising over the last 20 years. As part of this global warming, satellite data have detected a rise in temperature of the global ocean surface of about 1°C (1.8°F) over the past decade. Any rise in sea-surface temperature increases the rate of evaporation. And the increase in evaporation will raise the average atmospheric content of water vapor. What effect will this have on climate?

Water has several roles in global climate. First, in its vapor state it is one of the greenhouse gases. That is, it absorbs and emits longwave radiation, thus enhancing the warming effect of the atmosphere above the earth's surface. In fact, water is more important than CO_2 in creating the greenhouse effect. Thus, one result of an

Focus on Systems 6.2 • Feedback Loops in the Global Climate System

An important feature of many flow systems is the feedback loop, which we discussed in the Introduction. As we saw there, a feedback loop is a connection that either reduces or amplifies a change that occurs within a system. In the case of negative feedback, the change is reduced, while in the case of positive feedback, the change is amplified.

As an example of feedback, consider the linkage between global warming and atmospheric water in the form of water vapor, clouds, and precipitation, which is explained in the last section of this chapter's text. In this linkage, increased levels of atmospheric CO_2 and other greenhouse gases increase surface temperatures by enhancing the greenhouse effect. Increased surface temperatures in turn enhance evaporation, generating more water vapor in the atmo-

sphere, more clouds, and greater precipitation.

The linkages discussed in the text are diagrammed in the figure at right. Increased concentrations of CO_2 and other greenhouse gases boost the greenhouse effect, increasing planetary temperature and raising atmospheric specific humidity. Since water vapor is itself a greenhouse gas, the increase in specific humidity forms a positive feedback. Increased cloud cover also provides positive feedback, since liquid and solid water particles are more effective at greenhouse warming than water vapor alone. But increased clouds provide negative feedback as well when they reflect solar radiation back to space. Increased precipitation could provide a negative feedback if it resulted in greater snow and ice cover, which would reflect more solar radiation back to space. Note that these are

just a few of the feedback loops within the global climate system. Climatologists and global climate modelers have proposed and documented many others.

As you can see from this example, feedback loops can be quite important in understanding how systems behave. Systems with negative feedback loops will tend to be more stable, so that when changes occur, the system moderates them. Systems with positive feedback loops will tend to be more dynamic, since changes are amplified by the structure of the system. It's hard to tell at this time whether negative or positive feedback loops dominate the global climate system. However, considering the cost of changing climate patterns to human society, let's hope that future changes are moderated by negative feedback mechanisms, rather than accentuated by positive ones.

increase in global water vapor in the atmosphere should enhance warming.

Second, water vapor can condense or deposit, forming clouds. Will more clouds increase or decrease global temperatures? That question is still being debated by the scientists who study and model the atmosphere. Clouds can have two different effects on the surface radiation balance. As large, white bodies, they can reflect a large proportion of incoming shortwave radiation back to space, thus acting to cool global temperatures. But cloud droplets and ice particles also absorb longwave radiation from the ground and return that emission as counterradiation. This absorption is an important part of the greenhouse effect, and it is much stronger for water as cloud droplets or ice particles than as water vapor. Thus, clouds also act to warm global temperatures by enhancing longwave reradiation from the atmosphere to the surface.

Which effect, longwave warming or shortwave cooling, will dominate? The best information at present, obtained from satellite measurements, is that the average flow of shortwave energy reflected by clouds back to space is about 50 W/m² , while the greenhouse warming effect of clouds amounts to about 30 W/m². Thus,

the net effect of clouds is a cooling of the planet by an energy flow of about 20 W/m².

What will happen when surface temperature increases, water vapor in the atmosphere increases, and more clouds form? Computer models of global climate generally agree that longwave warming will be enhanced more than shortwave cooling. This means that the net effect of clouds will be a positive feedback that enhances the greenhouse effect and so accentuates the surface warming. After a small increase, the effect of clouds will still be to cool the planet but not at so great a rate. Shortwave cooling will still prevail but to a lesser degree. Another way to put it is to say that with global warming, the cooling effects of clouds on climate will be reduced somewhat, leading to temperatures that are even warmer. *Focus on Systems 6.2 • Feedback Loops in the Global Climate System* provides a more detailed look at the feedback loops between surface warming and clouds.

What about precipitation? This is the third role of water in global climate. With more water vapor and more clouds in the air, more precipitation should result. Think, however, about what might happen if precipitation increases in arctic and subarctic zones. In

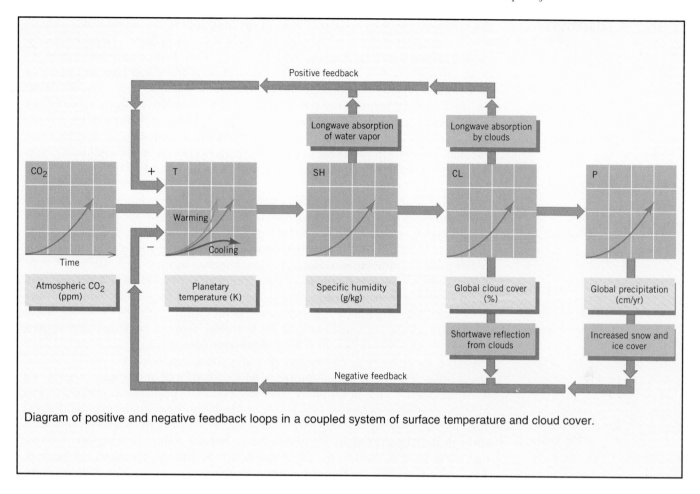

Diagram of positive and negative feedback loops in a coupled system of surface temperature and cloud cover.

this case, more of the earth's surface could be covered by snow and ice. Since snow is a good reflector of solar energy, this would increase the earth's albedo, thus tending to reduce global temperatures. Another effect might be to increase the depth of snow, thus tying up more water in snowpacks and reducing runoff to the oceans. Reducing runoff would reduce the rate at which sea level has been rising, presumably as a result of climate warming (see Chapter 17).

At this time, scientists are unsure how the global climate system will respond to global warming induced by the CO_2 increases predicted for the twenty-first century. Perhaps increased cloud cover and enhanced precipitation will reduce the global warming; perhaps not. As time goes by, however, our understanding of global climate and our ability to predict its changes are certain to increase.

With a knowledge of the global circulation patterns of the atmosphere and oceans, as well as an understanding of weather systems and how they produce precipitation, the stage is set for global climate—which is the topic of Chapters 7–9. As we will see, the annual cycles of temperature and precipitation that most regions experience are quite predictable, given the changes in wind patterns, air mass flows, and weather systems that occur with the seasons. The result will be a description of the world's climates that grows easily and naturally from the principles you have mastered in your study of physical geography thus far.

CHAPTER SUMMARY

A weather system is an organized state of the atmosphere associated with a characteristic weather pattern. Weather systems include wave cyclones, traveling anticyclones, tornadoes, easterly waves, weak equatorial lows, and tropical cyclones.

Air masses are distinguished by the latitudinal location and type of surface of their source regions. The boundaries between air masses are termed fronts. Wave cyclones typically form in the midlatitudes at the boundary between cool, dry air masses and warm, moist air masses. In the wave cyclone, a vast inspiraling motion produces cold and warm fronts, and eventually an occluded front. Precipitation normally occurs with each type of front. Tornadoes are very small, intense cyclones that occur as a part of thunderstorm activity. Their high winds can be very destructive.

Tropical weather systems include easterly waves and weak equatorial lows. Easterly waves occur when a weak low-pressure trough develops in the easterly wind circulation of the tropical zones, producing convergence, uplift, and shower activity. Weak equatorial lows occur near the intertropical convergence zone. In these areas of low pressure, convergence triggers abundant convectional precipitation.

Tropical cyclones can be the most powerful of all storms. They develop over very warm tropical oceans and can intensify to become vast inspiraling systems of very high winds with very low central pressures. As they move onto land, they can bring heavy surf and storm surges of very high waters. Tropical cyclones have caused great death and destruction in coastal regions.

Global air and ocean circulation provides the mechanism for poleward heat transport by which excess heat moves from the equatorial and tropical regions toward the poles. In the atmosphere, the heat is carried primarily in the movement of warm, moist air poleward, which releases its latent heat when precipitation occurs. In the oceans, a global circulation moves warm surface water northward through the Atlantic Ocean. Heated in the equatorial and tropical regions, the surface water loses its heat to the air in the North Atlantic and sinks to the bottom. These heat flows help make northern and southern climates warmer than we might expect based on solar heating alone.

Because global warming, produced by increasing CO_2 levels in the atmosphere, will increase the evaporation of surface water, atmospheric moisture levels will increase. This, in turn, could tend to increase temperatures still further through the effect of clouds. It could also tend to reduce temperatures by increasing the amount and duration of snow cover. Further research on the effect of global warming on climate is needed.

KEY TERMS

weather system	front	wave cyclone
cyclonic precipitation	cold front	tornado
cyclonic storm	warm front	tropical cyclone
air mass	occluded front	storm surge

REVIEW QUESTIONS

1. Define air mass. What two features are used to classify air masses?

2. Compare the characteristics and source regions for mP and cT air mass types.

3. Identify three types of fronts. Draw a cross section through each, showing the air masses involved, the contacts between them, and the direction of air mass motion.

4. What is a wave cyclone? How is it formed? Sketch two weather maps, showing a wave cyclone in open and occluded stages. Include isobars on your sketch. Identify the center of the cyclone as a low. Lightly shade areas where precipitation is likely to occur.

5. Describe a tornado. Where and under what conditions do tornadoes typically occur?

6. Identify three types of weather systems that bring rain in equatorial and tropical regions. Describe each system briefly.

7. Describe the structure of a tropical cyclone. What conditions are necessary for the development of a tropical cyclone? Give a typical path for the movement of a tropical cyclone in the northern hemisphere.

8. Why are tropical cyclones so dangerous?

9. How does the global circulation of the atmosphere and oceans provide poleward heat transport?

10. How does water, as vapor, clouds, and precipitation, influence global climate? How might water in these forms act to enhance or retard climatic warming?

PROBLEMS

Focus on Systems 6.1 • Hadley Cell Circulation as a Convection Loop Flow System

1. Sketch a cross section of Hadley cell circulation, using arrows to indicate airflow direction. At what latitudes are northerly, southerly, upward, and downward flows strongest?

Focus on Systems 6.2 • Feedback Loops in the Global Climate System

1. What is a feedback loop? Contrast the effects of negative and positive feedback loops on a system.

2. Provide an example of a system with a feedback loop and explain how the loop affects the system.

ESSAY QUESTIONS

1. Compare and contrast midlatitude and tropical weather systems. Be sure to include the following terms or concepts in your discussion: air mass, convectional precipitation, cyclonic precipitation, easterly wave, polar front, stable air, traveling anticyclone, tropical cyclone, unstable air, wave cyclone, and weak equatorial low.

2. Prepare a description of the annual weather patterns that are experienced through the year at your location. Refer to the general temperature and precipitation pattern as well as the types of weather systems that occur in each season.

Chapter 7

The Global Scope of Climate

Temperature Regimes

Global Precipitation
Seasonality of Precipitation

Climate Classification
Overview of the Climates

Dry and Moist Climates

Special Supplement • The Köppen Climate System

Focus on Systems 7.1 • Time Cycles of Climate

Working It Out 7.2 • Averaging in Time Cycles

The **climate** of a location is defined as the statistical collective of its weather conditions over a given period of time (normally several decades). In its most general sense, then, climate is "the synthesis of weather." To describe this weather, we could use many of the measures describing the state of the atmosphere that we have already encountered. These might include daily net radiation, barometric pressure, wind speed and direction, cloud cover and type, presence of fog, precipitation type and intensity, incidence of cyclones and anticyclones, frequency of frontal passages, and other such items of weather information. However, observations as detailed as these are not made regularly at most weather stations around the world. To study climate on a worldwide basis, we must turn to the two simple measurements that are made regularly at every weather station—temperature and precipitation. We will be concerned with both their average values and their variation within a period of time.

Note that temperature and precipitation are factors that strongly influence the natural vegetation of a region—for example, forests occur generally in moist regions, and grasslands in dry regions. The natural vegetation cover is often a distinctive feature of a climatic region and typically influences the human use of the area. Temperature and precipitation are also important factors in cultivation of crop plants—a necessary process for human survival. And the development of soils, as well as the types of processes that shape landforms, is partly dependent on temperature and precipitation. For these reasons, we will find that climates defined on the basis of temperature and precipitation also help set apart many features of the environment, not just climate alone. This is why the study of global climates is such an important part of physical geography.

A few simple principles discussed in earlier chapters are very helpful in understanding the global scope of climate. First, recall that two major factors influence the annual cycle of air temperature experienced at a station: latitude and coastal versus continental location.

• *Latitude.* The annual cycle of temperature at a station depends on its latitude. Near the equator, temperatures are warmer and the annual range is low. Toward the poles, temperatures are colder and the an-

---

Climate encompasses both means and extremes of weather. The Blizzard of '96, shown in this photo, brought New York City to a standstill.

Focus on Systems 7.1 • Time Cycles of Climate

Our Introduction presented the concept of a time cycle—a rhythmic change that affects natural flow systems. One major class of time cycles evident in climate consists of the astronomical cycles. The two most important ones are described in Chapter 1—the daily cycle of the earth's rotation on its axis, and the annual cycle of the earth's revolution around the sun. The daily cycle of rotation results in a daily temperature cycle, shown in (*a*), which is produced by hourly changes in insolation and surface energy balance described in Chapter 3. The annual cycle of revolution produces an annual cycle of daylight duration, shown in (*b*). Daylight duration is determined by latitude and solar declination, and is strongly related to the total daily energy flow received at the surface. In both curves, we've sketched the situation for a midlatitude continental location.

The standard climograph consists of annual cycles of temperature and precipitation averaged over several

or many consecutive years (see Figure 7.8). Each of the red dots showing the mean air temperature of a given month is itself the average value of all the monthly averages in the period of record. Suppose we were to plot the climographs of single years in sequence, as in (*c*), which shows a sketch for temperature. Each year's curve is different from those of the other years. The same would hold true for the precipitation bars on the climographs. Nevertheless, there is an annual rhythm in the climate data, even though the "beats" in this rhythm are far from uniform.

Let's turn now to long strings of annual data of air temperature, covering two to three centuries. A good example is Figure 3.26, showing the record of annual tree-ring growth. Here, the annual temperature value is reduced to a single number, represented by a single dot. The dots are then connected by short, straight lines, giving the sawtooth graph. Here you can see a major time cycle with a length of about

140 years. Superimposed is a smaller, shorter cycle that is on the order of perhaps one decade. Now examine Figure 3.24, a similar graph, in which annual values are shown by the yellow line. An added feature here is a smooth red curve—a five-year running average that helps to reveal significant cycles of more moderate magnitudes. The general, overall rise in values from left to right may be a much larger cycle of which we are seeing only one part. The interpretation of this large cycle is, of course, the subject of the great ongoing debate about global warming. We will return to time cycles again in Chapter 18 in the context of the Ice Age.

To sum it up, climate is always changing through time, showing rhythms of rise and fall of temperature and precipitation. The cycles we observe in nature are nested in a hierarchy of sizes—smaller ones within larger ones. Scientists can plot with some assurance the cycles of the past, but prediction of trends to come is filled with uncertainty.

nual range is greater. These effects are produced by the annual cycle of insolation, which varies with the seasons.

- *Coastal-continental location.* Coastal stations show a smaller annual variation in temperature, while the variation is larger for stations in continental interiors. This effect occurs because ocean surface temperatures vary less with the seasons than do temperatures of land surfaces.

Air temperature also has an important effect on precipitation. Recall this principle from Chapter 4:

- *Warm air can hold more moisture than cold air.* This means that colder regions generally have lower precipitation than warmer regions. Also, precipitation will tend to be greater during the warmer months of the temperature cycle.

Keeping these three key ideas in mind as you read this chapter and the next will help make the climates easy to understand and explain.

Another key idea associated with climate is the *time cycle*, which we discussed in the Introduction. The primary driving force for weather, as we have seen, is the flow of solar energy received by the earth and atmosphere. Since that energy flow varies on daily cycles with the planet's rotation and on annual cycles with the its revolution in orbit, it imposes these cycles on temperature and precipitation. There are other time cycles that appear in climate as well. *Focus on Systems 7.1 • Time Cycles of Climate* provides a closer look at time cycles in climate.

TEMPERATURE REGIMES

Let's look in more detail at the influence of latitude and location on the annual temperature cycle of a station. Figure 7.1 shows some typical patterns of mean monthly temperatures observed at stations around the globe. We can refer to these patterns as *temperature regimes*—distinctive types of annual temperature cycles

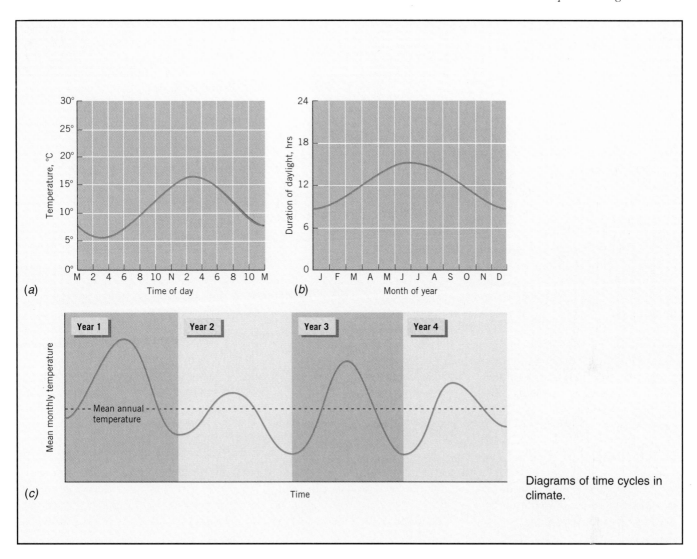

(a) Temperature, °C / Time of day

(b) Duration of daylight, hrs / Month of year

(c) Mean monthly temperature / Mean annual temperature / Time / Year 1 Year 2 Year 3 Year 4

Diagrams of time cycles in climate.

related to latitude and location. In the figure, each regime has been labeled according to its latitude zone: equatorial, tropical, midlatitude, and subarctic. Some labels also describe the location of the station in terms of its position on a landmass—"continental" for a continental interior location, and "west coast" or "marine" for a location close to the ocean.

The equatorial regime (Douala, Cameroon, 4° N) is uniformly very warm. Temperatures are close to 27°C (81°F) year-round. There are no temperature seasons, since insolation is nearly uniform throughout the year. In contrast, the tropical continental regime (In Salah, Algeria, 27° N) shows a very strong temperature cycle. Temperatures change from very hot when the sun is high, near one solstice, to mild at the opposite solstice. However, the situation is quite different at Walvis Bay, Southwest Africa (23° S), which is at nearly the same latitude and so has about the same insolation cycle as In Salah. At Walvis Bay, we find the tropical west coast regime, which has only a weak annual cycle and no extreme heat. The difference, of course, is due to the

moderating effects of the maritime location of Walvis Bay. This moderating effect persists poleward, as shown in the two temperature graphs for the midlatitude west coast regime—Monterey, California (36° N) and Sitka, Alaska (57° N).

In continental interiors, however, the annual temperature cycle remains strong. The midlatitude continental regime of Omaha, Nebraska (41° N), and the subarctic continental regime of Fort Vermillion, Alberta (58° N), show annual variations in mean monthly temperature of about 30°C (54°F) and 40°C (72°F), respectively. The ice sheet regime of Greenland (Eismitte, 71° N) is in a class by itself, with severe cold all year.

Other regimes can be identified, and, because they grade into one another, the list could be expanded indefinitely. However, what is important is that (1) annual variation in insolation, which is determined by latitude, provides the basic control on temperature patterns, and (2) the effect of location—maritime or continental—moderates that variation.

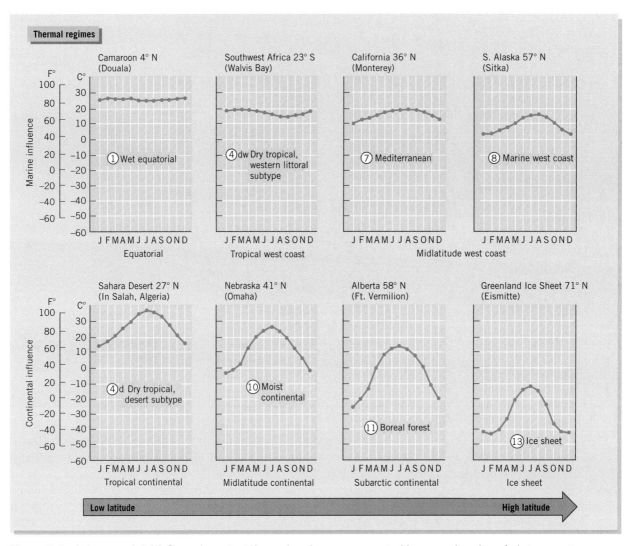

Figure 7.1 (*above and right*) Some important thermal regimes, represented by annual cycles of air temperature. (Based on the Goode Base Map.)

The temperature regimes shown in Figure 7.1 are defined by the variation in mean monthly temperature throughout the year. Monthly temperature is simply the average of daily temperatures during the month, where daily temperature is the average of the daily maximum and minimum. Mean monthly temperature is the monthly temperature for a particular month of the year averaged over a long period of record, usually several decades. *Working It Out 7.2 • Averaging in Time Cycles* demonstrates the power of averaging to reveal distinctive time cycles in data, such as the temperature regimes presented above.

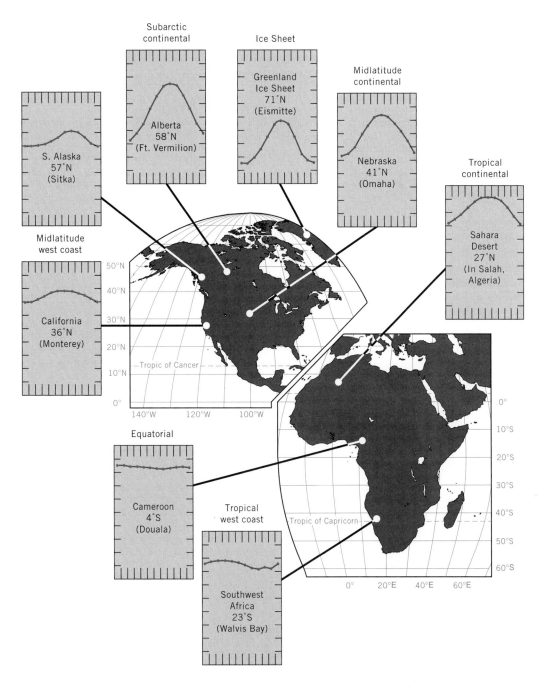

Subarctic
continental

Ice Sheet

Midlatitude
continental

Tropical
continental

S. Alaska
57°N
(Sitka)

Alberta
58°N
(Ft. Vermilion)

Greenland
Ice Sheet
71°N
(Eismitte)

Nebraska
41°N
(Omaha)

Sahara
Desert
27°N
(In Salah,
Algeria)

Midlatitude
west coast

California
36°N
(Monterey)

Equatorial

Cameroon
4°S
(Douala)

Tropical
west coast

Southwest
Africa
23°S
(Walvis Bay)

GLOBAL PRECIPITATION

Global precipitation patterns are largely determined by air masses and their movements, which in turn are produced by global air circulation patterns. Before taking a detailed look at global precipitation patterns, it will be helpful to examine the general patterns expected for a hypothetical supercontinent that has most of the features of the earth's continental assemblage but is simplified (Figure 7.2). The map recognizes and defines five classes of annual precipitation: wet, humid, subhumid, semiarid, and arid.

Beginning with the equatorial zone, the figure shows a wet band stretching across the continent. This band is produced by convectional precipitation in weak equatorial lows near the intertropical convergence, as described in Chapter 6. Note that the wet band widens and is extended poleward into the tropical zone along the continent's eastern coasts. This region is kept moist by the influence of the trade winds, which move warm, moist mT air masses and tropical cyclones westward onto the continental coast. Farther poleward, humid

conditions continue along the east coasts into the midlatitude zones. In these regions, subtropical high-pressure cells tend to move mT air masses from the east onto the continent in the summer (see Figure 5.15), while, in winter, wave cyclones bring cyclonic precipitation from the west (see Figure 6.10).

Another important feature of the hypothetical continent is the pattern of arid and semiarid regions that stretches from tropical west coasts to subtropical and midlatitude continental interiors. In the tropical and subtropical latitudes, the arid pattern is produced by dry, subsiding air in persistent subtropical high-pressure cells (Figure 5.15). The aridity continues eastward and poleward into semiarid continental interiors (for example, see Figure 7.3). These regions remain relatively dry, because they are far from source regions for moist air masses. Rainshadow effects provided by coastal mountain barriers are also important in maintaining inland aridity.

Yet another obvious feature of the supercontinent is the pair of wet bands along the west coasts of the midlatitude and subarctic zones. Figure 7.4 provides an ex-

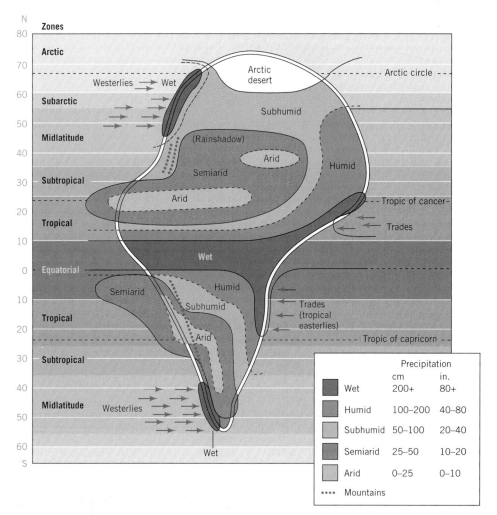

Figure 7.2 A schematic diagram of the distribution of annual precipitation over an idealized continent and adjoining oceans.

Figure 7.3 This vast sagebrush plain in Moffat County, Colorado, is typical of many parts of the semiarid continental interior region of North America.

Figure 7.4 West-coast midlatitude climates are moist and cool most of the year. The Irish coast, shown in this photo from the Dingle Peninsula, is an example.

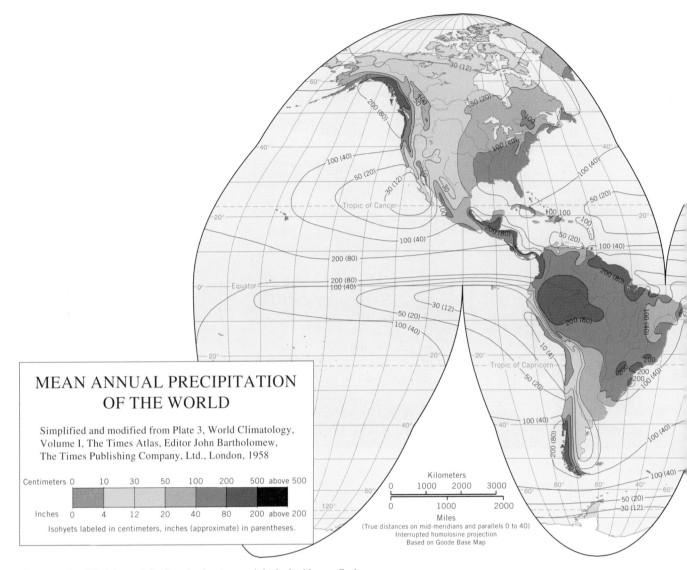

MEAN ANNUAL PRECIPITATION OF THE WORLD

Simplified and modified from Plate 3, World Climatology, Volume I, The Times Atlas, Editor John Bartholomew, The Times Publishing Company, Ltd., London, 1958

| Centimeters | 0 | 10 | 30 | 50 | 100 | 200 | 500 above 500 |

| Inches | 0 | 4 | 12 | 20 | 40 | 80 | 200 above 200 |

Isohyets labeled in centimeters, inches (approximate) in parentheses.

Kilometers: 0 1000 2000 3000
Miles: 0 1000 2000
(True distances on mid-meridians and parallels 0 to 40)
Interrupted homolosine projection
Based on Goode Base Map

Figure 7.5 World precipitation. Isohyets are labeled with cm (in.)

ample of the lush green landscapes found in these regions. The wet bands are produced by the eastward movement of moist mP air masses, typically as occluded wave cyclones, onto the continent. This motion is driven by the prevailing westerlies.

In the arctic zone, shown on the continent as arctic desert, precipitation remains low because air temperatures are low and only a small amount of moisture can be held in cold air.

These features are echoed in the actual pattern of global precipitation, shown in Figure 7.5. This map of mean annual precipitation shows **isohyets**—lines drawn through all points having the same annual precipitation. Using the same logic that we used to explain the precipitation patterns of the hypothetical continent, we can recognize seven global precipitation regions as follows. (Note that for regions where all or most of the precipitation is rain, we use the word "rainfall." For re-

gions where snow is a significant part of the annual total, we use the word "precipitation.") Table 7.1 summarizes the following observations.

1. Wet equatorial belt. This zone of heavy rainfall, over 200 cm (80 in.) annually, straddles the equator and includes the Amazon River basin in South America, the Congo River basin of equatorial Africa, much of the African coast from Nigeria west to Guinea, and the East Indies. Here the prevailing warm temperatures and high moisture content of the mE air masses favor abundant convectional rainfall. Thunderstorms are frequent year-round.

2. Trade-wind coasts. Narrow coastal belts of high rainfall, 150 to 200 cm (about 60 to 80 in.), and locally even more, extend from near the equator to latitudes of about 25° to 30° N and S on the eastern sides of every continent or large island. Examples include the eastern

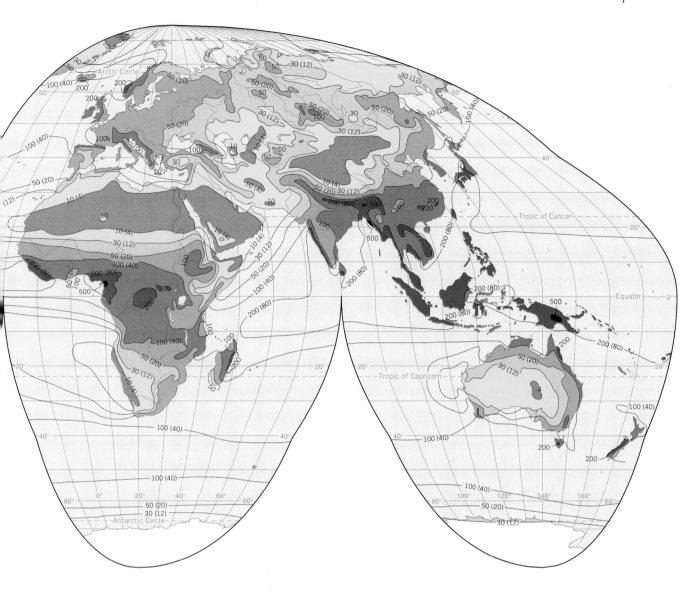

coast of Brazil, Central America, Madagascar, and northeastern Australia. The rainfall of these coasts is supplied by moist mT air masses from warm oceans, brought over the land by the trade winds. As they encounter coastal hills and mountains, these air masses produce heavy orographic rainfall.

3. Tropical deserts. In striking contrast to the wet equatorial belt astride the equator are the two zones of vast tropical deserts lying approximately on the tropics of cancer and capricorn. These are hot, barren deserts, with less than 25 cm (10 in.) of rainfall annually and in many places with less than 5 cm (2 in.). They are located under and are caused by the large, stationary subtropical cells of high pressure, in which the subsiding cT air mass is adiabatically warmed and dried. These deserts extend off the west coasts of the lands and out over the oceans. Rain here is largely convectional and extremely unreliable.

4. Midlatitude deserts and steppes. Farther northward, in the interiors of Asia and North America between lat. 30° and lat. 50°, are great deserts, as well as vast expanses of semiarid grasslands known as **steppes**. Annual precipitation ranges from less than 10 cm (4 in.) in the driest areas to 50 cm (20 in.) in the moister steppes. Dryness here results from remoteness from ocean sources of moisture.

Located in regions of prevailing westerly winds, these arid lands typically lie in the rainshadows of coastal mountains and highlands. For example, the Cordilleran Ranges of Oregon, Washington, British Columbia, and Alaska shield the interior of North America from moist mP air masses originating in the Pacific. Upon descending into the intermontane basins and interior plains, the mP air masses are warmed and dried. Similarly, mountains of Europe and the Scandinavian peninsula obstruct the flow of moist mP air

Working It Out 7.2 • Averaging in Time Cycles

The weather at most stations can vary widely from day to day, or even from hour to hour. For the study of climate, however, we are concerned with the weather in a statistical, or average, sense. To discover that average, it is often necessary to observe such variables as daily temperature and precipitation over periods of decades.

Part (*a*) of the figure at right shows monthly precipitation for a period of 20 years, from 1937 to 1994, as observed at St. Louis, Missouri. A first inspection of the graph shows that there is a rhythmic pattern to the precipitation, with sequences of wet months interspersed with sequences of drier months. This pattern matches that of the moist continental climate ⑩, with wet summers and dry winters. But beyond that general observation, it's hard to determine what the shape of the typical annual cycle of precipitation looks like for this station. In fact, it's difficult even to make a guess about which month is, on the average, the wettest or the driest.

To uncover the typical monthly cycle of precipitation, finding an average is necessary. The *average*, or *mean*, of a number of observations is simply the sum of the observations divided by the number of such observations. In commonly used algebraic notation, we can write

$$\bar{P} = \frac{1}{n}\sum_{i=1}^{n} P_i$$

where $\bar{P}$ (pronounced *P-bar*) is the mean precipitation for all observations in a particular month; P_i is the monthly precipitation value associated with the *i*th month; *n* is the number of months for which we have data; and the subscript *i* denotes the month, which ranges from *i*=1 to *n*. The Greek letter (capital) sigma, Σ, is used to denote the sum, with the lower and upper notations showing that the sum is to be formed for each observation in the sequence from *i*=1 to *n*. This is simply a rather formal way of saying, "Find the average by adding up all the values and then dividing the sum by the number of values."

The monthly averages for the 58 years of record (1937–1994) at this station are shown in graph (*b*) of the figure. Note that the monthly cycle has a fairly smooth shape. Maximum precipitation occurs in the spring months of April, May, and June, decreasing to a low point in January. By taking the average, the year-to-year variability is eliminated and the precipitation cycle appears.

Consider what the time cycle might look like with only a few years of record—say five years. By looking at graph (*a*), you can see that if we chose the five years from 1975 to 1979, we would get a different average than for the years 1981–1985. Part (*c*) of the figure compares the means of these two five-year sequences. Clearly, 1975–1979 is a sequence of drier years, while the 1981–1985 period is wetter. This is true for all months except January and March, which have more rainfall during the years 1975–1979 than 1981–1985.

In general, the longer the period of averaging, the closer the average will come to the true long-term average, and the smoother the cycle will be. Thus, averaging is a very useful tool for revealing patterns and time cycles in data sequences.

Table 7.1 World Precipitation Regions

Name	Latitude Range	Continental Location	Prevailing Air Mass	Annual Precipitation	
				Centimeters	*Inches*
1. Wet equatorial belt	10° N to 10° S	Interiors, coasts	mE	Over 200	Over 80
2. Trade-wind coasts (windward tropical coasts)	5–30° N and S	Narrow coastal zones	mT	Over 150	Over 60
3. Tropical deserts	10–35° N and S	Interiors, west coasts	cT	Under 25	Under 10
4. Midlatitude deserts and steppes	30–50° N and S	Interiors	cT, cP	10–50	4–20
5. Moist subtropical regions	25–45° N and S	Interiors, coasts	mT (summer)	100–150	40–60
6. Midlatitude west coasts	35–65° N and S	West coasts	mP	Over 100	Over 40
7. Arctic and polar deserts	60–90° N and S	Interiors, coasts	cP, cA	Under 30	Under 12

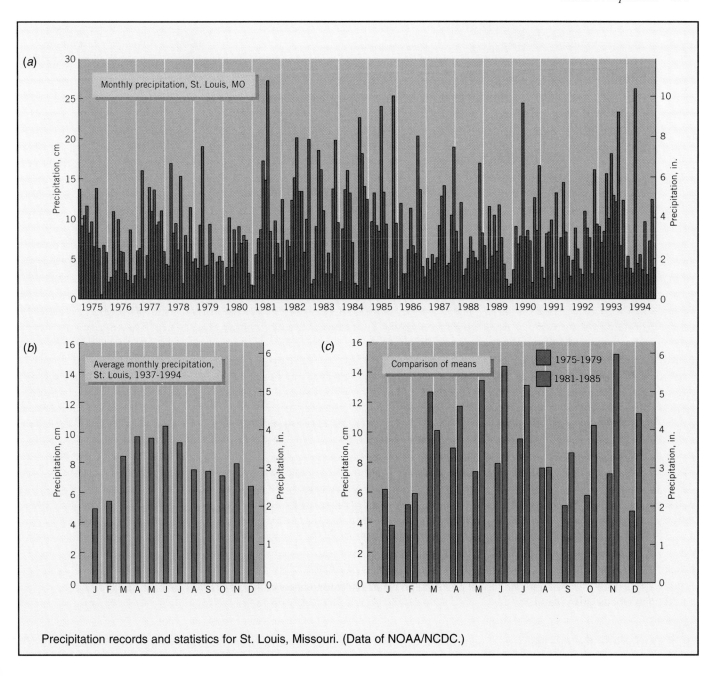

Precipitation records and statistics for St. Louis, Missouri. (Data of NOAA/NCDC.)

masses from the North Atlantic into western Asia. The great southern Asiatic ranges also prevent the entry of moist mT and mE air masses from the Indian Ocean.

The southern hemisphere has too little land in the midlatitudes to produce a true continental desert, but the dry steppes of Patagonia, lying on the lee side of the Andean chain, are roughly the counterpart of the North American deserts and steppes of Oregon and northern Nevada.

5. Moist subtropical regions. On the southeastern sides of the continents of North America and Asia, in lat. 25° to 45° N, are the moist subtropical regions, with 100 to 150 cm (about 40 to 60 in.) of rainfall annually. Smaller areas of the same kind are found in the southern hemisphere in Uruguay, Argentina, and southeastern Aus-

tralia. These regions are positioned on the moist western sides of the oceanic subtropical high-pressure centers. As a result, the lands receive moist mT air masses from the tropical ocean that are carried poleward over the adjoining land. Commonly, too, these areas receive heavy rains from tropical cyclones.

6. Midlatitude west coasts. Another distinctive wet location is on midlatitude west coasts of all continents and large islands lying between about 35° and 65° in the region of prevailing westerly winds. In these zones, abundant orographic precipitation occurs as a result of forced uplift of mP air masses. Where the coasts are mountainous, as in Alaska and British Columbia, southern Chile, Scotland, Norway, and South Island of New Zealand, the annual precipitation is over 200 cm (79

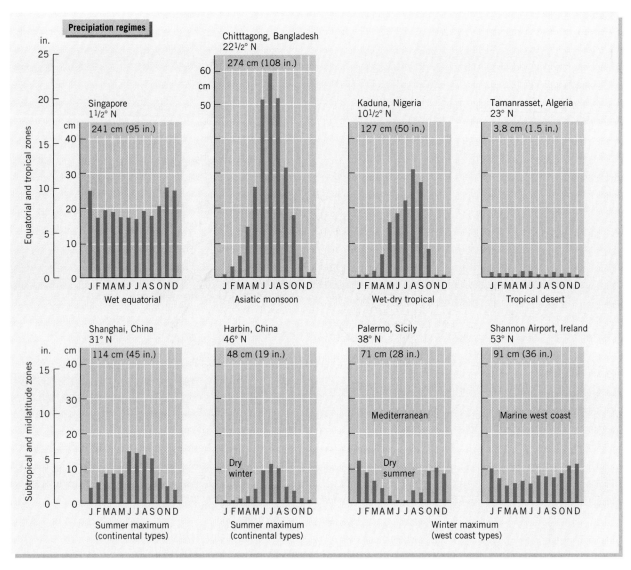

Figure 7.6 (*above and right*) Eight precipitation types selected to show various seasonal patterns. (Based on the Goode Base Map.)

in.). During the Ice Age, this precipitation fed alpine glaciers that descended to the coast, creating the picturesque deep bays (fiords) that are so typically a part of the scenery there.

7. Arctic and polar deserts. A seventh precipitation region is formed by the arctic and polar deserts. Northward of the 60th parallel, annual precipitation is largely under 30 cm (12 in.), except for the west coast belts. Cold cP and cA air masses cannot hold much moisture, and consequently, they do not yield large amounts of precipitation. At the same time, however, the relative humidity is high and evaporation rates are low.

Seasonality of Precipitation

Although total annual precipitation is a useful quantity in establishing the character of a climate type, it does not account for seasonality in precipitation. The varia-

tion in monthly precipitation through the annual cycle is a very important factor in climate description. If there is a pattern of alternating dry and wet seasons instead of a uniform distribution of precipitation throughout the year, we can expect that the natural vegetation, soils, crops, and human use of the land will all be different. It also makes a great deal of difference whether the wet season coincides with a season of higher temperatures or with a season of lower temperatures. If the warm season is also wet, growth of both native plants and crops will be enhanced. If the warm season is dry, the stress on growing plants will be great and irrigation will be required for crops.

The seasonality of precipitation can be covered largely by three types of monthly precipitation patterns: (1) uniformly distributed precipitation; (2) a precipitation maximum during the summer (or season of high sun) in which insolation is at its peak; and (3) a precip-

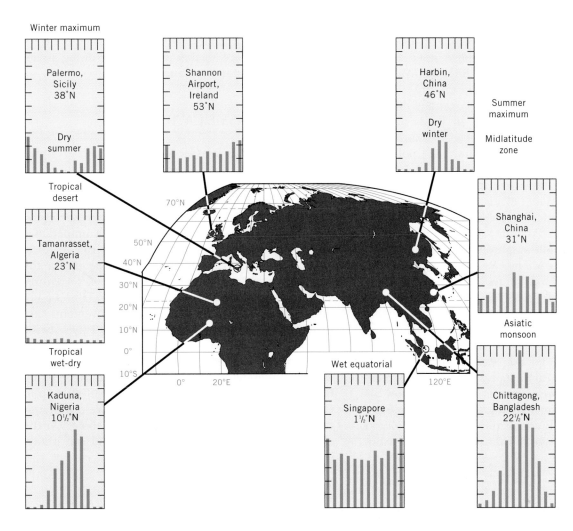

itation maximum during the winter or cooler season (season of low sun), when insolation is least. Note that the uniform type of pattern can include a wide range of possibilities from little or no precipitation in any month to abundant precipitation in all months.

Figure 7.6 shows a set of monthly precipitation diagrams selected to illustrate the major types that occur over the globe. Two of these show the uniformly distributed pattern described above—Singapore, a wet equatorial station near the equator (1½° N), and Tamanrasset, Algeria, a tropical desert station very near the tropic of cancer at 23° N. At Singapore, rainfall is abundant in all months, but some months have somewhat more than others. At Tamanrasset, there is so little rain in any month that it scarcely shows on the graph.

Chittagong, Bangaladesh (22½° N), and Kaduna, Nigeria (10½° N), both show patterns of the second

type—that is, a wet season at the time of high sun (summer solstice) and a dry season at the time of low sun (winter solstice). Chittagong is an Asian monsoon station, with a very large amount of precipitation falling during the high-sun season. Kaduna, an African station of the wet-dry tropical type, shows a similar pattern, but with about half the total annual precipitation. Both of these stations experience their wet season when the intertropical convergence zone (ITC) is nearby and their dry season when the ITC has retreated to the other hemisphere.

The summer precipitation maximum also occurs at higher latitudes on the eastern sides of continents. Shanghai, China (31° N), shows this pattern nicely in the subtropical zone. The same summer maximum persists into the midlatitudes. For example, Harbin, in eastern China (46° N), has a long, dry winter with a marked summer rain period.

In contrast to these patterns are cycles with a winter precipitation maximum. Palermo, Sicily (38° N) is an example of the Mediterranean type, named for its prevalence in the lands surrounding the Mediterranean Sea. This type experiences a very dry summer but has a moist winter. Southern and central California are also regions of this climate type. In Mediterranean climates, summer drought is produced by subtropical high-pressure cells, which intensify and move poleward during the high-sun season. They extend into the regions of Mediterranean climate, providing the hot dry weather of their associated cT air masses and blocking the passage of other, moister, air masses. In the low-sun season, the subtropical high-pressure cells move equatorward and weaken, allowing frontal and cyclonic precipitation to penetrate Mediterranean climate regions.

The dry-summer, moist-winter cycle is carried into higher midlatitudes along narrow strips of west coasts. Shannon Airport, Ireland (53° N) shows this marine west-coast type, although the difference between summer and winter rainfall is not as marked. Summers have less rainfall here for two reasons. First, the blocking effects of subtropical high-pressure cells tend to extend poleward into these regions, keeping moister air masses and cyclonic storms away. Second, the cyclonic storms that produce much of the winter precipitation are reduced in intensity during the high-sun season.

This occurs because the temperature and moisture contrasts between polar and arctic air masses and tropical air masses are weaker in summer, owing to increased high-latitude insolation.

CLIMATE CLASSIFICATION

Mean monthly values of air temperature and precipitation can describe the climate of a weather station and its nearby region quite accurately. To study climates from a global viewpoint, climatologists classify these values into a set of climate types. This requires developing a set of rules to use in examining monthly temperature and precipitation values. By applying the rules, the climatologist can use each station's data to determine the climate to which it belongs. This textbook recognizes 13 distinctive climate types that are designed to be understood and explained by air-mass movements and frontal zones. The rules that actually define these types, shown in Appendix IV, are based on an analysis of how the amount of moisture held in the soil varies throughout the year as determined by air temperature and rainfall. This is an important topic that we will take up in Chapter 19. The specific climate rules are not important to our discussion here, which is to show that the classification follows quite naturally from the under-

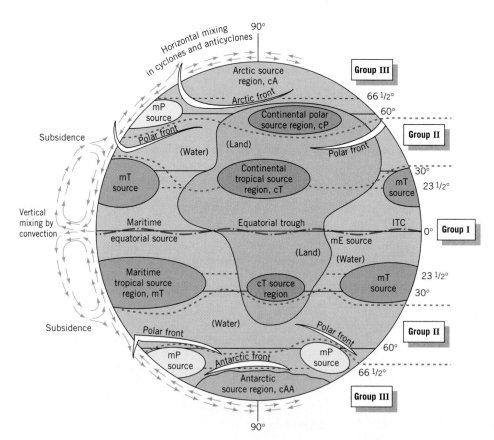

Figure 7.7 Using the map of air mass source regions, we can identify five global bands associated with three major climate groups. Within each group is a set of distinctive climates with unique characteristics that are explained by the movements of air masses and frontal zones.

standing of global temperature and precipitation processes that you have acquired in prior chapters.

Recall from Chapter 6 that air masses are classified according to the general latitude of their source regions and their surface type—land or ocean—within that region (see Figure 6.1). The latitude determines the temperature of the air-mass, which can also depend on the season. The kind of surface, land or ocean, controls the moisture content. Since the air-mass characteristics control the two most important climate variables—temperature and precipitation—we can explain climates using air masses as a guide.

We also know from Chapter 6 that frontal zones are regions in which air masses are in contact. When unlike air masses are in contact, cyclonic precipitation is likely to develop. However, the position of frontal zones changes with the seasons. For example, the polar-front zone lies generally across the middle latitudes of the United States in winter, but it retreats northward to Canada during the summer. The seasonal movements of frontal zones therefore influence annual cycles of temperature and precipitation.

Figure 7.7 shows our schematic diagram of air-mass source regions. We have subdivided this diagram into global bands that contain three broad groups of climates: low-latitude (Group I), midlatitude (Group II), and high-latitude (Group III). They are described briefly as follows.

- *Group I: Low-Latitude Climates.* The region of low-latitude climates (Group I) is dominated by the source regions of continental tropical (cT), maritime tropical (mT), and maritime equatorial (mE) air masses. These source regions are related to the three most obvious atmospheric features that occur within their latitude band—the two subtropical high-pressure belts and the equatorial trough at the intertropical convergence zone (ITC). Air of polar origin occasionally invades regions of low-latitude climates. Easterly waves and tropical cyclones are important in this climate group.
- *Group II: Midlatitude Climates.* The region of midlatitude climates (Group II) lies in the polar-front zone—a zone of intense interaction between unlike air masses. Here tropical air masses moving poleward and polar air masses moving equatorward are in conflict. Wave cyclones are normal features of the polar front, and this zone may contain as many as a dozen wave cyclones around the globe.
- *Group III: High-Latitude Climates.* The region of high-latitude climates (Group III) is dominated by polar and arctic (including antarctic) air masses. In the arctic belt of the 60th to 70th parallels, continental polar air masses meet arctic air masses along an *arctic-front zone*, creating a series of eastward-moving wave cyclones. In the southern hemisphere, there

are no source regions in the subantarctic belt for continental polar air—just a great single oceanic source region for maritime polar (mP) air masses. The pole-centered continent of Antarctica provides a single great source of the extremely cold, dry antarctic air mass (cAA). These two air masses interact along the *antarctic-front zone.*

Within each of these three climate groups are a number of climate types (or simply, climates)—four low-latitude climates (Group I), six midlatitude climates (Group II), and three high-latitude climates (Group III)—for a total of 13 climate types. In this textbook, the climates are numbered for ease in identification on maps and diagrams. We will refer to each climate by name, because the names describe the general nature of the climate and also suggest its global location. For convenience, we include the climate number next to its name in the text.

In presenting the climates, we will make use of a pictorial device called a **climograph**. It shows the annual cycles of monthly mean air temperature and monthly mean precipitation for a location, along with some other useful information. Figure 7.8 is an example of a

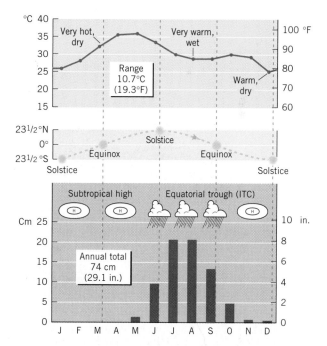

Figure 7.8 Climograph for Kayes, Mali, lat. 14° N. This station is located in western Africa, just south of the Sahara Desert. Here, the climate is nearly rainless for seven months of the year. During these months, dry subsiding air from subtropical high-pressure cells dominates. Rainfall occurs when the intertropical convergence zone moves northward and reaches the vicinity of Kayes in June or July. Temperatures are hottest in April and May, before the rainy season begins.

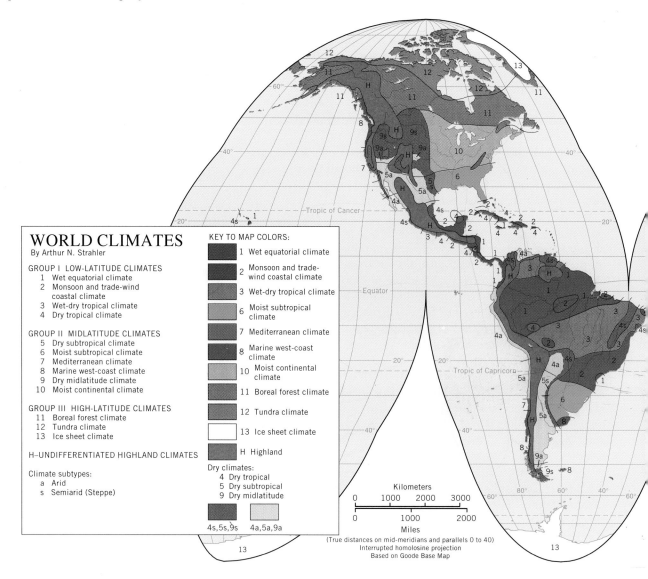

Figure 7.9 Climates of the world. Compiled from station data by A. N. Strahler.

climograph. The data plotted are for Kayes, a station in the African country of Mali, at 14° N. At the top of the climograph, the mean monthly temperature is plotted as a line graph. At the bottom, the mean monthly precipitation is shown as a bar graph. The annual range in temperature and the total annual precipitation are stated on every climograph as well. Most climographs also display dominant weather features, which are shown using picture symbols. For Kayes, the two dominant features are the subtropical high and equatorial trough (ITC). Many climographs also include a small graph of the sun's declination in order to help show when solstices and equinoxes occur.

The world map of climates, Figure 7.9, shows the actual distribution of climate types on the continents. This map, based on data collected at a large number of observing stations, is simplified because the climate boundaries are uncertain in many areas where observing stations are thinly distributed.

Overview of the Climates

Although Chapters 8 and 9 will provide a thorough examination of the 13 climates in our system, here is a quick and informal description of each.

Low-latitude Climates (Group I)

- *Wet equatorial* ①. Warm to hot with abundant rainfall, this is the steamy climate of the Amazon and Congo basins.
- *Monsoon and trade-wind coastal* ②. This warm to hot

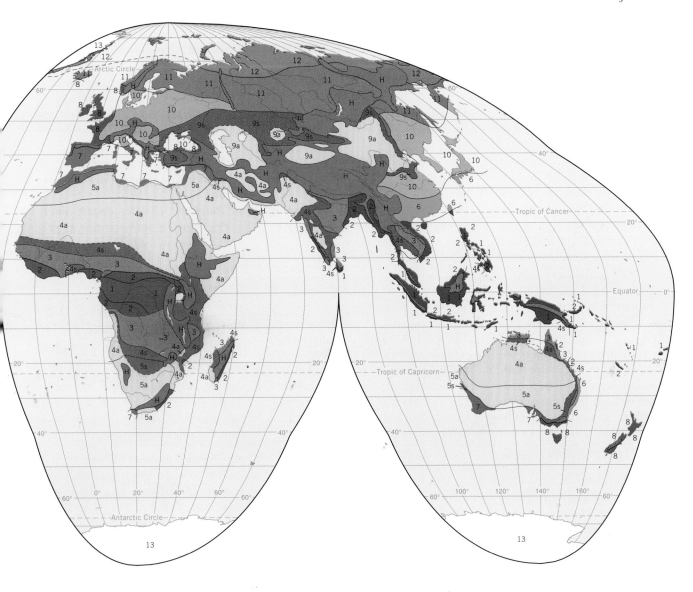

climate has a very wet rainy season. The climates of Vietnam and Bangladesh are good examples.

- *Wet-dry tropical* ③. A warm to hot climate with very distinct wet and dry seasons. Much of the Sahel region of Africa falls into this type.
- *Dry tropical* ④. The climate of the world's hottest deserts—extremely hot in the high-sun season, a little cooler in the low-sun season, with little or no rainfall. The Sahara Desert, Saudi Arabia, and the central Australian desert are of this climate.

Midlatitude Climates (Group II)

- *Dry subtropical* ⑤. Another desert climate, but not quite as hot as the dry tropical climate ④, since it is found farther poleward. This type includes the hottest part of the American southwest desert.
- *Moist subtropical* ⑥. The climate of the southeastern regions of the United States and China—hot and humid summers, with mild winters and ample rainfall year round.
- *Mediterranean* ⑦. Hot, dry summers and rainy winters mark this climate. Southern and central California, as well as the lands of the Mediterranean region—Spain, southern Italy, Greece, and the coastal regions of Lebanon and Israel—are prime examples.
- *Marine west-coast* ⑧. The climate of the Pacific northwest—coastal Oregon, Washington, and British Columbia. Warm summers and cool winters, with more rainfall in winter, are characteristics of this climate.
- *Dry midlatitude* ⑨. This dry climate is found in midlatitude continental interiors. The steppes of Central

Special Supplement • The Köppen Climate System

Air temperature and precipitation data have formed the basis for several climate classifications. One of the most important of these is the Köppen climate system, devised in 1918 by Dr. Vladimir Köppen of the University of Graz in Austria. For several decades, this system, with various later revisions, was the most widely used climate classification among geographers. Köppen was both a climatologist and plant geographer, so that his main interest lay in finding climate boundaries that coincided approximately with boundaries between major vegetation types.

Under the Köppen system, each climate is defined according to assigned values of temperature and precipitation, computed in terms of annual or monthly values. Any given station can be assigned to its particular climate group and subgroup solely on the basis of the records of temperature and precipitation at that place.

The Köppen system features a shorthand code of letters designating major climate groups, subgroups within the major groups, and further subdivisions to distinguish particular seasonal characteristics of temperature and precipitation. Five major climate groups are designated by capital letters as follows:

A *Tropical rainy climates*

Average temperature of every month is above 18°C (64.4°F). These climates have no winter season. Annual rainfall is large and exceeds annual evaporation.

B *Dry climates*

Evaporation exceeds precipitation on the average throughout the year. There is no water surplus; hence,

no permanent streams originate in B climate zones.

C *Mild, humid (mesothermal) climates*

The coldest month has an average temperature of under 18°C (64.4°F) but above –3°C (26.6°F); at least one month has an average temperature above 10°C (50°F). The C climates thus have both a summer and a winter.

D *Snowy-forest (microthermal) climates*

The coldest month has an average temperature of under –3°C (26.6°F). The average temperature of the warmest month is above 10°C (50°F). (Forest is not generally found where the warmest month is colder than 10°C (50°F).)

E *Polar climates*

The average temperature of the warmest month is below 10°C (50°F). These climates have no true summer.

Note that four of these five groups (*A*, *C*, *D*, and *E*) are defined by temperature averages, whereas one (*B*) is defined by precipitation-to-evaporation ratios. Groups *A*, *C*, and *D* have sufficient heat and precipitation for the growth of forest and woodland vegetation. Figure S7.1 shows the boundaries of the five major climate groups, and Figure S7.2 is a world map of Köppen climates.

Subgroups within the five major groups are designated by a second letter according to the following code.
S *Semiarid (steppe)*
W *Arid (desert)*
(The capital letters S and W are applied only to the dry B climates.)

f *Moist, adequate precipitation in all months, no dry season. This modifier is applied to A, C, and D groups.*
w *Dry season in the winter of the respective hemisphere (low-sun season).*
s *Dry season in the summer of the respective hemisphere (high-sun season).*
m *Rainforest climate, despite short, dry season in monsoon type of precipitation cycle. Applies only to A climates.*

From combinations of the two letter groups, 12 distinct climates emerge:

Af *Tropical rainforest climate*
The rainfall of the driest month is 6 cm (2.4 in.) or more.

Am *Monsoon variety of Af*
The rainfall of the driest month is less than 6 cm (2.4 in.). The dry season is strongly developed.

Aw *Tropical savanna climate*
At least one month has rainfall less than 6 cm (2.4 in.). The dry season is strongly developed.

Figure S7.3 shows the boundaries between *Af*, *Am*, and *Aw* climates as determined by both annual rainfall and rainfall of the driest month.

BS *Steppe climate*
A semiarid climate characterized by grasslands, it occupies an intermediate position between the desert climate (*BW*) and the more humid climates of the A, C, and D groups. Boundaries are determined by formulas given in Figure S7.4.

BW Desert climate
Desert has an arid climate with annual precipitation of usually less than 40 cm (15.7 in.). The boundary with the adjacent steppe climate (BS) is determined by formulas given in Figure 7.S4.

Cf *Mild humid climate with no dry season*

Precipitation of the driest month averages more than 3 cm (1.2 in.).

Cw *Mild humid climate with a dry winter*

The wettest month of summer has at least 10 times the precipitation of the driest month of winter. (Alternative definition: 70 percent or more of the mean annual precipitation falls in the warmer six months.)

Cs *Mild humid climate with a dry summer*

Precipitation of the driest month of summer is less than 3 cm (1.2 in.). Precipitation is at least three times as much as the driest month of summer. (Alternative definition: 70 percent or more of the mean annual precipitation falls in the six months of winter.)

Df *Snowy-forest climate with a moist winter*

No dry season.

Dw *Snowy-forest climate with a dry winter*

ET *Tundra climate*

The mean temperature of the warmest month is above 0°C (32°F) but below 10°C (50°F).

EF *Perpetual frost climate*

In this ice sheet climate, the mean monthly temperatures of all months are below 0°C (32°F).

To denote further variations in climate, Köppen added a third letter to the code group. The meanings are as follows:

a With hot summer; warmest month is over 22°C (71.6°F); C and D climates.

b With warm summer; warmest month is below 22°C (71.6°F); C and D climates.

c With cool, short summer; less than four months are over 10°C (50°F); C and D climates.

d With very cold winter; coldest month is below −38°C (−36.4°F); D climates only.

h Dry-hot; mean annual temperature is over 18°C (64.4°F); B climates only.

k Dry-cold; mean annual temperature is under 18°C (64.4°F); B climates only.

As an example of a complete Köppen climate code, *BWk* refers to a cool desert climate, and *Dfc* refers to a cold, snowy-forest climate with cool, short summer.

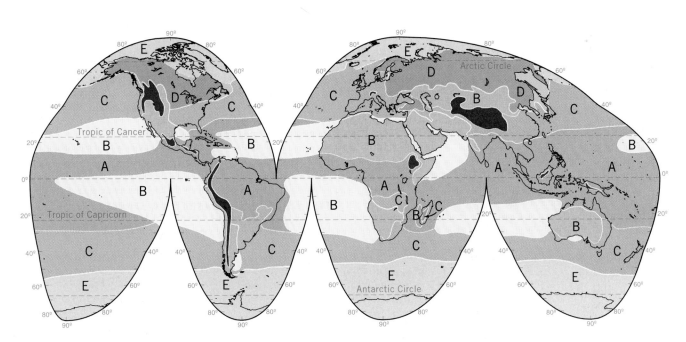

Figure S7.1 Highly generalized world map of major climate regions according to the Köppen classification. Highland areas are in black. (Based on Goode Base Map.)

Special Supplement • The Köppen Climate System (*continued*)

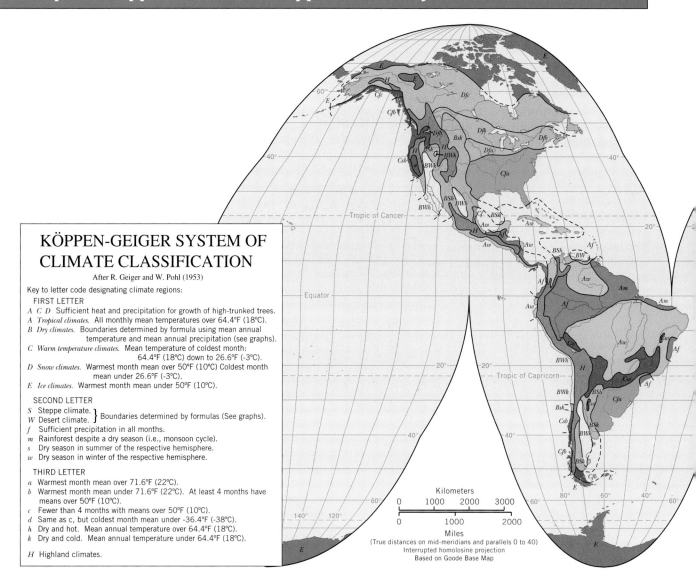

KÖPPEN-GEIGER SYSTEM OF CLIMATE CLASSIFICATION

After R. Geiger and W. Pohl (1953)

Key to letter code designating climate regions:

FIRST LETTER

A C D Sufficient heat and precipitation for growth of high-trunked trees.

A *Tropical climates.* All monthly mean temperatures over 64.4°F (18°C).

B *Dry climates.* Boundaries determined by formula using mean annual temperature and mean annual precipitation (see graphs).

C *Warm temperature climates.* Mean temperature of coldest month: 64.4°F (18°C) down to 26.6°F (-3°C).

D *Snow climates.* Warmest month mean over 50°F (10°C) Coldest month mean under 26.6°F (-3°C).

E *Ice climates.* Warmest month mean under 50°F (10°C).

SECOND LETTER

S Steppe climate. ⎫
W Desert climate. ⎬ Boundaries determined by formulas (See graphs).

f Sufficient precipitation in all months.

m Rainforest despite a dry season (i.e., monsoon cycle).

s Dry season in summer of the respective hemisphere.

w Dry season in winter of the respective hemisphere.

THIRD LETTER

a Warmest month mean over 71.6°F (22°C).

b Warmest month mean under 71.6°F (22°C). At least 4 months have means over 50°F (10°C).

c Fewer than 4 months with means over 50°F (10°C).

d Same as c, but coldest month mean under -36.4°F (-38°C).

h Dry and hot. Mean annual temperature over 64.4°F (18°C).

k Dry and cold. Mean annual temperature under 64.4°F (18°C).

H Highland climates.

Kilometers
0   1000   2000   3000

0   1000   2000
Miles
(True distances on mid-meridians and parallels 0 to 40)
Interrupted homolosine projection
Based on Goode Base Map

Figure S7.2   World map of climates according to the Köppen–Geiger–Pohl system.

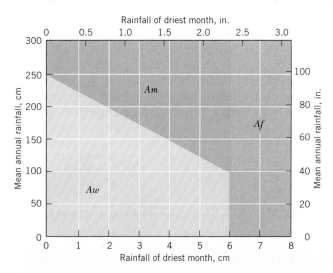

Figure S7.3   Boundaries of the A climates.

Figure S7.4   Boundaries of the B climates. Upper equations: metric system. Lower equations: English system.

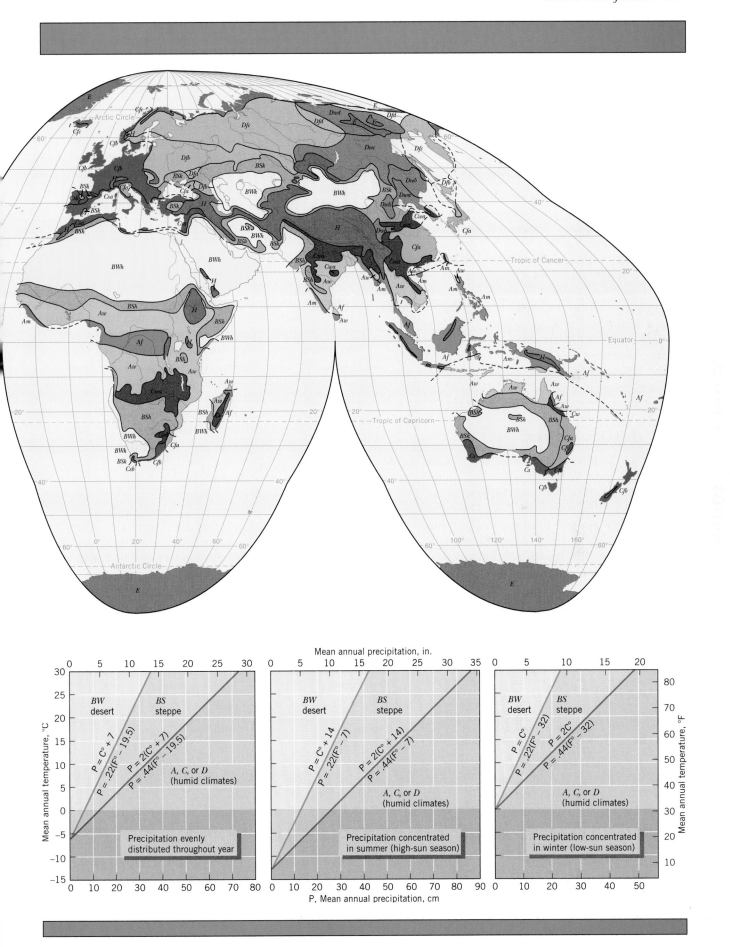

Asia and the Great Plains of North America are familiar locales with this climate—warm to hot in summer, cold in winter, and with low annual precipitation.

- *Moist continental* ⑩. This is the climate of the eastern United States and lower Canada—cold in winter, warm in summer, with ample precipitation through the year.

High-latitude Climates (Group III)

- *Boreal forest* ⑪. Short, cool summers and long, bitterly cold winters characterize this snowy climate. Northern Canada, Siberia, and central Alaska are regions of boreal forest climate.
- *Tundra* ⑫. Although this climate has a long, severe winter, temperatures on the tundra are somewhat moderated by proximity to the Arctic Ocean. This is the climate of the coastal arctic regions of Canada, Alaska, Siberia, and Scandinavia.
- *Ice sheet* ⑬. The bitterly cold temperatures of this climate, restricted to Greenland and Antarctica, can drop below –50°C (–58°F) during the sunless winter months. Even during the 24-hour days of summer, temperatures remain well below freezing.

The Köppen Climate System

The system of climates presented in this book and defined in Appendix IV is designed to flow quite naturally from the principles governing temperature and precipitation that we discussed earlier. An alternative classification is that devised by the Austrian climatologist Vladimir Köppen in 1918 and modified by R. Geiger and W. Pohl in 1953. It uses a system of letters to label climates. It is presented in a special supplement in this chapter.

DRY AND MOIST CLIMATES

All but 2 of the 13 climate types introduced in this chapter are classified as either dry climates or moist climates. Dry climates are those in which total annual evaporation of moisture from the soil and from plant foliage exceeds the annual precipitation by a wide margin. Generally speaking, the dry climates do not support permanently flowing streams. The soil is dry much of the year and the land surface is clothed with sparse plant cover—scattered grasses or shrubs—or simply lacks a plant cover. Moist climates are those with sufficient rainfall to maintain the soil in a moist condition through much of the year, and to sustain the year-round flow of the larger streams. Moist climates support forests or prairies of dense, tall grasses.

Within the dry climates there is a wide range of degree of aridity, ranging from very dry deserts nearly devoid of plant life to moister regions that support a partial cover of grasses or shrubs. To recognize this diversity, we will refer to two dry climate subtypes: (1) semiarid (or steppe), and (2) arid. The semiarid (steppe) subtype, designated by the letter *s*, has enough precipitation to support sparse grass- and shrubland. This subtype is found adjacent to moist climates. The arid subtype, designated by the letter *a*, ranges from extremely dry to transitional with semiarid.

Two of our 13 climates, as already noted, cannot be accurately described as either dry or moist climates. These are the wet-dry tropical ③ and Mediterranean ⑦ climate types. Instead, they show a seasonal alteration between a very wet season and a very dry season. This striking contrast in seasons gives a special character to the two climates, and thus we have singled them out for special recognition as wet-dry climates. In Figure 7.6,

Table 7.2 Moist, Dry, and Wet-Dry Climate Types

Climate Group	Climate Type		
	Moist	*Dry*	*Wet-Dry*
I: Low-latitude Climates	Wet equatorial ① Monsoon and trade-wind coastal ②	Dry tropical ④ (*s*, steppe; *a*, arid)	Wet-dry tropical ③
II: Midlatitude Climates	Moist subtropical ⑥ Marine west coast ⑧ Moist continental ⑩	Dry subtropical ⑤(*s*, *a*) Dry midlatitude ⑨ (*s*, *a*)	Mediterranean ⑦
III: High-latitude Climates	Boreal forest ⑪ Tundra ⑫	Ice sheet ⑬	

they are designated as having the wet-dry tropical and the Mediterranean precipitation patterns. Table 7.2 summarizes moist, dry, and wet-dry climates as they occur within the three main climate groups.

This introduction to global climate has stressed the relationship between climate and the factors that influence annual cycles of temperature and precipitation.

Temperature cycles may be uniform, seasonal (continental), or moderated by oceanic influences (marine). Precipitation may be uniform (ranging from scarce in all months to abundant in all months), may have a maximum at the time of high sun, or may have a maximum at the time of low sun. These cycles in turn are produced by the annual cycle of insolation as it varies with latitude and by the global patterns of atmospheric circulation and air mass movements. The next two chapters will examine the climates in more detail, including descriptions of key environmental characteristics that make each climate unique.

CHAPTER SUMMARY

Climate is the synthesis of weather of a region. Because temperature and precipitation are measured at many stations worldwide, we use the combined annual patterns of monthly averages of temperature and precipitation to assign climate types. Temperature regimes are typical patterns of annual variation in temperature. They depend on latitude, which determines the annual pattern of insolation, and on location—continental or maritime—which enhances or moderates the annual insolation cycle.

Global precipitation patterns are determined largely by air masses and their movements, which in turn are produced by global air circulation patterns. The main features of the global pattern of rainfall are (1) a wet equatorial belt produced by convectional precipitation around the ITC; (2) trade-wind coasts that receive moist flows of mT air from trade winds as well as tropical cyclones; (3) tropical deserts located under subtropical high-pressure cells; (4) midlatitude deserts and steppes, which are dry because they are far from maritime moisture sources; (5) moist subtropical regions that receive westward flows of moist mT air in the summer and eastward-moving cyclonic storms in winter; (6) midlatitude west coasts, which are subjected to eastward flows of mP air and occluded cyclones by prevailing westerly winds; and (7) polar and arctic deserts, where little precipitation falls because the air is too cold to hold much moisture. Annual patterns of precipitation fall into three patterns: uniform (ranging from abundant to scarce); high-sun maximum; and low-sun maximum.

There are three groups of climate types, arranged by latitude. Low-latitude climates (Group I) are dominated by mE, mT, and cT air masses, and are largely related to the global circulation patterns that produce the ITC, trade winds, and subtropical high pressure cells. They include wet equatorial ① (warm to hot with abundant rainfall); monsoon and trade-wind coastal ② (warm to hot with a very wet rainy season at high sun); wet-dry tropical ③ (warm to hot with very distinct wet and dry seasons); and dry tropical ④ (extremely hot in the high-sun season, a little cooler in the low-sun season, with little or no rainfall).

Midlatitude climates (Group II) lie in the polar-front zone and are strongly influenced by eastward-moving wave cyclones in which mT, mP, and cP air masses interact. They include dry subtropical ⑤ (a desert climate, not quite as hot as dry tropical ④); moist subtropical ⑥ (hot and humid summers, mild winters, ample rainfall); Mediterranean ⑦ (hot, dry summers, rainy winters); marine west-coast ⑧ (warm summers, cool winters, more rainfall in winter); and dry midlatitude ⑨ (warm in summer, cold in winter, low annual precipitation).

High-latitude climates (Group III) are dominated by polar and arctic (antarctic) air masses. Wave cyclones mixing mP and cP air masses along the arctic-front zone provide precipitation in this region. Climates of high latitudes include boreal forest ⑪ (short, cool summers, with bitterly cold, snowy winters); tundra ⑫ (very short or nonexistent summers with cold winters); and ice sheet ⑬ (bitterly cold, even in summer).

Dry climates are those in which precipitation is largely evaporated from soil surfaces and transpired by vegetation, so that permanent streams cannot be supported. Within dry climates, there are two subtypes: arid (driest) and semiarid, or steppe (a little wetter). In moist climates, precipitation exceeds evaporation and transpiration, providing for sustained year-round streamflow. In wet-dry climates, strong wet and dry seasons alternate.

KEY TERMS

climate	steppe	climograph
isohyet		

REVIEW QUESTIONS

1. Discuss the use of monthly records of average temperature and precipitation to characterize the climate of a region. Why are these measures useful?

2. Why are latitude and location (maritime or continental) important factors in determining the annual temperature cycle of a station?

3. How does air temperature, as a climatic variable, influence precipitation?

4. Describe three temperature regimes and explain how they are related to latitude and location.

5. Identify seven important features of the global map of precipitation and describe the factors that produce each.

6. The seasonality of precipitation at a station can generally be described as following one of three patterns. Identify these patterns, explaining how each can arise and providing an example.

7. What are the important global circulation patterns and air masses that influence low-latitude (Group I) climates? What are their effects?

8. What air masses and circulation patterns influence the midlatitude (Group II) climates and how?

9. Identify the air masses and frontal zones that are important in determining high-latitude climates (Group III) and explain their effects.

10. Name the climates by number (1–13) and group (I–III).

Focus on Systems 7.1 • Time Cycles of Climate

1. What are the two primary time cycles evident in climate data? How do they influence weather at a location?

2. Provide an example another time cycle evident in temperature records of past few centuries.

ESSAY QUESTION

1. Sketch a hypothetical supercontinent with the general shape of Eurasia merged with Africa. It should stretch from about 70° N to 40° S latitude. Add some north-south mountain ranges, positioned where you like. Then select four locations on the supercontinent, describing and explaining the annual cycles of temperature and precipitation you would expect at each.

PROBLEMS

Working It Out 7.2 • Averaging in Time Cycles

1. The table below shows monthly precipitation in cm at St. Louis for 1990–1994. Find the average for each month for this five-year period.

Yr	J	F	M	A	M	J	J	A	S	O	N	D
1990	3.6	9.0	6.8	7.8	24.4	7.7	8.5	7.2	2.0	12.6	8.5	16.6
1991	3.9	2.5	8.1	8.3	9.8	1.1	13.2	2.5	7.6	14.5	8.3	5.3
1992	2.8	4.8	8.8	6.2	3.7	3.0	10.9	8.8	7.6	3.1	16.1	9.3
1993	9.0	7.0	8.4	15.6	10.0	18.1	12.9	12.1	23.3	6.6	12.3	3.8
1994	5.3	3.8	3.2	26.2	4.4	5.5	3.6	9.6	3.0	7.2	12.4	3.9

2. Using graph paper, plot the monthly averages for 1990–1994 in a similar fashion to graph *b* of the figure. Then plot the 58-year monthly means, using the monthly values shown on graph (*b*). How do the two graphs compare?

Chapter 8

Low-Latitude Climates

With our introduction to global climates now complete in Chapter 7, we can now describe each of the climates and their associated environments in turn, beginning with the low-latitudes climates of this chapter. Keep in mind that the annual cycles of temperature and precipitation for each climate region can be easily related to the latitude and location (coastal or interior) of the region, and to the characteristic movements of air masses and fronts through the region as they are controlled by global-scale patterns of atmospheric circulation.

Our description of the environments associated with the climate types will stress the natural vegetation covers, agricultural practices, and other climate-related human activities that occur in the regions of each climate type. We will also touch on the nature of soils in some climate regions. Both soils and vegetation are covered more thoroughly in Part IV.

Zebras on the African savanna, in Kenya. Dry season.

LOW-LATITUDE CLIMATES

The *low-latitude climates* lie for the most part between the tropics of cancer and capricorn. In terms of world latitude zones, the low-latitude climates occupy all of the equatorial zone (10° N to 10° S), most of the tropical zone (10–15° N and S), and part of the subtropical zone. In terms of prevailing pressure and wind systems, the region of low-latitude climates includes the equatorial trough of the intertropical convergence zone (ITC), the belt of tropical easterlies (northeast and southeast trades), and large portions of the oceanic subtropical high-pressure belt (Figure 7.7).

Figure 8.1 shows climographs for the four low-latitude climates. These climates range from extremely moist—the wet equatorial climate ④—to extremely dry—the dry tropical climate ④. They also vary strongly in the seasonality of their rainfall. In the wet equatorial climate ①, rainfall is abundant all year round. But in the wet-dry tropical climate ③, rainfall is abundant for only part of the year. During the remainder of the year, little or no rain falls. The seasonal temperature cycle

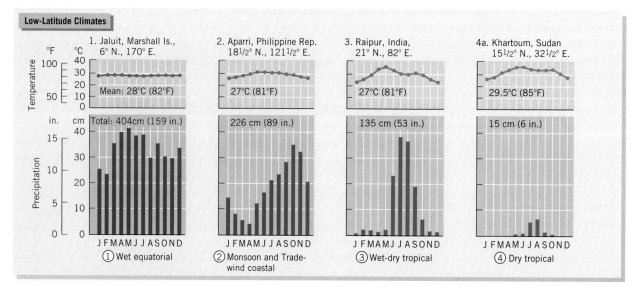

Figure 8.1 Climographs for the four low-latitude climates.

also varies among these climates. In the wet equatorial climate ①, temperatures are nearly uniform throughout the year. In the dry tropical climate ④, there is a strong annual temperature cycle. Table 8.1 summarizes some of the characteristics of these climates.

THE WET EQUATORIAL CLIMATE ① (Köppen: *Af*)

The **wet equatorial climate** ① is a climate of the intertropical convergence zone (ITC), which is nearby for most of the year. The climate is dominated by warm, moist maritime equatorial (mE) and maritime tropical (mT) air masses that yield heavy convectional rainfall. Rainfall is plentiful in all months, and the an-

nual total often exceeds 250 cm (about 100 in.). However, there is usually a seasonal pattern to the rainfall, so that rainfall is greater during some part of the year. This period of heavier rainfall occurs when the ITC migrates into the region. Remarkably uniform temperatures prevail throughout the year. Both mean monthly and mean annual temperatures are typically close to 27°C (81°F).

Figure 8.2 shows the world distribution of the wet equatorial climate ①. This climate is found in the latitude range 10° N to 10° S. Its major regions of occurrence include the Amazon lowland of South America, the Congo basin of equatorial Africa, and the East Indies, from Sumatra to New Guinea.

Figure 8.3 is a climograph for Iquitos, Peru (located in Figure 8.2), a typical wet equatorial station located

Figure 8.2 World map of wet equatorial ① and monsoon and trade-wind coastal climates ②. (Based on Goode Base Map.)

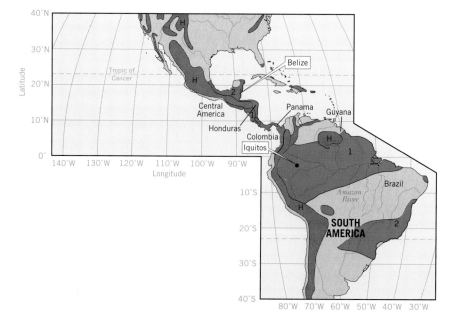

Table 8.1   Low-Latitude Climates

Climate	Temperature	Precipitation	Explanation
Wet equatorial ①	Uniform temperatures, mean near 27°C (81°F).	Abundant rainfall, all months, from mT and mE air masses. Annual total may exceed 250 cm (100 in.).	The ITC dominates this climate, with abundant convectional precipitation generated by convergence in weak equatorial lows. Rainfall is heaviest when the ITC is nearby.
Monsoon and trade-wind coastal ②	Temperatures show an annual cycle, with warmest temperatures in the high-sun season.	Abundant rainfall but with a strong seasonal pattern.	Trade-wind coastal: Rainfall from mE and mT air masses is heavy when the ITC is nearby, lighter when the ITC moves to the opposite hemisphere. Asian monsoon coasts: dry air flowing southwest in low-sun season alternates with moist oceanic air flowing northeast, producing a seasonal rainfall pattern on west coasts.
Wet-dry tropical ③	Marked temperature cycle, with hottest temperatures before the rainy season.	Wet high-sun season alternates with dry low-sun season.	Subtropical high pressure moves into this climate in the low-sun season, bringing very dry conditions. In the high-sun season, the ITC approaches and rainfall occurs. Asian monsoon climate: alternation of dry continental air in low-sun season with moist oceanic air in low-sun season with moist oceanic air in high-sun season brings a strong pattern of dry and wet seasons.
Dry tropical ④	Strong temperature cycle, with intense hot temperatures during high-sun season.	Low precipitation. Sometimes rainfall occurs when the ITC is near.	This climate is dominated by subtropical high pressure, which provides clear, stable air for much or all of year. Insolation is intense during high-sun period.

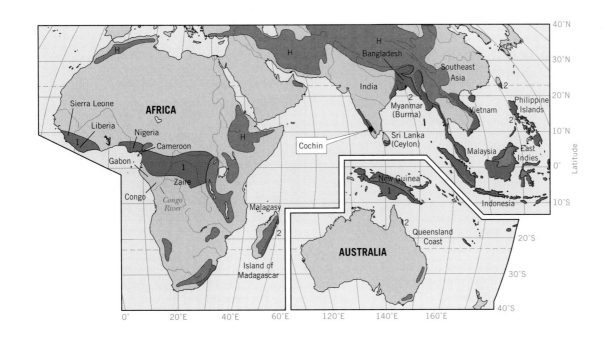

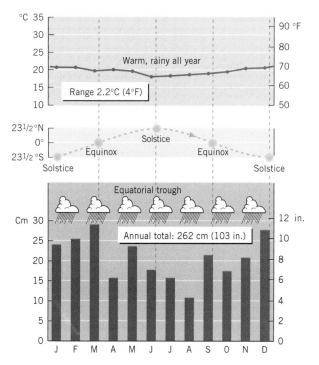

Figure 8.3 Wet equatorial climate ①. Iquitos, Peru, lat. 3° S, is located in the upper Amazon lowland, close to the equator. Temperatures differ very little from month to month, and there is abundant rainfall throughout the year.

As the name of the monsoon and trade-wind coastal climate ② suggests, this climate type is produced by two somewhat different situations. On trade-wind coasts, rainfall is produced by moisture-laden maritime tropical (mT) and maritime equatorial (mE) air masses. These are moved onshore onto narrow coastal zones by trade winds or by monsoon circulation patterns. As the warm, moist air passes over coastal hills and mountains, the orographic effect touches off convectional shower activity. Shower activity is also intensified by easterly waves, which are more frequent when the ITC is nearby. The east coasts of land masses experience this trade-wind effect because the trade winds blow from east to west. Trade-wind coasts are found along the east sides of Central and South America, the Caribbean Islands, Madagascar (Malagasy), Southeast Asia, the Philippines, and northeast Australia (see Figure 8.2).

The coastal precipitation effect also applies to the summer monsoon of Asia, when the monsoon circulation brings mT air onshore. However, the onshore monsoon winds blow from southwest to northeast, so it is the western coasts of land masses that are exposed to this moist airflow. Western India and Myanmar (formerly Burma) are examples. Moist air also penetrates well inland in Bangladesh, providing the very heavy monsoon rains for which the region is well known.

In central and western Africa, and southern Brazil, the monsoon pattern shifts the intertropical conver-

close to the equator in the broad, low basin of the upper Amazon River. Notice the very small annual range in temperature and the very large annual rainfall total. The monthly air temperatures are extremely uniform in the wet equatorial climate ①. Typically, mean monthly air temperature will range between 26° and 29°C (79° and 84°F) for stations at low elevation in the equatorial zone.

THE MONSOON AND TRADE-WIND COASTAL CLIMATE ②(Köppen: *Af, Am*)

Like the wet equatorial climate ①, the **monsoon and trade-wind coastal climate** ② has abundant rainfall. But unlike the wet equatorial climate ①, the rainfall of the monsoon and trade-wind coastal climate ② always shows a strong seasonal pattern. This seasonal pattern is due to the migration of the intertropical convergence zone (ITC). In the high-sun season ("summer," depending on the hemisphere), the ITC is nearby and monthly rainfall is greater. In the low-sun season, when the ITC has migrated to the other hemisphere, subtropical high pressure dominates and monthly rainfall is less. Figure 8.2 shows the global distribution of the monsoon and trade-wind coastal climate ②. The climate occurs over latitudes from 5° to 25° N and S.

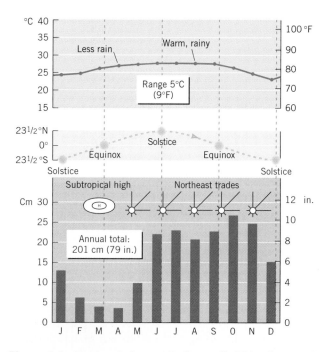

Figure 8.4 Trade-wind coastal climate ②. This climograph for Belize, a Central American east coast city at lat. 17° N, shows a marked season of low rainfall following the period of low sun. For the remainder of the year, precipitation is high, produced by warm, moist northeast trade winds.

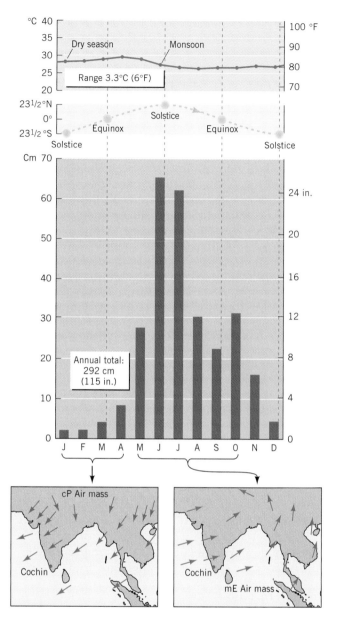

Figure 8.5 Monsoon coastal climate ②. Cochin, India, on a windward coast at lat. 10° N, shows an extreme peak of rainfall during the rainy monsoon, contrasting with a short dry season at time of low sun.

gence zone over 20° of latitude, or more (see Figure 5.16). Here, heavy rainfall occurs in the high-sun season, when the ITC is nearby. Drier conditions prevail in the low-sun season, when the ITC is far away.

Temperatures in the monsoon and trade-wind coastal climate ②, though warm throughout the year, also show an annual cycle. Warmest temperatures occur in the high-sun season, just before arrival of the ITC brings clouds and rain. Minimum temperatures occur at the time of low sun.

Figure 8.4 is a climograph for the city of Belize, in the Central American country of Belize (Figure 8.2). This east coast city, located at lat. 17° N, is exposed to the tropical easterly trade winds. Rainfall is abundant

from June through November, when the ITC is nearby. Easterly waves are common in this season, and an occasional tropical cyclone strikes the coast, bringing torrential rainfall. Following the December solstice, rainfall is greatly reduced, with minimum values in March and April. At this time, the ITC lies farthest away and the climate is dominated by subtropical high pressure. Air temperatures show an annual range of 5° C (9° F), with maximum in the high-sun months.

The Asiatic monsoon shows a similar pattern, but there is an extreme peak of rainfall during the high-sun period and a well-developed dry season with two or three months of only small rainfall amounts. The climograph for Cochin, India, provides an example (Figure 8.5). Located at lat. 10° N on the west coast of lower peninsular India (see Figure 8.2), Cochin receives the warm, moist southwest winds of the summer monsoon. In this season, monthly rainfall is extreme in both June and July. A strongly pronounced season of low rainfall occurs at the time of low sun—December through March. Air temperatures show only a very weak annual cycle, cooling a bit during the rains, so the annual range is small at this low latitude.

The Low-Latitude Rainforest Environment

Our first two climates—wet equatorial ① and monsoon and trade-wind coastal ②—are quite uniform in temperature and have a high annual rainfall. These factors create a special environment—the low-latitude rainforest environment (Figure 8.6). In the rainforest, streams

Figure 8.6 Rainforest of the western Amazon lowland, near Manaus, Brazil. The river is a tributary of the Amazon. Note the many different types of trees of varying shapes.

flow abundantly throughout most of the year, and river channels are lined along the banks with dense forest vegetation. Indigenous peoples of the rainforest once traveled the rivers in dugout canoes, although today most enjoy easier travel powered by outboard motor. Over a century ago, larger shallow-draft river craft turned the major waterways into the main arteries of trade, connecting towns and cities on the river banks. Aircraft have added a new dimension in mobility, criss-crossing the almost trackless green sea of forest to find landings in clearings or on broad reaches of the larger rivers.

In the low-latitude rainforest environment, the abundant rainfall and prevailing warm soil temperatures promote the decay and decomposition of rock to great depths, so that a thick soil layer is usually present. The soil is typically rich in oxides of iron and has a deep red color. This kind of soil has largely lost its ability to hold nutrient substances needed by plants such as the grasses and grain crops that are important in modern agriculture practiced in higher latitudes. However, many kinds of native plants have adapted to these soils. Most conspicuous of these are the great forest trees with broad leaves that comprise the majority of tree species within the evergreen rainforest. The key to the success of this kind of forest lies in the ability of the trees to quickly reuse (recycle) the essential plant nutrients released by the decay of fallen leaves and branches.

Low-latitude rainforests, unlike higher latitude forests, possess a great diversity of plant and animal species. The low-latitude rainforest can contain as many as 3000 different tree species in an area of only a few square kilometers, whereas the midlatitude forest possesses fewer than one-tenth that number. The num-

ber of types of animals found in the rainforest is also very large. A 16 km^2 (6 mi^2) area in Panama near the Canal, for example, includes about 20,000 species of insects.

Animal life of the rainforest is most abundant in the upper layers of the rainforest. Above the canopy, birds and bats are important predators and feed on insects above and within the topmost leaf canopy. Below this level live a wide variety of birds, mammals, reptiles, and invertebrates. These animals feed on the leaves, fruit, and nectar that are abundantly available in the main part of the canopy. Ranging between the canopy and ground are small climbing animals—monkeys, for example—that forage in both layers. At the surface are the large ground animals, including herbivores—plant-eating animals—that graze the low leaves and fallen fruits, and carnivores—meat eaters—that prey on the abundant animals.

Plant Products and Food Resources of the Rainforest

Many products of the rainforest are of economic value. Rainforest lumber, such as mahogany, ebony, or balsawood, is an important export. Quinine, cocaine, and other drugs come from the bark and leaves of tropical plants; cocoa comes from the seed kernel of the cacao plant. Natural rubber is made from the sap of the rubber tree. The tree comes from South America, where it was first exploited. Rubber trees also are widely distributed through the rainforest of Africa. Today, the principal production is from plantations in Indonesia, Malaysia, Thailand, Vietnam, and Sri Lanka (Ceylon).

An important class of food plants native to the wet low-latitude environment are starchy staples. Some of these are root crops while others are fruits. Manioc,

Figure 8.7 A coastal grove of coconut palms, City of Refuge, Kona Coast, Island of Hawaii. (A. N. Strahler.)

also known as cassava, has a tuberous root—something like a sweet potato—that reaches a length over 0.3 m (1 ft) and may weigh several kilograms. Yams, like maniocs, are large underground tubers. They are a major source of food in West Africa and the Caribbean. The taro plant possesses a starchy, enlarged, underground stem that has considerable food value. Visitors to Hawaii know taro through its transformation into poi, a fermented paste. Taro was imported into Africa from Southeast Asia and eventually reached the Caribbean region. Among the starchy fruits of the wet low-latitude environment are the banana and plantain. The banana was first cultivated for food in Southeast Asia, and then spread to Africa and the Americas. The plantain is a coarse variety of the banana that is starchy, has little sugar, and requires cooking.

Perhaps the plant that has been most important to humans in the low latitudes is the coconut palm. Besides being a staple food, it provides a multitude of useful products in the form of fiber and structural materials. The coconut palm flourishes on islands and coastal fringes of the wet equatorial and tropical climates (Figure 8.7). Copra, the dried meat of the coconut, is a valuable source of vegetable oil. Copra and coconut oil are major products of Indonesia, the Philippines, and New Guinea. Palm oil and palm kernels of other palm species are important products of the equatorial zone of West Africa and the Congo River basin.

THE WET-DRY TROPICAL CLIMATE ③ (Köppen: *Aw, Cwa*)

In the monsoon and trade-wind coastal climate ②, we noted that the movements of the intertropical convergence zone into and away from the climate region produce a seasonal cycle of rainfall and temperature. As we move farther poleward, this cycle becomes stronger, and the monsoon and trade-wind coastal climate ② grades into the **wet-dry tropical climate** ③.

The wet-dry tropical climate ③ is distinguished by a very dry season at low sun that alternates with a very wet season at high sun. During the low-sun season, when the equatorial trough is far away, dry continental tropical (cT) air masses prevail. In the high-sun season, when the ITC is nearby, moist maritime tropical (mT) and maritime equatorial (mE) air masses dominate. Cooler temperatures accompany the dry season but give way to a very hot period before the rains begin.

Figure 8.8 shows the global distribution of the wet-dry tropical climate ③. It is found at latitudes of 5° to 20° N and S in Africa and the Americas, and at 10° to 30° N in Asia. In Africa and South America, the climate occupies broad bands poleward of the wet equatorial and monsoon and trade-wind coastal climates. Because these regions are farther away from the ITC, less rainfall is triggered by the ITC during the rainy season, and

subtropical high pressure can dominate more strongly during the low-sun season. In central India and Indochina, the regions of wet-dry tropical climate ③ are somewhat protected by mountain barriers from the warm, moist mE and mT airflows provided by trade and monsoon winds. These barriers create a rainshadow effect, so that even less rainfall occurs during the rainy season and the dry season is drier still.

Figure 8.9 is a climograph for Timbo, Guinea, at lat. 10° N in West Africa (Figure 8.8). Here the rainy season begins just after the March equinox and reaches a peak in August, about two months following June solstice. At this time, the ITC has migrated to its most northerly position, and moist mE air masses flow into the region from the ocean lying to the south. Monthly rainfall then decreases as the low-sun season arrives and the ITC moves to the south. Three months—December through February—are practically rainless. During this season, subtropical high pressure dominates the climate, and stable, subsiding continental tropical (cT) air pervades the region. The temperature cycle is closely linked to both the solar cycle and the precipitation pattern. In February and March, insolation increases, and air temperature rises sharply. A brief hot season occurs. As soon as the rains set in, the effect of cloud cover and evaporation of rain causes the temperature to decline. By July, temperatures have resumed an even level.

The Savanna Environment

The native vegetation of the wet-dry tropical climate ③ must survive alternating seasons of very dry and very wet weather. Most plants enter a dormant phase during the dry period, then burst forth into leaf and bloom with the coming of the rains. For this reason, the native plant cover can be described as *rain-green vegetation.*

Rain-green vegetation consists of two basic types. First is *savanna woodland* (Figure 8.10). By *woodland* we mean an open forest in which trees are widely spaced apart. In the savanna woodland, coarse grasses occupy the open space between the trees, which are often coarse-barked and thorny. Large expanses of grassland may also be present (Figure 8.11). In the dry season, the grasses turn to straw, and many of the tree species shed their leaves to cope with the drought. Second, in the more arid parts of the climate region, small thorny trees and large shrubs form dense patches. This type of vegetation cover is referred to as *thorntree-tall-grass savanna.* (Further details are given in Chapter 21.) Because of the prevalence of the savanna vegetation in the wet-dry tropical climate ③, it can be identified as the savanna environment.

In the savanna environment, river channels that are not fed by nearby moist mountain regions are nearly or completely dry in the low-sun dry season. In the rainy season, these river channels become filled to their

Figure 8.8 World map of the wet-dry tropical climate ③. (Based on Goode Base Map.)

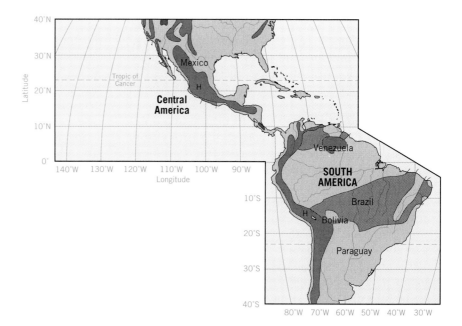

banks with swiftly flowing, turbid water. The rains are not reliable, and agriculture without irrigation is hazardous at best. When the rains fail, a devastating famine can ensue. The Sahel region of Africa, discussed later in this chapter, is a region well-known for such droughts and famines.

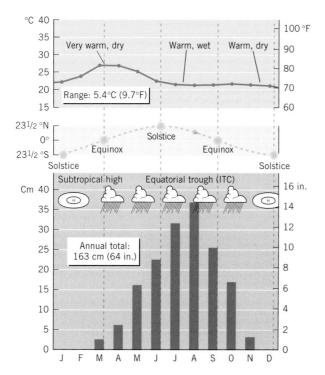

Figure 8.9 Wet-dry tropical climate ③. Timbo, Guinea, at lat. 10° N, is in West Africa. A long wet season at time of high sun alternates with an almost rainless dry season at time of low sun.

Soils of the savanna environment are similar in their physical characteristics and fertility to those of the rainforest environment—largely soils of low fertility and red color. However, substantial areas of the savanna environment have fertile soils developed and sustained by the slow infall of windblown dust from adjacent deserts. Equally important are highly fertile soils that occur along major through-flowing rivers. Annual flooding of these rivers leaves lowland deposits of fertile silt carried down from distant mountain ranges.

Animal Life of the African Savanna

The natural animal life of the savanna grasslands and woodlands is closely adapted to the vegetation and climate. These are the regions of the carnivorous game animals and the vast multitudes of grazing animals on which they prey (Figure 8.11). The savanna of Africa is the natural home of large herbivores, such as wildebeest, gazelle, deer, antelope, buffalo, rhinoceros, zebra, giraffe, and elephant. Their predators are the lion, leopard, hyena, wild dog, and jackal. Some of the herbivores depend on fleetness of foot to escape the predators. Others, such as the rhinoceros, buffalo, and elephant, defend themselves by their size, strength, or armor-thick hide. The giraffe is peculiarly well-adapted to savanna woodlands. Its long neck permits browsing on the higher foliage of the scattered trees.

The dry season brings a severe struggle for existence to animals of the African savanna. As streams and hollows dry up, the few muddy water holes must supply all drinking water. Danger of attack by carnivores is greatly increased.

The savanna ecosystem in Africa currently faces the prospect of widespread destruction. The primary threat

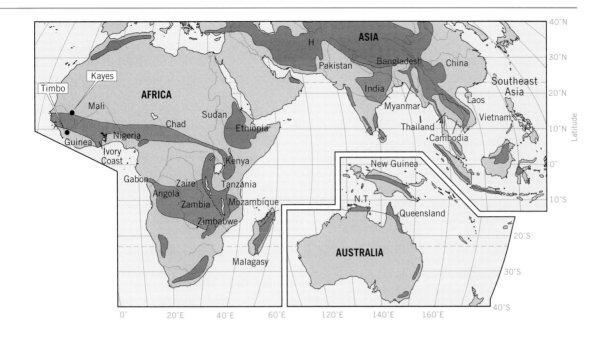

Figure 8.10 Savanna woodland, of the Serengeti Plains, Tanzania, East Africa. Acacia trees with flattened crowns remain green, although the coarse grasses have turned to straw in the dry season.

Figure 8.11 Vast herds of wildebeest and zebra graze the lush vegetation of the Serengeti Plains in Tanzania, still green from the rains. These animals migrate long distances in search of food and water.

is the loss of habitat that occurs as human activities claim ever-larger portions of the native savanna. Parks set aside to preserve the savanna ecosystem confine the grazing animals to a narrow range and prevent their seasonal migrations in search of food. In some instances, the growth in animal populations has been phenomenal because they have been protected from hunting. In confined preserves they rapidly consume all available vegetation in futile attempts to survive. Rapidly growing human populations are bringing increasing pressure on management agencies to allow encroachment on game preserves in order to expand cattle grazing and agriculture. Poaching on a large scale is now threatening the extinction of the elephant and rhinoceros in many areas.

Agricultural Resources of the Savanna Environment

The agricultural resources and practices of the savanna environment differ significantly between Asia and Africa. In Southeast Asia, the stronger Asiatic monsoon system is a key factor. There are also vastly larger human populations in the Asiatic monsoon nations than elsewhere in low latitudes, and these peoples have evolved their own unique culture patterns.

Of the staple foods grown widely in Southeast Asia, rice is dominant. About one-third of the human race subsists on rice, and most of this vast population is crowded into the arable lands of Southeast Asia. Intensive rice cultivation of this part of the world actually spans three climate types: monsoon and trade-wind coastal climate ②, wet-dry tropical climate ③, and moist subtropical climate ≈. Rice requires flooding of the ground at the time the seedling plants are cultivated, and this activity has been traditionally timed to coincide with the peak of the rainy monsoon season (Figure 8.12). The crop matures and is harvested in the dry season. Sugarcane is another important crop that grows rapidly during the rainy season and is harvested in the dry season.

Food and Woodland Resources of the African Savanna

Because the savanna environment in Africa is a transition zone between rainforest and desert environments, there is a corresponding gradation of plant resources and the ways in which they are used to support human life. In the moister zones, those of the savanna woodland with a long wet season, agriculture follows a pattern known as *bush-fallow farming*. This practice is related in some ways to the slash-and-burn agriculture of the rainforest. Trees are cut from a small area, piled up, and burned. The ash provides fertilizer for cultivated plots. After a few years, the land is allowed to revert to shrub and tree growth. Some of the same crops that are grown in the rainforest agriculture are grown in wetter areas of the bordering savanna. Crops include grains, yams, soybeans, and sugarcane. Tobacco and cotton are also grown.

In the drier savanna grassland and thorntree-savanna, where the wet season is short, the main subsistence crops of uplands are corn, millet (a kind of grain), sorghum, and peanuts. Sorghum—a tall, leafy plant somewhat resembling corn—is also an important food crop of the savanna environment. It is well-adapted to conditions of a short wet season and a long, hot dry season. The peanut is another major food crop of the savanna environment. Introduced from India, it is now grown in West Africa. Cotton, along with peanuts, is an important cash crop for export.

Besides the agricultural system of the permanent farmers, there exists a shifting cattle culture, in which large numbers of cattle are maintained as a display of wealth (Figure 8.13). The cattle provide food in the form of milk, butter, and blood. Following a seasonal pattern, the cattle are moved into the semidesert zone in the rainy season to graze on grasses. Then they are returned to the savanna grassland zone in the dry season, where they graze on fallow croplands and rely on water holes for survival.

The savanna woodland belt of Africa provides a number of other plant resources besides cultivated food

Figure 8.12 Much rice cultivation in the monsoon belt of Asia occurs in terraced fields such as these, on the island of Bali, Indonesia.

Figure 8.13 Masai herdsmen in southwestern Kenya, East Africa. Cattle represent a form of wealth and prestige; they also supply food in the form of milk and blood. The quality of these animals is secondary in importance to their numbers.

crops. Trees are cut for firewood, which is the only fuel for cooking available to most inhabitants. Trees also provide construction poles for dwellings. Among the savanna woodland trees that furnish important export products are the cashew-nut tree and the kapok tree. Gum arabic is taken from acacia trees. Most native trees of the savanna woodland have little commercial value as export lumber (an exception is black teak), but plantations of exotic tree species have been introduced in some areas. Both pine and eucalyptus have been successfully introduced as sources of pulpwood. These trees grow very rapidly, as compared with pulpwood trees in high latitudes.

EYE ON THE ENVIRONMENT:
Drought and Land Degradation in the African Sahel

The wet-dry tropical climate ③ is subject to years of devastating drought as well as to years of abnormally high rainfall that can result in severe floods. Climate records show that two or three successive years of abnormally low rainfall (a drought) typically alternate with several successive years of average or higher than average rainfall. This variability is a permanent feature of the wet-dry tropical climate ③, and the plants and animals inhabiting this region have adjusted to the natural variability in rainfall, with one exception: the human species.

The wet-dry tropical climate ③ of North Africa, including the adjacent semiarid southern belt of the dry tropical climate ④ to the north, provides a lesson on the human impact on a delicate ecological system. Countries of this perilous belt, called the *Sahel,* or *Sahelian zone,* are shown in Figure 8.14. From 1968 through

1974, all these countries were struck by a severe drought. Both nomadic cattle herders and grain farmers share this zone. During the drought, grain crops failed and foraging cattle could find no food to eat (Figure 8.15). In the worst stages of the Sahel drought, nomads were forced to sell their remaining cattle. Because the cattle were their sole means of subsistence, the nomads soon starved. Some 5 million cattle perished, and it has been estimated that 100,000 people died of starvation and disease in 1973 alone.

The Sahelian drought of 1968–1974 was associated with a special phenomenon, which at that time was called *desertification*—the permanent transformation of the land surface by human activities to resemble a desert, largely through the destruction of grasses, shrubs, and trees by grazing animals and fuelwood harvesting. That term has now been abandoned in favor of *land degradation.* Also visible are the effects of accelerated soil erosion, such as gullying of slopes and accumulations of sediment in stream channels. The removal of soil by wind is also intensified.

Periodic droughts throughout past decades are well documented in the Sahel, as they are in other world regions of wet-dry tropical climate ③. Figure 8.16 shows the percentage of departures from the long-term mean of each year's rainfall in the western Sahel from 1901 through 1991. Note the wide year-to-year variation. Since about 1950, the durations of periods of continuous departures both above and below the mean seem to have increased substantially. The period of sustained high-rainfall years in the 1950s contrasts sharply with a series of severe drought episodes starting in 1971. For an earlier record, scientists have examined fluctuations in the level of Lake Chad (Figure 8.14). In times of drought, the lake's shoreline retreats, while in times of abundant rainfall, it expands. The changes document

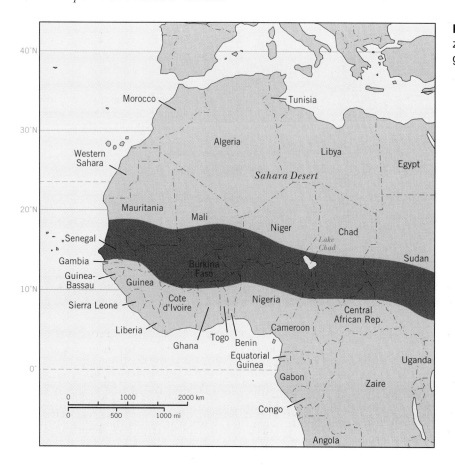

Figure 8.14 The Sahel, or Sahelian zone, shown in color, lies south of the great Sahara Desert of North Africa.

periods of rainfall deficiency and excess—1820–1840, below normal; 1870–1895, above normal; 1895–1920, below normal.

The sustained drought periods of the past show that droughts and wet periods are a normal phenomenon in the Sahel. In prior eras, however, the landscape was able to recover from the droughts during periods of abundant rainfall. The recent land degradation of the African Sahel is the direct result of greatly increased numbers of humans and their cattle. European managers of African colonies contributed to population increases by supplying food in time of famine, by reducing the death rate from disease, and by developing groundwater supplies for crop irrigation and livestock. With each succeeding drought, the increased population made a heavier impact on the vegetation and soil. Ultimately, land degradation became too sustained and too severe to permit recovery of the plant cover during

Figure 8.15 At the height of the Sahelian drought, vast numbers of cattle perished and even the goats were hard pressed to survive. Trampling of the dry ground prepared the region for devastating soil erosion in the rains that eventually ended the drought.

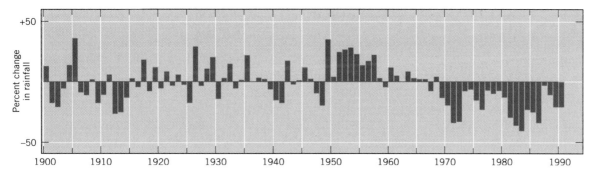

Figure 8.16 Rainfall fluctuations for stations in the western Sahel, 1901–1991, expressed as a percent departure from the long-term mean. (Courtesy of Sharon E. Nicholson, Department of Meteorology, Florida State University, Tallahassee.)

the rainy season or over a period of wetter years. As long as the population in the area remains high, the land degradation will be permanent.

The effect of an accelerated global warming in the coming decades on regions of wet-dry tropical climate ③ is difficult to predict. Some regions may experience greater swings between drought and surplus precipitation. At the same time, these regions may migrate poleward as a result of intensification of the Hadley cell circulation. Thus, the Sahel region may move northward, into the desert zone of the Sahara. However, such changes are highly speculative. At this time, there is no scientific consensus on how climatic change will affect the Sahel or other regions of wet-dry tropical climate ③.

The periodic droughts of the Sahel are an example of a multiyear climatic time cycle. *Working It Out 8.1 • Cycles of Rainfall in the Low Latitudes* presents some other examples of time cycles observed in low-latitude climates as well as some simple measures that describe their variability.

THE DRY TROPICAL CLIMATE ④
(Köppen: *BWh, BSh*)

The **dry tropical climate** ④ is found in the center and east sides of subtropical high-pressure cells. Here, air strongly subsides, warming adiabatically and inhibiting condensation. Rainfall is very rare and occurs only when unusual weather conditions move moist air into the region. Since skies are clear most of the time, the sun heats the surface intensely, keeping air temperatures high. During the high-sun period, heat is extreme. During the low-sun period, temperatures are cooler. Given the dry air and lack of cloud cover, the daily temperature range is very large.

The driest areas of the dry tropical climate ④ are near the tropics of cancer and capricorn. As we travel from the tropics toward the equator, we find that somewhat more rain falls. Continuing in this direction,

we encounter a short rainy season at the time of year when the ITC is near, and the climate grades into the wet-dry tropical ③ type.

Figure 8.17 shows the global distribution of the dry tropical climate ④. Nearly all of the dry tropical climate ④ areas lie in the latitude range 15° to 25° N and S. The largest region is the Sahara–Saudi Arabia–Iran–Thar desert belt of North Africa and southern Asia. This vast desert expanse includes some of the driest regions on earth. Another large region of dry tropical climate ④ is the desert of central Australia. The west coast of South America, including portions of Ecuador, Peru, and Chile, also exhibits the dry tropical climate ④. However, temperatures are moderated by the cool marine air layer that blankets the coast.

Figure 8.18 is a climograph for a dry tropical station in the heart of the North African desert. Wadi Halfa, Sudan (Figure 8.17), lies at lat. 22° N, almost on the tropic of cancer. The temperature record shows a strong annual cycle with a very hot period at the time of high sun, when three consecutive months average 32°C (90°F). Daytime maximum air temperatures are frequently between 43° and 48°C (about 110° to 120°F) in the warmer months. There is a comparatively cool season at the time of low sun, but the coolest month averages a mild 16°C (61°F) and freezing temperatures are rarely recorded. No rainfall bars are shown on the climograph because precipitation averages less than 0.25 cm (0.1 in.) in all months. Over a 39-year period, the maximum rainfall recorded in a 24-hour period at Wadi Halfa was only 0.75 cm (0.3 in.).

The world's dry climates consist largely of extremely arid deserts. In addition, there are broad zones at the margins of the desert that are best described as *semiarid*. These zones have a short wet season that supports the growth of grasses on which animals (both wild and domestic) graze. Geographers also call these semiarid regions **steppes**. Nomadic tribes and their herds of animals visit these areas during and after the brief moist period. In Figures 7.10 (world climate map) and 8.17 (world map of dry climates), the two subdivisions of dry

Working It Out 8.1 • Cycles of Rainfall in the Low Latitudes

The cycle of monthly precipitation shown on the climographs presented in Chapters 7–9 is basically controlled by the sun's declination cycle, which rhythmically varies the insolation received at a location according to its latitude and the time of the year. Through the mechanisms described in Chapters 4–7, this cycle produces an annual cycle of precipitation that is revealed by taking monthly averages over a long period of record. (See *Working It Out • Averaging in Time Cycles,* Chapter 7.)

Although this cycle is a strong one at most locations, there are also other cycles that influence rainfall over a sequence of years, rather than from month to month. However, they are not as strong or predictable as the annual cycle. But when we have several "wet" years followed by several "dry" years in a repeating sequence, another cycle may be at work.

Let's examine rainfall cycles that span several decades. These are shown in the three graphs at right. Each graph presents annual rainfall totals for a sequence of 46 years at three low-latitude stations: Padang, Sumatra, in the wet equatorial climate ①; Bombay, India, in the wet-dry tropical climate ③; and Abbassia, near Cairo, Egypt, which is in the dry tropical climate ④. (Notice the difference in vertical scale of the Abbassia record.) These data are selected for that month which, on the average, is the rainiest month at the recording station.

The data are presented in two ways: annual bars and a superimposed continuous curved line that smooths those annual values. From a close look at the graph, it seems that many of the peak bars are followed by either two or three lower bars before reversal to a higher bar. This suggests that there is a rainfall cycle with a length, or period, of between two and three years. The smoothed curve shows this cycle rather well for Padang and Bombay, especially in the early part of the sequence.

Another type of information about a cycle is its *amplitude.* For a smooth wavelike curve, amplitude is the difference in height between a crest and the adjacent trough. That difference is expressed here in centimeters of rainfall. Using a different vocabulary that better suits the description of climate, the varying amplitude of the curve is a measure of the *variability* of the precipitation. From an inspection of the amplitude of the cycles at Padang and Bombay, it appears that rainfall at Bombay has somewhat more variability.

One way to measure the variability is to examine the average deviation of each yearly value from the mean value. The deviation is simply the difference between a single value and the mean of all the values, taken without respect to sign. Thus, we can define the deviation as

$$D_i = \left| P_i - \overline{P} \right|$$

where D_i is the deviation of the *i*th observation, P_i; $\overline{P}$ is the mean (see *Working It Out • Averaging in Time Cycles,* Chapter 7); and the two vertical bars | | indicate absolute value. (The absolute value is simply the value taken without respect to sign. It is always positive.) The average deviation $\overline{D}$ is then

$$\overline{D} = \frac{1}{n} \sum_{i=1}^{n} D_i = \frac{1}{n} \sum_{i=1}^{n} \left| P_i - \overline{P} \right|$$

where there are *n* observations.

Calculation of the mean and average deviation for the three sets of observations yields the following results:

Station	Mean, cm	Average deviation, cm
Padang	51.4	12.9
Bombay	61.0	20.7
Abbassia	0.67	0.72

Looking first at the means, we see that Bombay has the largest value, with 61 cm (24 in.) of July rainfall. Padang is next, with an average of 51 cm (20 in.) of rainfall during November. Compared to these two stations, Abbassia is very dry, with an average of less than 1 centimeter of rainfall in its rainiest month, January. Considering the average deviations of these stations, we see that Bombay has the largest year-to-year fluctuation. Padang is next, followed by Abbassia with a very small value.

Note that if we look only at the average deviation, we would conclude that January rainfall is very constant at Abbassia. However, we

climates are distinguished with the letters *a* (arid) and *s* (semiarid).

An example of the semiarid dry tropical climate ④ is that of Kayes, Mali, which we presented in Chapter 7 as a sample climograph (Figure 7.8). Located in the Sahel region of Africa, this station has a distinct rainy season that occurs when the intertropical convergence moves north in the high-sun season. This precipitation pattern shows the semiarid subtype as a transition between the arid dry tropical climate ④ and the wet-dry tropical climate ③.

Much of the arid desert landscape consists of barren areas of drifting sand or sterile salt flats. However, in semiarid regions, thorny trees and shrubs are often abundant, since the climate includes a small amount of regular rainfall. Figure 8.19 shows a scene from the

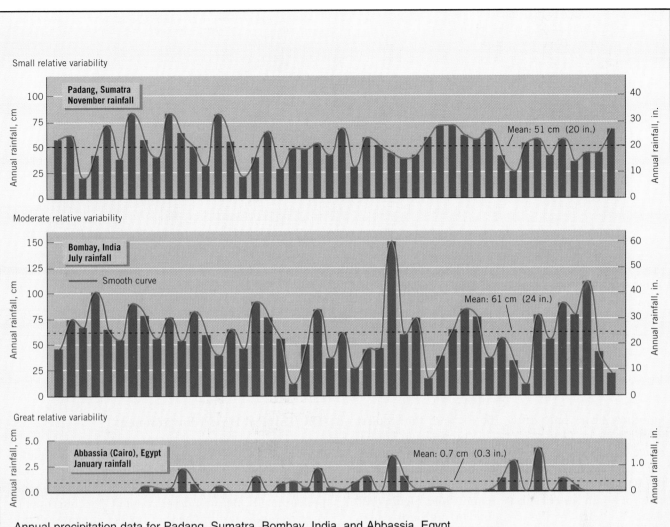

Annual precipitation data for Padang, Sumatra, Bombay, India, and Abbassia, Egypt.

can see from the graph that there are many years in which no rainfall at all occurs at Abbassia, and others where more than 3 cm falls. For the residents of this location, the variability is great, indeed! This suggests that we need to compare the average deviation with the mean as a measure of *relative variability*. Taking the ratio of the average deviation to the mean, or $\bar{D} \div \bar{P}$, we have

Station	Relative variability
Padang	12.9 ÷ 51.4 = 0.25
Bombay	20.7 ÷ 61.0 = 0.34
Abbassia	0.72 ÷ 0.67 = 1.07

Another way to put this is to observe that the average deviation is 25 percent of the mean for Padang, 34 percent for Bombay, and 107 percent for Abbassia. By this measure,

Abbassia's January rainfall is the most variable, while Padang's is the least variable.

This exercise not only presents a way to measure the variability of precipitation, but also demonstrates the large variability in rainfall that is characteristic of monthly rainfall in many of the low-latitude climates.

semiarid part of the Kalahari Desert in southwest Africa (Figure 8.17), featuring the strange-looking baobab tree.

An important variation of the dry tropical climate ④ occurs in narrow coastal zones along the western edge of continents. These regions are strongly influenced by cold ocean currents and the upwelling of deep, cold water, which occurs just offshore. The cool water mod-

erates coastal zone temperatures and reduces the seasonality of the temperature cycle. Figure 8.20 shows a climograph for Walvis Bay, a port city on the west coast of Namibia (Southwest Africa), at lat. 23° S (Figure 8.17). (The yearly cycle begins with July, because this is a southern hemisphere station.) For a location nearly on the tropic of capricorn (23 1/2° S), the monthly temperatures are remarkably cool, with the warmest

Figure 8.17 World map of the dry tropical ④, dry subtropical ⑤, and dry midlatitude ⑨ climates. (Based on Goode Base Map.)

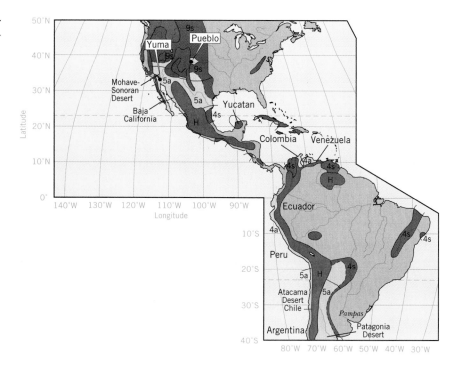

monthly mean only 19°C (66°F) and the coolest monthly mean 14°C (57°F). This provides an annual range of only 5°C (9°F). Coastal fog is a persistent feature of this climate. Another important occurrence of this western coastal desert subtype is along the western coast of South America in Peru and Chile. Figure 8.21 shows a stretch of this coastline in northern Chile, where it is known as the Atacama Desert.

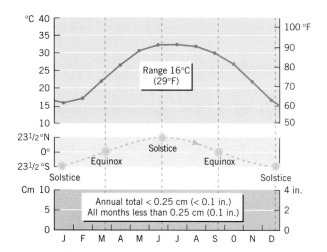

Figure 8.18 Dry tropical climate ④, dry desert. Wadi Halfa is a city on the Nile River in Sudan at lat. 22° N, close to the Egyptian border. Too little rain falls to be shown on the graph. Air temperatures are very high during the high-sun months.

The Tropical Desert Environment

The tropical deserts and their bordering semiarid zones comprise a global environmental region sustained by subsiding air masses of the continental high-pressure cells. Because desert rainfall is so infrequent, river channels and the beds of smaller streams are dry most of the time. However, a sudden and intense desert downpour can cause local flooding of brief duration that transports large amounts of silt, sand, gravel, and boulders. These events are termed *flash floods* (see Chapter 15). Major river channels often end in flat-floored basins having no outlet. Here clay and silt are deposited and accumulate, along with layers of soluble salts. Shallow salt lakes occupy some of these basins. The action of desert streams and the deposition of their salts by evaporation are covered in some detail in Chapters 13 and 15.

Large expanses of the very dry low-latitude desert appear to be entirely devoid of plant life except, perhaps, along the banks of some of the larger dry stream channels. Large barren areas may be covered with a layer of loosely fitted rock particles (sometimes called "desert pavement"), while loose drifting sand completely mantles other large areas. These kinds of surfaces are described in Chapter 17.

Many desert plants survive the harsh environment because they can quickly take advantage of a rare rainfall that may arrive only once in several years. These include fast-growing annuals, which survive the dry periods as dormant seeds, then sprout, grow, flower, and

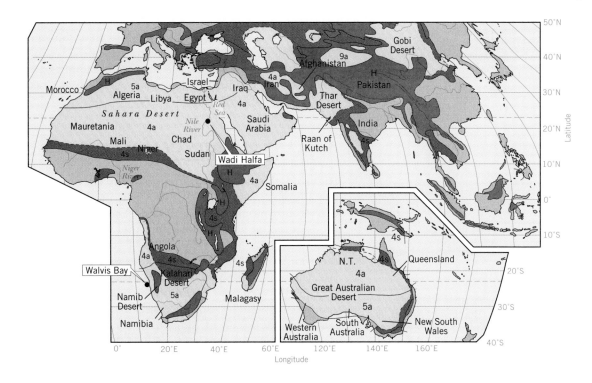

Figure 8.19 This strange-looking baobab tree is a common inhabitant of the thorntree semidesert of Botswana, in the Kalahari region of southern Africa.

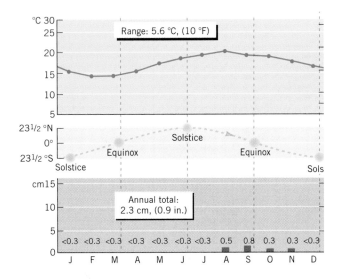

Figure 8.20 Dry tropical climate ④, western coastal desert subtype. Walvis Bay, Namibia (Southwest Africa), is a desert station on the west coast of Africa at lat. 23° S. Air temperatures are cool and remarkably uniform throughout the year.

set seed before soil moisture is exhausted. Hard-leaved or spiny shrubs, which resist water loss by transpiration, are also common. Some desert-dwellers are succulent plants and store water in their spongy tissues—cacti are an example. Still others, like the tamarisk (or salt cedar), grow along the dry beds of major watercourses and send their roots deep below the river bed to reach stores of water that infiltrated during river floods.

Figure 8.21 The Atacama Desert, on the coast of northern chile.

In various low places in the dry desert, water can be reached by digging or drilling wells that tap the groundwater zone, where porous rock material is saturated with fresh water. Where such water supplies are available, they can be used to irrigate agricultural plots, thus creating an *oasis* (Figure 8.22).

HIGHLAND CLIMATES OF LOW LATITUDES

Highland climates, shown in a gray tone on the world climate map, are cool to cold, usually moist, climates that occupy mountains and high plateaus. The higher the location, generally, the colder and wetter is its climate. Temperatures are lower since air temperatures in the free atmosphere normally decrease with altitude (see Chapter 3). Rainfall increases because orographic precipitation tends to be induced when air masses ascend to higher elevation (Chapter 4). Highland climates are not usually included in the broad schemes of climate classification. Many small highland areas are simply not shown on a world map.

The character of the climate of a given highland area usually is closely related to that of the climate of the surrounding lowland, particularly the form of the annual temperature cycle and the times of occurrence of wet and dry seasons. An example of this effect in the tropical zone is shown by climographs for two Indian stations in close geographical proximity (Figure 8.23). New Delhi, the capital city, lies in the Ganges lowland. Simla, a mountain refuge from the hot weather, is located nearby at about 2000 m (about 6500 ft) in altitude in the foothills of the Himalayas. When the hot-

Figure 8.22 This small farm surrounded by sand dunes is an oasis of irrigated crops. Gansu Province, China.

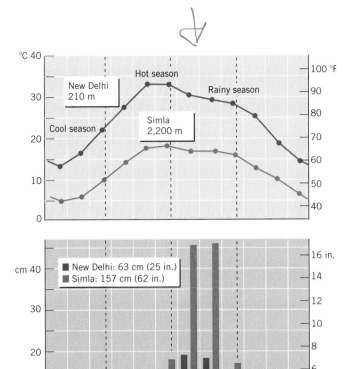

Figure 8.23 Climographs for New Delhi and Simla, both in northern India. Simla is a welcome refuge from the intense heat of the Gangetic Plain in May and June.

Figure 8.24 Potatoes are a mainstay of Peruvian agriculture in the high Andes. Here, a potato crop is being sorted.

season temperature averages over 32°C (90°F) in New Delhi, Simla is enjoying a pleasant 18°C (64°F), which is a full 14°C (25°F) cooler. Notice, however, that the two temperature cycles are quite similar in phase, with the minimum month being January for both. The annual rainfall cycles are also similar. New Delhi shows the typical rainfall pattern of the wet-dry tropical climate of Southeast Asia, with monsoon rains peaking in July and August. Simla has the same pattern, but the amounts are larger in every month, and the monsoon peak is very strong. Simla's annual total rainfall is well over twice that of New Delhi.

At high altitudes in tropical regions, native peoples have practiced agriculture in spite of a hostile environment that includes cold and intense solar radiation. The equatorial Andes provide an example. The high intermountain basins of the Bolivian Plateau, or Altiplano, are in the altitude range 3200 to 4300 m (about 10,500 to 14,000 ft) and are above the upper limits of the rainforest. Here, in small plots, corn, wheat, barley, and potatoes can be cultivated on a limited scale, and

nearby mountain pastures provide grazing for domesticated animals (Figure 8.24).

Low-latitude climates are a study in contrasts, from the wettest to the driest of climates and from the hottest to the most uniform of temperatures. They are dominated by two important features of atmospheric circulation—the intertropical convergence zone and the subtropical high-pressure cells. In our next chapter, we will turn to the climates of the midlatitudes and high latitudes. Although this group cannot claim the wettest and hottest of climates, it includes climates that are still very wet, very dry, and very hot, at least at certain times of the year. As we will see in the next chapter, the polar-front boundary and the contrast of warm, moist air masses of the subtropics with colder, drier air masses of the polar regions are the most important factors in influencing the temperature and precipitation cycles of midlatitude and high-latitude climates.

CHAPTER SUMMARY

Low-latitude climates are located largely between 30° N and S lat., and are controlled by the characteristics and annual movements of the ITC and the subtropical highs. These climates range from very moist to very dry and from quite uniform to very seasonal. In the wet equatorial climate ①, rainfall is abundant throughout the year because the ITC is always nearby. The annual temperature cycle is very weak, so that the daily range greatly exceeds the annual range. The monsoon and trade-wind coastal climate ② shows a dry period when the ITC has migrated toward the opposite tropic. With the return of the ITC, monsoon circulation and enhanced easterly waves provide increased rainfall in this climate type. Temperatures are highest in the dry weather before the onset of the wet season.

The equatorial climate ① and monsoon and trade-wind coastal climate ② provide the low-latitude rainforest environment, which is one of tall, broadleaved evergreen trees and great species diversity. The abundant rainfall yields freely flowing streams and rivers, and also acts to wash nutrients from the soil. The rainforest provides many products of economic value, including hardwoods, pharmaceuticals, and rubber. Manioc, yams, taro, banana, and coconut are native foodstuffs.

The wet-dry tropical climate ③ has a very dry period at the time of low sun and a wet season at the time of high sun, when the ITC is near. Temperatures peak strongly just before the onset of the wet season. Associated with this climate type is the savanna environment, including savanna woodland and thorntree-tall-grass savanna. The savanna is the home of the large grazing herbivores and carnivores that prey on them, but human pressure on the savanna environment is diminishing their habitat. In Asia, where the monsoon is strong, much of this climate zone is in rice cultivation. In Africa, the wetter savanna regions are in bush-fallow farming, yielding subsistence food crops and some exports. Cattle, raised for food, graze savanna grasses and fallow cropland. Extreme year-to-year variability in rainfall provides a constant threat of extreme drought, especially in the Sahel region of western and north-central Africa. The drought not only yields famine and disease, but also produces land degradation brought about by overgrazing and fuelwood harvesting.

The dry tropical climate ④ is dominated by subtropical high pressure and is often nearly rainless. Temperatures are very high during the high-sun season. While the semiarid subtype of this climate has a short wet season and supports a steppeland of sparse grasses, the arid subtype provides the tropical desert environment. In the desert, streams flow only after rare heavy rain showers, producing flash floods. Plants have adapted many strategies to survive the harsh environment, ranging from an annual habit to deep roots that tap groundwater. Human habitation is possible only where water is available.

Highland climates of low latitudes normally show a similar seasonality to lowland climates of nearby regions, but are cooler and wetter. Even on high plateaus well above the range of rainforest, agriculture is practiced.

KEY TERMS

wet equatorial climate ①
monsoon and trade-wind coastal climate ②
wet-dry tropical climate ③

dry tropical climate ④
steppes
highland climates

REVIEW QUESTIONS

1. Why is the annual temperature cycle of the wet equatorial climate ① so uniform?

2. Sketch the temperature and rainfall cycles for a typical station in the monsoon and trade-wind coastal climate ②. What factors contribute to the seasonality of the two cycles?

3. What are the main features of the low-latitude rainforest environment? Name some typical plant products of the rainforest. Which plants provide staple foods for inhabitants of the moist low-latitude environments?

4. The wet-dry tropical climate ③ has two distinct seasons. What factors produce the dry season? The wet season?

5. Describe the two main types of natural vegetation that occur in the savanna environment. Identify some typical animals of the African savanna.

6. How do native peoples of the African savanna utilize their environment?

7. What is meant by the term *land degradation*? Provide an example of a region in which land degradation has occurred.

8. Examine the Sahel rainfall graph carefully (Figure 8.16). Compare the pattern of rainfall fluctuations for the periods 1901–1950 and 1951–present. How do they differ?

9. Why are the dry tropical ④, dry subtropical ⑤, and dry midlatitude ⑨ climates dry? How do they differ in temperature and precipitation cycles?

10. How do the arid (a) and semiarid (s) subtypes of the dry climates differ?

11. Identify the key features of the tropical desert environment.

12. How do low-latitude highland climates differ from their counterparts at low elevation?

ESSAY QUESTIONS

1. The intertropical convergence zone (ITC) moves north and south with the seasons. Describe how this movement affects the four low-latitude climates.

2. Compare and contrast the low-latitude environments of Africa.

PROBLEMS

Working It Out 8.1 • Cycles of Rainfall in the Low Latitudes

1. The data below are measurements of total precipitation for November for the years 1985–1994 at San Juan, Puerto Rico. (November is the wettest month during this ten–year period.) Find the mean deviation of these data.

Year	*Precipitation, cm*
1985	11.5
1986	14.9
1987	19.0
1988	14.4
1989	12.6
1990	13.5
1991	15.6
1992	30.4
1993	11.0
1994	21.1

2. Calculate the relative variability by taking the ratio of the mean deviation to the mean. How does the result compare to those of the wettest months of Padang, Bombay, and Abbassia?

Chapter 9

Midlatitude and High-Latitude Climates

This chapter continues our examination of the global climates, focusing on midlatitude and high-latitude climates. As in Chapter 8, we will find that the temperature and precipitation cycles of the individual climates can be easily explained by the effects of latitude, position on the landmass, and movements of air masses and fronts.

MIDLATITUDE CLIMATES

The *midlatitude climates* almost fully occupy the land areas of the midlatitude zone and a large proportion of the subtropical latitude zone. Along the western fringe of Europe, they extend into the subarctic latitude zone as well, reaching to the 60th parallel. Unlike the low-lat-

An autumn scene from the moist continental climate ⑩ region. Fields, forests, vineyards, and a chateau in the distance provide an elegant and peaceful landscape. Jura region, eastern France.

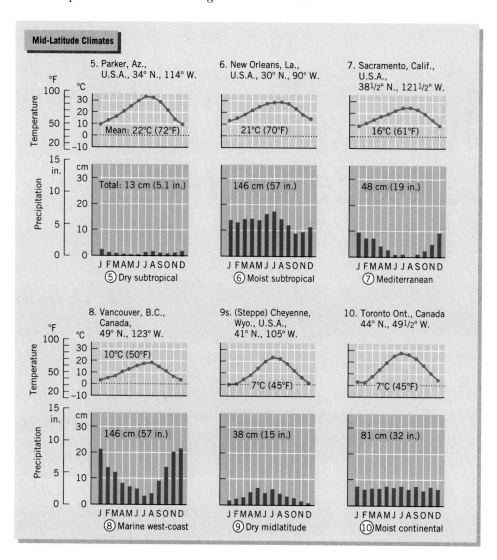

Figure 9.1 Climographs for the six midlatitude climates.

itude climates, which are about equally distributed between northern and southern hemispheres, nearly all of the midlatitude climate area is in the northern hemisphere. In the southern hemisphere, the land area poleward of the 40th parallel is so small that the climates are dominated by a great southern ocean and do not develop the continental characteristics of their counterparts in the northern hemisphere.

The midlatitude climates of the northern hemisphere lie in a broad zone of intense interaction between two groups of very unlike air masses (Figure 7.7). From the subtropical zone, tongues of maritime tropical (mT) air masses enter the midlatitude zone. There, they meet and conflict with tongues of maritime polar (mP) and continental polar (cP) air masses along the discontinuous and shifting polar-front zone.

In terms of prevailing pressure and wind systems, the midlatitude climates include the poleward halves of the great subtropical high-pressure systems and much of the belt of prevailing westerly winds. As a result, weather systems, such as traveling cyclones and their fronts, characteristically move from west to east. This

dominant global eastward airflow influences the distribution of climates from west to east across the North American and Eurasian continents.

The influence of traveling cyclones and the interaction between warm, moist air masses and cooler, drier air masses in the midlatitude climate zone makes for climates that are quite variable from day to day and month to month. *Working It Out 9.1 • Standard Deviation and Coefficient of Variation* uses monthly precipitation from stations in two different midlatitude climates to show how variability is measured quantitatively.

The six midlatitude climate types range from two that are very dry to three that are extremely moist (see Figure 9.1). The midlatitude climates span the range from those with strong wet and dry seasons—the Mediterranean climate ⑦—to those with precipitation that is more or less uniformly distributed through the year. Temperature cycles are also quite varied. Low annual ranges are seen along the windward west coasts. In contrast, annual ranges are large in the continental interiors. Table 9.1 summarizes the important features of these climates.

Table 9.1 Midlatitude Climates

Climate	Temperature	Precipitation	Explanation
Dry subtropical ⑤	Distinct cool or cold season at low-sun period.	Precipitation is low in nearly all months.	This climate lies poleward of the subtropical high-pressure cells and is dominated by dry cT air most of the year. Rainfall occurs when moist mT air reaches the region, either in summer monsoon flows or in winter frontal movements.
Moist subtropical ⑥	Temperatures show strong annual cycle, but with no winter month below freezing.	Abundant rainfall, cyclonic in winter and convectional in summer. Humidity generally high.	The flow of mT air from the west sides of subtropical high-pressure cells provides moist air most of the year. cP air may reach this region during the winter.
Mediterranean ⑦	Temperature range is moderate, with warm to hot summers and mild winters.	Unusual pattern of wet winter and dry summer. Overall, drier when nearer to subtropical high pressure.	The poleward migration of subtropical high-pressure cells moves clear, stable air cT into this region in the summer. In winter, cyclonic storms and polar frontal precipitation reach the area.
Marine west-coast ⑧	Temperature cycle is moderated by marine influence.	Abundant precipitation but with a winter maximum.	Moist mP air, moving inland from the ocean to the west, dominates this climate most of the year. In the summer, subtropical high pressure reaches these regions, reducing precipitation.
Dry midlatitude ⑨	Strong temperature cycle with large annual range. Summers warm to hot, winters cold to very cold.	Precipitation is low in all months but usually shows a summer maximum.	This climate is dry because of its interior location, far from mP source regions. In winter, cP dominates. In summer, a local dry continental air mass develops.
Moist continental ⑩	Summers warm, winters cold with three months below freezing. Very large annual temperature range	Ample precipitation, with a summer maximum.	This climate lies in the polar-front zone. In winter, cP air dominates, while mT invades frequently in summer. Precipitation is abundant, cyclonic in winter, and convectional in summer.

THE DRY SUBTROPICAL CLIMATE ⑤
(Köppen: *BWh, BWk, BSh, BSk*)

The **dry subtropical climate** ⑤ is simply a poleward extension of the dry tropical climate ④, caused by somewhat similar air-mass patterns. A point of difference is in the annual temperature range, which is greater for the dry subtropical climate ⑤. The lower latitude portions have a distinct cool season, and the higher latitude portions, a cold season. The cold season, which occurs at a time of low sun, is due in part to invasions of cold continental polar (cP) air masses from higher latitudes. Precipitation occurring in the low-sun season is produced by midlatitude cyclones that occasionally move into the subtropical zone. As in the dry tropical climate ④, both arid and semiarid subtypes are recognized.

Figure 8.20 shows the global distribution of the dry subtropical climate ⑤. A broad band of this climate type is found in North Africa, connecting with the Near East. Southern Africa and southern Australia also contain regions of dry subtropical climate ⑤ poleward of

Working It Out 9.1 • Standard Deviation and Coefficient of Variation

Although the mean, or average, of such variables as monthly precipitation or temperature is an important determiner of climate, the variability from one year to the next is also quite important. In *Working It Out 8.1 • Cycles of Rainfall in the Low Latitudes*, we introduced the average deviation as a useful measure of variation around a mean. We also provided the ratio of the average deviation to the mean as a measure of relative variability.

A more commonly used measure of variation is the *sample standard deviation*, which is defined for a sequence of n precipitation values P_i, $i = 1, \ldots, n$, in a sample as

$$s_p = \sqrt{\frac{1}{n} \sum_{i=1}^{n} D_i^{\,2}}$$

where s_p is the sample standard deviation of P, and $\overline{P}$ is the mean of the sample. D_i is the deviation of the ith observation from the sample mean, which was defined in Chapter 7 as $D_i = P_i - \overline{P}$. To find the sample standard deviation, simply take each deviation and square it. Then find the average of these squared values, and finally take the square root of that average. Note that any deviations that are negative will become positive after squaring. Thus, there is no need to take the absolute value of the deviation.

As an example of the sample standard deviation, examine the set of graphs shown to the right. Graphs (*a*) and (*b*) present monthly precipitation for two stations for a 10-year period from 1985 to 1994. San Jose, California, in the Mediterranean climate ⑦, has dry summers and wet winters and a low overall annual average precipitation of 36 cm (14 in.). Many months in the 10-year record are completely dry. Baton Rouge, Louisiana, in the moist subtropical climate ⑥, is a much wetter location, with an average annual precipitation of 173 cm (68 in.).

What about the variation in precipitation experienced at these stations? In a comparison of the two graphs, the Baton Rouge graph gives the impression of greater variation, largely because many months are much wetter than those of San Jose. Graph (*c*) compares the monthly sample standard deviations for each sample. In all months, the value is larger for Baton Rouge than for San Jose, in keeping with the impression gained from comparing graphs (*a*) and (*b*). The sample standard deviation for the annual totals, shown as the last pair of bars, is larger than the monthly values for both locations since the annual total itself is always larger than those of individual months.

The sample standard deviation does not measure the relative variation, however—that is, the variation with respect to the mean. Graph (*d*) shows the *coefficient of variation* for each location by month. The coefficient of variation for a sample is defined as the ratio of the standard deviation to the mean. That is,

$$CV_P = \frac{s_P}{\overline{P}}$$

where CV_P is the coefficient of variation for precipitation. The graph shows clearly that the relative variation of monthly rainfall at San Jose is larger than that of Baton Rouge in all months. Note that the coefficient of variation is highest in San Jose in the months of June, July and August. These are normally dry months, but in some years substantial rainfall occurs. Note also that the coefficients of variation for the annual totals (rightmost bars in graph *d*) are smaller than those of the monthly averages, showing that the relative variation of the total is less. This stands to reason, since a deficiency in one month can be made up in the next.

The mean, standard deviation, and coefficient of variation of a sample are values that give basic information about the sample and its variability. They are referred to as *sample statistics* and are widely used in many branches of science.

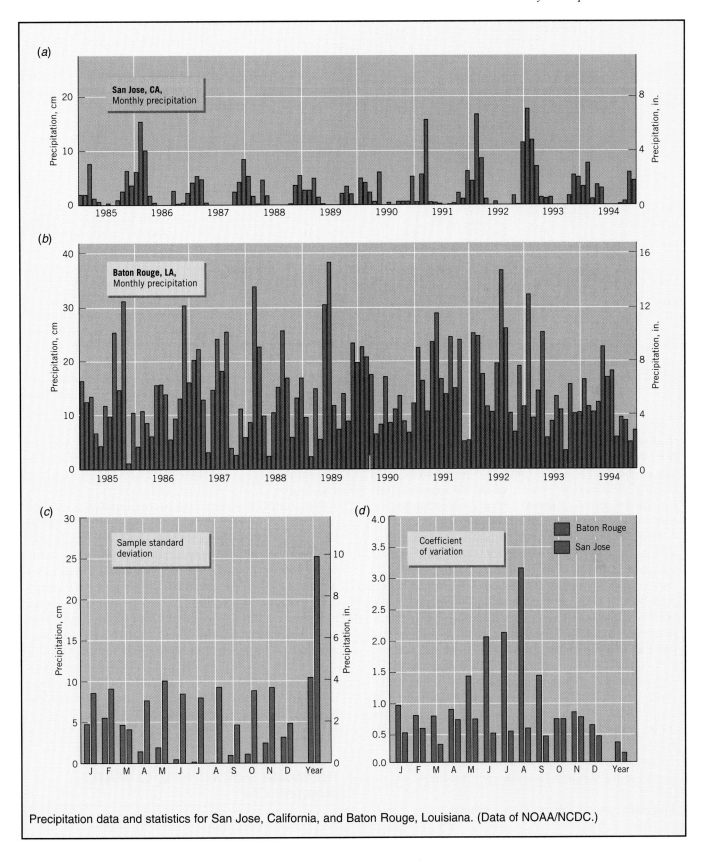

Precipitation data and statistics for San Jose, California, and Baton Rouge, Louisiana. (Data of NOAA/NCDC.)

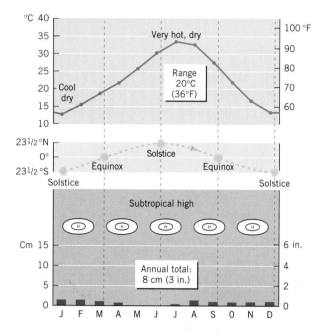

Figure 9.2 Dry subtropical climate ⑤. Yuma, Arizona, lat. 33° N, has a strong seasonal temperature cycle. Compare with Wadi Halfa (Figure 8.18).

the dry tropical climate ④. In South America, a band of dry subtropical climate ⑤ occupies Patagonia, a region east of the Andes in Argentina. In North America, the Mojave and Sonoran deserts of the American Southwest and northwest Mexico are of the dry subtropical type.

Figure 9.2 is a climograph for Yuma, Arizona, a city within the arid subtype of the dry subtropical climate ⑤, close to the Mexican border at lat. 33° N. The pattern of monthly temperatures shows a strong seasonal cycle with a dry, hot summer. A cold season brings monthly means as low as 13°C (55°F). Freezing temperatures—0°C (32°F) and below—can be expected at night in December and January. The annual range is 20°C (36°F). Precipitation, which totals about 8 cm (3 in.), is small in all months but has peaks in late winter and late summer. The August maximum is caused by the invasion of maritime tropical (mT) air masses, which bring thunderstorms to the region. Higher rainfalls from December through March are produced by midlatitude wave cyclones following a southerly path. Two months, May and June, are nearly rainless.

The Subtropical Desert Environment

The environment of the dry subtropical climate ⑤ is similar to that of the dry tropical climate ④ in that both are very dry. The boundary between these two climate types is gradational. But if we were to travel northward in the subtropical climate zone of North America, arriving at about 34° N in the interior Mojave Desert of

southeastern California, we would encounter environmental features significantly different from those of the low-latitude deserts of tropical Africa, Arabia, and northern Australia. Although the great summer heat of the low-elevation regions of the Mojave Desert is comparable to that experienced in the Sahara Desert, the low sun brings a winter season that is not found in the tropical deserts. In the Mojave Desert, cyclonic precipitation can occur in most months, including the cool low-sun months.

In the Mojave Desert (and adjacent Sonoran Desert), plants are often large and numerous, in some places giving the appearance of an open woodland. One example is the occurrence of forestlike stands of the tall, cylindrical saguaro cactus (see Figure 21.31). Another is the woodland of Joshua trees found in higher parts of the Mojave Desert (Figure 9.3). Other distinctive plants include the creosote bush and the ocotillo plant.

Both plants and animals of deserts are adapted to the dry environment. As in the dry tropical environment, many annual plants remain dormant as seeds during long dry periods, and then spring to life, flower, and bloom very quickly when rain falls. Certain invertebrate animals adopt the same life pattern. For example, the tiny brine shrimp of the North American desert may wait many years in dormancy until normally dry lakebeds fill with water, an event that occurs perhaps three or four times per century. The shrimp then emerge and complete their life cycles before the lake evaporates.

The human species adapts to the American subtropical desert environment by the application of technology. Irrigation projects import water in abundance, and growers lavishly apply it to croplands, where most of it is lost by intense evapotranspiration. Pipelines and highways provide fuels and building materials, machinery, and home appliances (the air conditioner). Transmission lines bring electricity from distant dams and coal-fired power plants. These human technological adaptations have run into serious environmental difficulties, one of which we will present in depth in Chapter 13. It is the deterioration of croplands by deposition of salts left by evaporation of irrigation waters and the accompanying waterlogging of the soil.

THE MOIST SUBTROPICAL CLIMATE ⑥ (Köppen: *Cfa*)

Recall that circulation around the subtropical high-pressure cells provides a flow of warm, moist air onto the eastern side of continents. This flow of maritime tropical (mT) air dominates the **moist subtropical climate** ⑥. Summer in this climate sees abundant rainfall, much of it convectional. Occasional tropical cyclones further enhance summer precipitation. In Southeast

Figure 9.3 A Mojave Desert landscape. The strange-looking Joshua tree, shown in the foreground, is abundant at higher elevations in the Mojave Desert.

Asia, this climate is characterized by a strong monsoon effect, with summer rainfall much increased above winter rainfall. Summer temperatures are warm, with persistent high humidity.

Winter precipitation in the moist subtropical climate ⑥ is also plentiful, produced in midlatitude cyclones. Invasions of continental polar (cP) air masses are frequent in winter, bringing spells of subfreezing weather. No winter month has a mean temperature below 0°C (32°F).

Figure 9.4 presents a global map of the moist subtropical climate ⑥. It is found on the eastern sides of continents in the latitude range 20° to 35° N and S. In South America, it includes parts of Uruguay, Brazil, and Argentina. In Australia, it consists of a narrow band between the eastern coastline and the eastern interior ranges. Southern China, Taiwan, and southernmost Japan are regions of the moist subtropical climate ⑥ in

Asia. In the United States, the moist subtropical climate ⑥ covers most of the Southeast, from the Carolinas to east Texas.

Figure 9.5 is a climograph for Charleston, South Carolina, located on the eastern seaboard at lat. 33° N. In this region, a marked summer maximum of precipitation is typical. Total annual rainfall is abundant—120 cm (47 in.)—and ample precipitation falls in every month. The annual temperature cycle is strongly developed, with a large annual range of 17°C (31°F). Winters are mild, with the January mean temperature well above the freezing mark.

The Moist Subtropical Forest Environment

The abundant rainfall of the moist subtropical climate ⑥ provides sufficient soil water for crops without irrigation in most years. Rivers and streams flow copiously

Figure 9.4 World map of the moist subtropical climate ⑥. (Based on Goode Base Map).

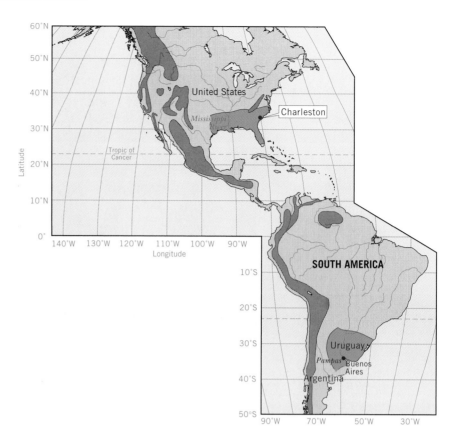

Figure 9.5 Moist subtropical climate ⑥. Charleston, South Carolina, lat. 33° N, has a mild winter and a warm summer. There is ample precipitation in all months but a definite summer maximum.

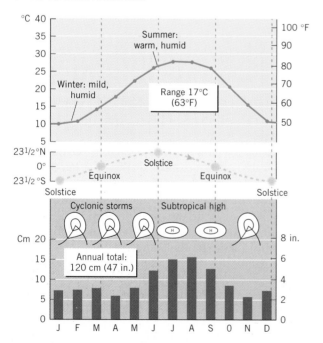

through much of the year. Flooding can be severe at times from tropical cyclones that come inland and produce torrential rains in the high-sun months. The large water surplus of the moist subtropical climate has important implications in terms of economic development. The large flows of rivers can furnish abundant freshwater resources for urbanization and industry without competition from irrigation demands. Evaporative losses from reservoirs are much less important than in arid lands. The maintenance of ample stream flows tends to reduce the dangers of severe water pollution and its adverse effects on ecosystems of streams and estuaries.

Much of the natural vegetation in the North American area of this moist climate type consists of *broadleaf deciduous forest*. (In a deciduous forest, leaf shedding occurs annually and the trees are bare throughout the winter season.) The oak tree is a typical representative of this forest. Broadleaf deciduous forest can be found on the interior uplands of the Carolinas and across Tennessee. Today, of course, large parts of this formerly forested zone have been replaced by agricultural croplands. An entirely different forest type occurs farther south and east on the low, sandy coastal plain of the Gulf states, on the Florida peninsula, and on the Atlantic coastal plain of Georgia and the Carolinas. It is the *southern pine forest*, which is well adapted to sandy

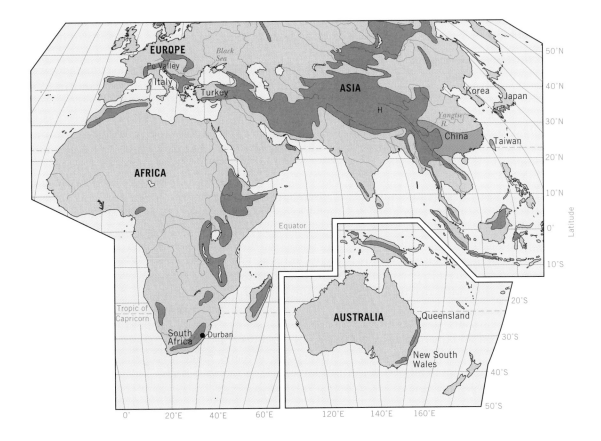

soils. (See world vegetation map, Figure 21.15.) Pines belong to the class of needleleaf trees, and they retain green foliage throughout the year.

Over a large part of southern China and the south island of Japan, the native vegetation was formerly a broadleaf forest of the evergreen type, in which the leaf canopy remains green throughout the year. It is called the *broadleaf evergreen forest* (Chapter 21). Today this forest is gone from China, but some small areas in Japan remain, allowing us to reconstruct its composition. It contained a large number of species of evergreen oaks, as well as species of the magnolia tree and the camphor tree, all peculiar to that part of the world. On the North American continent, in Louisiana and farther east on the Gulf coast, the evergreen broadleaf forest is represented by a few small areas of Evangeline oak and magnolia (Figure 9.6).

The warm summer climate and high rainfall of the moist subtropical environment tends to wash nutrients from the soil layer. As in the soils of the moist low-latitude climates, iron oxides accumulate in the soil, giving it colors ranging from yellow to red. For agricultural crops, especially grains or cereals, the soils of the moist subtropical environment rate as low in fertility and require large applications of fertilizers. Another unfavorable factor is the susceptibility of these soils to severe erosion and gullying when exposed by forest removal

Figure 9.6 Broadleaf evergreen forest of the Gulf coast region is represented here by Evangeline oaks bearing Spanish "moss," an epiphyte that forms long beardlike streamers. The ground beneath is maintained as a lawn. Evangeline State Park, Bayou Teche, Louisiana. (A. N. Strahler.)

and intensive cultivation. We will discuss these problems of soil fertility and erodibility in Chapter 19.

Agricultural Resources of the Moist Subtropical Forest Environment

In the two major regions of moist subtropical climate ⑥—North America and Southeast Asia—agricultural use of the land follows quite different patterns. The differences are partly historical and cultural, but also reflect the stronger monsoon effect in Asia, which causes a stronger concentration of precipitation in the summer. The human population is vastly denser in Southeast Asia than in the New World and subsists largely on rice, the dominant staple food crop. Two and even three rice crops are harvested annually in southern China. The rice crop is often followed by planting of wheat, winter legumes, peas, or green fertilizer crops. Except in the Mississippi delta region, little rice is produced in the southern United States. Both American and Asiatic regions produce sugarcane, peanuts, tobacco, and cotton, although not on an equal intensity in both regions. One striking difference is that tea is widely cultivated in Southeast Asia, but not at all in the southern United States. Corn is an important crop in the southern United States that is increasing in importance in southern China.

The potential of the moist subtropical climate to produce more food rests in more intensive land use instead of expansion of the area now under cultivation. The best land is already in use, and, in Southeast Asia, elaborate terrace systems have been used for centuries to allow farming of steep hill slopes. New genetic strains of rice and corn offer promise of increased yields when the necessary fertilizers are used. Japan and Taiwan apply very high levels of fertilizers and achieve high rice yields. The People's Republic of China is only now in the process of sharply increasing its production and use of fertilizers from comparatively low levels of the recent past. China is also developing independently some new high-yielding strains of rice and wheat.

In the southern United States, cattle production is another source of increased food production and makes use of soils too sandy for field crops. With soil water frequently replenished through the long, warm summer, pasture and range land can be continuously productive. Tree farming is also an important use of

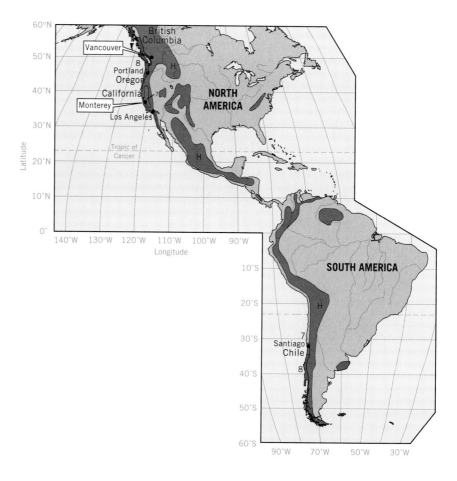

Figure 9.7 World map of the Mediterranean ⑦ and marine west coast ⑧ climates. (Based on Goode Base Map.)

sandy soils. Pines are well adapted to rapid growth on sandy soils and thrive where nutrient bases are in short supply.

THE MEDITERRANEAN CLIMATE ⑦
(Köppen: *Csa, Csb*)

The **Mediterranean climate** ⑦ is unique among the climate types because its annual precipitation cycle has a wet winter and a very dry summer. The reason for this precipitation cycle lies in the poleward movement of the subtropical high-pressure cells during the summer season. The Mediterranean climate ⑦ is located along the west coasts of continents, just poleward of the dry, eastern side of the subtropical high-pressure cells (Figures 5.13, 5.14). When the subtropical high-pressure cells move poleward in summer, they enter the region of this climate. Dry continental tropical (cT) air then dominates, producing the dry summer season. In winter, the moist mP air mass invades with cyclonic storms and generates ample rainfall.

In terms of total annual rainfall, the Mediterranean climate ⑦ spans a wide range from arid to humid, depending on location. Generally, the closer an area is to the tropics, the stronger the influence of subtropical high pressure will be, and thus the drier the climate. The temperature range is moderate, with warm to hot summers and mild winters. Coastal zones between lat. 30° and 35° N and S, such as southern California, show a smaller annual range, with very mild winters.

A global map of the Mediterranean climate ⑦ is shown in Figure 9.7. It is found in the latitude range 30° to 45° N and S. In the southern hemisphere, it occurs along the coast of Chile, in the Cape Town region of South Africa, and along the southern and western coasts of Australia. In North America, it is found in central and southern California. In Europe, this climate type surrounds the Mediterranean Sea, which gives the climate its distinctive name.

Figure 9.8 is a climograph for Monterey, California, a Pacific coastal city at lat. 36° N. The annual temperature cycle is very weak. The small annual range and cool summer reflect the strong control of the cold California current and its cool, marine air layer. Fogs are

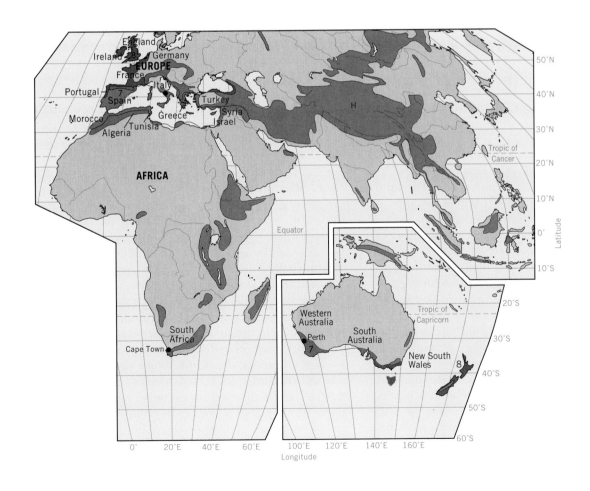

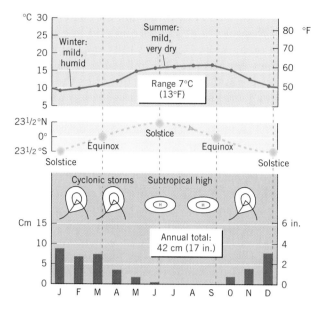

Figure 9.8 Mediterranean climate ⑦. Monterey, California, lat. 36° N, has a very weak annual temperature cycle because of its closeness to the Pacific Ocean. The summer is very dry.

Figure 9.9 This bark, stripped from the cork oak (*Quercus suber*), will be ground up and cemented into cork board and other structural products. Thick bark of choice quality is used for wine corks. Algeria, North Africa. (A. N. Strahler.)

frequent. This temperature regime is found only in a narrow coastal zone. Rainfall drops to nearly zero for four consecutive summer months but rises to substantial amounts in the rainy winter season. In California's Central Valley, the dry summers persist, but the annual temperature range is greater, so that daily high temperatures match those of the adjacent dry climates.

The Mediterranean Climate Environment

The Mediterranean climate ⑦ is a very attractive one for human habitation. The attraction lies in the thermal cycle of year-round pleasant temperatures, especially where moderated by coastal influences. The mild winters with abundant sunshine (despite periods of substantial rainfall) are a welcome refuge from the severe winters of the midlatitude continental interiors of Eurasia and North America. However, the low annual precipitation with dry summers makes fresh water scarce, requiring a large investment in technology to deliver enough water to support the heavy load of humanity.

Soil fertility in valley and lowland areas of the Mediterranean climate is naturally high, as it usually is in semiarid climates of the midlatitudes. Because overall rainfall is low, the nutrients that are essential to forage grasses and grains, fruit trees, vegetables, and many other varieties of plants are not washed out of the soil. The soils native to the Mediterranean environment belong to a special category of subclasses, explained in Chapter 20, and are not easily described in a few words.

The native vegetation of the Mediterranean climate environment is adapted to survival through the long

summer drought. Shrubs and trees are typically equipped with small, hard or thick leaves that resist water loss through transpiration. These plants are called *sclerophylls*; the prefix scler, from the Greek for "hard," is combined with phyllo, which is Greek for "leaf." Sclerophylls of the Mediterranean environment are typically evergreen, retaining their leaves through the entire yearly cycle. Examples are the evergreen oaks, of which there are several common species in California, and the cork oak of the Mediterranean lands (Figure 9.9). Another is the olive tree, native to the Mediterranean lands. In Australia, the thick-leaved eucalyptus tree is the dominant sclerophyll. Oak woodland of California bears a ground cover of grasses that turn to straw in the summer (Figure 9.10).

Another form of native vegetation in this environment is a cover of drought-resistant shrubs, including sclerophylls and spiny-leaved species. In the Mediterranean lands, this scrub vegetation goes under the name of maquis or garrigue. In California, where it is called *chaparral*, it clothes steep hill and mountain slopes too dry to support oak woodland or oak forest (Figure 9.11).

Wildfire is an integral part of the Mediterranean environment of California. Chaparral is extremely flam-

Figure 9.10 California oak woodland in summer. Blue oaks (*Quercus douglasii*) are scattered over a cover of dry grasses in the inner Coast Ranges, near Williams, California.

mable during the long summer fire season (Figure 5.18). Brushfires rage through chaparral and oak forests and leave the soil surface bare and unprotected. When torrential rains occur in winter, large quantities of coarse mineral debris are swept downslope and carried long distances by streams in flood. Mudflows and debris floods (usually called "mudslides" in the news media) are particularly destructive to communities on canyon floors.

Throughout the Mediterranean lands of Europe, North Africa, and the Near East, devastating soil ero-sion, induced by human activity over the past 2000 years or longer, has left its scars on the landscape. Many hillsides have been denuded of their soils and are now rocky and barren. Sediment, representing the displaced soil, has formed thick layers of sand and silt in adjacent valley floors.

Agriculture in the Mediterranean Environment

Lands bordering the Mediterranean Sea produce cereals—wheat, oats, and barley—where arable soils are ex-

Figure 9.11 California chaparral. Wiry shrubs cloak the steep slopes of a coast range near Acton, California.

Figure 9.12 Groves of lemon, orange, and avocado trees surround these homes and stables in Montecito, California. Chaparral covers the steep slopes of the Santa Ynez Mountains in the background. (A. N. Strahler.)

tensive enough to be cultivated. However, we usually think of that region as an important source of citrus fruits, grapes, and olives for European markets. Cork from the bark of the cork oak is also a product of economic value (Figure 9.9). In central and southern California, citrus, grapes, avocados, nuts (almond, walnut), and other fruits are grown extensively (Figure 9.12). Irrigated lowland soils are also highly productive of vegetable crops, such as carrots, lettuce, cauliflower, broccoli, artichokes, and strawberries, as well as sugar beets and forage crops (alfalfa). Cattle ranching and sheep grazing are of major importance on grassy hill slopes unsuited to field crops and orchards and on irrigated lowland pastures.

Because the Mediterranean environment is limited in global extent to comparatively small land areas, it offers little prospect for the expansion of croplands to provide major additions to the world's food supply. Irrigation is essential for high productivity, but there are hazards associated with heavy irrigation of lowland soils: salt accumulation and waterlogging (see Chapter 19). Urbanization and industrial development also face major problems of obtaining additional water supplies through importation over long distances by aqueduct. Nevertheless, the mild, sunny climate of southern California has proved a powerful population magnet, and water importation has already been developed on a mammoth scale. Expansion of suburban housing has, however, begun to take over rich, flat croplands, reducing the agricultural potential in a number of areas.

Rainfall in the Mediterranean climate ⑦ can be quite variable from year to year. Sometimes the weather patterns that provide winter precipitation fail to appear, or appear infrequently, leading to drought. In other years, precipitation may be much greater than

normal. *Focus on Systems 9.2 • California Rainfall Cycles and El Niño* provides an example—rainfall in Santa Barbara over a 22-year period—and examines its possible relation to such global climate events as El Niño cycles and volcanic eruptions.

THE MARINE WEST-COAST CLIMATE ⑧ (Köppen: *Cfb, Cfc*)

The **marine west-coast climate** ⑧ occupies midlatitude west coasts. These locations receive the prevailing westerlies from over a large ocean and experience frequent cyclonic storms involving cool, moist mP air masses. Where the coast is mountainous, the orographic effect causes a very large annual precipitation. In this moist climate, precipitation is plentiful in all months, but there is often a distinct winter maximum. In summer, subtropical high pressure extends poleward into the region, reducing rainfall. The annual temperature range is comparatively small for midlatitudes. The marine influence keeps winter temperatures mild, as compared with inland locations at equivalent latitudes.

The global map of the marine west-coast climate ⑧ (Figure 9.7) shows the areas in which this climate occurs. In North America, the climate occupies the western coast from Oregon to northern British Columbia. In Western Europe, the British Isles, Portugal, and much of France fall into the marine west-coast climate. New Zealand and the southern tip of Australia, as well as the island of Tasmania, are marine west-coast climate regions found in the southern hemisphere, as is the Chilean coast south of 35° S. The general latitude range of this climate is 35° to 60° N and S.

Figure 9.13 is a climograph for Vancouver, British

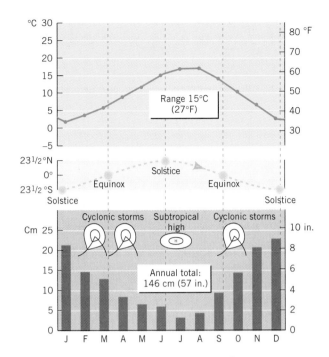

Figure 9.13 Marine west-coast climate ⑧. Vancouver, British Columbia, lat. 49° N, has a large annual total precipitation, but with greatly reduced amounts in the summer. The annual temperature range is small, and winters are very mild for this latitude.

Columbia, just north of the U.S.-Canadian border. The annual precipitation is very great, and most of it falls during the winter months. Notice the greatly reduced rainfall in the summer months. The temperature cycle shows a remarkably small range for this latitude. Even the winter months have averages above the freezing mark.

The Marine West-Coast Environment

Because of the abundant precipitation, lowland soils of marine west-coast climate ⑧ regions show a loss of nutrients, but in Europe have still retained moderate fertility. Applications of fertilizers and lime are needed for bountiful crop production, and in Europe these soils have been successfully cultivated for centuries. Much of the land surface within the marine west coast environment in northern Europe, British Columbia, southern Chile, and the South Island of New Zealand is on mountainous slopes that have been heavily scoured by the ice sheets and mountain glaciers of the recent Ice Age. Soils of these glaciated areas are extremely young and are poorly developed.

Forest is the native vegetation of this environment. Dense needleleaf forests of redwood, fir, cedar, hemlock, and spruce (Figure 9.14) flourish in the wet

Figure 9.14 Needleleaf forest of Sitka spruce and hemlock in the Hoh rainforest, Olympic National Park, Washington. Ferns and mosses provide a lush ground cover in this very wet environment.

Focus on Systems 9.2 • California Rainfall Cycles and El Niño

California's coastal zone from the San Francisco Bay area as far south as San Diego lies in a coastal subtype of the Mediterranean climate ⑦. It has a long, nearly rainless summer season, often cursed by disastrous wildfires. Relief from the summer drought comes in a few winter months that can bring heavy rainfalls, and with them disastrous stream floods and mudflows. Substantial amounts of rain can fall in November, or even earlier, but the months of heavy rainfall totals are December through March. In some years, heavy rainfall may continue well into April.

Our upper graph (*a*) shows the year's total rainfall for Santa Barbara, a coastal city lying about midway between San Francisco and San Diego, for the period 1963–1994. The rain gauge is positioned at low elevation not far from the coast. Values are shown for the water year. This period begins on October 1 and ends on September 30 in order to place all winter rainfall during the same recording period. The year-to-year relative variability is obviously great, with high peaks rising far above the mean value of 44.8 cm (17.6 in.). The time cycle of these annual accumulations seems to be irregular in both its period and its amplitude. (See *Working It Out 8.1 • Cycles of Rainfall in the Low Latitudes.*) Forecasting ahead into the winter rain season from one year to the next is a most unreliable undertaking, with no obvious repeating period to rely on.

On the other hand, rainfall forecasting for Santa Barbara could stand a good chance of being reasonably reliable if climatologists were to discover a phenomenon

with a similar pattern displayed several months in advance of the rain season. We would call it a "precursory event"—one that precedes and predicts an event to come. What precursory weather phenomenon might this be? Surely you can guess, for it is described in an earlier chapter. El Niño, in combination with La Niña, might serve as the precursory event. We can read these warning signals from a graph of the Southern Oscillation, which is under continuous observation. For the Southern Oscillation signal to be effective in forecasting, it must occur at least a few months in advance of the start of the rain season.

Our second graph (*b*) shows the recorded monthly values of the Southern Oscillation Index for the same period of years as the Santa Barbara rainfall graph. The index is the difference between the monthly barometric pressure at Darwin, Australia, and at the Island of Tahiti. Note that the *y*-axis is reversed, so that negative values of the index (*i.e.*, El Niños) can be correlated more easily with rainfall peaks. A persistent change in the upward (negative) direction signals El Niño, while such a change in the downward (positive) direction signals La Niña.

We have aligned our two graphs so that you can see whether there is any consistent association of the El Niño/La Niña pressure signals with strong highs and lows in the rainfall graph. The two wettest years—1978 and 1983—seem nicely related to El Niños. The Southern Oscillation Index begins dropping some months before the onset of the wet year, thus signaling the excess rain-

fall. However, the third wettest year, 1969, is not well signaled and is only associated with weak El Niño conditions. Furthermore, strong El Niños in 1965–1966 and 1987 are not associated with abnormally high rainfalls.

What about the relation between dry years and La Niña? Obvious La Niña events occur in 1971, 1974, 1975–1976, and 1989. While these do not seem strongly related to individual low-rainfall years, at least we can say that the annual rainfall during La Niña periods is always below the mean to some degree.

From these comparisons, we can see that while there is some relationship between Santa Barbara rainfall and the Southern Oscillation Index, it is not all that reliable. What other signals might predict a wet year for Santa Barbara? One possibility would be a rise in sea-surface temperature levels in the northern hemisphere of the eastern Pacific. Warmer sea-surface temperatures might add more moisture to maritime air masses that are drawn into occluded cyclones, which in turn provide much of the heaviest coastal rainfall to Southern California. Although there is no "sea-surface temperature index" for the northeastern Pacific to rely on, sea-surface temperature is now continuously monitored by satellite instruments and such an index might be developed.

Another possible controlling factor is volcanic eruptions. Marked on the upper graph are the eruptions of El Chichon (1982) and Pinatubo (1991). Both of these events propelled significant volumes of aerosols into the stratosphere, although the volume of Pinatubo aerosols was three times greater. (Recall from Chapter 3 that volcanic aerosols reflect more sunlight back to space, thus cooling the earth's surface and lower atmosphere, and so impacting global climate.) Whereas El Chichon has a perfect correspondence with a high rainfall at Santa Barbara in the following winter, Pinatubo shows no rainfall response in the first following year. Again we are frustrated!

Whatever the precursory event may prove to be, uncovering such teleconnections between weather and climate phenomena in different parts of the world is a challenging and fascinating pastime that can be of real value to human populations at the mercy of the vagaries of climate.

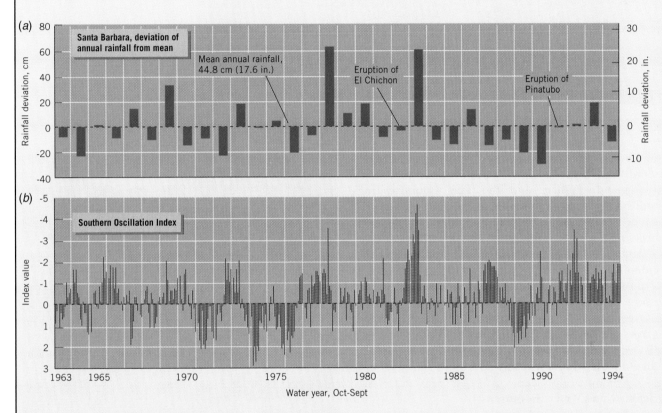

Rainfall in Santa Barbara compared with the Southern Oscillation Index. Note that the scale of the index is reversed for easier comparison with the rainfall data. (Data of NOAA/NCDC.)

Figure 9.15 This lush landscape near Devon, England, is located in the domain of the summergreen deciduous forest. After many centuries of human occupation and cultivation, the forest is now reduced to small patches and scattered individual trees in hollows and along fence rows.

mountainous areas of the northern Pacific coast. Under the lower precipitation regime of Ireland, southern England, France, and the Low Countries, a broadleaf deciduous forest was the native vegetation, but much of it disappeared many centuries ago under cultivation. Only scattered forest plots or groves remain (see Figure 9.15). Sometimes called "summergreen" deciduous forest, it is dominated by tall broadleaf trees that provide a continuous and dense canopy in summer but shed their leaves completely in winter. Dominant tree species of this forest type in Western Europe are oak and ash, with beech in the cooler and moister areas. Figure 21.16 is a map showing the distribution of midlatitude deciduous forests.

Agriculture and Water Resources

The marine west-coast environment of Western Europe and the British Isles has been intensively developed for centuries for such diverse uses as crop farming, dairying, orchards, and forest (Figure 9.15). It is an environment in most respects similar in agricultural character and productivity to the moist continental climate environment with which it merges on the east.

In North America, the mountainous terrain of the coastal belt offers only limited valley floors for agriculture, which is generally diversified farming. Forests are the primary plant resource, and here they constitute perhaps the greatest structural and pulpwood timber resource on earth. Douglas fir, western cedar, and western hemlock are the principal lumber trees of the Pacific Northwest (Figure 9.16). The same mountainous terrain that limits agriculture in the Pacific Northwest is a producer of enormous water surpluses that run to the sea in rivers. At various times in the past, schemes

Figure 9.16 The marine west-coast climate ⑧ of North America is the natural home of vast forests of evergreen needleleaf conifers. This photo shows clearcut patches of forest in British Columbia, on Vancouver Island.

have been devised to tap these surpluses and send the water south and east to drier regions, but so far no plan has been implemented.

THE DRY MIDLATITUDE CLIMATE ⑨
(Köppen: *BWk, BSk*)

The **dry midlatitude climate** ⑨ is limited almost exclusively to interior regions of North America and Eurasia, where it lies within the rainshadow of mountain ranges on the west or south. Maritime air masses are effectively blocked out much of the time, so that the continental polar (cP) air mass dominates the climate in winter. In summer, a dry continental air mass of local origin is dominant. Summer rainfall is mostly convectional and is caused by occasional invasions of maritime air masses. The annual temperature cycle is strongly developed, with a large annual range. Summers are warm to hot, but winters are cold to very cold.

The largest expanse of the dry midlatitude climate ⑨ is in Eurasia, stretching from the southern republics of the former Soviet Union to the Gobi Desert and northern China (Figure 8.20). In the central portions of this region lie true deserts of the arid climate subtype, with very low precipitation. Extensive areas of highlands occur here as well. In North America, the dry western interior regions, including the Great Basin, Columbia Plateau, and the Great Plains, are of the semiarid subtype. A small area of dry midlatitude climate ⑨ is found in southern Patagonia, near the tip of South America. The latitude range of this climate is 35° to 55° N.

Figure 9.17 is a climograph for Pueblo, Colorado, a semiarid station located at lat. 38° N, just east of the Rocky Mountains. Total annual precipitation is 31 cm (12 in.). Most of this precipitation is in the form of convectional summer rainfall, which occurs when moist maritime tropical (mT) air masses invade from the south and produce thunderstorms. In winter, snowfall is light and yields only small monthly precipitation averages. The temperature cycle has a large annual range, with warm summers and cold winters. January, the coldest winter month, has a mean temperature just below freezing.

The Dry Midlatitude Environment

The arid subtype of the dry midlatitude climate ⑨a is restricted to the driest of midlatitude continental interiors. As an extension of the arid subtype of the dry subtropical climate ⑤a, this cold desert environment supports only desert vegetation, or in North America, a cover of sagebrush and associated low woody shrubs.

Much more extensive is the steppe subtype ⑨s, which is found in large regions of interior North America and Central Asia. In this subtype, low annual precipitation combined with a strongly continental thermal regime has produced soils of high natural fertility that retain large supplies of the nutrient elements, including calcium, magnesium, potassium, and sodium. These soils are moderately to strongly alkaline, in contrast to soils of the moister midlatitude climates that are acid in chemical balance. (See Chapter 19 for details.)

Grasses thrive on such nutrient-rich soils in climates of low precipitation, and thus the native vegetation of the dry midlatitude steppe consists principally of hardy perennial short grasses capable of enduring severe summer drought. In North America, this cover is termed *short-grass prairie*, rather than steppe. Figure 21.26 maps the areas of short-grass prairie.

Agricultural Resources of the Short-Grass Prairie

Wheat is perhaps the most important single crop produced in unirrigated areas of short-grass prairie bordering on moister climate zones. One such important wheat-producing region lies in southern Alberta and Saskatchewan and in the northern border region of Montana. Another is the Palouse Hills region of southeastern Washington State and western Idaho (Figure 9.18). Here the crop is spring wheat, which is planted in the spring of the year. Using the water remaining in the soil from winter rains and snows, and the precipitation that falls in late spring and early summer, the crop is able to reach maturity for midsummer harvesting. In Ukraine, the country's rich wheat-growing area contin-

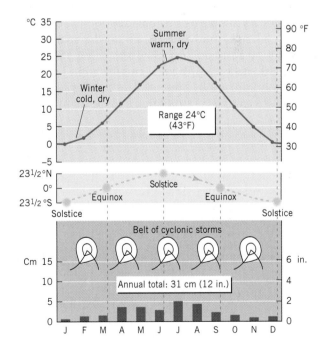

Figure 9.17 Dry midlatitude climate. Pueblo, Colorado, lat. 38° N, shows a marked summer maximum of rainfall in the summer months. Note also the cold, dry winter season. Figure 8.20 shows the location of this station.

Figure 9.18 Wheat farming in the Palouse Hills region of eastern Washington.

ues in a narrow zone far eastward across the steppes of Kazakhstan. In northern China, wheat is grown within a steppe region bordering the moist continental climate.

Wheat production of the midlatitude steppes is very much at the mercy of variations in seasonal rainfall. Good years and poor years follow cyclic variations. Soil water, not soil fertility, is the key to wheat production over these vast steppe lands. In the High Plains, there has been a great increase in recent decades in the use of groundwater pumped to the surface and distributed by irrigation systems. This groundwater source is rapidly being depleted and will ultimately fail.

Semiarid steppes form the great sheep and cattle ranges of the world. The steppes of Central Asia have for centuries supported a nomadic population whose sheep and goats find subsistence on the sparse grassland. On the vast expanses of the High Plains, the American bison lived in great numbers until being almost exterminated by hunters. The short-grass veldt of South Africa also supported much game at one time.

EYE ON THE ENVIRONMENT:

Drought and the Dust Bowl

Steppe grasses do not form a complete sod cover—loose, bare soil is exposed between the grass clumps. For this reason, overgrazing during a series of dry years can easily reduce the hold of grasses enough to permit destructive deflation (wind erosion), followed by water erosion and gullying.

On the Great Plains of Kansas, Oklahoma, Colorado, Texas, and New Mexico, deflation and soil drifting reached disastrous proportions during a series of drought years in the middle 1930s, following a great expansion of wheat cultivation. During the drought a sequence of exceptionally intense dust storms occurred (Figure 9.19). Within their formidable black clouds, visibility declined to nighttime darkness, even at noonday. The affected area, which also included part of the adjacent moist continental climate, became known as the *Dust Bowl.* Many centimeters of soil were removed from

Figure 9.20 World map of the moist continental climate ⑩. (Based on Goode Base Map.)

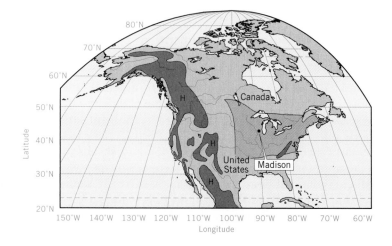

Figure 9.19 This swiftly moving black cloud signals the arrival of a stifling dust storm. Dust Bowl, Great Plains, ca. 1934.

fields and transported out of the region as suspended dust, while the coarser silt and sand particles accumulated in drifts along fence lines and around buildings. The combination of environmental degradation and repeated crop failures caused widespread abandonment of farms and a general exodus of farm families.

Among geographers who have studied the Dust Bowl phenomenon, there is a difference of opinion as to how great a role soil cultivation and livestock grazing played in inducing deflation. The drought was a natural event over which humans had no control, but it seems reasonable that the natural grassland would have sustained far less soil loss and drifting if it had not been destroyed by the plow.

Although we cannot prevent cyclic occurrences of drought over the Great Plains, measures can be taken to minimize the deflation and soil drifting occurring in

periods of dry soil conditions. Improved farming practices include use of deeply carved furrows that act as traps to soil movement. Leaving the stubble from harvested grain crops reduces deflation when land is lying fallow, and tree belts may have significant effect in reducing the intensity of wind at ground level.

THE MOIST CONTINENTAL CLIMATE ⑩ (Köppen: *Dfa, Dfb, Dwa, Dwb*)

The **moist continental climate** ⑩ is located in central and eastern parts of North America and Eurasia in the midlatitudes. This climate lies in the polar-front zone—the battleground of polar and tropical air masses. Seasonal temperature contrasts are strong, and day-to-day weather is highly variable. Ample precipitation

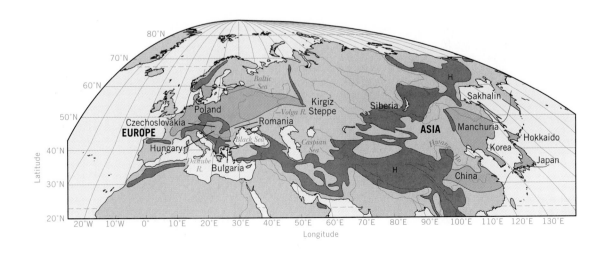

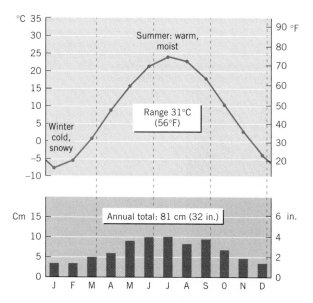

Figure 9.21 Moist continental climate ⑩. Madison, Wisconsin, lat. 43° N, has cold winters and warm summers, making the annual temperature range very large.

throughout the year is increased in summer by invading maritime tropical (mT) air masses. Cold winters are dominated by continental polar (cP) and continental arctic (cA) air masses from subarctic source regions.

In eastern Asia—China, Korea, and Japan—the seasonal precipitation pattern shows more summer rainfall and a drier winter than in North America. This is an effect of the monsoon circulation, which moves moist maritime tropical (mT) air across the eastern side of the continent in summer and dry continental polar southward through the region in winter. In Europe, the moist continental climate ⑩ lies in a higher latitude belt (45° to 60° N) and receives precipitation from mP air masses coming from the North Atlantic.

Figure 9.20 shows the global locations of the moist continental climate ⑩. It is restricted to the northern hemisphere, occurring in latitudes 30° to 55° N in North America and Asia, and in latitudes 45° to 60° N in Europe. In Asia, it is found in northern China, Korea, and Japan. Most of Central and Eastern Europe has a moist continental climate, as does most of the

Figure 9.22 The Mosel River in its winding, entrenched meandering gorge through the Rhineland-Pfalz province of western Germany. On the steep, undercut valley wall at the right are vineyards and forested land.

eastern half of the United States from Tennessee to the north, as well as the southernmost strip of eastern Canada.

Madison, Wisconsin, lat. 43° N, in the American Midwest (Figure 9.21), provides an example of the moist continental climate ⑩ (Figure 9.22). The annual temperature range is very large. Summers are warm, but winters are cold, with three consecutive monthly means well below freezing. Precipitation is ample in all months, and the annual total is large. There is a summer maximum of precipitation when the maritime tropical (mT) air mass invades, and thunderstorms are formed along moving cold fronts and squall lines. Much of the winter precipitation is in the form of snow, which remains on the ground for long periods.

The Moist Continental Forest and Prairie Environment

With ample precipitation throughout the year and only a short period of summer dryness, the moist continental climate supports forests as the native vegetation. Soils beneath these forests show the effects of the moist environment through the washing out of soil nutrients and other soil components and a strong tendency to soil acidity. These effects are most severe in the colder, more northerly parts of the climate zone, where strongly acidic soils are found on sandy surface layers. Here the evergreen needleleaf forest dominates—the pine forest of the Great Lakes region is an example. Throughout much of the northeastern United States and southeastern Canada, mixed coniferous and deciduous forest is the native type. It grades southward into broadleaf deciduous forest, which is found in a large area of the eastern United States. Here, soils have retained a high level of natural fertility suitable for crop cultivation. Deciduous forests remnants are also found in this climate in central and eastern Europe, and in a narrow belt penetrating far eastward into Siberia. Deciduous forests are also found in north-central China and in Korea.

An important environmental factor has been the activity of the great ice sheets that recently covered the northern parts of this climate region. The ice not only carved many landforms, but also left behind deposits of fresh soil materials in many places. These effects are explained in Chapter 18.

When traced westward into the continental interior of North America, the moist continental climate becomes progressively drier. This gradient influences both soils and vegetation. There is a large region of the Middle West, starting roughly in Illinois and continuing west through Iowa and into Nebraska, that formerly supported a natural cover of tall, dense grasses—*tall-grass prairie.* This kind of prairie once extended from about the United States–Canadian border southward to the Gulf coast, but today only a few small remnants can be found (see Figure 21.27).

Agricultural Resources of the Moist Continental Climate

Because of the availability of soil water through a warm summer growing season, the moist continental environment has an enormous potential for food production. Throughout Europe, large areas of the moist continental forest environment have been under field crops, pastures, and vineyards for centuries (Figure 9.22). The cooler, more northerly sections in North America and Europe support dairy farming on a large scale. A combination of acid soils and unfavorable glacial terrain in the form of bogs and lakes, rocky hills, and stony soils has deterred crop farming in many parts of these northern regions.

Farther south, plains formed on former lake floors and on undulating uplands are ideally suited to crop farming. Here, soils are of high fertility. Cereals grown extensively in North America and Europe include corn, wheat, rye (especially in Europe), oats, and barley. Corn is also an important crop in Hungary and Romania. Beet sugar is an important product of this environmental region in Europe, but not in North America. On the other hand, soybeans are intensively cultivated in the midwestern United States and in northern China and Manchuria, but very little in Europe. Rice is a dominant crop in both South Korea and Japan, much farther poleward than elsewhere in Asia. The rice seedlings can be planted in paddies flooded during the brief but copious rains of midsummer, and then harvested in the dry autumn. (Among geographers, this northern rice area is often included in the region called Monsoon Asia.) Agricultural productivity of the tall-grass prairie lands in the United States is now legendary under the name of the corn belt. Corn production is concentrated most heavily in the prairie plains of Illinois, Iowa, and eastern Nebraska. Wheat is also a major crop near the western limits of the tall-grass prairie in Kansas and Oklahoma.

HIGH-LATITUDE CLIMATES

By and large, the *high-latitude climates* are climates of the northern hemisphere (Figure 9.23). They occupy the northern subarctic and arctic latitude zones, but also extend southward into the midlatitude zone as far south as about the 47th parallel in eastern North America and eastern Asia. One of these, the ice sheet climate ⑬, is present in both hemispheres in the polar zones. Table 9.2 provides an overview of the three high-latitude climates.

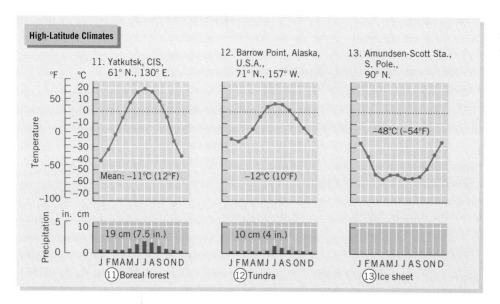

Figure 9.23 Climographs of the three high-latitude climates.

The high-latitude climates coincide closely with the belt of prevailing westerly winds that circles each pole. In the northern hemisphere, this circulation sweeps maritime polar (mP) air masses, formed over the northern oceans, into conflict with continental polar (cP) and continental arctic (cA) air masses on the continents. Rossby waves form in the westerly flow, bringing lobes of warmer, moister air poleward into the region in exchange for colder, drier air moved equatorward. The result of these processes is frequent wave cyclones, produced along a discontinuous and constantly fluctuating arctic-front zone. In summer, tongues of maritime tropical air masses (mT) reach the subarctic latitudes to interact with polar air masses and yield important amounts of precipitation.

THE BOREAL FOREST CLIMATE ⑪
(Köppen: *Dfc, Dfd, Dwc, Dwd*)

The **boreal forest climate** ⑪ is a continental climate with long, bitterly cold winters and short, cool sum-

mers. It occupies the source region for cP air masses, which are cold, dry, and stable in the winter. Invasions of the very cold cA air mass are common. The annual range of temperature is greater than that of any other climate and is greatest in Siberia. Precipitation increases substantially in summer, when maritime air masses penetrate the continent with traveling cyclones, but the total annual precipitation is small. Although much of the boreal forest climate is moist, large areas in western Canada and Siberia have low annual precipitation and are therefore cold and dry.

The global extent of the boreal forest climate ⑪ is presented in Figure 9.24. In North America, it stretches from central and western Alaska, across the Yukon and Northwest Territories to Labrador on the Atlantic coast. In Europe and Asia, it reaches from the Scandinavian Peninsula eastward across all of Siberia to the Pacific. In latitude, this climate type ranges from 50° to 70° N.

Figure 9.25 is a climograph for Fort Vermilion, Alberta, at lat. 58° N. The very great annual temperature range shown here is typical for North America. Monthly mean air temperatures are below freezing for

Figure 9.24 World map of the boreal forest climate ⑪. (Based on Goode Base Map.)

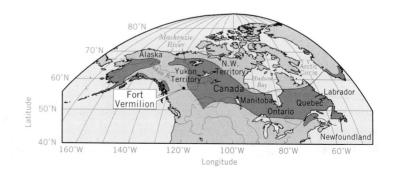

Table 9.2 High-Latitude Climates

Climate	Temperature	Precipitation	Explanation
Boreal forest ⑪	Short, cool summers and long, bitterly cold winters. Greatest annual range of all climates.	Annual precipitation small, falling mostly in summer months.	This climate falls in the source region for cold, dry stable cP air masses. Traveling cyclones, more frequent in summer, bring precipitation from mP air.
Tundra ⑫	No true summer but a short mild season. Otherwise, cold temperatures.	Annual precipitation small, falling mostly during mild season.	The coastal arctic fringes occupied by this climate are dominated by cP, mP, and cA air masses. The maritime influence keeps winter temperatures from falling to the extreme lows of interiors.
Ice sheet ⑬	All months below freezing, with lowest global temperatures on earth experienced during Antarctic winter.	Very low precipitation, but snow accumulates since temperatures are always below freezing.	Ice sheets are the source regions of cA and cAA air masses. Temperatures are intensely cold.

seven consecutive months. The summers are short and cool. Precipitation shows a marked annual cycle with a summer maximum, but the total annual precipitation is small. Although precipitation in winter is small, a snow cover remains over solidly frozen ground through the entire winter. On the same climograph, temperature data are shown for Yakutsk, a Siberian city at lat. 62° N. The enormous annual range is evident, as well as the extremely low means in winter months. January reaches a mean of about −42°C (−44°F). Except for the ice sheet interiors of Antarctica and Greenland, this region is the coldest on earth. Precipitation is not shown for Yakutsk, but the annual total is very small.

The Boreal Forest Environment

Land surface features of much of the region of boreal forest climate were shaped beneath the great ice sheets of the last ice age. Severe erosion by the moving ice exposed hard bedrock over vast areas and created numerous shallow rock basins. Bouldery rock debris mantles the rock surface in many places. Many of the

shallower rock basins have been filled by organic bog materials.

The dominant upland vegetation of the boreal forest climate ⑪ region is boreal forest, consisting of *needleleaf trees*. In North America and Europe, these are evergreen needleleaf trees, mostly pine, spruce, and fir. Figure 21.18 is a composite map showing the extent of the boreal forest. In central and eastern Siberia, the boreal forest is dominated by the larch, which sheds its needles in winter and is thus a deciduous tree. Associated with the needleleaf trees are stands of aspen, balsam poplar, willow, and birch. Along the northern fringe of boreal forest lies a zone of woodland in which low trees, such as black spruce, are spaced widely apart. The open areas are covered by a surface layer of lichens and mosses (Figure 9.26). This cold woodland is referred to as *taiga*.

Although the growing season in the boreal forest climate ⑪ is short, crop farming is still possible. It is largely limited to lands surrounding the Baltic Sea, bordering Finland and Sweden. Crops grown in this area include barley, oats, rye, and wheat. Along with dairying, these crops primarily supply food for subsistence.

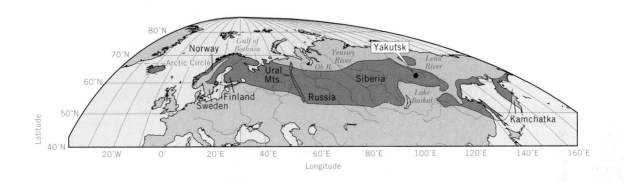

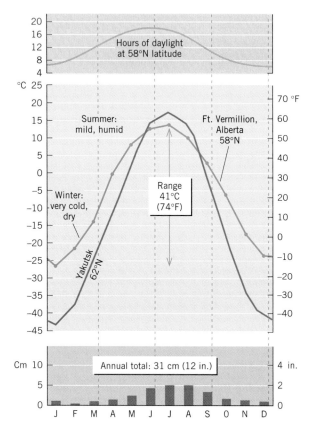

Figure 9.25 Boreal forest climate ⑪. Extreme winter cold and a very great annual range in temperature characterize the climate of Fort Vermilion, Alberta. The temperature range of Yakutsk, Siberia, is even greater.

The principal nonmineral economic products throughout the subarctic lands of eastern Canada are pulpwood and lumber from the needleleaf forests. Logs are carried down the principal rivers to pulp mills and lumber mills. Forests of pine and fir in Sweden, Finland, and northeast Russia are the primary plant resource. Their wood products are exported in the form of paper, pulp, cellulose, and construction lumber.

THE TUNDRA CLIMATE ⑫
(Köppen: *ET*)

The **tundra climate** ⑫ occupies arctic coastal fringes and is dominated by polar (cP, mP) and arctic (cA) air masses. Winters are long and severe. A moderating influence of the nearby ocean water prevents winter temperatures from falling to the extreme lows found in the continental interior. There is a very short mild season, which many climatologists do not recognize as a true summer.

The world map of the tundra climate ⑫ (Figure 9.27) shows the tundra ringing the Arctic Ocean and

extending across the island region of northern Canada. It includes the Alaskan north slope, the Hudson Bay region, and the Greenland coast in North America. In Eurasia, this climate type occupies the northernmost fringe of the Scandinavian Peninsula and Siberian coast. The Antarctic Peninsula (not shown in Figure 9.27) belongs to the tundra climate ⑫. The latitude range for this climate is 60° to 75° N and S, except for the northern coast of Greenland, where tundra occurs at latitudes greater than 80° N.

Figure 9.28 is a climograph for Upernivik, located on the west coast of Greenland at lat. 73° N. A short mild period, with above-freezing temperatures, is equivalent to a summer season in lower latitudes. The long winter is very cold, but the annual temperature range is not as large as that for the boreal forest climate to the south. Total annual precipitation is small. Increased precipitation beginning in July is explained by the melting of the sea-ice cover and a warming of ocean water temperatures. This increases the moisture content of the local air mass, allowing more precipitation.

Figure 9.26 Lichen woodland near Fort McKenzie, lat. 57° N, in northern Quebec. The trees are black spruce. Between the trees is a carpet of lichen.

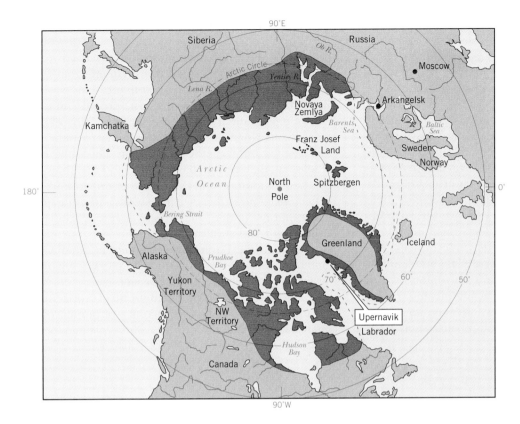

Figure 9.27 World map of the tundra climate ⑫.

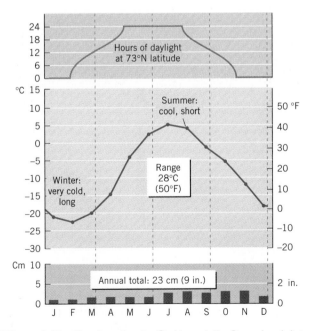

Figure 9.28 Tundra climate ⑫. Upernivik, Greenland, lat. 73° N, shows a smaller annual range than Fort Vermilion (Figure 9.25).

The Arctic Tundra Environment

The term *tundra* describes both an environmental region and a major class of vegetation (Figure 9.29). (An equivalent climatic environment—called *alpine tundra*—prevails in many global locations in high mountains above the timberline.) Soils of the arctic tundra are poorly developed and consist of freshly broken mineral particles and varying amounts of humus (finely divided, partially decomposed plant matter). Peat bogs are numerous. Because soil water is solidly and permanently frozen not far below the surface, the summer thaw brings a condition of water saturation to the soil.

Vegetation of the tundra consists of a cover of scattered grasses, sedges, and lichens, along with shrubs of willow. Vegetation is scarce on dry, exposed slopes and summits. Here the surface cover is often a rocky pavement of angular rock fragments. Trees exist in the tundra only as small, shrublike plants. They are stunted because of the seasonal damage to roots by freeze and thaw of the soil layer and to branches exposed to the abrading action of wind-driven snow. In some places a distinct tree line separates the forest and tundra. It coincides approximately with the 10°C (50°F) isotherm of the warmest month and has been used by geographers as a boundary between boreal forest and tundra.

The number of species in the tundra environment is small, but the abundance of individuals is high. Among the animals, vast herds of caribou in North America or reindeer (their Eurasian relatives) roam the tundra, lightly grazing the lichens and plants and moving constantly (Figure 9.29). A smaller number of musk-oxen are also consumers of the tundra vegetation. Wolves and wolverines, arctic foxes, and polar bears are predators. Among the smaller mammals, snowshoe rabbits and lemmings are important herbivores. Invertebrates are scarce in the tundra, except for a small number of insect species. Black flies, deerflies, mosquitoes, and "no-see-ums" (tiny biting midges) are all abundant and can make July on the tundra most uncomfortable for humans and animals. Reptiles and amphibians are also rare. The boggy tundra, however, offers an ideal summer environment for many migratory birds such as waterfowl, sandpipers, and plovers.

Arctic Permafrost

Because of the cold temperatures experienced in the tundra and northern boreal forest climate zones, the ground is typically frozen to great depth. This perennially frozen ground, or **permafrost**, prevails over the tundra region and a wide bordering area of boreal forest climate. Normally, a top layer of the ground will thaw each year during the mild season. This active layer of seasonal thaw is from 0.6 to 4 m (2 to 13 ft) thick, depending on latitude and the nature of the ground.

Continuous permafrost, which extends without gaps or interruptions under all surface features, coincides largely with the tundra climate, but also includes a large part of the boreal forest climate in Siberia. *Discontinuous permafrost*, which occurs in patches separated by frost-free zones under lakes and rivers, occupies much of the boreal forest climate zone of North America and Eurasia. Sporadic occurrence of permafrost in small patches extends into the southern limits of the boreal forest climate.

Figure 9.29 Caribou migration across the arctic tundra of northern Alaska.

The permafrost environment is in a stable, delicate equilibrium that can be easily disturbed. When human activities remove the cover of moss or peat and living plants that insulates the permafrost, the summer thaw is extended to a greater depth. Ice wedges or ice masses beneath the surface melt, causing the soil surface to sink. Meltwater mixes with the fine soil to form mud, which is then eroded and transported by water in streams. This activity is called *thermal erosion* and is reversed only with great difficulty.

To keep the heat from buildings on the tundra from melting the permafrost below, the buildings are placed on piles with an insulating air space below or built on a thick insulating pad of coarse gravel placed over the surface prior to construction. Steam and water lines that connect buildings are not placed underground, but are contained within above-ground heated, insulated tunnels to prevent thaw of the permafrost layer.

THE ICE SHEET CLIMATE ⑬
(Köppen: *EF*)

The **ice sheet climate** ⑬ coincides with the source regions of arctic (A) and antarctic (AA) air masses, situated on the vast, high ice sheets of Greenland and Antarctica and over polar sea ice of the Arctic Ocean. Mean annual temperature is much lower than that of any other climate, with no monthly mean above freezing. Strong temperature inversions develop over the ice sheets. In Antarctica and Greenland, the high surface altitude of the ice sheets intensifies the cold. Strong cyclones with blizzard winds are frequent. Precipitation, almost all occurring as snow, is very low, but accumulates because of the continuous cold. The latitude range for this climate is 65° to 90° N and S.

Figure 9.30 shows temperature graphs for several representative ice sheet stations. The graph for Eismitte, a research station on the Greenland ice cap, shows the northern hemisphere temperature cycle, whereas the other four examples are all from Antarctica. Temperatures in the interior of Antarctica have proved to be far lower than those at any other place on earth. A Russian meteorological station at Vostok, located about 1300 km (about 800 mi) from the south pole at an altitude of about 3500 m (11,500 ft), may be the world's coldest spot. Here a low of –88.3°C (–127°F) was observed in 1958. At the pole itself (Amundsen-Scott Station), July, August, and September of 1957 had averages of about –60°C (–76°F). Temperatures are considerably higher, month for month, at Little America in Antarctica because it is located close to the Ross Sea and is at a low altitude.

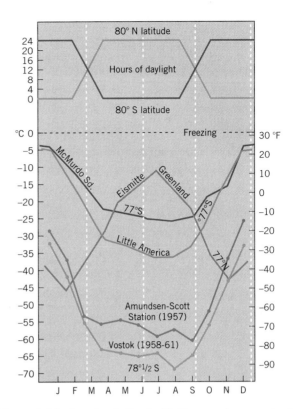

Figure 9.30   Ice sheet climate ⑬. Temperature graphs for five ice sheet stations.

The Ice Sheet Environment

Because of low monthly mean temperatures throughout the year over the ice sheets, this environment is devoid of vegetation and soils. The few species of animals found on the ice margins are associated with a marine habitat. In terms of habitation by humans, the ice sheet environment is extremely hostile because of extreme cold, high winds, and a total lack of food and fuel resources. Enormous expenditures of energy are required to import these necessities of life and to provide shelter. These efforts are justified for scientific research, but in the foreseeable future there is little prospect that this icy environment will provide useful supplies of energy or minerals.

The climates of the earth are remarkably diverse, ranging from the hot, humid wet equatorial climate ① at the equator to the bitterly cold and dry ice sheet climate ⑬ at the poles. Between these extremes are the other climates, each with distinctive features—such as dry summers or dry winters, or uniform or widely varying temperatures. The global environments associated with each climate type are also highly varied, from the lush, equatorial rainforest to the stunted willows of the tundra. Some of these environmental regions are more hospitable to human habitation and use than are others. As we have shown in this chapter and its predecessor, climate exerts strong controls on vegetation and soils, especially at the global level. We will return to a more detailed look at vegetation and soils in Part IV.

The next set of topics in our study in physical geography concerns the lithosphere—the realm of the solid earth, as we described in the Introduction. Our survey of the lithosphere, in *Part II: Systems and Cycles of the Solid Earth,* will begin with the nature of earth materials and then move to a description of how continents and ocean basins are formed and how they are continually changing, even today. Last, we turn to landforms occurring within continents that are produced by such lithospheric processes as volcanic activity and earthquake faulting.

CHAPTER SUMMARY

Midlatitude climates are quite varied, since they lie in a broad zone of intense interaction between tropical and polar air masses. The dry subtropical climate ⑤ is dominated by subtropical high pressure and resembles the dry tropical climate ④, but it has a larger annual temperature range and a distinct cool season. The Mojave and Sonoran Deserts, which occur in this climate type, are distinguished by a cover of specialized plants, such as the Joshua tree and saguaro cactus.

Abundant rainfall with a summer maximum is a characteristic of the moist subtropical climate ⑥. The temperature cycle of this type includes cool winters with spells of subfreezing weather and warm, humid summers. The natural vegetation cover is forest—broadleaved evergreen in most regions, with pine forest in regions of sandy soils. However, agriculture has largely replaced natural forests throughout this climate type. In Asia, rice culture is dominant, while in North America, sugarcane, peanuts, tobacco, and cotton are more typical.

The Mediterranean climate ⑦ is unique because its annual precipitation cycle has a wet winter and a very dry summer. The temperature range is moderate, with warm to hot summers and mild winters. The vegetation of the Mediterranean climate environment includes many evergreen sclerophylls, such as the cork oak, olive tree, and eucalyptus. Oak woodlands are also typical. Steep slopes are clothed with the drought-resistant shrubs of the chaparral. Because of a long history of human habitation, little natural vegetation remains in the Mediterranean climate environment of Europe.

The marine west-coast climate ⑧, like the Mediterranean climate ⑦, has a winter precipitation maximum. The marine influence keeps temperature mild with a narrowed annual range. Forest is the native vegetation of the marine west-coast environment—dense needleleaf forest on the northern Pacific coast, and broadleaf deciduous forest on the coast of Western Europe. In Europe, lands of this climate have been under intensive cultivation for many centuries, and little forest remains. In North America, this climate zone is a source of many diverse forest products.

The dry midlatitude climate ⑨ has both arid (⑨a) and semiarid (⑨s) subtypes. Both types have warm to hot summers and cold to very cold winters. The semiarid subtype typically has fertile soils with a sparse cover of grasses, which is termed short-grass prairie in North America and steppe in Eurasia. Wheat is the dominant crop, with cattle grazing an important commercial activity. Because rainfall is highly variable, drought is a recurring event. When combined with overgrazing, drought can create dust bowl conditions.

The moist continental climate ⑩ has ample precipitation with a summer maximum. Summers are warm and winters cold. In its northern regions, the moist continental environment sustains a cover of evergreen needleleaf forest, while farther south, the forest is broadleaf deciduous. In North America, a region of this climate in the Midwest once supported tall-grass prairie. This region is now the corn belt, an area of legendary crop productivity.

High-latitude climates have low precipitation since air temperatures are low. The precipitation generally occurs during the short warm period. The boreal forest climate ⑪ has long, bitterly cold winters. The boreal forest consists of needleleaf trees of pine, spruce, fir, and larch. Patches of deciduous aspen, poplar, willow, and birch also occur. Dairying, limited crop farming, and timber harvesting for pulp and lumber are economic activities of the boreal forest environment.

The tundra climate ⑫ occupies arctic coastal fringes. Because of the marine influence, winter temperatures are not as bitterly cold as those of the tundra climate ⑪. The tundra is a vegetation cover of scattered grasses, sedges, lichens, and dwarf shrubs, with patches of rock fragments devoid of plants. Although the environment is harsh, wildlife is abundant, especially in the warm months. Permafrost underlies much of the tundra, which presents special problems for construction of buildings and roads.

The ice sheet climate ⑬ is the coldest of all climates, with no monthly mean temperature above freezing. The severity of the climate prohibits nearly all human habitation.

KEY TERMS

dry subtropical climate ⑤
moist subtropical climate ⑥
Mediterranean climate ⑦
marine west-coast climate ⑧
dry midlatitude climate ⑨

moist continental climate ⑩
boreal forest climate ⑪
tundra climate ⑫
permafrost
ice sheet climate ⑬

REVIEW QUESTIONS

1. What climate type is associated with the Mojave Desert? Describe some of the features of the Mojave Desert environment.

2. Both the moist subtropical ⑥ and moist continental ⑩ climates are found on eastern sides of continents in the midlatitudes. What are the major factors that determine their temperature and precipitation cycles? How do these two climates differ?

3. What natural vegetation types are associated with the moist subtropical forest environment? How does human use of this environment differ between eastern North America and eastern Asia?

4. Both the Mediterranean ⑦ and marine west-coast ⑧ climates are found on the west coasts of continents. Why do they experience more precipitation in winter than in summer? How do the two climates differ?

5. Identify the characteristic features of the natural vegetation of the Mediterranean climate environment. Describe human use of this environment for agriculture and the limitations the climate places on these uses.

6. Contrast the natural vegetation types and human use of the land in the marine west-coast environments of North America and Europe.

7. The dry midlatitude environment is one of great agricultural importance. Why?

8. What environmental disaster occurred on the Great Plains of North America in the 1930s? Explain.

9. The moist continental climate ⑩ region supports at least two important types of forest. What are they?

10. Identify the key agricultural products of the moist continental climate region. What is the role of tall-grass prairie lands?

11. The subtropical high-pressure cells influence several climate types in the low latitudes and midlatitudes. Identify the climates and describe the effects of the subtropical high-pressure cells on them.

12. Both the boreal forest ⑪ and tundra ⑫ climate are climates of the northern regions, but the tundra is found fringing the Arctic Ocean and the boreal forest is located further inland. Compare these two climates from the viewpoint of coastal-continental effects.

13. The boreal forest environment includes vast areas of land that are little influenced by human activity. Describe the natural vegetation of the boreal forest environment.

14. What are the common plants of the tundra environment? What animals inhabit this climate zone?

15. Distinguish between two types of permafrost. What challenges does the permafrost environment provide for human habitation?

16. What is the coldest climate on earth? How is the annual temperature cycle of this climate related to the cycle of insolation?

Focus on Systems 9.2 • California Rainfall Cycles and El Niño

1. What is the relationship between annual rainfall at Santa Barbara and the Southern Oscillation Index?

2. Is the relation between annual rainfall and the Southern Oscillation Index reliable enough to predict high-rainfall years? Defend your answer by discussing the two graphs in detail.

ESSAY QUESTIONS

1. Discuss the role of the polar front and the air masses that come in conflict in the polar-front zone in the temperature and precipitation cycles of the midlatitude and high-latitude climates.

2. Compare and contrast the differing environments of the United States and Canada by constructing a transect linking five major cities in these countries and then describing the typical natural environments and human usage of the land you would encounter in traveling along the transect.

3. Suppose South America were turned over. That is, imagine that the continent was cut out and flipped over end-for-end so that the southern tip was at about 10° N latitude, and the northern end (Venezuela) was positioned at about 55° S. The Andean chain will still be on the west side, but the shape of the land mass will now be quite different. Sketch this continent and draw possible climate boundaries, using your knowledge of global air circulation patterns, frontal zones, and air mass movements.

PROBLEMS

Working It Out 9.1 • Standard Deviation and Coefficient of Variation

1. The data below are monthly rainfall values for Baton Rouge's wettest month (June) and driest month (September). For each month, find the sample standard deviation and the coefficient of variation. Which month has the larger standard deviation? Which has the highest relative variation?

Year	June	September
1985	11.8	14.7
1986	15.7	9.4
1987	24.2	3.9
1988	10.5	16.9
1989	38.4	14.0
1990	17.1	13.6
1991	16.7	15.0
1992	19.6	10.4
1993	8.9	3.4
1994	17.0	9.6

Part II:

Systems and Cycles of the Solid Earth

In Part II, we move from examining the flows of energy and matter at the earth's surface to the systems and cycles of the solid earth. These systems differ in two very important respects. One is their power sources—surface energy and matter flows are driven by solar radiation, while solid earth systems are driven by the earth's internal heat. Furthermore, these two great systems of energy and matter flow operate at very different time scales. For surface energy flow systems, the rhythms are daily and annual, responding to the rotation of the earth on its axis and the revolution of the earth around the sun. For earth systems, on the other hand, cycles are measured in thousnads to millions of years.

Chapter 10 begins Part II by discussing the basic materials of the solid earth—rock and minerals—and some principles of their formation. These are linked in the cycle of rock change, which describes how earth materials are constantly being cycled and recycled over geologic time. In Chapter 11, we introduce the grandest cycle of them all—that of plate tectonics, in which the earth's surface configuration of continents and ocean basins slowly changes as forces deep within the earth cause vast tectonic plates to converge, collide, split, and separate. The present pattern of plates and their motions explains the location of many geologic surface phenomena, such as earthquakes and volcanoes. These, and other manifestations of geologic activity such as folds and faults, are the subject of Chapter 12.

Physical geographers are concerned with the solid earth because the earth's outer layer serves as a platform, or base, for life on the lands. This platform provides the continental surfaces that are carved into landforms by moving water, wind, and glacial ice. Landforms, in turn, influence the distribution of ecosystems and exert strong controls over human occupation of the lands. Landforms made by water, wind, and ice are the subject of the chapters in Part III of this book.

Chapter 10

Earth Materials and the Cycle of Rock Change

We begin Part II, our study of the solid earth as a platform for both life and landform-making processes, with an examination of earth materials. These are the minerals and rocks that are found at or near the earth's surface. We will be concerned only with a few of a great number of minerals and rocks, focusing on those that are most important for a broad understanding of the continents, ocean basins, and their varied features. We will also examine the rock cycle, in which earth materials are constantly formed and re-formed in a cycle powered largely by heat from the earth's interior.

Layers of sedimentary rock are exposed in this sunrise view of the Grand Canyon, Arizona.

THE CRUST AND ITS COMPOSITION

The thin, outermost layer of our planet is the *earth's crust*. This mineral skin ranges from about 8 to 40 km (about 5 to 25 mi) thick and contains the continents and ocean basins. It is the source of soil on the lands, salts of the sea, gases of the atmosphere, and all the water of the oceans, atmosphere, and lands.

Figure 10.1 displays the eight most abundant elements of the earth's crust in terms of percentage by weight. Oxygen, the predominant element, accounts for a little less than half the total weight. Second is silicon, which accounts for a little more than a quarter. Together they account for 75 percent of the crust, by weight.

Aluminum accounts for approximately 8 percent and iron for about 5 percent of the earth's crust. These metals are very important to our industrial civilization,

Working It Out 10.1 • Radioactive Decay

The earth's interior is largely heated by the spontaneous decay of naturally occurring radioactive isotopes of certain elements. You may recall from basic chemistry that the properties of an element are determined by the number of positively charged particles, or *protons*, contained in its nucleus. The number of protons, called the *atomic number*, identifies the chemical element to which the nucleus belongs. For example, the element uranium has 92 protons. Nuclei of atoms also contain neutrons, which have nearly the same mass as protons but have no charge. The total of protons and neutrons within the nucleus of an atom is known as the *atomic mass number*.

Some elements are found in forms with different mass numbers. These forms are known as *isotopes*. For example, the most common isotope of uranium (U) is *uranium-238*, a form with an atomic mass of 238. Another form is *uranium-235*, which has an atomic mass number of 235. Chemists distinguish these

two forms by writing them as ^{238}U and ^{235}U.

A key to the understanding of radioactivity is that certain isotopes are *unstable*, meaning that the composition of the nucleus of the isotope can experience an irreversible change. This change process is known as *radioactive decay*. When a nucleus decays, it emits matter and energy. The energy is absorbed by the surrounding matter and thus ultimately takes the form of sensible heat. Furthermore, in some types of decay protons may be lost or gained. This means that an atom of one element may be transformed into an atom of another. For example, the uranium isotope ^{238}U decays to form ^{234}Th, an isotope of the element thorium. This new isotope is known as a *daughter product*. Often the new isotope created will be unstable as well and will decay into yet another isotope of a different element. As this process continues, the result is a *decay chain* of daughter products that eventually ends in the formation of a stable

isotope. For example, ^{238}U ultimately forms the stable isotope lead-206, ^{206}Pb.

The significance of radioactive decay is that it provides an internal source of heat for the earth—a source that accounts for the melting of solid rock to form magma and thus creates igneous rocks. This heating effect is described more fully in *Focus on Systems 10.2 • Powering the Cycle of Rock Change*. The decay of ^{238}U is only one of several radioactive decay chains that are important in heating the earth from within. Other chains begin with ^{235}U, thorium-232 (^{232}Th), and potassium-40 (^{40}K).

The time rate at which unstable isotopes decay spans a very large magnitude. Some isotopes decay in a matter of milliseconds, while others decay over a period of billions of years. The rate of decay of an unstable isotope is measured by its *half-life*—the period of time in which a number of atoms of the isotope will be reduced by half. For example, the half-life of ^{238}U is 4.46 billion

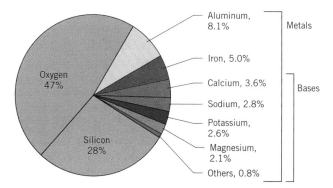

Figure 10.1 The eight most abundant elements in the earth's crust, measured by percentage of weight. Oxygen and silicon dominate, with aluminum and iron following.

and, fortunately, they are relatively abundant. Four other metallic elements—calcium, sodium, potassium, and magnesium—make up the remaining 12 percent. All four occur at about the same order of abundance (2 to 4 percent). These elements are essential for plant and animal life and are described in Chapter 20 as plant nutrients. Because of this role, they are also important determiners of soil fertility (Chapter 19).

Of the remaining dozens of chemical elements, a few are very important from the viewpoint of powering the cycles of the solid earth. These are the radioactive forms of elements that, through slow radioactive decay, provide a nearly infinite source of heat that seeps slowly outward from the earth's interior. Some of these are introduced in *Working It Out 10.1 • Radioactive Decay*.

years, meaning that one gram of ^{238}U will be reduced to one-half of a gram after that length of time. In another 4.46 billion years, this half-gram will be reduced to a quarter-gram, and so forth.

The graph to the right shows an example of the decay of potassium-40 (^{40}K). This unstable isotope has a half-life of 1.26 billion years. The *y*-axis shows the proportion of the original quantity remaining after the elapsed time shown on the *x*-axis. This curve has a *negative expo-nential form*—that is, a curve that de-creases as a negative exponential function of e, the base of natural logarithms. For this type of func-tion, we can write the proportion P remaining at time t as

$$P(t) = e^{-kt}$$

where k is a constant related to the half-life by the expression

$$k = \frac{0.693}{H}$$

Here, H is the half-life and 0.693 is the natural logarithm of 2. For the case of ^{40}K, $k = 0.693/1.26 = 0.550$ and therefore

$$P(t) = e^{-0.550t}$$

Since the half-life for ^{40}K is mea-sured in billions of years, t in this expression will also be in billions of years. For example, the proportion of ^{40}K atoms remaining after 500 million years (=0.5 billion years) will be

$$P(0.5) = e^{-0.550 \times 0.5} = e^{-0.275} = 0.760$$

When ^{40}K decays, one of two products is produced. In the first type of decay, calcium-40 (^{40}Ca) is created. In the second type, which occurs less frequently, argon-40 (^{40}Ar) is created. Both ^{40}Ca and ^{40}Ar are stable isotopes and un-dergo no further decay. The graph shows how these products will slowly build as they are formed by the decay of ^{40}K.

The phenomenon of radioactive decay is exploited in the application of one of the most important tools available to geologists—radiometric dating, which we explore further in *Working It Out 11.1* • *Radiometric Dating.*

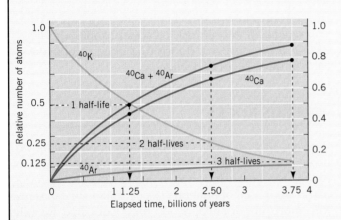

Exponential decay and growth curves for ^{40}K, ^{40}Ca, and ^{40}Ar. (Copyright © A. N. Strahler.)

Rocks and Minerals

The elements of the earth's crust occur in chemical compounds that we recognize as minerals. A **mineral** is a naturally occurring, inorganic substance that usually possesses a definite chemical composition and charac-teristic atomic structure. Most minerals have a crys-talline structure. Gemstones such as diamonds or ru-bies are examples, although their crystalline shape is enhanced by the stonecutter. Quartz is an example of a very common crystalline mineral. It usually occurs as a clear, six-sided prism (Figure 10.2).

Figure 10.2 These large crystals of quartz form six-sided, translucent columns.

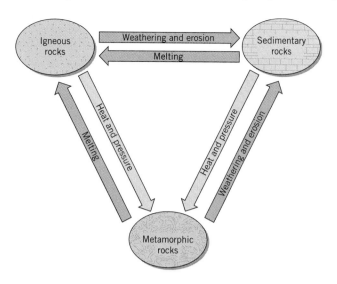

Figure 10.3 The cycle of rock change. In this simplified version of the cycle of rock change, the three classes of rock are transformed into one another by weathering and erosion, melting, and exposure to heat and pressure.

Minerals are combined into **rock**, which we can broadly define as an assemblage of minerals in the solid state. Rock comes in a very wide range of compositions, physical characteristics, and ages. A given variety of rock is usually composed of two or more minerals, and often many different minerals are present. However, a few rock varieties consist almost entirely of one mineral. Most rock of the earth's crust is extremely old by human standards, with the age of formation often dating back many millions of years. But rock is also being formed at this very hour as active volcanoes emit lava that solidifies on contact with the atmosphere.

Rocks of the earth's crust fall into three major classes. (1) **Igneous rocks** are solidified from mineral matter in a high-temperature molten state. (2) **Sedimentary rocks** are layered accumulations of mineral particles derived mostly by weathering and erosion of preexisting rocks. (3) **Metamorphic rocks** are formed from igneous or sedimentary rocks that have been physically or chemically changed, usually by application of heat and pressure during crustal mountain-making.

The three classes of rocks are constantly forming from one another in a continuous circuit—the **cycle of rock change**—through which the crustal minerals have been recycled during many millions of years of geologic time (Figure 10.3). In the process of melting, preexisting rock of any class is melted and then later cools to form igneous rock. In weathering and erosion, preexisting rock is broken down and accumulated in layers that become sedimentary rock. Heat and pressure convert igneous and sedimentary rocks to metamorphic rock. At the close of this chapter, after we have taken a more detailed look at rocks, minerals, and their formation processes, we will return to a more complete version of the cycle of rock change.

IGNEOUS ROCKS

Igneous rocks are formed when molten material moves from deep within the earth to a position within or atop the crust. There the molten material cools, forming rocks composed of mineral crystals. Table 10.1 presents the more important igneous rock types that we will refer to in this chapter.

Most igneous rock consists of *silicate minerals*, chemical compounds that contain silicon and oxygen atoms. Most of the silicate minerals also have one, two, or more of the metallic elements listed in Figure 10.1—that is, aluminum, iron, calcium, sodium, potassium, and magnesium. Although there are many silicate minerals, we will only be concerned with seven of them, which are shown in Figure 10.4.

Table 10.1   Some Common Igneous Rock Types

Subclass	Rock Type	Subclass	Rock Type	Composition
Intrusive (Cooling at depth, producing coarse crystal texture)	Granite	**Extrusive** (Cooling at the surface, producing fine crystal texture)	Rhyolite	Felsic minerals, typically quartz, feldspars, and mica
	Diorite		Andesite	Felsic minerals without quartz, usually including plagioclase feldspar and amphibole
	Gabbro		Basalt	Mafic minerals, typically plagioclase feldspar, pyroxene, and olivine
	Peridotite			An ultramafic rock of pyroxene and olivine

Among the most common minerals of all rock classes is **quartz**, which is silicon dioxide (SiO_2) (Figure 10.2). It is quite hard and resists chemical breakdown. Two silicate–aluminum minerals called *feldspars* follow. One type, *potash feldspar*, contains potassium as the dominant metal besides aluminum. A second type, *plagioclase feldspar*, is rich in sodium, calcium, or both. Quartz and feldspar form a silicate mineral group described as **felsic** ("fel" for feldspar; "si" for silicate).

Quartz and the feldspars are light in color (white, pink, or grayish) and lower in density than the other silicate minerals. By *density*, we mean the quantity of mat-

ter contained in a unit volume. Figure 10.5 shows the density of four earth materials: water, quartz, the mineral olivine (discussed later in this chapter), and pure iron.

The next three silicate minerals are actually mineral groups, with a number of mineral varieties in each group. They are the *mica*, *amphibole*, and *pyroxene* groups. All three are silicates containing aluminum, magnesium, iron, and potassium or calcium. The seventh mineral, *olivine*, is a silicate of only magnesium and iron that lacks aluminum. Altogether, these minerals are described as **mafic** ("ma" for magnesium; "f"

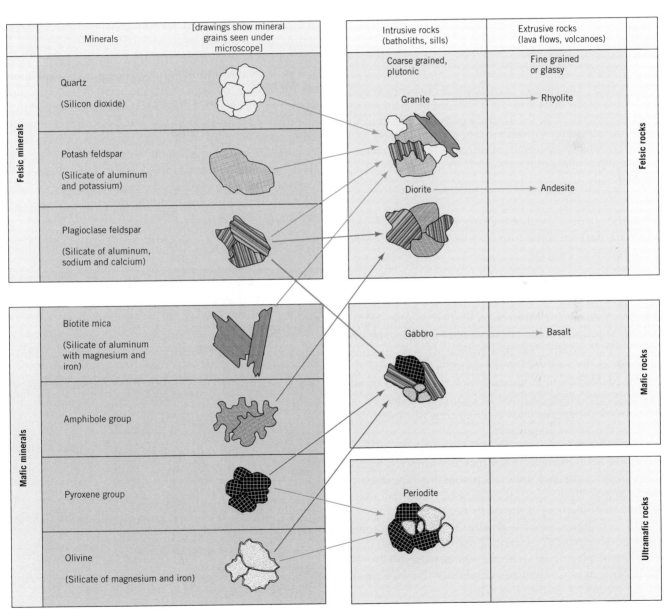

Figure 10.4 Silicate minerals and igneous rocks. Only the most important silicate mineral groups are listed, along with four common igneous rock types. The patterns shown for mineral grains indicate their general appearance through a microscope.

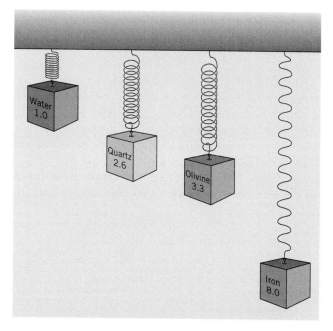

Figure 10.5 The concept of density is illustrated by several cubes of the same size, but of different materials, hung from coil springs under the influence of gravity. The stretching of the coil spring is proportional to the density of each material, shown in thousands of kilograms per cubic meter. Quartz and olivine are common minerals. (Copyright © A. N. Strahler.)

from the chemical symbol for iron, Fe). The mafic minerals are dark in color (usually black) and are denser than the felsic minerals.

Common Igneous Rocks

Igneous rocks solidify from rock in a hot, molten state, known as **magma**. From pockets a few kilometers below the earth's surface, magma makes its way upward through older solid rock and eventually solidifies as igneous rock. No single igneous rock is made up of all seven silicate minerals listed in Figure 10.4. Instead, a given rock variety contains three or four of those minerals as the major ingredients.

The column in the center of Figure 10.4 shows four common igneous rocks. Each rock is connected by arrows to the principal minerals it contains. These four rocks are carefully selected to be used in our later explanation of features of the earth's crust.

The first igneous rock is **granite**. The bulk of granite consists of quartz (27 percent), potash feldspar (40 percent), and plagioclase feldspar (15 percent). The remainder is mostly biotite and amphibole. Because most of the volume of granite is of felsic minerals, we classify granite as a *felsic igneous rock*. Granite is a mixture of white, grayish or pinkish, and black grains, but the overall appearance is a light gray or pink color (Figure 10.6).

Diorite, the second igneous rock on the list, lacks quartz. It consists largely of plagioclase feldspar (60 percent) and secondary amounts of amphibole and pyroxene. Diorite is a light-colored felsic rock that is only slightly denser than granite.

The third igneous rock is *gabbro*, in which the major mineral is *pyroxene* (60 percent). A substantial amount of plagioclase feldspar (20 to 40 percent) is present, and, in addition, there may be some olivine (0 to 20 percent). Since the mafic minerals pyroxene and olivine are dominant, gabbro is classed as a *mafic igneous rock*. It is dark in color and denser than the felsic rocks.

The fourth igneous rock, *peridotite*, is dominated by olivine (60 percent). The rest is mostly pyroxene (40 percent). Peridotite is classed as an *ultramafic igneous rock*, denser even than the mafic types. The mineral grains in igneous rocks are very tightly interlocked, and the rock is normally very strong.

These four common varieties of igneous rock show an increasing range of density, from felsic, through mafic, to ultramafic types. This arrangement is duplicated on a grand scale in the principal rock layers that comprise the solid earth, with the least dense layer (mostly felsic rocks) near the surface and the densest layer (ultramafic rocks) deep in the earth's interior. We will stress this layered arrangement again in describing the earth's crust and the deeper interior zones in Chapter 11.

Figure 10.6 This freshly broken piece of granite shows white grains of quartz, pinkish grains of feldspar, and dark grains of biotite crystals.

neous rocks cool very slowly and, as a result, develop large mineral crystals—that is, they are *coarse-textured*. A good example is the granite pictured in Figure 10.6.

In an intrusive igneous rock, individual mineral crystals are visible with the unaided eye or with the help of a simple magnifying lens. In an extrusive rock, which cools very rapidly, the individual crystals are very small. They can only be seen through a microscope. Rocks with such small crystals are termed *fine-textured*. If the lava contains dissolved gases, it can cool to form a rock with a frothy, bubble-filled texture, called *scoria* (Figure 10.8, *right*). Sometimes a lava cools to form a shiny natural volcanic glass (Figure 10.8, *left*). Most lava solidifies simply as a dense, uniform rock with a dark, dull surface.

Since the outward appearance of intrusive and extrusive rocks formed from the same magma are so different, they are named differently. The igneous rock types we have discussed so far—granite, diorite, gabbro, and peridotite—are intrusive rocks. Except for peridotite, all have counterparts as extrusive rocks. They are named in the right column of Figure 10.4. Each one has the same mineral composition as the plutonic rock named at the left. *Rhyolite* is the name for lava of the same composition as granite; **andesite** is lava with the mineral composition of diorite; and **basalt** is lava of the composition of gabbro. Andesite and basalt are the two most common types of lavas. Rhyolite and andesite are pale grayish or pink in color, whereas basalt is black. Lava flows, along with particles of solidified lava blown explosively from narrow vents, often accumulate as isolated hills or mountains that we recognize as volcanoes. They are described in Chapter 12.

A body of intrusive igneous rock is called a **pluton**. Granite typically accumulates in enormous plutons, called *batholiths*. Figure 10.9 shows the relationship of a batholith to the overlying rock. As the hot fluid magma rises, it melts and incorporates the older rock lying above it. A single batholith extends down several kilometers and may occupy an area of several thousand square kilometers.

Figure 10.7 This fresh black lava shows a glassy surface and flow structures. Chain of Craters Road, Hawaii Volcanoes National Park, Hawaii.

Intrusive and Extrusive Igneous Rocks

Magma that solidifies below the earth's surface and remains surrounded by older, preexisting rock is called **intrusive igneous rock**. The process itself is *intrusion*. Where magma reaches the surface, it emerges as **lava**, which solidifies to form **extrusive igneous rock** (Figure 10.7). The process is one of *extrusion*.

Although both intrusive rock and extrusive rock can solidify from the same original body of magma, their outward appearances are quite different when you compare freshly broken samples of each. Intrusive ig-

Figure 10.8 (*left*) Obsidian, or volcanic glass. The smooth, glassy appearance is acquired when a gas-free lava cools very rapidly. This sample shows red and black streaks, caused by minor variations in composition. (*right*) A specimen of scoria, a form of lava containing many small holes and cavities produced by gas bubbles.

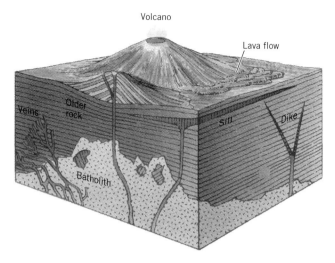

Figure 10.9 This block diagram illustrates various forms of intrusive igneous rock plutons as well as an extrusive lava flow.

Figure 10.9 shows two other common forms of plutons. One is a *sill*, a platelike layer formed when magma forces its way between two preexisting rock layers. In the example shown, the sill has lifted the overlying rock layers to make room. A second kind of pluton is the *dike*, a wall-like body formed when a vertical rock fracture is forced open by magma. Commonly, these vertical fractures conduct magma to the land surface in the process of extrusion. Figure 10.10 shows a dike of mafic rock cutting across layers of older rock. The dike rock is fine-textured because of its rapid cooling. Magma entering small, irregular, branching fractures in the surrounding rock solidifies in a branching network of thin *veins*.

Chemical Alteration of Igneous Rocks

The minerals in igneous rocks are formed at high temperatures, and often at high pressures, as magma cools. When igneous rocks are exposed at or near the earth's surface, the conditions are quite different. Temperatures and pressures are low. Also, the rocks are exposed to soil water and groundwater solutions that contain dissolved oxygen and carbon dioxide. In this new environment, the minerals within an igneous rock may no longer be stable. Instead, most of these minerals undergo a slow chemical change that weakens their structure. Chemical change in response to this alien environment is called **mineral alteration**. It is a process of weathering, which we will discuss further in Chapter 13.

Weathering also includes the physical forces of disintegration that break up igneous rock into small fragments and separate the component minerals, grain by grain. This breakup, or fragmentation, is essential for the chemical reactions of mineral alteration. The reason is that fragmentation results in a great increase in mineral surface area exposed to chemically active solutions. (We will take up the processes of physical disintegration of rocks in Chapter 13.)

Oxygen dissolved in soil or groundwater can oxidize minerals. *Oxidation* occurs when oxygen is added in a chemical reaction. Oxidation is the normal fate of most silicate minerals exposed at the surface. With oxidation, the silicate minerals are converted to *oxides*, in which silicon and the metallic elements—such as calcium, magnesium, and iron—each bond completely with oxygen. Oxides are very stable. Quartz, with the composition silicon dioxide (SiO_2), is a mineral oxide that also occurs naturally in granite. Because it is stable, quartz is very long-lasting. As we will see shortly, it is a major constituent of sedimentary rocks.

Water combines with some silicate minerals in a reaction known as *hydrolysis*. This process is not merely a soaking or wetting of the mineral, but a true chemical change that produces a different mineral compound. The products of hydrolysis are stable and long-lasting, as are the products of oxidation.

Some of the alteration products of silicate minerals are clay minerals, described in Chapter 19. A **clay mineral** is one that has plastic properties when moist, because it consists of very small, thin flakes that become lubricated by layers of water molecules. Clay minerals formed by mineral alteration are abundant in common types of sedimentary rocks.

Figure 10.10 A dike of mafic igneous rock with nearly vertical parallel sides, cutting across flat-lying sedimentary rock layers. Arrows mark the contact between igneous rock and sedimentary rock. Spanish Peaks region, Colorado.

Table 10.2 Some Common Sedimentary Rock Types

Subclass	*Rock Type*	*Composition*
Clastic (Composed of rock and/or mineral fragments)	Sandstone	Cemented sand grains
	Conglomerate	Sandstone containing pebbles of hard rock
	Mudstone	Silt and clay, with some sand
	Claystone	Clay
	Shale	Clay, breaking easily into flat flakes and plates
Chemically precipitated (Formed by chemical precipitation from sea water or salty inland lakes)	Limestone	Calcium carbonate, formed by precipitation on sea or lake floors
	Dolomite	Magnesium and calcium carbonates, similar to limestone
	Chert	Silica, a noncrystalline form of quartz
	Evaporites	Minerals formed by evaporation of salty solutions in shallow inland lakes or coastal lagoons
Organic (Formed from organic material)	Coal	Rock formed from peat or other organic deposits; may be burned as a mineral fuel
	Petroleum	Liquid hydrocarbon found in sedimentary deposits; not a true rock but a mineral fuel
	Natural gas	Gaseous hydrocarbon found in sedimentary deposits; not a true rock but a mineral fuel

When carbon dioxide dissolves in water, a weak acid—*carbonic acid*—is formed. Carbonic acid can dissolve certain minerals, especially calcium carbonate. In addition, where decaying vegetation is present, soil water contains many complex organic acids that are capable of reacting with minerals. Certain common minerals, such as rock salt (sodium chloride), dissolve directly in water, but simple solution is not particularly effective for the silicate minerals.

SEDIMENTS AND SEDIMENTARY ROCKS

We can now turn to the second great rock class, the *sedimentary rocks*. The mineral particles in sedimentary rocks can be derived from preexisting rock of any of the three rock classes as well as from newly formed organic matter. However, igneous rock is the most important original source of the inorganic mineral matter that makes up sedimentary rock. For example, a granite can weather to yield grains of quartz and particles of clay minerals derived from feldspars, thus contributing sand and clay to a sedimentary rock. Sedimentary rocks include rock types with a wide range of physical and chemical properties. We will only touch on a few of the most important kinds of sedimentary rocks, which are shown in Table 10.2.

In the process of mineral alteration, solid rock is weakened, softened, and fragmented, yielding particles of many sizes and mineral compositions. When transported by a fluid medium—air, water, or ice—these particles are known collectively as **sediment**. Used in its broadest sense, sediment includes both inorganic and organic matter. Dissolved mineral matter in solution must also be included.

Streams and rivers carry sediment to lower land levels, where sediment can accumulate. The most favorable sites of sediment accumulation are shallow sea floors bordering continents. But sediments also accumulate in inland valleys, lakes, and marshes. Thick accumulations of sediment may become deeply buried under newer (younger) sediments. Wind and glacial ice also transport sediment, but not necessarily to lower elevations or to places suitable for accumulation. Over long spans of time, the sediments can undergo physical or chemical changes, becoming compacted and hardened to form sedimentary rock.

There are three major classes of sediment. First is **clastic sediment**, which consists of inorganic rock and mineral fragments. Examples are the materials in a sand bar of a river bed, or on a sandy ocean beach. Second is **chemically precipitated sediment**, which consists of inorganic mineral compounds precipitated from a saltwater solution or as hard parts of organisms. In the process of chemical precipitation, ions in solution combine to form solid mineral matter separate from the solution. A layer of rock salt, such as that found in dry lake beds in arid regions, is an example. A third class is **organic sediment** which consists of the tissues of plants and animals, accumulated and preserved after the death of the organism. An example is a layer of peat in a bog or marsh.

Figure 10.11 Eroded horizontal beds of sedimentary rock, Lost City, King Canyon, Northwest Territories, Australia. The intense red color of these rocks is produced by their high iron oxide content.

Sediment accumulates in more-or-less horizontal layers, called **strata**, or simply "beds" (Figure 10.11). Individual strata are separated from those below and above by surfaces called stratification planes or bedding planes. These separation surfaces allow one layer to be easily removed from the next. Strata of widely different compositions can occur alternately, one above the next.

Clastic Sedimentary Rocks

Clastic sediments are derived from one or more of the rock groups—igneous, sedimentary, metamorphic—and thus include a very wide range of minerals for sedimentary rock formation. Silicate minerals are the most important, both in original form and as altered by oxidation and hydrolysis. Quartz and feldspar usually dominate. Because quartz is hard and is immune to alteration, it is the most important single component of the clastic sediments (Figure 10.12). Second in abundance are fragments of unaltered fine-grained parent rocks, such as tiny pieces of lava rock. Feldspar and mica are also commonly present. Clay minerals are major constituents of the finest clastic sediments.

The range of particle sizes in a clastic sediment determines how easily and how far the particles are transported by water currents. The finer the particles, the more easily they are held suspended in the fluid. On the other hand, the coarser particles tend to settle to the bottom of the fluid layer. In this way, a separation of size grades, called *sorting*, occurs. Sorting determines the texture of the sediment deposit and of the sedimentary rock derived from that sediment. The finest clay particles do not settle out unless they are made to clot together into larger clumps. This clotting process, called flocculation, normally occurs when river water carrying clay mixes with the saltwater of the ocean.

When sediments accumulate in thick sequences, the lower strata are exposed to the pressure produced by the weight of the sediments above them. This pressure compacts the sediments, squeezing out excess water. Cementation occurs as dissolved minerals recrystallize where grains touch and in the spaces between mineral particles. Silicon dioxide (quartz, SiO_2) is very slightly soluble in water, and so the cement is often a form of quartz, called *silica*, which lacks a true crystalline form. Calcium carbonate ($CaCO_3$) is another common material that cements clastic sedimentary rocks. Compaction and cementation produce sedimentary rock.

The important varieties of clastic sedimentary rock are distinguished by the size of their particles. They in-

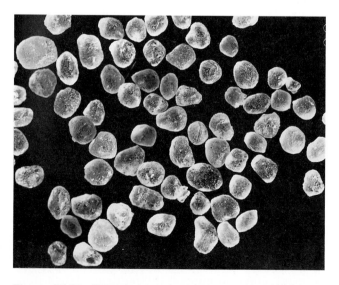

Figure 10.12 Rounded quartz grains from an ancient sandstone. The grains average about 1 mm (0.039 in.) in diameter.

Figure 10.13 An erosional remnant of massive sandstone in Utah. The thin neck of sandstone has been eroded from both sides to form a natural arch.

clude sandstone, conglomerate, mudstone, claystone, and shale. **Sandstone** is formed from fine to coarse sand (Figure 10.13). The cement may be silica or calcium carbonate. The sand grains are commonly of quartz, such as those shown in Figure 10.12. (Refer to Figure 10.2 for the names and diameters of the various grades of sediment particles.) Sandstone containing numerous rounded pebbles of hard rock is called *conglomerate* (Figure 10.14).

A mixture of water with particles of silt and clay, along with some sand grains, is called *mud*. The sedimentary rock hardened from such a mixture is called *mudstone*. Compacted and hardened clay layers become *claystone*. Sedimentary rocks of mud composition are commonly layered in such a way that they easily break apart into small flakes and plates. The rock is then described as being *fissile* and is given the name **shale**. Shale, the most abundant of all sedimentary rocks, is formed largely of clay minerals. The compaction of the mud to form mudstone and shale involves a considerable loss of volume as water is driven out of the clay.

Chemically Precipitated Sedimentary Rocks

Under favorable conditions, mineral compounds are deposited from the salt solutions of sea water and of salty inland lakes in desert climates. One of the most common sedimentary rocks formed by chemical precipitation is **limestone**, composed largely of the mineral calcite. *Calcite* is calcium carbonate ($CaCO_3$). Marine limestones—limestone strata formed on the sea floor—accumulated in thick layers in many ancient seaways in past geologic eras (Figure 10.15). A closely related rock is *dolomite*, composed of calcium-magnesium carbonate. Limestone and dolomite are grouped together as the *carbonate rocks*. They are dense rocks, ranging in color from white to pale gray or even black.

Sea water also yields sedimentary layers of silica in a hard, noncrystalline form called *chert*. Chert is a variety of sedimentary rock, but it also commonly occurs combined with limestone (cherty limestone).

Shallow water bodies acquire a very high level of salinity where evaporation is sustained and intense. One type of shallow water body is a bay or estuary in a coastal desert region. Another type is the salty lake of inland desert basins (see Figure 14.24). Sedimentary minerals and rocks deposited from such concentrated solutions are called **evaporites**. Ordinary rock salt, the

Figure 10.14 A piece of quartzitic conglomerate, cut through and polished, reveals rounded pebbles of quartz (clear and milky colors) and chert (grayish). It is about 12 cm (4.7 in.) in diameter.

Figure 10.15 The Etretât cliffs of Normandy, France, are composed of soft limestone, known as chalk.

mineral halite (sodium chloride), has accumulated in this way in thick sedimentary rock layers. These layers are mined as major commercial sources of halite.

EYE ON THE ENVIRONMENT:
Hydrocarbon Compounds in Sedimentary Rocks

Hydrocarbon compounds (compounds of carbon, hydrogen, and oxygen) form a most important type of organic sediment—one on which human society increasingly depends for fuel. These substances occur both as solids (peat and coal) and as liquids and gases (petroleum and natural gas). Only coal qualifies physically as a rock. *Peat* is a soft, fibrous substance of brown to black color. (See Figure 19.20.) It accumulates in a bog environment where the continual presence of water inhibits the decay of plant remains.

At various times and places in the geologic past, plant remains accumulated on a large scale, accompanied by sinking of the area and burial of the compacted organic matter under thick layers of inorganic clastic sediments. *Coal* is the end result of this process (Figure 10.16). Individual coal seams are interbedded with shale, sandstone, and limestone strata.

Petroleum (or *crude oil*, as the liquid form is often called) includes many hydrocarbon compounds. *Nat-*

ural gas, which is found in close association with accumulations of liquid petroleum, is a mixture of gases. The principal gas is methane (marsh gas, CH_4). Geologists generally agree that petroleum and natural gas are of organic origin, but the nature of the process is not fully understood. They are not classed as minerals, but rather are separately classed as mineral fuels.

Natural gas and petroleum commonly occupy open interconnected pores in a thick sedimentary rock layer—a porous sandstone, for example. The simplest arrangement of strata favorable to trapping petroleum and natural gas is an uparching of the type shown in Figure 10.17. Shale forms an impervious cap rock. A porous sandstone beneath the cap rock serves as a reservoir. Natural gas occupies the highest position, with the oil below it.

Yet another form of occurrence of hydrocarbon fuels is *bitumen*, a variety of petroleum that behaves much as a solid, although it is actually a highly viscous liquid. Bitumen goes by other common names, such as tar, asphalt, or pitch. In some localities, bitumen occupies pore spaces in layers of sand or porous sandstone. It remains immobile in the enclosing sand and will flow only when heated. Outcrops of *bituminous sand* (*oil sand*) exposed to the sun will show bleeding of the bitumen. Perhaps the best known of the great bituminous sand deposits are those occurring in Alberta, Canada. Where exposed along the banks of the

Figure 10.16 A coal seam near Sheridan, Wyoming, being strip mined by heavy equipment.

Athabasca River, the oil sand is extracted from surface mines.

Hydrocarbon compounds in sedimentary rocks are important because they provide an energy resource on which modern human civilization depends. These **fossil fuels**, as they are called collectively, have required millions of years to accumulate. However, they are being consumed at a very rapid rate by our industrial society. These fuels are nonrenewable resources. Once they are gone, there will be no more, because the quantity produced by geologic processes even in a thousand years is scarcely measurable in comparison to the quantity stored through geologic time. 🍂

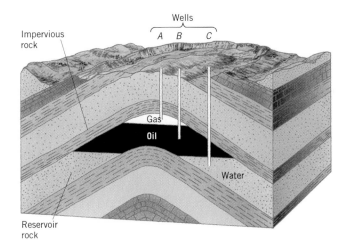

Figure 10.17 Idealized cross section of an oil pool on a dome structure in sedimentary strata. Well *A* will draw gas; well *B* will draw oil; and well *C* will draw water. The cap rock is shale; the reservoir rock is sandstone. (Copyright © A. N. Strahler.)

METAMORPHIC ROCKS

Any type of igneous or sedimentary rock may be altered by the tremendous pressures and high temperatures that accompany the mountain-building processes of the earth's crust. The result is a rock so changed in texture and structure as to be reclassified as **metamorphic rock**. Mineral components of the parent rock are, in many cases, reconstituted into different mineral varieties. Recrystallization of the original minerals can also occur. Our discussion of metamorphic rocks will mention only five common types—slate, schist, quartzite, marble, and gneiss (Table 10.3).

Table 10.3 Some Common Metamorphic Rock Types

Rock Type	Description
Slate	Shale exposed to heat and pressure that splits into hard flat plates
Schist	Shale exposed to intense heat and pressure that shows evidence of shearing
Quartzite	Sandstone that is "welded" by a silica cement into a very hard rock of solid quartz
Marble	Limestone exposed to heat and pressure, resulting in larger, more uniform crystals
Gneiss	Rock resulting from the exposure of clastic sedimentary or intrusive igneous rocks to heat and pressure

Figure 10.18 This freshly exposed schist shows shiny surfaces along which shearing has occurred. Known as a "greenschist," it represents the end product of deep deformation of black mud once deposited on the floor of an ancient oceanic trench.

Slate is formed from shale that is heated and compressed by mountain-making forces. This fine-textured rock splits neatly into thin plates, which are familiar as roofing shingles and as patio flagstones. With applica-

Figure 10.19 A surf-washed rock surface exposing banded gneiss. Pemaquid Point, Maine.

tion of increased heat and pressure, slate changes into **schist**, representing the most advanced stage of metamorphism. Schist has a structure called foliation, consisting of thin but rough, irregularly curved planes of parting in the rock (Figure 10.18). These are evidence of *shearing*—a compressional stress that pushes the layers sideways, much as a deck of cards can be pushed into a fan with the sweep of a palm. Schist is set apart from slate by the coarse texture of the mineral grains, the abundance of mica, and occasionally the presence of scattered large crystals of newly formed minerals, such as garnet.

The metamorphic equivalent of conglomerate, sandstone, and siltstone is **quartzite**, which is formed by the addition of silica to fill completely the open spaces between grains. This process is carried out by the slow movement of undergroundwaters carrying silicate into the rock, where it is deposited.

Limestone, after undergoing metamorphism, becomes *marble*, a rock of sugary texture when freshly broken. During the process of internal shearing, calcite in the limestone reforms into larger, more uniform crystals than before. Bedding planes are obscured, and masses of mineral impurities are drawn out into swirling bands.

Finally, the important metamorphic rock **gneiss** may be formed either from intrusive igneous rocks or from clastic sedimentary rocks that have been in close contact with intrusive magmas. A single description will not fit all gneisses because they vary considerably in appearance, mineral composition, and structure. One variety of gneiss is strongly banded into light and dark layers or lenses (Figure 10.19), which may be bent into wavy folds. These bands have differing mineral compositions. They are thought to be relics of sedimentary strata, such as shale and sandstone, to which new mineral matter has been added from nearby intrusive magmas.

THE CYCLE OF ROCK CHANGE

The processes that form rocks, when taken together, constitute a single system that cycles and recycles earth materials over geologic time from one form to another. The *cycle of rock change*, a concept that we introduced toward the beginning of this chapter, describes this system. Figure 10.20 shows a more complete version of this cycle. There are two environments—a surface environment of low pressures and temperatures and a deep environment of high pressures and temperatures. The surface environment is the site of rock alteration and sediment deposition. In this environment, igneous, sedimentary, and metamorphic rocks are uplifted and exposed to air and water. Their minerals are altered chemically and broken free from the parent rock, yielding sediment. The sediment accumulates in basins, where deeply buried sediment layers are compressed and cemented into sedimentary rock.

Sedimentary rock, entering the deep environment, is heated by the slow radioactive decay of elements and comes into a zone of high confining pressure. Here it is transformed into metamorphic rock. Pockets of magma are formed in the deep environment and move upward, melting and incorporating surrounding rock as they rise. Upon reaching a higher level, magma cools and solidifies, becoming intrusive igneous rock, or it may emerge at the surface to form extrusive igneous rock. And thus the cycle is completed.

The cycle of rock change has been active since our planet became solid and internally stable, continuously forming and reforming rocks of all three major classes. Not even the oldest dated igneous and metamorphic rocks thus far found are the "original" rocks of the earth's crust, for they were recycled eons ago.

The loops in the cycle of rock change are powered by a number of sources, ranging from solar energy to radiogenic heat. *Focus on Systems 10.2 • Powering the Cycle of Rock Change* describes these in more detail.

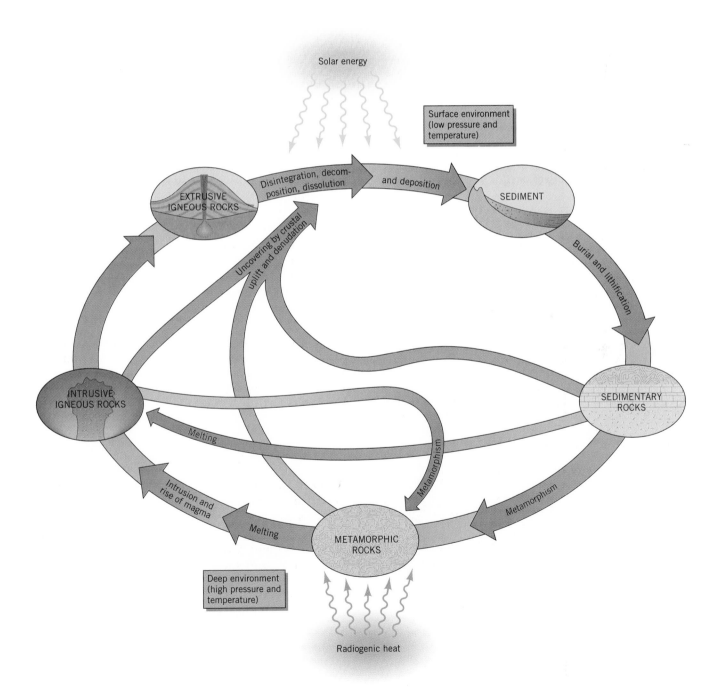

Figure 10.20 The cycle of rock change.

Focus on Systems 10.2 • Powering the Cycle of Rock Change

In the cycle of rock change, the materials of the lithosphere are constantly being formed and transformed in both their physical and mineral composition. Deep within the crust, rocks and sediments are compressed, consolidated, baked, sheared, and sometimes melted. They eventually emerge at the surface either by extrusion, in the case of lava, or by the stripping off of the overlying rock layers by erosion. Once at the surface, the rocks are physically and chemically weathered into sediment. The sediment then moves to low places where it accumulates and can be buried, thus completing the cycle.

What energy sources power the rock cycle? Let's examine the underground part of the cycle first.

For this part of the cycle, the main power source is *radiogenic heat*. This term refers to heat that is slowly released by the radioactive decay of unstable isotopes that were originally formed early in the history of the solar system. (See *Working It Out 10.1 • Radioactive Decay* for more details.) Isotopes of uranium (^{238}U, ^{235}U), thorium (^{232}Th), and potassium (^{40}K), along with the daughter products generated by their decay, are the source of nearly all of this heating.

The upper part of the figure below plots temperature with depth below the earth's surface. Two curves are shown—one for temperatures beneath continents and one for temperatures beneath ocean basins. Although the curves are slightly different, they show that the temperature increases rapidly at first and then increases only slowly with depth. Most of the radiogenic heat is liberated in the rock beneath the continents, within the uppermost 100 km (about 60 mi) or so. This observation fits with chemical analyses of continental rocks, which show that isotopes of uranium, thorium, and potassium are most abundant in the upper layers of continents. Radiogenic heat is sufficient to keep earth layers below the crust close to the melting point, and thus provides the power source for the heating by which metamorphic and igneous rocks are formed from preexisting rocks.

A secondary power source is gravity. As sedimentary layers are buried

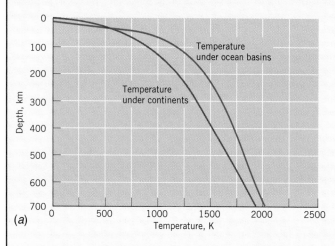

(a)

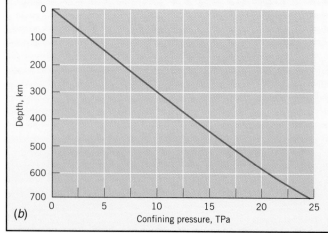

(b)

Change in temperature (*a*) and pressure (*b*) with depth from the earth's surface. In (*b*), units are terapascals, TPa (1 TPa = 10 kilobars). (Copyright © A. N. Strahler.)

more and more deeply, the pressure on the layers increases. The lower graph plots pressure with depth below the earth's surface. This pressure acts to bring mineral grains or plates into very close contact, where they can bond to one another. Water, often containing dissolved silica or calcium carbonate, is forced out of the sediments while leaving deposits of these minerals to cement the grains in place. Thus, the rock density increases with time.

Gravity and radiogenic heat are power sources that alter rock composition and structure. But what causes the motions of rocks, in which sediments and preexisting rocks are buried deep within the earth and then later brought to the surface? These motions are part of the plate tectonic cycle, which we will examine in the following chapter.

The above-ground portion of the cycle of rock change is driven by several power sources, shown in the figure below. Of these, the sun and gravity are the most important. Solar energy, and the unequal heating of the earth's surface that it provides, is the primary energy source for the motions of the atmosphere and oceans, as we have seen in prior chapters. A lesser, but still important, power source for atmospheric and oceanic movements is the earth's rotation, which transfers heat energy through friction to the fluid atmosphere and ocean that it contacts. The tidal forces of the sun and moon also act to keep the ocean and atmosphere in motion.

The circulation of the atmosphere and oceans produces moving fluids—air, water, and glacial ice—that transport weathered rock particles while reducing them to ever-finer sizes. In this way, the particles are also directly exposed to oxygen and water, allowing alteration of their chemical composition. The action of gravity on the particles eventually brings them to sedimentary basins, where they accumulate and the underground portion of the cycle of rock change begins. The figure below shows these relationships among the flows of energy that power the cycle of rock change.

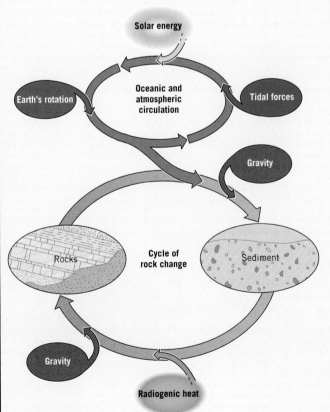

Flows of energy that drive the cycle of rock change.

This chapter has described the minerals and rocks of the earth's surface and the processes of their formation. As we look further into the earth's outermost layers in the next chapter, we will see that these processes do not occur everywhere. Instead, there is a grand plan that organizes the formation and destruction of rocks and distributes the processes of the cycle of rock change in a geographic pattern. This pattern is controlled by the pattern in which the solid earth's brittle outer layer is fractured into great plates that split and separate and also converge and collide. The grand plan is plate tectonics, the scheme for understanding the dynamics of the earth's crust over millions of years of geologic time.

CHAPTER SUMMARY

The elements oxygen and silicon dominate the earth's crust. Metallic elements, which include aluminum, iron, and the base elements, account for nearly all the remainder. These elements are found in minerals—naturally occurring, inorganic substances. Each has an individual chemical composition and atomic structure.

Silicate minerals make up the bulk of igneous rocks. They contain silicon and oxygen together with some of the metallic elements. There are three broad classes of igneous rocks, depending on their mineral content. Felsic rocks contain mostly felsic minerals and are least dense. Mafic rocks, containing mostly mafic minerals, are denser. Ultramafic rocks are most dense. Because felsic rocks are least dense, they are generally found in the upper layers of the earth's crust. Mafic and ultramafic rocks are more abundant in the layers below. If magma erupts on the surface to cool rapidly as lava, the rocks formed are extrusive and have a fine crystal texture. If the magma cools slowly below the surface as a pluton, the rocks are intrusive and the crystals are larger. Granite (felsic, intrusive), andesite (felsic, extrusive), and basalt (mafic, extrusive) are three very common igneous rock types.

Most silicate minerals found in igneous rocks undergo mineral alteration when exposed to air and moisture at the earth's surface. Mineral alteration occurs through oxidation, hydrolysis, or solution. Clay minerals are commonly produced by mineral alteration.

Sedimentary rocks are formed in layers, or strata. Clastic sedimentary rocks are composed of fragments of rocks and minerals that usually accumulate on ocean floors. As the layers are buried more and more deeply, water is pressed out and particles are cemented together. Sandstone and shale are common examples. Chemical precipitation also produces sedimentary rocks, such as limestone. Coal, petroleum, and natural gas are hydrocarbon compounds occurring in sedimentary rocks that are used as mineral fuels.

Metamorphic rocks are formed when igneous or sedimentary rocks are exposed to heat and pressure. Shale is altered to schist, sandstones become quartzite, and intrusive igneous rocks or clastic sediments are metamorphosed into gneiss.

In the cycle of rock change, rocks are exposed at the earth's surface, and their minerals are broken free and altered to form sediment. The sediment accumulates in basins, where the layers are compressed and cemented into sedimentary rock. Deep within the earth, the heat of radioactive decay melts preexisting rock into magma, which can move upward into the crust to form igneous rocks that cool at or below the surface. Rocks deep in the crust are exposed to the heat and pressure, forming metamorphic rock. Mountain-building forces move deep igneous, sedimentary, and metamorphic rocks upward to the surface, providing new material for surface alteration and breakup and completing the cycle.

KEY TERMS

mineral	quartz	lava
rock	felsic	extrusive igneous rock
igneous rocks	mafic	andesite
sedimentary rocks	magma	basalt
metamorphic rocks	granite	pluton
cycle of rock change	intrusive igneous rock	mineral alteration

clay mineral
sediment
clastic sediment
chemically precipitated
  sediment
organic sediment

strata
sandstone
shale
limestone
evaporites
fossil fuels

metamorphic rock
schist
quartzite
gneiss

REVIEW QUESTIONS

1. What is the earth's crust? What elements are most abundant in the crust?

2. Define the terms *mineral* and *rock*. Name the three major classes of rocks.

3. What are silicate minerals? Describe two classes of silicate minerals.

4. Name four types of igneous rocks and arrange them in order of density.

5. How do igneous rocks differ when magma cools (a) at depth, and (b) at the surface?

6. Sketch a cross section of the earth showing the following features: batholith, sill, dike, veins, lava, and volcano.

7. How are igneous rocks chemically altered? Identify and describe three processes of chemical alteration.

8. What is sediment? Define and describe three types of sediments.

9. Describe two processes that produce sedimentary rocks, and identify at least three important varieties of clastic sedimentary rocks.

10. How are sedimentary rocks formed by chemical precipitation?

11. What types of sedimentary deposits consist of hydrocarbon compounds? How are they formed?

12. What are metamorphic rocks? Describe at least three types of metamorphic rocks and how they are formed.

13. Sketch the cycle of rock change and describe the processes that act within it to form igneous, sedimentary, and metamorphic rocks.

Focus on Systems 10.2 • Powering the Cycle of Rock Change
1. How do pressure and temperature change with depth below the earth's surface? Sketch a graph of temperature change with depth and explain its shape.

2. Identify and describe the sources of energy that power the cycle of rock change.

ESSAY QUESTION

1. A granite is exposed at the earth's surface, high in the Sierra Nevada mountain range. Describe how mineral grains from this granite might be released, altered, and eventually become incorporated in a sedimentary rock. Trace the route and processes that would incorporate the same grains in a metamorphic rock.

PROBLEMS

Working It Out 10.1 • Radioactive Decay
1. The half-life of potassium-40 (^{40}K) is 1.26 billion years. What percentage of a quantity of ^{40}K will remain after 1 billion years? 3 billion years? (*Hint:* Check your work by referring to the figure in the box.)

2. The half-life of ^{232}Th is 14 billion years. What percentage of a quantity of pure ^{232}Th will remain after 5 billion years? 10 billion years? 15 billion years?

3. The unstable isotope carbon-14 (^{14}C) has a half-life of 5715 years. How many years will be required for the proportion of a pure sample of ^{14}C to be reduced to 0.1 (10 percent)?

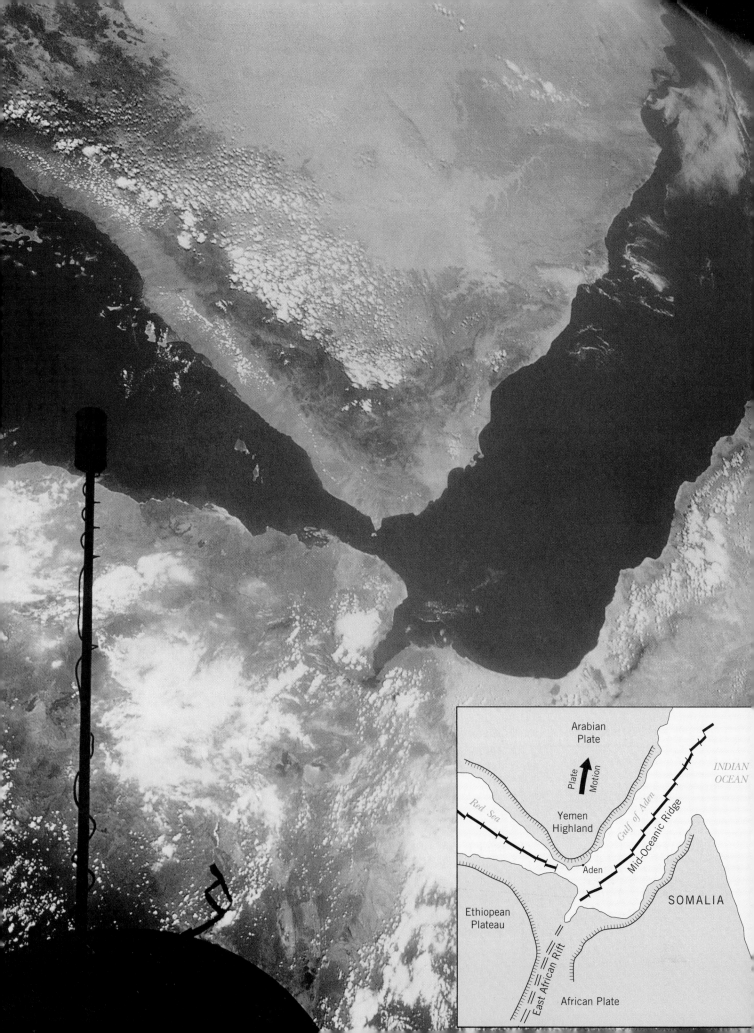

Arabian
Plate

Plate
Motion

Red Sea

Yemen
Highland

INDIAN
OCEAN

Gulf of Aden

Mid-Oceanic Ridge

Aden

SOMALIA

Ethiopean
Plateau

East African Rift

African Plate

The Lithosphere and the Tectonic System

O n the globes and maps we've seen since early childhood, the outline of each continent is so unique that we would never mistake one continent for

Astronauts of the Gemini XI mission took this photograph of the Red Sea and Gulf of Aden as their space vehicle passed eastward over northeastern Africa. The view is toward the north, but directions and distances are severely distorted. Cloud formations cover much of the high plateaus. We are seeing an early stage of the separation of the Arabian plate from the African plate. The first new ocean floor to appear was the Gulf of Aden—a branch of the Indian Ocean seen at the far right. Visualize how the Yemen Highland would have fit into the Somalian coastline. More recently, the Red Sea graben to the west began to open. The great East African Rift Valley enters the scene from the south, so we have a triple-plate junction in process of formation. Some time in the far distant future a new "Somalian Plate" will require recognition.

another. But why are no two continents even closely alike? As we will see in this chapter, the continents have had a long history. In fact, in each of the continents some regions of metamorphic rocks date back more than 2 billion years. As part of that history, the continents have been fractured and split apart, as well as pushed together and joined.

Perhaps you've visited an old New England farmhouse that was constructed over hundreds of years—first the small two-room house, to which the kitchen shed was added at the back, and then the parlor wing at the side. Later the roof was raised for a second story, the tool house and barn were joined to the main house, the carriage house was built, and so it went. The earth's continents have that kind of history. They are composed of huge masses of continental crust that have been assembled at different times in each continent's history. The theory describing the motions and changes through time of the continents and ocean basins, and the processes that fracture and fuse them, is called *plate tectonics.*

The forces that move continents and cause them to collide are powered by energy sources deep within the earth. These forces are not influenced by the surface patterns of temperature, winds, precipitation, vegetation, or soils. Today we find volcanoes erupting in the cold desert of Antarctica as well as near the equator in African savannas. An alpine mountain range has been pushed up in the cold subarctic zone of Alaska, where it runs east-west. Yet another range lies astride the equator in South America and runs north-south. Both ranges lie in belts of crustal collision where many strong earthquakes occur.

Although internal crustal processes operate independently, the processes that govern climate, vegetation, and soils are dependent on the major relief features and earth materials provided by the crustal processes. Thus, an understanding of plate tectonics is important to our understanding of the global patterns of the earth's landscapes—including its climate, soils, vegetation, and, ultimately, human activity.

In this chapter we will survey the major geologic features of our planet, starting with the layered structure of its deep interior. We will then examine the outermost layer, or crust, and compare the crust of the continents with the crust of the ocean basins. Lastly, we will turn to plate tectonics and describe how plate movements have created broad regions of igneous, sedimentary, and metamorphic rocks. The motions and histo-

ries of these vast rock plates, splitting and separating to form ocean basins, closing and colliding to form mountain ranges, can be taken as an overarching framework to organize a great body of geologic and geographic knowledge. That is why this revolutionary framework—plate tectonic theory—ranks with the theory of evolution as one of the great milestones in scientific study of the earth.

THE STRUCTURE OF THE EARTH

What lies deep within the earth? From studies of earthquake waves, reflected from deep earth layers, scientists have discovered that our earth is far from uniform from its outer crust to its center. Instead, it consists of a central core with several layers, or shells, surrounding it. The densest matter is at the center, and each layer above it is increasingly less dense. We will begin our examination of the earth's inner structure at the center and then work outward.

The Inner Structure and Crust

Figure 11.1 is a cutaway diagram of the earth showing its interior. The earth as a whole is an almost spherical body approximately 6400 km (about 4000 mi) in radius. The center is occupied by the **core**, which is about 3500 km (about 2200 mi) in radius. Because of the sudden change in behavior of earthquake waves upon reaching the core, scientists have concluded that the

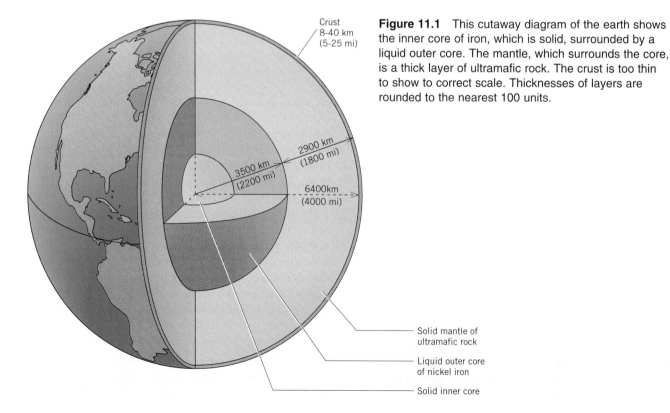

Figure 11.1 This cutaway diagram of the earth shows the inner core of iron, which is solid, surrounded by a liquid outer core. The mantle, which surrounds the core, is a thick layer of ultramafic rock. The crust is too thin to show to correct scale. Thicknesses of layers are rounded to the nearest 100 units.

outer core has the properties of a liquid. However, the innermost part of the core is in the solid state. Based on earthquake waves (and other kinds of data), it has long been inferred that the core consists mostly of iron, with some nickel. The core is very hot—its temperature ranges from about 2800°C to 3100°C (about 5100°F to 5600°F).

Enclosing the metallic core is the **mantle**, a rock shell about 2900 km (about 1800 mi) thick, composed of ultramafic mineral matter. Judging from the behavior of earthquake waves, we can conclude that the mantle is composed of mafic minerals similar to olivine (a silicate of iron and magnesium). Thus, the mantle rock may resemble the ultramafic igneous rock peridotite, which is found exposed here and there on the continental surface. Temperatures in the mantle range from about

(a)

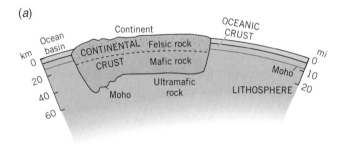

(b)

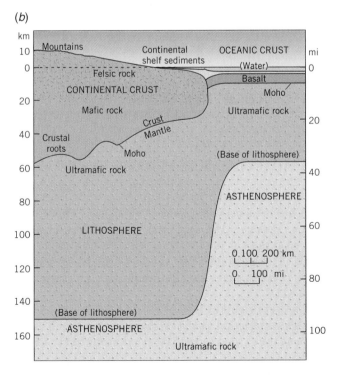

Figure 11.2 (*a*) Idealized cross section of the earth's crust and upper mantle. (*b*) Details of the crust and mantle at the edge of a continent, including the types of rocks found there. Also shown are the lithosphere and asthenosphere. (Copyright © A. N. Strahler.)

2800°C (about 5100°F) near the core to about 1800°C (about 3300°F) near the crust.

The outermost and thinnest of the earth shells is the **crust**, a layer normally about 8 to 40 km (about 5 to 25 mi) thick (Figure 11.2). It is formed largely of igneous rock, but it also contains substantial proportions of metamorphic rock and a comparatively thin upper layer of sedimentary rock. The base of the crust, where it contacts the mantle, is sharply defined. This contact is detected by the way in which earthquake waves abruptly change velocity at that level. The boundary surface between crust and mantle is called the *Moho*, a simplification of the name of the seismologist, Andrija Mohorovicic, who discovered it in 1909.

The continental crust is quite different from the crust beneath the oceans. From their study of earthquake waves, geologists have concluded that the **continental crust** consists of two continuous zones—a lower, continuous rock zone of mafic composition, which is more dense, and an upper zone of felsic rock, which is less dense (Figure 11.2). Because the felsic portion has a chemical composition similar to that of granite, it is commonly described as being *granitic rock*. Much of the granitic rock is metamorphic rock. There is no sharply defined separation between the felsic and mafic zones.

The crust of the ocean basins is sharply different from continental crust. **Oceanic crust** consists almost entirely of the mafic rocks basalt and gabbro. Basalt, as lava, forms an upper zone, whereas gabbro, an intrusive rock of the same composition, lies beneath the basalt.

Now, add to the above description the fact that the crust is much thicker beneath the continents than beneath the ocean floors, as Figure 11.2 shows. While 35 km (22 mi) is a good average figure for crustal thickness beneath the continents, 7 km (4 mi) is a typical figure for thickness of the basalt/gabbro crust beneath the deep ocean floors. The differences in both thickness and rock composition between continental and oceanic crust are attributable to formative processes that have created the crust, which we discuss later in this chapter.

THE LITHOSPHERE

Geologists use the term **lithosphere** to mean an outer earth zone, or shell, of rigid, brittle rock. This usage of the term is different from that found in our Introduction, which identified the lithosphere as one of the four great realms of earth—atmosphere, hydrosphere, lithosphere, and biosphere. The word does not carry any special meaning in terms of chemical makeup of the rock, since the lithosphere can include felsic, mafic, and ultramafic rocks.

The lithosphere is a much thicker zone than the crust, as shown in Figure 11.2. It includes not only the crust, but also the cooler, upper part of the mantle that is composed of brittle rock. Since mantle rock beneath the brittle lithosphere is highly heated, it is in a plastic physical state. You can think of it as being much like white-hot iron on the verge of melting and capable of being shaped with little difficulty. In this state, iron is easily pressed into a form or drawn out into wire. Using the same analogy, we can say that the lithosphere resembles cold cast iron that responds to strong twisting or sharp bending by breaking abruptly ("snapping") along sharp fractures.

Some tens of kilometers deep in the earth, the brittle condition of the lithospheric rock gives way gradually to a plastic, or "soft," layer named the **asthenosphere** (Figure 11.2). (This word is derived from the Greek root *asthenes* meaning "weak.") However, at still greater depth in the mantle, the strength of the rock material again increases. Thus, the asthenosphere is a soft layer sandwiched between the "hard" lithosphere above and a "strong" mantle rock layer below. In terms of states of matter, the asthenosphere is not a liquid, even though

its temperature reaches 1400°C (about 2600°F). Its melting point is raised by the immense pressure from the weight of overlying rocks.

As shown in Figure 11.2, the lithosphere ranges in thickness from 60 to 150 km (37 to 93 mi). It is thickest under the continents and thinnest under the ocean basins. Figure 11.3 idealizes the lithosphere and asthenosphere as two simple global layers of uniform thickness. We use the thickness of the oceanic lithosphere (60 km, or 37 mi) in this diagram. We present the layers this way only to show their true-scale dimensions in relation to the entire earth. The asthenosphere extends down to a depth of at least 300 km (186 mi), but both upper and lower boundaries are actually gradational. Under the ocean floors, the weakest portion of the asthenosphere lies at a depth of roughly 200 km (about 125 mi). Here the ultramafic mantle rock is close to its melting point.

The rigid, brittle lithosphere (or "hard shell") can move bodily over the soft, plastic asthenosphere. The yielding of the asthenosphere to allow this "gliding" motion is distributed through a thickness of many tens of kilometers. Also, the lithospheric shell is broken into large pieces called **lithospheric plates**. A single plate can be as large as a continent and can move independently of the plates that surround it. Like great slabs of floating ice on the polar sea, lithospheric plates can separate from one another at one location, while elsewhere they may collide in crushing impacts that raise great ridges. Along these collision ridges, one plate can be found diving down beneath the edge of its neighbor. These varied sorts of plate movements are the primary focus of this chapter.

THE GEOLOGIC TIME SCALE

To place the great movements of lithospheric plates in their correct positions in historical sequence, we will need to refer to some major units in the scale of geologic time. Table 11.1 lists the major geologic time divisions. All time older than 570 million years (m.y.) before the present is *Precambrian time.* Three *eras* of time follow: *Paleozoic, Mesozoic,* and *Cenozoic.* These eras saw the evolution of life-forms in the oceans and on the lands. The geologic eras are subdivided into *periods.* Their names, ages, and durations are also given in Table 11.1.

The Cenozoic Era is particularly important in terms of the continental surfaces, because nearly all landscape features seen today have been produced in the 65 million years since that era began. The Cenozoic Era is comparatively short in duration, scarcely more than the average duration of a single period in older eras. It is subdivided directly into seven lesser time units called *epochs.* (Note that the terms *Tertiary Period* and *Quaternary Period* are sometimes applied to the Cenozoic

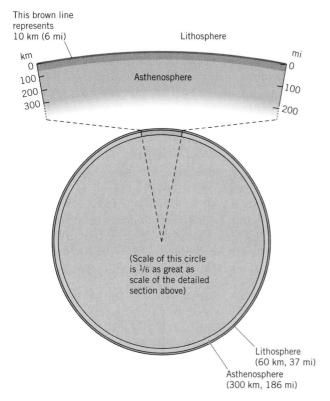

Figure 11.3 The lithosphere and asthenosphere drawn to true scale. The diagram illustrates the extreme thinness of the mobile lithospheric plates that move over the asthenosphere. The black line at the top of the upper diagram is scaled to represent a thickness of 10 km (6 mi). It will accommodate about 98 percent of the earth's surface features, from ocean floors to high mountains and plateaus. Only a few lofty mountains would project above the black line, and only a few deep ocean trenches would project below the line.

Table 11.1 Geologic Time Scale

Era	Period	Epoch	Duration (millions of years)	Age (millions of years)	Orogenies
Cenozoic		Holocene	(10,000 yr)		
		Pleistocene	2	2	
		Pliocene	3	5	
		Miocene	19	24	
		Oligocene	13	37	
		Eocene	21	58	
		Paleocene	8	66	Cordilleran
Mesozoic	Cretaceous		78	144	
	Jurassic		64	208	
	Triassic		37	245	Allegheny, or Hercynian
Paleozoic	Permian		41	286	
	Carboniferous		74	360	
	Devonian		48	408	Caledonian
	Silurian		30	438	
	Ordovician		67	505	
	Cambrian		65	570	

Precambrian Time (Extends to oldest known rocks, about 4 billion years)
Age of earth as a planet: 4.6 to 4.7 billion years.
Age of universe: 17 to 18 billion years.

epochs Paleocene through Pliocene, and Pleistocene, respectively.) Details of the Pleistocene and Holocene epochs are given in Chapter 18.

The human genus *Homo* evolved during the late Pliocene Epoch and throughout the Pleistocene Epoch. As you can see, the period of human occupation of the earth's surface—a few million years at best—is but a fleeting moment in the vast duration of our planet's history.

Geologists use a variety of techniques to establish the ages of rocks within the geologic time scale. One of the most important tools is *radiometric dating*, which establishes the age of certain minerals within rocks using principles of radioactive decay. This process is described in more detail in *Working It Out 11.1. • Radiometric Dating.*

CONTINENTS AND OCEAN BASINS

The major relief features of the earth are the continents and ocean basins (Figure 11.4). Detailed global maps show that about 29 percent of the earth's surface is land and 71 percent oceans. If the seas were to drain away, however, we would see that broad areas lying close to the continental shores are actually covered by

shallow water, less than 150 m (500 ft) deep. From these relatively shallow continental shelves, the ocean floor drops rapidly to depths of thousands of meters. In that respect, the ocean basins seem to be more than "brim full" of water. That is, the oceans have spread over the margins of ground that would otherwise be assigned to the continents. If the ocean level were to drop by 150 m (about 500 ft), the shelves would be exposed, adding about 6 percent to the area of the continents. Then, the surface area of continents would be

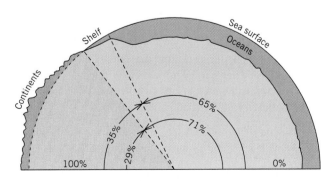

Figure 11.4 Actual global percentages of land and ocean areas compared with percentages if sea level were to drop 180 m (about 600 ft), exposing the continental shelves.

Working It Out 11.1 • Radiometric Dating

How old is the earth? How old are the oldest rocks? These questions have interested geologists ever since systematic study of the earth began. However, it was not until the early part of the twentieth century, with the discovery of radioactive decay, that scientists uncovered a principle that would provide accurate answers to these questions. And it was not until the middle of the century, when technology provided methods to measure minute amounts of trace elements in rocks, that it became possible to apply this principle to the practical problem of determining the ages of rocks.

As noted in *Working It Out 10.1 • Radioactive Decay*, certain isotopes of elements undergo a spontaneous decay process in which their atomic composition changes. Recall from that feature that ^{238}U decays to form the daughter product ^{234}Th. The half-life for this decay is 4.47×10^9 years. However, ^{234}Th is itself unstable and decays to form another isotope, which is also unstable, and decays to form yet another isotope, and so on, until the stable isotope lead-206 (^{206}Pb) is ultimately formed. Recall also that another naturally occurring isotope of uranium, ^{235}U, has a similar decay sequence and yields the stable isotope lead-207 (^{207}Pb) with a half-life of 7.04×10^8 years.

Suppose that an initial quantity of ^{238}U is included in the formation of an igneous rock early in the earth's history. Locked in place within the crystalline structure of minerals within the rock, the atoms of ^{238}U will slowly transform themselves to ^{206}Pb at a rate that depends only on the half-life of ^{238}U. This is the principle behind *radiometric dating*—using the ratios of particular elemental isotopes to determine the time of formation or metamorphism of a rock.

How exactly does radiometric dating work? Recall from *Working It Out 10.1* that

$$P(t) = e^{-kt}$$

where $P(t)$ is the proportion of an unstable isotope remaining after time t, and $k = 0.693/H$, with H the half-life of the isotope. From this relation, we can derive the formula

$$t = \frac{1}{k}\ln\left[\frac{D}{M}+1\right]$$

which gives the time of decay, t, as a function of k and D/M, the ratio of daughter to mother atoms in the sample. Suppose, for example, that a chemical analysis of a mineral shows that for every two atoms of ^{238}U, one atom of ^{206}Pb is present. Then $D/M = 1/2 = 0.5$, and since $k = 0.693/4.47 = 0.155$ using units of billions of years (1 b.y. $= 10^9$ yr), we have

$$t = \frac{1}{0.155}\ln[0.5+1] = \frac{0.405}{0.155} = 2.61 \text{ b.y.}$$

The practice of radioactive dating goes well beyond this simple example, but the principle is basically the same—that by analyzing the concentration of certain radioactive isotopics in certain minerals carefully and exactly, it is possible to date them with certainty.

How old are the oldest rocks? Among the oldest rocks known are ancient shield rocks from Greenland, which have been dated at 3.80 b.y. Even older are some shield rocks from Antarctica, which date to 3.93 b.y. But the oldest rocks so far discovered are ancient gneisses in northwestern Canada. They contain zircon crystals with ages of 3.96 b.y.

How old is the earth? During the earliest eras of the earth's history, continental crust was being vigorously formed and reformed, so rocks and minerals from these early times are long gone. However, some types of meteorites have radiometric ages of about 4.6 b.y., and these also have the same mixture of naturally occurring lead isotopes as the earth's rocks. This latter fact leads geochemists to suspect strongly that these meteorites were formed from the same primordial matter as the earth itself, thus dating the formation of the earth at about 4.6 billion years ago.

---

increased to 35 percent, and the ocean basin area decreased to 65 percent. These revised figures represent the true relative proportions of continents and oceans. This same ratio—35 to 65—also applies to the global ratio of continental crust to oceanic crust.

Relief Features of the Continents

Broadly viewed, the continental masses consist of two basic subdivisions: (1) active belts of mountain-making and (2) inactive regions of old, stable rock. The moun-

tain ranges in the active belts grow through one of two very different geologic processes. First is *volcanism*, the formation of massive accumulations of volcanic rock by extrusion of magma. Many lofty mountain ranges consist of chains of volcanoes built of extrusive igneous rocks. The second mountain-building process is *tectonic activity*, the breaking and bending of the earth's crust under internal earth forces. This tectonic activity usually occurs when great lithospheric plates come together in titanic collisions. (We will explain this topic later in this chapter.) Crustal masses that are raised by tectonic activity form mountains and plateaus. Masses that are lowered form crustal depressions. In some instances, volcanism and tectonic activity have combined to produce a mountain range. Landforms produced by volcanic and tectonic activity are the subject of Chapter 12.

Alpine Chains

Active mountain-making belts are narrow zones that are usually found along continental margins. These belts are sometimes referred to as *alpine chains* because they are characterized by high, rugged mountains, such as the Alps of Central Europe. These mountain belts were formed in the Cenozoic Era by volcanism, or tectonic activity, or a combination of both. Alpine mountain-building continues even today in many places.

The alpine chains are characterized by broadly curved patterns on the world map (Figure 11.5). Each curved section of an alpine chain is referred to as a **mountain arc**. The arcs are linked in sequence to form the two principal mountain belts. One is the *circum-Pacific belt*, which rings the Pacific Ocean basin. In North and South America, this belt is largely on the conti-

nents and includes the Andes and Cordilleran ranges. In the western part of the Pacific basin, the mountain arcs lie well offshore from the continents and take the form of **island arcs**. Partly submerged, they join the Aleutians, Kurils, Japan, and the Philippines, as well as many small islands. These island arcs are the result of volcanic activity. Between the larger islands, the arcs are represented by volcanoes rising above the sea as small, isolated islands.

The second chain of major mountain arcs forms the *Eurasian-Indonesian belt*, shown in Figure 11.5. It starts in the west at the Atlas Mountains of North Africa and continues through the European Alps and the ranges of the Near East and Iran to join the Himalayas. The belt then continues through Southeast Asia into Indonesia, where it abruptly meets the circum-Pacific belt. Later we will return to these active belts of mountain-making and explain them in terms of lithospheric plate motions.

Continental Shields

Belts of recent and active mountain-making account for only a small portion of the continental crust. The

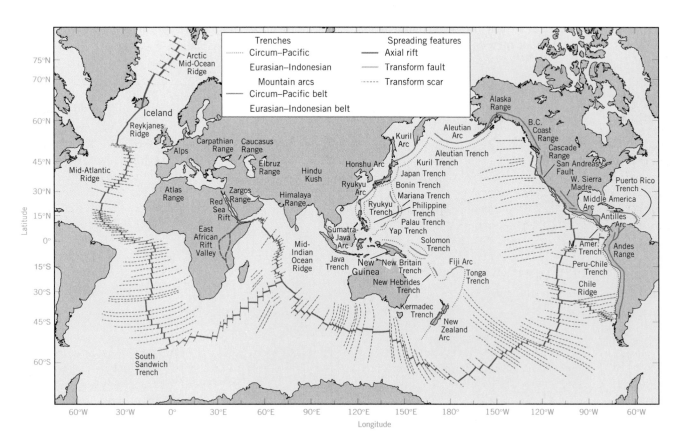

Figure 11.5 Principal mountain arcs, island arcs, and trenches of the world and the midoceanic ridge. (Midoceanic ridge map copyright © A. N. Strahler.)

remainder consists of comparatively inactive regions under which lie much older rock. Within these stable regions we recognize two structural types of crust—shields and mountain roots. **Continental shields** are low-lying continental surfaces beneath which lie igneous and metamorphic rocks in a complex arrangement. Figure 11.6 is a very generalized map showing the shield areas of the continents. Two classes of shield are shown: exposed shields and covered shields. In the *exposed shields,* very old rocks lie at the surface. They are mostly of Precambrian age and have had a very complex geologic history. An example of an exposed shield is the Canadian Shield of North America. Exposed shields are also extensive in Scandinavia, South America, Africa, peninsular India, and Australia.

The exposed shields are largely regions of low hills and low plateaus, although there are some exceptions where large crustal blocks have been recently uplifted. Many thousands of meters of rock have been eroded from these shields during their continuous exposure throughout the past half-billion or more years.

Large areas of the continental shields are covered by younger sedimentary layers, ranging in age from Paleozoic through Cenozoic eras. These strata accumulated at times when the shields subsided and were inundated by shallow seas. Marine sediments were laid down on the ancient shield rocks in thicknesses ranging from hundreds to thousands of meters. These shield areas were then broadly arched and again became land surfaces. Erosion has since removed large sections of the sedimentary cover, but it still remains intact over vast areas. We refer to such areas as *covered shields* in order to distinguish them from the exposed shields in which the Precambrian rocks lie bare. The covered shields are shown in Figure 11.6.

Some core areas of the shields are composed of rock as old as early Precambrian time, dating back to a time period called the *Archean Eon,* 2.5 to 3.5 billion years ago. On our map, these ancient areas are shown encircled by bold lines. The ancient cores are exposed in some areas but covered in others.

Ancient Mountain Roots

Remains of older mountain belts lie within the shields in many places. These *mountain roots* are mostly formed of Paleozoic and early Mesozoic sedimentary rocks that have been intensely bent and folded, and in most locations changed into metamorphic rocks—slate, schist, and quartzite, for example. Thousands of meters of overlying rocks have been removed from these old tectonic belts, so that only the lowermost structures remain. Roots appear as chains of long, narrow ridges, rarely rising over a thousand meters above sea level. (Landforms of mountain roots are described in Chapter 16.)

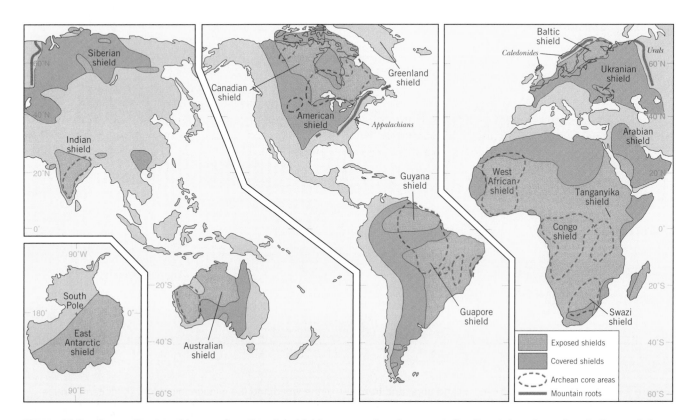

Figure 11.6 Generalized world map of continental shields, exposed and covered. Continental centers of early Precambrian age (Archean) lie with the areas encircled by broken red line. Heavy brown lines show mountain roots of Caledonian and Hercynian orogenies. (Based in part on data of R. E. Murphy, P. M. Hurley, and others. Copyright © A. N. Strahler.)

One important system of mountain roots was formed in the Paleozoic Era, during a great collision between two enormous lithospheric plates that took place about 400 million years ago. What were then high alpine mountain chains have since been worn down to belts of subdued mountains and hills. Today these roots, called Caledonides, form a highland belt across the northern British Isles and Scandinavia. They are also present in the Maritime Provinces of eastern Canada and in New England. A second, but younger, root system was formed during another great collision of plates near the close of the Paleozoic Era, about 250 million years ago. In North America, this highland system is represented by the Appalachian Mountains. The Caledonides and Appalachians are shown as mountain roots in Figure 11.6.

Relief Features of the Ocean Basins

Crustal rock of the ocean floors consists almost entirely of basalt, which is covered over large areas by a comparatively thin accumulation of sediments. Age determinations of the basalt and its sediment cover show that the oceanic crust is quite young, geologically speaking. Much of that crust was formed during the Cenozoic Era and is less than 60 million years old. Over some large areas, however, the rock is of Mesozoic age, mostly from 65 to about 135 million years old. When we consider that the great bulk of the continental crust is of Precambrian age—mostly over 1 billion years old—the young age of the oceanic crust is quite remarkable. However, we will soon see how plate tectonic theory explains this young age.

The Midoceanic Ridge

Figure 11.7 shows the important relief features of ocean basins. A typical ocean basin is characterized by a central ridge structure that divides the basin in about half. The *midoceanic ridge* consists of submarine hills that rise gradually to a rugged central zone. Precisely in the center of the ridge, at its highest point, is the *axial rift*, which is a narrow, trenchlike feature. The location and form of this rift suggest that the crust is being pulled apart along the line of the rift.

The midoceanic ridge and its principal branches can be traced through the ocean basins for a total distance of about 60,000 km (about 37,000 mi). Figure 11.5 shows the extent of the ridge. Beginning in the Arctic Ocean, it then divides the Atlantic Ocean basin from Iceland to the South Atlantic. Turning east, it enters the Indian Ocean, where one branch penetrates Africa. The other branch continues east between Australia and Antarctica, and then swings across the South Pacific. Nearing South America, it turns north and reaches North America at the head of the Gulf of California.

The Ocean Basin Floor

On either side of the midoceanic ridge are broad, deep plains and hill belts that belong to the *ocean basin floor* (Figure 11.7). Their average depth below sea level is about 5000 m (about 16,000 ft). The flat surfaces are called *abyssal plains*. They are smoothly coated with fine sediment that has settled slowly and evenly from ocean water above.

Many details of the ocean basins and their submarine landforms are shown in Figure 11.8. This image of the ocean floor was constructed from data of a U.S. Navy satellite that measured the surface height of the ocean very precisely using radar. Because small variations in gravity caused by undersea ridges and valleys produce corresponding small variations in the ocean height, it is possible to use the surface height data to infer the undersea topography.

The right page of the figure shows the North Atlantic basin. The prominent central feature is the Mid-Atlantic Ridge. The left page shows the deep trenches of

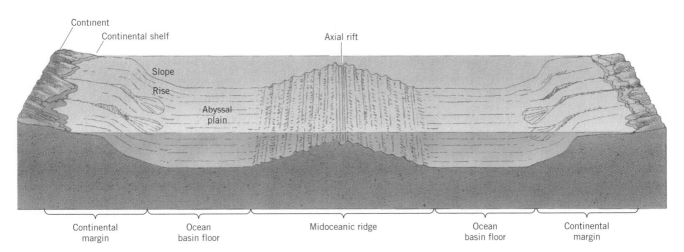

Figure 11.7 This schematic block diagram shows the main features of ocean basins. It applies particularly well to the North and South Atlantic oceans.

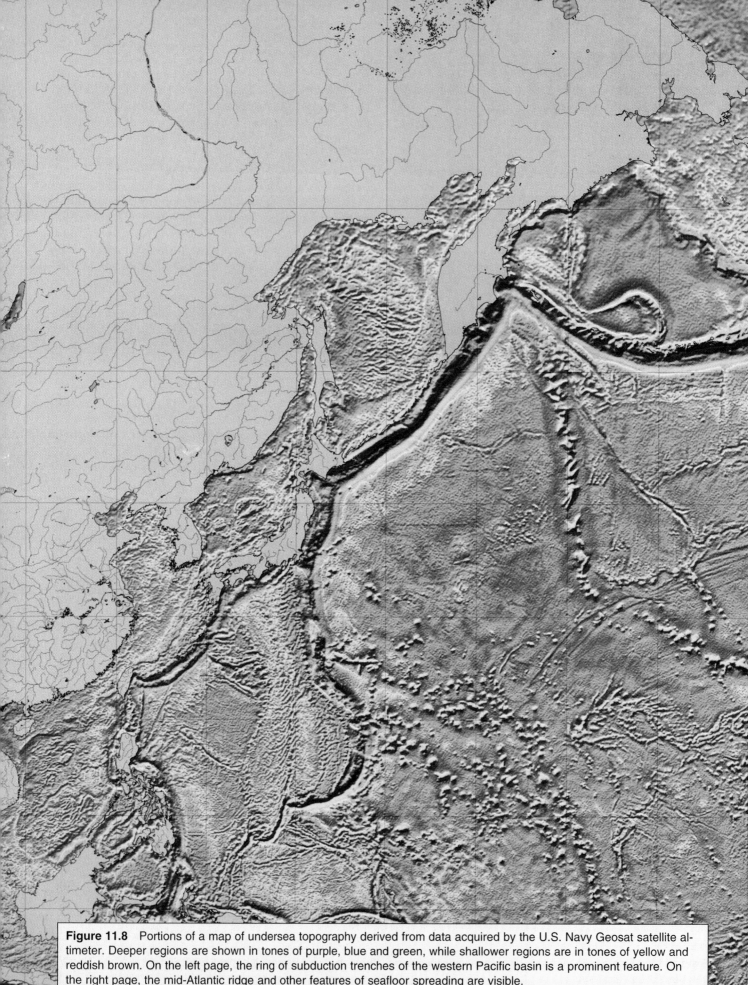

Figure 11.8 Portions of a map of undersea topography derived from data acquired by the U.S. Navy Geosat satellite altimeter. Deeper regions are shown in tones of purple, blue and green, while shallower regions are in tones of yellow and reddish brown. On the left page, the ring of subduction trenches of the western Pacific basin is a prominent feature. On the right page, the mid-Atlantic ridge and other features of seafloor spreading are visible. (Copyright © 1995, David T. Sandwell. Used by permission.)

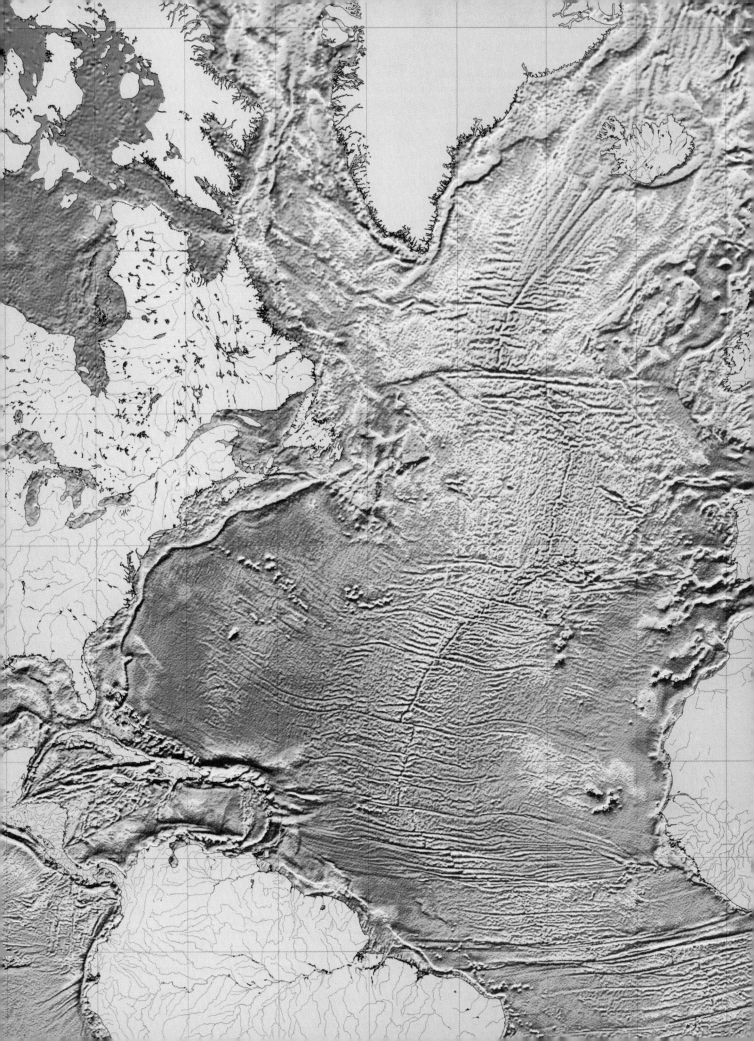

the western Pacific that mark the positions of subduction arcs—features of plate motions that we will discuss shortly.

Continental Margins

The *continental margin*, labeled in Figure 11.7 as the third form element of the typical ocean basin, can be defined as the narrow zone in which oceanic lithosphere is in contact with continental lithosphere (see Figure 11.2*b*). Thus, the continental margin is a feature shared by the continent and its adjacent ocean basin.

As the continental margin is approached, the ocean floor begins to slope gradually upward, forming the *continental rise* (Figure 11.7). The floor then steepens greatly on the *continental slope*. At the top of this slope we arrive at the brink of the *continental shelf*, a gently sloping platform some 120 to 160 km (about 75 to 100 mi) wide. Water depth is about 150 m (about 500 ft) at the outer edge of the shelf.

The symmetrical model illustrated in Figure 11.7 is nicely shown in the North Atlantic and South Atlantic Ocean basins. It also applies rather well to the Indian Ocean and Antarctic Ocean basins. The margins of these symmetrical basins are described as *passive continental margins*, which means they have not been subjected to Cenozoic tectonic and volcanic activity. Both continental and oceanic lithosphere at a passive continental margin are part of the same lithospheric plate and move together, away from the axial rift.

Great thicknesses of sedimentary strata, derived from the continents, have accumulated along passive continental margins. The strata range in age from the late Mesozoic (Jurassic, Cretaceous) through the Cenozoic. Near the continent, the shelf strata form a wedge-shaped deposit, thickening oceanward. A block dia-

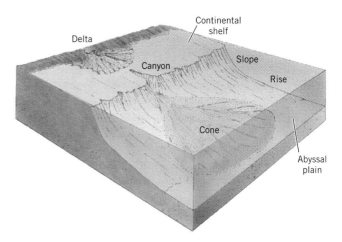

Figure 11.10 Block diagram of a deep-sea cone and its relationship to the continental shelf. (Copyright © A. N. Strahler.)

gram, Figure 11.9, shows details of this deposit. The sediments have been brought from the land by rivers and spread over the shallow sea floor by currents. Below the continental rise and its adjacent abyssal plain is another thick sediment deposit. It is formed of deep-sea sediments carried down the continental slope by swift muddy currents. The weight of these sediments causes them to subside, depressing the oceanic crust at its margin with the continental crust.

Deltas built by rivers contribute a great deal of the shelf sediment. Tonguelike turbid currents thickened with fine silt and clay from these deltas carve submarine canyons into the continental shelf. Where these currents emerge from the canyons, they deposit their sediment and build *deep-sea cones* (Figure 11.10).

The Pacific Ocean basin differs from our ideal model in the layout of its margins. Although it has a mid-oceanic ridge with ocean basin floors on either side, it has quite different continental margins. They are characterized by mountain arcs or island arcs with deep offshore *oceanic trenches*. Geologists refer to these trenched ocean basin limits as *active continental margins*.

The locations of the major trenches are shown in Figure 11.5. Trench floors can reach depths of nearly 11,000 m (about 36,000 ft), although most range from 7,000 to 10,000 m (about 23,000 to 33,000 ft) (see also Figure 11.8). Many lines of scientific evidence show that the oceanic crust is bent down sharply to form these trenches and that they mark the boundary between two lithospheric plates that are being brought together.

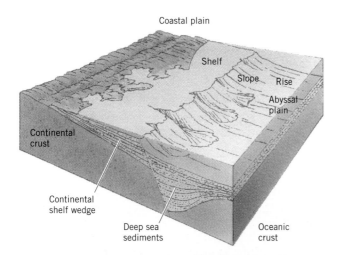

Figure 11.9 This block diagram shows an inner wedge of sediments beneath the continental shelf and an outer wedge of deep-sea sediments beneath the continental rise and abyssal plain.

PLATE TECTONICS

Both crustal spreading along the axial rift of the mid-oceanic ridge and crustal downbending beneath oceanic trenches involve the entire thickness of lithospheric plates. The general theory of lithospheric plates with their relative motions and boundary interactions is called **plate tectonics**. **Tectonics** is a noun meaning "the study of tectonic activity." As defined earlier, tectonic activity refers to all forms of breaking and bending of the entire lithosphere, including the crust.

Figure 11.11 shows the major features of plate interactions. The vertical dimension of the block diagram (*a*) is greatly exaggerated, as are the landforms. A true-scale cross section (*b*) shows the correct relationships between crust and lithosphere, but surface relief features can scarcely be shown. Diagram (*c*) is a sketch of plate motions on a spherical earth.

There are two very different kinds of lithospheric plates (see Figure 11.11*b*). Plates that lie beneath the ocean basins consist of *oceanic lithosphere*. The diagrams show two plates, plates X and Y, both made up of oceanic lithosphere, which is comparatively thin (about 50 km or 30 mi thick). Plate Z, bearing thick continental crust, is made up of *continental lithosphere*, which is much thicker (about 150 km or 95 mi).

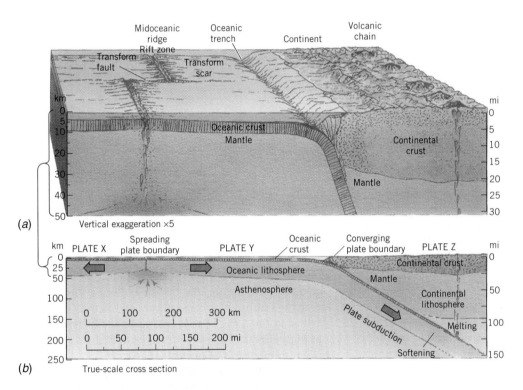

(a)

(b)

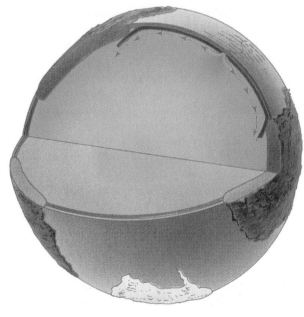

(c)

Figure 11.11 Schematic cross sections showing some of the important elements of plate tectonics. Diagram (*a*) is greatly exaggerated in vertical scale, and emphasizes surface and crustal features. Only the uppermost 50 km (31 mi) is shown. Diagram (*b*) is drawn to true scale and shows conditions to a depth of 250 km (155 mi). Here the actual relationships between lithospheric plates can be examined, but surface features are too small to be shown. Diagram (*c*) is a pictorial rendition of plates on a spherical earth and is not to scale. (Copyright © A. N. Strahler.)

An important scientific principle applies to the relative surface heights of the two kinds of lithosphere. The lithosphere can be thought of as "floating" on the soft asthenosphere. Consider two blocks of wood, one thicker than the other, floating in a pan of water. The surface of the thick block will ride higher above the water surface than that of the thin block. This principle explains why the continental surfaces rise so high above the ocean floors.

Plate Motions and Interactions

As shown in Figure 11.11, plates X and Y are pulling apart along their common boundary, which lies along the axis of a midoceanic ridge. This pulling apart tends to create a gaping crack in the crust, but magma continually rises from the mantle beneath to fill it. The magma appears as basaltic lava in the floor of the rift and quickly congeals. At greater depth under the rift, magma solidifies into gabbro, an intrusive rock of the same composition as basalt. Together, the basalt and gabbro continually form new oceanic crust. This type of boundary between plates is termed a *spreading boundary*.

At the right, the oceanic lithosphere of plate Y is moving toward the thick mass of continental lithosphere that comprises plate Z. Where these two plates collide, they form a *converging boundary*. Because the oceanic plate is comparatively thin and dense, in contrast to the thick, buoyant continental plate, the oceanic lithosphere bends down and plunges into the soft layer, or asthenosphere. The process of downplunging of one plate beneath another is called **subduction**.

The leading edge of the descending plate is cooler than the surrounding asthenosphere—sufficiently cooler, in fact, that this descending slab of brittle rock is denser than the surrounding soft hot rock. Consequently, once subduction has begun, the slab "sinks under its own weight," so to speak. However, the slab is

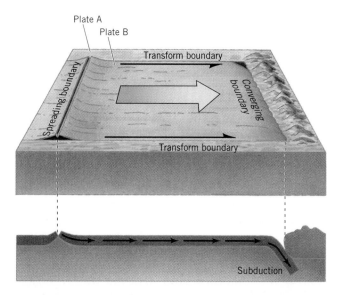

Figure 11.13 A schematic diagram of a single rectangular lithospheric plate with two transform boundaries. (Copyright © A. N. Strahler.)

gradually heated by the surrounding hot rock and thus eventually softens. The underportion, which is mantle rock in composition, simply reverts to mantle rock as it softens. The thin upper crust, formed of less dense mineral matter, can melt and become magma. This magma tends to rise because it is less dense than the surrounding material. Figure 11.11 shows some magma pockets formed from the upper edge of the slab. They are pictured as rising like hot-air balloons through the overlying continental lithosphere. When they reach the earth's surface, they form a volcano chain lying about parallel with the deep oceanic trench that marks the line of descent of the oceanic plate.

Viewing plate Y as a unit in Figure 11.11, we see that this single lithospheric plate is simultaneously undergoing *accretion* (growth by addition) and *consumption* (by softening and melting). If rates of accretion and consumption are equal, the plate will maintain its overall size. If consumption is slower, the plate will expand. If accretion is slower, the plate will shrink.

We have yet to consider a third type of lithospheric plate boundary. Two lithospheric plates may be in contact along a common boundary on which one plate merely slides past the other with no motion that would cause the plates either to separate or to converge (Figure 11.12). This is a *transform boundary*. The plane along which motion occurs is a nearly vertical fracture extending down through the entire lithosphere, and it is called a *transform fault*. A *fault* is a rock plane along which there is motion of the rock mass on one side with respect to that on the other. (More about faults will appear in Chapter 12.) Transform boundaries are often associated with midoceanic ridges and are shown in Figures 11.8 and 11.11.

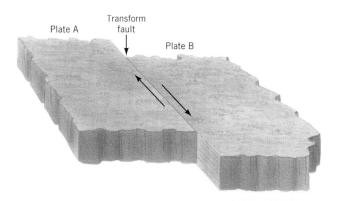

Figure 11.12 A transform fault involves the horizontal motion of two adjacent lithospheric plates, one sliding past the other. (Copyright © by A. N. Strahler.)

Plate Boundaries Summarized

In summary, there are three major kinds of active plate boundaries:

- *Spreading boundaries.* New lithosphere is being formed by accretion. Example: Sea-floor spreading along the axial rift.
- *Converging boundaries.* Subduction is in progress, and lithosphere is being consumed. Example: Active continental margin.
- *Transform boundaries.* Plates are gliding past one another on a transform fault. Example: Transform boundary associated with the midoceanic ridge.

Let us put these three boundaries into a pattern to include an entire lithospheric plate (Figure 11.13). Visualize the sunroof of an automobile in which a portion of the roof slides to open. Just as the sunroof can move by sliding past the fixed portion of the roof at its sides, so a lithospheric plate can move by sliding past other plates on transform faults. Where the sunroof opens, the situation is similar to a spreading boundary. Where the sunroof slides under the rear part of the roof, the situation is similar to a converging boundary. Boundaries of lithospheric plates can be curved as well as straight, and individual plates can pivot as they move. There are many geometric variations in the shapes and motions of individual plates.

The motions of lithospheric plates follow a time cycle in which a single supercontinent is split apart and then joined back together. Named the Wilson Cycle, it is described in more detail in *Focus on Systems 11.2 • The Wilson Cycle and Supercontinents.*

The Global System of Lithospheric Plates

The global system of lithospheric plates consists of six great plates. These are listed in Table 11.2 and are shown on a world map in Figure 11.14. Several lesser plates and subplates are also recognized. They range in size from intermediate to comparatively small. Plate boundaries are shown by symbols, explained in the key accompanying the map.

The great Pacific plate occupies much of the Pacific Ocean basin and consists almost entirely of oceanic lithosphere. Its relative motion is northwesterly, so that

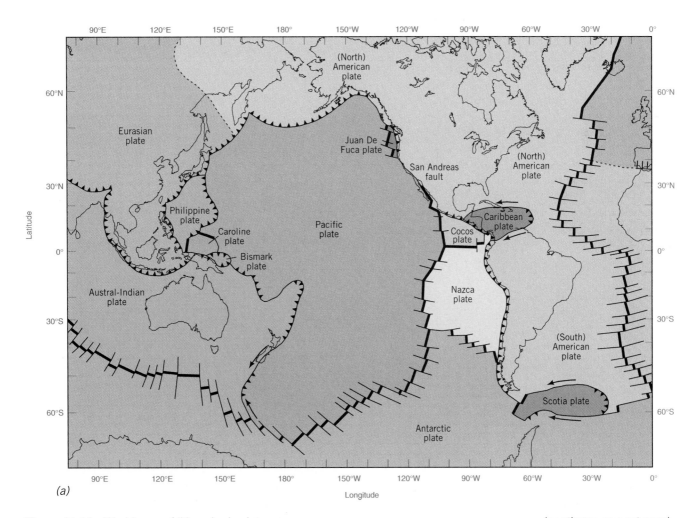

(a)

Figure 11.14 World map of lithospheric plates.

(continues on next page)

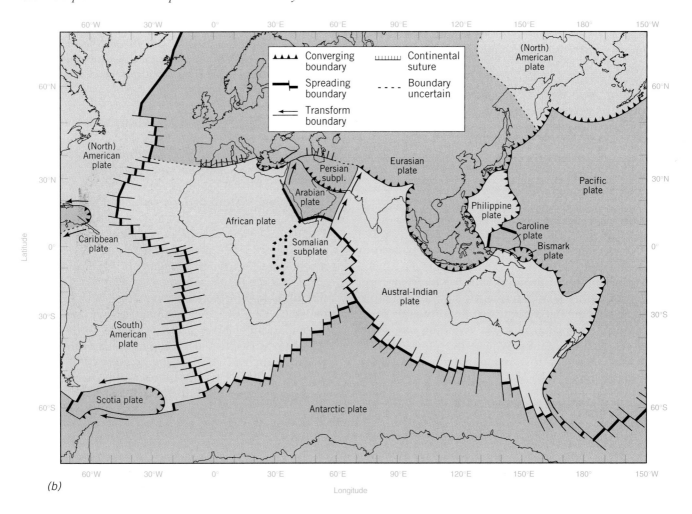

(b)

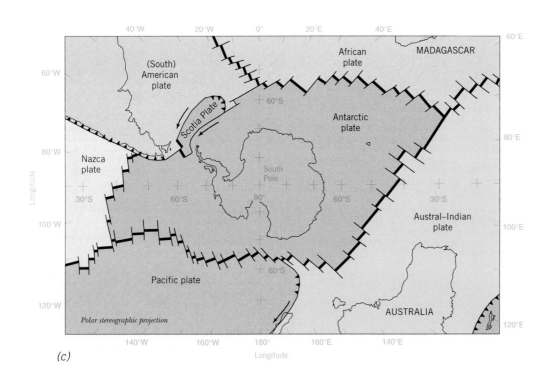

(c)

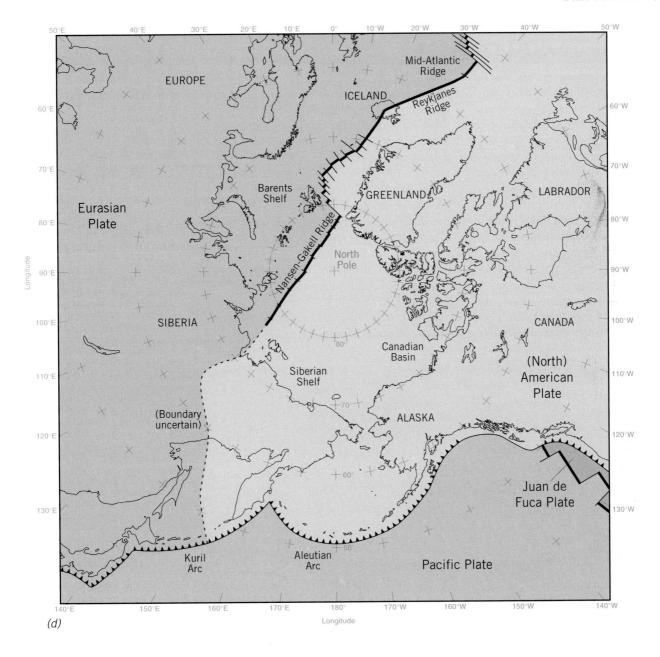

(d)

Table 11.2   The Lithospheric Plates

Great plates	Lesser plates
Pacific	Nazca
American (North, South)	Cocos
Eurasian	Philippine
Persian subplate	Caribbean
African	Arabian
Somalian subplate	Juan de Fuca
Austal-Indian	Caroline
Antarctic	Bismark
	Scotia

Focus on Systems 11.2 • The Wilson Cycle and Supercontinents

When geoscientists think about time cycles, they have in mind spans of time measuring hundreds of millions of years from beginning to end of just a single cycle. Recognition of a grand cycle of tectonic events bears the name of a distinguished Canadian geophysicist—Professor J. Tuzo Wilson of the University of Toronto. He made important new discoveries and interpretations that helped lead to a full-blown theory of plate tectonics. All of the parts were finally there—continental rupture to start the opening of a new ocean basin, plate subduction to generate island arcs,

and collisions to weld continents together. Using all of these separate lithospheric activities, Wilson forged a logical time chain of plate tectonic events, and it quickly gained wide acceptance as the *Wilson cycle.*

The Wilson cycle can best be appreciated by use of a cartoonlike presentation of the essential elements of plate tectonics, using a few symbolic icons. In our illustration at below, these elements are shown arranged throughout six stages and are listed here with references to our text diagrams and examples found on the planet today. This

ideal cycle requires some 300 to 500 million years to complete.

- *Stage 1.* Embryonic ocean basin. (See also Figure 11.22*b.*) The Red Sea separating the Arabian Peninsula from Africa is an active example.
- *Stage 2.* Young ocean basin. The Labrador basin, a branch of the North Atlantic lying between Labrador and Greenland, is a fair example of this stage.
- *Stage 3.* Old ocean basin. (Figure 11.22*c.*) Includes all of the vast expanse of the North and South Atlantic oceans and the Antarctic Ocean. Passive margin sedimentary wedges have become wide and thick.
- *Stage 4a.* Closing of the ocean basin begins with formation of new subduction boundaries. A brief substage.
- *Stage 4b.* Island arcs have risen and grown into great volcanic island chains. (Figure 11.18*a.*) These are found surrounding the Pacific plate, with the Aleutian arc as a fine example.
- *Stage 5.* Closing continues. Formation of new subduction margins close to the continents is followed by arc–continent collisions. (Figure 11.18*c.*) The Japanese Islands represent this stage.
- *Stage 6.* The ocean basin has finally closed with a collision orogen, forming a continental suture. (Figure 11.21*c.*) The Himalayan orogen is a fine recent example, with activity continuing today.

Note that the continent of the final stage is much wider than the conti-

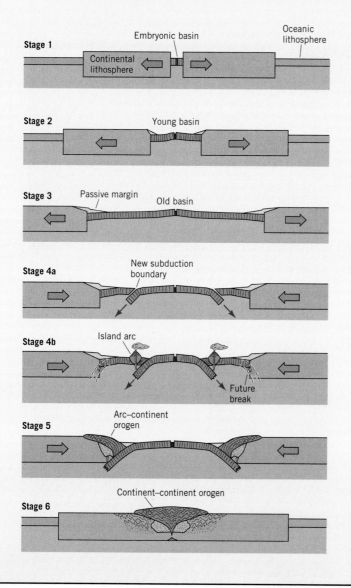

Schematic diagram depicting the six stages of the Wilson cycle. The diagrams are not to true scale. (From *Plate Tectonics,* copyright © 1997 by Arthur N. Strahler. Used by permission.)

nent of the first stage, because of the great added quantity of continental lithosphere formed during the intervening collisions. This is how continents grow through geologic time.

Our stage diagrams show a single continent splitting into two continents, then rejoining to form one. Suppose that over the entire globe several continents are splitting apart and reclosing at about the same time. Expanding on this idea, imagine that these fragments became detached from a single world continent that can be called a *supercontinent*.

Several powerful lines of evidence show that a supercontinent actually came into existence, starting about 200 million years ago.

Called *Pangea*, it is described in our chapter text. Good evidence has now been found that an earlier supercontinent, dubbed *Rodinia*, was fully formed about 700 million years ago. It consisted of early representatives of the same continents that later made up Pangea. Rodinia broke apart, and its fragments were carried away in different directions. Then they reversed their motions and headed back toward a common center, where many continent–continent collisions bonded them together by sutures to comprise Pangea. Some interesting evidence has also pointed to the former existence of a supercontinent even older than Rodinia.

The illustration below presents the supercontinent cycle as a loop

with three stages. Analogous to Stages 1–3 of the Wilson cycle is the dispersal phase, followed by a convergence phase that corresponds to Wilson stages 4 through 6. The cycle ends in a complete new supercontinent. Many new collision orogens are formed between the original continental fragments. Given, say, 3000 million years ago (Middle Archean time) as the first occurrence of a full-blown supercontinent cycle, there could have been some 6 to 10 such cycles. The hypothesis of a time cycle of supercontinents, repeating the Wilson cycle over and over again, now holds its place as the basic theme of the geologic evolution of our planet.

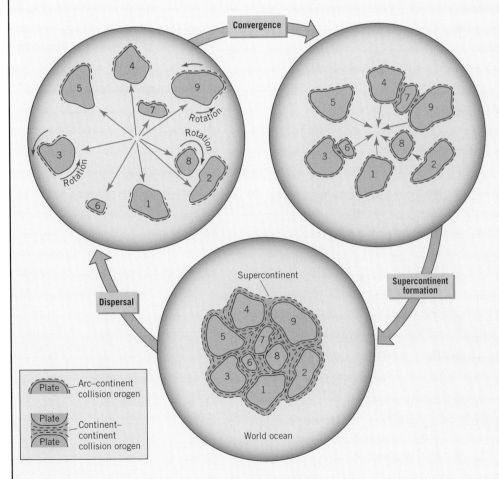

Schematic diagram of the supercontinent cycle in three phases. A pattern of wiggly dashes shows collision orogens that accumulate around and between the original continents. (From *Plate Tectonics*, copyright © 1997 by Arthur N. Strahler. Used by permission.)

it has a subduction boundary along most of the western and northern edge. The eastern and southern edge is mostly a spreading boundary. A sliver of continental lithosphere is included and makes up the coastal portion of California and all of Baja California. The California portion is bounded by an active transform fault (the San Andreas Fault).

The American plate includes most of the continental lithosphere of North and South America as well as the entire oceanic lithosphere lying west of the midoceanic ridge that divides the Atlantic Ocean basin down the middle. For the most part, the western edge of the American plate is a subduction boundary, with oceanic lithosphere diving beneath the continental lithosphere. The eastern edge is a spreading boundary. (Some classifications recognize separate North American and South American plates.)

The Eurasian plate is mostly continental lithosphere, but it is fringed on the west and north by a belt of oceanic lithosphere. The African plate has a central core of continental lithosphere nearly surrounded by oceanic lithosphere.

The Austral-Indian plate takes the form of a long rectangle. It is mostly oceanic lithosphere but contains two cores of continental lithosphere—Australia and peninsular India. Recent evidence shows that these two con-

tinental masses are moving independently and may actually be considered to be parts of separate plates. The Antarctic plate has an elliptical shape and is almost completely enclosed by a spreading plate boundary. This means that the other plates are moving away from the pole. The continent of Antarctica forms a central core of continental lithosphere completely surrounded by oceanic lithosphere.

Of the nine lesser plates, the Nazca and Cocos plates of the eastern Pacific are rather simple fragments of oceanic lithosphere bounded by the Pacific midoceanic spreading boundary on the west and by a subduction boundary on the east. The Philippine plate is noteworthy as having subduction boundaries on both east and west edges. Two small but distinct lesser plates—Caroline and Bismark—lie to the southeast of the Philippine plate. The Arabian plate has two transform fault boundaries, and its relative motion is northeasterly. The Caribbean plate also has important transform fault boundaries. The tiny Juan de Fuca plate is steadily diminishing in size and will eventually disappear by subduction beneath the American plate. Similarly, the Scotia plate is being consumed by the American and Antarctic plates.

Figure 11.15 is a schematic circular cross section of the lithosphere along a great circle in low latitudes. It

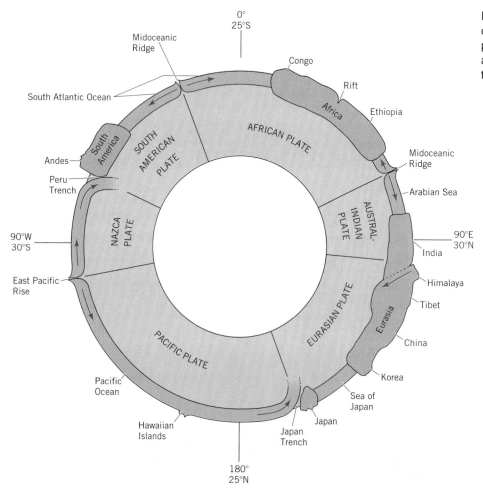

Figure 11.15 Schematic circular cross section of the major plates on a great circle tilted about 30 degrees with respect to the equator. (A. N. Strahler.)

shows several of the great plates and their boundaries. The great circle is tilted by about 30° with respect to the equator. The section crosses the African plate, heads northeast across the Eurasian plate through the Himalayas to Japan and Korea, then dips southeast across the Pacific plate, cutting across the South American plate, and finally returns to Africa. Three spreading boundaries at midoceanic ridges are encountered, with two subduction zones (the Japan and Peru trenches) and a continent-continent collision, where the Austral-Indian plate dives under the Eurasian plate.

SUBDUCTION TECTONICS

Converging plate boundaries, with subduction in progress, are zones of intense tectonic and volcanic activity. The narrow zone of a continent that lies above a plate undergoing subduction is therefore an active continental margin. Figure 11.16 shows some details of the geologic processes that are associated with plate subduction. Two diagrams are used. Part (*a*) is exaggerated to show crustal and surface details, and part (*b*) is drawn to true scale to show the lithospheric plates.

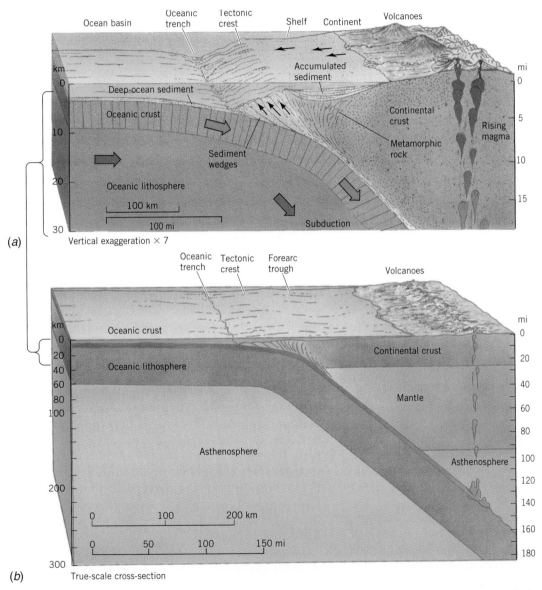

Figure 11.16 Some typical features of an active subduction zone. Diagram (*a*) uses a great vertical exaggeration to show surface and crustal details. Sediments scraped off the moving plate form tilted wedges that accumulate in a rising tectonic mass. Between the tectonic crest and the mainland is a shallow trough in which sediment brought from the land is accumulating. Metamorphic rock is forming above the descending plate. Magma rising from the top of the descending plate reaches the surface to build a chain of volcanoes. Diagram (*b*) is a true-scale cross section showing the entire thickness of the lithospheric plates. (Copyright © A. N. Strahler.)

The trench axis represents the line of meeting of sediment coming from two sources. Carried along on the moving oceanic plate is deep-ocean sediment—fine clay and ooze—that has settled to the ocean floor. From the continent comes terrestrial sediment in the form of sand and mud brought by streams to the shore and then swept into deep water by currents. In the bottom of the trench, both types of sediment are intensely deformed and are carried downward with the moving plate. The deformed sediment is then shaped into wedges that ride, one over the other, on steep fault planes. The wedges accumulate into an *accretionary prism* in which metamorphism takes place. In this way, the continental margin is built outward, and a new continental crust of metamorphic rock is formed.

The accretionary prism is of relatively low density and tends to rise, forming a *tectonic crest.* The tectonic crest is shown to be submerged in the figure, but in some cases it forms an island chain paralleling the coast—that is, a *tectonic arc.* Between the tectonic crest and the mainland is a shallow trough, the *forearc trough.* This trough traps a great deal of terrestrial sediment, which accumulates in a basinlike structure. The bottom of the forearc trough continually subsides under the load of the added sediment. In some cases, the sea floor of the trough is flat and shallow, forming a type of continental shelf. Sediment carried across the shelf moves down the steep outer slope of the accretionary prism in tonguelike flows of turbidity currents.

The lower diagram of Figure 11.16 shows the descending lithospheric plate entering the asthenosphere. Intense heating of the upper surface of the plate melts the oceanic crust, forming basaltic magma. As this magma rises, it is changed in chemical composition at the base of the crust and becomes andesite magma. The rising andesite magma reaches the surface to form volcanoes of andesite lava, such as those we see in the Andes of South America.

Orogeny

Generally speaking, high-standing mountain masses (other than volcanic mountains) are elevated by one of two basic tectonic processes: *compression* and *extension.* Compressional tectonic activity—"squeezing together" or "crushing"—acts at converging plate boundaries. Extensional tectonic activity—"pulling apart"—occurs where oceanic plates are separating or where a continental plate is undergoing breakup into fragments. We turn first to compressional tectonic processes.

Alpine mountain chains typically consist of intensely deformed strata of marine origin. The strata are tightly compressed into wavelike structures, called *folds.* As compression continues, the folds are first overturned and then become *recumbent* as they are further overturned upon themselves (Figure 11.17). Accompanying the folding is a form of faulting in which slices of rock

move over the underlying rock on fault surfaces of low inclination. These are *overthrust faults.* Individual rock slices, called *thrust sheets,* are carried many tens of kilometers over the underlying rock. In the European Alps, thrust sheets of this kind were named *nappes* (from the French word meaning "cover sheet" or "tablecloth"). Nappes may be thrust one over the other to form a great pile. The entire deformed rock mass produced by such compressional mountain-making is called an *orogen,* and the event that produced it is an *orogeny.*

Thrust sheets are known to have moved nearly horizontally for distances of several tens of kilometers. As one theory of thrust-sheet development, geologists postulate that at the time the sheets were in motion, the thrust planes actually sloped downward in the direction of motion of the sheet. You might visualize the thrust sheets as sliding "downhill" under the force of gravity. The phenomenon is known as *gravity gliding.* The presence of groundwater under great pressure beneath the thrust sheet is thought to have reduced the friction so much that the enormous rock layer could glide easily for a long distance.

Orogens and Collisions

Visualize, now, a situation in which two masses of continental crust converge along a subduction boundary. Ultimately, the two masses must collide because the impacting masses are too thick and too buoyant to allow either mass to slip under the other. The result is an orogeny in which various kinds of crustal rocks are crumpled into folds and sliced into nappes. This process has been called "telescoping" after the behavior of a folding telescope that collapses from a long tube into a short cylinder. Collision permanently unites the two masses, terminating further tectonic activity along that collision zone.

Two types of orogens are recognized. One type results from the impact of a relatively small mass of high-standing continental lithosphere against a full-sized mass of continental lithosphere. Following this type of collision, the small mass is firmly welded to the continent and will become a permanent part of the continental shield. The second type of orogen results from the collision of two very large bodies of continental lithosphere (full-sized plates), permanently uniting them and terminating further tectonic activity along that collision zone.

To prepare for the first type of orogeny, a simple mechanical model will be useful. Imagine we are in a custom bakery that makes cakes on special order. As each cake is completed, it is placed on a continually moving conveyer belt that transports it to the packaging department. Here, the cake slides off onto a short table surface at the level of the belt. The table will accommodate only a few cakes, so that if the attendant leaves the station for too long, disaster soon sets in. As another cake arrives, it slams into those already on the table, crushing them severely and compacting them into a single mass.

The conveyer belt in our analogy represents oceanic lithosphere disappearing into a subduction zone. Each cake represents one of several possible crustal features that can project above the otherwise uniform abyssal surface. Small projecting objects, such as seamounts, usually cause no problem—they may be knocked off at the base by impact with the overlying plate to become incorporated into the accretionary prism. The problem is with much larger crustal protrusions that are too massive and too firmly rooted to be subducted.

What kinds of traveling crustal masses are capable of colliding with a large continental mass and adhering to it? One is the volcanic island arc—for example, the Kuril arc or the Lesser Antilles arc. An entirely different class of objects consists of small fragments or bits of continental crust called *microcontinents*. Some of the older island arcs with a long history of accretion qualify as microcontinents—Honshu, the Philippines, or Hispaniola, for example. In other cases, a fragment of continent has been pulled away from the mainland on a widening rift that develops into a backarc basin. These islands of continental crust are thus surrounded by oceanic crust.

The objects described above can arrive at the active margin of a large body of continental lithosphere as an oceanic plate moves toward a subduction boundary adjacent to continental lithosphere. Over millions of years, the impacting object can travel thousands of kilometers.

Arc–Continent Collisions

The impact of an island arc is called an *arc–continent collision*. It is illustrated by a set of diagrams in Figure 11.18. The setting, shown in (*a*), is one of a passive continental margin (*right*) closing with a volcanic arc (*left*). In this case, the subduction is occurring along the volcanic arc, thus narrowing the expanse of oceanic lithosphere separating the two features. An accretionary prism is growing larger as the volcanic arc increases in size and depth. In a later stage (*b*), the intervening ocean is completely closed, and the accretionary prism is forced to ride up the continental margin, gliding on a basal thrust fault. As shown in (*c*), impact of the volcanic arc causes the overthrusting to telescope the strata of the passive margin, extending the low-angle thrust slice far over the continental shield. This kind of thrust plane is called a *décollement*, for it is located at the interface of the strata and the ancient crystalline basement rocks beneath. Rocks of the volcanic arc are also severely deformed and converted into metamorphic rock.

Notice also that in the last stage, a new subduction boundary has come into existence on the oceanward

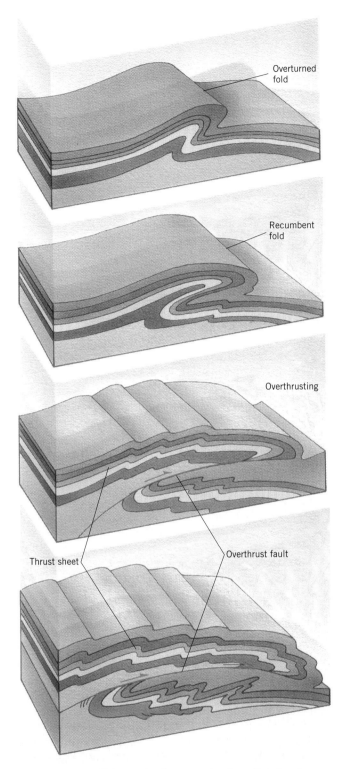

Figure 11.17 Schematic diagrams to show the development of a recumbent fold, broken by a low-angle overthrust fault to produce a thrust sheet, or *nappe*, in alpine structure. (Based on diagrams by A. Heim, 1922, *Geologie der Schweiz*, vol. II–1, Tauschnitz, Leipzig.)

Labels in figure: Overturned fold; Recumbent fold; Overthrusting; Thrust sheet; Overthrust fault

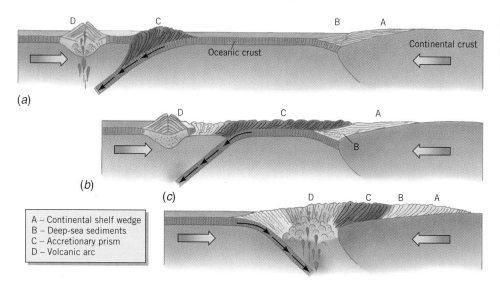

Figure 11.18 Schematic cross section of an arc–continent collision. Not shown to true scale. (Drawn by A. N. Strahler.)

A – Continental shelf wedge
B – Deep-sea sediments
C – Accretionary prism
D – Volcanic arc

side of the former volcanic arc. Magmas rising from the downplunging plate penetrate the metamorphic zone of the orogen. These give rise to bodies of felsic magma that rise even higher and become granitic batholiths. Magma also reaches the surface to produce outpourings of andesitic and rhyolitic lava. The process of intrusion by magma greatly thickens the crust beneath the orogen, making it a permanent addition to the continent.

Examples of orogens resulting from arc–continent collisions can be found on both eastern and western sides of North America. The relationships shown in Figure 11.18 are patterned in a general way after Paleozoic arc–continent collisions that affected the Appalachian and Ouachita mountain belts. Another example is the Cordilleran Orogeny that (as the name indicates) built the Cordilleran Ranges of western North America. In the northern Rocky Mountains of Montana, Alberta, and British Columbia, the overthrust zone is remarkably displayed in the high, glaciated mountains of Waterton-Glacier International Peace Park, Banff Park, and Jasper Park.

Accreted Terranes of Western North America

In recent years, geologists who have been involved in mapping the bedrock features of western North America from Mexico to Alaska have come to recognize that it consists of a mosaic of crustal patches called *terranes.* A particular terrane is quite distinct geologically from those that surround it, in that it has its own special rock type or suite of component rock types. If we think of this assemblage of terranes as a mosaic of irregularly shaped tiles, we see that whereas most of the tiles are unique, others are duplicated in widely separated places. Figure 11.19 is a map of these terranes. There are at least 50 of them, and each is separated from its neighbors by a fault contact. In many cases, the fault is

similar to a transform fault and trends more or less parallel with the continental margin.

Each terrane has a name. For example, the terrane called "Wrangellia," shown on the map by special color, occurs in five separate patches. It is a terrane consisting of basaltic island-arc volcanics and sedimentary rocks of types that include both cherts of deep-sea origin and limestones of shallow-depth origin. Using paleomagnetic methods, geologists have been able to establish the original global location of certain of the terranes. Much to their initial surprise, Wrangellia was found to have come from a location at lat. 10° and as far distant as a spot equivalent to where New Guinea is now located—traveling since then a distance of nearly 10,000 km (about 6000 mi)! It occupied that location in Triassic time, when a thickness of about 3 km (about 10,000 ft) of basaltic island-arc lavas was deposited.

Each of the many terranes is now regarded as a former microcontinent—using that term in a general sense. Obviously, a terrane can only travel by being carried along with the lithospheric plate in which it is embedded. These bits of continental lithosphere were embedded in oceanic lithosphere, which moved toward North America, eventually bringing each microcontinent to a location close to a subduction boundary at the active western margin of North America. Here a collision took place, and the microcontinent was welded to the continent. Many such collisions took place, eventually forming the mosaic of terranes.

Once the microcontinents were added to the continent, most were sheared into two or more small fragments by transcurrent faults trending about parallel with the continental margin. Faults of this type form in continental crust when the subducting lithospheric plate approaches at an oblique angle to the subduction boundary. The San Andreas Fault is an example of this type of fault, which is discussed in more detail in Chapter 12. Contained within the moving slices bounded by

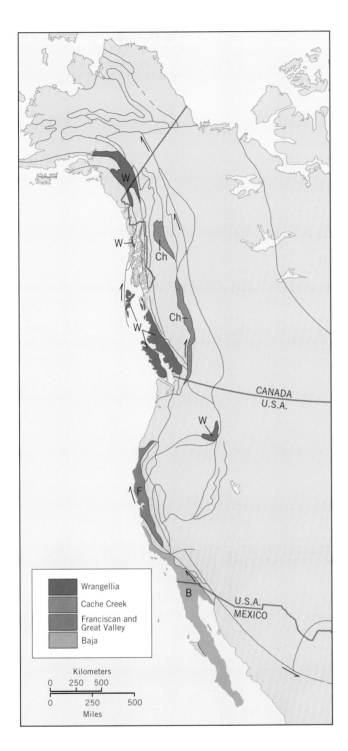

Figure 11.19 Map of microcontinent terranes of western North America. Fragments of four representative terranes are shown in color. (Based on data of U.S. Geological Survey.)

Wrangellia

Cache Creek

Franciscan and Great Valley

Baja

Kilometers
0   250   500

0   250   500
Miles

meters across, the tectonic effect would perhaps have been localized. In the case of an impacting volcanic island arc thousands of kilometers long, a major orogeny would have ensued that strongly affected thousands of kilometers along the continental margin.

Continent–Continent Collisions

The second important type of collision occurs when two continents collide. Continent–continent collisions occurred in the Cenozoic Era along a great tectonic line that marks the southern boundary of the Eurasian plate (Figure 11.20). The line begins with the Atlas Mountains of North Africa, and it runs across the Aegean Sea region into western Turkey. Beyond a major gap in Turkey, the line takes up again in the Zagros Mountains of Iran. Jumping another gap in southeastern Iran and Pakistan, the collision line sets in again in the great Himalayan Range.

Each segment of this collision zone represents the collision of a different north-moving plate against the single and relatively immobile Eurasian plate. A European segment containing the Alps was formed when the African plate collided with the Eurasian plate in the Mediterranean region. A Persian segment resulted from the collision of the Arabian plate with the Eurasian plate. A Himalayan segment represents the collision of the Indian continental portion of the Austral-Indian plate with the Eurasian plate.

Figure 11.21 is a series of cross sections in which the tectonic events of a typical continent–continent collision are reconstructed. Diagram *a* shows a passive margin at the left and an active subduction margin at the right. As the ocean between the converging continents is eliminated, a succession of overlapping thrust faults cuts through the oceanic crust in diagram *b.* The thrust slices ride up, one over the other, telescoping the oceanic crust and the sediments above it. As the slices become more and more tightly squeezed, they are forced upward. The upper part of each thrust sheet assumes a horizontal attitude to form a nappe in diagram *c,* which then glides forward under gravity on a low downgrade. A mass of metamorphic rock is formed between the joined continental plates, welding them together. This new rock mass is the **continental suture**. It is a distinctive type of orogen. Our world map of plates, Figure 11.14, uses a special symbol for these sutures.

Continent–continent collisions have occurred many times since the late Precambrian time. Several ancient

transcurrent faults, terrane fragments became separated and distributed up and down the entire continental margin. This is what seems to have happened in the case of Wrangellia. The accretion of microcontinents to form a mosaic pattern in the continental crust may have also been one of the modes by which the shields of Precambrian age were constructed.

Whether the arrival and impact of a microcontinent was capable of generating a full-blown orogeny is questionable. For a small microcontinent, a few tens of kilo-

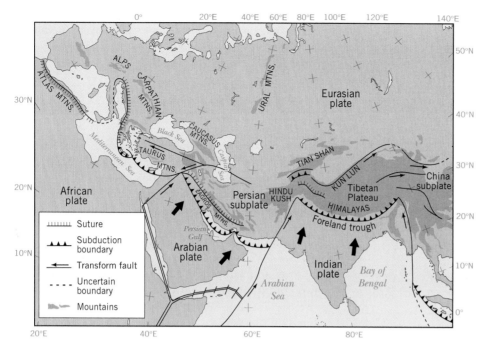

Figure 11.20 Sketch map of Eurasian collision segments.

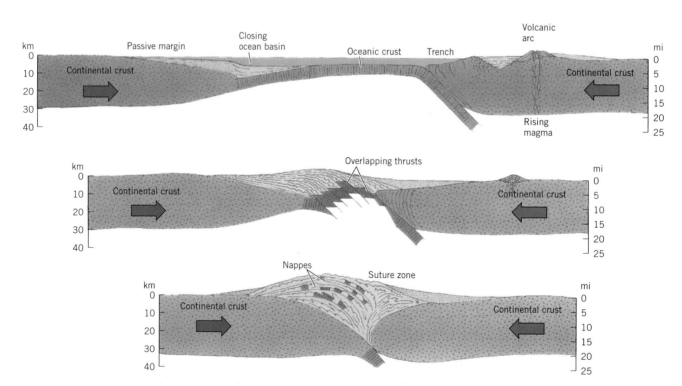

Figure 11.21 Schematic cross sections showing continent–continent collision and the formation of a suture zone with nappes. (Copyright © A. N. Strahler.)

sutures have been identified in the continental shields. The Ural Mountains, which divide Europe from Asia, are one such suture, formed near the end of the Paleozoic Era.

CONTINENTAL RUPTURE AND NEW OCEAN BASINS

We have already noted that the continental margins bordering the Atlantic Ocean basin on both its eastern and western sides are very different from the active margin of a subduction zone. At present, the Atlantic margins have no important tectonic activity and are passive continental margins. Even so, they represent the contact between continental lithosphere and oceanic lithosphere, with continental crust meeting oceanic crust, as shown in Figure 11.9. Passive margins are formed when a single plate of continental lithosphere is rifted apart. This process is called *continental rupture.*

Figure 11.22 uses three schematic block diagrams to show how continental rupture takes place and leads to the development of passive continental margins. At first, the crust is both lifted and stretched apart as the lithospheric plate is arched upward. The crust fractures and moves along faults—upthrown blocks form mountains, while downdropped blocks form basins.

Eventually a long, narrow valley, called a *rift valley,* appears (block *a*). The widening crack in its center is continually filled in with magma rising from the mantle below. The magma solidifies to form new crust in the floor of the rift valley. Crustal blocks on either side slip down along a succession of steep faults, creating a mountainous landscape. The Rift Valley system of East Africa—described in more detail in Chapter 12—is a notable example of this stage of continental rupture. As separation continues, a narrow ocean appears, with a spreading plate boundary running down its center (block *b*). Plate accretion takes place in the central rift to produce new oceanic crust and lithosphere.

The Red Sea is a narrow ocean formed by a continental rupture. Its straight coasts mark the edges of the rupture The widening of such an ocean basin can continue until a large ocean has formed and the continents are widely separated (Figure 11.22, block *c*).

During the process of opening of an ocean basin, the spreading boundary develops a series of offsets, one of which is shown in the upper left-hand portion of part (*a*) of Figure 11.11. The offset ends of the axial rift are connected by an active transform fault. As spreading continues, a scarlike feature is formed on the ocean floor as an extension of the transform fault. These *transform scars* take the form of narrow ridges or clifflike features and may extend for hundreds of kilometers

across the ocean floors. The scars are not, for the most part, associated with active faults.

What is the power source for these slow motions of vast lithospheric plates? Geophysicists generally agree that the heat produced by radioactive decay of unstable isotopes in the crust and mantle powers plate motions. *Focus on Systems 11.3 • The Power Source for Plate Movements* provides more information.

CONTINENTS OF THE PAST

Although modern plate tectonic theory is only a few decades old, the concept of a breakup of an early supercontinent into fragments that drifted apart dates back to the nineteenth century and beyond. Almost as soon as good navigational charts became available to

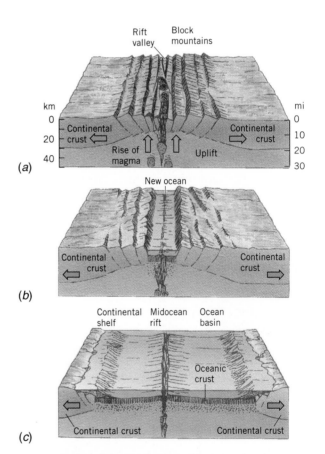

Figure 11.22 Schematic block diagrams showing stages in continental rupture and the opening up of a new ocean basin. The vertical scale is greatly exaggerated to emphasize surface features. (*a*) The crust is uplifted and stretched apart, causing it to break into blocks that become tilted on faults. (*b*) A narrow ocean is formed, floored by new oceanic crust. (*c*) The ocean basin widens, while the passive continental margins subside and receive sediments from the continents. (Copyright © A. N. Strahler.)

Focus on Systems 11.3 • The Power Source for Plate Movements

The system of lithospheric plates in motion represents a huge flow system of dense mineral matter. Its operation requires enormous power within an internal energy system. It is generally agreed that the source of this energy system lies in the phenomenon of radioactivity. (See *Working It Out 10.1 • Radioactive Decay*, and *Focus on Systems 10.3 • Powering the Cycle of Rock Change*.)

Radioactive elements in the crust and upper mantle constantly give off heat in the spontaneous process of radioactive decay. As the temperature of mantle rock rises, the rock expands. Recall that in the case of the atmosphere, upward motion of warmer, less dense air takes place by convection. It is thought that, in a somewhat similar type of convection, mantle material rises steadily beneath spreading plate boundaries. Geologists do not yet have a full understanding of how this rise of heated rock causes plates to move. The favored hypothesis states that, as the rising mantle lifts the

oceanic lithospheric plate to a higher elevation, the plate tends to move horizontally away from the spreading axis under the influence of gravity (see broad arrows on accompanying figure at right). At the same time, of course, the plate undergoes cooling, becomes denser, and sinks steadily lower.

At the far edge of the plate, subduction occurs because the oceanic plate is colder and denser than the asthenosphere through which it sinks. The motion of the downgoing oceanic plate exerts a drag on the surrounding asthenosphere. This drag on the overlying asthenosphere sets in motion a horizontal flow current moving toward the sinking slab. The current then turns abruptly down to drag against the upper surface of the slab. At this downturn the mantle beneath the continental lithosphere becomes highly heated and causes the upper surface of the slab to melt. Here, magma is released and rises in bubblelike bodies to penetrate the

thick continental lithosphere. Upon cooling, this magma forms plutons. Magma that reaches the surface builds volcanoes. The lower end of the descending slab is also heated and softened, and is carried slowly away in the mantle at great depth, moving toward the spreading boundary, where some of it can perhaps be recycled.

So we have here a power system that generates a vertical loop of material flow bringing hot asthenosphere near the earth's surface, where heat is passed up into the oceanic and atmospheric layers. It is a one-way power system that depends on a radiogenic heat source that is being slowly depleted. As time passes, the plate tectonic system will lose power and steadily slow its rates of plate motion. Looming in the far, far distant future is a "heat death" of the earth's tectonic system, when there will be no more great earthquakes and no more violent volcanic eruptions.

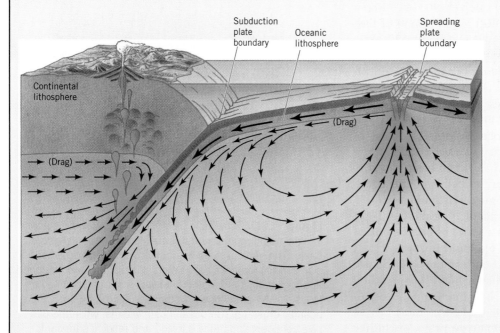

A schematic diagram showing a system of mantle flow (thin arrows) that transports heat to the earth's surface where it is lost through volcanic action. (Copyright © A. N. Strahler.)

show the continental outlines, geographers became intrigued with the close correspondence in outline between the eastern coast of South America and the western coastline of Africa. Credit for the first full-scale scientific hypothesis of the breakup of a single large continent belongs to Alfred Wegener, a German meteorologist and geophysicist, who offered geologic evidence as early as 1915 that the continents had once been united (Figure 11.23). He reconstructed a supercontinent named *Pangea*, which existed intact as early as about 300 million years age, in the Carboniferous Period.

A storm of controversy followed, and many American geologists denounced Wegener's "continental drift" hypothesis. However, he had some loyal supporters in Europe, South Africa, and Australia. Several lines of hard scientific evidence presented by Wegener strongly favored the former existence of Pangea. The prominent geographer and climatologist Vladimir Köppen collaborated with Wegener in introducing favorable evidence from distribution patterns of fossil and present-day plant species. But the physical process of separation of the continents offered by Wegener was weak and was strongly criticized on valid physical grounds. He had proposed that a continental layer of less dense rock had moved like a great floating "raft" through a "sea" of denser oceanic crustal rock. Geologists could show by use of established principles of physics that this proposed mechanism was impossible. In 1930, Wegener perished of cold and exhaustion while on an expedition into the Greenland Ice Sheet. There followed three decades in which only a small minority of scientists promoted his hypothesis.

In the 1960s, however, seismologists showed beyond doubt that thick lithospheric plates are in motion, both along the midoceanic ridge and beneath the deep offshore trenches of the continents. Other geophysicists used magnetic data "frozen" into crustal rock to conclude that the continents had moved great distances apart. Within only a few years, Wegener's scenario was validated, but only by applying a mechanism never dreamed of in his time.

Yes, the continents are moving today. Data from orbiting satellites have shown that rates of separation, or of convergence, between two plates are on the order of 5 to 10 cm (about 2 to 4 in.) per year, or 50 to 100 km (about 30 to 60 mi) per million years. At that rate, global geography must have been very different in past geologic eras than it is today. Many continental riftings and many plate collisions have taken place over the past 2 billion years. Single continents have fragmented into smaller ones, while at other times, small continents have merged to form large ones. (See *Focus on Systems 11.2 • The Wilson Cycle and Supercontinents.*)

A brief look at the changing continental arrangements and locations over the past 250 m.y. is well worthwhile in our study of physical geography because those changes brought with them changes in climate, soils, and vegetation on each continent as the continents moved across the parallels of latitude. Geography of the geologic past bears the name "paleogeography." If you decide to become a paleogeographer, you will need to carry with you into the past all you have learned about the physical geography of the present.

Wegener had made a crude map showing stages in the breakup of Pangea, but his timetable had been in part incorrect. Modern reconstructions of the global arrangements of past continents have been available since the mid-1960s. Global maps drawn for each geologic period have been repeatedly revised in the light of new evidence, but differences of interpretation affect only the minor details.

Figure 11.24 shows stages in the separations and travels of our continents, starting in the Permian Period of the Mesozoic Era, about 250 m.y. ago. In the first map, Pangea lies astride the equator and extends nearly to

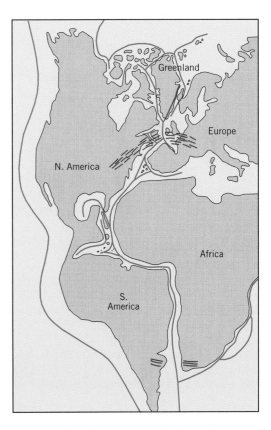

Figure 11.23 Alfred Wegener's 1915 map fitting together the continents that today border the Atlantic Ocean basin. The sets of short lines show the fit of Paleozoic tectonic structures between Europe and North America and between southernmost Africa and South America. (From A. Wegener, 1915, *De Entstehung der Kontinente und Ozeane*, F. Vieweg, Braunschweig.)

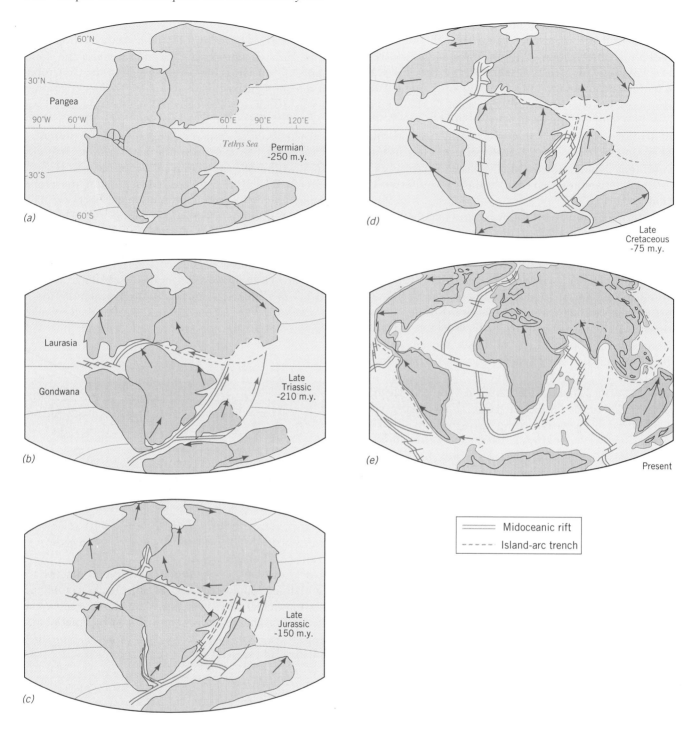

Figure 11.24 The breakup of Pangea is shown in five stages. Inferred motion of lithospheric plates is indicated by arrows. (Redrawn and simplified from maps by R. S. Dietz and J. C. Holden, *Journal of Geophysical Research*, vol. 75, pp. 4943–4951, Figures 2 to 6. Copyrighted by the American Geophysical Union. Used by permission.)

the two poles. Regions that are now North America and western Eurasia lie in the northern hemisphere. Jointly they are called *Laurasia*. Regions that are now South America, Africa, Antarctica, Australia, New Zealand, Madagascar, and peninsular India lie south of the equator. Jointly, they go by the name *Gondwana*. Subsequent maps (*b* through *d*) show the breaking apart and

dispersal of the Laurasia and Gondwana plates to yield their modern components and locations, as seen in map (*e*).

Note in particular that North America traveled from a low-latitude location into high latitudes, finally closing off the Arctic Ocean as a largely landlocked sea. This change may have been a major factor in bringing on the ice age in late Cenozoic time. (See Chapter 18.) Another traveler was the Indian Peninsula, which started out from a near-subarctic southerly location in Permian time and streaked northeast across the Thethys Sea to collide with Asia in the northern tropical savanna zone. India's bedrock still bears the grooves and scratches of a great glaciation in Permian time, when it was nearer to the south pole. Looking ahead, we may safely predict that the Atlantic Ocean basins will become much wider, and the westward motion of the Americas will cause a reduction in the width of the Pacific basin.

Over the long spans of geologic time, our planet's surface is shaped and reshaped endlessly. Oceans open and close, mountains rise and fall. Rocks are folded, faulted, and fractured. Arcs of volcanoes spew lava to build chains of lofty peaks. Deep trenches consume sediments that are carried beneath adjacent continents. These activities are byproducts of the motions of lithospheric plates and are described by plate tectonics.

For geographers, the great value of plate tectonics is that it provides a grand scheme for understanding the nature and distribution of the largest and most obvious features of our planet's surface—its continents and ocean basins and their major relief features. However, large features, such as mountain ranges, are made up of many smaller features—individual peaks, for example. In the next chapter, we will look at the continental surface in more detail, examining the volcanic and tectonic landforms that result when volcanoes erupt and rock layers are folded and faulted. As we will see, the formation of these landforms is often marked by powerful earthquakes—phenomena that can produce major impacts indeed on modern technological society.

CHAPTER SUMMARY

At the center of the earth lies the core—a dense mass of liquid iron and nickel that is solid at the very center. Enclosing the metallic core is the mantle, composed of ultramafic rock. The outermost layer is the crust. Continental crust consists of two zones—a lighter zone of felsic rocks atop a denser zone of mafic rocks. Oceanic crust consists only of denser, mafic rocks. The lithosphere, the outermost shell of rigid, brittle rock, includes the crust and an upper layer of the mantle. Below the lithosphere is the asthenosphere, a region of the mantle in which mantle rock is soft, or plastic.

Geologists trace the history of the earth through the geologic time scale. Precambrian time includes the earth's earliest history. It is followed by three major divisions—the Paleozoic, Mesozoic, and Cenozoic eras.

Continental masses consist of active belts of mountain-making and inactive regions of old, stable rock. Mountain-building occurs by volcanism and tectonic activity. Alpine chains occur in two principal mountain belts—the circum-Pacific and Eurasian-Indonesian belts. Continental shields are regions of low-lying igneous and metamorphic rocks. They may be exposed or covered by layers of sedimentary rocks. Ancient mountain roots lie within some shield regions.

The ocean basins are marked by a midoceanic ridge with its central axial rift. This ridge occurs at the site of crustal spreading. Most of the ocean floor is abyssal plain, covered by fine sediment. As passive continental margins are approached, the continental rise, slope, and shelf are encountered. At active continental margins, deep oceanic trenches lie offshore.

Continental lithosphere includes the thicker, lighter continental crust and a rigid layer of mantle rock beneath. Oceanic lithosphere is comprised of the thinner, denser oceanic crust and rigid mantle below. The lithosphere is fractured and broken into a set of plates, large and small, that move with respect to each other. Where plates move apart, a spreading boundary occurs. At converging boundaries, plates collide. At transform boundaries, plates move past one another on a transform fault. There are six major lithospheric plates.

When oceanic lithosphere and continental lithosphere collide, the denser oceanic lithosphere plunges beneath the continental lithospheric plate, a process called subduction. A trench marks the site of downplunging. Some subducted oceanic crust melts and rises to the surface, producing volcanoes near the continent's edge. Sediments carried on the converging oceanic plate accumulate at the continental margin, where they are folded and faulted and may be thrust horizontally over the rocks of the continent.

Orogens are the deformed rock masses that result when two masses of continental lithospheric material are sutured together at a converging boundary. Island arcs, microcontinents, or full-sized continents may collide with the margin of a large continent. Arc–continent collisions produce an orogen of intensely deformed arc rocks thrust over continental rock. Microcontinents become joined to the continent as accreted terranes. Continent–continent collisions form a distinctive type of orogen—the continental suture. In continental rupture, extensional tectonic forces move a continental plate in opposite directions, creating a rift valley. Eventually the rift valley widens and opens to the ocean, and new oceanic crust forms as spreading continues.

Plate movements are thought to be powered by convection currents in the plastic mantle rock of the asthenosphere. During the Permian Period, the continents were joined in a single, large supercontinent—Pangea—that broke apart, leading eventually to the present arrangement of continents and ocean basins.

KEY TERMS

core	lithosphere	continental shields
mantle	asthenosphere	plate tectonics
crust	lithospheric plates	tectonics
continental crust	mountain arc	subduction
oceanic crust	island arcs	continental suture

REVIEW QUESTIONS

1. Describe the earth's inner structure, from the center outward. What types of crust are present? How are they different?

2. How do geologists use the term *lithosphere*? What layer underlies the lithosphere, and what are its properties? Define the term *lithospheric plate*.

3. More recent geologic time is divided into three eras. Name them in order from oldest to youngest. How do geologists use the terms *period* and *epoch*? What age is applied to time before the earliest era?

4. What proportion of the earth's surface is in ocean? in land? Do these proportions reflect the true proportions of continents and oceans? If not, why not?

5. What are the two basic subdivisions of continental masses?

6. What term is attached to belts of active mountain-making? What are the two basic processes by which mountain belts are constructed? Provide examples of mountain arcs and island arcs.

7. What is a continental shield? How old are continental shields? What two types of shields are recognized?

8. Sketch a cross section of an ocean basin with passive continental margins. Label the following features: midoceanic ridge, axial rift, abyssal plain, continental rise, continental slope, and continental shelf.

9. Identify and describe two types of lithospheric plates. Sketch a cross section showing a collision between the two types. Label the following features: oceanic crust, continental crust, mantle, oceanic trench, and rising magma. Indicate where subduction is occurring.

10. Name the six great lithospheric plates. Identify an example of a spreading boundary by general geographic location and the plates involved. Do the same for a converging boundary.

11. Describe the process of subduction as it occurs at a converging boundary of continental and oceanic lithospheric plates. How is the continental margin ex-

tended? Identify the terms *tectonic crest, tectonic arc,* and *forearc trough.* How is subduction related to volcanic activity?

12. Describe how compressional mountain-building produces folds, faults, overthrust faults, and thrust sheets (nappes). How do the thrust sheets move long distances?

13. What are the essential features of an arc–continent collision? Provide a series of sketches showing the process.

14. How are accreted terranes formed? Describe Wrangellia as an example of a microcontinent that has been accreted onto a large continental mass.

15. Describe the formation of a continental suture, keying your description to a sketch of a continent–continent collision. Provide a present-day example where a continental suture is being formed, and give an example of an ancient continental suture.

16. How does continental rupture produce passive continental margins? Describe the process of rupturing and its various stages.

17. What are transform faults? Where do they occur?

18. What forces are thought to power plate tectonics? Explain.

19. Briefly summarize the history of our continents that geologists have reconstructed beginning with the Permian Period.

Focus on Systems 11.2 • The Wilson Cycle and Supercontinents

1. What is the Wilson cycle of plate tectonics? What is the net effect of a Wilson cycle on the continental mass?

2. Using a sketch, describe the cycle by which supercontinents are formed and reformed.

Focus on Systems 11.3 • The Power Source for Plate Movements

1. How is the principle of convection thought to be related to the power source for plate tectonic motions?

2. What role does gravity play in the motion of lithospheric plates?

ESSAY QUESTIONS

1. Figure 11.14 is a Mercator map of lithospheric plates, whereas Figure 11.15 presents a cross section of plates on a great circle. Construct a similar cross section of plates on the 30° S parallel of latitude. As in Figure 11.15, label the plates and major geographic features. Refer to Figure 11.5 for arcs, trenches, and midocean ridges.

2. Suppose astronomers discover a new planet that, like earth, has continents and oceans. They dispatch a reconnaissance satellite to photograph the new planet. What features would you look for, and why, to detect past and present plate tectonic activity on the new planet?

PROBLEMS

Working It Out 11.1 • Radiometric Dating

1. A geochemist analyzes a sample of zircon obtained from an igneous rock and observes that the ratio of ^{206}Pb to ^{238}U atoms is 0.448. Using the formula, what is the age of the sample?

2. The geochemist also determines the atomic ratio of ^{207}Pb to ^{235}U as 9.56 for the same sample. For this decay process, the half-life is $H = 7.04 \times 10^8$ years. Calculate k and then determine the age from the D/M ratio of 9.56. Is it consistent with the age obtained from the ^{206}Pb/^{238}U ratio?

Chapter 12

Volcanic and Tectonic Landforms

The two previous chapters in Part II—on earth materials and plate tectonics—have now set the stage for the study of volcanic and tectonic landforms, which is the subject of this chapter. These features of the landscape are created by internal earth forces. Upwelling magma spews forth explosively from vents and fissures, creating steep-sided volcanoes or flooding vast areas with molten rock. Compressional stresses generated by plate motions fold flat-lying rocks into wavelike forms, producing long ridges from upfolds and long valleys from downfolds. Earthquake faults mark the lines on which rock layers break and move past one another, producing huge uplifted mountain blocks next to deep, downdropped valleys.

By looking in detail at landforms created by volcanic and tectonic activity, we are moving down from the global perspective of enormous lithospheric plates that collide to form vast mountain arcs or separate to form yawning ocean basins. We are now "zooming in" to examine landscape features that are actually small enough to view from a single vantage point. As we will see, the global perspective of plate tectonics helps explain how these landscape features are created and where they occur.

LANDFORMS

Landforms are the surface features of the land—for example, mountain peaks, cliffs, canyons, plains, beaches, and sand dunes. Landforms are created by many processes, and much of the remainder of this book describes landforms and how they are produced. **Geomorphology** is the scientific study of the processes that

An active Costa Rican stratovolcano.

shape landforms. In this chapter, we will examine landforms produced directly by volcanic and tectonic processes.

The shapes of continental surfaces reflect the "balance of power," so to speak, between two sets of forces. Internal earth forces act to move crustal materials upward through volcanic and tectonic processes, thus creating new rock at the earth's surface. In opposition are processes and forces that act to lower continental surfaces by removing and transporting mineral matter through the action of running water, waves and currents, glacial ice, and wind. We can refer to these processes of land sculpture collectively as **denudation**.

Seen in this perspective, landforms in general fall into two basic groups—initial landforms and sequential landforms. *Initial landforms* are produced directly by volcanic and tectonic activity. They include volcanoes and lava flows, as well as downdropped rift valleys and elevated mountain blocks in zones of recent crustal deformation. Figure 12.1*a* shows a mountain block uplifted by crustal activity. The energy for lifting molten rock and rigid crustal masses to produce the initial landforms has an internal heat source. As we explained in Chapter 11, this heat energy is generated largely by natural radioactivity in rock of the earth's crust and mantle. It is the fundamental energy source for the motions of lithospheric plates.

Landforms shaped by processes and agents of denudation belong to the group of *sequential landforms.*

The word "sequential" means that they follow in sequence after the initial landforms have been created, and a crustal mass has been raised to an elevated position. Figure 12.1*b* shows an uplifted crustal block (an initial landform) that has been attacked by agents of denudation and carved up into a large number of sequential landforms.

You can think of any landscape as representing the existing stage in a great contest of opposing forces. As lithospheric plates collide or pull apart, internal earth forces periodically elevate parts of the crust to create initial landforms. The external agents of denudation—running water, waves, wind, and glacial ice—persistently wear these masses down and carve them into vast numbers of smaller sequential landforms.

Today we can observe the many stages of this endless struggle between internal and external forces by traveling to different parts of the globe. Where we find high alpine mountains and volcanic chains, internal earth forces have recently dominated the contest. In the rolling low plains of the continental interiors, the agents of denudation have won a temporary victory. At other locations, we can find many intermediate stages. Because the internal earth forces act repeatedly and violently, new initial landforms keep coming into existence as old ones are subdued.

VOLCANIC ACTIVITY

We have already identified *volcanism*, or volcanic activity, as one of the forms of mountain-building. The extrusion of magma builds landforms, and these can accumulate in a single area both as volcanoes and as thick lava flows. Through these volcanic activities, imposing mountain ranges are constructed.

A **volcano** is a conical or dome-shaped initial landform built by the emission of lava and its contained gases from a constricted vent in the earth's surface (Figure 12.2). The magma rises in a narrow, pipelike conduit from a magma reservoir lying beneath. Upon reaching the surface, igneous material may pour out in tonguelike lava flows. Magma may also be ejected in the form of solid fragments driven skyward under pressure of confined gases. Ejected fragments, ranging in size from boulders to fine dust, are collectively called *tephra*. Forms and dimensions of a volcano are quite varied, depending on the type of lava and the presence or absence of tephra.

Stratovolcanoes

The nature of volcanic eruption, whether explosive or quiet, depends on the type of magma. Recall from Chapter 10 that there are two main types of igneous

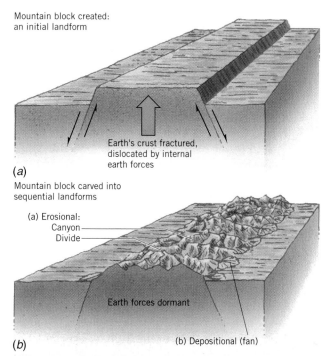

Mountain block created: an initial landform

Earth's crust fractured, dislocated by internal earth forces

(*a*)

Mountain block carved into sequential landforms

(a) Erosional:
Canyon
Divide

Earth forces dormant

(b) Depositional (fan)

(*b*)

Figure 12.1 Initial and sequential landforms. (Drawn by A. N. Strahler)

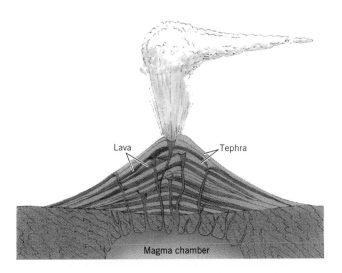

Figure 12.2 Idealized cross section of a stratovolcano with feeders from magma chamber beneath. The steep-sided cone is built up from layers of lava and tephra. (Copyright © A. N. Strahler.)

Figure 12.3 Mount St. Helens, a stratovolcano of the Cascade Range in southwestern Washington, violently erupted on the morning of May 18, 1980, emitting a great cloud of condensed steam, heated gases, and ash from the summit crater. Within a few minutes, the plume had risen to a height of 20 km (12 mi), and its contents were being carried eastward by stratospheric winds. The eruption was initiated by explosive demolition of the northern portion of the cone, which is concealed from this viewpoint.

rocks: felsic and mafic. The felsic lavas (rhyolite and andesite) have a high degree of viscosity—that is, they are thick and gummy, and resist flow. So, volcanoes of felsic composition typically have steep slopes, and lava usually does not flow long distances from the volcano's vent. In addition, felsic lavas usually hold large amounts of gas under high pressure. As a result, these lavas can produce explosive eruptions. The eruption of Mount St. Helens in 1980 is an example of an explosive eruption of felsic lavas (Figure 12.3).

Tall, steep-sided volcanic cones are produced by felsic lavas. These cones usually steepen toward the summit, where a bowl-shaped depression—the *crater*—is located. When the volcano erupts, tephra falls on the area surrounding the crater and contributes to the structure of the cone. *Volcanic bombs* are also included in the tephra. These solidified masses of lava range up to the size of large boulders and fall close to the crater. Very fine volcanic dust can rise high into the troposphere and stratosphere, traveling hundreds or thousands of kilometers before settling to the earth's surface (Figure 12.3).

The interlayering of ash strata and sluggish streams of felsic lava produces a **stratovolcano**. Lofty, conical stratovolcanoes are well known for their scenic beauty. Fine examples are Mount Hood and Mount St. Helens in the Cascade Range (Figure 12.3), Fujiyama in Japan,

Figure 12.4 Mount Mayon, in southeastern Luzon, the Philippines, is often considered the world's most nearly perfect composite volcanic cone. It is an active volcano, and its summit rises to an altitude of nearly 2400 m (about 8000 ft). A series of eruptions began here in February 1993.

Figure 12.5 Crater Lake, Oregon, is surrounded by the high, steep wall of a great caldera. Wizard Island (center foreground) is an almost perfect basaltic cinder cone with basalt lava flows. It was built on the floor of the caldera after the major explosive activity had ceased.

Mount Mayon in the Philippines (Figure 12.4), and Mount Shishaldin in the Aleutian Islands.

Another important form of emission from explosive stratovolcanoes is a cloud of white-hot gases and fine ash. This intensely hot cloud, or "glowing avalanche," travels rapidly down the flank of the volcanic cone, searing everything in its path. On the Caribbean island of Martinique, in 1902, a glowing cloud issued without warning from Mount Pelée. It swept down on the city of St. Pierre, destroying the city and killing all but two of its 30,000 inhabitants.

Calderas

One of the most catastrophic of natural phenomena is a volcanic explosion so violent that it destroys the entire central portion of the volcano. Vast quantities of ash and dust are emitted and fill the atmosphere for many hundreds of square kilometers around the volcano. There remains only a great central depression named a *caldera*. Although some of the upper part of the volcano is blown outward in fragments, most of it settles back into the cavity formed beneath the volcano by the explosion.

Krakatoa, a volcanic island in Indonesia, exploded in 1883, leaving a huge caldera. It is estimated that 75 cu km (18 cu mi) of rock was blown out of the crater during the explosion. Great seismic sea waves generated by the explosion killed many thousands of persons living in low coastal areas of Sumatra and Java.

A classic example of a caldera produced in prehistoric times is Crater Lake, Oregon (Figure 12.5). The former volcano, named Mount Mazama, is estimated to have risen 1200 m (about 4000 ft) higher than the present caldera rim. The great explosion and collapse occurred about 6600 years ago.

Stratovolcanoes and Subduction Arcs

Most of the world's active stratovolcanoes lie within the circum-Pacific belt. Here, subduction of the Pacific, Nazca, Cocos, and Juan de Fuca plates is active. In Chapter 11, we explained how andesitic magmas rise beneath volcanic arcs of active continental margins and island arcs (see Figure 11.5). One good example is the volcanic arc of Sumatra and Java, lying over the subduction zone between the Australian plate and the Eurasian plate. Another is the Aleutian volcanic arc, located where the Pacific plate dives beneath the North American plate. The Cascade Mountains of northern California, Oregon, and Washington form a similar chain. Important segments of the Andes Mountains in South America consist of stratovolcanoes.

Shield Volcanoes

In contrast to thick, gassy felsic lava, mafic lava (basalt) is often highly fluid. It typically has a low viscosity and holds little gas. As a result, eruptions of basaltic lava are usually quiet, and the lava can travel long distances to spread out in thin layers. Typically, then, large basaltic volcanoes are broadly rounded domes with gentle slopes. Hawaiian volcanoes are of this type.

Volcanic activity involving basaltic lavas can occur in the midst of oceanic or continental lithospheric plates. Here, in isolated spots far from active plate boundaries, basaltic lavas are created by *mantle plumes*—isolated columns of molten material rising slowly within the asthenosphere. Directly above a mantle plume, crustal basalt is heated to the point of melting and produces a magma pocket. The site of rising magma is called a *hot spot*. Magma of basaltic composition makes its way through the overlying lithosphere to emerge at the surface as lava.

Where mantle plumes create hot spots in the oceanic lithosphere, the emerging basalt builds a class of initial landforms known as **shield volcanoes**. These broad basaltic domes are constructed on the deep ocean floor, far from plate boundaries, and may be built high enough to rise above sea level as volcanic islands. As a lithospheric plate drifts slowly over a mantle plume beneath, a succession of shield volcanoes is formed. Thus, a chain of basaltic volcanic islands comes into existence. Several such chains exist in the Pacific Ocean basin, including the Midway Islands and the Hawaiian group.

A few basaltic volcanoes also occur along the mid-oceanic ridge, where sea-floor spreading is in progress. Perhaps the outstanding example is Iceland, in the North Atlantic Ocean. Iceland is constructed entirely of basalt. Basaltic flows are superimposed on older basaltic rocks as dikes and sills formed by magma emerging from deep within the spreading rift. Mount Hekla, an active volcano on Iceland, is a shield volcano somewhat similar to those of Hawaii. Other islands consisting of basaltic volcanoes located along or close to the axis of the Mid-Atlantic Ridge are the Azores, Ascension, and Tristan da Cunha.

Hawaiian Volcanoes

The shield volcanoes of the Hawaiian Islands are characterized by gently rising, smooth slopes that flatten near the top, producing a broad-topped volcano (Figure 12.6). Domes on the island of Hawaii rise to summit elevations of about 4000 m (about 13,000 ft) above sea level. Including the basal portion lying below sea level, they are more than twice that high. In width they range from 16 to 80 km (10 to 50 mi) at sea level and up to 160 km (about 100 mi) wide at the submerged base. The basalt lava of the Hawaiian volcanoes is highly fluid and travels far down the gentle slopes. Most of the lava flows issue from fissures (long, gaping cracks) on the flanks of the volcano.

Hawaiian lava domes have a wide, steep-sided central depression that may be 3 km (2 mi) or more wide and

Figure 12.6 Basaltic shield volcanoes of Hawaii. At lower left is the now-cold Halemaumau fire pit, formed in the floor of the central depression of Kilauea volcano. On the distant skyline is the snow-capped summit of Mauna Kea volcano, its elevation over 4000 m (about 13,000 ft).

several hundred meters deep. These large depressions are a type of collapse caldera. Molten basalt is sometimes seen in the floors of deep pit craters that occur on the floor of the central depression or elsewhere over the surface of the lava dome.

The chain of Hawaiian volcanoes were created by the motion of the Pacific plate over a "hotspot"—a plume of upwelling basaltic magma deep within the mantle. With this mechanism, each volcano is created by the rising of hotspot magma, but then eventually the volcano moves away from the hotspot, becoming extinct and subject to erosion. Wind, water, waves, and subsidence eventually reduce the volcano to a sunken island platform. *Focus on Systems 12.1 • The Life Cycle of a Volcano* describes this process in more detail.

Flood Basalts

Where a mantle plume lies beneath a continental lithospheric plate, the hotspot may generate enormous volumes of basaltic lava that emerge from numerous vents and fissures and accumulate layer upon layer. The basalt may ultimately attain a thickness of thousands of meters and cover thousands of square kilometers. These accumulations are called *flood basalts*.

An important American example is found in the Columbia Plateau region of southeastern Washington, northeastern Oregon, and westernmost Idaho. Here, basalts of Cenozoic age cover an area of about 130,000 km^2 (about 50,000 mi^2)—nearly the same area as the state of New York. Individual basalt flows, exposed in the walls of river gorges, are expressed as cliffs in which vertical joint columns are conspicuous (Figure 12.7).

Cinder Cones

Associated with flood basalts, shield volcanoes, and scattered occurrences of basaltic lava flows is a small volcano known as a *cinder cone* (Figure 12.8). Cinder cones form when frothy basalt magma is ejected under high pressure from a narrow vent, producing tephra. The rain of tephra accumulates around the vent to form a roughly circular hill with a central crater. Cinder cones rarely grow to heights of more than a few hundred meters. An exceptionally fine example of a cinder cone is Wizard Island, built on the floor of Crater Lake long after the caldera was formed (Figure 12.5).

Hot Springs and Geysers

Where hot rock material is near the earth's surface, it can heat nearby groundwater to high temperatures. When the groundwater reaches the surface, it provides

Figure 12.7 Basalt lava flows exposed in cliffs bordering the Columbia River in Washington. Each set of cliffs is a major lava flow. In cooling, the lava acquires a vertical structure that makes it crack into tall columns.

Figure 12.8 This young cinder cone is surrounded by rough-surfaced basalt lava flows. Lava Beds National Monument, northern California.

Focus on Systems 12.1 • The Life Cycle of a Volcano

The time cycles we have examined so far in earlier chapters are of the repeating type. That is, each individual cycle is followed by the next, in an endless procession. The astronomical cycles of earth–sun relationships are examples, generating repeating annual cycles of such variable quantities as daily air temperature or monthly rainfall amounts.

In this Focus on Systems feature, we turn to what can be called a *life cycle*. This type of a cycle does not repeat, but rather consists of a continuous progression that can be seen as having a number of stages from a beginning to an end. The higher organisms have life cycles, expressed in the various stages of growth, development, and aging from conception to death.

Nonorganic life cycles can also be identified in nature. An example is the life cycle of a volcano. A volcano can appear without warning at a location where no other volcanic activity has been seen. The volcanic cone grows to a great mountain, then seemingly becomes dormant, and may end its life in a gigantic explosion that leaves a deep caldera.

The life cycle of a basaltic shield volcano that grows upward from the abyssal ocean floor is particularly interesting. This growth may take it from a depth greater than 5000 m (about 16,000 ft) below the sea surface to a height greater than 4000 m (about 13,000 ft) above the surface, thus building a huge mass of igneous rock atop the sea floor. A unique feature of this type of volcano is that it rests on an oceanic lithosphere that is in steady horizontal motion over a soft asthenosphere. You might say that the volcano rides on a conveyer belt.

The chain of Hawaiian volcanic domes is the best example of this type of volcano. However, this chain lies far away from the spreading boundary of the Pacific plate, which would be a natural source of magma. What, then, is the magma source for these islands? Based on geophysical evidence, geologists have concluded that there is a *hotspot* beneath the islands—a persistent source of heat providing pulses or bubbles of rising magma that built the Hawaiian chain, island by island, as the Pacific plate moved over the hotspot. The figure below diagrams this process.

The life cycle of these basaltic shield volcanoes was worked out by two geologists who became leading authorities on the Hawaiian volcanoes, Gordon Macdonald and Agatin Abbot. The typical stages of the life cycle are shown in figure at right. In the first stage (1), magma rising from the hotspot on the ocean floor forms a low basaltic lava dome. After the dome is constructed to full height (2), a central caldera is formed (3). There follows a post-caldera stage (4), in which large cinder cones fill the caldera and renew some of the original mass, but then dormancy sets in and erosional processes lower the mountain as waves cut back the coast (5). Fringing coral reefs then become broader. The volcano is now fully extinct, and the oceanic crust is steadily subsiding. Erosional beveling is completed at the atoll stage (6), when a thick layer of reef corals and related lagoon sediments has formed. In a final stage (7), continued crustal subsidence drowns the reef corals, and only a sunken island, called a *guyot*, remains.

It is now agreed that the hotspot that generated the entire chain of islands is produced by a *mantle*

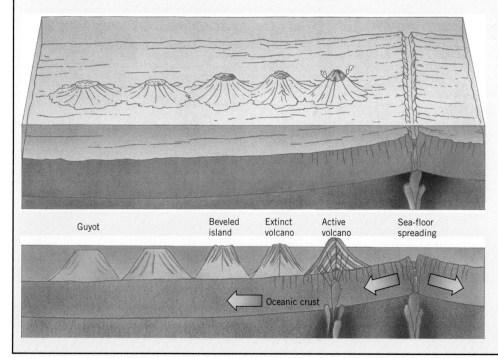

Guyot   Beveled island   Extinct volcano   Active volcano   Sea-floor spreading

Oceanic crust

A cartoon showing a chain of volcanoes formed by an oceanic plate moving over a hotspot. (Copyright © by A. N. Strahler.)

plume, arising far down in the asthenosphere. As the hot mantle rock rises, magma forms in bodies that melt their way through the lithosphere and reach the sea floor. Each major pulse of the plume sets off a volcanic cycle. As the volcano moves away from the location of the deep plume, it undergoes the middle and late stages (5–7) of the cycle. This has produced a long trail of sunken islands and guyots, shown on the map below. (See Figure 11.8 as well.) The Hawaiian trail, trending northwestward, is 2400 km (about 1500 mi) long and includes a sharp bend to the north caused by a sudden change of direction of the Pacific plate. This distant leg consists of the Emperor Seamounts. Several other long trails of volcanic seamounts cross the Pacific Ocean basin. They, too follow parallel paths that reveal the plate motion.

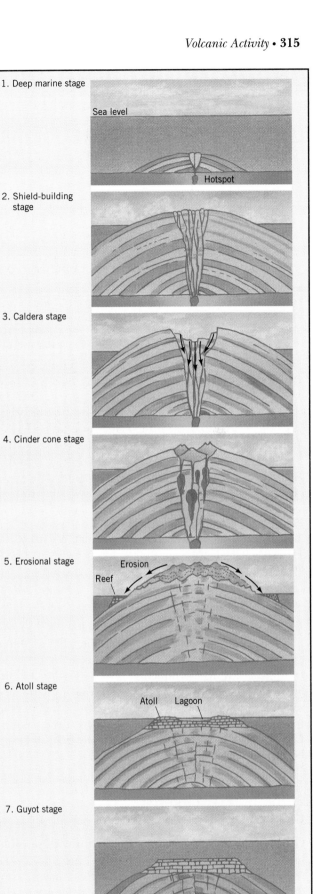

1. Deep marine stage

2. Shield-building stage

3. Caldera stage

4. Cinder cone stage

5. Erosional stage

6. Atoll stage

7. Guyot stage

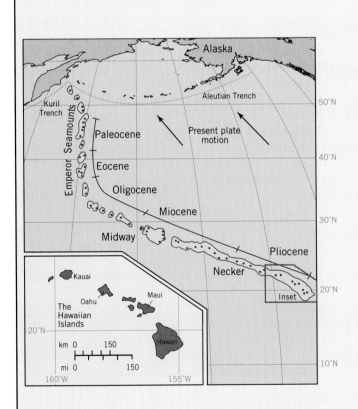

Sketch map of the Hawaiian seamount chain in the northwest Pacific Ocean basin. Dots are summits; the enclosing colored area marks the base of the volcano at the ocean floor. (Copyright © by A. N. Strahler.)

Life cycle of a typical Hawaiian volcano. (Adapted by permission from Macdonald and Abbott, 1970, *Volcanoes in the Sea: The Geology of Hawaii*, Honolulu, University of Hawaii Press, p. 138, Figure 92.)

Figure 12.9 Mammoth Hot Springs, an example of geothermal activity in Yellowstone National Park, Wyoming. Terraces ringed by mineral deposits hold steaming pools of hot water.

Figure 12.10 Old Faithful Geyser in Yellowstone National Park, Wyoming.

hot springs at temperatures not far below the boiling point of water (Figure 12.9). At some places, jetlike emissions of steam and hot water occur at intervals from small vents—producing *geysers* (Figure 12.10). Since the water that emerges from hot springs and geysers is largely groundwater that has been heated in contact with hot rock, this water is recycled surface water. Little, if any, is water that was originally held in rising bodies of magma.

EYE ON THE ENVIRONMENT:
Geothermal Energy Sources

Geothermal energy is energy in the form of sensible heat that originates within the earth's crust and makes its way to the surface by conduction. Heat may be conducted upward through solid rock or carried up by circulating groundwater that is heated at depth and returns to the surface. Concentrated geothermal heat sources are usually associated with igneous activity, but there also exist deep zones of heated rock and groundwater that are not directly related to igneous activity.

Observations made in deep mines and boreholes show that the temperature of rock increases steadily with depth. Although the rate of increase falls off quite rapidly with increasing depth, temperatures attain very high values in the upper mantle, where rock is close to its melting point. Heat within the earth's crust and mantle is produced largely by radioactive decay. (See *Focus on Systems 10.3 • Powering the Cycle of Rock Change.*) Slow as this internal heat production is, it scarcely diminishes with time and the basic energy resource it provides can be regarded as limitless on a human scale.

It might seem simple enough to obtain all our energy needs by drilling deep holes at any desired location into the crust and letting the hot rock turn injected fresh water into steam, which we could use to generate electricity as our primary energy resource. Unfortunately, at the depths usually required to furnish the needed heat intensity, crustal rock tends to close any cavity or opening by rupture and slow flowage. This phenomenon would either prevent the holes from being drilled or would close them in short order. Generally, then, we must look for geothermal localities, where special conditions have caused hot rock and hot groundwater to lie within striking distance of conventional drilling methods. Areas of hot springs and geysers are primary candidates.

Natural hot water and steam localities were the first type of geothermal energy source to be developed and at present account for nearly all production of electrical power. Wells are drilled to tap the hot water, which flashes into steam under reduced atmospheric pressure as it reaches the surface. The steam is separated from the water in large towers and fed into generating turbines to produce electricity (Figure 12.11). The hot water is usually released into surface stream flow, where it may create a pollution problem. The larger steam fields have sufficient energy to generate at least 15 megawatts of electric power, and a few can generate 200 megawatts or more.

In certain areas, the intrusion of magma has been sufficiently recent that solid igneous rock of a batholith is still very hot in a depth range of perhaps 2 to 5 km (about 1 to 3 mi). At this depth, the rock is strongly compressed and contains little, if any, groundwater. Rock in this zone may be as hot as 300°C (about 575°F) and could supply an enormous quantity of heat energy. The planned development of this resource includes drilling into the hot zone and then shattering the surrounding rock by hydrofracture—a method using water under pressure that is widely used in petroleum development. Surface water would then be pumped down one well into the fracture zone and heated water pumped up another well. Although some experiments have been conducted, this heat source has not yet been exploited operationally. 🐾

EYE ON THE ENVIRONMENT:
Volcanic Eruptions as Environmental Hazards

The eruptions of volcanoes and lava flows are environmental hazards of the severest sort, often taking a heavy toll of plant and animal life and devastating human habitations. What natural phenomenon can compare with the Mount Pelée disaster in which thousands of lives were snuffed out in seconds? Perhaps only an earthquake or storm surge of a tropical cyclone is equally disastrous.

Wholesale loss of life and destruction of towns and cities are frequent in the history of peoples who live near active volcanoes. Loss occurs principally from sweeping clouds of incandescent gases that descend the volcano slopes like great avalanches, from lava flows whose relentless advance engulfs whole cities, from the descent of showers of ash, cinders, and bombs, and from violent earthquakes associated with volcanic activity. For habitations along low-lying coasts there is the additional peril of great seismic sea waves, generated elsewhere by explosive destruction of volcanoes.

In 1985, an explosive eruption of Ruiz Volcano in the Colombian Andes caused the rapid melting of ice and snow in the summit area. Mixing with volcanic ash, the water formed a variety of mudflow known as a *lahar*. Rushing downslope at speeds up to 145 km (90 mi) per hour, the lahar became channeled into a valley on the lower slopes, where it engulfed a town and killed more than 20,000 persons.

Despite their potential for destructive activity, volcanoes are a valuable natural resource in terms of recreation and tourism. Few landscapes can rival in beauty the mountainous landscapes of volcanic origin. National parks have been made of Mount Rainier, Mount Lassen, and Crater Lake in the Cascade Range, a mountain mass largely of volcanic construction. Hawaii Volcanoes National Park recognizes the natural beauty of Mauna Loa and Kilauea. Their breathtaking displays of molten lava are a living textbook of igneous processes. 🐾

Figure 12.11 An electricity-generating plant at The Geysers, California. Steam pipes in the foreground lead to the plant. After use in generating turbines, the steam is condensed in large cylindrical towers.

LANDFORMS OF TECTONIC ACTIVITY

Recall from Chapter 11, our introduction to global plate tectonics, that there are basically two different forms of tectonic activity: compressional and extensional. Along converging lithospheric plate boundaries, tectonic activity is primarily compression, illustrated schematically in Figure 12.12. In subduction zones, sedimentary layers of the ocean floor are subject to compression within a trench as the descending plate forces them against the overlying plate. In continental collision, compression is of the severest kind.

In zones of rifting of continental plates, explained in Chapter 11, the brittle continental crust is pulled apart and yields by faulting. This motion is described as extensional. In the simple model shown in Figure 12.12, rifting is expressed in a pair of opposite-facing faults. The crustal block between them moves down to form a depressed area.

Fold Belts

When continental collision begins to take place, broad wedges of strata of a passive continental margin come under strong forces of compression (Figure 11.16). The strata, which were originally more or less flat-lying, experience **folding**, as shown in Figure 12.12. The wavelike shapes imposed on the strata consist of alternating archlike upfolds, called *anticlines*, and troughlike downfolds, called *synclines*. Thus, the initial landform associated with an anticline is a broadly rounded mountain ridge, and the landform corresponding to a syncline is an elongate, open valley.

An example of open folds of comparatively young geologic age that has long attracted the interest of geographers is the Jura Mountains of France and Switzerland. Figure 12.13 is a block diagram of a small portion of that fold belt. The rock strata are mostly limestone layers and were capable of being deformed by bending with little brittle fracturing. Folding occurred in late Cenozoic (Miocene) time. Notice that each mountain crest is associated with the axis of an anticline, while each valley lies over the axis of a syncline. Some of the anticlinal arches have been partially removed by erosion processes. The rock structure can be seen clearly in the walls of the winding gorge of a major river that crosses the area. The Jura folds lie just to the north of the main collision orogen of the Alps. Because of their location near a mountain mass, they are called *foreland folds*.

Another example of a belt of foreland folds produced during continental collision is the Zagros Mountains of southwestern Iran. Seen from an orbiting space vehicle, the Zagros folds resemble an army of caterpillars crawling in parallel tracks (Figure 12.14). The folds

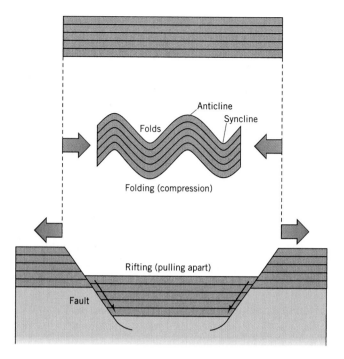

Figure 12.12 Two basic forms of tectonic activity that produce initial landforms. Flat-lying rock layers may be compressed to form folds or may be pulled apart to form faults by rifting.

were formed during late Cenozoic time when the Arabian plate smashed into the Eurasian plate, closing an ocean basin that formerly separated the two continental masses. Flat-lying strata on the passive margin of the Arabian plate were crumpled into open folds over a belt more than 200 km (about 125 mi) wide.

Faults and Fault Landforms

A **fault** in the brittle rocks of the earth's crust occurs when rocks suddenly yield to unequal stresses by fracturing. Faulting is accompanied by a displacement—a slipping motion—along the plane of breakage, or *fault plane*. Faults are often of great horizontal extent, so that the surface trace, or fault line, can sometimes be followed along the ground for many kilometers. Most major faults extend down into the crust for at least several kilometers.

Faulting occurs in sudden slippage movements that generate earthquakes. A single fault movement may result in slippage of as little as a centimeter or as much as 15 m (about 50 ft). Successive movements may occur many years or decades apart, even several centuries apart. Over long time spans, the accumulated displacements can amount to tens or hundreds of kilometers.

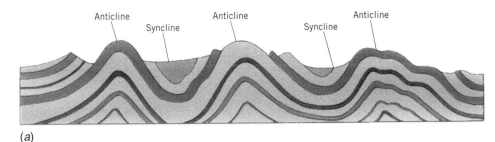

Anticline   Syncline   Anticline   Syncline   Anticline

(a)

Figure 12.13 Anticlines and synclines of the Jura Mountains, France and Switzerland. A cross section shows the folds and ground surface (a). The landscape developed on the folds is shown in the block diagram (b). (After E. Raisz.)

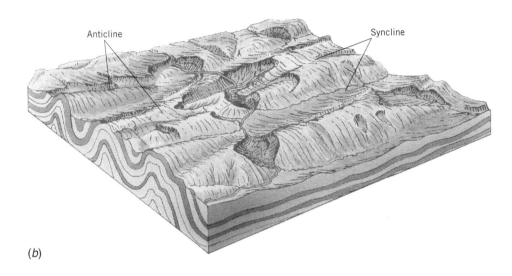

Anticline   Syncline

(b)

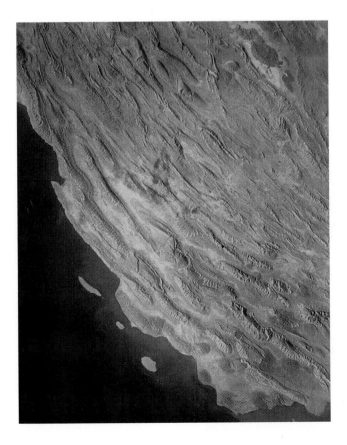

In some places, clearly recognizable sedimentary rock layers are offset on opposite sides of a fault, allowing the total amount of displacement to be measured accurately.

Normal Faults

One common type of fault associated with crustal rifting is the **normal fault** (Figure 12.15a). The plane of slippage, or fault plane, is steeply inclined. The crust on one side is raised, or upthrown, relative to the other, which is downthrown. A normal fault results in a steep, straight, clifflike feature called a *fault scarp*. Fault scarps range in height from a few meters to a few hundred meters (Figure 12.16). Their length is usually measur-

Figure 12.14 The Zagros Mountains of Iran, photographed by astronauts of the Gemini XII orbiting space vehicle in 1966. The view is toward the northwest and shows a portion of the Persian Gulf. The elongate ridges are anticlines, elevated by folding which began late in the Cenozoic Era. They are partly eroded by streams.

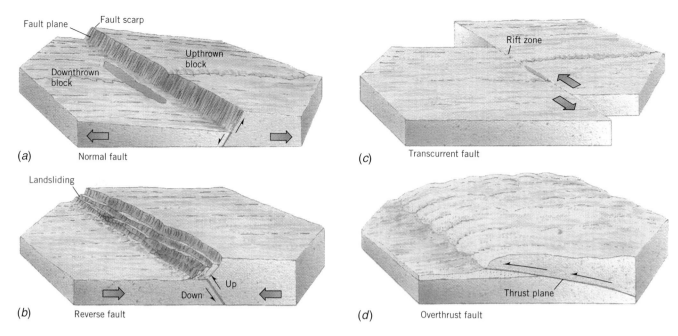

(a) Normal fault

(b) Reverse fault

(c) Transcurrent fault

(d) Overthrust fault

Figure 12.15 Four types of faults.

able in kilometers. In some cases they attain lengths as great as 300 km (about 200 mi).

Normal faults are not usually isolated features. Commonly, they occur in multiple arrangements, often as a set of parallel faults. This gives rise to a grain or pattern of rock structure and topography. A narrow block dropped down between two normal faults is a *graben* (Figure 12.17). A narrow block elevated between two

normal faults is a *horst*. Grabens make conspicuous topographic trenches, with straight, parallel walls. Horsts make blocklike plateaus or mountains, often with a flat top but steep, straight sides.

In rifted zones of the continents, regions where normal faulting occurs on a grand scale, mountain masses called *block mountains* are produced. The up-faulted mountain blocks can be described as either tilted or

Figure 12.16 This fault scarp was formed during the Hebgen Lake, Montana, earthquake of 1959. In a single instant, a displacement of 6 m (20 ft) took place on a normal fault.

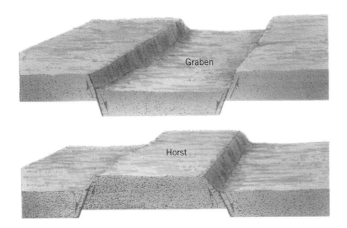

Figure 12.17 Initial landforms of normal faulting. A graben is a downdropped block, often forming a long, narrow valley. A horst is an upthrown block, forming a plateau, mesa, or mountain. (A. N. Strahler)

lifted (Figure 12.18). A tilted block has one steep face—the fault scarp—and one gently sloping side. A lifted block, which is a type of horst, is bounded by steep fault scarps on both sides.

Transcurrent Faults

Recall that lithospheric plates slide past one another along major transform faults and that these features comprise one type of lithospheric plate boundary. Long before the principles of plate tectonics became known, geologists referred to such faults as **transcurrent faults** (Figure 12.15*c*), or sometimes as *strike-slip faults*. In a transcurrent fault, the movement is predominantly horizontal. Where the land surface is nearly flat, no scarp, or a very low one at most, results. Only a thin fault line is traceable across the surface. In some places a narrow trench, or rift, marks the fault.

The best known of the active transcurrent faults is the great San Andreas Fault, which can be followed for a distance of about 1000 km (about 600 mi) from the Gulf of California to Cape Mendocino, a location well north of the San Francisco area, where it heads out to sea. It is a transform fault that marks the active boundary between the Pacific plate and the North American plate (see Figure 11.14). The Pacific plate is moving to-ward the northwest, which means that a great portion of the state of California and all of Lower (Baja) California are moving bodily northwest with respect to the North American mainland.

Throughout many kilometers of its length, the San Andreas Fault appears as a straight, narrow scar. In some places this scar is a trenchlike feature, and elsewhere it is a low scarp (Figure 12.19). Frequently, a stream valley takes an abrupt jog—for example, first right, then left—when crossing the fault line. This offset in the stream's course shows that many meters of movement have occurred in fairly recent time.

Reverse and Overthrust Faults

In a *reverse fault*, the inclination of the fault plane is such that one side rides up over the other and a crustal shortening occurs (Figure 12.15*b*). Reverse faults produce fault scarps similar to those of normal faults, but the possibility of landsliding is greater because an overhanging scarp tends to be formed. The San Fernando, California, earthquake of 1971 was generated by slippage on a reverse fault.

The *low-angle overthrust fault* (Figure 12.15*d*) involves predominantly horizontal movement. One slice of rock rides over the adjacent ground surface. A thrust slice

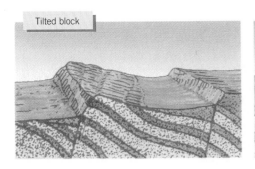
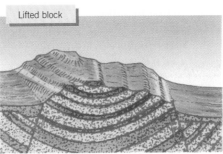

Figure 12.18 Fault block mountains may be of tilted type (*left*) or lifted type (*right*). (After W. M. Davis.)

Figure 12.19 The San Andreas Fault in southern California. The fault is marked by a narrow trough. The fault is slightly offset in the middle distance.

may be up to 50 km wide (about 30 mi). The evolution of low-angle thrust faults was explained in Chapter 11 and illustrated in Figure 11.15.

The Rift Valley System of East Africa

Rifting of continental lithosphere is the very first stage in the splitting apart of a continent to form a new ocean basin. The process is beautifully illustrated by the East African rift valley system. This region has attracted the attention of geologists since the early 1900s. They gave the name *rift valley* to what is basically a graben, but with a more complex history that includes the building of volcanoes on the graben floor.

Figure 12.20 is a sketch map of the East African rift valley system, which extends from the Red Sea southward about 3000 km (about 1900 mi). Along this axis, the earth's crust is being lifted and spread apart in a long, ridgelike swell. The rift valley system consists of a number of grabenlike troughs. Each is a separate rift valley ranging in width from about 30 to 60 km (about 20 to 40 mi). Major rivers and several long, deep lakes—Lake Nyasa and Lake Rudolph, for example—occupy some of the valley floors. Two great stratovolcanoes have been built close to the rift valley east of Lake

Victoria. One is Mount Kilimanjaro, whose summit rises to over 6000 m (about 19,000 ft). The other, Mount Kenya, is only a little lower and lies right on the equator.

EARTHQUAKES

You have probably seen television news accounts of disastrous earthquakes and their destructive effects (Figure 12.21). Californians know about severe earthquakes from first-hand experience, but several other areas in North America have also experienced strong earthquakes, and a few of these have been very severe. An **earthquake** is a motion of the ground surface, ranging from a faint tremor to a wild motion capable of shaking buildings apart.

The earthquake is a form of energy of wave motion transmitted through the surface layer of the earth.

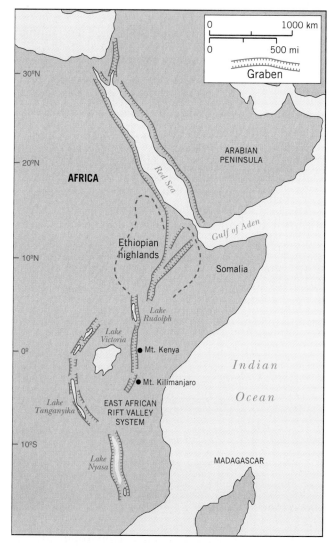

Figure 12.20 A sketch map of the East African rift valley system and the Red Sea to the north.

Figure 12.21   Earthquake devastation in Mexico City following the earthquake of September 1985.

Waves move outward in widening circles from a point of sudden energy release, called the *focus.* Like ripples produced when a pebble is thrown into a quiet pond, these seismic waves gradually lose energy as they travel outward in all directions. (The term *seismic* means "pertaining to earthquakes.")

Most earthquakes are produced by sudden slip movements along faults. They occur when rock on both sides of the fault is slowly bent over many years by tectonic forces. Energy accumulates in the bent rock, just as it does in a bent archer's bow. When a critical point is reached, the strain is relieved by slippage on the fault, and the rocks on opposite sides of the fault move in different directions. A large quantity of energy is instantaneously released in the form of seismic waves, which shake the ground. In the case of a transcurrent fault, on which movement is in a horizontal direction, slow bending of the rock that precedes the shock takes place over many decades. Sometimes a slow, steady displacement known as *fault creep* occurs, which tends to reduce the accumulation of stored energy.

The devastating San Francisco earthquake of 1906 resulted from slippage along the San Andreas Fault, which is dominantly a transcurrent fault. This fault also passes about 60 km (about 40 mi) inland of Los Angeles, placing the densely populated metropolitan Los Angeles region in great jeopardy. Associated with the San Andreas Fault are several important parallel and branching transcurrent faults, all of which are capable of generating severe earthquakes.

Earthquakes can also be produced by volcanic activity, as when magma rises or recedes within a volcanic chamber.

A scale of earthquake magnitudes was devised in 1935 by the distinguished seismologist Charles F. Richter. Now called the *Richter scale,* it describes the quantity of energy released by a single earthquake. Scale numbers range from 0 to 9, but there is really no upper limit other than nature's own energy release limit. For each whole unit of increase (say, from 5.0 to 6.0), the quantity of energy release increases by a factor of 32. The Chilean earthquake of 1960 is the largest observed to date. It earned a rating of 8.3 on the Richter scale, but that value was later updated to 9.5. The great San Francisco earthquake of 1906 is now rated as magnitude 7.9. *Working It Out 10.2* • *The Richter Scale* provides more information about the Richter scale and the energy released by some great earthquakes.

Earthquakes and Plate Tectonics

Seismic activity—the repeated occurrence of earthquakes—occurs primarily near lithospheric plate boundaries. Figure 12.22, which shows the location of all earthquake centers during a typical seven-year period, clearly reveals this pattern. The greatest intensity of seismic activity is found along converging plate boundaries where oceanic plates are undergoing subduction. Strong pressures build up at the downslanting contact of the two plates, and these are relieved by sudden fault slippages that generate earthquakes of large magnitude. This mechanism explains the great earthquakes experienced in Japan, Alaska, Central America, Chile, and other narrow zones close to trenches and volcanic arcs of the Pacific Ocean basin.

Good examples can be cited from the Pacific coast of Mexico and Central America, where the subduction boundary of the Cocos Plate lies close to the shoreline. The great earthquake that devastated Mexico City in 1986 was centered in the deep trench offshore. Two great shocks in close succession, the first of magnitude 8.1 and a second of 7.5, damaged cities along the coasts of the Mexican states of Michoacán and Guerrero. Although Mexico City lies inland about 300 km (about 185 mi) distant from the earthquake epicenters, it experienced intense ground shaking of the underlying saturated clay formations, with the resulting death toll of some 10,000 persons (Figure 12.21).

Transform boundaries that cut through the continental lithosphere are also sites of intense seismic activity, with moderate to strong earthquakes. The most familiar example is the San Andreas Fault, discussed further in *Eye on the Environment: Earthquakes Along the San Andreas Fault.*

Spreading boundaries are the location of a third class of narrow zones of seismic activity related to lithospheric plates. Most of these boundaries are identified with the midoceanic ridge and its branches. For the most part, earthquakes in this class are limited to moderate intensities.

Earthquakes also occur at scattered locations over the continental plates, far from active plate boundaries. In many cases, no active fault is visible, and the geologic cause of the earthquake is uncertain. For example, the great New Madrid earthquake of 1811 was centered in the Mississippi River floodplain in Missouri. It produced three great shocks in close succession, rated from 8.1 to 8.3 on the Richter scale.

Seismic Sea Waves

An important environmental hazard often associated with a major earthquake centered on a subduction plate boundary is the *seismic sea wave*, or **tsunami**, as it is known to the Japanese. A train of these water waves is

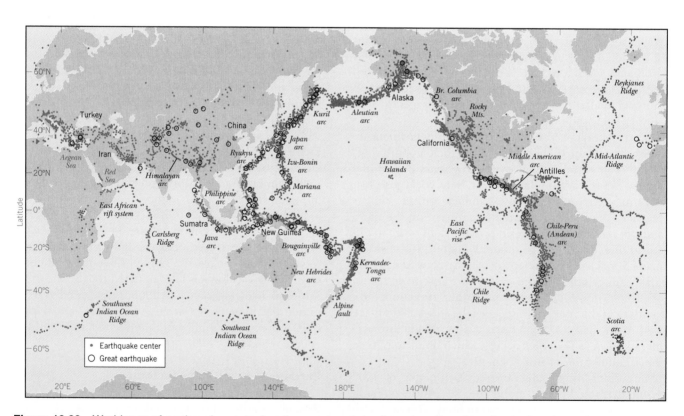

Figure 12.22   World map of earthquake center locations and centers of great earthquakes. Center locations of all earthquakes originating at depths of 0 to 100 km (62 mi) during a six-year period are shown by red dots. Each dot represents a single location or a cluster of centers. Black circles identify centers of earthquakes of Richter magnitude 8.0 or greater during an 80-year period. The map clearly shows the pattern of earthquakes occurring at subduction boundaries. (Compiled by A. N. Strahler from data of U.S. government. Copyright © A. N. Strahler.)

often generated in the ocean by a sudden movement of the sea floor at a point near the earthquake source. The waves travel over the ocean in ever-widening circles, but they are not perceptible at sea in deep water. (Seismic sea waves are sometimes referred to as "tidal waves," but since they have nothing to do with the tides, the name is quite misleading.)

When a tsunami arrives at a distant coastline, the effect is to cause a rise of water level. Normal wind-driven waves, superimposed on the heightened water level, attack places inland that are normally above their reach. For example, in this century several destructive seismic sea waves in the Pacific Ocean attacked ground as high as 10 m (30 ft) above normal high-tide level, causing widespread destruction and many deaths by drowning in low-lying coastal areas. It is thought that the coastal flooding that occurred in Japan in 1703, with an estimated loss of life of 100,000 persons, may have been caused by seismic sea waves.

EYE ON THE ENVIRONMENT:
Earthquakes along the San Andreas Fault

More than nine decades have passed since the great San Francisco earthquake of 1906 was generated by movement on the San Andreas Fault. The maximum horizontal displacement of the ground was about 6 m (20 ft). Since then, this sector of the fault has been locked—that is, the rocks on the two sides of the fault have been held together without sudden slippage (Figure 12.23). In the meantime, the two lithospheric plates that meet along the fault have been moving steadily with respect to one another. This means that a huge amount of unrelieved strain energy has already accumulated in the crustal rock on either side of the fault.

On October 17, 1989, the San Francisco Bay area was severely jolted by an earthquake with a Richter magnitude of 7.1. The earthquake's epicenter was located

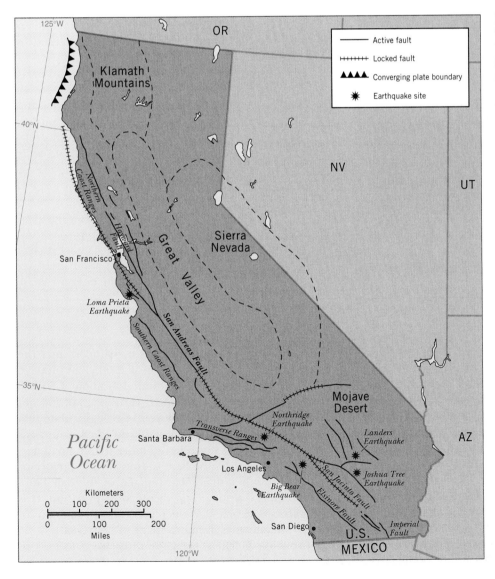

Figure 12.23 A sketch map of the San Andreas Fault and Transverse Ranges showing the locked sections of the fault alternating with active sections. (Based on data of the U.S. Geological Survey.)

Working It Out 12.2 • The Richter Scale

The magnitude of earthquakes is most commonly assessed by the *Richter scale*, which was devised by the seismologist Charles F. Richter in 1935, and then modified in 1956 by Richter and his colleague, Beno Gutenberg. This scale can be re- lated to the energy released at the earthquake center, and thus can be used as an estimate of the severity of a particular earthquake. The scale has neither a fixed maximum nor a minimum, but earthquakes rated as high as 8.4 on the Richter scale have been measured thus far. Earth- quakes of magnitude 2.0 are the smallest normally detected by human senses, but instruments can detect quakes as small as –3.0 on the scale. The Richter scale is loga- rithmic and is based on the ampli- tude of the largest earthquake waves measured for a particular earthquake. For each integer in- crease in the Richter scale, the am- plitude of the earthquake wave in- creases by a factor of 10.

How is the Richter scale related to the amount of energy released by an earthquake? The Richter scale is proportional to the amplitude of the largest wave, but that is not a di- rect measure of energy released. However, there is a relationship be- tween the Richter scale number and energy release that is used by geophysicists:

$$\log_{10} E = 4.8 + 1.5M$$

where E is the energy in joules and M is the Richter scale magnitude. The figure at the left plots this rela- tionship using a logarithmic scale for energy release. By taking the ex- ponential of both sides, we can write this expression as

$$E = 10^{(4.8 + 1.5M)} = 10^{4.8} \times 10^{1.5M}$$

From this form, we can see that if the Richter scale number increases by one unit, then the energy release

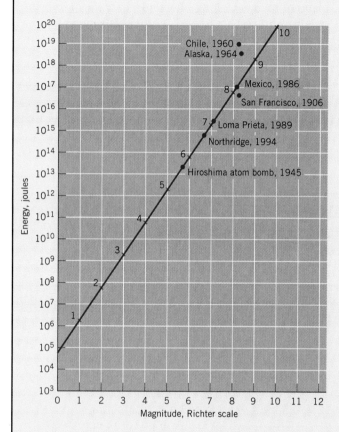

The Richter scale and energy released by earthquakes. (Copyright © by A. N. Strahler.)

near Loma Prieta peak, about 80 km (50 mi) southeast of San Francisco, at a point only 12 km (7 mi) from the city of Santa Cruz, on Monterey Bay. The city of Santa Cruz suffered severe structural damage to older build- ings. Destructive ground shaking in the distant San Francisco Bay area proved surprisingly severe. Build- ings, bridges, and viaducts on landfills were particularly hard hit (Figure 12.24). Altogether, 62 lives were lost in this earthquake, and the damage was estimated to be about $6 billion. In comparison, the 1906 earthquake took a toll of 700 lives and property damage equivalent to 20 billion 1987 dollars.

The displacement that caused the Loma Prieta earth- quake occurred deep beneath the surface not far from the San Andreas Fault, which has not slipped since the great San Francisco earthquake of 1906. The slippage on the Loma Prieta fault amounted to about 1.8 m (6 ft) horizontally and 1.2 m (4 ft) vertically, but did not break the ground surface above it. Geologists state that the Loma Prieta slippage, though near the San Andreas Fault, probably has not relieved more than a small por- tion of the strain on the San Andreas. While the occur- rence of another major earthquake in the San Fran- cisco region cannot be predicted with precision, it is inevitable. As each decade passes, the probability of that event becomes greater.

Surprising as it may seem, you may need to look to Japan for a scenario that the citizens of the San Fran-

will increase by a factor of $10^{1.5 \times 1}$, or 31.6. This means that for each unit increase in the Richter scale, say from 4.0 to 5.0, about 32 times as much energy is released. For a two-unit increase, say from 4.0 to 6.0, the energy released will be $10^{1.5 \times 2} = 10^3 = 1000$ times as large!

News media accounts of earthquakes often get this wrong and state that an increase of one unit in the Richter scale produces a tenfold increase in energy released. These accounts confuse the amplitude of the earthquake wave, which does increase tenfold per scale unit, with

the energy release, which increases by the factor of 31.6.

The relationship between energy release and Richter scale magnitude shown above works best for small and medium earthquakes. For large earthquakes, however, the energy release is underestimated. In 1977, the seismologist Hiroo Kanamori of the California Institute of Technology developed a method to calculate earthquake magnitude which more accurately reflects the energy release. After examining the seismograms of great earthquakes of the twentieth century, he assigned larger magnitudes to many of them. Some of these are shown in the table. When they are plotted on the graph according to their Richter magnitude, they fall above the line, showing that they release more energy than expected by their Richter rating.

However they are rated, great earthquakes are among the most powerful natural phenomena known. The Alaskan earthquake of 1964 released a quantity of energy equal to that of about 200,000 atomic bombs of the size exploded at Hiroshima in 1945. As human habitation of large cities near active faults continues to increase, the potential for death and destruction from such powerful earthquakes increases as well.

Richter Magnitude and Energy Release

Magnitude, Richter Scale*	Energy Release (joules)	Comment
2.0	6×10^7	Smallest quake normally detected by humans.
2.5–3.0	10^8–10^9	Quake can be felt if it is nearby. About 100,000 shallow quakes of this magnitude per year.
4.5	4×10^{11}	Can cause local damage.
5.7	2×10^{13}	Energy released by Hiroshima atom bomb.
6.0	6×10^{13}	Destructive in a limited area. About 100 shallow quakes per year of this magnitude.
6.7	7×10^{14}	Northridge earthquake of 1994.
7.0	2×10^{15}	Rated a major earthquake above this magnitude. Quake can be recorded over whole earth. About 14 per year this great or greater.
7.1	3×10^{15}	Loma Prieta earthquake of 1989.
8.25 (7.9)	4.5×10^{16}	San Francisco earthquake of 1906.
8.1	10^{17}	Mexican earthquake of 1986.
8.4 (9.2)	4×10^{18}	Alaskan earthquake of 1964.
8.3 (9.5)	10^{19}	Chilean earthquake of 1960, near the border of Equador.

* () indicates magnitude as adjusted by Kanamori.

cisco Bay area could experience when a branch of the San Andreas Fault lets go. The Hyogo-ken Nanbu earthquake that devastated the city of Kobe in January of 1995 (Figure 12.25) occurred on a short side branch of a major strike-slip fault quite similar tectonically to the San Andreas. A similar side branch of the main San Andreas Fault, called the Hayward Fault, runs through the East Bay region of San Francisco. Seismologists estimate that a slip on the Hayward Fault would generate a quake of magnitude 7.0—about the same as the Kobe quake—and has a 28-percent probability of occurring by the year 2018. Although the nation of Japan prides itself on its earthquake preparedness, the Kobe catastrophe left 5000 dead, 300,000 homeless, and did prop-

erty damage estimated to have been ten times that of the Northridge earthquake of 1994, which was the most costly earthquake in American history.

Along the southern California portion of the San Andreas Fault, a recent estimate placed at about 50 percent the likelihood that a very large earthquake will occur within the next 30 years. In 1992 three severe earthquakes occurred in close succession along local faults a short distance north of the San Andreas Fault within the area labeled the southern active area on the map (Figure 12.23). The second of these, the Landers earthquake, a powerful 7.5 on the Richter scale, occurred on a strike-slip fault trending north-northwest. It caused a 80-km (50-mi) rupture across the Mojave

Figure 12.24 A collapsed section of the double-decked Nimitz Freeway (Interstate 880) in Oakland, California. Shaking by the Loma Prieta earthquake caused the upper level of the freeway to collapse on the lower deck, crushing at least 39 persons in their cars.

Desert. These three events have led to speculation that the likelihood of a major slip on the nearby San Andreas Fault in the near future has substantially increased.

For residents of the Los Angeles area, an additional serious threat lies in the large number of active faults close at hand. Movements on these local faults have produced more than 40 damaging earthquakes since 1800, including the Long Beach earthquakes of the 1930s and the San Fernando earthquake of 1971. The San Fernando earthquake measured 6.6 on the Richter scale and produced severe structural damage near the earthquake center. The brief but intense primary shock and aftershocks that followed damaged beyond repair many older structures built of unreinforced brick masonry. In 1987 an earthquake of magnitude 6.1 struck the vicinity of Pasadena and Whittier, located within about 20 km (12 mi) of downtown Los Angeles. Known as the Whittier Narrows earthquake, it was generated along a local fault system that had not previously shown significant seismic activity.

One of the most damaging earthquakes in recent years occurred along a similar local fault system—the Northridge earthquake of 1994. Centered near Northridge, a San Fernando Valley suburb of Los Angeles, the quake occurred early on the morning of January 17 and was assigned a Richter scale rating of 6.7. It produced the strongest ground motions ever recorded in an urban setting in North America and the greatest financial losses from a natural disaster in the United States since the great San Francisco earthquake of 1906.

Northridge lies in an east-west trending fault-rich valley close to the base of the lofty San Gabriel Mountains. This range is an uplifted fault-block that has produced earthquakes at many locations along the base of its steep southern flank, including the equally severe San Fernando earthquake of 1971. These active earthquake faults belong to a tectonic belt known as the Transverse Ranges, which runs more or less east-west through the

Figure 12.25 This area within the city of Kobe, Japan, was reduced to rubble by the earthquake of January 1995 and the fires that followed in its aftermath.

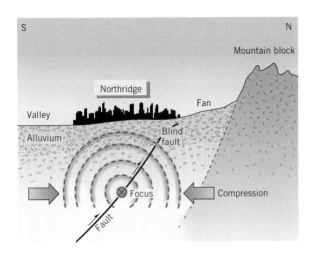

Figure 12.26 Geologic setting of the Northridge earthquake.

Los Angeles basin and heads out across the Pacific coastline beyond Santa Barbara (see Figure 12.23).

The Northridge and San Fernando earthquake faults are of the reverse type (see Figure 12.15*b*). Figure 12.26 shows how it works. Large arrows pointing toward each other indicate that along the Transverse Ranges the earth's crust is under a strong, north-south, horizontal compressional force. The fault plane slants steeply upward. When the fault suddenly starts to slip at the focus, located about 20 km below the valley surface, the upper block delivers a powerful upward punch that sharply lifts the ground surface. As with a boxer's "uppercut," what counts is the acceleration of that first upward slip motion. Seismic waves also spread out radially from the focus, generating strong ground shaking with horizontal motions. Notice that the displacement along the earthquake fault did not extend upward to the ground surface—that is, it is a "blind fault." Nevertheless, disturbances of the ground surface occurred over a wide area.

Northridge had more severe structural damage than San Fernando, although both were Richter 6.7 energy releases. In the Northridge case, the focus lay directly beneath the city, whereas the focus of the San Fernando earthquake was located some distance back under the mountain block. Northridge casualties included 33 persons dead as a direct result of the earthquake, more than 7000 injuries treated at local hospitals, and over 20,000 persons left homeless. Total financial losses were estimated as high as $20 billion. Sections of three freeways were closed, including some of the busiest highways in the country.

A slip along the San Andreas Fault, some 50 km (31 mi) to the north of the densely populated region of Los Angeles, will release an enormously larger quantity of energy than local earthquakes, such as the Northridge or San Fernando earthquakes. On the other hand, the destructive effects of a San Andreas earthquake in downtown Los Angeles will be somewhat moderated by the greater travel distance. Although the intensity of ground shaking might not be much different from that of the San Fernando earthquake, for example, it will last much longer and cover a much wider area of the Los Angeles region. The potential for damage and loss of life is enormous.

In the last three chapters, we have surveyed the composition, structure, and geologic activity of the earth's crust. We began with a study of the rocks and minerals that make up the earth's crust and core. We saw how the rock cycle describes a continuous cycling and recycling of rocks and minerals that has occurred over some 3 billion years or more of geologic time. In the second chapter, we developed plate tectonics as the mechanism powering the rock cycle. The global pattern of plate tectonics also explains the geographical distribution of mountain ranges, ocean basins, and continental shields occurring on the earth's surface. In this chapter, we described the landforms that result directly from volcanic and tectonic activity, which occur primarily at the boundaries of spreading or colliding lithospheric plates.

With this survey of the earth's crust and the geologic processes that shape it now completed, we can turn to other landform-creating processes in the following chapters. First will be the processes of weathering, which breaks rock into small particles, and mass wasting, which moves them downhill as large and small masses under the influence of gravity. Next, we turn to running water in three chapters that first describe the behavior of rivers and streams, then their work in shaping landforms, and lastly how running water dissects rock layers to reveal their inner structures. In our two concluding chapters, we will examine landforms created by waves, wind, and glacial ice.

CHAPTER SUMMARY

Landforms are the surface features of the land, and geomorphology is the scientific study of landforms. Initial landforms are shaped by volcanic and tectonic activity, while sequential landforms are sculpted by agents of denudation, including running water, waves, wind, and glacial ice.

Volcanoes are landforms marking the eruption of lava at the earth's surface. Stratovolcanoes, formed by the emission of thick, gassy, felsic lavas, have steep slopes and tend to explosive eruptions that can form calderas. Most active stratovolcanoes lie along the Pacific rim, where subduction of oceanic lithospheric plates is occurring. At hotspots, rising mantle material provides mafic magma that erupts as basaltic lavas. Because these lavas are more fluid and contain little gas, they form broadly rounded shield volcanoes. Hotspots occurring beneath continental crust can also provide vast areas of flood basalts. Some basaltic volcanoes occur along the midoceanic ridge.

The two forms of tectonic activity are compressional and extensional. Compressional activity occurs at lithospheric plate collisions. At first, the compression produces folding—anticlines (upfolds) and synclines (downfolds). If compression continues, folds may be overturned, and eventually overthrust faulting can occur. Extensional activity occurs where lithospheric plates are spreading apart, generating normal faults. These can produce upthrown and downdropped blocks that are sometimes as large as mountain ranges or rift valleys. Transcurrent faults occur where two rock masses move horizontally past each other.

Earthquakes occur when rock layers, bent by tectonic activity, suddenly fracture and move. The sudden motion at the fault produces earthquake waves that shake and move the ground surface in the adjacent region. The energy released by an earthquake is measured by the Richter scale. Large earthquakes occurring near developed areas can cause great damage. Most severe earthquakes occur near plate collision boundaries. The San Andreas Fault is a major transcurrent fault located near two great urban areas—Los Angeles and San Francisco. The potential for a severe earthquake on this fault is high, and the probability of a major earth movement increases every year.

KEY TERMS

landform	stratovolcano	normal fault
geomorphology	shield volcano	transcurrent fault
denudation	folding	earthquake
volcano	fault	tsunami

REVIEW QUESTIONS

1. Distinguish between initial and sequential landforms. How do they represent the balance of power between internal earth forces and external forces of denudation agents?

2. What is a stratovolcano? What is its characteristic shape, and why does that shape occur? Where do stratovolcanoes generally occur and why?

3. What is a shield volcano? How is it distinguished from a stratovolcano? Where are shield volcanoes found, and why? Give an example of a shield volcano. How are flood basalts related to shield volcanoes?

4. How can volcanic eruptions become natural disasters? Be specific about the types of volcanic events that can devastate habitations and extinguish nearby populations.

5. What types of landforms characterize foreland fold belts? Where do these belts occur, and how are they formed?

6. Sketch a cross section through a normal fault, labeling the fault plane, upthrown side, downthrown side, and fault scarp.

7. Briefly describe the rift valley system of East Africa as an example of normal faulting.

8. How does a transcurrent fault differ from a normal fault? What landforms are expected along a transcurrent fault? How are transcurrent faults related to plate tectonic movements?

9. What is an earthquake, and how does it arise? How are the locations of earthquakes related to plate tectonics?

10. Describe the tsunami, including its origin and effects.

11. Briefly summarize the geography and recent history of the San Andreas Fault system in California. What are the prospects for future earthquakes along the San Andreas Fault?

Focus on Systems 12.1 • The Life Cycle of a Volcano

1. How does a life cycle contrast with a repeating time cycle?

2. Describe the stages in the life cycle of a basaltic shield volcano of the Hawaiian type.

3. What is a hotspot? What produces it? How is it related to the life cycle?

ESSAY QUESTIONS

1. Write a fictional news account of a volcanic eruption. Select a type of volcano—composite or shield—and a plausible location. Describe the eruption and its effects as it was witnessed by observers. Make up any details you need, but be sure they are scientifically correct.

2. How are mountains formed? Provide the plate tectonic setting for mountain formation, and then describe how specific types of mountain landforms arise.

PROBLEMS

Working It Out 12.2 • The Richter Scale

1. After measuring the amplitude and distance of an earthquake on a seismogram, a geophysicist determines that it rates 5.2 on the Richter scale. How much energy will the earthquake release?

2. Suppose that the energy of a magnitude 6.0 earthquake is released evenly over a 60-second period. Recalling that one watt (W) is equal to a joule per second (J/sec), how many 100-watt light bulbs would be powered by this rate of energy flow during that minute? The average U.S. electric power consumption rate is about 3×10^{12} W. What fraction of the U.S. power consumption could be sustained by this energy flow rate?

3. After recalculation, a geophysicist revises the magnitude of an earthquake upward from 7.1 to 7.3 on the Richter scale. By what proportion has the estimate of the energy released changed?

Part III

Systems of Landform Evolution

Physical geography is focused on the life layer—the zone of interactions among the atmosphere, hydrosphere, lithosphere, and biosphere in which we live, breathe, and carry out our daily lives. Part I approached the life layer from the viewpoint of energy and matter flow systems involving the atmosphere and surface. These systems are largely driven by solar power and so operate on time scales of hours to days to years. Part II examined the matter and flow systems of the solid earth. Here the power source is largely the earth's internal heat, generated by radioactive decay of unstable isotopes. These operate primarily on time scales of million of years, with the exception of some forms of volcanic activity.

Part III, Systems of Landform Evolution, returns our focus to the life layer. It examines the matter and energy flow systems that shape the surface of the land. These systems operate largely on time scales intermediate between those of the atmosphere and solid earth—on the order of hundreds to thousands to a few millions of years.

Systems of landform evolution are powered mostly by gravity—the constant attraction of the earth's mass for solids (rock particles and soil) and fluids (water and glacial ice) that moves them downhill, creating landforms in the process. But gravity only releases potential energy that has been stored in positioning these solids and fluids above a base level, such as the ocean. The potential energy is ultimately derived from two sources—solar energy, which places liquid and solid water atop the continents through precipitation, and the earth's internal heat, which generates the earth forces that uplift masses of rock and soil above base level.

Chapter 13 begins Part III by examining the processes of weathering that break up rock material in place, and then turns to movements of masses of weathered material under gravity—landslides and earthflows, for example. Chapter 14 focuses on water at or near the land surface, in lakes, rivers, and as groundwater. It is followed by Chapter 15, which moves from the study of water as surface water and groundwater to consider its role as a landform-creating agent of erosion and deposition. In Chapter 16 we show how fluvial action interacts with rock structures—folds, faults, and the like—to produce distinctive configurations of landforms related to the structures. Chapter 17 turns to two other landform-creating agents powered not by gravity, but by atmospheric motion—waves and wind. Chapter 18 concludes Part III by examining landforms created by glacial ice, both as high mountain glaciers and as vast continental ice sheets.

Parts I, II, and III focus on the life layer from the viewpoints of the atmosphere, hydrosphere, and lithosphere. In Part IV, we will complete our study of physical geography by focusing on the life layer from the viewpoint of the biosphere.

Chapter 13

Weathering and Mass Wasting

Now that we have completed our study of the earth's crust—including its mineral composition, its moving lithospheric plates, and its tectonic and volcanic landforms—we can focus on the shallow life layer itself. At this sensitive interface, the external processes of denudation carve sequential landforms from the earth materials uplifted by the earth's internal processes. Our

These flutes and grooves in limestone rock, near Vinales, in Pinar Del Rio, Cuba, illustrate solution weathering in a warm, moist climate.

investigation of what happens to rock upon exposure at the surface began in Chapter 10 with a description of mineral alteration of rock and the production of sediment, which is an essential part of the cycle of rock transformation. In this chapter, we look further at the softening and breakup of rock, termed weathering, and how the resulting particles move downhill under the force of gravity, a process termed mass wasting.

Weathering is the general term applied to the combined action of all processes that cause rock to disintegrate physically and decompose chemically because of exposure near the earth's surface. In Chapter 10, we described the processes of mineral alteration (oxidation, hydrolysis, and carbonic acid solution) that transform minerals from types that were stable when the rocks were formed to types that are now stable at surface temperatures and pressures. These processes are part of *chemical weathering*, in which rock particles are altered chemically at the surface. A second type of weathering is *physical weathering*, in which rock particles are fractured and broken apart. Weathering also leads to a number of distinctive landforms, which we will describe in this chapter.

This chapter also includes a related process that creates landforms. It is the spontaneous downhill movement of soil, regolith, and rock under the influence of gravity, but without the action of moving water, air, or ice. This downhill movement is referred to as **mass wasting**. Movement of a mass of soil or rock to lower levels takes place when the internal strength declines to a critical point below which the force of gravity cannot be resisted. This failure of strength under the ever-present force of gravity takes many forms and scales, and we will see that human activity causes or aggravates several forms of mass wasting.

We conclude our chapter with an examination of a suite of special landforms and geomorphic processes that are found in the arctic lands. They are created primarily by the freezing and thawing of water, acting in concert with gravity.

SLOPES AND REGOLITH

Mass wasting occurs on slopes. As used in physical geography, the term *slope* designates a small strip or patch of the land surface that is inclined from the horizontal. Thus, we speak of "mountain slopes," "hill slopes," or "valley-side slopes" to describe some of the inclined ground surfaces that we might encounter in traveling across a landscape. Slopes guide the downhill flow of surface water and fit together to form drainage systems within which surface-water flow converges into stream

channels. Nearly all natural surfaces slope to some degree. Very few are perfectly horizontal or vertical.

As described in Chapter 9, most slopes are mantled with *regolith* (Figure 13.1), a surface layer of weathered rock particles. The regolith grades downward into solid, unaltered rock, known simply as **bedrock**. Regolith, in turn, provides the source for **sediment**, consisting of detached mineral particles that are transported and deposited in a fluid medium. This fluid may be water, air, or even glacial ice. As we saw in Chapter 9, both regolith and sediment comprise parent materials for the formation of the true soil, which is a surface layer capable of supporting the growth of plants.

Figure 13.1 shows a typical hill slope that forms one wall of the valley of a small stream. Soil and regolith blanket the bedrock, except in a few places where the bedrock is particularly hard and projects in the form of *outcrops*. *Residual regolith* is derived directly from the rock beneath and moves very slowly down the slope toward the stream. Beneath the valley bottom are layers of transported regolith, called **alluvium**, which is sediment transported and deposited by the stream. This sediment had its source in regolith prepared on hill slopes many kilometers or miles upstream. All accumulations of sediment on the land surface, whether deposited by streams, waves and currents, wind, or glacial ice, can be designated *transported regolith*, in contrast to residual regolith.

The thickness of soil and regolith is quite variable. Although the soil is rarely more than 1 or 2 m thick (about 3 to 6 ft), residual regolith on decayed and fragmented rock may extend down to depths of 5 to 100 m (about 15 to 300 ft), or more at some locations. On the other hand, soil or regolith, or both, may be missing. In some places, everything is stripped off down to the bedrock, which then appears at the surface as an outcrop. In other places, following cultivation or forest fires, the fertile soil is partly or entirely eroded away, and severe erosion exposes the regolith.

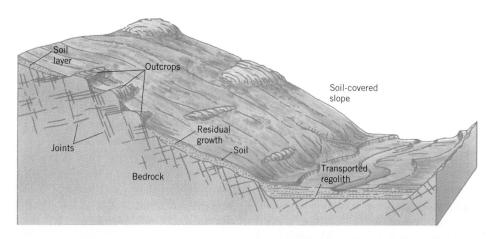

Figure 13.1 Soil, regolith, and outcrops on a hill slope. Alluvium, a form of transported regolith, lies in the floor of an adjacent stream valley.

PHYSICAL WEATHERING

Physical weathering produces fine particles of regolith from massive rock by the action of forces strong enough to fracture the rock. The physical weathering processes discussed in this chapter include frost action, salt-crystal growth, unloading, expansion–contraction, and wedging by plant roots.

Frost Action

One of the most important physical weathering processes in cold climates is *frost action*, the repeated growth and melting of ice crystals in the pore spaces of soil and in rock fractures. This activity can rupture even extremely hard rocks. Frost action produces a number of conspicuous effects and forms in all climates that have cold winters. Features caused by frost action and the buildup of ice below the surface are particularly visible in the tundra climate of arctic coastal fringes and islands, and above timberline in high mountains.

Almost everywhere, bedrock is cut through by systems of fractures called *joints* (Figure 13.1). Rarely is igneous rock free of numerous joints that permit the entry of water. In sedimentary rocks, the planes of stratification, or bedding planes, comprise a natural set of planes of weakness cut at right angles by sets of joints. Obviously, comparatively weak stresses can separate joint blocks, whereas strong stresses are required to make fresh fractures through solid rock. The process can be called *block separation* (Figure 13.2).

As coarse-grained igneous rock becomes weakened by chemical decomposition, water is able to penetrate the contact surfaces between mineral grains. In this lo-

cation, the water can freeze and exert strong forces to separate the grains. This form of breakup is termed *granular disintegration* (Figure 13.2). The product is a fine gravel or coarse sand in which each grain consists of a single mineral particle separated from the others along its original crystal or grain boundaries.

The effects of frost action can be seen in all climates having a winter season with many alternations of freeze and thaw. Where bedrock is exposed on knolls and mountain summits, joint blocks are pried apart by water that freezes in joint cracks (Figure 13.3). Under the most favorable conditions, seen on high mountain summits and in the arctic tundra, large angular rock fragments accumulate in a layer that completely blankets the bedrock beneath. The German name *felsenmeer* ("rock sea") has been given to such expanses of broken rock (Figure 13.4).

Frost action on cliffs of bare rocks in high mountains detaches rock fragments that fall to the cliff base.

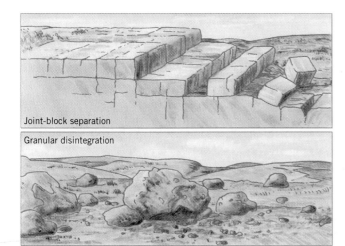

Figure 13.2 Two common forms of bedrock disintegration. (Drawn by A. N. Strahler.)

Figure 13.3 Frost-shattered blocks, Sierra Nevada, California. Ice-crystal growth within the joint planes of rock can cause the rock to split apart. This boulder is an example.

Figure 13.4 Frost-shattered blocks of quartzite on the summit of the Snowy Range, Wyoming, at an elevation of about 3700 m (about 12,000 ft).

Where production of fragments is rapid, they accumulate to form *talus slopes*. Most cliffs are notched by narrow ravines that funnel the rock fragments into separate tracks. Each track, or chute, feeds a growing, conelike talus body. Talus cones are arranged side by side along the base of the cliff (Figure 13.5). Fresh talus slopes are unstable, so that the disturbance created by walking across the slope or dropping a large rock fragment from the cliff above will easily set off a sliding or rolling motion within the surface layer of fragments.

In fine-textured soils and sediments, composed largely of silt and clay, freezing of soil water takes place in horizontal layers or lens-shaped bodies. As these ice layers thicken, the overlying soil layer is heaved upward. Prolonged frost heaving can produce minor irregularities and small mounds on the soil surface. Where a rock fragment lies at the surface, perpendicular ice needles can grow beneath the fragment and raise it above the surface (Figure 13.6). The same process acting on a rock fragment below the soil surface will eventually bring the fragment to the surface.

Frost action is a dominant process in arctic and high-mountain tundra environments, where it is a factor in

Figure 13.5 Talus cones near the shore of Lake Louise in the Canadian Rockies.

Figure 13.6 Needle ice growth. At night, water in the soil freezes at the surface, creating ice needles that can lift particles of soil or move larger stones.

the formation of a wide variety of unique soil structures and landforms. These we investigate in a section later in this chapter.

Salt-Crystal Growth

Closely related to the growth of ice crystals is the weathering process of rock disintegration by growth of salt crystals in rock pores. This process, *salt-crystal growth*, operates extensively in dry climates and is responsible for many of the niches, shallow caves, rock arches, and pits seen in sandstone formations. During long drought periods, groundwater is drawn to the surface of the rock by capillary force. As evaporation of the water takes place in the porous outer zone of the sandstone, tiny crystals of salts such as halite (sodium chloride), calcite (calcium carbonate), or gypsum (calcium sulfate) are left behind. The growth force of these crystals produces grain-by-grain breakup of the sandstone, which crumbles into a sand and is swept away by wind and rain.

Especially susceptible are zones of rock lying close to the base of a cliff, because there the groundwater seeps outward to reach the rock surface (Figure 13.7). In the southwestern United States, many of the deep niches or cavelike recesses formed in this way were occupied by Native Americans. Their cliff dwellings gave them protection from the elements and safety from armed attack (Figure 13.8).

Salt crystallization also damages masonry buildings as well as concrete sidewalks and streets. Brick and concrete in contact with moist soil are highly susceptible to grain-by-grain disintegration from salt crystallization. The salt crystals can be seen as a soft, white, fibrous layer on damp basement floors and walls. The deicing salts spread on streets and highways can be quite de-

structive. Sodium chloride (rock salt), widely used for this purpose, is particularly damaging to concrete pavements and walks, curbstones, and other exposed masonry structures.

Unloading

A widespread process of rock disruption related to physical weathering results from *unloading*, the relief of confining pressure of overlying rock. Unloading occurs

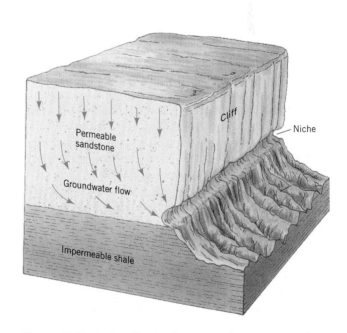

Figure 13.7 In dry climates there is a slow seepage of water from the cliff base. Salt-crystal growth separates the grains of permeable sandstone, breaking them loose and creating a niche. (After A. N. Strahler.)

Figure 13.8 The White House Ruin, a former habitation of Native Americans, occupies a large niche in sandstone in the lower wall of Canyon de Chelly, Arizona.

as rock is brought near the surface by erosion of overlying layers. Rock formed at great depth beneath the earth's surface (particularly igneous and metamorphic rock) is in a slightly compressed state because of the confining pressure of overlying rock. As the rock above is slowly worn away, the pressure is reduced, and the rock expands slightly in volume. Expansion causes thick shells of rock to break free from the parent mass below. The new surfaces of fracture are a form of jointing called *sheeting structure*. The rock sheets show best in massive rocks such as granite and marble.

Where sheeting structure has formed over the top of a single large body of massive rock, an *exfoliation dome* is produced (Figure 13.9). Domes are among the largest of the landforms shaped primarily by weathering. In Yosemite Valley, California, where domes are spectacularly displayed, the individual rock sheets may be as thick as 15 m (50 ft).

Other Physical Weathering Processes

Most rock-forming minerals expand when heated and contract when cooled. Where rock surfaces are exposed daily to the intense heating of the sun alternating with nightly cooling, the resulting expansion and contraction exerts powerful disruptive forces on the rock. Although first-hand evidence is lacking, it seems likely that daily temperature changes can cause the breakup of a surface layer of rock already weakened by other agents of weathering.

Another mechanism of rock breakup is the growth of plant roots, which can wedge joint blocks apart. You may have seen a tree whose lower trunk and roots are firmly wedged between two great joint blocks of massive rock. Whether the tree has actually been able to spread the blocks farther apart or has merely occupied the available space is sometimes open to question. However, it is certain that pressure exerted by growth of tiny rootlets in joint fractures causes the loosening of countless small rock scales and grains.

Figure 13.9 North Dome and Royal Arches, Yosemite National Park, California. Thick exfoliation rock layers are visible in the lower part of the photo.

CHEMICAL WEATHERING AND ITS LANDFORMS

We investigated **chemical weathering** processes in Chapter 10 under the heading of mineral alteration. Recall that the dominant processes of chemical change affecting silicate minerals are oxidation, hydrolysis, and carbonic acid action. While feldspars and the mafic minerals are very susceptible to chemical change, quartz is a highly stable mineral, almost immune to decay.

Hydrolysis and Oxidation

Decomposition by hydrolysis and oxidation changes strong rock into very weak regolith. This change allows erosion to operate with great effectiveness wherever the regolith is exposed. Weakness of the regolith also makes it susceptible to natural forms of mass wasting. In warm, humid climates of the equatorial, tropical, and subtropical zones, hydrolysis and oxidation often result in the decay of igneous and metamorphic rocks to depths as great as 100 m (about 300 ft). To the construction engineer, deeply weathered rock is of major concern in the building of highways, dams, or other heavy structures. Although the regolith is soft and can often be removed by power shovels with little blasting, foundations on regolith can fail under heavy loads.

This regolith also has undesirable plastic properties because of its high content of clay minerals.

The hydrolysis of exposed granite surfaces is accompanied by the grain-by-grain breakup of the rock. This process creates many interesting boulder and pinnacle forms by the rounding of angular joint blocks (Figure 13.10). These forms are particularly conspicuous in arid regions. In most deserts there is ample moisture for hydrolysis to act, given sufficient time. The products of grain-by-grain breakup form a fine desert gravel, which consists largely of quartz and partially decomposed feldspar crystals.

Carbonic Acid Action

Atmospheric carbon dioxide is dissolved in all surface waters of the lands, including rainwater, soil water, and stream water. The solution of carbon dioxide in water produces a weak acid, called *carbonic acid*, which can dissolve some types of minerals. Carbonate sedimentary rocks, such as limestone and marble, are particularly susceptible to the acid action. In this process, the mineral calcium carbonate is dissolved and carried away in solution in stream water.

Carbonic acid reaction with limestone produces many interesting surface forms, mostly of small dimensions. Outcrops of limestone typically show cupping, rilling, grooving, and fluting in intricate designs (Fig-

Figure 13.10 Large joint blocks of granite are gradually rounded into smooth forms by grain-by-grain disintegration in a desert environment. Alabama Hills, Owens Valley, California.

Figure 13.11 This outcrop of pure limestone shows rills and cups formed by solution weathering. County Clare, Ireland.

ure 13.11). In a few places, the scale of deep grooves and high wall-like rock fins reaches proportions that keep people and animals from passing through. Dissolution of limestone by carbonic acid in groundwater can produce underground caverns as well as distinctive landscapes that are formed when underground caverns collapse. These landforms and landscapes will be described in Chapter 14.

In urban areas, air pollution by sulfur and nitrogen oxides commonly occurs. When these gases dissolve in rainwater, the result is acid precipitation (see Chapter 4). The acids rapidly dissolve limestone and chemically weather other types of building stones. The result can be very damaging to stone sculptures, building decorations, and tombstones (Figure 13.12).

In the wet low-latitude climates, mafic rock, particularly basaltic lava, undergoes rapid removal under attack by soil acids and produces landforms quite similar to those formed by carbonic acid on massive limestones in the moist climates of higher latitudes. The effects of

solution removal of basaltic lava are displayed in spectacular grooves, fins, and spires on the walls of deep alcoves in part of the Hawaiian Islands (Figure 13.13).

MASS WASTING

Everywhere on the earth's surface, gravity pulls continuously downward on all materials. Bedrock is usually so strong and well supported that it remains fixed in place. However, when a mountain slope becomes too steep, bedrock masses break free and fall or slide to new positions of rest. In cases where huge masses of bedrock are involved, the result can be catastrophic in loss to life and property in towns and villages in the path of the slide. Such slides are a major form of environmental hazard in mountainous regions. They are but one form of *mass wasting*, which we defined earlier as spontaneous downhill movement of soil, rock, and regolith under the influence of gravity.

Because soil, regolith, and many forms of sediment are held together poorly, they are much more susceptible to movement under the force of gravity than hard, massive bedrock. Abundant evidence shows that on most slopes at least a small amount of downhill movement is going on constantly. Although much of this motion is imperceptible, the regolith sometimes slides or flows rapidly.

Soil Creep

On almost any soil-covered slope, you can find evidence of extremely slow downhill movement of soil and regolith, a process called **soil creep**. Figure 13.14 shows some of the evidence that the process is going on. Joint blocks of distinctive rock types are found moved far downslope from the outcrop. In some layered rocks such as shales or slates, edges of the strata seem to "bend" in the downhill direction (Figure 13.15). This is not true plastic bending, but is the result of the slight movement of many rock pieces on small joint cracks. Creep causes fence posts and utility poles to lean downslope and even shift measurably out of line. Roadside retaining walls can buckle and break under the pressure of soil creep.

What causes soil creep? Alternate drying and wetting of the soil, growth of frost needles, heating and cooling of the soil, trampling and burrowing by animals, and shaking by earthquakes all produce some disturbance of the soil and regolith. Because gravity exerts a downhill pull on every such rearrangement, the particles very gradually work their way downslope.

Earthflows

In regions of humid climate, a mass of water-saturated soil, regolith, or weak shale may move down a steep

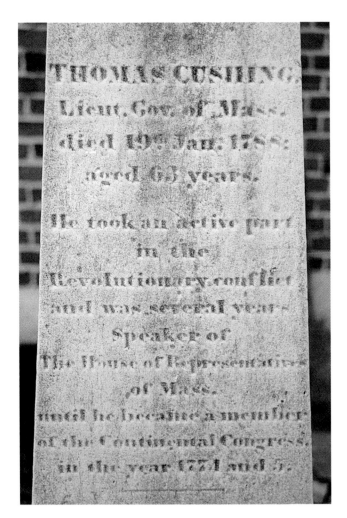

Figure 13.12 Weathered tombstones from a burying ground in Boston, Massachusetts. The marker on the left, carved in marble, has been strongly weathered, weakening the lettering. The marker on the right, made of slate, is much more resistant to erosion.

Figure 13.13 The steep walls of many narrow coastal ravines are deeply grooved by chemical weathering of the basaltic lava on the Napili Coast, Island of Hawaii, Hawaii.

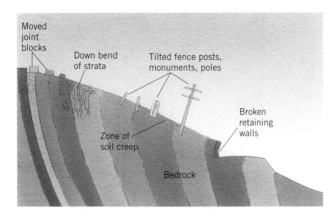

Figure 13.14 The slow, downhill creep of soil and regolith shows up in many ways on a hillside. (After C.F.S. Sharpe.)

slope during a period of a few hours in the form of an **earthflow**. Figure 13.16 is a sketch of an earthflow showing how the material slumps away from the top, leaving a steplike terrace bounded by a curved, wall-like scarp. The saturated material flows sluggishly to form a bulging toe.

Shallow earthflows, affecting only the soil and regolith, are common on sod-covered and forested slopes that have been saturated by heavy rains. An earthflow may affect a few square meters, or it may cover an area of several hectares. If the bedrock of a mountainous region is rich in clay (derived from shale or deeply weathered volcanic rocks), earthflows sometimes involve millions of metric tons of bedrock moving by plastic flowage like a great mass of thick mud.

Earthflows are a common cause of blockage of highways and railroad lines, usually during periods of heavy rains. Generally, the rate of flowage is slow, so that the flows are not a threat to life. Property damage to buildings, pavements, and utility lines is often severe where construction has taken place on unstable soil slopes.

EYE ON THE ENVIRONMENT:
Environmental Impact of Earthflows

Examples of both large and small earthflows induced or aggravated by human activities are found in the Palos Verdes Hills of Los Angeles County, California (Figure 13.17). These movements occur in shales that tend to become plastic when water is added. The upper part of the earthflow undergoes a subsiding motion with backward rotation of the down-sinking mass, as illustrated in Figure 13.16. The interior and lower parts of the mass move by slow flowage, and a toe of extruded flowage material may be formed.

Largest of the earthflows in the Palos Verdes Hill area was the Portuguese Bend "landslide" that affected an area of about 160 hectares (about 400 acres). The total movement over a three-year period was about 20 m (about 70 ft). Damage to residential and other structures totaled some $10 million. The most interesting observation, from the point of view of assessing human impact on the environment, is that the slide has been attributed by geologists to infiltration of water from septic tanks and from irrigation water applied to lawns and gardens. A discharge of over 115,000 liters (about 30,000 gallons) of water per day from some 150 homes

Figure 13.15 Slow, downhill creep of regolith on this mountainside near Downieville, California, has caused vertical rock layers to seem to "bend over."

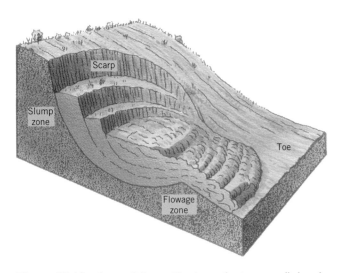

Figure 13.16 An earthflow with slump features well developed in the upper part. Flowage has produced a bulging toe. (After A. N. Strahler.)

is believed to have sufficiently weakened the shale beneath it to start and sustain the flowage.

One special form of earthflow has proved to be a major environmental hazard in parts of Norway and Sweden and along the St. Lawrence River and its tributaries in Quebec Province of Canada. This type of flowage involves horizontally layered clays, sands, and silts of late Pleistocene age that form low, flat-topped

terraces adjacent to rivers and lakes. Figure 13.18 provides a block diagram of a classic example that occurred in 1898 near St. Thuribe, Quebec. Over a large area, which may be 600 to 900 m (about 2000 to 3000 ft) across, a layer of silt and sand 6 to 12 m (about 20 to 40 ft) thick begins to move toward the river, sliding on a layer of soft clay that has spontaneously turned into a near-liquid state. The moving mass also settles downward and breaks into steplike masses. Carrying along houses or farms, the layer ultimately reaches the river, into which it pours as a great disordered mass of mud.

A particularly spectacular example was the Nicolet, Quebec, earthflow of 1955, which carried a large piece of the town into the Nicolet River (Figure 13.19). Fortunately, only three lives were lost, but the damage to buildings and a bridge ran into millions of dollars.

This type earthflow is caused by *quick clays*—clays that spontaneously change from a solid condition to a near-liquid condition when subjected to a shock or disturbance. Quick clays are thought to have formed in the shallow waters of saltwater bays during the late Pleistocene. When deposited, the thin plates of clay in these layers have a "house of cards" structure with a large proportion of water-filled void space between plates. The saltwater provides positively and negatively charged atoms that help to bind the structure and give it strength. But with the regional uplift that often occurs after ice sheets melt away, the quick clay is elevated above sea level, and fresh groundwater replaces the saltwater. A mechanical shock then causes the house of

Figure 13.17 This aerial view shows houses on Point Fermun, in Palos Verdes, California, disintegrating as they slide downward toward the sea.

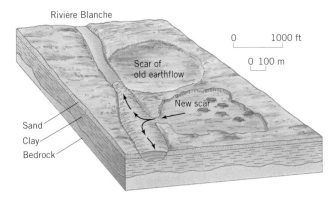

Figure 13.18 Block diagram of the 1898 earthflow near St. Thuribe, Quebec. (After C.F.S. Sharpe, *Landslides and Related Phenomena*, Columbia University Press, New York.)

Figure 13.19 The Nicolet, Quebec, earthflow of 1955.

cards structure to collapse. Because such a large proportion of water (from 45 to 80 percent by volume) is present, the clay-water mixture behaves as a liquid, with almost no strength remaining. 🐾

Mudflows

One of the most spectacular forms of mass wasting and a potentially serious environmental hazard is the **mudflow**. This mud stream of fluid consistency pours swiftly down canyons in mountainous regions (Figure 13.20). In deserts, where vegetation does not protect the

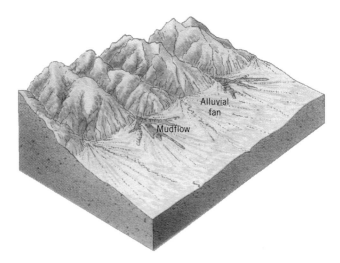

Figure 13.20 Thin, streamlike mudflows issue occasionally from canyon mouths in arid regions. The mud spreads out on the piedmont slopes below in long, narrow tongues. (Drawn by A. N. Strahler.)

mountain soils, local thunderstorms produce rain much faster than it can be absorbed by the soil. As the water runs down the slopes, it forms a thin mud, which flows down to the canyon floors. Following stream courses, the mud continues to flow until it becomes so thick it must stop. Great boulders are carried along, buoyed up in the mud. Roads, bridges, and houses in the canyon floor are engulfed and destroyed. Where the mudflow emerges from the canyon and spreads across an alluvial fan, severe property damage and even loss of life may be the result (Figure 13.21).

As explained in Chapter 12, mudflows that occur on the slopes of erupting volcanoes are called *lahars*. Freshly fallen volcanic ash and dust are turned into mud by heavy rains or melting snows and flow down the slopes of the volcano. Herculaneum, a city at the base of Mount Vesuvius, was destroyed by a mudflow during the eruption of A.D. 79. At the same time, the neighboring city of Pompeii was buried under volcanic ash.

Mudflows vary in consistency, from a mixture like concrete emerging from a mixing truck to consistencies similar to that of turbid river floodwaters. The watery type of mudflow is called a *debris flood* or *debris flow* in the western United States, and particularly in southern California, where it occurs commonly and with disastrous effects.

Landslides

A **landslide** is the rapid sliding of large masses of bedrock. Wherever mountain slopes are steep, there is a possibility of large, disastrous landslides (Figure 13.22). In Switzerland, Norway, or the Canadian Rock-

Figure 13.21 This mudflow, carrying numerous large boulders, issued from a steep mountain canyon in the Wasatch Mountains, Utah.

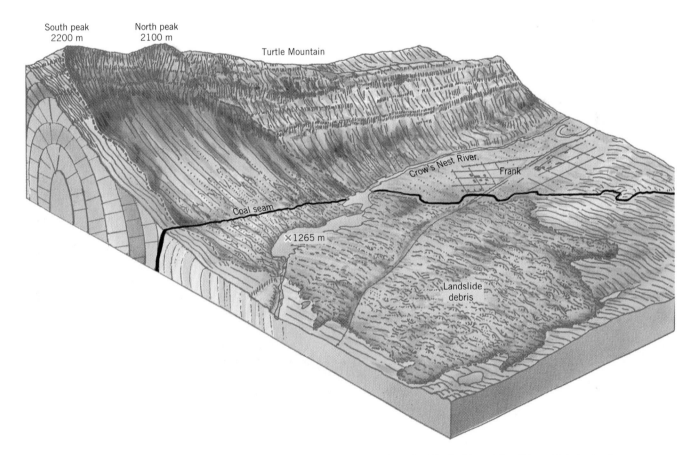

Figure 13.22 A classic example of an enormous, disastrous landslide is the Turtle Mountain slide, which took place near Frank, Alberta. A huge mass of limestone slid from the face of Turtle Mountain between South and North peaks, descended to the valley, then continued up the low slope of the opposite valley side until it came to rest as a great sheet of bouldery rock debris. (Data from Canadian Geological Survey, Department of Mines. Drawn by A. N. Strahler. Copyright © A. N. Strahler.)

Working It Out 13.1 • The Power of Gravity

Directly or indirectly, gravity powers many of the processes that shape the landscape, from the erosion of running water in carving landforms (Chapter 15) to the processes of mass wasting—earth flows, mudflows, solifluction, and landslides—that are the subject of this chapter. By acting to move mineral matter and water downhill, gravity does work over a span of time and is thus a source of power.

Let's briefly review gravitation, gravity, force, work, and power. *Gravitation* is the simultaneous attraction that occurs between two physical bodies. The strength of the attraction depends on the mass of each body and the distance between the two bodies. The attraction of the earth for an object near the earth's surface is called *gravity*. It is described as an acceleration that acts on the mass of the object to produce a force according to the simple relationship

$$F = m \times g$$

where F is the force, m is the mass of the object, and g is the acceleration due to gravity (about 9.8 m/s²). If the mass is measured in kg and the acceleration is measured in m/s², then the force is in units of newtons (N).

Suppose now that gravity actually does some work—that is, it moves a mass through a distance. The work is measured as force times distance:

$$W = F \times d$$

where W is the work and d is the distance. If the force is measured in newtons and the distance in meters, then the work done is measured in joules (J).

Power describes the time rate at which work is accomplished—that is, work per unit time, or

$$P = \frac{W}{T}$$

where P is the power, W is the work, and T is the time. If work is measured in joules and time is measured in seconds, then power is in unit of watts (W). Recall from *Focus on Systems 2.1 • Forms of Energy* that work and energy have the same units, and that energy is defined as the ability to do work. Thus, power is both a measure of a rate of energy flow and of a rate at which work is being done.

As an example of the power of gravity, let's take the Madison Slide—a catastrophic landslide described in the text. In this event, a rock body with a volume of about 28 million cubic meters (28×10^6 m³) moved a vertical distance downward of about 500 m into the valley of the Madison River, attaining a velocity of about 150 km/hr. Propelled by its own inertia, the flow then moved uphill on the far side of the river valley by a vertical distance of about 120 m. In this slide, the potential energy of position of the mass above the valley bottom was converted to kinetic energy during the slide, and the kinetic energy was

Figure 13.23 Seen from the air, the Madison Slide forms a great dam of rubble across the Madison River Canyon.

dissipated as heat produced by friction during the sliding motion.

Note that the slide moved material downhill through a vertical distance of 500 m, converting potential energy to kinetic energy and friction, then uphill a vertical distance of 120 m. On the uphill leg, some kinetic energy was converted back to potential energy as the mass moved upward against the acceleration of gravity. So the net release of potential energy was produced by a vertical distance change of 500 − 120 = 380 m.

How much energy was released? Or, alternatively, how much work was done? From the formulas above, we can see that this requires knowing the mass, acceleration, and distance. Of these, only the mass is unknown. However, we know the volume of the landslide— 28×10^6 m^3—so what is missing is the density—the mass of the rock per unit of volume. Let's assume that the rock has the density of granite, which is about 2700 kg/m^3

(2.7×10^3 kg/m^3) (Figure 10.5). Then we have

$$m = 28 \times 10^6 \, \text{m}^3 \times \frac{2.7 \times 10^3 \, \text{kg}}{\text{m}^3}$$
$$= 7.56 \times 10^{10} \, \text{kg}$$

Gravity will act on this mass with a force

$$F = m \times g = 7.56 \times 10^{10} \, \text{kg} \times \frac{9.8 \, \text{m}}{\text{s}^2}$$
$$= 7.41 \times 10^{11} \, \text{N}$$

This force is applied through a net distance of 380 m, so the work done is

$$W = F \times d = 7.41 \times 10^{11} \, \text{N} \times 380 \, \text{m}$$
$$= 2.82 \times 10^{14} \, \text{J}$$

This work was entirely converted to heat through the friction of the slide.

The amount of work done is rather similar to the amount of electrical energy consumed in a day by the United States, which is about 2.33×10^{14} J/day. Thus, the energy released by the Madison Slide, if converted completely to electric

power, could satisfy U.S. electrical needs for slightly more than one day.

To calculate the power expended by gravity in the slide, we need to know how long the motion took. With some simple assumptions about the angles of the mountain slopes and the magnitude of friction in the earthflow, a reasonable guess for the time required is about 70 seconds. The power applied by gravity in the downward motion would then be

$$P = \frac{W}{T} = \frac{2.82 \times 10^{14} \, \text{J}}{70 \, \text{s}} = 4.02 \times 10^{12} \, \text{W}$$

Expressed in watts, the average rate of U.S. electric power consumption is about 2.70×10^9 W. The rate of energy release in the Madison Slide is therefore about 1500 times greater than the rate at which the entire U.S. consumes electricity. This large number demonstrates the awesome power of gravity to shape and reshape the landscape.

---

ies, for example, villages built on the floors of steep-sided valleys have been destroyed and their inhabitants killed by the sliding of millions of cubic meters of rock set loose without warning.

Severe earthquakes in mountainous regions are a major cause of landslides and earthflows. An example is the Madison Canyon landslide caused by the Hebgen Lake earthquake of 1959 in Montana (Figure 13.23). The Madison Slide involved the motion of 28 million cu m (about 100 million cu ft) of rock which formed a part of the south wall of the canyon of the Madison River. The rock mass, which measured over 600 m (about 2000 ft) in length and 300 m (about 1000 ft) in height, descended half a kilometer (0.3 mi) to the Madison River at a speed later estimated to have exceeded 150 km/hr (about 90 mi/hr). The rock mass quickly disintegrated into boulders and pulverized rock, which crossed the canyon floor. Momentum of the moving mass carried the debris over 120 m (about 400 ft) in vertical distance up the opposite canyon wall. Acting as a huge natural dam, the slide debris blocked the Madison River and a new lake quickly began to

form. Within three weeks' time, it was 60 m (about 200 ft) deep. The lake outlet has since been stabilized and the lake, now named Earthquake Lake, is permanent. At least 26 persons died beneath the slide, and their bodies have never been recovered.

The amazing speed and freedom with which rock-slide rubble travels down a mountainside has been explained through the presence of a layer of compressed air trapped between the slide and the ground surface. The air layer reduces frictional resistance and may keep the rubble from making contact with the surface beneath.

A landslide like the Madison Slide releases a large amount of gravitational energy. Some simple calculations indicate that the energy released by the Madison Slide within the minute or two of its occurrence was more than enough to equal American electrical energy consumption for a day. *Working It Out 13.1 • The Power of Gravity* documents the enormous power of gravity and shows how this estimate was made.

Aside from occasional local catastrophes, landslides have rather limited environmental influence because they occur only sporadically and usually in thinly populated mountainous regions. Small slides can, however, repeatedly block or break important mountain highways or railway lines.

Induced Mass Wasting

Human activities can induce mass wasting in forms ranging from mudflow and earthflow to landslide. These activities include (1) piling up of waste soil and rock into unstable accumulations that can move spontaneously, and (2) removal of support by undermining natural masses of soil, regolith, and bedrock. We can refer to mass movements produced by human activities as *induced mass wasting*.

In Los Angeles County, California, real estate development has been carried out on very steep hillsides and mountainsides by the process of bulldozing roads and homesites out of the deep regolith. The excavated regolith is pushed out into adjacent embankments where its instability poses a threat to slopes and stream channels below. When saturated by heavy winter rains, these embankments can give way, producing earthflows, mudflows, and debris floods that travel far down the canyon floors and spread out on the alluvial fan surfaces below, burying streets and yards in bouldery mud.

Many debris floods of this area are also produced by heavy rains falling on mountain slopes denuded of vegetation by fire in the preceding dry season. Some of these fires are set carelessly or deliberately by humans.

Industrial societies now possess enormous machine power and explosives capable of moving great masses of regolith and bedrock from one place to another. One use of this technology is the extraction of mineral resources. Another is the movement of earth in the construction of highway grades, airfields, building foundations, dams, canals, and various other large structures. Both activities involve removal of earth materials, a process that destroys the preexisting ecosystems and habitats of plants and animals. When the materials are used to build up new land on adjacent surfaces, ecosystems and habitats are also destroyed—by burial. What distinguishes artificial forms of mass wasting from the natural forms is that machinery is used to raise earth materials against the force of gravity. Explosives used in blasting can produce disruptive forces many times more powerful than the natural forces of physical weathering.

Scarification is a general term for excavations and other land disturbances produced to extract mineral resources. It includes the accumulation of rock waste as spoil or tailings. Among the forms of scarification are open-pit mines, strip mines, quarries for structural materials, borrow pits along highway grades, sand and gravel pits, clay pits, phosphate pits, scars from hydraulic mining, and stream gravel deposits reworked by dredging.

PROCESSES AND LANDFORMS OF THE ARCTIC AND ALPINE TUNDRA

The landscape of the treeless arctic and alpine tundra environment, as we saw in Chapter 9, shows many distinctive effects of weathering and mass wasting under a regime of severe winter cold. In the tundra, soil water is solidly frozen throughout a winter of many months duration. During the short summer season, however, surface thawing occurs, leaving the soil saturated and vulnerable to mass wasting and water erosion. With the return of cold temperatures, the freezing of soil water exerts a strong mechanical influence on the surface layer, creating a number of distinctive landforms.

The tundra environment can be thought of as a unique natural system. Today it is often found in close proximity to glacial ice—near tongue-like alpine glaciers and along the fringes of ice sheets. This unique location has long been described by the adjective **periglacial** (*peri*, meaning "near"). Here, intense frost action creates a distinctive set of landforms and landforming processes that we may term the *periglacial system*. This system is driven by a very large annual range in temperature, which creates a strong annual cycle in the flow of energy into and out of the surface layer of the ground. These cycles provide the mechanism for the movement of soil particles individually and collectively by changing water from liquid to solid and back to liquid again, thus creating the distinctive processes that operate in the periglacial environment. We preface our discussion of periglacial processes with a look at permafrost.

Arctic Permafrost

Ground that is perennially (year-round) below the freezing point of fresh water (0°C; 32°F) is called **permafrost**. By "ground" we mean to include mineral matter falling within the full range of particle sizes described in Chapter 10—that is, clay, silt, sand, pebbles, and boulders. The term *permafrost* refers only to the ground temperature. Thus, solid bedrock perennially below the freezing point is also included in permafrost. Water is commonly present in pore spaces in the ground and in the frozen state is known as **ground ice**.

Permafrost prevails over most of the tundra climate region and in a wide bordering area of boreal forest climate. A distinctive feature of permafrost terrains is a zone of seasonal thaw, occurring at the ground surface, that is called the *active layer*. It is from 0.5 to 4 m thick (about 1.6 to 13 ft), depending on the latitude and na-

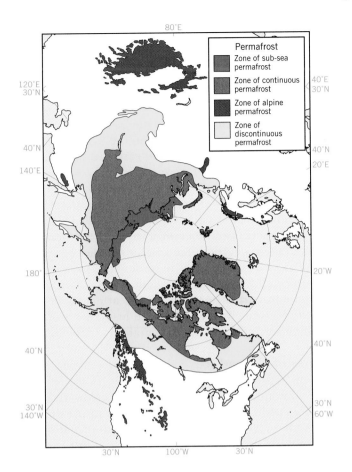

Figure 13.24 Distribution of permafrost in the northern hemisphere. (Adapted from Troy L. Péwé, *Geotimes*, vol. 29, no. 2, p. 11, copyright, 1984, by the American Geological Institute. Used by permission.)

ture of the ground. The upper surface of the perennially frozen zone is called the *permafrost table*.

The distribution of permafrost in the northern hemisphere is shown in Figure 13.24. Cross sections of the permafrost zone in Alaska and Asia are shown in Figure 13.25. Four zones are recognized. *Continuous permafrost*, which extends without gaps or interruptions under all surface features, coincides largely with the tundra climate, but also includes a large part of the boreal forest climate in Siberia. *Discontinuous permafrost*, which occurs in patches separated by frost-free zones under lakes and rivers, occupies much of the boreal forest climate zone of North America and Eurasia. Sporadic occurrence of permafrost in small patches extends into the southern limits of the boreal forest climate. A third zone—*sub-sea permafrost*—lies beneath sea level in a shallow offshore zone. On our map you will find it beneath the Arctic sea offshore of the Asian coast and the coasts of Alaska, the Yukon, and the Northwest Territories in North America.

Permafrost reaches to a depth of 600 to 1000 m (about 2000 to 3000 ft) in the continuous zone near lat.

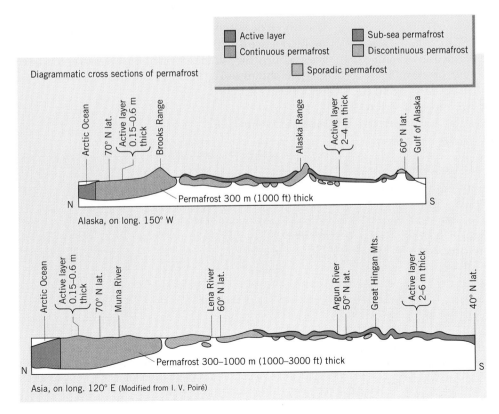

Figure 13.25 North-south cross sections of permafrost in Alaska and Asia. (From Robert F. Black, "Permafrost," Chapter 14 of P. D. Trask's *Applied Sedimentation*. Copyright © by John Wiley & Sons. Reprinted by permission of John Wiley & Sons, Inc.)

Figure 13.26 Exposed by river erosion, this great wedge of solid ice fills a vertical crack in organic-rich floodplain silt along the Yukon River, near Galena in western Alaska.

70° N. Much of this permanently frozen zone is an inheritance from the more severe conditions of the last ice age, but some permafrost bodies may be growing under existing climate conditions.

Permafrost is an example of an energy flow system in which the depth of the active layer and the base of the permafrost are determined by the balance of heat energy flowing into the ground from the surface and the heat energy upwelling from the earth's interior. *Focus on Systems 13.2 • Permafrost as an Energy Flow System* describes this energy flow system in more detail.

Even in the areas of deepest continuous permafrost there exist isolated pockets that never fall below the freezing point. The Siberian word *talik* has been applied to these pockets. One kind of talik lies beneath a deep lake that freezes over in winter and thaws in summer. The talik may extend to a depth of 100 m in the continuous permafrost zone. The heat necessary for a talik to be formed and maintained may come from circulating surface water that penetrates to these depths.

Forms of Ground Ice

The amount of ground ice present below the permafrost table varies greatly from place to place. Near the surface, it can take the form of a body of almost 100 percent ice. At increasing depth, the percentage of ice becomes lower and approaches a zero value at great depth.

A common form of ground ice is the **ice wedge**. In silty alluvium, such as that formed on river floodplains and delta plains in the arctic environment, ice accumulates in vertical wedge-forms in deep cracks in the sediment (Figure 13.26). Ice wedges are thought to originate as shrinkage cracks formed during extreme winter cold. As shown in Figure 13.27, surface water enters the

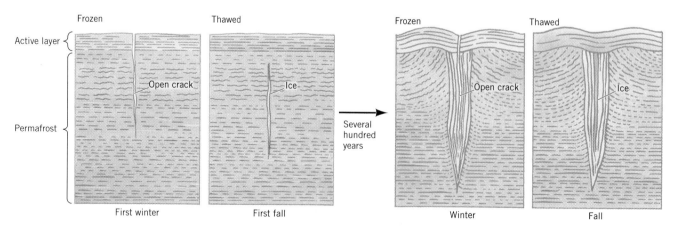

Figure 13.27 Formation of an ice wedge. In the beginning stage (*left*), an open crack appears during the winter and is filled with meltwater, which freezes into ice during the following autumn. After several hundred winters (*right*), the ice wedge has grown and continues to grow as the same seasonal sequence is repeated. (Adapted by permission from A. H. Lachenbruch in Rhodes W. Fairbridge, Ed., *The Encyclopedia of Geomorphology,* Van Nostrand Reinhold, N.Y. N.Y.)

Figure 13.28 These polygons are formed by the growth of ice wedges. On the Alaskan north slope, near the border of Alaska and Yukon Territory, Canada.

cracks during the spring melt and becomes frozen. Repeated cracking and the addition of new ice causes the ice wedge to thicken until it becomes as wide as 3 m (about 10 ft) and as deep as 30 m (about 50 ft). Ice wedges are typically interconnected into a system of polygons, called *ice-wedge polygons* (Figure 13.28).

Another remarkable ice-formed feature of the arctic tundra is a conspicuous conical mound, call a *pingo* (Figure 13.29). The pingo has a core of ice and grows in height as more ice accumulates, forcing up the overlying sediment. In extreme cases, pingos reach a height of 50 m (164 ft) and a basal diameter as great as 600 m (about 2000 ft).

Patterned Ground and Solifluction

In areas of coarse-textured regolith consisting of rock particles in a wide range of sizes, the annual cycle of thawing and freezing causes the coarsest fragments—pebbles and cobbles—to move horizontally as well as vertically and thus become sorted out from the finer particles. This type of sorting produces ringlike arrangements of coarse fragments. The linking of adjacent rings produces a netlike pattern to form a system of *stone polygons* (also called "stone rings" or "stone nets") (Figure 13.30). On steep slopes, the process of soil creep causes the polygons to be elongated in the downslope direction and to become transformed into parallel stone stripes (Figure 13.31). Both stone polygons and ice-wedge polygons belong to a class of features called **patterned ground**.

A special variety of earthflow characteristic of arctic permafrost tundra regions is **solifluction** (from Latin

Figure 13.29 This large pingo rises above the flat plain of the McKenzie River delta, N.W.T., Canada.

Focus on Systems 13.2 • Permafrost as an Energy Flow System

The upper few hundred meters of the solid earth receive flows of heat from two directions—downward from the atmosphere (or ocean) above and upward from the ground below. The flow of energy into and out of the ground at the surface depends on the surface energy balance. In winter, when the air is, on average, colder than the ground, heat energy will flow out of the ground to the atmosphere. In summer, when the air is, on average, warmer, heat energy will flow into the ground from the atmosphere. (*Focus on Systems 3.2 • The Surface Energy Balance Equation* describes the surface energy balance in detail.)

The surface flow of heat energy into and out of the ground occurs mostly by conduction. Because the ground, as a dense substance, offers a strong resistance to heat penetration, the ground temperature changes most rapidly close to the surface. At increasing depth in the ground, the temperature changes much more slowly. Even so, given long periods of time, the temperature of a layer tens of meters deep in the ground can still show seasonal fluctuations.

Permafrost is defined as ground that is perennially at or below 0°C. Permafrost occurs in regions where the mean annual surface temperature is below freezing. Over very long periods of time (hundreds to thousands of years), the ground at a depth of 15 to 20 m tends to take on the average annual air temperature

and can thus acquire a year-round temperature well below 0°C.

The figure to the right shows this effect for a mean annual air temperature of –10°C. In the winter (*left*), air temperatures drop and heat energy flows out of the ground. In this example, we assume that the minimum annual air temperature during the winter is –25°C. The blue curve shows the average temperature profile at that time. Ground temperature assumes the air temperature at the surface and then warms with depth until about 15 m.

Summer conditions are shown on the right side of the graph. Here, the maximum monthly surface temperature is +5°C, and heat energy flows into the ground. The heat flow raises the ground temperature most near the surface, as shown on the red temperature plot. Note that the uppermost layer of the ground is heated above freezing. This, of course, is the active layer of permafrost that freezes and thaws each year. As before, the temperature remains unchanged below about 15 m.

In contrast to the seasonally varying flow of heat near the surface is the steady flow of heat upward from the earth's interior. As we saw in *Focus on Systems 10.2 • Powering the Rock Cycle*, crustal temperature increases with increasing depth due to the constant flow of heat from the interior. This increase is shown by a straight slanting line on the lower graph of the figure. The tem-

perature gradient varies somewhat, but a typical value is about +3°C per 100 m of ground depth. Since the temperature increases, eventually it exceeds 0°C. This defines the base of the permafrost layer. It is reached, in this example, at a depth of about 375 m.

Suppose that the mean annual temperature is somewhat warmer—say, –5°C instead of –10°C. In this case, we would expect the active layer to be thicker, as summer temperatures would be warmer and the soil would thaw more deeply. We would also expect the bottom of the permafrost to be encountered at a level that is not quite as deep. Thus, both the thickness of the active layer and the depth of permafrost are related to the mean annual surface temperature—the colder the temperature, the thinner is the active layer and the deeper is the permafrost base.

Permafrost is another example of an energy flow system in which inflows and outflows of energy tend to reach a balance, or equilibrium, over time. However, because heat flows so slowly through the ground, it takes a long time for permafrost to reach an equilibrium when a climate change occurs—perhaps thousands of years in some cases. Researchers have speculated that permafrost bodies in regions where mean annual surface temperature is now above 0°C may be relics of the Ice Age, slowly wasting away but still in place after 12,000 years of postglacial climate.

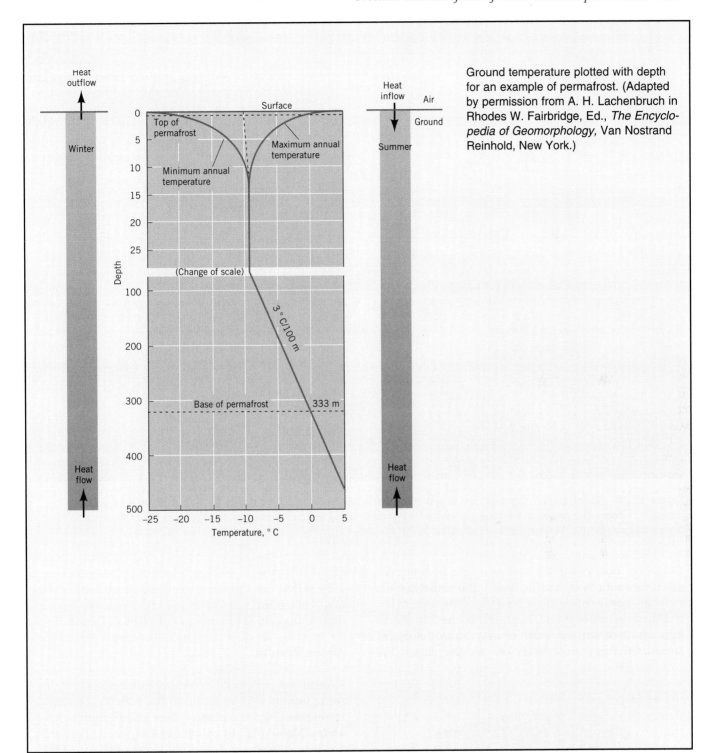

Ground temperature plotted with depth for an example of permafrost. (Adapted by permission from A. H. Lachenbruch in Rhodes W. Fairbridge, Ed., *The Encyclopedia of Geomorphology,* Van Nostrand Reinhold, New York.)

Figure 13.30   Sorted circles of gravel form a system of netlike stone rings on this nearly flat land surface where water drainage is poor. The circles in the foreground are 3 to 4 m (10 to 13 ft) across; the gravel ridges are 20 to 30 cm (8 to 12 in.) high. Broggerhalvoya, western Spitsbergen, latitude 78° N.

words meaning "soil" and "to flow"). It is active in early summer, when thawing has penetrated the upper few tenths of meters of the surface. At this time, the soil is fully saturated with water that cannot escape downward because of the permafrost below. Moving almost im-

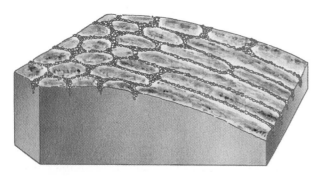

Figure 13.31   Stone rings grading downslope into stone stripes. (Adapted from C.F.S. Sharpe, *Landslides and Related Phenomena*, p. 37, Figure 5. New York, Columbia University Press. Used by permission of the publisher.)

perceptibly, this saturated soil is deformed into *solifluction terraces* and *solifluction lobes* that give the tundra slope a stepped appearance (Figure 13.32).

Alpine Tundra

Most of the periglacial processes and forms of the low arctic tundra are also found in the high alpine tundra throughout high mountains of the middle and high latitudes. Major regions of *alpine permafrost* in the northern hemisphere are shown on our map, Figure 13.24. In the alpine tundra, there is a dominance of steep mountainsides with large exposures of hard bedrock that has been strongly abraded by glacial ice. Postglacial talus slopes and cones formed of large angular blocks are conspicuously developed. Patterned ground and solifluction terraces occupy relatively small valley floors where slopes are low and tend to accumulate finer sediment.

Alpine permafrost is restricted to elevations at which the mean annual temperature is below freezing. The general elevation at which mean annual temperatures

Figure 13.32 Solifluction lobes in the Richardson Mountains, Northwest Territories, Canada. Bulging masses of water-saturated regolith have slowly moved downslope, overriding the ground surface below while bearing intact their covers of plants and soil.

fall to 0°C (32°F) will, of course, be related to latitude. In subtropical latitudes, elevations of 4000 m (about 13,000 ft) or greater are required, while permafrost can be encountered at sea level at latitudes above 70°. Figure 13.33 is a schematic graph showing the boundary between continuous and discontinuous permafrost as it varies with elevation and latitude. Whereas the boundary at lat. 40° lies at about 3500 m (about 11,500 ft), it has fallen to 2000 m (about 6600 ft) at lat. 50°, and by lat. 68° has descended close to sea level.

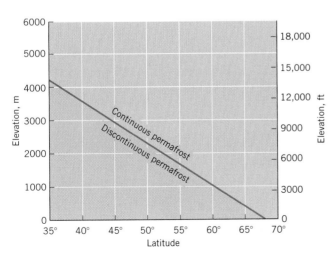

Figure 13.33 This graph shows how the altitude of the boundary between continuous and discontinuous alpine permafrost changes with the latitude.

In this chapter, we have examined two related topics—weathering and mass wasting. In the weathering process, rock near the surface is broken up into smaller fragments and often altered in chemical composition. In the mass wasting process, weathered rock and soil move downhill in slow to sudden mass movements. The landforms of mass wasting are produced by gravity acting directly on soil and regolith. Gravity also powers another landform-producing agent—running water—which we take up in the next three chapters. The first deals with water in the hydrologic cycle, in soil, and in streams. The second deals specifically with how streams and rivers erode regolith and deposit sediment to create landforms. The third describes how stream erosion strips away rock layers of different resistance, providing large landforms that reveal underlying rock structures.

CHAPTER SUMMARY

Weathering is the action of processes that cause rock to disintegrate and decompose because of surface exposure. Mass wasting is the spontaneous downhill motion of soil, rock, or regolith under gravity. Regolith, the surface layer of weathered rock particles, may be residual or transported.

Physical weathering produces regolith from solid rock by breaking bedrock into pieces. Frost action breaks rock apart by the repeated growth and melting of ice crystals in rock fractures and joints, as well as between individual mineral grains. In mountainous regions of vigorous frost, fields of angular blocks accumulate as felsenmeers. Slopes of rock fragments form talus cones. In soils and sediments, needle ice and ice lenses push rock and soil fragments upward. Salt-crystal growth in dry climates breaks individual grains of rock free, and can damage brick and concrete. Unloading of the weight of overlying rock layers can cause some types of rock to expand and break loose into thick shells, producing exfoliation domes. Daily temperature cycles in arid environments are thought to cause rock breakup. Wedging by plant roots also forces rock masses apart.

Chemical weathering results from mineral alteration. Igneous and metamorphic rocks can decay to great depths through hydrolysis and oxidation, producing a regolith that is often rich in clay minerals. Carbonic acid reacts with and dissolves limestone. In warm, humid environments, basaltic lavas can also show features of solution weathering.

Soil creep is a process of mass wasting in which soil moves down slopes almost imperceptibly under the influence of gravity. In an earthflow, water-saturated soil or regolith slowly flows downhill. Quick clays, which are unstable and can liquify when exposed to a shock, have produced earthflows in previously glaciated regions. A mudflow is much swifter than an earthflow. It follows stream courses, becoming thicker as it descends. A landslide is a rapid sliding of large masses of bedrock, sometimes triggered by an earthquake.

Scarification of land by human activities, such as mining of coal or ores, can heap up soil and regolith into unstable masses that produce earthflows or mudflows. Mass wasting can also be caused by removal of support, undermining the natural support of soil or regolith. These actions are termed induced mass wasting.

The tundra environment is dominated by the periglacial system of distinctive landforms and processes related to freezing and thawing of water in the active layer of permafrost. Continuous permafrost and sub-sea permafrost occur at the highest latitudes, flanked by a band of discontinuous permafrost that is transitional to warmer regions. Pingos and ice wedges are forms of ground ice found in permafrost terrains. Patterned ground occurs when ice growth and melting in freeze–thaw cycles create stone polygons and ice wedge polygons. During the brief summer thaw, saturated soils flow to form solifluction terraces and lobes. Alpine tundra and permafrost extend to lower latitudes in high-mountain environments.

KEY TERMS

weathering	mass wasting	permafrost
mass wasting	soil creep	ground ice
bedrock	earthflow	ice wedge
sediment	mudflow	patterned ground
alluvium	landslide	solifluction
physical weathering	scarification	
chemical weathering	periglacial	

REVIEW QUESTIONS

1. What is meant by the term *weathering*? What types of weathering are recognized?

2. Define the terms *regolith*, *bedrock*, *sediment*, and *alluvium.* Sketch a cross section through a part of the landscape showing these features and label them on the sketch.

3. How does frost action break up rock? Describe some landforms created by frost action and how they are formed.

4. How does salt-crystal growth break up rock? Give an example of a landform that arises from salt-crystal growth.

5. What is an exfoliation dome, and how does it arise? Provide an example.

6. Name three types of chemical weathering. Describe how limestone is often altered by a chemical weathering process.

7. Define mass wasting and identify the processes it includes.

8. What is soil creep, and how does it arise?

9. What is an earthflow? What features distinguish it as a landform?

10. Contrast earthflows and mudflows, providing an example of each.

11. Define the term *landslide*. How does a landslide differ from an earthflow?

12. Define and describe induced mass wasting. Provide some examples.

13. Explain the term *scarification*. Provide an example of an activity that produces scarification.

14. What is meant by the term *periglacial*?

15. Define and describe permafrost and some of its features, including ground ice, active layer, and permafrost table.

16. Identify and describe two forms of ground ice.

17. What is patterned ground? Identify two types of patterned ground.

Focus on Systems 13.2 • Permafrost as an Energy Flow System

1. Contrast the energy flows reaching the ground through the surface above and rising from below.

2. How does the profile of temperature with depth change from winter to summer in a permafrost region?

3. How is mean annual surface temperature related to the thickness of the active layer and the depth of permafrost?

ESSAY QUESTIONS

1. A landscape includes a range of lofty mountains elevated above a dry desert plain. Describe the processes of weathering and mass wasting that might be found on this landscape and identify their location.

2. Imagine yourself as the newly appointed director of public safety and disaster planning for your state. One of your first jobs is to identify locations where human populations are threatened by potential disasters, including those of mass wasting. Where would you look for mass wasting hazards and why? In preparing your answer, you may want to consult maps of your state.

PROBLEMS

Working It Out 13.1 • The Power of Gravity

1. According to the account, the slide reached a velocity of at least 150 km/hr. What was the kinetic energy of the slide at that point? Recall from *Focus on Systems 2.1 • Forms of Energy* that the formula for kinetic energy is $E = \frac{1}{2} mv^2$, where m is the mass (kg) and v is the velocity (m/s). How does this value compare to the total energy released in the slide?

2. Physics predicts the velocity, v (m/s), of a falling body in the absence of friction to be $v = \sqrt{2gd}$, where g (m/s^2) is the acceleration of gravity and d (m) is the distance of fall from rest. What is the velocity of an object falling without friction for a distance of 500 m? How does this compare with the velocity observed for the Madison Slide? Why are the two values different?

Chapter 14

The Cycling of Water on the Continents

Before we proceed with our survey of landforms and the processes that create them, we need to look more closely at the universal environmental agent that plays the dominant role in shaping the landscape—water. In this chapter we will focus first on two parts of the hydrologic cycle—the parts that involve water at the land surface and water that lies within the ground.

Recall from Chapter 4 that fresh water on the continents in surface and subsurface water constitutes only about 3 percent of the hydrosphere's total water. Most of this fresh water is locked into ice sheets and mountain glaciers. Groundwater accounts for a little more than half of 1 percent, and fresh water in lakes, streams, and rivers constitutes only three-hundredths of 1 percent of the total water. Although very small proportionally, the fresh, liquid water at the land surface is a vital part of the environment of plants and animals dwelling on continents.

In Chapter 4, we discussed atmospheric moisture and precipitation, describing the part of the hydrologic cycle in which water evaporates from ocean and land surfaces, condenses, or deposits, and then precipitates as rain or snow. What happens then to this precipitation? Figure 14.1 provides the answer. As the diagram

---

Detifoss Falls, Iceland, demonstrates the power of running water.

shows, a portion of the precipitation returns directly to the atmosphere through evaporation from the soil. Another portion travels downward, moving through the soil under the force of gravity to become part of the underlying groundwater body. Following underground flow paths, this subsurface water eventually emerges to become surface water, or it may emerge directly in the shore zone of the ocean. A third portion flows over the ground surface as runoff to lower levels. As it travels, the water flow becomes collected into streams, which eventually conduct the running water to the ocean.

In this chapter, we will trace the parts of the hydrologic cycle that include both the subsurface and surface pathways of water flow. The study of these flows is part of the science of *hydrology*, which is the study of water as a complex but unified system on the earth.

Figure 14.2 shows what happens to water from precipitation as it first reaches the land surface. Most soil surfaces in their undisturbed, natural states are capable of absorbing the water from light or moderate rains by **infiltration**. In this process, water enters the small natural passageways between irregularly shaped soil particles, as well as the larger openings in the soil surface.

These openings are formed by the borings of worms and animals, earth cracks produced by soil drying, cavities left from decay of plant roots, or spaces made by the growth and melting of frost crystals. A mat of decaying leaves and stems breaks the force of falling water drops and helps to keep these openings clear.

The precipitation that infiltrates the soil is temporarily held in the soil layer as soil water, occupying the *soil water belt* (Figure 14.2). Water within this belt can be returned to the surface and then to the atmosphere through a process that combines two components— evaporation and transpiration. As we explained in Chapter 9, these two forms of water vapor transport are combined in the term *evapotranspiration*.

When rain falls too rapidly to be passed downward through soil openings, **runoff** occurs and a surface water layer runs over the surface and down the direction of ground slope. This surface runoff is called **overland flow**. In periods of heavy, prolonged rain or rapid snowmelt, streams are fed directly by overland flow.

Overland flow also occurs when rainfall or snowmelt provides water to a soil that is already saturated. Since soil openings and pores are already filled, water can in-

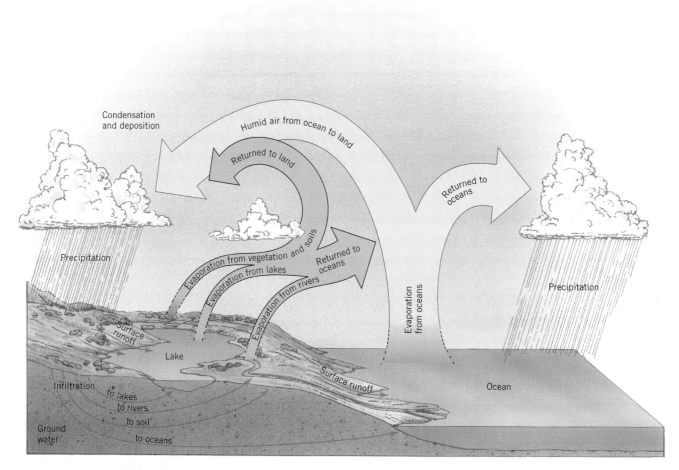

Figure 14.1 The hydrologic cycle traces the various paths of water from oceans, through the atmosphere, to land, and its return to the oceans.

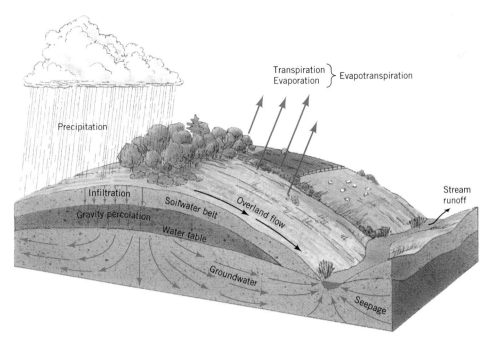

Figure 14.2 Paths of precipitation falling on the land. Some is returned to the atmosphere through evapotranspiration. Some runs off the soil surface. The remainder sinks into the soil water belt, where it is accessible to plants. A portion of the infiltrating water passes through the soil water belt and percolates down to the groundwater zone.

filtrate the soil only as quickly as it can drain down to deeper layers. Under these conditions, nearly all of the precipitation will run off.

Runoff as overland flow moves surface particles from hills to valleys, and so it is an agent that shapes landforms. Because runoff supplies water to streams and rivers, it also allows rivers to cut canyons and gorges, and carry sediment to the ocean—but we are getting ahead of our story. We'll return to landforms carved by running water in Chapter 15.

GROUNDWATER

As shown in Figure 14.2, water derived from precipitation can continue to flow downward beyond the soil water belt. This slow downward flow under the influence of gravity is termed *percolation*. Eventually, the percolating water reaches groundwater. **Ground water** is the part of the subsurface water that fully saturates the pore spaces in bedrock, regolith, or soil, and so occupies the *saturated zone* (Figure 14.3). The **water table**

Figure 14.3 Zones of subsurface water.

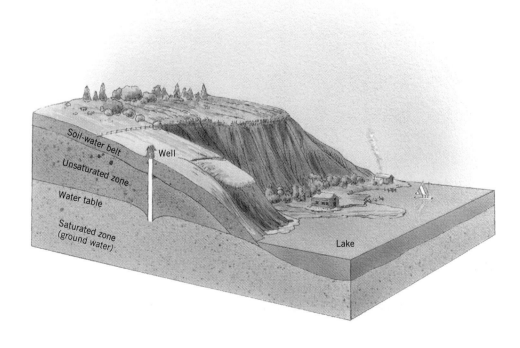

marks the top of this zone. Above it is the *unsaturated zone*, in which water does not fully saturate the pores. Here, water is held by capillary tension—as thin films of water adhering to mineral surfaces. This zone also includes the soil water belt.

Groundwater moves slowly in deep paths, eventually emerging by seepage into streams, ponds, lakes, and marshes. In these places the land surface dips below the water table. Streams that flow throughout the year—perennial streams—derive much of their water from groundwater seepage.

The Water Table

Where there are many wells in an area, the position of the water table can be mapped in detail (Figure 14.4). This is done by plotting the water heights and noting the trend of change in elevation from one well to the other. The water table is highest under the highest areas of land surface—hilltops and divides. The water table declines in elevation toward the valleys, where it appears at the surface close to streams, lakes, or marshes.

The reason for this water table configuration is that water percolating down through the unsaturated zone tends to raise the water table, while seepage into lakes, streams, and marshes tends to draw off groundwater and to lower its level. Because groundwater moves extremely slowly, a difference in water table level is built up and maintained between high and low points on the water table. In periods of high precipitation, the water table rises under divide areas. In periods of water deficit, or during a drought, the water table falls (Figure 14.4).

Figure 14.4 also shows paths of groundwater flow. Water that enters the hillside midway between the hilltop and the stream flows rather directly toward the stream. Water reaching the water table midway between streams, however, flows almost straight down to great depths before recurving and rising upward again. Progress along these deep paths is incredibly slow, while flow near the surface is much faster. The most rapid flow is close to the stream, where the arrows converge. Over time, the level of the water table tends to remain stable, and the flow of water released to streams and lakes must balance the flow of water percolating down into the water table.

LIMESTONE SOLUTION BY GROUNDWATER

We saw in Chapter 13 that limestone at the surface is slowly dissolved by carbonic acid action in moist climates, producing lowland areas. The slow flow of groundwater in the saturated zone can also dissolve limestone below the surface, producing deep underground caverns. These features can then collapse, producing a sinking of the ground above and the development of a unique type of landscape.

Limestone Caverns

You are probably familiar with such famous caverns as Mammoth Cave or Carlsbad Caverns. Millions of Americans have visited these natural wonders. *Limestone caverns* are interconnected subterranean cavities in bedrock formed by the corrosive action of circulating groundwater on limestone. Figure 14.5 shows how caverns develop. As shown in the upper diagram, the action of carbonic acid is particularly concentrated in the saturated zone just below the water table. This removal process forms many kinds of underground "land-

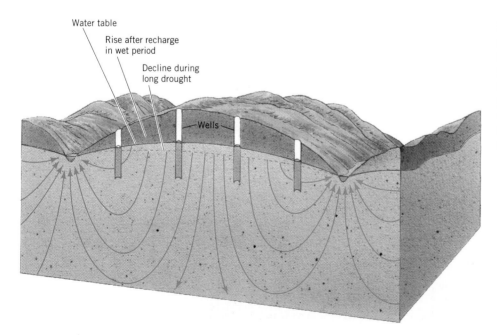

Water table
Rise after recharge in wet period
Decline during long drought
Wells

Figure 14.4 The configuration of the water table surface conforms broadly with the land surface above it. It varies in response to prolonged wet and dry periods. Groundwater flow paths circulate water to deep levels in a very slow motion and eventually feed streams by seepage.

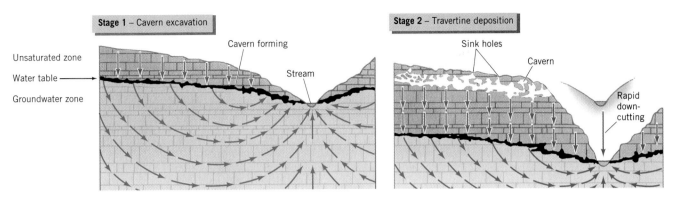

Figure 14.5 Cavern development in the groundwater zone, followed by travertine deposition in the unsaturated zone. (Copyright © A. N. Strahler.)

forms," such as tortuous tubes and tunnels, great open chambers, and tall chimneys. Subterranean streams can be found flowing in the lowermost tunnels, and these carry the products of solution to emerge along the banks of surface streams and rivers.

In a later stage, shown in the lower diagram, the stream has deepened its valley, and the water table has been correspondingly lowered to a new position. The cavern system previously formed is now in the unsaturated zone. Deposition of carbonate matter, known as *travertine*, takes place on exposed rock surfaces in the caverns. Encrustations of travertine take many beautiful forms—stalactites, stalagmites, columns, drip curtains, and terraces (Figure 14.6).

Figure 14.6 Travertine deposits in the Papoose Room of Carlsbad Caverns include stalactites (slender rods hanging from the ceiling) and sturdy columns.

Figure 14.7   Sinkholes in limestone, near Roswell, New Mexico.

Karst Landscapes

Where limestone solution is very active, we find a landscape with many unique landforms. This is especially true along the Dalmatian coastal area of Croatia, where the landscape is called *karst*. Geographers apply that term to the topography of any limestone area where sinkholes are numerous and small surface streams are nonexistent. A sinkhole is a surface depression in a region of cavernous limestone (Figure 14.7). Some sinkholes are filled with soil washed from nearby hillsides, while others are steep-sided, deep holes.

Development of a karst landscape is shown in Figure 14.8. In an early stage, funnel-like sinkholes are numerous. Later, the caverns collapse, leaving open, flat-

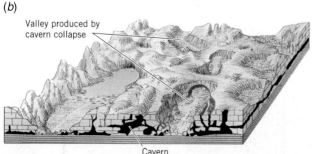

Figure 14.8   Features of a karst landscape. (*a*) Rainfall enters the cavern system through sinkholes in the limestone. (*b*) Extensive collapse of caverns reveals surface streams flowing on shale beds beneath the limestone. Some parts of the flat-floored valleys can be cultivated. (Drawn by Erwin Raisz. Copyright © A. N. Strahler.)

floored valleys. Examples of some important regions of karst or karstlike topography are the Mammoth Cave region of Kentucky, the Yucatan Peninsula, and parts of Cuba and Puerto Rico. In regions such as southern China and west Malaysia, the karst landscape is dominated by steep-sided, conical limestone hills or towers, 100 to 500 m (about 300 to 1500 ft) high (Figure 14.9). These hills are often riddled with caverns and passageways.

EYE ON THE ENVIRONMENT:
PROBLEMS OF GROUNDWATER MANAGEMENT

Rapid withdrawal of groundwater has seriously impacted the environment in many places. Increased urban populations and industrial developments require larger water supplies—needs that cannot always be met by constructing new surface-water reservoirs. To fill these needs, vast numbers of wells using powerful pumps draw great volumes of groundwater to the surface, greatly altering nature's balance of groundwater recharge and discharge.

In dry climates, agriculture is often heavily dependent on irrigation water from pumped wells—espe-

cially since major river systems are likely to be already fully utilized for irrigation. Wells are also convenient water sources. They can be drilled within the limits of a given agricultural or industrial property and can provide immediate supplies of water without any need to construct expensive canals or aqueducts.

In earlier times, the small well that supplied the domestic and livestock needs of a home or farmstead was actually dug by hand and sometimes lined with masonry. By contrast, a modern well supplying irrigation and industrial water is drilled by powerful machinery that can bore a hole 40 cm (16 in.) or more in diameter to depths of 300 m (about 1000 ft) or more. Drilled wells are sealed off by metal casings that exclude impure near-surface water and prevent clogging of the tube by caving of the walls. Near the lower end of the hole, in the groundwater zone, the casing is perforated to admit the water. The yields of single wells range from as low as a few hundred liters or gallons per day in a domestic well to many millions of liters or gallons per day for large industrial or irrigation wells.

Water Table Depletion

As water is pumped from a well, the level of water in the well drops. At the same time, the surrounding water

Figure 14.9   Tower karst near Guilin (Kweilin), Guanxi Province, southern China at lat. 25° N. White limestone can be seen exposed in the nearly vertical sides of the towers.

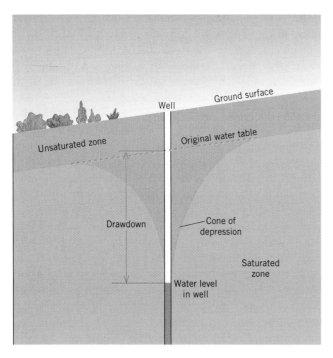

Figure 14.10 Drawdown and cone of depression in a pumped well. As the well draws water, the water table is depressed in a cone shape centered on the well.

table is lowered in the shape of a downward-pointing cone, termed the *cone of depression* (Figure 14.10). The difference in height between the cone tip and the original water table is the *drawdown*. The cone of depression may extend out as far as 16 km (10 mi) or more from a well where heavy pumping is continued. Where many wells are in operation, their intersecting cones will produce a general lowering of the water table.

Water table depletion often greatly exceeds recharge—the rate at which infiltrating water moves downward to the saturated zone. In arid regions, much of the groundwater for irrigation is drawn from wells driven into thick sands and gravels. These deposits are often recharged by the seasonal flow of streams that head high in adjacent mountains. Fanning out across the dry lowlands, the streams lose water, which sinks into the sands and gravels and eventually percolates to the water table below. The extraction of groundwater by pumping can greatly exceed this recharge by stream flow, lowering the water table. Deeper wells and more powerful pumps are then required. The result is exhaustion of a natural resource that is not renewable except over long periods of time.

Ground Subsidence

Another important environmental effect of excessive groundwater withdrawal is subsidence of the ground surface. Venice, Italy, provides a dramatic example of this side effect. Venice was built in the eleventh century A.D. on low-lying islands in a coastal lagoon, sheltered from the ocean by a barrier beach. Underlying the area are some 1000 m (about 3300 ft) of layers of sand, gravel, clay, and silt, with some layers of peat. Compaction of these soft layers has been going on gradually for centuries under the heavy load of city buildings. However, groundwater withdrawal, which has been greatly accelerated in recent decades, has aggravated the condition.

Figure 14.11 Land subsidence has subjected Venice to episodes of flooding by waters of the Adriatic Sea. Here, high water has flooded the Piazza San Marco and an outdoor café.

Many ancient buildings in Venice now rest at lower levels and have suffered severe damage as a result of flooding during winter storms on the adjacent Adriatic Sea (Figure 14.11). Sea level is normally raised by the effects of coastal storms, and when high tides occur at the same time, water rises even higher. The problem of flooding during storms is aggravated by the fact that many of the canals of Venice receive raw sewage, so that the floodwater is contaminated.

Most of the subsidence in recent decades has been attributed to withdrawals of large amounts of groundwater from industrial wells at Porto Marghere, the modern port of Venice, located a few kilometers distant on the mainland shore. This pumping has now been greatly curtailed, reducing the rate of subsidence to a very small natural rate (about 1 mm per year). However, the threat of flooding and damage to churches and other buildings of great historical value remains. Flood control now depends on the construction of seawalls and floodgates on the barrier beach that lies between Venice and the open ocean.

Groundwater withdrawal has affected several regions in California, where groundwater for irrigation has been pumped from basins filled with alluvial (stream and lake-deposited) sediments. Water table levels in these basins have dropped over 30 m (98 ft), with a maximum drop of 120 to 150 m (394 to 492 ft) being recorded in one locality of California's San Joaquin Valley. In the Los Baños-Kettleman City area, ground subsidence of as much as 4.2 to 4.8 m (14 to 16 ft) was measured at some locations in a 35-year period.

Another important area of ground subsidence accompanying water withdrawal is beneath Houston, Texas, where the ground surface has subsided from 0.3 to 1 m (1 to 3 ft) in a metropolitan area 50 km (31 mi) across. Damage has resulted to buildings, pavements, airport runways, and other structures.

Perhaps the most celebrated case of ground subsidence is that affecting Mexico City. Carefully measured ground subsidence has ranged from about 4 to 7 m (13 to 23 ft). The subsidence has resulted from the withdrawal of groundwater from an aquifer system beneath the city and has caused many serious engineering prob-

lems. The volume of clay beds overlying the aquifer has contracted greatly as water has been drained out. To combat the ground subsidence, recharge wells were drilled to inject water into the aquifer. In addition, new water supplies from sources outside the city area were developed to replace local groundwater use.

Contamination of Groundwater

Another major environmental problem related to groundwater withdrawal is contamination of wells by pollutants that infiltrate the ground and reach the water table. Both solid and liquid wastes are responsible. Disposal of solid wastes poses a major environmental problem in the United States because our advanced industrial economy provides an endless source of garbage and trash. Traditionally, these waste products were trucked to the town dump and burned there in continually smoldering fires that emitted foul smoke and gases. The partially consumed residual waste was then buried under earth.

In recent decades, a major effort has been made to improve solid-waste disposal methods. One method is high-temperature incineration, but it often leads to air pollution. Another is the sanitary landfill method in which waste is not allowed to burn. Instead, layers of waste are continually buried, usually by sand or clay available on the landfill site. The waste is thus situated in the unsaturated zone. Here it can react with rainwater that infiltrates the ground surface. This water picks up a wide variety of chemical compounds from the waste body and carries them down to the water table (Figure 14.12).

Once in the water table, the pollutants follow the flow paths of the groundwater. As the arrows in the figure indicate, the polluted water may flow toward a supply well, which is drawing in groundwater from a large radius. Once the polluted water has reached the well, the water becomes unfit for human consumption. Polluted water may also move toward a nearby valley, causing pollution of the stream flowing there (left side of Figure 14.12).

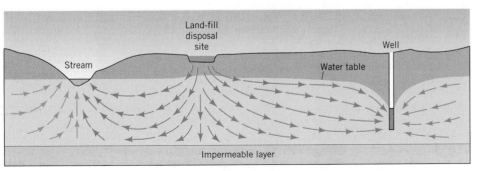

Figure 14.12 Polluted water, leached from a waste-disposal site, moves toward a supply well (*right*) and a stream (*left*). (Copyright © A. N. Strahler.)

SURFACE WATER

So far, we have examined how water moves below the land surface. Now, we turn to tracing the flow paths of surplus water that runs off the land surface and ultimately reaches the sea. Here, we will be concerned primarily with rivers and streams. (In general usage, we speak of "rivers" as large watercourses and "streams" as smaller ones. However, the word "stream" is also used as a scientific term designating the channeled flow of surface water of any amount.)

Overland Flow and Stream Flow

As we saw earlier in this chapter, runoff that flows down the slopes of the land in broadly distributed sheets is overland flow. We can distinguish overland flow from *stream flow*, in which the water occupies a narrow channel confined by lateral banks. Overland flow can take several forms. It may be a continuous thin film, called *sheet flow*, where the soil or rock surface is smooth (Figure 14.13). Where the ground is rough or pitted, flow may take the form of a series of tiny rivulets connecting one water-filled hollow with another. On a grass-covered slope, overland flow is subdivided into countless tiny threads of water, passing around the stems. Even in a heavy and prolonged rain, you might not notice overland flow in progress on a sloping lawn. On heavily forested slopes, overland flow may pass entirely concealed beneath a thick mat of decaying leaves.

Overland flow eventually contributes to a stream, which is a much deeper, more concentrated form of runoff. We can define a **stream** as a long, narrow body of flowing water occupying a trenchlike depression, or channel, and moving to lower levels under the force of gravity. The **channel** of a stream is a narrow trough, shaped by the forces of flowing water to be most effective in moving the quantities of water and sediment supplied to the stream (Figure 14.14). Channels may be so narrow that a person can jump across them, or, in the case of the Mississippi River, as wide as 1.5 km (about 1 mi).

As a stream flows under the influence of gravity, the water encounters resistance—a form of friction—with the channel walls. As a result, water close to the bed and banks moves slowly and water in the deepest and most centrally located zone flows fastest. If the channel is straight and symmetrical, the single line of maximum velocity is located in midstream. If the stream curves, the maximum velocity is found toward the bank on the outside of the curve.

Actually, the arrows in Figure 14.14 only show average velocity. In all but the most sluggish streams, the water is affected by *turbulence*, a system of countless eddies that are continually forming and dissolving. If we follow a particular water molecule, it will travel a highly irregular, corkscrew path as it is swept downstream. Motions include upward, downward, and sideward directions. Only if we measure the water velocity at a certain fixed point for a long period of time, say, several minutes, will the average motion at that point be downstream and in a line parallel with the surface and bed.

Stream Discharge

Stream flow at a given location is measured by its **discharge**, which is defined as the volume of water per

Figure 14.13 Overland flow taking the form of a thin sheet of water covers the nearly flat plain in the middle distance. This water is converging into stream flow in a narrow, steep-sided gully (*left*). The photograph was taken shortly after a summer thunderstorm had deluged the area. The locality, near Raton, New Mexico, shows steppe grassland vegetation.

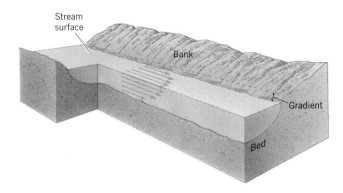

Figure 14.14 Stream flow within a channel is most rapid near the center.

unit time passing through a cross section of the stream at that location. It is measured in cubic meters (cubic feet) per second. The cross-sectional area and average velocity of a stream can change within a short distance, even though the stream discharge does not change (Figure 14.15). These changes occur because of changes in the *gradient* of the stream channel. The gradient is the rate of fall in elevation of the stream surface in the downstream direction (Figure 14.14).

When the gradient is steep, the force of gravity will act more strongly and flow velocity will be greater. When the stream channel has a gentle gradient, the velocity will be slower. Figure 14.15 shows a short stretch of stream throughout which the discharge remains constant. In stretches of rapids, where the stream flows swiftly, the stream channel will be shallow and narrow. In pools, where the stream flows more slowly, the stream channel will be wider and deeper to maintain the same discharge. Sequences of pools and rapids can be found along streams of all sizes.

The discharges of streams and rivers change from day to day, and records of daily and flood discharges of major streams and rivers are important information. They are used in planning the development and distribution of surface waters, as well as in designing flood-protection structures and predicting floods as they progress down a river system. An important activity of the U.S. Geological Survey is the measurement, or gauging, of stream discharge in the United States. In cooperation with states and municipalities, this organization maintains over 6000 gauging stations on principal streams and their tributaries.

Figure 14.16 is a map showing the relative discharge of major rivers of the United States. The mighty Mississippi with its tributaries dwarfs all other North America rivers. Two other discharges are of major proportions—the Columbia River, draining a large segment of the Rocky Mountains in southwestern Canada and the northwestern United States, and the Great Lakes, discharging through the St. Lawrence River. The Colorado River, a much smaller stream, crosses a vast semiarid and arid region in which little tributary flow is added to its snowmelt source high in the Rocky Mountains.

As the figure shows, the discharge of major rivers increases downstream as a natural consequence of the way streams and rivers combine to deliver runoff and sediment to the oceans. The gradient also changes in a downstream direction. The general rule is the larger the cross-sectional area of the stream, the lower the gradient. Great rivers, such as the Mississippi and Amazon, have gradients so low that they can be described as "flat." For example, the water surface of the lower Mississippi River falls in elevation about 3 cm for each kilometer of downstream distance (1.9 in. per mi).

Rivers with headwaters in high mountains have characteristics that are especially desirable for utilizing river flow as irrigation and for preventing floods. The higher ranges serve as snow storage areas, storing the winter

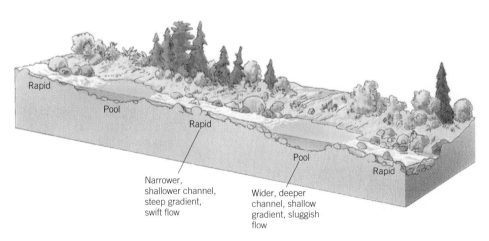

Figure 14.15 In order to preserve a uniform discharge, a stream becomes wider and deeper in pools, where flow velocity is slower. In rapids, flow is swifter, and the stream becomes narrower and shallower.

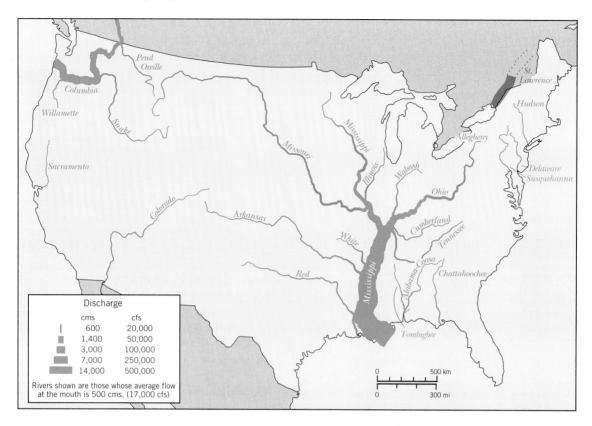

Figure 14.16 This schematic map shows the relative magnitude of United States rivers. Width of the river as drawn is proportional to mean annual discharge. (After U.S. Geological Survey.)

and spring precipitation until early or midsummer, when it is released slowly through melting. As melting proceeds to successively higher levels, the meltwater is supplied to the river. In this way, a continuous river flow is maintained. Among the snow-fed rivers of the western United States are the Columbia, Snake, Missouri, Arkansas, and Colorado.

What powers streams? The answer, of course, is grav-

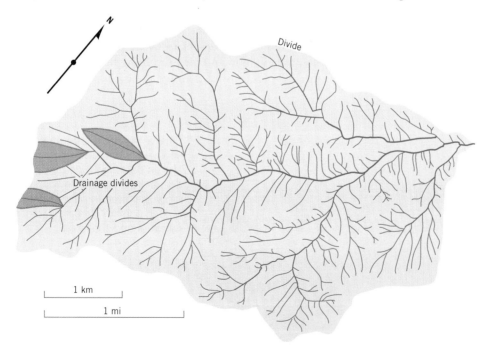

Figure 14.17 Channel network of a small drainage basin. Drainage divides mark the boundaries of stream basins. (Data of U.S. Geological Survey and Mark A. Melton.)

ity. This pervasive force moves water downhill and is opposed by the friction of the water with the bed and banks of the channel. *Focus on Systems 14.1 • Energy in Stream Flow* explains how gravitational power is related to the gradient of the stream and how that power is dissipated in stream flow.

Drainage Systems

As runoff moves to lower and lower levels and eventually to the sea, it becomes organized into a **drainage system**. The system consists of a branched network of stream channels, as well as the sloping ground surfaces that contribute overland flow to those channels. Between the channels on the ridges are *drainage divides*, which mark the boundary between slopes that contribute water to different streams or drainage systems (Figure 14.17). The entire system is bounded by a drainage divide that outlines a more-or-less pear-shaped **drainage basin**.

A typical stream network within a drainage basin is shown in Figure 14.17. Each fingertip tributary receives runoff from a small area of land surface surrounding the channel. The entire surface within the outer divide of the drainage basin constitutes the watershed for overland flow. The drainage system provides a converging mechanism that funnels overland flow and smaller streams into larger ones.

STREAM FLOW

The discharge of a stream will increase in response to a period of heavy rainfall or snowmelt. However, the response is delayed, for the movement of water into stream channels takes time. The length of delay depends on a number of factors. The most important factor is the size of the drainage basin feeding the stream above the place where the gauging station is located.

The relationship between stream discharge and precipitation is best studied by means of a simple graph, called a *hydrograph*. Figure 14.18 is a hydrograph for a drainage basin about 800 km^2 (300 mi^2) in area located in Ohio within the moist continental climate ⑩. The graph presents data for a two-day summer storm. Rainfall is shown by a bar graph giving the number of centimeters of precipitation in each two-hour period. Also plotted on the graph (smooth line) is the discharge of Sugar Creek, the trunk stream of the drainage basin. The average total rainfall over the watershed of Sugar Creek was about 15 cm (6 in.). About half of this amount passed down the stream within three days' time. Some rainfall was held in the soil as soil water, some evaporated, and some infiltrated to the water table to be held in long-term storage as groundwater.

Studying the rainfall and runoff graphs in Figure 14.18, we see that prior to the onset of the storm, Sugar Creek was carrying a small discharge. This flow, being supplied by the seepage of groundwater into the channel, is termed *base flow*. After the heavy rainfall began, several hours elapsed before the stream gauge at the basin mouth began to show a rise in discharge. This interval, called the *lag time*, indicates that the branching system of channels was acting as a temporary reservoir. The channels were at first receiving inflow more rapidly than it could be passed down the channel system to the stream gauge.

Lag time is measured as the difference between the center of mass of precipitation (CMP) and the center of mass of runoff (CMR), as labeled in Figure 14.18. If you imagine the set of precipitation columns as weights positioned on a bar at the base of the graph, the balance point of the bar and weights would be the center of mass.

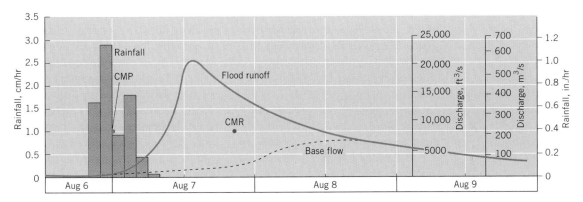

Figure 14.18 Four days of precipitation and stream flow at Sugar Creek, Ohio, following a heavy rainstorm in August. (After Hoyt and Langbein, *Floods*, copyright © Princeton University Press. Used by permission.)

Focus on Systems 14.1 • Energy in Stream Flow

Watching a flowing stream from a perch on its bank can be a fascinating experience. If it's a small stream, you may hear splashing and gurgling as the water flows around obstacles in its bed. If it's a large, deep stream—what we usually call a river—it may be flowing silently, but with large eddies that roil the surface as they move by. In either case, energy is constantly being dissipated as the river flows. That is our theme here.

First, we need to examine briefly the geometric elements of a stream channel—what hydraulic engineers call the "hydraulic geometry." These elements are shown in the diagram at right. Notice first the velocity arrows that show the flow pattern at the stream surface. Velocity is nearly zero at either bank and then increases as the center line of the channel is approached. Shown below the surface is a vertical cross section along the center line of the channel. Velocity increases from the bottom of the channel to the water surface.

The discharge of a stream depends on the mean water velocity and the cross-sectional area of the stream. For example, if the mean velocity of a stream passing a certain point on the bank is 2 m/s and the cross-sectional area of the stream at that point is 20 m², then in one second $2 \times 20 = 40$ m³ will flow past the point. This simple relation is expressed by the equation

$$Q = A \times V$$

where Q is the discharge, A is the cross-sectional area, and V is the mean velocity.

Another important variable describing a stream at a particular point is the slope of the stream (S). This value measures the inclination, or gradient, of the entire channel in the downstream direction. As you might expect, the slope is related to the velocity of the stream. If the slope is steep, the water will flow more quickly and the velocity will be greater.

How are these quantities related from point to point along the stream? Imagine walking some distance along a stream. As you watch the stream, you will encounter pools, where the slope of the stream is low and the water moves slowly. Between these you will find stretches of rapids, where the slope is steeper and the water moves more swiftly. This situation is shown in the lower part of the figure.

Suppose that no new tributaries enter the stream in the distance that you traverse. In this case, Q will be constant through this stretch. If the slope steepens, velocity will increase, and we can see from the equation $Q = A \times V$ that if the velocity increases, then the cross-sectional area of the stream must decrease if Q is unchanged. In this way, we see that in pools, the slope will be low, velocity will be low, and cross-sectional area will be large. In the rapids, the situation is reversed—slope is steep, velocity is high, and cross-sectional area is small.

Let's turn now to the flow of energy accompanying stream flow. The driving force for stream flow is, of course, gravity. As water moves downslope in a stream channel, its potential energy is converted to kinetic energy of mass in motion, and the kinetic energy in turn is changed into heat by the friction of flow.

Every molecule of water in a stream tends to be drawn vertically down under the force of gravity, so that the water exerts a pressure—a force per unit area—on the enclosing channel walls. This force increases with depth. A component of that force acts in the downstream direction, parallel with the streambed, and tends to cause the flow of one water layer over the next layer in a type of motion known as *shear*. These layers can be regarded as thin sheets of molecules, each sheet slipping over the one below it, much as playing cards slip over one another when the deck is gently pushed across a table top. The layer immediately in contact with the solid streambed does not slip, but each higher layer slips over the one below, so that the velocity (V) of forward motion increases upward from the bed. The same effect applies to the steep sides of the channel banks. This type of flow is known as *laminar flow*. Moving away from the bed and banks, however, this smoothly layered form of shear gives way to *turbulent flow* in which shear occurs in eddies of a wide range of sizes, intensities, and orientations.

Shear of both types meets with internal resistance, which requires the conversion of much of the potential energy into sensible heat; this in turn raises the temperature of the water. The heat is eventually lost by conduction, radiation, or evaporation to the air above, and by conduction into the solid channel walls. As you might imagine, turbulent flow creates much more friction than laminar flow, so that much more energy is dissipated when water flow is turbulent.

How does the energy dissipated in stream flow differ in pools and rapids? In pools, slope is low, flow velocity is low, and less energy is dissipated as water moves a given distance downstream. In rapids, the slope is steeper, velocity is greater, and the energy dissipated will be much higher, for a given distance downstream.

You may be curious about just how much heat is released in the friction of flow of a stream. Since the potential energy released will be converted to heat, we only need to calculate the potential energy loss, which we can easily do using the same principles as in *Working It Out 13.1 • The Power of Gravity*, where we calculated the energy released by a landslide.

Suppose a cubic meter of stream water flows a horizontal distance of 1 km while descending 10 meters. This amounts to a slope of $10/1000 = 0.01$ or 1 percent. The force of gravity acting on the cubic meter is $F = mg = 10^3$ kg $\times$ 9.8 m/s^2 $= 9.8 \times 10^3$ N. The potential energy

released in falling 10 meters is then $F \times d = 9.8 \times 10^3$ N $\times$ 10 m $= 9.8 \times 10^4$ J. Let's further suppose that the mean velocity of the water flow is 1 m/s. Then the water will cover the distance of 1 km $= 10^3$ m in 10^3 s. The power produced is then energy per unit time, or 9.8×10^4 J $\div 10^3$ s $= 98$ W. That is, stream flow will release a flow of energy equal to 98 watts per cubic meter, per degree of slope, per meter per second of velocity.

For example, Rock Creek near Red Lodge, Montana, discharges

about 17 m^3/s at a velocity of about 0.85 m/s when it completely fills its channel, and its slope there is about .021 or 2.1 percent. Using the factor of 98 watts per cubic meter, per degree of slope, per meter per second of velocity, we find that this stream flow will release about 2980 W of power, which is enough to meet the demands of a few suburban households.

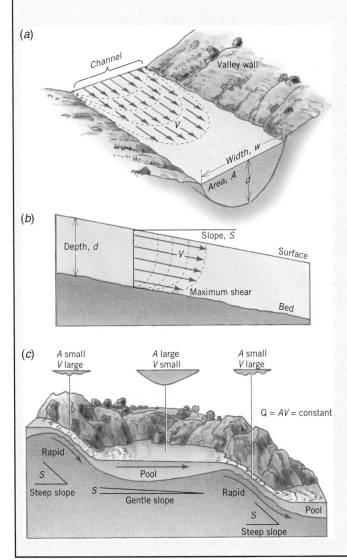

(a)

(b)

(c)

Characteristics of stream flow. The velocity of flow is greatest in the middle (*a*) and at the top (*b*) of the stream. (*c*) Mean velocity, cross-sectional area, and slope change in the pools and rapids of a stream section of uniform discharge. (*b*, Copyright © A. N. Strahler.)

In the Sugar Creek example, the lag time was about 18 hours, with the peak flow reached almost 24 hours after the rainfall began. Note also that the stream's discharge rose much more abruptly than it fell. In general, the larger a watershed, the longer is the lag time between peak rainfall and peak discharge, and the more gradual is the rate of decline of discharge after the peak has passed. Another typical feature of a flood hydrograph is the slow but distinct rise in the amount of discharge contributed by base flow.

EYE ON THE ENVIRONMENT:
How Urbanization Affects Stream Flow

The growth of cities affects the flow of small streams in two ways. First, an increasing percentage of the surface becomes impervious to infiltration as it is covered by buildings, driveways, walks, pavements, and parking lots. In a closely builtup residential area with small lot sizes, the percentage of impervious surface may run as high as 80 percent.

An increase in proportion of impervious surface increases overland flow generally from the urbanized area. This change acts to increase the frequency and height of flood peaks during heavy storms for small watersheds lying largely within the urban area. There is also a reduction of recharge to the groundwater body beneath, and this reduction, in turn, decreases the base-flow contribution to channels in the same area. Thus, the full range of stream discharges, from low stages in dry periods to flood stages, is made greater by urbanization.

A second change caused by urbanization is brought about by the introduction of storm sewers that quickly carry storm runoff from paved areas directly to stream channels for discharge. Thus, runoff travel time to channels is shortened while the proportion of runoff is increased by the expansion in impervious surfaces. The two changes together act to reduce the lag time of urban streams and increase their peak discharge levels. Many rapidly expanding suburban communities are finding that low-lying, formerly flood-free, residential areas now experience periodic flooding as a result of stream urbanization.

The Annual Flow Cycle of a Large River

In regions of humid climates, where the water table is high and normally intersects the important stream channels, the hydrographs of larger streams will show clearly the effects of two sources of water—base flow and overland flow. Figure 14.19 is a hydrograph of the Chattahoochee River in Georgia, a fairly large river that drains a watershed of 8700 k^2 (3350 mi^2), much of it in the southern Appalachian Mountains. The sharp, abrupt fluctuations in discharge are produced by overland flow following rain periods of one to three days' duration. These fluctuations are each similar to the hydrograph of Figure 14.18, except that here they are shown much compressed by the time scale.

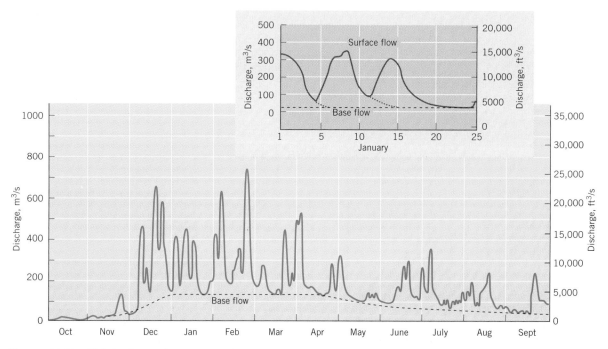

Figure 14.19 This hydrograph shows the fluctuating discharge of the Chattahoochee River, Georgia, throughout a typical year. The high peaks are caused by runoff from streams that produced heavy overland flow. (Data of U.S. Geological Survey.)

Figure 14.20 Flooding of the Susquehanna River in West Nanticoke, Pennsylvania, during the disasterous flood of January 1996.

After each rain period, the discharge falls off rapidly, but if another storm occurs within a few days, the discharge rises to another peak. The enlarged inset graph in Figure 14.19 shows details for the month of January. Here, three storms occur in rapid succession. Runoff drops between each storm, but not completely to the level of base flow. When a long period intervenes between storms, the discharge falls to a low value, the base flow, at which it levels off.

Throughout the year the base flow, which represents groundwater inflow into the stream, undergoes a marked annual cycle. During winter and early spring, water table levels are raised, and the rate of inflow into streams is increased. For the Chattahoochee River, the rate of base flow during January, February, March, and April holds uniform at about 100 m^3/s (about 3500 ft^3/s). The base flow begins to decline in spring, as heavy evapotranspiration losses reduce soil water and therefore cut off the recharge of groundwater. The decline continues through the summer, reaching a low of about 30 m^3/s (about 1000 ft^3/s) by the end of October.

River Floods

You've probably been exposed to enough media coverage of river floods to have a good idea of what floodwaters look like and what kind of damage is caused by flood erosion and deposition of silt and clay. Even so, it is not easy to define the term **flood**. Perhaps it is enough to say that a flood condition exists when the discharge of a river cannot be accommodated within its normal channel. As a result, the water spreads over the adjoining ground, which is normally cropland or forest. Sometimes, however, the ground is occupied by houses, factories, or transportation corridors.

Most rivers of humid climates have a *floodplain*, a broad belt of low, flat ground bordering the channel on one or both sides that is flooded by stream waters about once a year. This flood usually occurs in the season when abundant surface runoff combines with the effects of a high water table to supply more runoff than can be carried in the channel. This annual inundation is considered a flood, even though its occurrence is expected and does not prevent the cultivation of crops after the flood has subsided. The seasonal inundation does not interfere with the growth of dense forests, which are widely distributed over low, marshy floodplains in all humid regions of the world. Still higher discharges of water, the rare and disastrous floods that may occur as rarely as once in 30 or 50 years, inundate ground lying well above the floodplain (Figure 14.20).

For practical purposes, the National Weather Service, which provides a flood-warning service, designates a particular river surface height, or *stage*, at a given place as the *flood stage*. Above this critical level, inundation of the floodplain will occur.

Flash floods are characteristic of streams draining small watersheds with steep slopes. These streams have short lag times—perhaps only an hour or two—so when an intense rainfall event occurs, the stream rises very quickly to a high level. The flood arrives as a swiftly moving wall of turbulent water, sweeping away build-

Working It Out 14.2 • Magnitude and Frequency of Flooding

It is a simple fact of nature that more extreme events happen less frequently. For example, hundreds or thousands of small earthquakes may occur within a region during the period of a century, but only a few earthquakes are really large. In other words, the greater the magnitude of an event, the lesser the frequency at which it recurs.

The figure at the right shows this principle for the flood frequency of the Clearwater River at Kamaiah, Idaho. Each dot on the graph plots the maximum discharge of the river recorded within a particular year—that is, the peak flow, or largest flood, within the year. The numbers on the bottom horizontal scale tell the probability that the given discharge will be equaled or exceeded in a given year. For example, a discharge associated with the value of 20 percent is interpreted to mean "a discharge of this magnitude (about 2000 m³/s) can be expected to be equaled or exceeded in 20 out of 100 years." The numbers on the top horizontal scale show the recurrence interval (or return period). This value is simply the probability percentage divided into 100. For ex-

ample, the return period for the probability of 20 percent is 100/20 = 5 years.

Note that the scale on the horizontal axis is not uniform. For this type of graph, the spacing is adjusted to follow a mathematical probability function such that the points will tend to plot on a straight line. The data for the Clearwater River follow the straight line fairly well.

As an example of using this graph, suppose that we would like to know how often a flow of 1500 m³/s is likely to occur. Reading across from the 1500 m³/s value on the vertical axis, we intersect the straight line at about 50 percent. This means that in 50 years out of 100, a flow of 1500 m³/s will be equaled or exceeded. Reading from the top of the graph, we note that the recurrence interval for this event is 2 years. For a higher flow, say 2500 m³/s, we can read about 6 percent probability for a recurrence interval of about 17 years.

How is the recurrence interval determined for a particular flood? The procedure is quite simple. First, assemble the maximum dis-

charges for each year in a list. Next, reorder the values from greatest to least, assigning a rank of 1 to the largest value, 2 to the next largest, and so on. Then determine the recurrence interval for each flow from its rank, using the formula

$$I = \frac{N+1}{R}$$

where I is the recurrence interval in years, N is the number of years in the record, and R is the rank of the value. Suppose that a flow is the largest in a set of 44 years of observations. Then its recurrence interval is $I = (N + 1)/R = (44 + 1)/1 = 45/1 = 45$ years. Suppose the flow is the fifth largest value. Its recurrence interval would then be $I = (44 + 1)/5 = 45/5 = 9$ years.

Keep in mind that the recurrence interval is only a way of expressing a probability. If a community experiences a "20-year flood" in one year, it does not mean that it will be 20 years until another flood of this magnitude comes along. There is nothing to prevent two 20-year floods from occurring in successive years. It's just not very likely.

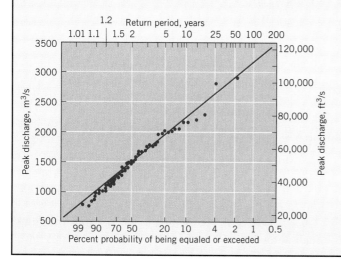

Flood frequency data for the Clearwater River at Kamaiah, Idaho. Each dot is a measured maximum yearly discharge in a 53-year record. (Data of U.S. Geological Survey from R. K. Linsley, M. A. Kohler, and J. L. Paulhus, *Hydrology for Engineers*, second edition, McGraw-Hill, New York.)

ings and vehicles in its path. In heavily forested watersheds, as in the Appalachians, the water carries little sediment load and is highly erosive. In arid western watersheds, great quantities of coarse rock debris are swept into the main channel and travel with the floodwater. Because flash floods often occur too quickly to warn affected populations, they can cause significant loss of life.

Flood Prediction

The magnitude of a flood is usually measured by the peak discharge or highest stage of a river during the period of flooding. Large floods occur less frequently than smaller ones—that is, the greater the discharge or the higher the stage, the less likely is the flood. *Working It Out 14.2 • Magnitude and Frequency of Flooding* presents a graph of flood frequency for a river and shows you

how to determine the recurrence interval for various discharges given a record of peak annual discharges.

Another tool used to present the flood history of a river is the flood expectancy graph. Figure 14.21 shows expectancy graphs for two rivers. The meaning of the bar symbols is explained in the key. The Mississippi River at Vicksburg illustrates a great river responding largely to spring floods so as to yield a simple annual cycle. The Sacramento River shows the effects of the Mediterranean climate, with its winter wet season and long severe summer drought. Winter floods are caused by torrential rainstorms and snowmelt in the mountain watersheds of the Sierra Nevada and southern Cascades. By midsummer, river flow has shrunk to a very low stage.

The National Weather Service operates a River and Flood Forecasting Service through 85 offices located at strategic points along major river systems of the United States. When a flood threatens, forecasters analyze pre-

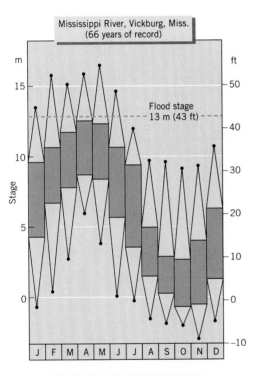

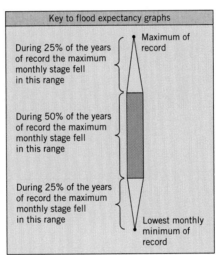

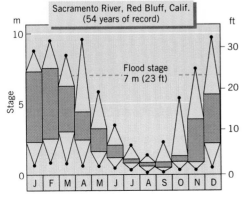

Figure 14.21 Maximum monthly stages of the Mississippi River at Vicksburg, Mississippi, and the Sacramento River at Red Bluff, California.

cipitation patterns and the progress of high waters moving downstream. Examining the flood history of the rivers and streams concerned, they develop specific flood forecasts. These are delivered to communities within the associated district, which usually covers one or more large watersheds. Flood warnings are publicized by every possible means. Close cooperation is maintained with various agencies to plan evacuation of threatened areas and the removal or protection of property.

EYE ON THE ENVIRONMENT:
The Mississippi Flood of 1993

One of the most serious natural disasters of the late twentieth century was the Mississippi flood of 1993. This extreme event occurred in midsummer, although the usual time for the flooding of the Mississippi is in spring (see Figure 14.21). The 1993 flood was triggered by a succession of unprecedented rainfalls in the upper Mississippi basin. In June of that year, rainfall totaled over 30 cm (12 in.) for large areas of the region. Southern Minnesota and western Wisconsin received even larger amounts. Monthly rainfall totals were the highest on many records dating back more than 100 years. The heavy rainfall continued into July and was concentrated in Iowa, Illinois, and Missouri. Inundated by huge volumes of rain, the vast wet landscape came to resemble a sixth Great Lake when viewed from the air.

As the water drained from the landscape, so the Mississippi and its tributaries rose, creating a flood crest of swiftly moving turbid water that worked its way slowly downstream through July and early August (Figure 14.22). At nearly every community along the length of the Mississippi and Missouri rivers, citizens fought to raise and reinforce the levees that protected their lands, homes, and businesses from the ravages of the flood. Where levees failed to keep them in check, the

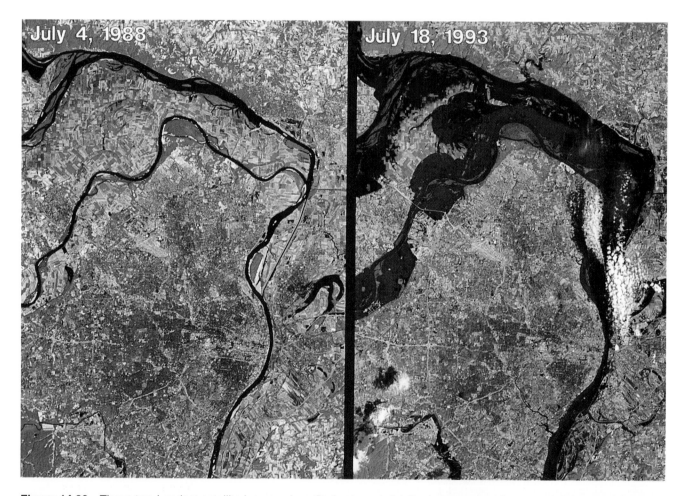

Figure 14.22   These two Landsat satellite images show St. Louis and vicinity during a normal year and during the flood of 1993. In these images, vegetation appears green, and urban areas appear in pink and purple tones. Some clouds appear in the 1993 image. The Mississippi River (*topmost*) joins the Missouri River to the north of the city. In 1993, the two rivers left their banks to spread into the bottomlands of their floodplains. The floodwaters crested at an even higher level at St. Louis on August 1, 1993.

rivers filled their broad floodplains from bluff to bluff. By the time it was over, the waters had crested at 14.8 m (49.4 ft) at St. Louis, nearly 6 m (20 ft) above flood stage.

The Mississippi River Flood of 1993 was an environmental disaster of the first magnitude. An area twice the size of New Jersey was flooded, taking 50 lives and driving nearly 70,000 people from their homes. Property damage, including loss of agricultural crops, was estimated at $12 billion. Of the region's 1400 levees, at least 800 were breached or overtopped. Some flood engineers classed the event as a 500-year flood, meaning that only once in a 500-year interval is a flood of this magnitude likely to occur. However, this does not mean that a similar flood, or an even larger one, could not occur in the near future. For the people living on its banks and bottomlands, the mighty Mississippi is a sleeping giant that can awaken with devastating consequences.

LAKES

A **lake** is a water body that has an upper surface exposed to the atmosphere and no appreciable gradient. The term *lake* includes a wide range of water bodies. Ponds (which are small, usually shallow water bodies), marshes, and swamps with standing water can all be included under the definition of a lake. Lakes receive water input from streams, overland flow, and groundwater, and so are included as parts of drainage systems. Many lakes lose water at an outlet, where water drains over a dam (natural or constructed) to become an outflowing stream. Lakes also lose water by evaporation. Lakes, like streams, are landscape features but are not usually considered to be landforms.

Lakes are quite important from the human viewpoint. They are frequently used as sources of fresh water, and they also support ecosystems that provide food for humans. Where dammed to a high level above the outlet stream, they can provide hydroelectric power as well. Lakes and ponds are also important recreation sites and sources of natural beauty.

Where lakes are not naturally present in the valley bottoms of drainage systems, we create lakes as needed by placing dams across the stream channels. Many regions that formerly had almost no natural lakes are now abundantly supplied. Some are small ponds made to serve ranches and farms, while others cover hundreds of square kilometers. In some areas, the number of artificial lakes is large enough to have significant effects on the region's hydrologic cycle.

Basins occupied by lakes show a wide range of origins as well as a vast range in dimensions. Lake basins, like stream channels, are true landforms. Basins are created by a number of geologic processes. For example, the tectonic process of crustal faulting creates many large, deep lakes. Lava flows often form a dam in a river valley, causing water to back up as a lake. Landslides suddenly create lakes, as we saw in the case of the Madison Slide (Chapter 13).

An important point about lakes in general is that they are short-lived features on the geologic time scale. Lakes disappear by one of two processes, or a combination of both. First, lakes that have stream outlets will be gradually drained as the outlets are eroded to lower levels. Where a strong bedrock threshold underlies the outlet, erosion will be slow but nevertheless certain. Second, lakes accumulate inorganic sediment carried by streams entering the lake and organic matter produced by plants within the lake. Eventually, they fill up, forming a boggy wetland with little or no free water surface.

Lakes can also disappear when climate changes. If precipitation is reduced within a region, or temperatures and net radiation increase, evaporation can exceed input and the lake will dry up. Many former lakes of the southwestern United States flourished in moister periods of glacial advance during the Ice Age. Today, they have shrunk greatly or have disappeared entirely under the present arid regime.

In moist climates, the water level of lakes and ponds coincides closely with the water table in the surrounding area. Seepage of groundwater into the lake, as well as direct runoff of precipitation, maintains these free water surfaces permanently throughout the year. Examples of such freshwater ponds are found widely distributed in glaciated regions of North America and Europe. Here, plains of glacial sand and gravel contain natural pits and hollows left by the melting of stagnant ice masses that were buried in the sand and gravel deposits (see Chapter 18). Figure 14.23 is a block diagram

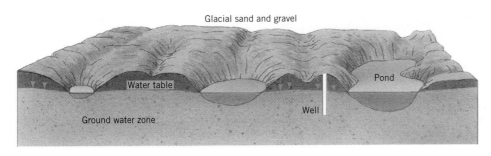

Figure 14.23 Freshwater ponds in sandy glacial deposits on Cape Cod, Massachusetts. (Redrawn from *A Geologist's View of Cape Cod*, Doubleday & Co., New York. Copyright © A. N. Strahler. Used by permission of Doubleday & Co.)

showing small freshwater ponds on Cape Cod. The surface elevation of these ponds coincides closely with the level of the surrounding water table.

Many former freshwater water table ponds have become partially or entirely filled by organic matter from the growth and decay of water-loving plants. The ultimate result is a bog with its surface close to the water table. Freshwater marshes and swamps, in which water stands at or close to the ground surface over a broad area, also represent the appearance of the water table at the surface. Such areas of poor surface drainage are included under the name of *wetlands*.

Saline Lakes and Salt Flats

Lakes with no surface outlet are characteristic of arid regions. Here, the average rate of water loss by evaporation balances the average rate of stream inflow. If the rate of inflow increases, the lake level will rise. At the same time, the lake surface will increase in area, allowing a greater rate of evaporation. A new balance can then be achieved. Similarly, if the region becomes more arid, reducing input and increasing evaporation, the water level will fall to a lower level.

Lakes without outlets often show salt buildup. Dissolved solids are brought into the lake by streams—usu-ally streams that head in distant highlands where a water surplus exists. Since evaporation removes only pure water, the salts remain behind and the salinity of the water slowly increases. Salinity, or degree of "saltiness," refers to the abundance of certain common ions in the water. Eventually, salinity levels reach a point where salts are precipitated as solids (Figure 14.24). (See the section "Evaporites" in Chapter 10.)

Sometimes the surfaces of such lakes lie below sea level. An example is the Dead Sea, with a surface elevation of –396 m (–1299 ft). The largest of all lakes, the Caspian Sea, has a surface elevation of –25 m (82 ft). Both of these large lakes are saline.

In regions where climatic conditions consistently favor evaporation over input, the lake may be absent. Instead, a shallow basin covered with salt deposits (see Figure 15.31), otherwise known as a *salt flat* or dry lake, will occur. In Chapter 15 we will describe dry lakebeds as landforms. On rare occasions, these flats are covered by a shallow layer of water, brought by flooding streams heading in adjacent highlands.

A number of desert salts are of economic value and have been profitably extracted from salt flats. In shallow coastal estuaries in the desert climate, sea salt for human consumption is commercially harvested by allowing it to evaporate in shallow basins. One well-

Figure 14.24 These salt encrustations at the edge of Great Salt Lake, Utah, were formed when the lake level dropped during a dry period.

known source of this salt is the Rann of Kutch, a coastal lowland in the tropical desert of westernmost India, close to Pakistan. Here the evaporation of shallow water of the Arabian Sea has long provided a major source of salt for inhabitants of the interior.

EYE ON THE ENVIRONMENT:
The Aral Sea—A Dying Saline Lake

The Aral Sea, located in the former Soviet republics of Kazakhstan and Uzbekistan, is a prime example of an inland saline lake that has been severely impacted by human activity. Sustained for centuries by runoff from the Syr Dar'ya and Amy Dar'ya rivers, the lake's inflow was severely limited by diversion of most of the water of the two rivers. The water was used to irrigate a vast development of cotton agriculture that began in earnest in the 1960s and continues to the present.

Once the world's fourth largest lake, the Aral Sea has lost more than two-thirds of its volume since 1960. In shrinking, its salinity has increased from 10 grams per liter to more 30. Its shoreline has receded rapidly, exposing the former lakebed, which became encrusted with salts. The vast salt-covered plain that remained has provided a source of wind-blown salt and mineral particles that continues to damage agricultural fields and natural landscapes within a wide area to the southwest of the lake. Of 24 native fish species in the lake, only 4 remain and the commercial catch of fish, once as much as 10 percent of the total catch of the Soviet Union, has been reduced to zero.

The future of the lake appears grim indeed. Reduction of inflow to near-zero values is foreseen, and the salinity of the remaining lake will be about the same as that of the ocean. Without the sacrifice of agricultural production for water to fill the lake, or the import of more water from vast distances to the north, there is little that can be done to save the lake.

DESERT IRRIGATION

Human interaction with the tropical desert environment is as old as civilization itself. Two of the earliest sites of civilization—Egypt and Mesopotamia—lie in the tropical deserts. Successful occupation of these deserts requires irrigation with large supplies of water from nondesert sources. For Egypt and Mesopotamia, the water sources of ancient times were the rivers that cross the desert but derive their flow from regions that have a water surplus. These are referred to as *exotic rivers* because their flows are derived from an outside region.

Irrigation systems in arid lands divert the discharge of an exotic river such as the Nile, Indus, Jordan, or Colorado into a distribution system that allows the water to infiltrate the soil of areas under crop cultivation. Ultimately, such irrigation projects can suffer from two undesirable side effects: salinization and waterlogging of the soil.

Salinization occurs when salts build up in the soil to levels that inhibit plant growth. This happens because an irrigated area within a desert loses large amounts of soil water through evapotranspiration. Salts contained in the irrigation water remain in the soil and increase to high concentrations. Salinization may be prevented or cured by flushing the soil salts downward to lower levels by the use of more water. This remedy requires greater water use than for crop growth alone. In addition, new drainage systems must be installed to dispose of the excess saltwater.

Waterlogging occurs when irrigation with large volumes of water causes a rise in the water table, bringing the zone of saturation close to the surface. Most food crops cannot grow in perpetually saturated soils. When the water table rises to the point at which upward movement under capillary action can bring water to the surface, evaporation is increased and salinization is intensified.

Agricultural areas of major salinization include the Indus River Valley in Pakistan, the Euphrates Valley in Syria, the Nile Delta of Egypt, and the wheat belt of western Australia. In the United States, extensive regions of heavily salinized agriculture are found in the San Joaquin and Imperial valleys of California. Other areas of salinization occur throughout the entire semiarid and arid regions of the western United States.

EYE ON THE ENVIRONMENT:
POLLUTION OF SURFACE WATER

Streams, lakes, bogs, and marshes are specialized habitats of plants and animals. Their ecosystems are particularly sensitive to changes induced by human activity in the water balance and in water chemistry. Our industrial society not only makes radical physical changes in water flow by construction of engineering works (dams, irrigation systems, canals, dredged channels), but also pollutes and contaminates our surface waters with a large variety of wastes.

The sources of water pollutants are many and varied. Some industrial plants dispose of toxic metals and organic compounds by discharging them directly into streams and lakes. Many communities still discharge untreated or partly treated sewage wastes into surface waters. In urban and suburban areas, pollutant matter entering streams and lakes includes deicing salt and lawn conditioners (lime and fertilizers). These ions can also contaminate groundwater. In agricultural regions, important sources of pollutants are fertilizers and the body wastes of livestock. Mining and processing of mineral deposits are major sources of water pollution. In addition to chemical pollution, there is thermal pollution from discharge of heated water from electric power-generating plants. The possibility of contamination by radioactive substances released from nuclear power and processing plants also exists.

Among the common chemical pollutants of both surface water and groundwater are sulfate, chloride, sodium, nitrate, phosphate, and calcium ions. (Recall from simple chemistry that an ion is the charged form of a molecule or an atom, and that many chemical compounds dissolve in water by forming ions. Chapter 19 provides more details.) Sulfate ions enter runoff both by fallout from polluted urban air and as sewage effluent. Chloride and sodium ions are contributed both by fallout from polluted air and by deicing salts used on highways. In some locations, community water supplies located close to highways have become polluted from deicing salts. Important sources of nitrate ions are fertilizers and sewage effluent. Excessive concentrations of nitrate in freshwater supplies are highly toxic, and, at the same time, their removal is difficult and expensive. Phosphate ions are contributed in part by fertilizers and by detergents in sewage effluent.

Phosphate and nitrate are plant nutrients and can lead to excessive growth of algae and other aquatic plants in streams and lakes. Applied to lakes, this process is known as *eutrophication*, which is often described as the "aging" of a lake. In eutrophication, the accumulation of nutrients stimulates plant growth, producing a large supply of dead organic matter in the lake. Microorganisms break down this organic matter but require oxygen in the process. However, oxygen dissolves only slightly in water, and so it is normally present only in low concentrations. The added burden of oxygen use by the decomposers reduces the oxygen level to the point where other organisms, such as desirable types of fish, cannot survive. After a few years of nutrient pollution, the lake can take on the characteristics of a shallow pond that results when a lake is slowly filled with sediment and organic matter over thousands of years by natural "aging" processes.

A particular form of chemical pollution of surface water goes under the name of *acid mine drainage*. It is an important form of environmental degradation in parts of Appalachia where abandoned coal mines and strip-mine workings are concentrated. Groundwater emerges from abandoned mines and as soil water percolating through strip-mine waste banks. This water contains sulfuric acid and various salts of metals, particularly of iron. Acid of this origin in stream waters can have adverse effects on animal life. In sufficient concentrations, it is lethal to certain species of fish and has at times caused massive fish kills.

Toxic metals, among them mercury, along with pesticides and a host of other industrial chemicals, are introduced into streams and lakes in quantities that are locally damaging or lethal to plant and animal communities. In addition, sewage introduces live bacteria and viruses that are classed as biological pollutants. These pose a threat to the health of humans and animals.

Thermal pollution is a term applied generally to the discharge of heat into the environment from combustion of fuels and from the conversion of nuclear energy into electric power. We have described thermal pollution of the atmosphere and its effects in Chapter 4. Thermal pollution of water is different in its environmental effects because it takes the form of heavy discharges of heated water locally into streams, estuaries, and lakes. The thermal environmental impact may thus be quite drastic in a small area.

This chapter has focused on water, including the precipitation that runs off the land and flows into the sea or accumulates in inland basins. Think for a moment about the drainage system of a stream that conducts this flow of runoff. The smallest streams catch runoff from slopes, carrying the runoff into larger streams with which they join. The larger streams, in turn, receive runoff from their side slopes and also pass their flow on to still larger streams. However, this cannot happen unless the gradients of slopes and streams are adjusted so that water keeps flowing downhill. This means that the landscape is shaped and organized into landforms that are an essential part of the drainage system. The shaping of landforms within the drainage system occurs as running water erodes the landscape, which is the subject of the next chapter.

CHAPTER SUMMARY

The fresh water of the lands accounts for only a small fraction of the earth's water. Since it is produced by precipitation over land, it depends on the continued operation of the hydrologic cycle for its existence. The soil layer plays a key role in determining the fate of precipitation by diverting it in three ways: to the atmosphere as evapotranspiration, to groundwater through percolation, and to streams and rivers as runoff.

Groundwater occupies the pore spaces in rock and regolith. The water table marks the upper surface of the saturated zone of groundwater, where pores are completely full of water. Groundwater moves in slow paths deep underground, recharging rivers, streams, ponds, and lakes by upward seepage and thus contributing to runoff. Solution of limestone by groundwater can produce limestone caverns and generate karst landscapes.

Wells draw down the water table and, in some regions, lower the water table more quickly than it can be recharged. Groundwater contamination can occur when precipitation percolates through contaminated soils or waste materials. Landfills and dumps are common sources of groundwater contaminants.

Runoff includes overland flow, moving as a sheet across the land surface, and flow in streams and rivers, which is confined to a channel. Rivers and streams are organized into a drainage network that moves runoff from slopes into channels and from smaller channels into larger ones. The discharge of a stream measures the flow rate of water moving past a given location. Discharge increases downstream as tributary streams add more runoff.

The hydrograph plots the discharge of a stream at a location through time. Since it takes time for water to move down slopes and into progressively larger stream channels, peak discharge differs from peak precipitation by a lag time. The larger the stream, the longer the lag time and the more gradual the rate of decline of discharge after the peak. Because urbanization typically involves covering ground surfaces with impervious materials, urban streams exhibit shorter lag times and higher peak discharges. Annual hydrographs of streams from humid regions show an annual cycle of base flow on which are superimposed discharge peaks related to individual rainfall episodes.

Floods occur when river discharge increases, and the flow can no longer be contained within the river's usual channel. Water spreads over the floodplain, inundating low fields and forests adjacent to the channel. When discharge is high, floodwaters can rise beyond normal levels to inundate nearby areas of development, causing damage and sometimes taking lives. Flash floods occur in small, steep watersheds and can be highly destructive. The Mississippi Flood of 1993, occurring during the summer rather than during its usual time, the spring, is an example of an extreme flood event that caused widespread destruction and damage.

Lakes are especially important parts of the drainage system because they are sources of fresh water. They are also used for recreation and, in many cases, can supply hydroelectric power. Where lakes occur in inland basins, they are often saline. When climate changes, such lakes can dry up, creating salt flats.

Irrigation is the diversion of fresh water from streams and rivers to supply the water needs of crops. In desert regions, where irrigation is most needed, problems of salinization and waterlogging can occur, reducing productivity and eventually creating unusable land.

Water pollution arises from many sources, including industrial sites, sewage treatment plants, agricultural activities, mining, and processing of mineral deposits. Sulfate, nitrate, phosphate, chloride, sodium, and calcium ions are frequent contaminants. Toxic metals, pesticides, and industrial chemicals are also hazards.

KEY TERMS

infiltration
runoff
overland flow
groundwater

water table
stream
channel
discharge

drainage system
drainage basin
flood
lake

REVIEW QUESTIONS

1. What happens to precipitation falling on soil? What processes are involved?

2. How and under what conditions does precipitation reach groundwater?

3. How and where does groundwater flow? Sketch a diagram showing the flow paths of groundwater.

4. How do caverns come to be formed in limestone? Describe the key features of a karst landscape.

5. How do wells affect the water table? What happens when pumping exceeds recharge?

6. What is land subsidence and how does it occur? Provide an example.

7. How is groundwater contaminated? Describe how a well might become contaminated by a nearby landfill dump.

8. Define discharge (of a stream) and the two quantities that determine it. How does discharge vary in a downstream direction? How does gradient vary in a downstream direction?

9. What is a drainage system? How are slopes and streams arranged in a drainage basin?

10. Sketch a hydrograph for a stream showing a peak discharge in response to a rainfall event. Indicate base flow.

11. Define the term *flood*. What is the floodplain? What factors are used in forecasting floods?

12. Why was the Mississippi Flood of 1993 so unusual?

13. How are lakes defined? What are some of their characteristics? What factors influence the size of lakes?

14. Describe some of the problems that can arise in long-continued irrigation of desert areas.

15. Identify common surface-water pollutants and their sources.

Focus on Systems 14.1 • Energy in Stream Flow

1. How are the cross-sectional area and velocity of a stream related to the flow?

2. How do cross-sectional area, velocity, and slope change in pools and rapids?

3. What two types of flow occur in a stream, and how do they differ?

4. How does the energy dissipated by stream flow differ in pools and rapids?

ESSAY QUESTIONS

1. A thundershower causes heavy rain to fall in a small region near the headwaters of a major river system. Describe the flow paths of that water as it returns to the atmosphere and ocean. What human activities influence the flows? In what ways?

2. Imagine yourself a recently elected mayor of a small city located on the banks of a large river. What issues might you be concerned with that involve the river? In developing your answer, choose and specify some characteristics for this city—such as its population, its industries, its sewage systems, and the present uses of the river for water supply or recreation.

PROBLEMS

Working It Out 14.2 • Magnitude and Frequency of Flooding

1. The data below are peak flows for the West River, near Newfane, Vermont, for 33 years (1929–1961). (Following 1961, the river was controlled by a reservoir upstream.) Rank the data from greatest to least and determine the recurrence interval for each flow using the formula. If two values are tied, use the average of the two ranks in determining the recurrence interval for the flow. What are the magnitudes of floods with recurrence intervals closest to 1, 2, 5, 10, and 35 years? (*Hint:* Many word processing programs and spreadsheets have the ability to sort data and thus can be used to order the list quickly and easily.)

Year	Flow (m^3/s)	Year	Flow (m^3/s)	Year	Flow (m^3/s)
1929	234	1940	292	1951	312
1930	113	1941	119	1952	309
1931	217	1942	229	1953	337
1932	184	1943	234	1954	139
1933	245	1944	266	1955	279
1934	209	1945	210	1956	337
1935	225	1946	181	1957	164
1936	677	1947	357	1958	264
1937	216	1948	419	1959	172
1938	487	1949	711	1960	351
1939	193	1950	278	1961	194

(Data from U.S. Geological Survey.)

2. Note that the percent probability that a flow will be equaled or exceeded is simply $100/I$, where I is the recurrence interval. What is the percent probability that a flood flow in any one year will exceed 250 m^3/s? 500 m^3/s? What flow will be equaled or exceeded in 25 percent of all years?

Chapter 15

Fluvial Processes and Landforms

Running water is a powerful land-forming agent. Flowing as a sheet across a land surface, running water picks up particles and moves them downslope into a stream channel. When rainfall is heavy, streams and rivers swell, lifting large volumes of sediment and carrying them downstream. In this way, running water erodes mountains and hills, carves valleys, and deposits sediment. This chapter describes the work of running water and the landforms it shapes.

Running water is one of four flowing substances that erode, transport, and deposit mineral and organic matter. The other three are waves, glacial ice, and wind. These four fluid agents carry out the processes of *denudation*, which we discussed in Chapter 12. Recall that denudation is the total action of all processes by which the exposed rocks of the continents are worn away and the resulting sediments are transported to the sea or closed inland basins.

Denudation is an overall lowering of the land surface. Denudation processes, if left unchecked to operate over geologic time, will reduce a continent to a nearly featureless, sea-level surface. However, that fate has always been averted, since plate tectonic activity has kept continental crust elevated well above the ocean basins. The result is that running water, waves, glacial ice, and wind have always had plenty of raw material available to create the many landforms that we see around us.

Thanks to denudation processes and plate tectonics, the land environments of life have always been in constant change, even as plants and animals have undergone their evolutionary development. Wind, water, waves, and ice have produced, maintained, and

---

Sinuous meanders of a Tambopata River tributary, Peru.

changed a wide variety of landforms, and these have been the habitats for evolving life-forms. In turn, the life-forms have become adapted to those habitats and have diversified to a degree that matches the diversity of the landforms themselves.

FLUVIAL PROCESSES AND LANDFORMS

Landforms shaped by running water are conveniently described as **fluvial landforms** in order to distinguish them from landforms made by the other flowing agents—glacial ice, wind, and waves. Fluvial landforms are shaped by the **fluvial processes** of overland flow and stream flow, which we described in Chapter 14. Weathering and the slower forms of mass wasting (Chapter 13), such as soil creep, operate hand in hand with overland flow, providing the rock and mineral fragments that are carried into stream systems.

Fluvial landforms and fluvial processes dominate the continental land surfaces the world over. Throughout geologic history, glacial ice has been present only in continental areas located in mid- and high latitudes and in high mountains. Landforms made by wind action are found only in very small parts of the continental surfaces. And landforms made by waves and currents are restricted to the narrow contact zone between oceans and continents. That is why, in terms of area, the fluvial landforms dominate the environment of terrestrial life.

Areas of fluvial landforms are also the major source areas of human food resources through the practice of agriculture. Except for areas in the northern hemisphere that were formerly occupied by glacial ice, most land areas used in crop cultivation or for grazing have been shaped by fluvial processes.

Erosional and Depositional Landforms

All agents of denudation perform the geological activities of erosion, transportation, and deposition. Consequently, there are two major groups of landforms—erosional landforms and depositional landforms. When an initial landform, such as an uplifted crustal block, is created, it is attacked by the processes of denudation and especially by fluvial action. Valleys are formed where rock is eroded away by fluvial agents. Between the valleys are ridges, hills, or mountain summits, representing the remaining parts of the crustal block that is as yet uncarved by running water. These sequential landforms, shaped by progressive removal of the bedrock mass, are *erosional landforms.*

Fragments of soil, regolith, and bedrock that are removed from the parent rock mass are transported by the fluid agent and deposited elsewhere to make an entirely different set of surface features—the depositional landforms. Figure 15.1 illustrates the two groups of

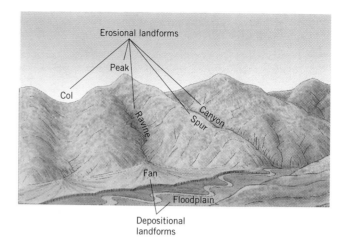

Figure 15.1 Erosional and depositional landforms. (A. N. Strahler)

landforms as produced by fluvial processes. The ravine, canyon, peak, spur, and col are erosional landforms. The fan, built of rock fragments below the mouth of the ravine, is a depositional landform. The floodplain, built of material transported by a stream, is also a *depositional landform.*

SLOPE EROSION

Fluvial action starts on the uplands as *soil erosion.* By exerting a dragging force over the soil surface, overland flow picks up particles of mineral matter ranging in size from fine colloidal clay to coarse sand or even gravel. The size grade selected depends on the speed of the current and the degree to which the particles are bound by plant rootlets or held down by a mat of leaves. Added to this solid matter is dissolved mineral matter in the form of ions produced by acid reactions or direct solution.

This ongoing removal of soil is part of the natural geological process of denudation. It occurs everywhere that precipitation falls on land. Under stable natural conditions in a humid climate, the erosion rate is slow enough that a soil with distinct horizons is formed and maintained. Each year a small amount of soil is washed away, while a small amount of solid rock material becomes altered to new regolith and soil. These conditions also enable plant communities to maintain themselves in a stable equilibrium. Soil scientists refer to this state of activity as the *geologic norm.*

EYE ON THE ENVIRONMENT:
Accelerated Erosion

In contrast, the rate of soil erosion may be enormously speeded up by human activities or by rare natural erosional events to produce a state of *accelerated erosion.*

What happens then is that the soil is removed much faster than it can be formed, and the uppermost soil horizons are progressively exposed. Accelerated erosion arises most commonly when the plant cover and the physical state of the ground surface change. Destruction of vegetation by the clearing of land for cultivation or by forest fires sets the stage for a series of drastic changes. No foliage remains to intercept rain, and the protection of a ground cover of fallen leaves and stems is removed. Consequently, the raindrops fall directly on the mineral soil.

The direct force of falling drops on bare soil causes a geyserlike splashing in which soil particles are lifted and then dropped into new positions. This process is termed *splash erosion* (Figure 15.2). Soil scientists estimate that a torrential rainstorm has the ability to disturb as much as 225 metric tons of soil per hectare (about 100 U.S. tons per acre). On a sloping ground surface, splash erosion shifts the soil slowly downhill. An even more important effect is to cause the soil surface to become much less able to absorb water. This occurs because the natural soil openings become sealed by particles shifted by raindrop splash. Reduced infiltration, in turn, permits a much greater depth of overland flow to occur from a given amount of rain. So, the rate of soil erosion is intensified.

Another effect of the destruction of vegetation is to reduce greatly the resistance of the ground surface to the force of erosion under overland flow. On a slope covered by grass sod, even a deep layer of overland flow causes little soil erosion. This protective action is present because the energy of the moving water is dissipated in friction with the grass stems, which are tough and elastic. On a heavily forested slope, the surface layer of leaves, twigs, roots, and even fallen tree trunks take up the force of overland flow. Without such a cover, the eroding force is applied directly to the bare soil surface, easily dislodging the grains and sweeping them downslope.

We can get a good appreciation of the contrast between normal and accelerated erosion rates by comparing the quantity of sediment derived from cultivated surfaces with that derived from naturally forested or reforested surfaces. The comparison is made within a single region in which climate, soil, and topography are fairly uniform. *Sediment yield* is a technical term for the quantity of sediment removed by overland flow from a unit area of ground surface in a given unit of time. Yearly sediment yield is stated in metric tons per hectare, or tons per acre.

Figure 15.3 presents annual average sediment yield and runoff by overland flow from several types of up-

Figure 15.2 A large raindrop (*above*) lands on a wet soil surface, producing a miniature crater (*below*). Grains of clay and silt are thrown into the air, and the soil surface is disturbed.

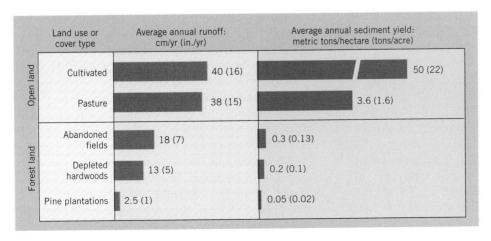

Figure 15.3 This bar graph shows that both runoff and sediment yield are much greater for open land than for land covered by shrubs and forest. (Data of S. J. Ursic, U.S. Department of Agriculture)

Figure 15.4 Deep branching gullies have carved up an overgrazed pasture near Shawnee, Oklahoma. Contour terracing and check dams have halted the headward growth of the gullies.

land surfaces in northern Mississippi. Notice that both surface runoff and sediment yield decrease greatly with increased effectiveness of the protective vegetative cover. Sediment yield from cultivated land undergoing accelerated erosion is over ten times greater than that from pasture and about one thousand times greater than that from pine plantation land. 🍂

Sheet Erosion and Rilling

Accelerated soil erosion is a constant problem in cultivated regions with a substantial water surplus. When the natural cover of forest or prairie grasslands is first removed and the soil is plowed for cultivation, little erosion will occur until the action of rain splash has broken down the soil aggregates and sealed the larger openings. Then, however, overland flow begins to remove the soil in rather uniform thin layers, a process termed *sheet erosion*. Because of seasonal cultivation, the effects of sheet erosion are often little noticed until the upper horizons of the soil are removed or greatly thinned.

Where land slopes are steep, runoff from torrential rains produces a more destructive activity, *rill erosion*, in which many closely spaced channels are scored into the soil and regolith. If the rills are not destroyed by soil tillage, they may soon begin to join together into still larger channels. These deepen rapidly and soon become *gullies*—steep-walled, canyonlike trenches whose upper ends grow progressively upslope (Figure 15.4). Ultimately, a rugged, barren topography results from accelerated soil erosion that is allowed to proceed unchecked.

Colluvium and Alluvium

With an increased depth of overland flow and no vegetation cover to absorb the eroding force, soil particles are easily picked up and moved downslope. Eventually, they reach the base of the slope, where the surface slope becomes more gentle and meets the valley bottom. There the particles come to rest and accumulate in a thickening layer termed *colluvium*. Because this deposit is built by overland flow, it has a sheetlike distribution and may be little noticed, except where it eventually buries fence posts or tree trunks.

If not deposited as colluvium, sediment carried by overland flow eventually reaches a stream in the adjacent valley floor. Once in the stream, it is carried farther downvalley and may accumulate as alluvium in layers on the valley floor. The term **alluvium** is used to describe any stream-laid sediment deposit. In a region of accelerated erosion, deposition of alluvium can bury fertile floodplain soil under infertile, sandy layers.

Coarse alluvium chokes the channels of small streams and can cause the water to flood broadly over the valley bottoms.

Slope Erosion in Semiarid and Arid Environments

Thus far, we have discussed slope erosion in moist climates with a natural vegetation of forest or a dense prairie grassland. Conditions are quite different in a midlatitude semiarid climate with summer drought. Here, the natural plant cover consists of short-grass prairie (steppe). Although it is sparse and provides a rather poor ground cover of plant litter, the grass cover is normally strong enough that a slow pace of erosion can be sustained. Much the same conditions are also found in the tropical savanna grasslands.

In these semiarid environments, however, the natural equilibrium is highly sensitive and easily upset. Depletion of the plant cover by fires or the grazing of herds of domesticated animals can easily set off rapid erosion. These sensitive, marginal environments require cautious use, because they lack the potential to recover rapidly from accelerated erosion once it has begun.

Erosion at a very high rate by overland flow is actually a natural process in certain favorable locations in semiarid and arid lands. Here, the erosion produces *badlands*. Badlands are underlain by clay formations, which are easily eroded by overland flow. Erosion rates are too fast to permit plants to take hold, and no soil can develop. A maze of small stream channels is developed, and ground slopes are very steep (Figure 15.5).

One well-known area of badlands in the semiarid short-grass prairie is the Big Badlands of South Dakota, along the White River. Badlands such as these are self-sustaining and have been in existence on continents throughout much of geologic time. Badlands can also result from poor agricultural processes, especially when the vegetation cover of clay formations is disturbed by plowing or overgrazing.

THE WORK OF STREAMS

The work of streams consists of three closely related activities—erosion, transportation, and deposition. **Stream erosion** is the progressive removal of mineral material from the floor and sides of the channel, whether bedrock or regolith. **Stream transportation** consists of movement of the eroded particles dragged over the streambed, suspended in the body of the stream, or held in solution as ions. **Stream deposition** is the accumulation of transported particles on the streambed and floodplain, or on the floor of a standing body of water into which the stream empties. Erosion cannot occur without some transportation taking place, and the transported particles must eventually come to rest. Thus, erosion, transportation, and deposition are simply three phases of a single activity.

Stream Erosion

Streams erode in various ways, depending on the nature of the channel materials and the tools with which the current is armed. The force of the flowing water not only sets up a dragging action on the bed and banks, but also causes particles to impact the bed and banks. Dragging and impact can easily erode alluvial materials, such as gravel, sand, silt, and clay. This form

Figure 15.5 Badlands at Zabriskie Point, Death Valley National Monument, California.

Figure 15.6 These potholes in lava bedrock attest to the abrasion that takes place on the bed of a swift mountain stream. McCloud River, California.

smaller grains to produce a wide assortment of grain sizes. This process of mechanical wear is called *abrasion*. It is the principal means of erosion in bedrock that is too strong to be affected by simple hydraulic action. A striking example of abrasion is the erosion of a *pothole*. This occurs when a shallow depression in the bedrock of a streambed acquires one or several grinding stones, which are spun around and around by the flowing water and carve a deep depression (Figure 15.6).

Finally, the chemical processes of rock weathering—acid reactions and solution—are effective in removing rock from the stream channel. The process is called *corrosion*. Effects of corrosion are conspicuous in limestone, which develops cupped and fluted surfaces.

Stream Transportation

The solid matter carried by a stream is the **stream load**. It is carried in three forms (Figure 15.7). Dissolved matter is transported invisibly in the form of chemical ions. All streams carry some dissolved ions resulting from mineral alteration. Sand, gravel, and larger particles move as *bed load* close to the channel floor by rolling or sliding. Clay and silt are carried in *suspension*—that is, they are held up in the water by the upward elements of flow in turbulent eddies in the stream. This fraction of the transported matter is the *suspended load* (Figure 15.8). Of the three forms, suspended load is generally the largest.

A large river such as the Mississippi carries as much as 90 percent of its load in suspension. Most of the suspended load comes from its great western tributary, the Missouri River, which is fed from semiarid lands, including the Dakota Badlands. (Figure 14.16 shows the Mississippi River system and its major branches.)

of erosion, called *hydraulic action*, can excavate enormous quantities in a short time. The undermining of the banks causes large masses of alluvium to slump into the river, where the particles are quickly separated and become part of the stream's load. This process of bank caving is an important source of sediment during high river stages and floods.

Where rock particles carried by the swift current strike against bedrock channel walls, chips of rock are detached. The large, strong fragments become rounded as they travel. The rolling of cobbles and boulders over the streambed further crushes and grinds the

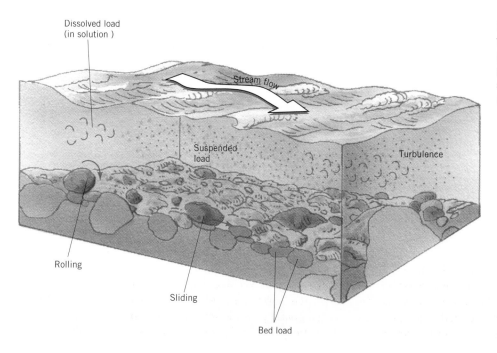

Figure 15.7 Streams carry their load as dissolved, suspended, and bed load. Suspended load is kept in suspension by turbulence. Bed load moves by sliding or rolling.

Figure 15.8 This turbulent stream in Madagascar carries a heavy load of suspended sediment. Gullying erodes the denuded hills in the foreground.

The Yellow (Huang) River of China heads the world list in annual suspended sediment load, and its watershed sediment yield is one of the highest known for a large river basin. This is because much of its basin consists of cultivated upland surfaces of wind-deposited silt that is very easily eroded. (See Chapter 17 and Figure 17.31.) In addition, much of the river's upper watershed is in a semiarid climate with dry winters. Vegetation is sparse, and the runoff from heavy summer rains sweeps up a large amount of sediment.

Capacity of a Stream to Transport Load

The maximum solid load of debris that can be carried by a stream at a given discharge is a measure of the *stream capacity*. This load is usually measured in units of metric tons per day passing downstream at a given location. Total solid load includes both bed load and suspended load.

A stream's capacity to carry suspended load increases sharply with an increase in the stream's velocity, because the swifter the current, the more intense is the turbulence. The capacity to move bed load also increases with velocity—the faster water motion produces

a stronger dragging force against the bed. In fact, the capacity to move bed load increases according to the third to fourth power of the velocity. In other words, when a stream's velocity is doubled in flood, its ability to transport bed load is increased from eight to sixteen times. Thus, most of the conspicuous changes in the channel of a stream occur in a flood stage. *Working It Out 15.1 • River Discharge and Suspended Sediment* shows how the suspended sediment load of a stream increases with discharge.

When water flow increases, a stream in a channel flowing over thick layers of silt, sand, and gravel will easily widen and deepen its channel. When the flow slackens, the stream will deposit material in the bed, filling the channel again. Where a stream flows in a channel of hard bedrock, the channel cannot be quickly deepened in response to rising waters and may change very little during a single flood. Such conditions exist in streams that occupy deep canyons and have steep gradients.

STREAM GRADATION

A main stream, fully developed within its drainage basin, has normally undergone thousands of years of adjustment of its channel. It can discharge not only the surplus runoff produced by the basin, but also the solid load that the tributary channels supply. This transport capability is related to the gradient of the channel (discussed in Chapter 14). The steeper the gradient, the higher the stream velocity and the greater the ability of the stream to carry sediment. The gradient of a stream adjusts over time to achieve an average balanced state of operation, year in and year out and from decade to decade. In this equilibrium condition, the stream is referred to as a **graded stream**.

To develop the concept of a graded stream system, we will investigate the changes that take place along a stretch of stream that is initially very poorly adjusted to the transport of its load. Such an ungraded channel is illustrated by an example in Figure 15.9. The side of the block shows a series of *stream profiles*—plots of elevation of the stream with distance from the sea. The starting profile (1) is produced by crustal uplift in a series of fault steps, bringing to view a land surface that was formerly beneath the ocean and exposing it to fluvial processes for the first time. Overland flow collects in shallow depressions, which fill and overflow from higher to lower levels. In this way, a through-flowing channel originates and begins to conduct runoff to the sea. As time passes, the landscape is slowly eroded by fluvial action, and the stream profile is smoothed out into a uniform curve, shown as profile line 3. The profile has now been graded. From that point on into time, this *graded profile* is steadily lowered in elevation as the landscape is further eroded (curves 4 through 6).

Working It Out 15.1 • River Discharge and Suspended Sediment

As you might imagine, a river carries more sediment in times of flood than during periods of normal flow. This happens for a number of reasons.

- First, as the discharge increases, more water is available to carry sediment downstream.
- Second, as discharge increases, so does the velocity of the stream flow. And as the velocity of the water flow increases, so does the intensity of the turbulence that keeps suspended sediment in motion. Thus, more sediment can be carried in a cubic meter of swiftly moving water.
- Third, as the water rises and increases in velocity, the force of the stream flow on the bed increases. The stream then erodes sediment in its bed, scouring and deepening the channel, a process that also increases the sediment available for transportation.
- Fourth, the heavy precipitation that causes a flood generates overland flow that drags more sediment into upstream river channels. Thus, more sediment is available for transport.

The figure to the right is a graph plotting discharge against the suspended sediment load for the Powder River at Arvada, Wyoming. The discharge is measured in cubic meters per second (m³/s), and the suspended sediment load in metric tons per day (t/d). The points are individual measurements collected over a period of time at a single location. The graph easily shows that as discharge increases, so does suspended sediment. For example, when the discharge goes from 1 to 10 m³/s, the suspended sediment load increases from about 200 to 10,000 t/d. Thus, when discharge increases by a factor of 10, sediment load increases by a factor of about 50.

This type of increase is exponential—that is, it fits a function of the form

$$y = ax^b$$

where $b > 1$. (*Working It Out 3.3 • Exponential Growth* described exponential functions in more detail.) On normal (arithmetic) graph paper, this function plots as an ascending curve. On a log-log plot, however, it is a straight line. In the log-log plot, the scale positions a value based on the logarithm of the value rather than the value itself. (Note that the logarithm used here is the common logarithm, which uses the base 10.) The right and upper axes on the graph demonstrate this point. They show the logarithm of the values changing on a uniform scale from –2 to +2 for discharge, and from 0 to 6 for suspended sediment load, while the log axes change by powers of 10.

This formulation works because, if we take the log of both sides of the exponential function $y = ax^b$, we have

$$\log y = \log a + b \log x$$

By substituting y' for $\log y$, x' for $\log x$, and a' for $\log a$, we have

$$y' = a' + bx'$$

which is the equation for a straight line. Thus, by plotting the logs of x and y values, rather than the values themselves, we get a straight line. Using the log scale does this conversion automatically.

In the case of the relation between suspended sediment load and discharge for the Powder River at Arvada, Wyoming, the measurements come close to fitting a line with the equation

$$S = 178Q^{1.75}$$

where S is the suspended sediment load (t/d) and Q is the discharge (m³/s). This can also be written in log-log form, as a straight-line function:

$$\log S = \log(178) + 1.75 \log Q$$
$$= 2.25 + 1.75 \log Q$$

The exponential increase in suspended load with increasing dis-

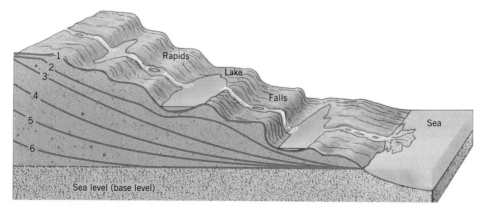

Figure 15.9 Schematic diagram of gradation of a stream. Originally, the channel consists of a succession of lakes, falls, and rapids. (Copyright © A. N. Strahler.)

charge is characteristic of nearly all rivers. In fact, the rate of increase for many rivers is actually larger than for the Powder River at Arvada—values of 2–3 are typical for the power to which Q is raised.

The important point to carry home from this examination of the behavior of rivers is that transportation of suspended sediment increases very rapidly with discharge. Therefore, most of the transportation work of rivers is carried out in times of high flow. Many times more sediment may be carried downstream during a single small flood than over many months of normal river flow. And a large flood may move vast volumes of sediment that has been accumulating in the channel and floodplain in wait for a rare and extreme event.

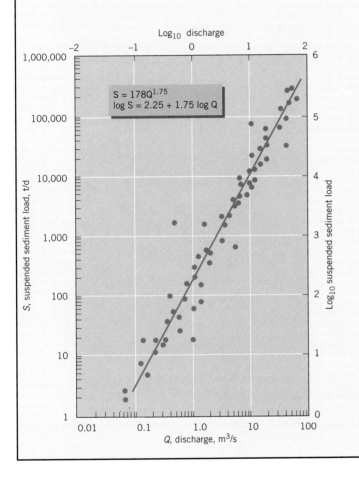

Increase of suspended sediment load with increase in discharge plotted on logarithmic scales. The dots are individual observations scattered about the line of best fit. Data are for the Powder River at Arvada, Wyoming. (Data of U.S. Geological Survey.)

Landscape Evolution of a Graded Stream

Early Stages

Figure 15.10 illustrates the stream gradation process in a series of block diagrams. In (*a*), we see *waterfalls* and *rapids*, which are simply portions of the channel with steep gradients. Flow velocity at these points is greatly increased, and abrasion of bedrock is therefore most intense. As a result, the falls are cut back and the rapids are trenched. At the same time, the ponded stretches of the stream are first filled by sediment and are later lowered in level as the lake outlets are cut down. In time, the lakes disappear and the falls are transformed into rapids.

In the early stages of gradation and tributary extension, the capacity of the stream exceeds the load supplied to it, so that little or no alluvium accumulates in the channels. Abrasion continues to deepen the major channels, with the result that they come to occupy steep-walled *gorges* or *canyons* (Figure 15.11). Weathering and mass wasting of these rock walls contribute an increasing supply of rock debris to the channels. Also on the increase is debris shed from land surfaces that contribute overland flow to the newly developed branches.

(a)

(b)

(c)

(d)

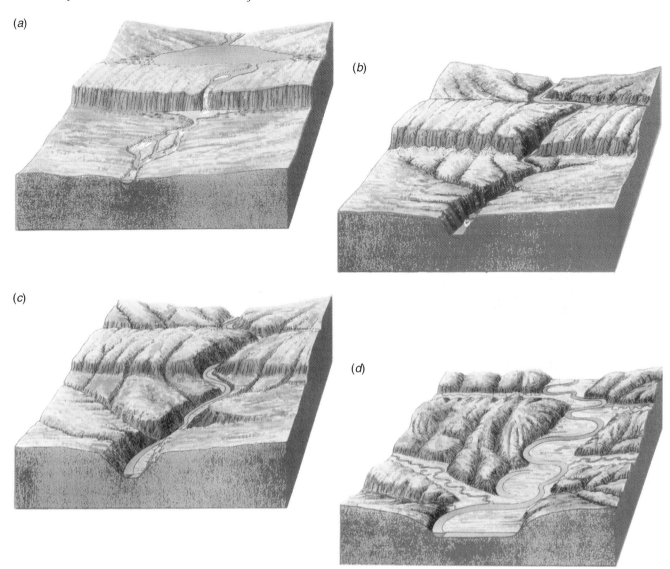

Figure 15.10 Evolution of a stream and its valley. (Drawn by E. Raisz. Copyright © A. N. Strahler.) (a) Stream established on a land surface dominated by landforms of recent tectonic activity. (b) Gradation in progress. The lakes and marshes drained. The gorge is deepening, and tributary valleys are extending.(c) Graded profile attained. Floodplain development is beginning, and the widening of the valley is in progress.(d) Floodplain widened to accommodate meanders. Floodplains now extend up tributary valleys.

The erosion of rapids reduces the gradient to a slope angle that more closely approximates the average gradient of that section of the stream (Figure 15.10b). At the same time, branches of the main stream are being extended into higher parts of the original landmass. These carve out many new small drainage basins. Thus, the original tectonic landscape is transformed into a complete fluvial landform system.

Attaining a Graded Condition

As landscape change continues, a gradual decrease in a stream's capacity to move bed load results from the gradual reduction in the channel gradient. Simultane-

ously, the load supplied to the stream from the entire upstream area is on the increase. So, a time will come when the supply of load exactly matches the stream's capacity to transport it. Figure 15.10c shows the landscape at this point in time. Now, all the major streams have achieved the graded condition and possess graded profiles that descend smoothly and uniformly.

The first indication that a stream has attained a graded condition is the beginning of floodplain development. On the outside of each bend, the channel shifts laterally (sidewise) to make a curve of larger radius (Figure 15.12). This forced change requires the stream to undercut and steepen the valley wall. On the inside of the bend, alluvium accumulates as a long,

Figure 15.11 The Grand Canyon of the Yellowstone River has been carved in volcanic rock (rhyolite) of the Yellowstone Plateau. Yellowstone National Park, Wyoming.

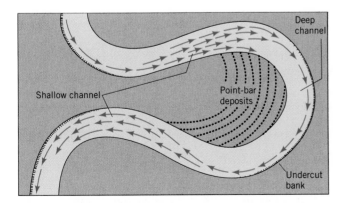

Figure 15.12 Idealized map and cross-profile of a meander bend of a large alluvial river, such as the lower Mississippi. Arrows show the position of the swiftest current. (Copyright © A. N. Strahler.)

curving deposit of sediment—termed a *point bar*. Widening of the bar deposit produces a crescent-shaped area of low ground, which is the first stage in floodplain development. This stage is illustrated in Figure 15.10*c*. As lateral cutting continues, the floodplain strips are widened, and the channel develops sweeping bends. These winding river bends are called **alluvial meanders** (*d*). In this way, the floodplain is widened into a continuous belt of flat land between steep valley walls. Figure 15.13 illustrates these changes.

Floodplain development reduces the frequency with which the river attacks and undermines the adjacent valley wall. Weathering, mass wasting, and overland flow can then act to reduce the steepness of the valley-side slopes (Figure 15.13). As a result, in a humid climate, the gorgelike aspect of the valley gradually disap-

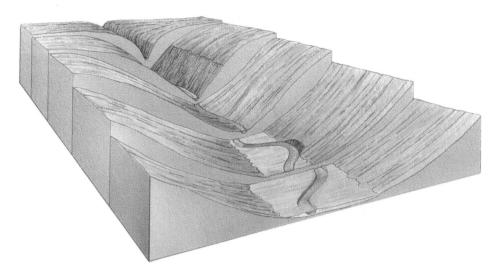

Figure 15.13 Following stream gradation, the valley walls become gentler in slope, and the bedrock is covered by soil and weathered rock. (After W. M. Davis. Copyright © A. N. Strahler.)

pears and eventually gives way to an open valley with soil-covered slopes protected by a dense plant cover. The time required for the stream to reach a graded condition and erode a broad valley will be a few tens of millions of years.

The graded stream network has the property that the gradient and channel form of each portion of the stream network is adjusted to carry the sediment load it receives. The gradient and channel form varies, however, from the small streams at the headwaters of the network to the broad river channels found farther downstream.

One way to study the properties of streams is to organize them by stream order, in which stream segments are ordered from those draining the smallest of slopes to those formed far downstream from the progressive joining of many upstream segments. Figure 15.14 illustratestime concept. The smallest tributaries are first–order streams, which receive overland flow from within their first-order basin. When two first–order streams join, the result is a second–order stream, and so forth. As you might well imagine, many stream properties increase with stream order. Two principal ones are drainage area and discharge. We will encounter others in chapter 16.

Great Waterfalls

Although small, high waterfalls are common features of alpine mountains carved by glacial erosion (Chapter 18), large waterfalls on major rivers are comparatively rare the world over. Faulting and dislocation of large crustal blocks have caused spectacular waterfalls on several east African rivers (Figure 15.15). As we explained in Chapter 12, this is the Rift Valley region, where block faulting has been taking place.

Another class of large waterfalls involves new river channels resulting from glacial activity in the Pleistocene Epoch. Erosion and deposition by large moving ice sheets greatly disrupted drainage patterns in north-

ern continental regions, creating lakes and causing river courses to be shifted to new locations. Niagara Falls is a good example. The outlet from Lake Erie into Lake Ontario, the Niagara River, is situated over a gently inclined layer of limestone, beneath which lies easily eroded shale (Figure 15.16). The fall is maintained by continual undermining of the limestone by erosion of the less resistant shale at the base of the fall (Figure 15.17). The height of the falls is now 52 m (171

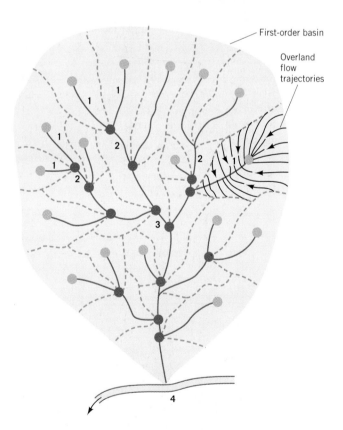

Figure 15.14 Schematic map of a third–order drainage basin showing both the stream channel system and the drainage basin network. (Copyright © 1996 by A. N. Strahler.)

Figure 15.15 Victoria Falls is located on the Zambesi River, which forms the border between Zambia and Zimbabwe in southern Africa. The water falls 128 m (420 ft) to the bottom of a deep, narrow gorge excavated along a fault line.

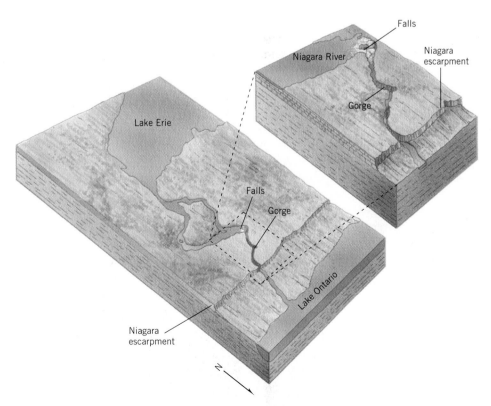

Figure 15.16 A bird's-eye view of the Niagara River with its falls and gorge carved in strata of the Niagara Escarpment. The view is toward the southwest from a point over Lake Ontario. (After a sketch by G. K. Gilbert. Redrawn from A. N. Strahler. Copyright © A. N. Strahler.)

Limestone

Shale

Figure 15.17 Niagara Falls is formed where the river passes over the eroded edge of a massive limestone layer. Continual undermining of weak shales at the base keeps the fall steep. (Drawn by E. Raisz. Copyright © A. N. Strahler.)

ft), and its discharge is about 17,000 m³ per second (about 200,000 ft³/s). The drop of Niagara Falls is utilized for the production of hydroelectric power by the Niagara Power Project. Water is withdrawn upstream from the falls and is carried in tunnels to generating plants located about 6 km (4 mi) downstream from the falls.

EYE ON THE ENVIRONMENT:
Dams and Resources

Because most large rivers of steep gradient do not have falls, dams are necessary to create the vertical drop required to spin the turbines of electric power generators. Hydroelectric power is cheap, nonpolluting, and renewable, and large dams provide fresh water for urban use and irrigation. However, lakes behind dams drown river valleys and can inundate the gorges, rapids, and waterfalls of major rivers. Scenic and recreational resources such as these can be lost. White-water boating and rafting, for example, are now popular sports on our wild rivers. The Grand Canyon of the Colorado River, probably more than any single product of fluvial processes, demonstrates the scenic and recreational value of a great river gorge.

Dam construction also destroys ecosystems adapted to the river environment. In addition, deposition of sediment behind the dam rapidly reduces the holding capacity of the lake and within a century or so may fill the lake basin. This reduces the ability of the lake to provide consistent supplies of water and hydroelectric power. It is small wonder, then, that new dam projects can meet with stiff opposition from concerned local citizens' groups and national environmental organizations. 🐾

Evolution of a Fluvial Landscape

With the concept of a graded stream in mind, this is a good place to examine some of the broader aspects of fluvial denudation. Just as a stream undergoes a life cycle of change, an entire landmass has its stages of evolution.

Consider a landscape made up of many drainage basins and their branching stream networks. The region is rugged, with steep mountainsides and high, narrow crests (Figure 15.18a). The total volume of rock lying above sea level and available for removal by fluvial action is called the *landmass*. Rock debris is transported out of each drainage basin by the main graded streams at the same average rate as the debris is being contributed from the land surfaces within the basin. Eventually, the export of debris lowers the land surface generally, and the average altitude of the land surface steadily declines. This decline must be accompanied by a reduction in the average gradients of all streams (b).

As time passes, the streams and valley-side slopes of the drainage basins undergo gradual change to lower gradients. In theory, the ultimate goal of the denudation process is to reduce the landmass to a featureless plain at sea level. In this process, a sea-level surface imagined to lie beneath the entire landmass represents the lower limiting level, or *base level*, of the fluvial denudation (labeled in Figures 15.9 and 15.18). But because the rate of denudation becomes progressively slower, the land surface approaches the base-level surface of zero elevation at a slower and slower pace. Under this scenario, the ultimate goal can never be reached. Instead, after the passage of some millions of years, the land surface is reduced to a gently rolling surface of low elevation, called a **peneplain** (c). You can think of this strange term as meaning an "almost-plain."

Production of a peneplain requires a high degree of crustal and sea-level stability for a period of many millions of years. One region that has been cited as a possible example of a contemporary peneplain is the Amazon–Orinoco basin of South America. This vast region is a stable continental shield of ancient rock.

Uplifted peneplains that are now high-standing land surfaces are numerous within the continental shields. These regions are characterized by an upland surface of nearly uniform elevation. The upland is trenched by stream valleys graded with respect to a lower base level.

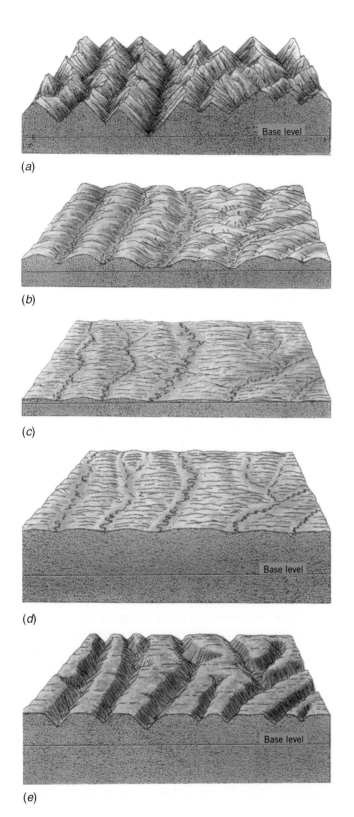

(a)

(b)

(c)

(d)

(e)

Figure 15.18 Reduction of a landmass by fluvial erosion. (a) In early stages, relief is great, slopes are steep, and the rate of erosion is rapid. (b) In an advanced stage, relief is greatly reduced, slopes are gentle, and rate of erosion is slow. Soils are thick over the broadly rounded hill summits.(c) After many millions of years of fluvial denudation, a peneplain is formed. Slopes are very gentle, and the landscape is an undulating plain. Floodplains are broad, and the stream gradients are extremely low. All of the land surface lies close to base level.(d) The peneplain is uplifted.(e) Streams trench a new system of deep valleys in the phase of landmass rejuvenation. (Drawn by A. N. Strahler.)

called *landmass rejuvenation*. Landscapes in various stages of rejuvenation are recognized throughout the continental shield area of North America. With the passage of many millions of years, the landscape will be carved into the rugged stage shown in (a), and the later stages of (b) and (c) will follow.

Aggradation and Alluvial Terraces

A graded stream, delicately adjusted to its supply of water and rock waste from upstream sources, is highly sensitive to changes in those inputs. Changes in climate or vegetation cover bring changes in discharge and load at downstream points, and these changes in turn require channel readjustments. One kind of change is the buildup of alluvium in the valley floors.

Consider first what happens when bed load increases, exceeding the transporting capacity of the stream. Along a section of channel where the excess load is introduced, the coarse sediment accumulates on the streambed in the form of bars of sand, gravel, and pebbles. These deposits raise the elevation of the streambed, a process called *aggradation*. As more bed materials accumulate, the stream channel gradient is steepened and flow velocity increases. This increase enables bed materials to be dragged downstream and spread over the channel floor at more and more distant downstream sections. In this way, sediment introduced at the head of a stream will be gradually spread along all the length of the stream.

Aggradation typically changes the channel cross section from a narrow and deep form to a wide and shallow one. Because bars are continually being formed, the flow is divided into multiple threads. These rejoin and subdivide repeatedly to give a typical *braided stream* (Figure 15.19). The coarse channel deposits spread across the former floodplain, burying fine-textured alluvium under the coarse material.

What processes cause alluvium to build up on valley floors and induce stream aggradation? One way is for alluvium to accumulate as a result of accelerated soil erosion, which we have already discussed. Other causes

Figure 15.18d shows the peneplain of (c) uplifted to an elevation of several hundred meters. The base level is now far below the land surface, and a large thickness of new landmass has been created. Soon, however, streams begin to trench the landmass and to carve deep, steep-walled valleys, shown in (e). This process is

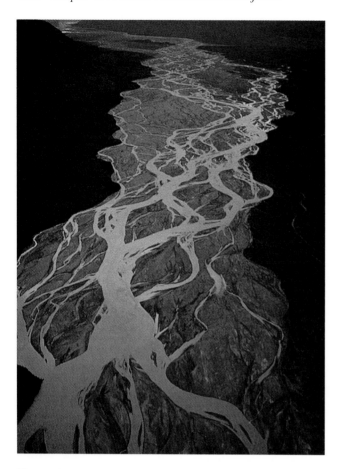

Figure 15.19 The braided channel of the Chitina River, Wrangell Mountains, Alaska, shows many separate channels separating and converging on a floodplain filled with glacial debris.

for aggradation are related to major changes in global climate, such as the onset of an ice age. Figure 15.19 illustrates one natural cause of aggradation—glacial activity—that has been of major importance in stream systems of North America and Eurasia during the recent Ice Age. In our photo example, a modern valley glacier has provided a large quantity of coarse rock debris at the head of the valley. Valley aggradation of a similar kind was widespread in a broad zone near the edges of the great ice sheets of the Pleistocene Epoch. The accumulated alluvium filled most valleys to depths of several tens of meters. Figure 15.20*a* shows a valley filled in this manner by an aggrading stream. The case could represent any one of a large number of valleys in New England or the Middle West.

Suppose, next, that the source of bed load is cut off or greatly diminished. In the case illustrated in Figure 15.20, the ice sheets have disappeared from the distant headwater areas and, with them, the supplies of coarse rock debris. Reforestation of the landscape restores a protective cover to valley-side and hill slopes of the region, keeping coarse mineral particles from entering

the stream through overland flow. Now the streams have abundant water discharges but little bed load. In other words, they are operating below their transporting capacity. The result is channel scour. The channel form becomes both deeper and narrower, and it also begins to develop meanders.

Gradually, the stream profile level is lowered, in a process called *degradation*. Because the stream is very close to being in the graded condition at all times, its dominant activity is lateral (sidewise) cutting by growth of meander bends, as shown in Figure 15.20*b*. By this process the valley alluvium is gradually excavated and carried downstream, but it cannot all be removed because the channel encounters hard bedrock masses in many places. These obstructions prevent cutting away of more alluvium. Consequently, as shown in (*c*), step-like alluvial surfaces remain on both sides of the valley. The treads of these steps are called *alluvial terraces*.

Alluvial terraces have always attracted human settlement because of the advantages over both the valley-bottom floodplain—which is subject to annual flooding—and the hill slopes beyond—which may be too steep and rocky to cultivate. Besides, terraces are easily tilled and make prime agricultural land (Figure 15.21). Towns are also easily laid out on the flat ground of a terrace. Roads and railroads can be constructed along the terrace surfaces with little difficulty.

Alluvial Rivers and Their Floodplains

We turn now to the graded river with its floodplain. As time passes, the floodplain is widened, so that broad areas of floodplain lie on both sides of the river channel. Civil engineers have given the name **alluvial river** to a large river of very low channel gradient. It flows on a thick floodplain accumulation of alluvium constructed by the river itself in earlier stages of its activity. Characteristically, an alluvial river experiences overbank floods each year or two. These occur during the season of large water surplus over the watershed. Overbank flooding of an alluvial river normally inundates part or all of a floodplain that is bounded on either side by rising steep slopes, called *bluffs*.

Typical landforms of an alluvial river and its floodplain are illustrated in Figure 15.22. Dominating the floodplain is the meandering river channel itself and abandoned stretches of former channels. Meanders develop narrow necks, which are cut through, thus shortening the river course and leaving a meander loop abandoned. This event is called a *cutoff*. It is quickly followed by deposition of silt and sand across the ends of the abandoned channel, producing an *oxbow lake*. The oxbow lake is gradually filled in with fine sediment brought in during high floods and with organic matter produced by aquatic plants. Eventually, the oxbows are converted into swamps or marshes, but their identity is retained indefinitely (Figure 15.23).

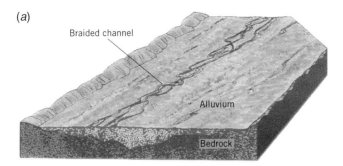

(a)

Braided channel

Alluvium

Bedrock

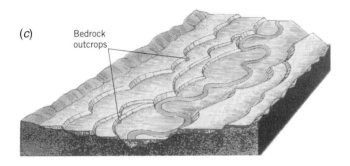

(b)

Alluvial terrace

Meander

Bluff

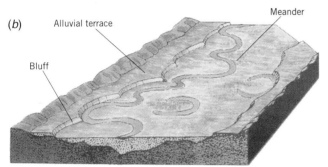

(c)

Bedrock outcrops

Figure 15.20 Alluvial terraces form when a graded stream slowly cuts away the alluvial fill in its valley. (Drawn by A. N. Strahler.)

Figure 15.21 Alluvial terraces of the Rakaia River gorge on the South Island of New Zealand. The flat terrace surface in the foreground is used as pasture for sheep. Two higher terrace levels can be seen at the left.

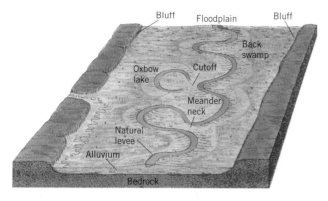

Figure 15.22 Floodplain landforms of an alluvial river. (Drawn by A. N. Strahler.)

During periods of overbank flooding, when the entire floodplain is inundated, water spreads out from the main channel over adjacent floodplain deposits (Figure 15.24). As the spreading water slackens, sand and silt are deposited in a zone adjacent to the channel.

Figure 15.23 This aerial photo of the Mississippi River floodplain shows three ox-bow lakes — former river bends that were cut off as the river shifted its course. Vegetation appears bright red in this color-infrared photo.

The result is an embankment of higher land on either side of the channel known as a **natural levee**. Because deposition is heavier closest to the channel and decreases away from the channel, the levee surface slopes away from the channel (Figure 15.22). Between the levees and the bluffs is lower ground, called the *backswamp*. In Figure 15.24, the small settlements are located on the higher ground of the levee, next to the river, while the agricultural fields occupy the backswamp area.

Overbank flooding not only results in the deposition of a thin layer of silt on the floodplain, but also brings an infusion of dissolved mineral substances that enter the soil. As a result of the resupply of nutrients, floodplain soils retain their remarkable fertility, even though they are located in regions of rainfall surplus from which these nutrients are normally leached away.

Entrenched Meanders

What happens when a broadly meandering river is uplifted in the process of stream rejuvenation? The uplift increases the river's gradient, so that it cuts downward into the bedrock below. This forms a steep-walled inner gorge. On either side lies the former floodplain, now a flat terrace high above river level. Any river deposits left on the terrace are rapidly stripped off by runoff, since no new sediment is accumulating.

Rejuvenation may cause the meanders to become impressed into the bedrock and give the inner gorge a meandering pattern. These sinuous bends are termed *entrenched meanders* to distinguish them from the floodplain meanders of an alluvial river (Figure 15.25). Although entrenched meanders are not free to shift about as floodplain meanders do, they can enlarge slowly so as to produce cutoffs. Cutoff of an entrenched meander leaves a high, round hill separated from the valley wall by the deep abandoned river channel and the shortened river course (Figure 15.25). As you might guess, such hills formed ideal natural fortifications. Many European fortresses of the Middle Ages were built on such cutoff meander spurs. Under unusual circumstances, where the bedrock includes a strong, massive sandstone formation, meander cutoff leaves a *natural bridge* formed by the narrow meander neck.

FLUVIAL PROCESSES IN AN ARID CLIMATE

Desert regions look strikingly different from humid regions, in both vegetation and landforms. Obviously, the lower precipitation makes the difference. Vegetation is sparse or absent, and land surfaces are mantled with mineral material—sand, gravel, rock fragments, or bedrock itself.

Figure 15.24 This river floodplain in Bangladesh is largely under water during a 1973 flood. Villages occupy the higher ground of the natural levees bordering the river channel.

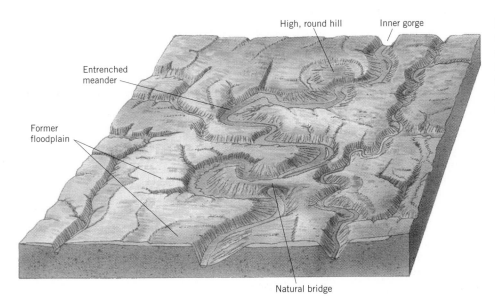

Natural bridge

Figure 15.25 Rejuvenation of a meandering stream has produced entrenched meanders. One meander neck has been cut through, forming a natural bridge. (Drawn by E. Raisz. Copyright © A. N. Strahler.)

Figure 15.26   A flash flood has filled this desert channel in the Tucson Mountains of Arizona with raging, turbid waters. A distant thunderstorm produced the runoff.

Although deserts have low precipitation, rain falls in dry climates as well as in moist, and most landforms of desert regions are formed by running water. A particular locality in a dry desert may experience heavy rain only once in several years. But when rain does fall, stream channels carry water and perform important work as agents of erosion, transportation, and deposition. Fluvial processes are especially effective in shaping desert landforms because of the sparse vegetation cover. The few small plants that survive offer little or no protection to soil or bedrock. Without a thick vegetative cover to protect the ground and hold back the swift downslope flow of water, large quantities of coarse rock debris are swept into the streams. A dry channel is transformed in a few minutes into a raging flood of muddy water heavily charged with rock fragments (Figure 15.26).

An important contrast between regions of arid and humid climates lies in the way in which the water enters and leaves a stream channel (Figure 15.27). In a humid region with a high water table sloping toward a stream channel, groundwater moves steadily toward the channel and seeps into the streambed, producing permanent (perennial) streams. In arid regions, the water table normally lies far below the channel floor. Where a stream flows across a plain of gravel and sand, water is lost from the channel by seepage. Loss of discharge by

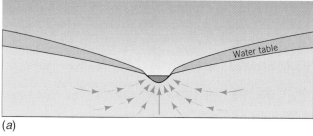

(a)

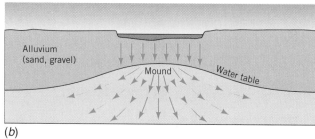

(b)

Figure 15.27   In humid regions (a), a stream channel receives groundwater through seepage. In arid regions (b), stream water seeps out of the channel and into the water table below. (Copyright © A. N. Strahler.)

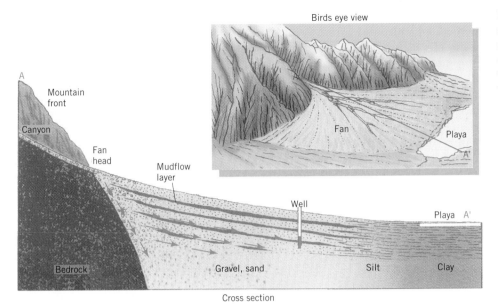

Figure 15.28 Features of an alluvial fan. A cross section shows mudflow layers interbedded with sand layers, providing water (arrows) for a well in the fan. (Copyright © A. N. Strahler.)

seepage and evaporation strongly depletes the flow of streams in alluvium-filled valleys of arid regions. As a result, aggradation occurs and braided channels are common. Streams of desert regions are often short and end in alluvial deposits or on the floors of shallow, dry lakes.

Alluvial Fans

One very common landform built by braided, aggrading streams is the **alluvial fan**, a low cone of alluvial sands and gravels resembling in outline an open Japanese fan (Figure 15.28). The apex, or central point of the fan, lies at the mouth of a canyon or ravine. The fan is built out on an adjacent plain. Alluvial fans are of many sizes. In fact, some desert fans are many kilometers across (Figure 15.29).

Fans are built by streams carrying heavy loads of coarse rock waste from a mountain or an upland region. The braided channel shifts constantly, but its position is firmly fixed at the canyon mouth. The lower part of the channel, below the apex, sweeps back and forth. This activity accounts for the semicircular fan form and the downward slope in all radial directions away from the apex.

Figure 15.29 Great alluvial fans extend out upon the floor of Death Valley. The canyons from which they originate are carved deeply into a great uplifted fault block.

Large, complex alluvial fans also include mudflows (Figure 15.28). Mud layers are interbedded with sand and gravel layers. Water infiltrates the fan at its head, making its way to lower levels along sand layers. The mudflow layers serve as barriers to groundwater movement. The trapped groundwater is under pressure from water higher in the fan apex. When a well is drilled into the lower slopes of the fan, water rises spontaneously as artesian flow. (See Chapter 16 and Figure 16.8.)

Alluvial fans are the primary sites of groundwater reservoirs in the southwestern United States. In many fan areas, sustained heavy pumping of these reserves for irrigation has lowered the water table severely. The rate of recharge is extremely slow in comparison. At some locations, recharge is increased by building water-spreading structures and infiltrating basins on the fan surfaces. A serious side effect of excessive groundwater withdrawal is subsidence (sinking) of the land surface.

The Landscape of Mountainous Deserts

Where tectonic activity has recently produced block faulting in an area of continental desert, the assemblage of fluvial landforms is particularly diverse. The basin-and-range region of the western United States is such an area. It includes large parts of Nevada and Utah, southeastern California, southern Arizona and New Mexico, and adjacent parts of Mexico. The uplifted and tilted blocks, separated by downdropped tectonic basins, provide an environment for the development of spectacular erosional and depositional fluvial landforms.

Figure 15.30 demonstrates some landscape features of mountainous deserts. Figure 15.30a shows two uplifted fault blocks with a downdropped block between them. Although denudation acts on the uplifted blocks as they are being raised, we have shown them as very little modified at the time tectonic activity has ceased. At first, the faces of the fault block are extremely steep. They are scored with deep ravines, and talus blocks form cones at the bases of the blocks.

Figure 15.30b shows a later stage of denudation in which streams have carved up the mountain blocks into a rugged landscape of deep canyons and high divides. Rock waste furnished by these steep mountain slopes is carried from the mouths of canyons to form numerous large alluvial fans. The fan deposits form a continuous apron extending far out into the basins.

In the centers of desert basins lie the saline lakes and dry lake basins mentioned in Chapter 14. Accumulation of fine sediment and precipitated salts produces an extremely flat basin floor, called a **playa** in the southwestern United States and in Mexico. Salt flats are found where an evaporite layer forms the surface. In some playas, shallow water forms a salt lake.

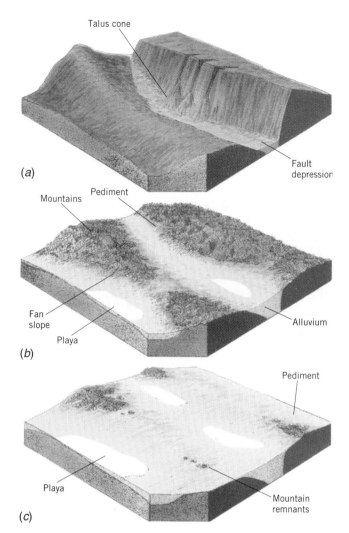

Figure 15.30 Idealized diagrams of landforms of the mountainous deserts of the southwestern United States. (*a*) Initial uplift of two blocks, with a downdropped valley between them. (*b*) Stage of rapid filling of tectonic basins with debris from high, rugged mountain blocks. (*c*) Advanced stage with small mountain remnants and broad playa and fan slopes. (Drawn by A. N. Strahler.)

Figure 15.31 is an air photograph of a mountainous desert landscape in the Death Valley region of eastern California. Three environmental zones can be seen— the rugged mountain masses dissected into canyons with steep rocky walls, a zone of alluvial fans along the mountain front, and the white playa occupying the central part of the basin.

In this type of mountainous desert, fluvial processes are limited to local transport of rock particles from a mountain range to the nearest adjacent basin, which receives all the sediment. The basin gradually fills as the mountains diminish in elevation. Because there is no outflow to the sea, the concept of a base level of denudation has no meaning. For that same reason, a peneplain does not represent the next-to-the-last stage of denudation, as it does in the humid environment.

Figure 15.31 Racetrack Playa, a flat, white plain, is surrounded by alluvial fans and rugged mountains. This desert valley lies in the northern part of the Panamint Range, which is not far west of Death Valley, California. In the distance rises the steep eastern face of the Inyo Mountains, a great fault block.

Each arid basin becomes a closed system as far as transport of sediment is involved. Only the hydrologic system is open, with water entering as precipitation and leaving as evaporation.

As the desert mountain masses are lowered in height and reduced in extent, a gently sloping rock floor, called a *pediment*, develops close to the receding mountain front (Figure 15.32). As the remaining mountains shrink further in size, the pediment surfaces expand to form wide rock platforms thinly veneered with alluvium. This advanced stage is shown in Figure 15.30*c*. The desert land surface produced in an advanced stage of fluvial activity is an undulating plain consisting largely of areas of pediment surrounded by areas of alluvial fan and playa surfaces.

Running water, the primary agent of denudation, does not act equally on all types of rocks. Some rocks are more resistant to erosion, while others are less so. As a result, fluvial erosion can create unique and interesting landscapes that reveal rock structures of various kinds. This will be the subject of our next chapter.

Figure 15.32 A pediment formerly carved in granite at the foot of a desert mountain slope is now seen in profile as a straight, sloping line. Later erosion has exposed the bedrock, and weathering has formed the joint blocks into boulders (intermediate distance). Little Dragoon Mountains, near Adams Peak, about 20 km (about 12 mi) northeast of Benson, Arizona.

CHAPTER SUMMARY

This chapter has covered the landforms and land-forming processes of running water, one of the four active agents of denudation. Like the other agents, running water erodes, transports, and deposits rock material, forming both erosional and depositional landforms.

The work of running water begins on slopes, producing colluvium where overland flow moves soil particles downslope, and producing alluvium when the particles enter stream channels and are later deposited. In most natural landscapes, soil erosion and soil formation rates are more or less equal, a condition known as the geologic norm. Badlands are an exception in which natural erosion rates are very high.

The work of streams includes stream erosion and stream transportation. Where stream channels are carved into soft materials, large amounts of sediment can be obtained by hydraulic action. Where stream channels flow on bedrock, channels are deepened only by the abrasion of bed and banks by mineral particles, large and small. Both the suspended load and bed load of rivers increase greatly as velocity increases. Velocity, in turn, depends on gradient.

Over time, streams tend to a graded condition, in which their gradients are adjusted to move the average amount of water and sediment supplied to them by slopes. Lakes and waterfalls, created by tectonic, volcanic, or glacial activity, are short-lived events, geologically speaking, that give way to a smooth, graded stream profile. Grade is maintained as landscapes are eroded toward base level. If the land surface is tectonically stable for a very long period, a peneplain can form. When provided with a sudden inflow of rock material, as, for example, by glacial action, streams build up their beds by aggradation. When that inflow ceases, streams resume downcutting, leaving behind alluvial terraces.

Large rivers with low gradients that move large quantities of sediment are termed alluvial rivers. The meandering of these rivers forms cutoff meanders, oxbow lakes, and other typical landforms. Alluvial rivers are sites of intense human activity. Their fertile floodplains yield agricultural crops and provide easy transportation paths. When a region containing a meandering alluvial river is uplifted, the rejuvenation can produce entrenched meanders.

Although rainfall is scarce in deserts, running water is very effective there in producing fluvial landforms. Desert streams, subject to flash flooding, build alluvial fans at the mouths of canyons. Water sinks into the fan deposits, creating local groundwater reservoirs. Eventually, desert mountains are worn down into gently sloping pediments. Fine sediments and salts, carried by streams, accumulate in playas, from which water evaporates, leaving sediment and salt behind.

KEY TERMS

fluvial landforms	stream deposition	alluvial river
fluvial processes	stream load	natural levee
alluvium	graded stream	alluvial fan
stream erosion	alluvial meander	playa
stream transportation	peneplain	

REVIEW QUESTIONS

1. List and briefly identify the four flowing substances that serve as agents of denudation.

2. Compare erosional and depositional landforms. Sketch an example of each type.

3. Describe the process of slope erosion. What is meant by the geologic norm?

4. Contrast the two terms *colluvium* and *alluvium*. Where on a landscape would you look to find each one?

5. What special conditions are required for badlands to form?

6. When and how does sheet erosion occur? How does it lead to rill erosion and gullying?

7. In what ways do streams erode their bed and banks?

8. What is stream load? Identify its three components. In what form do large rivers carry most of their load?

9. How is velocity related to the ability of a stream to move sediment downstream?

10. What is a graded stream? Sketch the profile of a graded stream.

11. Describe the evolution of a fluvial landscape. What is meant by rejuvenation?

12. How does stream degradation produce alluvial terraces?

13. Define the term *alluvial river*. Identify some characteristic landforms of alluvial rivers. Why are alluvial rivers important to human civilization?

14. Sketch the floodplain of a graded, meandering river. Identify key landforms on the sketch. How do they form?

15. Why is fluvial action so effective in arid climates, considering that rainfall is scarce? How do streams in arid climates differ from streams of moist climates?

16. Describe the evolution of the landscape in a mountainous desert. Use the terms *alluvial fan*, *playa*, and *pediment* in your answer.

ESSAY QUESTIONS

1. A river originates high in the Rocky Mountains, crosses the high plains, flows through the agricultural regions of the Midwest, and finally reaches the sea. Describe the fluvial processes and landforms you might expect to find on a journey along the river from its headwaters to the ocean.

2. What would be the effects of climate change on a fluvial system? Choose either the effects of cooler temperatures and higher precipitation in a mountainous desert, or warmer temperatures and lower precipitation in a humid agricultural region.

PROBLEMS

Working It Out 15.1 • River Discharge and Suspended Sediment

1. What suspended load would you project for the Powder River at Arvada when the discharge is 50 m^3/s? 5 m^3/s? (*Hint:* Confirm your answers by examining the graph.)

2. If the discharge of the Powder River at Arvada doubles, what is the effect on its suspended sediment load?

3. A hydrologist measures discharge and sediment load for a river and plots the points as logarithms on ordinary graph paper. She fits a line to the points with the following equation:

$$\log S = 2.50 + 2.12 \log Q$$

Rewrite this equation as an exponential of the form $S = aQ^b$.

Landforms and Rock Structure

Denudation acts to wear down all rock masses exposed at the land surface. However, different types of rocks are worn down at different rates. Some are easily eroded, while others are highly resistant to erosion. Generally, weak rocks will underlie valleys, and strong rocks will underlie hills, ridges, and uplands. This means that the pattern of landforms on a landscape can reveal the sequence of rock layers underneath it. The sequence, in turn, is determined by the type of rock structure present—for example, a series of folds, or perhaps a dome, in a set of rock layers. Thus, there is often a direct relationship between landforms and rock structure.

Over the world's vast land area, many types of rock and rock structures appear. Repeated episodes of uplift, followed by periods of denudation, can bring to the surface rocks and rock structures formed deep within the crust. In this way, even ancient mountain roots, formed during continental collisions, may eventually appear at the surface and be subjected to denudation. Batholiths and other plutons—igneous rock structures produced by magmas that cool at depth—can similarly be exposed.

As we will see in this chapter, rock structure controls not only the locations of uplands and lowlands, but also the placement of streams and the shapes and heights of the intervening divides. A distinctive assemblage of landforms and stream patterns develops on each of the major types of crustal structures. Recall that in Chapter 12 we classified all landforms as either initial or sequential in origin. Initial landforms are produced directly by volcanic activity, folding, and faulting. De-

---

Hogback ridges are formed when erosion reveals resistant beds in a warped sequence of sedimentary rock layers.

nudation soon converts these initial landforms into sequential landforms, which are controlled in shape, size, and arrangement by the underlying rocks and their structure. The subject of this chapter is therefore the sequential landforms that arise through erosion of rock structures.

ROCK STRUCTURE AS A LANDFORM CONTROL

As denudation takes place, landscape features develop according to patterns of bedrock composition and structure. Figure 16.1 shows an arrangement of five layers of sedimentary rock, together with a mass of much older igneous rock. The sediments were originally deposited in horizontal layers atop the igneous rock, but the block of igneous and sedimentary rock has been bent and tilted in an ancient orogeny. The diagram shows the usual landform shape of each rock type and whether it forms valleys or mountains. The cross section on the front of the block diagram shows conventional symbols used by geologists for each rock type—for example, fine dots for sandstone or "bricks" for limestone.

As indicated by the diagram, shale is a weak rock that is easily eroded and forms the low valley floors of the region. Limestone, subjected to solution by carbonic acid in rainwater and surface water, also forms valleys in humid climates. In arid climates, on the other hand, limestone is a resistant rock and usually stands high to form ridges and cliffs. Sandstone and conglomerate are typically resistant to denudation, and they form ridges or uplands. As a group, the igneous rocks are also resistant—they typically form uplands or mountains rising above adjacent areas of shale and limestone. Although metamorphic rocks in general are more resistant to denudation, individual metamorphic rock types—marble or schist, for example—vary in their resistance.

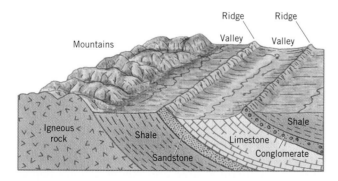

Figure 16.1 Landforms evolve through the slow erosional removal of weaker rock, leaving the more resistant rock standing as ridges or mountains. (Drawn by A. N. Strahler.)

Figure 16.2 provides a preview of the important rock structures and their accompanying landforms that we will discuss in this chapter. In these sketches, resistant beds form ridges or belts of hills, while weaker rocks form lowlands or valleys. The upper four sketches show landforms and rock structures of sedimentary rock layers. The structures include horizontal layers, gently sloping layers, and layers that are folded or warped into domes. Other structures shown are fault blocks, plutons, metamorphic belts, and eroded volcanoes.

Strike and Dip

The surfaces of most rock layers are not flat but are tipped away from the horizontal. In addition, internal planes—such as joints—occur within rock layers and are likely to be tilted. We need a system of geometry to enable us to measure and describe the position in which these natural planes are held and to indicate them on maps. Examples of such planes include the bedding layers of sedimentary strata, the sides of a dike, and the joints in granite.

The tilt and orientation of a natural rock plane are measured with reference to a horizontal plane. Figure 16.3 shows a water surface resting against tilted sedimentary strata. Since it is horizontal, we can use the water surface as the horizontal plane. The acute angle formed between the rock plane and the horizontal plane is termed the *dip*. The amount of dip is stated in degrees, ranging from 0 for a horizontal rock plane to 90 for a vertical rock plane. Next, we examine the line of intersection between the inclined rock plane and the horizontal water plane. This horizontal line will have an orientation with respect to the compass. This orientation is termed the *strike*. In Figure 16.3, the strike is along a north-south line.

LANDFORMS OF HORIZONTAL STRATA AND COASTAL PLAINS

Extensive areas of the ancient continental shields are covered by thick sequences of horizontal sedimentary rock layers. At various times in the 600 million years following the end of Precambrian time, these strata were deposited in shallow inland seas. Following crustal uplift, with little disturbance other than minor crustal warping or faulting, these areas became continental surfaces undergoing denudation.

Arid Regions

In arid climates, where vegetation is sparse and the action of overland flow especially effective, sharply defined landforms develop on horizontal sedimentary strata (Figure 16.4). The normal sequence of land-

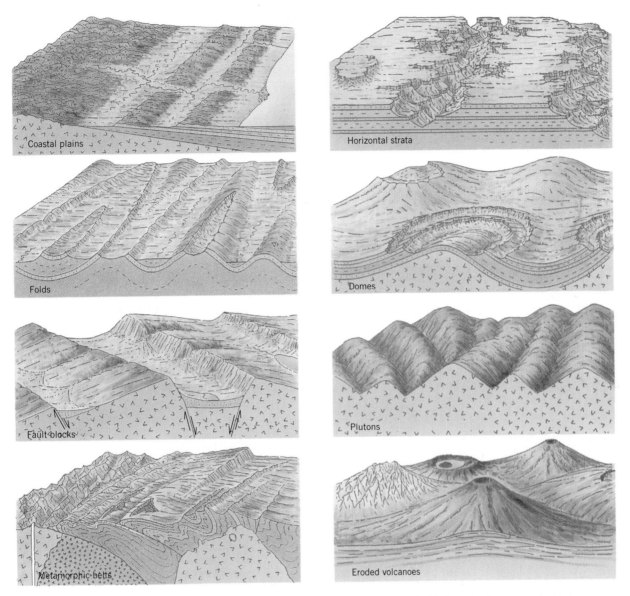

Figure 16.2 Landforms associated with various rock structures. (Drawn by A. N. Strahler.)

Figure 16.3 Strike and dip. (Drawn by A. N. Strahler.)

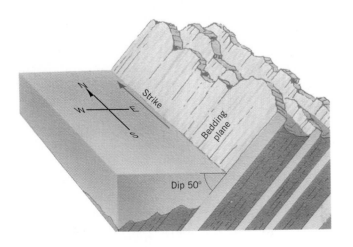

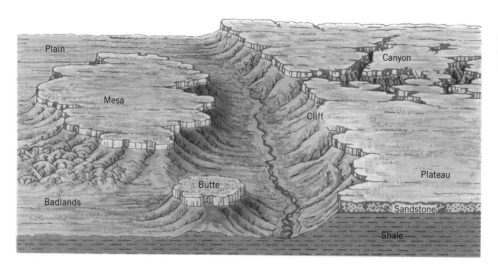

Figure 16.4 In arid climates, distinctive erosional landforms develop in horizontal strata. (Drawn by A. N. Strahler.)

forms is a sheer rock wall, called a *cliff*, which develops at the edge of a resistant rock layer. At the base of the cliff is an inclined slope, which flattens out into a plain beyond.

In these arid regions, erosion strips away successive rock layers, leaving behind a broad platform capped by hard rock layers. This platform is usually called a **plateau**. Cliffs retreat as near-vertical surfaces because the weak clay or shale formations exposed at the cliff base are rapidly washed away by storm runoff and channel erosion. When undermined, the rock in the upper cliff face repeatedly breaks away along vertical fractures.

Cliff retreat produces a **mesa**, which is a table-topped plateau bordered on all sides by cliffs. Mesas represent the remnants of a formerly extensive layer of resistant rock. As a mesa is reduced in area by retreat of the rimming cliffs, it maintains its flat top. Eventually, it becomes a small steep-sided hill known as a **butte**. Further erosion may produce a single, tall column before the landform is totally consumed.

In the walls of the great canyons of the Colorado Plateau region, these landforms are wonderfully displayed. Figure 16.5 shows a view of the Grand Canyon. The flat surface of the canyon rim marks the edge of the vast plateau surrounding the canyon. Within the

Figure 16.5 Bright Angel Canyon, Grand Canyon National Park, Arizona. Note the "stairstep" landscape in which resistant sandstones and limestones form flat-topped benches and mesas, while weaker shales form slopes connecting them.

Figure 16.6 This dendritic drainage pattern is from an area of horizontal strata in the Appalachian Plateau of northern Kentucky. (Derived from data of the U.S. Geological Survey.)

canyon, resistant rock layers crop out at different levels, producing a series of steps. Cliffs of resistant rock form the risers, while weaker rocks form the sloping treads below. Benches or platforms develop atop resistant strata, extending the treads and producing a series of levels that break the descent into the canyon. Small mesas and buttes remain within the canyon, isolated by erosion of surrounding rock strata.

Drainage Patterns on Horizontal Strata

Imagine making a map of a region that showed only its stream channels. This map would show the stream pattern, or **drainage pattern**, of the region. The drainage pattern that develops on each landmass type is often related to the underlying rock type and structure.

Regions of horizontal strata show broadly branching stream networks formed into a *dendritic drainage pattern.* Figure 16.6 is a map of the stream channels in a small area within the Appalachian Plateau region of Kentucky. The smaller streams in this pattern take a great variety of compass directions. Because the rock layers lie flat, no one direction is favored.

Drainage patterns of stream networks, like the dendritic pattern in Figure 16.6, have a number of interesting properties that geomorphologists have studied extensively. *Working It Out 16.1 • Properties of Stream Networks* describes two of these properties—the number and mean length of streams of increasing order.

Coastal Plains

Coastal plains are found along passive continental margins that are largely free of tectonic activity. They are underlain by nearly horizontal strata that slope gently toward the ocean. Figure 16.7*a* shows a coastal zone that has recently emerged from beneath the sea. For-

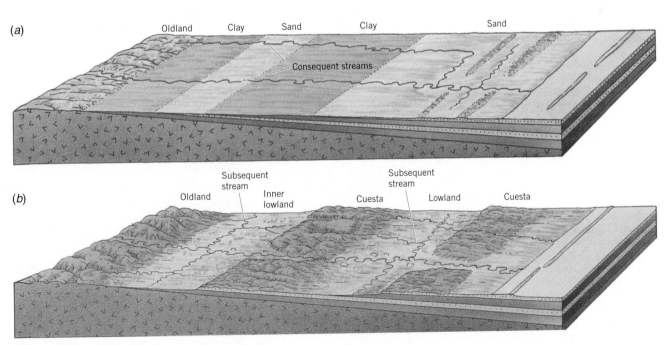

Figure 16.7 Development of a broad coastal plain. (*a*) Early stage—plain recently emerged. (*b*) Advanced stage—cuestas and lowlands developed. (Drawn by A. N. Strahler.)

Working it Out 16.1 • Properties of Stream Networks

Fluvial landforms are produced by the erosion and deposition of streams that are connected into networks. For this reason, the properties of stream networks have always been of interest to geomorphologists studying the landform-making process. Many of these properties are studied with respect to stream order. Recall from Chapter 15 that the finest stream segments are of first order. When two first-order streams join, the downstream segment is of second order. Third-order stream segments are formed by the junction of second-order segments, and so forth.

Two interesting properties of stream networks are the number of segments of each order and the average length of the segments of each order. As you might imagine, there are many more segments of first order than of second order, and of second order than of third order, and so on. Thus, the number of segments decreases with increasing stream order. As to average length, we find that first-order stream segments are shortest, while second-order segments are longer, third-order are longer still, and so forth. This observation stands to

reason, since there will be fewer streams of higher order. Thus, a higher-order stream segment will have to flow for a longer distance before being joined by another segment of the same order.

The table below provides some data from the Allegheny River, in McKean County, Pennsylvania. The Allegheny ultimately joins the Ohio River, but for this study, only that portion upstream from a particular point on a seventh-order segment of the Allegheny was examined. To provide the information shown in the table, detailed topographic maps were carefully studied over a large area. Each stream segment was identified, and its length was measured. The results were tabulated by stream order.

Looking first at the number of segments, note that the value drops very quickly as order increases. However, the ratio of number of segments of one order to the next seems roughly constant. This ratio is called the *bifurcation ratio* and is shown in the third column. For example, the ratio of first- to second-order segment number is $5966/1529 = 3.9$, meaning that each second-order segment branches to

form 3.9 first-order segments, on the average. Looking at the values in the column, we can see that the bifurcation ratio varies from 3.0 to 5.7; the average value is 4.37. In algebraic notation, we can write the bifurcation ratio, R_b as

$$R_b = \frac{N_u}{N_{u+1}}$$

where N_u is the number of segments of stream order u and N_{u+1} is the number of segments of stream order $u + 1$.

Part (*a*) of the figure plots the number of segments against stream order for the Allegheny. This graph is *semilogarithmic* (*semilog*, for short)—that is, the *x*-axis is linear, while the *y*-axis is logarithmic. Note that each point falls very close to a straight line. This line follows an equation of the form

$$N_u = R_b^{(k-u)} \qquad (1)$$

Here, k is the order of the largest stream in the network being considered—7 in the case of the Allegheny River drainage network in Table A. We can then write the equation for the Allegheny as

$$N_u = 4.37^{(7-u)}$$

using the mean bifurcation ratio, 4.37, for R_b. With this formula, the number of segments expected for stream orders 1 to 7 will then be 4.37^6, 4.37^5, 4.37^4, 4.37^3, 4.37^2, 4.37^1, 4.37^0, or 6964, 1594, 365, 83, 19, 4, 1.

The figure also shows data for a fifth-order stream network in the Big Badlands region of South Dakota. For this stream network, the average bifurcation ratio is 3.47. The Big Badlands region differs considerably from the environment of the Allegheny River in that a given area of easily eroded badland terrain has many more stream channels (see Figure 15.5). However, the lines plotted for Big Badlands and the Allegheny River are close to par-

Allegheny River Drainage Basin Characteristics

Stream Order u	Number of Segments N_u	Bifurcation Ratio R_b	Mean Segment Length (km) L_u	Cumulative Mean Segment Length (km) L_u^*	Length Ratio R_L
1	5966		0.15	0.15	
		3.9			3.2
2	1529		0.48	0.63	
		4.0			2.7
3	378		1.29	1.9	
		5.7			3.1
4	68		4.0	5.9	
		5.3			2.8
5	13		11.3	17.2	
		4.3			2.8
6	3		32.2	49.4	
		3.0			
7	1		13+		
		$\bar{R}_b = 4.37$			$\bar{R}_L = 2.92$

allel, indicating that they have similar average bifurcation ratios. In fact, bifurcation ratios for all natural stream networks tend to range between 3 and 5, no matter what type of substrate is being subjected to fluvial action. This remarkable observation was first made by the hydraulic engineer Robert E. Horton, and thus equation (1) above is sometimes referred to as Horton's law of stream numbers.

Turning now to average stream length, we anticipated that the average length of stream segments increases with stream order. The table confirms this fact, with mean lengths for the Allegheny ranging from 0.15 km for first-order segments to 32.2 km for sixth-order segments. As in the case of the bifurcation ratio for stream segment numbers, we can define a length ratio R_L for mean length of segments from two adjacent orders as

$$R_L = \frac{L_{u+1}}{L_u}$$

where L_u is the mean length of segments of order u. As the table shows, values of R_L are rather similar to values of R_b. For the Allegheny data, the average length ratio is $R_L = 2.92$.

Like the number of stream segments, the progression of lengths also closely follows a straight line on a semilog plot, provided that it is plotted as the cumulative mean length. This value, shown in the fifth column of the table, results when mean length values are added cumulatively. That is, the second value is the sum of the first two, the third value is the sum of the first three values, and so on. Stating this algebraically,

$$L_u^* = \sum_{i=1}^{u} L_u$$

where L_u^* is the cumulative mean length of segments of order u, and i is a whole number counting from 1 to u.

The equation for the straight line that connects values of L_u^* on a semilog graph is

$$L_u^* = L_1 R_L^{(u-1)} \qquad (2)$$

where L_1 is the mean length of first-order segments. For the case of the Allegheny, the table shows that $L_1 = 0.15$ km. Using the average length ratio $R_L = 2.92$, we have

$$L_u^* = 0.15 \times [2.92^{(u-1)}]$$

Part (b) of the figure plots the cumulative mean lengths for the first six orders of the Allegheny stream network. The points lie very close to a straight line. Also shown are data from a fourth-order stream network at Fern Canyon, California. The fact that both lines are nearly parallel shows that cumulative mean length tends to increase by the same factor in spite of differences in climate, substrate, and rock type. Equation (2) was also derived by Horton and is referred to as Horton's law of stream lengths.

Number and mean length of stream segments are just two interesting properties of stream networks studied by geomorphologists. Many other properties have been studied as well. If you continue the study of physical geography, you will learn more of the interesting behavior of streams.

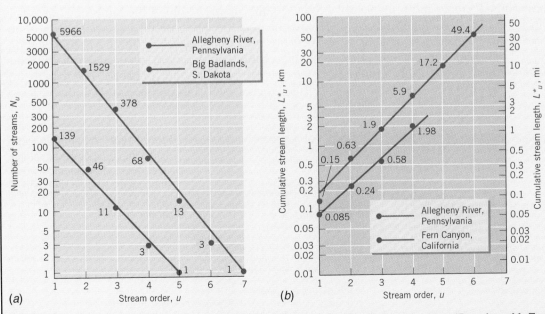

Numbers of segments (a) and mean segment length (b) plotted against stream order. (Data from M. E. Morisawa, K. G. Smith, and J. C. Maxwell.)

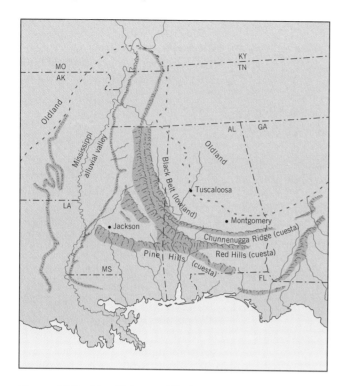

Figure 16.8 The Alabama–Mississippi coastal plain is belted by a series of sandy cuestas and shale lowlands. (After A. K. Lobeck.)

merly, this zone was a shallow continental shelf that accumulated successive layers of sediment brought from the land and distributed by currents. On the newly formed land surface, streams flow directly seaward, down the gentle slope. A stream of this origin is a *consequent stream*. It is a stream whose course is controlled by the initial slope of a new land surface. Consequent streams form on many kinds of initial landforms, such as volcanoes, fault blocks, or beds of drained lakes.

In an advanced stage of coastal-plain development, a new series of streams and topographic features has developed (Figure 16.7*b*). Where more easily eroded strata (usually clay or shale) are exposed, denudation is

rapid, making *lowlands*. Between them rise broad belts of hills called **cuestas**. Cuestas are commonly underlain by layers of sand, sandstone, limestone, or chalk that dip away from the cuesta at a low angle. The lowland lying between the area of older rock—the oldland—and the first cuesta is called the *inner lowland*.

Streams that develop along the trend of the lowlands, parallel with the shoreline, are of a class known as *subsequent streams*. They take their position along any belt or zone of weak rock, and therefore follow closely the pattern of rock exposure. Subsequent streams occur in many regions, and we will mention them again in the discussion of folds, domes, and faults.

The drainage lines on a fully dissected coastal plain combine to form a *trellis drainage pattern*. In this type of pattern, a main stream has tributaries that are arranged at right angles. You can see this pattern in the streams of Figure 16.7*b*. The subsequent streams trend at about right angles to the consequent streams.

The coastal plain of the United States is a major geographical region, ranging in width from 160 to 500 km (about 100 to 300 mi) and extending for 3000 km (about 2000 mi) along the Atlantic and Gulf coasts. In Alabama and Mississippi (Figure 16.8), the coastal plain shows many of the features of Figure 16.7. In this region, the coastal plain sweeps far inland, as the oldland boundary shows. Cuestas and lowlands run in belts curving to follow the ancient coast. The cuestas, named "hills" or "ridges," are underlain by sandy formations and support pine forests. Limestone forms lowlands, such as the Black Belt in Alabama. This belt is named for its dark, fertile soils.

An interesting feature of coastal plain strata is the occurrence of **artesian wells**. As Figure 16.9 shows, cuestas of porous sandstone absorb precipitation, and the water moves down the dip of the sandstone layer. This water-bearing formation is called an *aquifer*. Beds of impervious rock, called *aquicludes*, lie above and below the aquifer. Far out on the coastal plain, at depth, the water is under strong confining pressure. When wells are drilled there, water rises spontaneously to the surface.

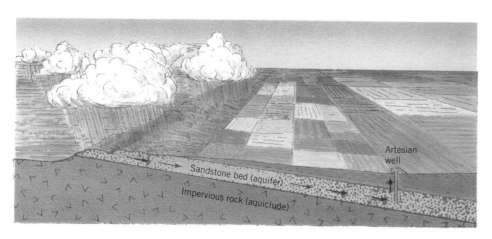

Figure 16.9 An artesian well drilled into a dipping sandstone layer overlain by an impervious shale layer. (Drawn by Erwin Raisz. Copyright © by A. N. Strahler.)

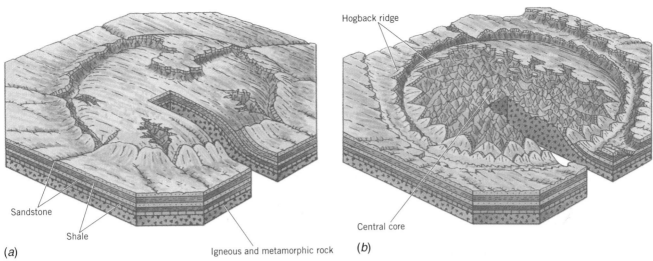

Figure 16.10   Erosion of sedimentary strata from the summit of a dome structure. (*a*) The strata are partially eroded, forming an encircling hogback ridge. (*b*) The strata are eroded from the center of dome, revealing a core of older igneous or metamorphic rock. (Drawn by A. N. Strahler.)

LANDFORMS OF WARPED ROCK LAYERS

Although sedimentary rocks are horizontal or gently sloping in some regions, in other areas rock layers may be upwarped or downwarped. In some cases, these form domes and basins. In others, they form anticlines and synclines—wavelike forms of crests and troughs—as we noted in Chapter 12.

Sedimentary Domes

A distinctive landmass type is the **sedimentary dome**, a circular or oval structure in which strata have been forced upward into a domed shape. Sedimentary domes occur in various places within the covered shield areas of the continents. Igneous intrusions at great depth may have been responsible for some of these uplifts. For others, upthrusting on deep faults may have been the cause.

Erosion features of a sedimentary dome are illustrated in the two block diagrams of Figure 16.10. Strata are first removed from the summit region of the dome, exposing older strata beneath. Eroded edges of steeply dipping strata form sharp-crested sawtooth ridges called *hogbacks* (Figure 16.11). When the last of the

Figure 16.11   Hogbacks of sandstone lie along the eastern base of the Colorado Front Range, which is a large domed structure. The view is southward toward the city of Boulder. The rugged forested terrain in the distance is developed on igneous and metamorphic rock exposed in the core of the uplift.

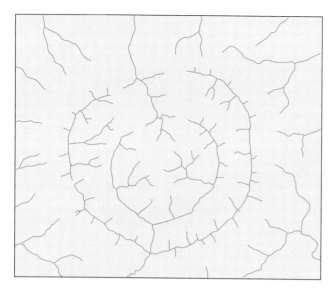

Figure 16.12 The drainage pattern on an eroded dome combines annular and radial elements.

strata have been removed, the ancient shield rock is exposed in the central core of the dome, which then develops a mountainous terrain.

The stream network on a deeply eroded dome shows dominant subsequent streams forming a circular system, called an annular drainage pattern (Figure 16.12). ("Annular" means "ring-shaped.") The shorter tributaries make a radial arrangement. The total pattern resembles a trellis pattern bent into a circular form.

The Black Hills Dome

A classic example of a large and rather complex sedimentary dome is the Black Hills dome of western South Dakota and eastern Wyoming (Figure 16.13). The Red Valley is continuously developed around the entire dome and is underlain by a weak shale, which is easily washed away. On the outer side of the Red Valley is a high, sharp hogback of Dakota sandstone, known as Hogback Ridge, rising some 150 m (about 500 ft) above the level of the Red Valley. Farther out toward the margins of the dome, the strata are less steeply inclined and form a series of cuestas.

The eastern central part of the Black Hills consists of a mountainous core of intrusive and metamorphic rocks. These mountains are richly forested, while the surrounding valleys are beautiful open parks. Thus, the region is very attractive as a summer resort area. In the northern part of the central core, there are valuable ore deposits. At Lead is the fabulous Homestake Mine, one of the world's richest gold-producing mines. The

west-central part of the Black Hills consists of a limestone plateau deeply carved by streams. The plateau represents one of the last remaining sedimentary rock layers to be stripped from the core of the dome.

Fold Belts

According to the model of mountain-making developed in Chapter 11, strata of the continental margins are deformed into folds along narrow belts during continental collision. As we explained in Chapter 12, a troughlike downbend of strata is a *syncline,* and the archlike upbend next to it is an *anticline.* In a belt of folded strata, synclines alternate with anticlines, like a succession of wave troughs and wave crests on the ocean surface.

The two block diagrams of Figure 16.14 show some of the distinctive landforms resulting from fluvial denudation of a belt of folded strata. Deep erosion of these simple, open folds in this landmass type produces a **ridge-and-valley landscape**. Weaker formations such as shale and limestone are eroded away, leaving hard strata, such as sandstone or quartzite, to stand in bold

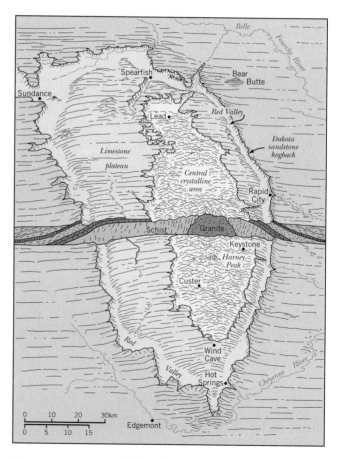

Figure 16.13 The Black Hills consist of a broad, flat-topped dome deeply eroded to expose a core of igneous and metamorphic rocks. (Drawn by A. N. Strahler.)

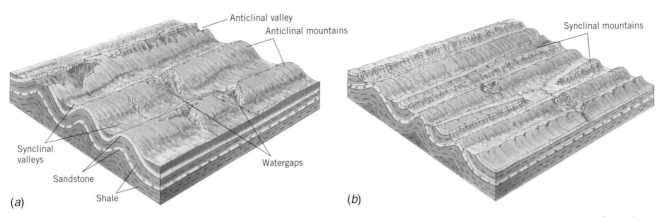

Figure 16.14 Stages in the erosional development of folded strata. (*a*) Erosion exposes a highly resistant layer of sandstone or quartzite, which controls much of the ridge-and-valley landscape. (*b*) Continued erosion partly removes the resistant formation but reveals another below it. (Drawn by A. N. Strahler.)

relief as long, narrow ridges (Figure 16.15). Note that anticlines, or upwarps, are not always ridges. If a resistant rock type at the center of the anticline is eroded through to reveal softer rocks underneath, an *anticlinal valley* may form. An *anticlinal valley* is shown in Figure 16.14*a*. A *synclinal mountain* is also possible. It occurs when a resistant rock type is exposed at the center of a syncline, and the rock stands up as a ridge (Figure 16.14*b*).

On eroded folds, the stream network is distorted into a trellis drainage pattern (Figure 16.16). The principal elements of this pattern are long, parallel subsequent streams occupying the narrow valleys of weak rock. In a few places, a larger stream cuts across a ridge in a deep, steep-walled *water gap*.

The folds illustrated in Figure 16.14 are continuous and even-crested. They produce ridges that are approximately parallel in trend and continue for great dis-

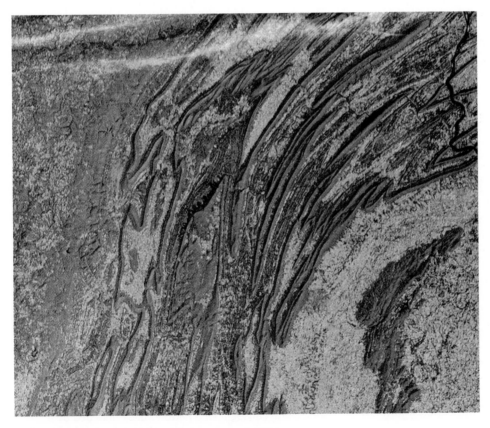

Figure 16.15 Like an old wooden plank deeply etched by drifting sand to reveal the grain, the surface of south-central Pennsylvania shows zigzag ridges formed by bands of hard quartzite. Strata were crumpled into folds during a collision of continents that took place over 200 million years ago to produce the Appalachians. In this September image, the red color depicts vegetation cover, while the blue colors indicate mainly agricultural fields. The cloud bands are formed by condensation from passing aircraft. Note the shadows that the cloud bands cast on the ground.

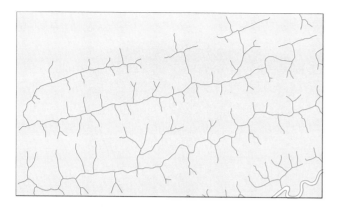

Figure 16.16 A trellis drainage pattern on deeply eroded folds in the Central Appalachians.

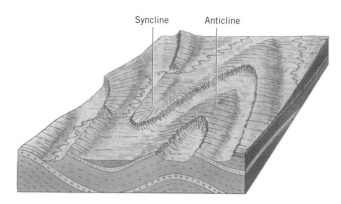

Figure 16.17 Folds with crests that plunge downward give rise to zigzag ridges following erosion. (Drawn by Erwin Raisz. Copyright © by A. N. Strahler.)

tances. In some fold regions, however, the folds are not continuous and level-crested. Instead, the fold crests rise or descend from place to place. A descending fold crest is said to "plunge." Plunging folds give rise to a zigzag line of ridges (Figure 16.17). Excellent examples of plunging folds can be seen in Figure 16.15.

LANDFORMS DEVELOPED ON OTHER LANDMASS TYPES

We now turn to the landforms developed on a few remaining landmass types. First are erosional forms that develop on a landmass exposed to faulting. Then we discuss the landforms that develop on metamorphic rocks and on exposed igneous batholiths. Last are the landforms of eroded volcanoes.

Erosion Forms on Fault Structures

In Chapter 12, we found that active normal faulting produces a sharp surface break called a *fault scarp* (see Figure 12.16). Repeated faulting may produce a great rock cliff hundreds of meters high. Erosion quickly modifies a fault scarp, but, because the fault plane extends hundreds of meters down into the bedrock, its effects on erosional landforms persist for long spans of geologic time. Figure 16.18 shows both the original fault scarp (above) and its later landform expression (below), known as a *fault-line scarp*. Even though the cover of sedimentary strata has been completely removed, exposing the ancient shield rock, the fault continues to produce a landform. Because the fault plane is a zone of weak rock that has been crushed during faulting, it is occupied by a subsequent stream. A scarp persists along the upthrown side. Such scarps on ancient fault lines are numerous in the exposed and covered continental shields.

Figure 16.19 shows some erosional features of a large, tilted fault block. The freshly uplifted block has a steep mountain face, but it is rapidly dissected into deep canyons. The upper mountain face becomes less steep as it is removed, and rock debris accumulates in the form of alluvial fans adjacent to the fault block. Vestiges of the fault plane are preserved in a line of triangular facets. Each facet represents the snubbed end of a ridge between two canyon mouths.

In addition to creating faults, tectonic activity is also capable of uplifting landmasses well above base level. The landmass is often elevated much more rapidly than it can be eroded away, producing a plateau or mountain range that can persist for millions of years. *Focus on*

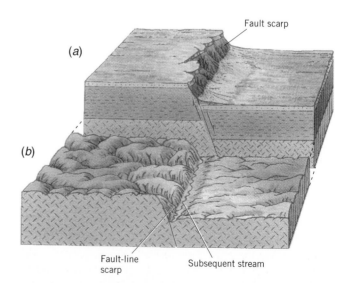

Figure 16.18 (a) A recently formed fault scarp. (b) Despite continental denudation over several million years, the fault plane causes a long narrow valley bounded by a scarp. (Drawn by A. N. Strahler. Copyright © A. N. Strahler.)

Figure 16.19 The near-vertical faces of the foreground peaks in this photo mark a fault scarp along which the rocks have been uplifted by many successive motions. Talus slopes of rock debris line the base of the scarp. Eastern Sierra Nevada, California.

Systems 16.2 • A Model Denudation System explores rates of uplift and erosion in more detail.

Metamorphic Belts

Where strata have been tightly folded and altered into metamorphic rocks during continental collision, denudation eventually develops a landscape with a strong grain of ridges and valleys. These features lack the sharpness and straightness of ridges and valleys in belts of open folds. Even so, the principle that governs the development of ridges and valleys in open folds applies to metamorphic landscapes as well—namely, that rocks resistant to denudation form highlands and ridges, while rocks that are more susceptible to denudation form lowlands and valleys.

Figure 16.20 shows typical erosional forms associated with parallel belts of metamorphic rocks, such as schist, slate, quartzite, and marble. Marble forms valleys, while slate and schist make hill belts. Quartzite stands out boldly and may produce conspicuous narrow hogbacks. Areas of gneiss form highlands.

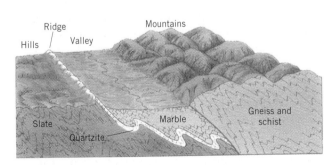

Figure 16.20 Metamorphic rocks form long, narrow parallel belts of valleys and mountains. (Drawn by A. N. Strahler.)

Focus on Systems 16.2 • A Model Denudation System

As landmasses are eroded by fluvial action, streams move mineral matter downhill from uplands to lowlands and ultimately to the ocean or a closed inland basin. As we saw in *Focus on Systems 14.1 • Energy in Stream Flow*, this motion is powered by gravity. When erosion strips upland surfaces, it also lightens the upland landmass. In Chapter 11, we established the principle that the earth's outer shell of hard, brittle rock—the lithosphere—rests on the asthenosphere, a plastic rock layer below. Geologists have shown that the lithosphere literally floats on the plastic asthenosphere, much as an iceberg floats on denser sea water. In the case of the iceberg, if you were to remove the part showing above the water, the iceberg would rise and bring more ice

above the water level. If you placed an added load of ice on the iceberg, it would sink lower in the water but would come to a new position of rest with its summit somewhat higher than before. This flotation principle, as applied to the lithosphere, is called *isostasy*.

When bedrock is eroded from a continent, the lithosphere rises beneath it according to isostacy. However, the rise is somewhat less than the thickness of material removed. When the eroded sediment accumulates in undersea layers on the continental margin, the added load causes the lithosphere to sink lower, but again, the amount of sinking is somewhat less than the thickness of sediment accumulating. A rising or sinking of the crust in response to denudation or sediment accumula-

tion is referred to as *isostatic compensation.*

Isostatic compensation applies to the erosion of landmasses lifted above a base level of erosion. What is responsible for the initial lifting? In many cases, it is the tectonic activity associated with lithospheric plate interactions. During arc-continent and continent-continent collisions, thickening of the continental lithosphere may occur rapidly, producing a high alpine chain—the Himalayas or Andes, for example.

The rate of uplift by tectonic and isostatic processes can be much greater than the rate of denudation by fluvial erosion. Based on direct geologic evidence, rates of recent tectonic uplift in various parts of the world are on the order of 4 to 12 m per 1000 years (about 15 to 40 ft/1000 yr). Rates of denudation are much lower. In rugged terrain, observed rates are on the order of 1 to 1.5 m/1000 yr (about 3 to 5 ft/1000 yr).

The rate of denudation will be strongly influenced by the average summit elevation of the landmass. The higher a mountain block rises, the steeper will be the gradients, on the average, of streams carving into that landmass. As we saw in Chapter 15, both the intensity of stream abrasion and the ability of a stream to transport load increase strongly as channel gradients increase. Valley wall slopes are also steeper, so mass wasting and erosion by overland flow will be more intense. As gradients diminish with time, however, the rate of denudation will also diminish.

Average denudation rates over thousands of years are determined by measuring sediment accumulation in lowland areas. If we had been able to monitor surface elevations over geologic time, we would have recorded the effect of isostatic compensation. For example, if ero-

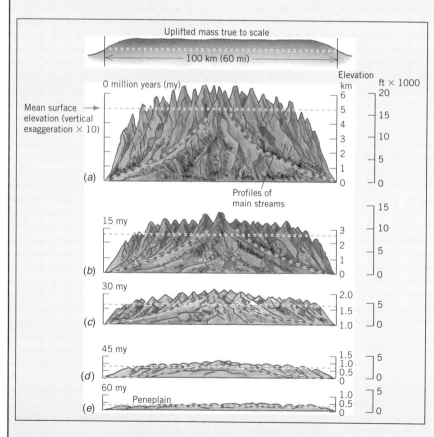

Figure 1   Schematic diagram of landmass denuation. In this model, the average surface elevation is reduced by one-half every 15 million years. (Copyright © A. N. Strahler.)

sion strips 1 m of sediment from an uplifted landmass, isostatic compensation might uplift the mass by 0.8 m in response, leaving a net reduction of elevation of only 0.2 m. Thus, the landmass as a whole will lower at a much slower rate than the rate at which it is stripped by erosion.

Figure 1 diagrams the erosion of an uplifted landmass. Here we assume that a large crustal mass is raised as a single block from a previous position close to sea level to a mean surface elevation of 5000 m (about 16,000 ft). The uplifted block has a very broadly domed summit and steep sides, and is bordered by low areas, at or below sea level, that serve as receiving sites for sediment carried out from the mountain block by streams. During the uplift, streams are active, and so the landmass shown in stage *a* has already been carved into a maze of steep-walled gorges that are organized into a fluvial system of steep-gradient streams. In stage *b*, after 15 m.y., the average elevation is reduced by half, to 2500 m (about 8000 ft). Slopes and stream gradients are still steep, but have declined significantly. In stage *c*, at 30 m.y., the elevation is reduced by half again, to 1250 m (about 4000 ft). By now the slopes are much more gentle, as are stream gradients. Stage *d* marks another half-reduction, as does stage *e*. At this time (60 m.y. after initial uplift), the landmass has been eroded to a low-lying peneplain with a mean surface elevation of about 300 m (about 1000 ft).

This description follows a simple model for landmass denudation—exponential decay. This is the same model that describes the decay of radioactive isotopes, which we presented in *Working It Out 10.1 • Radioactive Decay*. Figure 2 shows how the mean elevation decreases with time following the exponential decay model. For each 15 m.y. period, the mean landmass elevation is reduced by half. The net denudation rate also drops by half with each 15 m.y. interval. Thus, the landmass is eroded at an ever-slower rate as it approaches a peneplain.

The initial rate of denudation, 231 m/m.y. (758 ft/m.y.), translates to 0.231 m/1000 yr (0.758 ft/1000 yr). Note that this rate is the net lowering rate, which includes the effect of isostatic compensation. The true rate at which rock material is stripped from the landmass will be much greater. Assuming that isostatic compensation recovers 80 percent of the lost elevation, the erosion rate will be five times greater. Thus, the initial erosion rate in our example would be about 1.15 m/1000 yr (3.79 m/1000 yr), which agrees well with the rates of 1.0–1.5 m/1000 yr (about 3–5 ft/yr) observed for mountainous terrain. As noted, this rate will decrease as the landmass is eroded.

Note also that tectonic processes may interrupt the slow denudation of the landmass at any time. The landmass may be uplifted, or rejuvenated, as shown in Figure 15.18. Or, it may be downdropped and subjected to sedimentation from surrounding highland areas, or even lowered below sea level. Furthermore, climate may change, causing erosion rates to change. In fact, there are few places on earth where the simple model presented above will actually apply. However, it establishes the principles of uplift, isostatic compensation, and denudation that are always in operation, even though a particular landscape may have a complex history.

What powers the denudation of the landscape? Tectonic processes, including isostatic rebound, provide gravitational potential energy as an input when the landmass is lifted above base level. Solar energy also powers the process by providing the precipitation that drives the fluvial processes of erosion.

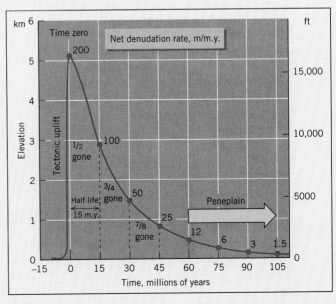

Figure 2 Graph of decrease in average surface elevation with time, as shown in Figure 2. (Copyright © A. N. Strahler.)

Parts of New England, particularly the Taconic and Green Mountains of New Hampshire and Vermont, illustrate the landforms eroded on an ancient metamorphic belt. The larger valleys trend north and south and are underlain by marble. These are flanked by ridges of gneiss, schist, slate, or quartzite.

Exposed Batholiths and Monadnocks

Recall from Chapter 12 that batholiths—huge plutons of intrusive igneous rock—are formed deep below the earth's surface. Some are eventually uncovered by erosion and appear at the surface. Because batholiths are typically composed of resistant rock, they are eroded into hilly or mountainous uplands.

Batholiths of granitic composition are a major ingredient in the mosaic of ancient rocks comprising the continental shields. A good example is the Idaho batholith, a granite mass exposed over an area of about 40,000 sq km (about 16,000 sq mi)—a region almost as large as New Hampshire and Vermont combined. Another example is the Sierra Nevada batholith of California, which makes up most of the high central part of that great mountain block.

A dendritic drainage pattern is often well developed on an eroded batholith (Figure 16.21). Note how similar this pattern is to that on horizontal strata (Figure 16.6).

Small bodies of granite, representing domelike projections of batholiths that lie below, are often found surrounded by ancient metamorphic rocks into which the granite was intruded (Figure 16.22). The name **monadnock** has been given to an isolated mountain or

hill that rises conspicuously above a peneplain. A monadnock develops because the rock within the monadnock is much more resistant to denudation processes than the bedrock of the surrounding region. The name is taken from Mount Monadnock in southern New Hampshire.

Deeply Eroded Volcanoes

Recall from Chapter 12 that volcanoes can be divided into two types—stratovolcanoes and shield volcanoes. Stratovolcanoes are typically constructed by explosive eruptions accumulating of layers of thick, gummy andesite lava flows and airborne tephra, while shield volcanoes are built primarily of thin, runny flows of basalt. Thus, these two types of volcanoes produce different types of landforms as they erode away.

Figure 16.23 shows successive stages in the erosion of stratovolcanoes, lava flows, and a caldera. Shown in block (*a*) are active volcanoes in the process of building. These are initial landforms. Lava flows issuing from the volcanoes have spread down into a stream valley, following the downward grade of the valley and forming a lake behind the lava dam.

In block (*b*) some changes have taken place. The most conspicuous change is the destruction of the largest volcano to produce a caldera. A lake occupies the caldera, and a small cone has been built inside. One of the other volcanoes, formed earlier, has become extinct. It has been dissected by streams and has lost its smooth conical form. Smaller, neighboring volcanoes are still active, and the contrast in form is marked.

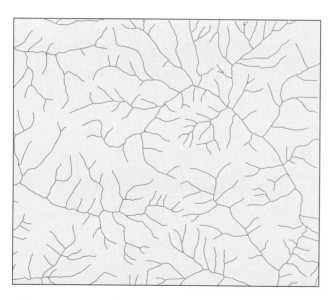

Figure 16.21 This dendritic drainage pattern is developed on the dissected Idaho batholith.

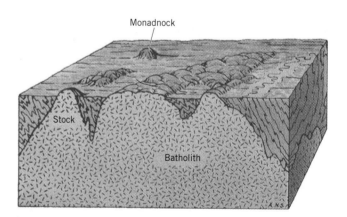

Figure 16.22 Batholiths appear at the land surface only after long-continued erosion has removed thousands of meters of overlying rocks. Small projections of the granite intrusion appear first and are surrounded by older rock. (Drawn by A. N. Strahler.)

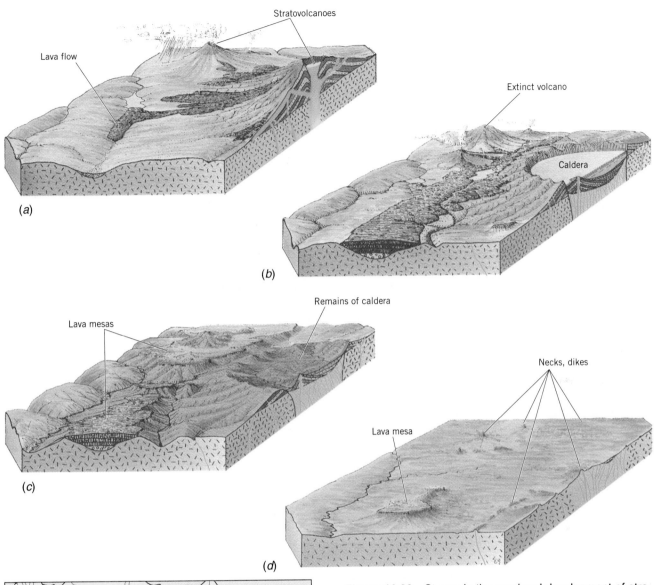

Figure 16.23   Stages in the erosional development of stratovolcanoes and lava flows. (Drawn by Erwin Raisz. Copyright © by A. N. Strahler.)

Figure 16.24   Radial drainage patterns of volcanoes in the East Indies. The letter "C" shows the location of a crater.

The system of streams on a dissected volcano cone is a *radial drainage pattern.* Because these streams take their position on a slope of an initial land surface, they are of the consequent variety. It is often possible to recognize volcanoes from a drainage map alone because of the perfection of the radial pattern (Figure 16.24). A good example of a partly dissected volcano is Mount Shasta in California (Figure 16.25).

In Figure 16.23c all volcanoes are extinct and have been deeply eroded. The caldera lake has been drained, and the rim has been worn to a low, circular ridge. The lava flows have been able to resist erosion far better than the rock of the surrounding area and have come to stand as mesas high above the general level of the region.

Block (*d*) of Figure 16.23 shows an advanced stage of erosion of stratovolcanoes. All that remains now is a small, sharp peak, called a *volcanic neck*—which is the remains of lava that solidified in the pipe of the volcano. Radiating from it are wall-like dikes, formed of magma, which filled radial fractures around the base of the volcano. Perhaps the finest illustration of a volcanic neck with radial dikes is Ship Rock, New Mexico (Figure 16.26).

Shield volcanoes show erosion features that are quite different from those of stratovolcanoes. Figure 16.27a shows the first stage of erosion of Hawaiian shield volcanoes, beginning in the upper diagram with the active volcano and its central depression. These are initial landforms. Radial consequent streams cut deep

Figure 16.25 Mount Shasta in the Cascade Range, northern California, is a partly dissected composite volcano. It has been eroded by streams and small alpine glaciers. Part way up the right-hand side is a more recent subsidiary volcanic cone, called Shastina, with a sharp crater rim.

canyons into the flanks of the extinct shield volcano (*b*), and these canyons are opened out into deep, steep-walled amphitheaters such as Waimea Canyon, on the island of Kauai (Figure 16.28). In the last stages (*c*), the

Figure 16.26 Ship Rock, New Mexico, is a volcanic neck enclosed by a weak shale formation. The peak rises about 520 m (about 1700 ft) above the surrounding plain. In the foreground and to the left, wall-like dikes extend far out from the central peak.

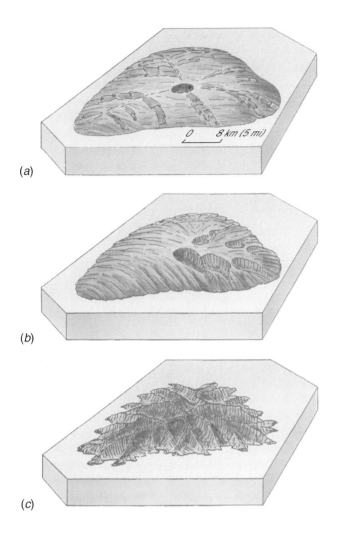

(a)

(b)

(c)

Figure 16.27 Shield volcanoes in various stages of erosion make up the Hawaiian Islands. (*a*) Newly formed dome with central depression. (*b*) Early stage of erosion with deeply eroded valley heads. (*c*) Advanced erosion stage with steep slopes and mountainous relief. (Drawn by A. N. Strahler.)

original surface of the shield volcano is entirely obliterated, leaving a rugged mountain mass made up of sharp-crested divides and deep canyons.

With this chapter, we conclude our series of three chapters on running water as a landforming-making agent. Because the landscapes of most regions on the earth's land surface are produced by fluvial processes acting on differing rock types, running water is by far the most important agent in creating the variety of landforms that we see around us. The three remaining agents of denudation are waves, wind, and glacial ice. These are the subjects of the final two chapters of Part III. In the first, we describe how waves shape the landforms of beaches and coasts, and how wind shapes the dunes of coasts and deserts. In the second, we relate how glacial ice scrapes, pushes, and deposits rock materials to form the distinctive landforms of glaciated regions.

Figure 16.28 Waimaea Canyon, over 760 m (about 2500 ft) deep, has been eroded into the flank of an extinct shield volcano on Kauai, Hawaii. Gently sloping layers of basaltic lava flows are exposed in the canyon walls.

CHAPTER SUMMARY

This chapter has added a new dimension to the realm of landforms produced by fluvial denudation—the variety and complexity introduced by differences in rock composition and crustal structure. Because the various kinds of rocks offer different degrees of resistance to the forces of denudation, they exert a controlling influence on the shapes of landforms. We have seen, too, that streams carving up a landmass are controlled to a high degree by the structure of the rock on which they act. In this way, distinctive drainage patterns evolve.

In arid regions of horizontal strata, resistant rock layers produce vertical cliffs separated by gentler slopes on less resistant rocks. The drainage pattern is dendritic. Where strata are gently dipping, as on a coastal plain, the more resistant rock layers stand out as cuestas, interspersed with lowland valleys on weaker rocks. Consequent streams cut across the cuestas toward the ocean, and subsequent streams follow the valleys. This forms a trellis drainage pattern.

Where rock layers are uparched into a dome, erosion produces a circular arrangement of rock layers outward from the center of the dome. Resistant strata form hogbacks, and weaker rocks form lowlands. Igneous rocks are often revealed in the center. The drainage pattern is annular. In fold belts, the sequence of synclines and anticlines brings a linear pattern of rock layers to the surface. Resistant strata form ridges, and weaker strata form valleys. Like the coastal plain, the drainage pattern will be trellised.

Faulting provides an initial surface along which rock layers are moved—the fault scarp. This feature can persist as a fault-line scarp long after the initial scarp is gone. In metamorphic belts, weak layers of shale, slate, and marble underlie valleys, while gneiss and schist form uplands. Quartzite stands out as a ridge-former. Exposed batholiths are often composed of uniform, resistant igneous rock. They erode to form a dendritic drainage pattern. Monadnocks of intrusive igneous rock stand up above a plain of weaker rocks.

Stratovolcanoes produce lava flows that initially follow valleys, but are highly resistant to erosion. After erosion of the surrounding area, they can remain as highlands, lava ridges, or mesas. At the last stages of erosion, all that remains of stratovolcanoes are necks and dikes. Hawaiian shield volcanoes are eroded by streams that form deeply carved valleys with steeply sloping heads. Eventually, these merge to produce a landscape of steep slopes and knifelike ridges.

KEY TERMS

plateau	coastal plain	ridge-and-valley land-
mesa	cuesta	scape
butte	artesian well	monadnock
drainage pattern	sedimentary dome	

REVIEW QUESTIONS

1. Why is there often a direct relationship between landforms and rock structure? How are geologic structures formed deep within the earth exposed at the surface?

2. Which of the following types of rocks—shale, limestone, sandstone, conglomerate, igneous rocks—tends to form lowlands? uplands?

3. How are the tilt and orientation of a natural rock plane measured?

4. Identify and sketch a typical landform of flat-lying rock layers found in an arid region.

5. How are coastal plains formed? Identify and describe two landforms found on coastal plains. What drainage pattern is typical of coastal plains?

6. What types of landforms are associated with sedimentary domes? How are they formed? What type of drainage pattern would you expect for a dome and why? Provide an example of an eroded sedimentary dome and describe it briefly.

7. How does a ridge-and-valley landscape arise? Explain the formation of the ridges and valleys. What type of drainage pattern is found on this landscape? Where in the United States can you find a ridge-and-valley landscape?

8. What type of landform(s) might you expect to find in the presence of a fault? Why?

9. Which of the following types of rocks—schist, slate, quartzite, marble, gneiss— tends to form lowlands? uplands?

10. What types of landform and drainage patterns develop on batholiths? Provide a sketch of a batholith. What is the difference between a batholith and a monadnock?

11. Describe and sketch the stages through which a landscape of stratovolcanoes is eroded. What distinctive landforms remain as the last remnants?

12. Do shield volcanoes erode differently from stratovolcanoes? How so?

Focus on Systems 16.2 • A Model Denudation System

1. What is isostasy? Explain isostatic compensation and why it occurs.

2. How do rates of uplift and denudation compare? How is the net rate of denudation affected by isostatic compensation?

3. What principle explains how the mean elevation of a landmass and the net rate of its denudation change with time?

ESSAY QUESTIONS

1. Imagine the following sequence of sedimentary strata—sandstone, shale, limestone, and shale. What landforms would you expect to develop in this structure if the sequence of beds is (*a*) flat-lying in an arid landscape; (*b*) slightly tilted as in a coastal plain; (*c*) folded into a syncline and an anticline in a fold belt; (*d*) fractured and displaced by a normal fault? Use sketches in your answer.

2. A region of ancient mountain roots, now exposed at the surface, includes a central core of plutonic rocks surrounded on either side by belts of metamorphic rocks. What landforms would you expect for this landscape and why?

PROBLEMS

Working It Out 16.1 • Properties of Stream Networks

1. A sixth-order stream network has an average bifurcation ratio of 3.5. How many stream segments would you expect for streams of order 1–6?

2. A fourth-order stream network has an average mean segment length ratio of 3.2, and the mean length of first-order stream segments is 0.12 km. What are the expected cumulative mean lengths for stream segments of order 2–4? Note that the (noncumulative) mean length for order u is simply the difference between the cumulative length at order u and the cumulative length at order u-1. That is, $L_u = L_u^* - L_{u-1}^*$. Using this relation, derive the expected (noncumulative) mean lengths for segments of orders 2–4.

Chapter 17

The Work of Waves and Wind

 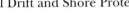
In Chapter 5, we saw how the rotation of the earth on its axis, combined with unequal solar heating of the earth's surface, produces a system of global winds in which air moves as a fluid across the earth's surface. This motion of air produces a frictional drag that can move surface materials. When winds blow over broad expanses of water, waves are generated. These waves then expend their energy at coastlines, eroding rock, moving sediment, and creating landforms. When winds blow over expanses of soil or alluvium that are not protected by vegetation, fine particles can be picked up and carried long distances before coming to rest. Coarser particles, such as sand, can also be moved to build landforms, such as sand dunes. Because wave action on beaches provides abundant sand, we often find landforms of both wind and waves near coastlines.

This chapter describes the processes of erosion and deposition, as well as the landforms, that result when wind power moves surface materials, either directly as wind or indirectly as waves. Unlike fluvial action or

---

Wave-cut stacks at Port Campbell National Park, Victoria, Australia.

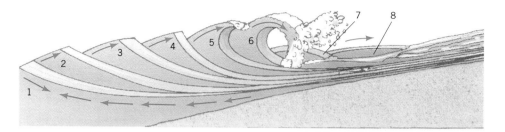

Figure 17.1 A breaking wave. As the wave approaches the beach (1–3), it steepens (4–5) and finally falls forward (6–7), rushing up the beach slope (8). (After W. M. Davis.)

Figure 17.2 Mohegan Bluffs, Block Island, Rhode Island. This marine scarp of Ice Age sediments is being rapidly eroded, threatening historic Southeast Lighthouse. In 1993 the lighthouse was moved to a safer spot 73.5 m (245 ft) away.

mass wasting, wind and breaking waves can move materials uphill, against the force of gravity. For this reason, wind power is a unique land-forming agent. Another agent that acts in coastal regions is the tide—the rhythmic rise and fall in sea level that is generated by the gravitational attraction of the sun and moon. This subject is also covered in our chapter.

Throughout this chapter we will use the term **shoreline** to mean the shifting line of contact between water and land. The broader term **coastline**, or simply **coast**, refers to a zone in which coastal processes operate or have a strong influence. The coastline includes the shallow water zone in which waves perform their work, as well as beaches and cliffs shaped by waves, and coastal dunes. Also often present are **bays**—bodies of water that are sheltered by the configuration of the coast from strong wave action. Where a river empties into an ocean bay, the bay is termed an **estuary**. In an estuary, fresh and ocean water mix, creating a unique habitat for many plants and animals which is neither fresh water nor ocean.

THE WORK OF WAVES

The most important agent shaping coastal landforms is wave action. The energy of waves is expended primarily in the constant churning of mineral particles and water as waves break at the shore. This churning erodes shoreline materials, moving the shoreline landward. But, as we will see shortly, the action of waves and currents can also move sediment along the shoreline for long distances. This activity can build beaches outward as well as form barrier islands just off shore.

Waves travel across the deep ocean with little loss of energy. When waves reach shallow water, the drag of the bottom slows and steepens the wave until it falls and collapses as a *breaker* (Figure 17.1). Many tons of water surge forward, riding up the beach slope.

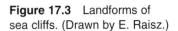

Figure 17.3 Landforms of sea cliffs. (Drawn by E. Raisz.)

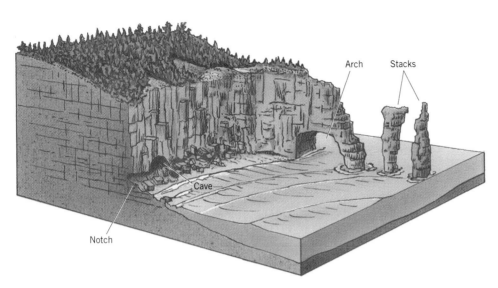

Where weak or soft materials—various kinds of regolith, such as alluvium—make up the coastline, the force of the forward-moving water alone easily cuts into the coastline. Here, erosion is rapid, and the shoreline may recede rapidly. Under these conditions, a steep bank—a *marine scarp*—is the typical coastal landform (Figure 17.2). It retreats steadily under attack of storm waves.

Marine Cliffs

Where a **marine cliff** lies within reach of the moving water, it is impacted with enormous force. Rock fragments of all sizes, from sand to cobbles, are carried by the surging water and thrust against bedrock of the cliff. The impact breaks away new rock fragments, and the cliff is undercut at the base. In this way, the cliff erodes shoreward, maintaining its form as it retreats. The retreat of a marine cliff formed of hard bedrock is exceedingly slow, when judged in terms of a human life span.

Figure 17.3 illustrates some details of a typical marine cliff. A deep basal indentation, the *wave-cut notch*, marks the line of most intense wave erosion. The waves find points of weakness in the bedrock and penetrate deeply to form crevices and *sea caves*. Where a more resistant rock mass projects seaward, it may be cut through to form a picturesque *sea arch*. After an arch collapses, a rock column, known as a *stack*, remains. Eventually, the stack is toppled by wave action and is leveled. As the sea cliff retreats landward, continued wave abrasion forms an *abrasion platform* (Figure 17.4).

Figure 17.4 A wave-cut abrasion platform on the central coast of California, Montana d'Oro State Park.

This sloping rock floor continues to be eroded and widened by abrasion beneath the breakers. If a beach is present, it is little more than a thin layer of gravel and cobblestones atop the abrasion platform.

Beaches

Where sand is in abundant supply, it accumulates as a thick, wedge-shaped deposit, or **beach**. Beaches absorb the energy of breaking waves. During short periods of storm activity, the beach is cut back, and sand is carried offshore a short distance by the heavy wave action. However, the sand is slowly returned to the beach during long periods when waves are weak. In this way, a beach may retain a fairly stable but alternating configuration over many years' time.

Beaches are shaped by alternate landward and seaward currents of water generated by breaking waves. After a breaker has collapsed, a foamy, turbulent sheet of water rides up the beach slope. This *swash* is a powerful surge that causes a landward movement of sand and gravel on the beach. When the force of the swash has been spent against the slope of the beach, a return flow, or *backwash*, pours down the beach (Figure 17.5*a*). This "undercurrent" or "undertow," as it is popularly called, can be strong enough to sweep unwary bathers off their feet and carry them seaward beneath the next oncoming breaker.

Littoral Drift

The unceasing shifting of beach materials with swash and backwash of breaking waves also results in a sidewise movement known as *beach drift* (Figure 17.5*a*). Wave fronts usually approach the shore at less than a right angle, so that the swash and its burden of sand ride obliquely up the beach. After the wave has spent its energy, the backwash flows down the slope of the beach in the most direct downhill direction. The particles are dragged directly seaward and come to rest at a position to one side of the starting place. This movement is repeated many times and individual rock particles travel long distances along the shore. Multiplied many thousands of times to include the numberless particles of

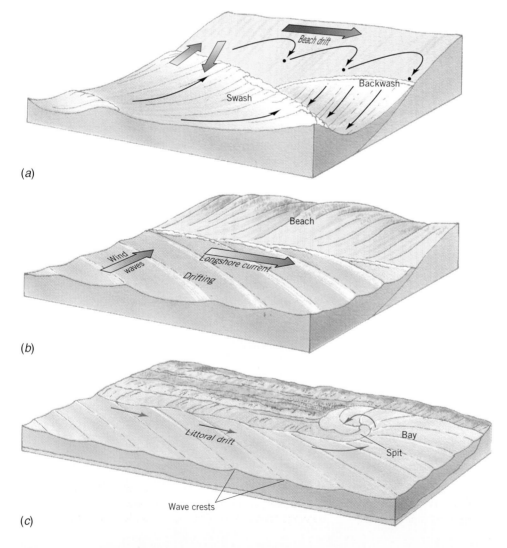

(a)

(b)

(c)

Figure 17.5 How waves move sediment by littoral drift. (*a*) Swash and backwash move particles along the beach in beach drift. (*b*) Waves set up a longshore current that move particles by longshore drift. (*c*) Littoral drift, produced by these two processes, creates a sandspit.

the beach, beach drift becomes a very significant form of sediment transport.

Sediment in the shore zone is also moved along the beach in a related, but different, process. When waves approach a shoreline at an angle to the beach, a current is set up parallel to the shore in a direction away from the wind. This is known as a *longshore current* (Figure 17.5*b*). When wave and wind conditions are favorable, this current is capable of carrying sand along the sea bottom. The process is called *longshore drift*. Beach drift and longshore drift, acting together, move particles in the same direction for a given set of onshore winds. The total process is called **littoral drift** (Figure 17.5*c*). ("Littoral" means "pertaining to a coast or shore.")

Littoral drift operates to shape shorelines in two quite different situations. Where the shoreline is straight or broadly curved for many kilometers at a stretch, littoral drift moves the sand along the beach in one direction for a given set of prevailing winds. This situation is shown in Figure 17.5*c*. Where a bay exists, the sand is carried out into open water as a long finger, or *sandspit* (Figure 17.6). As the sandspit grows, it forms a barrier, called a *bar*, across the mouth of the bay.

A second situation is shown in Figure 17.7. Here the coastline consists of prominent headlands, projecting seaward, and deep bays. Approaching wave fronts slow when the water becomes shallow, and this slowing effect causes the wave front to wrap around the headland. High, wave-cut cliffs develop shoreward of an abrasion platform. Sediment from the eroding cliffs is carried by littoral drift along the sides of the bay, converging on the head of the bay. The result is a crescent-shaped beach, often called a *pocket beach*.

Beach erosion and deposition is one part of a larger system involving the movement of sediment provided by streams through the shallow water zone by littoral drift. In *Focus on Systems 17.1 • The Coastal Sediment Cell as a Matter Flow System*, we provide a systems view of this sediment transport.

EYE ON THE ENVIRONMENT:
Littoral Drift and Shore Protection

When sand arrives at a particular section of the beach more rapidly than it is carried away, the beach is widened and built oceanward. This change is called **progradation** (building out). When sand leaves a section of beach more rapidly than it is brought in, the beach is narrowed and the shoreline moves landward. This change is called **retrogradation** (cutting back).

Along stretches of shoreline affected by retrogradation, the beach may be seriously depleted or even entirely disappear, destroying valuable shore property. In some circumstances, structures can be installed that will cause progradation, and so build a broad, protective beach. This is done by installing groins at close in-

Figure 17.6 This white sandspit is growing in a direction toward the observer. Monomoy Point, Cape Cod, Massachusetts.

tervals along the beach. A *groin* is simply a wall or embankment built at right angles to the shoreline. It may be constructed of large rock masses, of concrete, or of wooden pilings. The groins act to trap sediment moving along the shore as littoral drift (Figure 17.8).

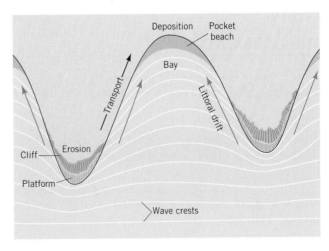

Figure 17.7 On an embayed coast, sediment is carried from eroding headlands to the bayheads, where pocket beaches are formed. (Copyright © A. N. Strahler.)

Figure 17.8 Erosion by severe storms during the winter of 1993 has carved out an inlet in this barrier beach on the south shore of Long Island, New York. A system of groins has trapped sand, protecting the far stretch of beach. In the foreground, the beach has receded well inland of the houses that were once located on its edge. (© Cliff de Bear/Newsday)

In some cases, the source of beach sand is sediment delivered to the coast by a river. Construction of dams far upstream on the river may drastically reduce the sediment load of the river, cutting off the source of sand for littoral drift. Retrogradation can then occur on a long stretch of shoreline.

TIDAL CURRENTS

Most marine coastlines are influenced by the *ocean tide*, a rhythmic rise and fall of sea level under the influence of changing attractive forces of moon and sun on the rotating earth. Where tides are great, the effects of changing water level and the tidal currents thus set in motion are of major importance in shaping coastal landforms.

The tidal rise and fall of water level is graphically represented by the *tide curve*. Figure 17.9 is a tide curve for Boston Harbor covering a day's time. The water reached its maximum height, or high water, at 3.6 m (11.8 ft) on the tide staff, and then fell to its minimum height, or low water, at 0.8 m (2.6 ft), occurring about 6¼ hours later. A second high water occurred about

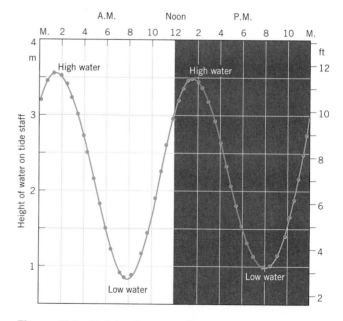

Figure 17.9 Height of water at Boston Harbor measured every half hour.

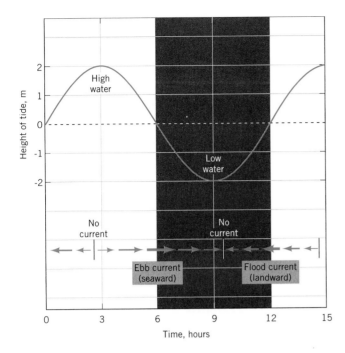

Figure 17.10 The ebb current flows seaward as the tide level falls. The flood current flows landward as the tide level rises. Tidal currents are strongest when water level is changing most rapidly, in the middle of the cycle.

$12^1/_2$ hours after the previous high water, completing a single tidal cycle. In this example, the tidal range, or difference between heights of successive high and low waters, is 2.8 m (9.2 ft).

In bays and estuaries, the changing tide sets in motion currents of water known as *tidal currents*. The relationships between tidal currents and the tide curve are shown in Figure 17.10. When the tide begins to fall, an *ebb current* sets in. This flow ceases about the time when the tide is at its lowest point. As the tide begins to rise, a landward current, the *flood current*, begins to flow.

Tidal Current Deposits

Ebb and flood currents generated by tides perform several important functions along a shoreline. First, the currents that flow in and out of bays through narrow inlets are very swift and can scour the inlet strongly. This keeps the inlet open, despite the tendency of shore-drifting processes to close the inlet with sand.

Second, tidal currents carry large amounts of fine silt and clay in suspension. This fine sediment is derived from streams that enter the bays or from bottom muds agitated by storm wave action. It settles to the floors of the bays and estuaries, where it accumulates in layers and gradually fills the bays. Much organic matter is present in this sediment.

In time, tidal sediments fill the bays and produce mud flats, which are barren expanses of silt and clay.

They are exposed at low tide but covered at high tide. Next, a growth of salt-tolerant plants takes hold on the mud flat. The plant stems trap more sediment, and the flat is built up to approximately the level of high tide, becoming a *salt marsh* (Figure 17.11). A thick layer of peat is eventually formed at the surface. Tidal currents maintain their flow through the salt marsh by means of a highly complex network of winding tidal streams.

TYPES OF COASTLINES

The world's coastlines present a number of different coastline types. Each type is unique because of the distinctive landmass against which the ocean water has come to rest. One group of coastline types derives its qualities from **submergence**, the partial drowning of a coast by a rise of sea level or a sinking of the crust. Another group derives its qualities from **emergence**, the exposure of submarine landforms by a falling of sea level or a rising of the crust. Another group of coastline types results when new land is built out into the ocean by volcanoes and lava flows, by the growth of river deltas, or by the growth of coral reefs.

A few important types of coastlines are illustrated in Figure 17.12. The first two are the result of submergence. The *ria coast* is a deeply embayed coast resulting from submergence of a landmass dissected by streams. The *fiord coast* is deeply indented by steep-

Figure 17.11 A tideland salt marsh. Chincoteague National Wildlife Refuge, Virginia.

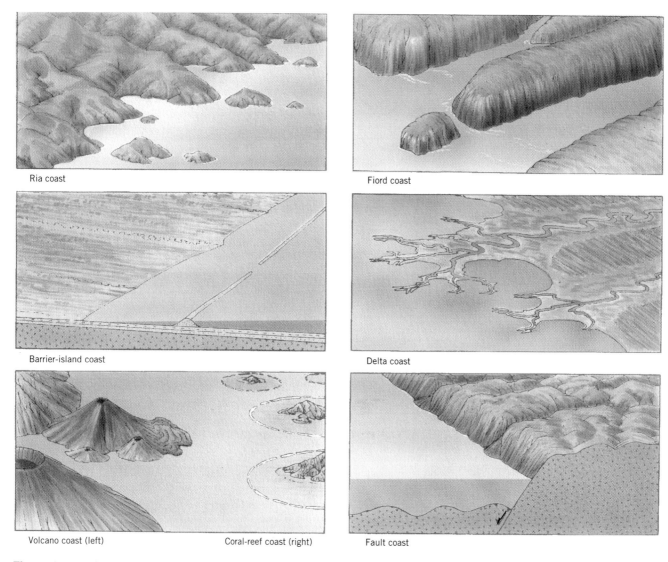

Ria coast

Fiord coast

Barrier-island coast

Delta coast

Volcano coast (left)   Coral-reef coast (right)

Fault coast

Figure 17.12   Seven common kinds of coastlines are illustrated here. These examples have been selected to illustrate a wide range in coastal features. (Drawn by A. N. Strahler.)

walled fiords, which are submerged glacial troughs (discussed further in Chapter 18). The *barrier-island coast* is associated with a recently emerged coastal plain. Here, the offshore slope is very gentle, and a barrier island of sand, lying a short distance from the coast, is created by wave action. Large rivers build elaborate deltas, producing *delta coasts*. The *volcano coast* is formed by the eruption of volcanoes and lava flows, partly constructed below water level. Reef-building corals create new land and make a *coral-reef coast*. Down-faulting of the coastal margin of a continent can allow the shoreline to come to rest against a fault scarp, producing a *fault coast*.

Shorelines of Submergence

Shorelines of submergence include ria coasts and fiord coasts. The ria coast, which takes its name from the

Spanish word for estuary, *ria*, has many offshore islands. A ria coast is formed when a rise of sea level or a crustal sinking (or both) brings the shoreline to rest against the sides of valleys previously carved by streams (Figure 17.13*a*). Wave attack forms cliffs on the exposed seaward sides of islands and headlands (*b*). Sediment produced by wave action accumulates in the form of beaches along the cliffed headlands and at the heads of bays. This sediment is carried by littoral drift and is often built into sandspits across bay mouths and as connecting links between islands and mainland (*c*). Eventually, the sandspits seal off the bays, forming estuaries (*d*). If sea level remains at the same height with respect to the land for a long time, the coast may evolve into a cliffed shoreline of narrow beaches (*e*).

The fiord coast is similar to the ria coast. However, the submerged valleys were carved by flowing glaciers instead of streams. As a result, the valleys are deep, with straight, steep sides. Because sediment rapidly sinks

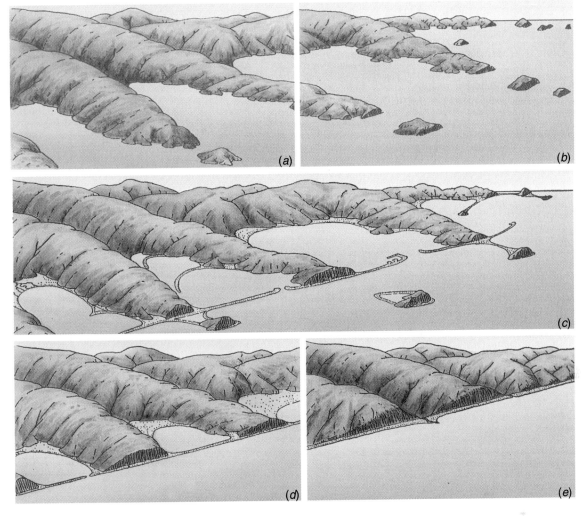

Figure 17.13   Stages in the evolution of a ria coastline. (Drawn by A. N. Strahler.)

into the deep water, beaches are rare. Fiords are described further in Chapter 18 (see Figure 18.7).

Many glaciated coastlines, such as those of New England and the Canadian Maritime Provinces, are shorelines of submergence. During the Ice Age, the weight of ice sheets depressed the crust of these regions, and they are still slowly rising now that the ice sheets are gone. Meanwhile, ocean waves and currents are rapidly eroding their rocky shorelines, creating bays, bars, and estuaries as shown in Figure 17.13.

Barrier-Island Coasts

In contrast to the bold relief and deeply embayed outlines of coastlines of submergence are low-lying coasts from which the land slopes gently beneath the sea. The coastal plain of the Atlantic and Gulf coasts of the United States presents a particularly fine example of such a gently sloping surface. As we explained in Chapter 16, this coastal plain is a belt of relatively young sedimentary strata, formerly accumulated beneath the sea

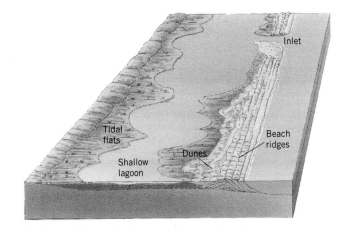

Figure 17.14   A barrier island is separated from the mainland by a wide lagoon. Sediments fill the lagoon, while dune ridges advance over the tidal flats. An inlet allows tidal flows to pass in and out of the lagoon. (Drawn by A. N. Strahler.)

Focus on Systems 17.1 • The Coastal Sediment Cell as a Matter Flow System

Regular beachgoers are familiar with the changes in beaches that occur from day to day and season to season. During languid summer days of light winds and waves, the beach builds outward toward the ocean and establishes a steep face exposed to the rise and fall of the waves with the tide. Afternoon sea breezes move sand grains along the beach and landward into the dunes nearby.

With the storms and surf of winter, the waves work the beach intensively, scouring sand away from the beach and building offshore bars. Waves break farther out, and their swash traverses a long span of beach. At high tides, storm waves attack coastal cliffs, undercutting and releasing sediment or loose fragments of bedrock. Beyond the breakers, a strong littoral drift current, powered by winter's larger waves, moves large quantities of sediment parallel to the shore.

If you walk the beach for some

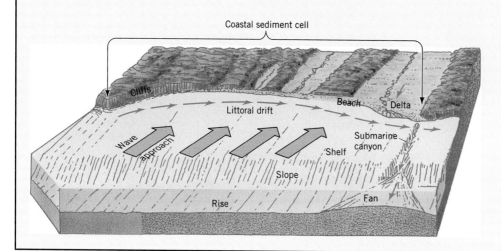

Pictorial diagram of a coastal sediment cell typical of the coast of southern California. (Drawn by A. N. Strahler. Copyright © A. N. Strahler.)

as deposits on the continental shelf. During the latter part of the Cenozoic Era and into recent time, the coastal plain emerged from the ocean as a result of repeated crustal uplifts.

Along much of the Atlantic Gulf coast there exist *barrier islands,* low ridges of sand built by waves and further increased in height by the growth of sand dunes (Figure 17.14). Behind the barrier island lies a *lagoon.* It is a broad expanse of shallow water in places largely filled with tidal deposits.

A characteristic feature of barrier islands is the presence of gaps, known as *tidal inlets.* Strong currents flow alternately landward and seaward through these gaps as the tide rises and falls. In heavy storms, the barrier island may be breached by new inlets (Figure 17.8). After that occurs, tidal current scour will tend to keep a new inlet open. In some cases, the inlet is closed later by shore drifting of beach sand. Perhaps the finest example of a barrier-island coast is the Gulf coast of Texas (Figure 17.15).

Figure 17.15 A Landsat image of the Texas barrier-island coast. Corpus Christi is located on the round bay just below and to the left of center. Red colors on the Landsat image indicate vegetation. Bright white areas are dunes or beach sand.

distance, you will also notice changes. The beach will be very narrow at headlands and promontories, where wave action is intense and littoral drift quickly moves sediment away. In sheltered locations, littoral drift provides sediment that accumulates in pocket beaches. At the mouth of a stream or river, the beach broadens in response to an influx of sediment, and as you walk in the direction of littoral drift, the beach continues to be wider for some distance beyond the river mouth.

What you cannot see, however, is the presence of submarine canyons crossing the continental shelf and extending out to great depth. Sediment moving offshore in the littoral drift zone is carried into these canyons, where currents of turbid, sediment-laden water carry the sediment to undersea fans deposited along the continental rise (see Figure 11.10). Many submarine canyons were cut by streams during the Ice Age, when sea level was 100 meters or more below present levels.

The movement of sediment along a typical stretch of the North American Pacific Coast can be described as an open matter flow system, which we term here a coastal sediment cell. The figure at the left is a sketch of a coastal sediment cell. Wave approach at an angle to the beach generates a littoral drift that moves sediment parallel to the beach. Cliff erosion provides a source of sediment sufficient for narrow beaches to form in the direction of the drift current. Streams provide the major portion of the sediment, however, with very little derived from cliff erosion in most locations. Beach sediment forms a storage pool that is increased in summer and reduced in winter. Coupled to the beach is another pool of sediment in storage—sand in dunes that are built by wind and eroded by waves in intense storms. The system output is a flow of sediment through the submarine canyon to the continental rise.

The matter flow system of the coastal sediment cell underscores the importance of the flow of sediment from rivers in nourishing beaches. Many coastal communities have experienced beach recession because the rivers that feed their coastlines have been dammed, trapping sediment upstream and keeping it out of the flow system. With less sediment flowing through the coastal system as a whole, less is available for storage on the beach, and rock cliffs are more rapidly cut back. The California coast, with so many of its rivers dammed for water supply, provides many good examples of this accelerated wastage of cliffs and the loss of increasing numbers of oceanside homes each year.

Delta and Volcano Coasts

The deposit of clay, silt, and sand made by a stream or river where it flows into a body of standing water is known as a **delta**. Deposition is caused by rapid reduction in velocity of the current as it pushes out into the standing water. Typically, the river channel divides and subdivides into lesser channels called *distributaries*. The coarser sand and silt particles settle out first, while the fine clays continue out farthest and eventually come to rest in fairly deep water. Contact of fresh with saltwater causes the finest clay particles to clot together and form larger particles that settle to the sea floor.

Deltas show a wide variety of outlines. The Nile delta has the basic triangular shape of the Greek letter delta. In outline, it resembles an alluvial fan. The Mississippi delta has a different shape. Long, branching fingers grow far out into the Gulf of Mexico at the ends of the distributaries, giving the impression of a bird's foot. A satellite image of the delta, Figure 17.16, shows the great quantity of suspended sediment—clay and fine silt—being discharged by the river into the Gulf. It amounts to about 1 million metric tons per day.

Delta growth is often rapid, ranging from 3 m (about 10 ft) per year for the Nile to 60 m (about 200 ft) per year for the Mississippi delta. Some cities and towns that were at river mouths several hundred years ago are today several kilometers inland.

Volcano coasts arise where volcanic deposits—lava and ash—flow from active volcanoes into the ocean. Low cliffs occur when wave action erodes the fresh deposits. Beaches are typically narrow, steep, and composed of fine particles of the extrusive rock.

Coral-Reef Coasts

Coral-reef coasts are unique in that the addition of new land is made by organisms—corals and algae. Growing together, these organisms secrete rocklike deposits of mineral carbonate, called **coral reefs**. As coral colonies die, new ones are built on them, accumulating as limestone. Coral fragments are torn free by wave attack, and the pulverized fragments accumulate as sand beaches.

Coral-reef coasts occur in warm, tropical and equatorial waters between the limits of lat. 30° N and 25° S. Water temperatures above 20°C (68°F) are necessary for dense growth of coral reefs. Reef corals live near the water surface. The sea water must be free of suspended sediment and well aerated for vigorous coral growth. For this reason, corals thrive in positions exposed to wave attack from the open sea. Because muddy water prevents coral growth, reefs are missing opposite the

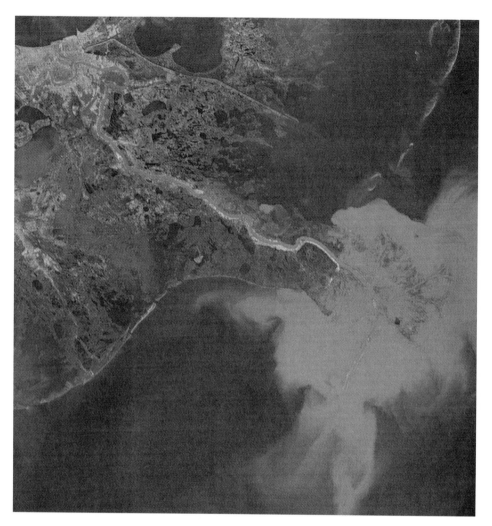

Figure 17.16 The Mississippi delta recorded from Landsat orbiting satellite. The natural levees of the bird-foot delta appear as lacelike filaments in a great pool of turbid river water. New Orleans can be seen at the upper left, occupying the region between the natural levees of the Mississippi River and the southern shore of Lake Pontchartrain.

Figure 17.17 The Island of Moorea and its coral reefs, Society Islands, South Pacific Ocean. The island is a deeply dissected volcano with a history of submergence. Tahiti lies in the background.

mouths of muddy streams. Coral reefs are remarkably flat on top. They are exposed at low tide and covered at high tide.

There are three distinctive types of coral reefs—fringing reefs, barrier reefs, and atolls. *Fringing reefs* are built as platforms attached to shore (Figure 17.17). They are widest in front of headlands where wave attack is strongest. *Barrier reefs* lie out from shore and are separated from the mainland by a lagoon (Figure 17.17). Narrow gaps occur at intervals in barrier reefs. Through these openings, excess water from breaking waves is returned from the lagoon to the open sea.

Atolls are more or less circular coral reefs enclosing a lagoon but have no land inside. On large atolls, parts of the reef have been built up by wave action and wind to form low island chains connected by the reef. Most atolls are built on a foundation of volcanic rock that has subsided below sea level.

Raised Shorelines and Marine Terraces

The active life of a shoreline is sometimes cut short by a sudden rise of the coast. When this tectonic event occurs, a *raised shoreline* is formed. If present, the marine cliff and abrasion platform are abruptly raised above the level of wave action. The former abrasion platform has now become a *marine terrace* (Figure 17.18). Of course, fluvial denudation acts to erode the terrace as soon as it is formed. The terrace may also undergo partial burial under alluvial fan deposits.

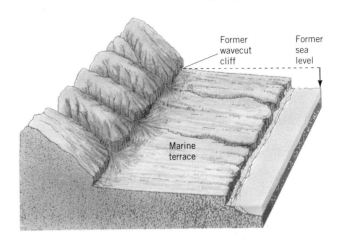

Figure 17.18 A raised shoreline becomes a cliff parallel with the newer, lower shoreline. The former abrasion platform is now a marine terrace. (Drawn by A. N. Strahler.)

Raised shorelines are common along the continental and island coasts of the Pacific Ocean, because here tectonic processes are active along the mountain and island arcs. Repeated uplifts result in a series of raised shorelines in a steplike arrangement. Fine examples of these multiple marine terraces are seen on the western slope of San Clemente Island, off the California coast (Figure 17.19).

Figure 17.19 Marine terraces on the western slope of San Clemente Island, off the southern California coast. More than 20 terraces have been identified in this series. The highest has an elevation of about 400 m (about 1300 ft).

EYE ON THE ENVIRONMENT:
THE THREAT OF RISING SEA LEVEL FROM GLOBAL WARMING

As we have seen in previous chapters, the earth's surface temperature has slowly increased over the past century or so and appears to be increasing at present. Among the possible effects of climatic warming is the rise of sea level, brought about by the melting of glacial ice and thermal expansion of sea water. What would be the global impact of a sea-level rise of several meters? Low-lying areas—such as tidal estuaries and deltas—would be flooded. Coastal cities would have to be defended from wave attack. Barrier islands would be swept landward or overtopped. Atolls and other low islands would disappear. The threat of a rise in sea level is real, but like other possible impacts of global warming, it is far from certain how long it will take sea level to rise and how great the effect will be.

A review of the scientific principles involved here may be helpful. First, these changes in sea level are *eustatic*—that is, they refer to worldwide sea-level change. Local relative changes in sea level resulting from the uplift or downsinking of the earth's crust are tectonic effects and not included in eustatic change. Eustatic change can result from a change either in water volume of the world ocean (as by melting of glaciers) or in the water-holding capacity of the ocean basins. The second possibility relates to plate tectonics and is on a longer time scale than that involved in global warming.

Climate affects the water volume of the oceans in two ways. First, a volume change results from water temperature increase. Global warming will raise the temperature of the uppermost layer of the oceans—at most a few hundred meters thick—and the resulting thermal expansion of that water will cause a small but significant eustatic rise.

Second, global warming may increase the rate of glacial melting, causing mountain glaciers and ice sheets to shrink in volume. Meltwater would then increase the volume of the world ocean. Keep in mind, however, that, by increasing precipitation, global warming may also promote the growth of glaciers. Thus, we must consider the relative strength of the opposite processes of melting and growth. Large swings in world sea level have taken place throughout the past 3 million years as ice sheets have cyclically expanded and contracted. (In Chapter 18 we will return to a discussion of the mechanism of possible sea-level rise by ice melting.)

There is strong evidence to suggest that a slow eustatic rise of sea level has accompanied the overall global temperature rise that started in the mid-1880s. Scientists examining tide gauge records have concluded that in the 80-year period 1900–1980, the total eustatic rise has been about 80 mm (about 3 in.). About half of this increase is attributed to thermal expansion. The average rate of rise was about 1.2 mm/yr (about 0.05 in./yr) through the early and middle 1900s. Recently, the rate has doubled to about 2.4 mm/yr (about 0.1 in./yr). Although this accelerating rate of rise seems to fit the effects of observed global warming, keep in mind that the linkage between sea-level rise and global warming is still uncertain.

Projections of the rate of sea-level rise into the next century vary widely. For example, computer modeling evaluated by the National Academy of Sciences in 1989 showed estimates of sea-level increase by the year 2100 from as small as 0.3 m (about 1 ft) to as large as 2 m (about 7 ft). Using a rise of 1 m (3.28 ft) by the middle or late 2000s, various national and international agencies have estimated the impacts, both physical and financial, on major coastal cities of the world. Large river deltas, including both the cities built on them and their reclaimed tidal estuaries and marshes, would face catastrophic damage through storm surf and storm surges in concert with high tides. The costs of building protective barriers to meet this threat are enormously high, and the barriers may not be effective in all situations. Along cliffed coasts under direct attack by waves, retrogradation will be intensified, resulting in massive losses of shorefront properties. Some island nations would lose most or all of their land. The Maldives, a chain of atolls in the Indian Ocean with a population of about a quarter of a million people, is an example.

WIND ACTION

Transportation and deposition of sand by wind is an important process in shaping certain coastal landforms. We have already mentioned coastal sand dunes, which are derived from beach sand. In the remainder of this chapter, we investigate the transport of mineral particles by wind and the shaping of dune forms. Our discussion also provides information about dune forms far from the coast—in desert environments, where the lack of vegetation cover allows dunes to develop if a source of abundant sand particles is present.

Wind blowing over the land surface is one of the active agents of landform development. Ordinarily, wind is not strong enough to dislodge mineral matter from the surfaces of unweathered rock, or from moist, clay-rich soils, or from soils bound by a dense plant cover. Instead, the action of wind in eroding and transporting sediment is limited to land surfaces where small mineral and organic particles are in a loose, dry state. These areas are typically deserts and semiarid lands (steppes). An exception is the coastal environment, where beaches provide abundant supplies of loose sand. In this environment, wind action shapes coastal dunes, even where the climate is humid and the land surface inland from the coast is well protected by a plant cover.

Erosion by Wind

Wind performs two kinds of erosional work: abrasion and deflation. Loose particles lying on the ground surface may be lifted into the air or rolled along the ground by wind action. In the process of *wind abrasion*, wind drives sand and dust particles against an exposed rock or soil surface. This causes the surface to be worn away by the impact of the particles. Abrasion requires cutting tools—mineral particles—carried by the wind, while deflation is accomplished by air currents alone.

The sandblasting action of wind abrasion against exposed rock surfaces is limited to the basal meter or two of a rock mass that rises above a flat plain. This height is the limit to which sand grains can rise high into the air. Wind abrasion produces pits, grooves, and hollows in the rock. Wooden utility poles on windswept sandy plains are quickly cut through at the base unless a protective metal sheathing or heap of large stones is placed around the base.

The removal of loose particles from the ground is termed **deflation**. Deflation acts on loose soil or sediment. Dry river courses, beaches, and areas of recently formed glacial deposits are susceptible to deflation. In dry climates, much of the ground surface is subject to deflation because the soil or rock is largely bare of vegetation.

Wind is selective in its deflational action. The finest particles, those of clay and silt sizes, are lifted and raised into the air—sometimes to a height of a thousand meters (about 3300 ft) or more. Sand grains are moved only when winds are at least moderately strong and usually travel within a meter or two (about 3 to 6 ft) of the ground. Gravel fragments and rounded pebbles can be rolled or pushed over flat ground by strong winds, but they do not travel far. They become easily lodged in hollows or between other large grains. Consequently, where a mixture of size of particles is present on the ground, the finer sized particles are removed and the coarser particles remain behind.

A landform produced by deflation is a shallow depression called a **blowout**. The size of the depression may range from a few meters (10 to 20 ft) to a kilometer (0.6 mi) or more in diameter, although it is usually only a few meters deep. Blowouts form in plains regions of dry climate. Any small depression in the surface of the plain, especially where the grass cover has been broken or disturbed, can form a blowout. Rains fill the depression and create a shallow pond or lake. As the water evaporates, the mud bottom dries out and cracks, leaving small scales or pellets of dried mud. These particles are lifted out by the wind.

Deflation is also active in semidesert and desert regions. In the southwestern United States, playas often occupy large areas on the flat floors of tectonic basins (see Figure 15.31). Deflation has reduced many playas several meters in elevation.

Rainbeat, overland flow, and deflation may be active for a long period on the gently sloping surface of a desert alluvial fan or alluvial terrace. These processes remove fine particles, leaving coarser, heavier materials behind. As a result, rock fragments ranging in size from pebbles to small boulders become concentrated into a surface layer known as a **desert pavement** (Figure 17.20). The large fragments become closely fitted together, concealing the smaller particles—grains of

Figure 17.20 This desert pavement is formed of closely fitted rock fragments. Lying on the surface are fine examples of wind-faceted rocks, which attain their unusual shapes by long-continued sandblasting.

sand, silt, and clay—that remain beneath. The pavement acts as an armor that effectively protects the finer particles from rapid removal by deflation. However, the pavement is easily disturbed by the wheels of trucks and motorcycles, exposing the finer particles and allowing severe deflation and water erosion to follow.

Dust Storms

Strong, turbulent winds blowing over barren surfaces lift great quantities of fine dust into the air, forming a dense, high cloud called a **dust storm**. In semiarid grasslands, a dust storm is generated where ground surfaces have been stripped of protective vegetation cover by cultivation or grazing. Strong winds cause soil particles and coarse sand grains to hop along the ground. This motion breaks down the soil particles and disturbs more soil. With each impact, fine dust is released that can be carried upward by turbulent winds.

A dust storm approaches as a dark cloud extending from the ground surface to heights of several thousand meters (Figure 17.21). Typically, the advancing cloud wall represents a rapidly moving cold front. Within the dust cloud there is deep gloom or even total darkness. Visibility is cut to a few meters, and a fine choking dust penetrates everywhere.

SAND DUNES

A **sand dune** is any hill of loose sand shaped by the wind. Active dunes constantly change form under wind currents. Dunes form where there is a source of sand—

for example, a sandstone formation that weathers easily to release individual grains, or perhaps a beach supplied with abundant sand from a nearby river mouth. Dunes must be free of a vegetation cover in order to form and move. They become inactive when stabilized by a vegetation cover, or when patterns of wind or sand sources change.

Dune sand is most commonly composed of the mineral quartz, which is extremely hard and largely immune to chemical decay. The grains are beautifully rounded by abrasion (see Figure 10.12). Figure 17.22 shows how sand grains are moved by strong winds—in long, low leaps, bouncing after impact with other grains. Rebounding grains rarely rise more than half a centimeter above the dune surface. Grains struck by bouncing grains are pushed forward, and, in this way, the surface sand layer creeps downwind. This type of hopping, bouncing movement is termed *saltation*.

Types of Sand Dunes

One common type of sand dune is an isolated heap of free sand called a *barchan*, or *crescentic dune*. This type of dune has the outline of a crescent, and the points of the crescent are directed downwind (Figure 17.23). On the upwind side of the crest, the sand slope is gentle and smoothly rounded. On the downwind side of the dune, within the crescent, is a steep dune slope, the *slip face*. This face maintains a more or less constant angle from the horizontal (Figure 17.24), which is known as the *angle of repose*. The slip face is oversteepened slightly by the accumulation of individual wind-carried sand grains until it becomes unstable and the outermost

Figure 17.21 An approaching dust storm in eastern Kenya.

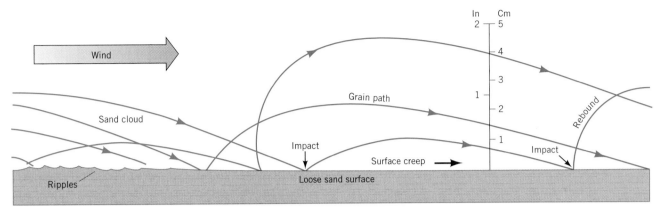

Figure 17.22   Sand particles travel in a series of long leaps. (After R. A. Bagnold.)

layer of sand on the face slips down the dune slope, restoring the angle of repose. For loose sand, this angle is about 35°. *Working It Out 17.2 • Angle of Repose of Dune Sands* describes this process and develops a simple statistical test to determine whether two different samples of sand have different repose angles.

Barchan dunes usually rest on a flat, pebble-covered ground surface. The life of a barchan dune may begin as a sand drift in the lee of some obstacle, such as a small hill, rock, or clump of brush. Once a sufficient mass of sand has formed, it begins to move downwind, taking on the crescent form. For this reason, the dunes are usually arranged in chains extending downwind from the sand source.

Where sand is so abundant that it completely covers the solid ground, dunes take the form of wavelike ridges separated by troughlike furrows. These dunes are called *transverse dunes* because, like ocean waves,

their crests trend at right angles to the direction of the dominant wind (Figure 17.25). The entire area may be called a *sand sea*, because it resembles a storm-tossed sea suddenly frozen to immobility. The sand ridges have sharp crests and are asymmetrical, the gentle slope being on the windward and the steep slip face on the lee side. Deep depressions lie between the dune ridges. Sand seas require enormous quantities of sand, supplied by material weathered from sandstone formations or from sands in nearby alluvial plains. Transverse dune belts also form adjacent to beaches that supply abundant sand and have strong onshore winds.

Wind is a major agent of landscape development in the Sahara Desert. Enormous quantities of reddish dune sand have been derived from weathering of sandstone formations. The sand is formed into a great sand sea, called an *erg*. Elsewhere, there are vast flat-surfaced sheets of sand that are armored by a layer of pebbles that forms a desert pavement. A surface of this kind in the Sahara is called a *reg*.

Some of the Saharan dunes are elaborate in shape. For example, the *star dune* (heaped dune), is a large hill of sand whose base resembles a many-pointed star in plan. The Arabian star dunes remain fixed in position and have served for centuries as reliable landmarks for desert travelers. The star dune also occurs in the deserts of the border region between the United States and Mexico.

Another group of dunes belongs to a family in which the curve of the dune crest is bowed outward in the downwind direction. (This curvature is the opposite of the barchan dune.) These are termed *parabolic dunes*. A common type of parabolic dune is the *coastal blowout dune*, formed adjacent to beaches. Here, large supplies of sand are available, and the sand is blown landward by prevailing winds (Figure 17.26a). A saucer-shaped depression is formed by deflation, and the sand is heaped in a curving ridge resembling a horseshoe in plan. On the landward side is a steep slip face that advances over the lower ground and buries forests, killing

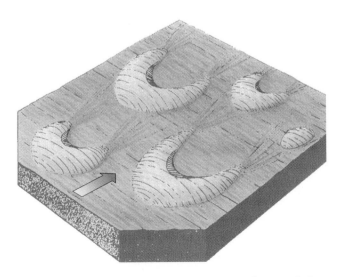

Figure 17.23   Barchan dunes. The arrow indicates wind direction. (Drawn by A. N. Strahler.)

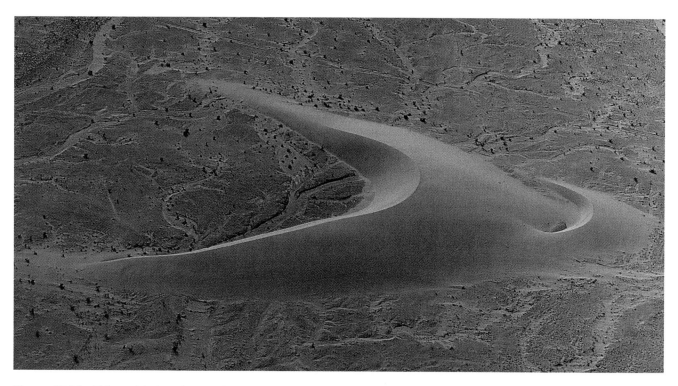

Figure 17.24 This aerial view shows a large barchan dune moving from right to left. At its apex is a smaller barchan dune that is overtaking it. Note the network of dry stream channels. Obviously, these dunes have migrated across this channel network since the last water flood occurred. Salton Sea region, California.

Figure 17.25 Transverse dunes of a sand sea, near Yuma, Arizona. The view is eastward; prevailing winds are northerly (from the left side of the photo).

the trees (Figure 17.27). Coastal blowout dunes are well displayed along the southern and eastern shore of Lake Michigan. Dunes of the southern shore have been protected for public use as the Indiana Dunes State Park.

On semiarid plains, where vegetation is sparse and winds are strong, groups of parabolic blowout dunes develop to the lee of shallow deflation hollows (Figure 17.26b). Sand is caught by low bushes and accumulates on a broad, low ridge. These dunes have no steep slip faces and may remain relatively immobile. In some cases, the dune ridge migrates downwind, drawing the dune into a long, narrow form with parallel sides resembling a hairpin in outline (Figure 17.26c).

Another class of dunes, described as *longitudinal dunes*, consists of long, narrow ridges oriented parallel with the direction of the prevailing wind (Figure 17.28). These dune ridges may be many kilometers long and cover vast areas of tropical and subtropical deserts in Africa and Australia.

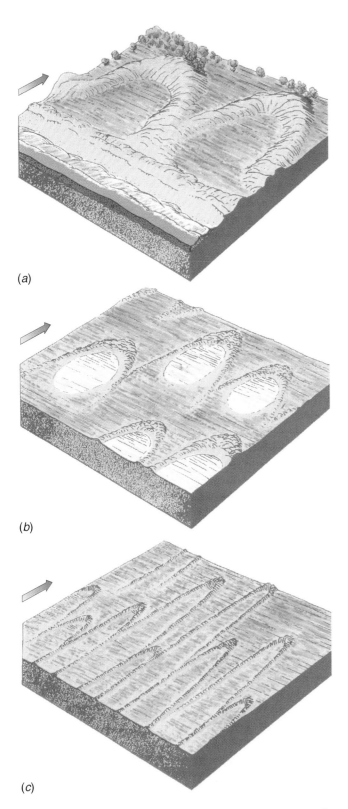

(a)

(b)

(c)

Figure 17.27   This coastal blowout dune is advancing over a coniferous forest, with the slip face gradually burying the tree trunks. Pacific coast, near Florence, Oregon.

Figure 17.26   Three types of parabolic dunes. The prevailing wind direction is the same for all three types. (*a*) Coastal blowout dunes. (*b*) Parabolic dunes on a semiarid plain. (*c*) Parabolic dunes drawn out into hairpin forms. (Drawn by A. N. Strahler.)

Figure 17.28   Longitudinal dunes run parallel to the direction of the wind. (Drawn by A. N. Strahler.)

Working It Out 17.2 • Angle of Repose of Dune Sands

In North Africa during World War II, German panzer tank forces under General Rommel were driving the British forces eastward. Battles were being fought on difficult desert terrain that included desert pavement as well as large areas of active sand dunes. Sudden, violent sandstorms that might rage for many hours severely tried both men and machines. A British Royal Army Engineer, Major R. A. Bagnold, was sent to Egypt to study this phenomenon. He had just completed in England a laboratory study of the motions of sand grains driven by strong winds. His treatise on dune dynamics, published in 1941 and reprinted in 1954, remains a classic—perhaps familiar to Allied officers who directed Desert Storm four decades later.

We reproduce here one of Bagnold's illustrations, showing how a new sand dune with a broadly rounded surface is transformed into a barchan dune (Figure 1). Leaping sand grains (Figure 17.25) transfer the sand downwind to form a high, sharp crest. From this crest, the grains leap out on to the steep slip face of the dune, where they come to rest. This activity tends to steepen the slope of the slip face, so that there comes a point when a thin layer of sand grains slides down to the base. In this way the slope is alternately diminished and steepened. Our question now is: What is

the maximum slope angle that the grains can hold before the next slip occurs? Does this angle—called the *angle of repose*—vary according to the particle size or to mineral composition of the sand grains? Or perhaps to some other factor?

To investigate this question, we perform a lab experiment that makes use of two sand samples. One, called "Ottawa sand," consists of beautifully rounded grains of pure quartz. In fact, they were deposited as dune sands back in Paleozoic time and are now being quarried for industrial uses. The sand has been screened so as to include only a very small range of diameters. Take a good look at these grains—they are shown in Chapter 10 as Figure 10.12. Our second sample consists of a common kind of beach sand scooped up at low tide. Both samples have been passed through a series of screens to eliminate grains larger than 2 mm and smaller than 1 mm, leaving coarse sand.

To carry out our experiment, we use a small box with vertical glass sides and a removable end piece (Figure 2) . Sand is poured gently into the box until it is almost full. The end plate is then pulled out, allowing the sand to pour out into a container below. The angle of the sand surface can now be measured through the side glass. Each reading is observed to the nearest one-

half degree. Ten such trials were made with the Ottawa sand and ten with the beach sand. The bar graph (Figure 3) shows how frequently each slope angle occurred for each type of sand. There is a clear difference between the two samples, for a wide gap separates the two samples.

Why does the beach sand show the higher mean? Because both samples were carefully screened, difference in size of particles is probably not a factor. But there are other differences. Because the beach sand has been worked by waves as well as wind, it may include grains that are more angular and thus able to resist movement better. The beach sand may also contain mineral particles other than quartz—heavy minerals, perhaps—that have higher density and would move farther downslope before stopping. Also, some of these other minerals might come in sharp-edged grains that serve to pack the mass more tightly together. Only a careful microscopic analysis would tell for sure.

Our two sets of data offer an opportunity to apply some simple mathematical statistics as an example of how physical geographers use statistics to ask and answer questions about data. Consider the bar graph for a moment. Do you suppose it's possible that both types of sand actually do have the same angle of repose, but that by some

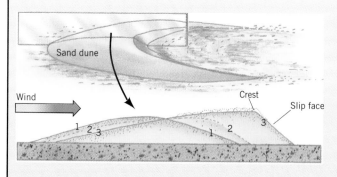

Figure 1 Growth of a dune and development of a slip face. (After R. A. Bagnold, *The Physics of Blown Sand and Desert Dunes,* London, Methuen, 1941.)

accident of chance the data show they are separated? Statisticians have developed a test to see whether this might be the case—a *t*-test of two means. In this test, a sample statistic, *t*, is calculated. The greater the value of *t*, the more likely it is that the two means are different.

The formula for the *t*-statistic is

$$t = \frac{\left| \overline{X}_1 - \overline{X}_2 \right|}{s_{\overline{X}_1 - \overline{X}_2}} \qquad (1)$$

The numerator of this expression is the absolute value (value without respect to sign) of the difference between the means of the two samples, here designated by $\overline{X}_1$ and $\overline{X}_2$. The greater is the absolute difference between the means, the larger will be the value of *t*. The denominator, $s_{\overline{X}_1 - \overline{X}_2}$, is an estimate of the standard deviation of this difference, based on the standard deviations of both samples. It is calculated from the individual sample standard deviations as shown in

Working It Out 9.1 • Standard Deviation and Coefficient of Variation. The formula for this type of standard deviation is

$$s_{\overline{X}_1 - \overline{X}_2} = \sqrt{\frac{s_1^2 + s_2^2}{n}} \qquad (2)$$

where s_1 and s_2 are the sample standard deviations for the two groups. From the formula for *t* (1), we see that if the sample values have a small standard deviation, that will make the standard deviation of the difference small as well, and that in turn will make *t* larger.

For the comparison between Ottawa sand and beach sand, we have $\overline{X}_1 = 32.6$ and $s_1 = 0.32$, and $\overline{X}_2 = 35.1$ and $s_2 = 0.55$. Finding the standard deviation of the difference from (2),

$$s_{\overline{X}_1 - \overline{X}_2} = \sqrt{\frac{(0.32)^2 + (0.55)^2}{10}}$$
$$= \sqrt{\frac{0.102 + 0.303}{10}}$$
$$= \sqrt{0.041} = 0.20$$

Then, applying this to the calculation of *t* in (1), we have

$$t = \frac{\left| \overline{X}_1 - \overline{X}_2 \right|}{s_{\overline{X}_2 - \overline{X}_2}} = \frac{|32.6 - 35.1|}{0.20} = \frac{2.5}{0.20}$$
$$= 12.5$$

Consulting a statistical table, we note that if *t* is greater than 3.610 for the case of two samples of size 10, there is less than 1 chance in 1000 that the means are the same but the difference has arisen by chance alone. Thus, we can conclude with confidence that the angles of repose of the two sand samples are different.

The *t*-test of two means is just one of many types of statistical tests that geographers use in examining data and drawing conclusions. If you continue your study of physical geography, you will undoubtedly learn more about such statistics and how they are used in geography and studies of the environment.

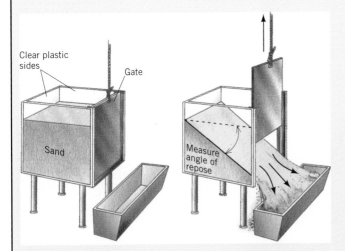

Figure 2 A simple experiment to measure the angle of repose of sand.

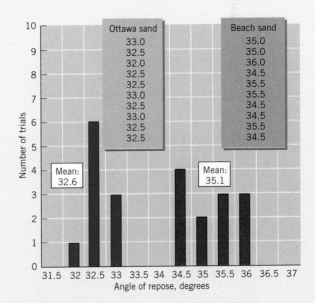

Figure 3 Frequency graph counting the number of occurrences of repose angles observed for Ottawa and beach sand. (Copyright © A. N. Strahler.)

Coastal Foredunes

Landward of sand beaches, we usually find a narrow belt of dunes in the form of irregularly shaped hills and depressions. These are the *foredunes*. They normally bear a cover of beachgrass and a few other species of plants capable of survival in the severe environment (Figure 17.29). On coastal foredunes, the sparse cover of beachgrass and other small plants acts as a baffle to trap sand moving landward from the adjacent beach. As a result, the foredune ridge is built upward to become a barrier standing several meters above high-tide level.

Foredunes form a protective barrier for tidal lands that often lie on the landward side of a beach ridge or barrier island. In a severe storm, the swash of storm waves cuts away the upper part of the beach. Although the foredune barrier may then be attacked by wave action and partly cut away, it will not usually yield. Between storms, the beach is rebuilt, and, in due time, wind action restores the dune ridge, if plants are maintained.

If the plant cover of the dune ridge is reduced by vehicular and foot traffic, a blowout will rapidly develop. The new cavity can extend as a trench across the dune ridge. With the onset of a storm that brings high water levels and intense wave action, swash is funneled through the gap and spreads out on the tidal marsh or tidal lagoon behind the ridge. Sand swept through the gap is spread over the tidal deposits. If eroded, the gap can become a new tidal inlet for ocean water to reach the bay beyond the beach. For many coastal communities of the eastern United States seaboard, the breaching of a dune ridge with its accompanying overwash may bring unwanted change to the tidal marsh or estuary.

Loess

In several large midlatitude areas of the world, the surface is covered by deposits of wind-transported silt, which has settled out from dust storms over many thousands of years. This material is known as **loess**. (The pronunciation of this German word is somewhere between "lerse" and "luss.") It generally has a uniform yellowish to buff color and lacks any visible layering. Loess tends to break away along vertical cliffs wherever it is exposed by the cutting of a stream or grading of a roadway. It is also very easily eroded by running water and is subject to rapid gullying when the vegetation cover that protects it is broken. Because it is easily excavated, loess has been widely used for cave dwellings both in China and in Central Europe.

The thickest deposits of loess are in northern China, where a layer over 30 m (about 100 ft) thick is common and a maximum thickness of 100 m (about 300 ft) has been measured. It covers many hundreds of square kilometers and appears to have been brought as dust from the interior of Asia. Loess deposits are also of major importance in the United States, Central Europe, Central Asia, and Argentina.

In the United States, thick loess deposits lie in the Missouri-Mississippi Valley (Figure 17.30). Large areas of the prairie plains region of Indiana, Illinois, Iowa, Missouri, Nebraska, and Kansas are underlain by loess ranging in thickness from 1 to 30 m (about 3 to 100 ft). Extensive deposits also occur in Tennessee and Mississippi in areas bordering the lower Mississippi River floodplain. Still other loess deposits are in the Palouse region of northeast Washington and western Idaho.

The American and European loess deposits are directly related to the continental glaciers of the Pleistocene Epoch. At the time when the ice covered much

Figure 17.29 Beachgrass thriving on coastal foredunes has trapped drifting sand to produce a dune ridge. Queen's County, Prince Edward Island, Canada.

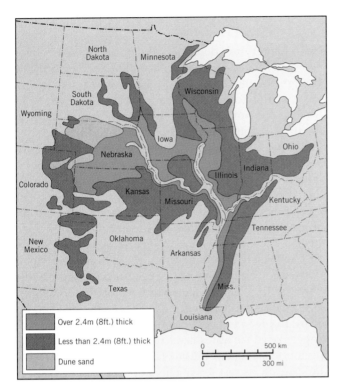

Figure 17.30 Map of loess distribution in the central United States. (Data from *Map of Pleistocene Eolian Deposits of the United States*, Geological Society of America.)

bare ground, picking up silt from the floodplains of braided streams that discharged the meltwater from the ice. This dust settled on the ground between streams, gradually building up a smooth, level ground surface. The loess is particularly thick along the eastern sides of the valleys because of prevailing westerly winds. It is well exposed along the bluffs of most streams flowing through these regions today.

Loess is of major importance in world agricultural resources. Loess forms the parent matter of rich black soils (Mollisols, Chapter 19) especially suited to cultivation of grains. The highly productive plains of southern Russia, the Argentine pampa, and the rich grain region of north China are underlain by loess. In the United States, corn is extensively cultivated on the loess plains in Kansas, Iowa, and Illinois, where rainfall is sufficient. Wheat is grown farther west on loess plains of Kansas and Nebraska and in the Palouse region of eastern Washington.

The thick loess deposit covering a large area of north-central China in the province of Shanxi and adjacent provinces poses a difficult problem of severe soil erosion. Although the loess is capable of standing in vertical walls, it also succumbs to deep gullying during the period of torrential summer rains. From the steep walls of these great scars, fine sediment is swept into streams and carried into tributaries of the Huang He (Yellow River). The Chinese government has implemented an intensive program of slope stabilization by using artificial contour terraces (seen in Figure 17.31) in combination with tree planting. Valley bottoms have

of North America and Europe, a generally dry winter climate prevailed in the land bordering the ice sheets. Strong winds blew southward and eastward over the

Figure 17.31 A succession of contour terraces, cut into thick loess and covered by tree plantings, is designed to prevent deep gullying and loss of soil. Arched entrances to cave dwellings can be seen at lower left and lower right. Grain fields on the flat valley floor occupy surfaces of thick sediment trapped behind dams. Shanxi Province, near Xian (Sian), People's Republic of China.

been dammed so as to trap the silt to form flat patches of land suitable for cultivation.

EYE ON THE ENVIRONMENT: Induced Deflation

Induced deflation is a frequent occurrence when short-grass prairie in a semiarid region is cultivated without irrigation. Plowing disturbs the natural soil surface and grass cover, and in drought years, when vegetation dies out, the unprotected soil is easily eroded by wind action. Much of the Great Plains region of the United States has suffered such deflation, experiencing dust storms generated by turbulent winds. Strong cold fronts frequently sweep over this area and lift dust high into the troposphere at times when soil moisture is low. The "Dust Bowl" of the 1930s is an example (see Chapter 9).

Human activities in very dry, hot deserts contribute measurably to the raising of high dust clouds. In the desert of northwest India and Pakistan (the Thar Desert bordering the Indus River), the continued trampling of fine-textured soils by hooves of grazing animals and by human feet produces a dust cloud that hangs over the region for long periods. It extends to a height of 9 km (about 30,000 ft).

This chapter has described the processes and landforms associated with wind action, either directly or through the medium of wind-driven ocean waves. The power source for wind action is, of course, the sun. By heating the earth's surface in a nonuniform pattern, the flow of solar energy produces pressure gradients that cause wind. The sun also powers the last active landform-making agent of erosion, transportation, and deposition on our list—glacial ice. By evaporating water from the oceans and returning that water to the lands as snow, solar power creates the bodies of solid ice that we distinguish as mountain glaciers and ice sheets.

Compared to wind and water, glacial ice moves much more slowly but is far steadier in its motion. Like a vast conveyor belt, glacial ice moves sediment forward relentlessly, depositing the sediment at the ice margin, where the ice melts. By plowing its way over the landscape, glacial ice also shapes the local terrain—bulldozing loose rock from hillsides and plastering sediments underneath its vast bulk. This slow but steady action is very different from that of water, wind, and waves, and produces a set of landforms that is the subject of our next text chapter.

CHAPTER SUMMARY

This chapter has described the landforms of waves and wind, both of which are indirectly powered by the earth's rotation and the unequal heating of its surface by the sun. Waves act at the shoreline—the boundary between water and land. Waves expend their energy as breakers, which erode hard rock into marine cliffs and create marine scarps in softer materials.

Beaches, usually formed of sand, are shaped by the swash and backwash of waves, which continually work and rework beach sediment. Wave action produces littoral drift, which moves sediment parallel to the beach. This sediment accumulates in bars and sandspits, which further extend the beach. Depending on the nature of longshore currents and the availability of sediment, shorelines can experience progradation or retrogradation.

Tidal forces cause sea level to rise and fall rhythmically, and this change of level produces tidal currents in bays and estuaries. Tidal flows redistribute fine sediments within bays and estuaries, which can accumulate with the help of vegetation to form salt marshes. These are sometimes reclaimed to form new agricultural land.

Coastlines of submergence result when coastal lands sink below sea level or sea level rises rapidly. Scenic ria and fiord coasts are examples. Coastlines of emergence include barrier-island coasts and delta coasts. Coral-reef coasts occur in regions of warm tropical and equatorial waters. Along some coasts, rapid uplift has occurred, creating raised shorelines and marine terraces.

Global sea level has varied widely in the past and is presently increasing slowly. A possible explanation is that global warming is increasing the volume of sea water by thermal expansion and by causing glacial ice on lands to melt. Future rises may be very costly to human society as coasts erode, islands are submerged, and coastal cites are subjected to frequent flooding.

Wind is a landform-creating agent that acts by moving sediment. Deflation occurs when wind removes mineral particles—especially clay and silt, which can be carried long distances. Deflation creates blowouts in semidesert regions and lowers playa surfaces in deserts. In arid regions, deflation produces dust storms.

Sand dunes form when a source, such as a sandstone outcrop or a beach, provides abundant sand that can be moved by wind action. Barchan dunes are arranged individually or in chains leading away from the sand source. Transverse dunes form a sand sea of frozen "wave" forms arranged perpendicular to the wind direction. Parabolic dunes are arc-shaped—coastal blowout dunes are an example. Longitudinal dunes parallel the wind direction and cover vast desert areas. Coastal foredunes are stabilized by dune grass and help protect the coast against storm wave action.

Loess is a surface deposit of fine, wind-transported silt. It can be quite thick, and it typically forms vertical banks. Loess is very easily eroded by water and wind. In eastern Asia, the silt forming the loess was transported by winds from extensive interior deserts located to the north and west. In Europe and North America, the silt was derived from fresh glacial deposits during the Pleistocene Epoch. Human activities can hasten the action of deflation by breaking protective surface covers of vegetation and desert pavement.

KEY TERMS

shoreline
coastline
coast
bay
estuary
marine cliff
beach

littoral drift
progradation
retrogradation
submergence
emergence
delta
coral reef

atoll
deflation
blowout
desert pavement
dust storm
sand dune
loess

REVIEW QUESTIONS

1. What is the energy source for wind and wave action?
2. What landforms can be found in areas where bedrock meets the sea?
3. What is littoral drift, and how is it produced by wave action?
4. Identify progradation and retrogradation. How can human activity influence retrogradation?
5. How are salt marshes formed? How can they be reclaimed for agricultural use?
6. What key features identify a coastline of submergence? Identify and compare the two types of coastlines of submergence.
7. Under what conditions do barrier-island coasts form? What are the typical features of this type of coastline? Provide and sketch an example of a barrier-island coast.
8. Describe the features of delta coasts and their formation. Sketch and compare the Mississippi and Nile deltas.

9. What conditions are necessary for the development of coral reefs? Identify three types of coral-reef coastlines.

10. How are marine terraces formed?

11. How might global climate change bring about a rise in sea level?

12. Is a sea-level rise occurring at present? What are the projections for the future, and what are the implications of the projections?

13. What is deflation, and what landforms does it produce? What role does the dust storm play in deflation?

14. How do sand dunes form? Describe and compare barchan dunes, transverse dunes, star dunes, coastal blowout dunes, parabolic dunes, and longitudinal dunes.

15. What is the role of coastal dunes in beach preservation? How are coastal dunes influenced by human activity? What problems can result?

16. Define the term *loess*. What is the source of loess, and how are loess deposits formed?

Focus On Systems 17.1 • The Coastal Sediment Cell as a Matter Flow System

1. What are the typical changes from summer to winter that you might expect for midlatitude beaches?

2. Identify and describe the pathways of flow in a coastal sediment cell, including inputs, outputs, and locations of sediment in storage.

3. How does the damming of coastal rivers affect the coastal sediment cell and the storage of sediment in beaches?

ESSAY QUESTIONS

1. Consult an atlas to identify a good example of each of the following types of coastlines: ria coast, fiord coast, barrier-island coast, delta coast, coral-reef coast, and fault coast. For each example, provide a brief description of the key features you used to identify the coastline type.

2. Wind action moves sand close to the ground in a bouncing motion, whereas silt and clay are lifted and carried longer distances. Compare landforms and deposits that result from wind transportation of sand with those that result from wind transportation of silt and finer particles.

PROBLEMS

Working It Out 17.2 • Angle of Repose of Dune Sands

1. A geomorphologist is investigating the effect of grain size on the angle of repose. She obtains two samples of small faceted glass beads (rhinestones) with diameters of 3 mm and 4 mm from a jewelry manufacturer. Using these beads, she conducts 20 angle of repose trials for each size with the following results:

Angle of repose, degrees

Size	$\bar{X}$	s	32.0	32.5	33.0	33.5	34.0	34.5	35.0	35.5	36.0
3 mm	34.4	0.91			2	4	3	3	5	1	2
4 mm	33.3	0.72	1	4	7	2	4	2			

Plot the data on a graph following the example of the graph in *Working It Out 17.2.* Do the two samples appear to be as well separated as those of Ottawa and beach sand?

2. Calculate the *t*-statistic for the difference in the two means. Compare the result with the values in the table below, shown for two samples of size 20 each. Can you conclude that the size has an effect on the angle of repose?

t value	Chance
>2.42	1 in 100
>2.70	1 in 200
>3.31	1 in 1000

Chapter 18

Glacier Systems and the Ice Age

In this chapter we turn to the last of the active agents that create landforms—glacial ice. Not long ago, during the Ice Age, much of northern North America and Eurasia was covered by massive sheets of glacial ice. As a result, glacial ice has played a dominant role in shaping landforms of large areas in midlatitude and subarctic zones. Yet, glacial ice still exists today in two great accumulations of continental dimensions—the Greenland and Antarctic Ice Sheets—and in many smaller masses in high mountains.

The glacial ice sheets of Greenland and Antarctica strongly influence the radiation and heat balance of the globe. Because of their intense whiteness, they reflect much of the solar radiation they receive. Their in-

tensely cold surface air temperatures contrast with temperatures at more equatorial latitudes. This temperature difference helps drive the system of meridional heat transport that we described in Chapters 2 and 4. In addition, these enormous ice accumulations represent water in storage in the solid state. They figure as a major component of the global water balance. When the volume of glacial ice increases, as during an ice age, sea levels must fall. When ice sheets melt away, sea level rises. Today's coastal environments evolved during the rising sea level that followed the melting of the last ice sheets of the Ice Age.

GLACIERS

Most of us know ice only as a brittle, crystalline solid because we are accustomed to seeing it in small quanti-

Shorefast sea ice at Barrow, Alaska.

ties. Where a great thickness of ice exists, the pressure on the ice at the bottom makes the ice lose its rigidity. That is, the ice becomes plastic. This allows the ice mass to flow in response to gravity, slowly spreading out over a larger area or moving downhill. On steep mountain slopes, the ice can also move by sliding. Movement is the key characteristic of a *glacier*, defined as any large natural accumulation of land ice affected by present or past motion.

Glacial ice accumulates when the average snowfall of the winter exceeds the amount of snow that is lost in summer by ablation. The term **ablation** means the loss of snow and ice by evaporation and melting. When winter snowfall exceeds summer ablation, a layer of snow is added each year to what has already accumulated. As the snow compacts by surface melting and refreezing, it turns into a granular ice and is then compressed by overlying layers into hard crystalline ice. When the ice mass is so thick that the lower layers become plastic, outward or downhill flow starts, and the ice mass is now an active glacier.

Glacial ice forms where temperatures are low and snowfall is high. These conditions can occur both at high elevations and at high latitudes. In mountains, glacial ice can form even in tropical and equatorial zones if the elevation is high enough to keep average annual temperatures below freezing. Orographic precipitation encourages the growth of glacial ice. In high mountains, glaciers flow from small high-elevation collecting grounds down to lower elevations, where tem-

peratures are warmer. Here the ice disappears by ablation. Typically, mountain glaciers are long and narrow because they occupy former stream valleys. These **alpine glaciers** are a distinctive type (Figure 18.1).

In arctic and polar regions, prevailing temperatures are low enough that snow can accumulate over broad areas, eventually forming a vast layer of glacial ice. Accumulation starts on uplands that intercept heavy snowfall. The uplands become buried under enormous volumes of ice, which can reach a thickness of several thousand meters. The ice then spreads outward, over surrounding lowlands, and covers all landforms it encounters. This extensive type of ice mass is called an **ice sheet**. As already noted, ice sheets exist today in Greenland and Antarctica.

Glacial ice normally contains abundant rock fragments ranging from huge angular boulders to pulverized rock flour. Some of this material is eroded from the rock floor on which the ice moves. In alpine glaciers, rock debris is also derived from material that slides or falls from valley walls onto the ice.

Glaciers are capable of eroding and depositing great quantities of sediment. *Glacial abrasion* is a glacial erosion process caused by rock fragments that are held within the ice and scrape and grind against bedrock (Figure 18.2). Erosion also occurs by *plucking*, as moving ice lifts out blocks of bedrock that have been loosened by the freezing and expansion of water in joint fractures. Abrasion and plucking act to smooth the bed of a glacier as glacial flow continues through time.

Figure 18.1 Aerial view of the Grand Plateau Glacier, St. Elias Mountains, Glacier Bay National Park. The dark stripes are medial moraines.

Rock debris brought into a glacier is eventually deposited at the lower end of a glacier, where the ice melts. Both erosion and deposition result in distinctive glacial landforms.

ALPINE GLACIERS

Figure 18.3 illustrates a number of features of alpine glaciers. The illustration shows a simple glacier occupying a sloping valley between steep rock walls. Snow collects at the upper end in a bowl-shaped depression, the **cirque**. The upper end lies in a zone of accumulation. Layers of snow in the process of compaction and recrystallization are called *firn*.

The smooth firn field is slightly bowl-shaped in profile. Flowage in the glacial ice beneath the firn carries the ice downvalley out of the cirque. The rate of ice flow is accelerated at a steep rock step, where deep crevasses (gaping fractures) mark an ice fall. The lower part of the glacier lies in the zone of ablation. In this area, the rate of ice wastage is rapid, and old ice is exposed at the glacier surface. This surface may be quite rough, with deep crevasses. At its lower end, or terminus, the glacier carries abundant rock debris. As the downward-flowing ice melts, the debris accumulates.

Although the uppermost layer of a glacier is brittle and fractures readily into crevasses, the ice beneath behaves as a plastic substance and moves by slow flowage (Figure 18.4). Like stream flow, glacier flow is most rapid far from the glacier's bed—near the midline and toward the top of the glacier's surface. Alpine glaciers also move by basal sliding. In this process, the ice slides downhill, lubricated by meltwater and mud at its base.

A glacier establishes a dynamic balance in which the rate of accumulation at the upper end balances the rate of ablation at the lower end. This balance is easily upset by changes in the average annual rates of accumulation or ablation, causing the glacier's terminus to move forward or melt.

Glacial flow is usually very slow. It amounts to a few centimeters per day for large ice sheets and the more sluggish alpine glaciers, but as fast as several meters per day for an active alpine glacier. However, some alpine glaciers experience episodes of very rapid movement, termed *surges*. A surging glacier may travel downvalley at speeds of more than 60 m (about 200 ft) per day for several months. The reasons for surging are not well understood, but probably involve mechanisms that increase the amount of meltwater beneath the ice, enhancing basal sliding. Most glaciers do not experience surging.

An alpine glacier is a good example of a flow system of matter and energy. Matter flows as ice and rock debris move downhill under the power of gravity. Potential energy is converted to kinetic energy in the motion

Figure 18.2 This grooved and polished surface, now partly eroded, marks the former path of glacial ice. Cathedral Lakes, Yosemite National Park, California.

of flow, then dissipated as heat through friction. More details are provided in *Focus on Systems 18.1 • A Glacier as a Flow System of Matter and Energy*.

Landforms Made by Alpine Glaciers

Landforms made by alpine glaciers are shown in a series of diagrams in Figure 18.5. Mountains are eroded and shaped by glaciers, and after the glaciers melt, the remaining landforms are exposed to view. Diagram *a* shows a region sculptured entirely by weathering, mass wasting, and streams. The mountains have a smooth, rounded appearance. Soil and regolith are thick.

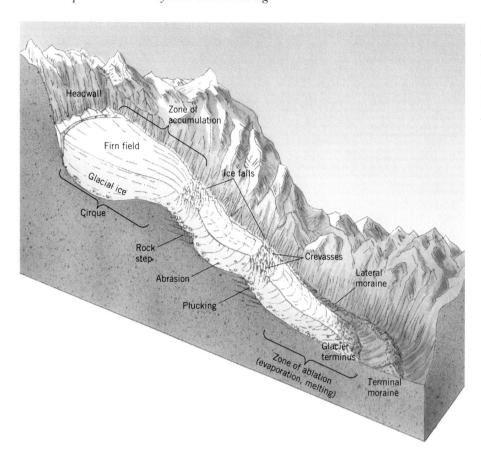

Figure 18.3 Cross section of an alpine glacier. Ice accumulates in the glacial cirque, then flows downhill, abrading and plucking the bedrock. Glacial debris accumulates at the glacier terminus. (After A. N. Strahler.)

Imagine now that a climatic change results in the accumulation of snow in the heads of the higher valleys. An early stage of glaciation is shown at the right side of diagram *b*, where snow is collecting and cirques are being carved by the grinding motion of the ice. Deepening of the cirques is aided by intensive frost shattering of the bedrock near the masses of compacted snow.

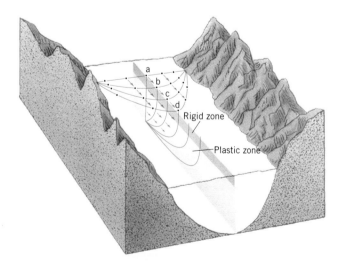

Figure 18.4 Motion of glacial ice. Ice moves most rapidly on the glacier's surface at its midline. Movement is slowest near the bed, where the ice contacts bedrock or sediment.

At a later stage (center), glaciers have filled the valleys and are integrated into a system of tributaries that feed a trunk glacier. Tributary glaciers join the main glacier smoothly. The cirques grow steadily larger. Their rough, steep walls soon replace the smooth, rounded slopes of the original mountain mass. Where two cirque walls intersect from opposite sides, a jagged, knifelike ridge, called an *arête*, is formed. Where three or more cirques grow together, a sharp-pointed peak is formed. Such peaks are called *horns*. On the left side of the diagram, a portion of the landscape is drawn as unglaciated.

A ridge or pile of rock debris left by glacial action that marks the edge of a glacier is termed a **moraine**. A *lateral moraine* is a debris ridge formed along the edge of the ice adjacent to the trough wall (Figure 18.3). Where two ice streams join, this marginal debris is dragged along to form a narrow band riding on the ice in midstream (Figures 18.5*b*, 18.1), called a *medial moraine*. At the terminus of a glacier, rock debris accumulates in a *terminal moraine*, an embankment curving across the valley floor and bending upvalley along each wall of the trough (Figure 18.6).

Glacial Troughs and Fiords

Glacier flow constantly deepens and widens its rock channel, so that after the ice has finally melted, a deep,

(a)

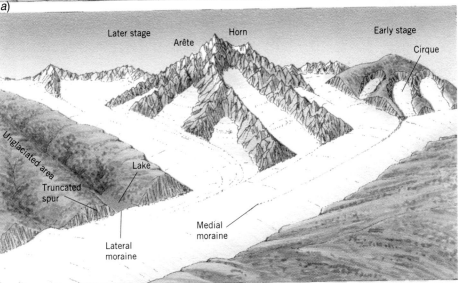

Later stage   Horn   Early stage
            Arête   Cirque

Unglaciated area

Truncated spur   Lake

                    Lateral moraine   Medial moraine

(b)

Tarns

Hanging valley   Glacial trough

(c)

Figure 18.5 Landforms produced by alpine glaciers. (*a*) Before glaciation sets in, the region has smoothly rounded divides and narrow, V-shaped stream valleys. (*b*) After glaciation has been in progress for thousands of years, new erosional forms are developed. (*c*) With the disappearance of the ice, a system of glacial troughs is exposed. (Drawn by A. N. Strahler.)

Figure 18.6 A terminal moraine, shaped like the bow of a great canoe, lies at the mouth of a deep glacial trough on the east face of the Sierra Nevada. Cirques can be seen in the distance. Near Lee Vining, California.

Figure 18.7 Development of a glacial trough. (*a*) During maximum glaciation, the U-shaped trough is filled by ice to the level of the small tributaries. (*b*) After glaciation, the trough floor may be occupied by a stream and lakes. (*c*) If the main stream is heavily loaded, it may fill the trough with alluvium. (*d*) Should the glacial trough have been deepened below sea level, it will be occupied by an arm of the sea, or fiord. (Drawn by E. Raisz.)

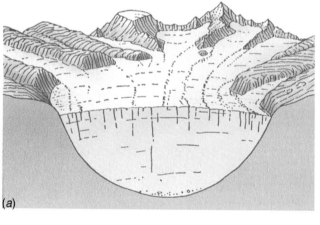

(*a*)

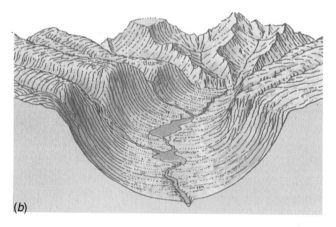

(*b*)

(*c*)

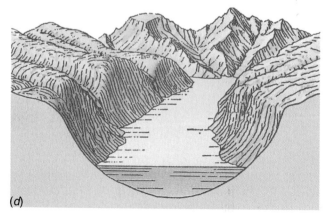

(*d*)

Figure 18.8 Geirangerfjord, Norway, is a deeply carved glacial trough occupied by an arm of the sea.

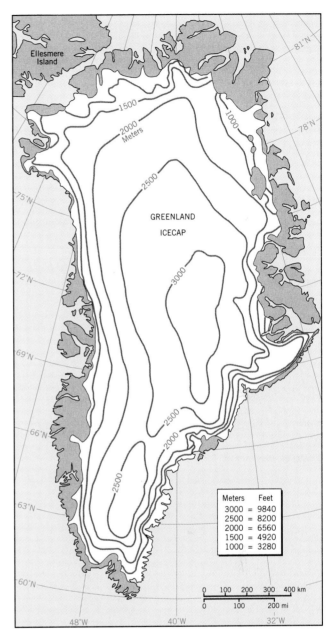

Figure 18.9 The Greenland Ice Sheet. Contours show elevations of the ice sheet surface. (Based on data of R. F. Flint, *Glacial and Pleistocene Geology*, John Wiley & Sons, New York.)

steep-walled **glacial trough** remains (Figure 18.5*c*). The trough typically has a U-shape in cross-profile (Figure 18.7). Tributary glaciers also carve U-shaped troughs, but they are smaller in cross section and less deeply eroded by their smaller glaciers. Because the floors of these troughs lie high above the level of the main trough, they are called *hanging valleys*. Streams later occupy the abandoned valleys, providing a scenic waterfall that cascades over the lip of the hanging valley to the main trough below. High up in the smaller troughs, the bedrock is unevenly excavated, so that the floors of troughs and cirques contain rock basins and rock steps. The rock basins are occupied by small lakes, called *tarns* (Figure 18.5*c*). Major troughs sometimes hold large, elongated trough lakes.

Many large glacial troughs now are filled with alluvium and have flat floors. Aggrading streams that issued from the receding ice front were heavily laden with rock fragments so that the deposit of alluvium extended far downvalley. Figure 18.7 shows a comparison between a trough with little or no fill (*b*) and another with an alluvial-filled bottom (*c*).

When the floor of a trough open to the sea lies below sea level, the sea water enters as the ice front recedes. The result is a deep, narrow estuary known as a **fiord** (Figure 18.7*d*). Fiords are opening up today along the Alaskan coast, where some glaciers are melting back rapidly and ocean waters are filling their troughs. Fiords are found largely along mountainous coasts between lat. 50° and 70° N and S (Figure 18.8). On these coasts, glaciers were nourished by heavy orographic snowfall, associated with the marine west-coast climate ⑧.

Focus on Systems 18.1 • A Glacier as a Flow System of Matter and Energy

A glacier provides a good example of a system of coupled matter and energy flow. For this analysis, we will consider an alpine glacier flowing from its head in high mountain cirques to its terminus in the valley below.

Part (*a*) of the figure below sketches the pathways of matter flow. In the zone of accumulation, snowfall provides input to the glacier as the snow becomes compacted and recrystallized to solid ice. During warm summer periods, some

snow is lost to melting, evaporation, and sublimation, but the annual balance is on the side of accumulation.

As the glacier reaches lower elevations, loss of water from the glacier by evaporation, sublimation,

Matter and energy flow in an alpine glacier system. (Copyright © A. N. Strahler.)

Matter flow system

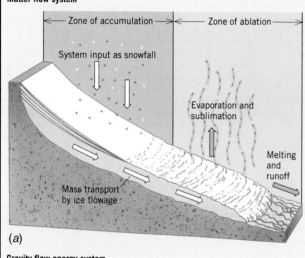

(*a*)

Gravity flow energy system

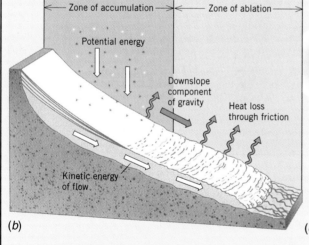

(*b*)

Thermal flow energy system

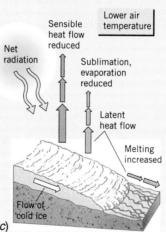

(*c*)

ICE SHEETS OF THE PRESENT

In contrast to alpine glaciers are the enormous ice sheets of Antarctica and Greenland. These are huge plates of ice, thousands of meters thick in the central areas, resting on land masses of subcontinental size. The Greenland Ice Sheet has an area of 1.7 million sq km (about 670,000 sq mi) and occupies about seven-

eighths of the entire island of Greenland (Figure 18.9). Only a narrow, mountainous coastal strip of land is exposed. The Antarctic Ice Sheet covers 13 million sq km (about 5 million sq mi) (Figure 18.10). Both ice sheets are developed on large, elevated land masses in high latitudes. No ice sheet exists near the north pole, which is positioned in the vast Arctic Ocean. Ice there occurs only as floating sea ice.

and melting increases. The balance shifts to a net loss of ice over the year, and the glacier enters the zone of ablation. With further descent, the rate of loss increases with increasing mean annual temperature and the glacier's cross section shrinks rapidly. At the terminus, no more ice remains.

Over a period of years, the glacier flow system reaches a steady state in which the excess accumulation of water as ice in the zone of accumulation is balanced by the loss of water in the zone of ablation. Within the glacier, there is a continuous downhill flow of ice, and each year an increment equal to the amount of new ice formed moves across the boundary between the zone of accumulation and zone of ablation. The shape and appearance of the glacier remains unchanged, although there may be some small variation from year to year.

Consider what would happen if temperatures cooled and snowfall increased. In that event, more ice would form annually in the zone of accumulation. The boundary between the zone of accumulation and zone of ablation would move to a lower elevation. The glacier would take longer to melt at lower elevations, so the terminus would extend farther down the valley. Eventually, a new steady state would be reached in which a larger, thicker glacier terminates farther from its source. If temperatures warmed, the changes would reverse, producing a new steady state in which a thinner glacier terminates at a higher elevation.

As the glacier flows downhill, it also transports rock debris. Scouring its bed, it breaks off rock fragments and creates a valley with a distinctive U-shape. The glacier also receives material from sideslopes as rock fragments fall onto the surface of the glacier. This debris, ranging from large blocks to fine rock flour, is transported to the lowlands below and shaped into different types of depositional landforms. (These are discussed in the main portion of this chapter.) Note that the transport of debris by the glacier can be treated as a separate open flow system powered by gravity.

Let's turn now to the flows of energy coupled to the matter flow system. We can consider these as two subsystems—a gravity flow subsystem (part *b* of the figure) and a thermal flow subsystem (part *c*). The power sources of the gravity flow subsystem are solar energy and gravity. Solar energy flow provides the water at high elevations through precipitation, thus contributing potential energy as an input. Gravity powers the downhill motion of the ice. In this motion, potential energy is converted to kinetic energy, and the kinetic energy is dissipated as heat in the friction of glacial flow. Gravity also powers the landform-making processes of glacial erosion, transportation, and deposition.

The thermal flow subsystem, shown in part *c*, is easier to understand if we think of cold as a negative quantity, as the absence of heat. The flow of glacial ice, then, serves to export "cold" from higher elevations to lower ones. How does this occur? When ice descends to low elevations, its surface is subjected to shortwave heat flows from solar radiation. This input provides a positive net radiation balance during the day. The energy flow goes largely to melting the ice, which means that less energy is available to warm the overlying air through sensible heat transfer. Also, less energy remains to sublimate the snow and evaporate the meltwater, thus reducing the flow of latent heat to the atmosphere. The effect is that more incoming energy is converted to latent heat of melting, leaving less available to warm the air. This makes the local climate colder.

These three interlinked systems demonstrate the important point that real energy and matter flow systems can be complex when examined in detail. Analysis of energy and matter budgets of systems thus requires careful study if the analysis is to include all important factors.

The surface of the Greenland Ice Sheet has the form of a very broad, smooth dome. Underneath the ice sheet, the rock floor lies near or slightly below sea level under the central region but is higher near the edges. The Antarctic Ice Sheet is thicker than the Greenland Ice Sheet—as much as 4000 m (about 13,000 ft) at maximum. At some locations, ice sheets extend long tongues, called outlet glaciers, to reach the sea at the heads of fiords. From the floating edge of the glacier, huge masses of ice break off and drift out to open sea with tidal currents to become icebergs. An important glacial feature of Antarctica is the presence of great plates of floating glacial ice, called *ice shelves* (Figure 18.10). Ice shelves are fed by the ice sheet, but they also accumulate new ice through the compaction of snow.

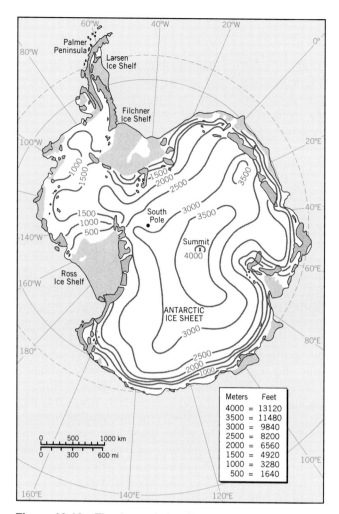

Figure 18.10   The Antarctic Ice Sheet and its ice shelves. Contours show elevations of the ice sheet surface. (Based on data of American Geophysical Union.)

SEA ICE AND ICEBERGS

Free-floating ice on the sea surface is of two types—sea ice and icebergs. *Sea ice* (Figure 18.11) is formed by direct freezing of ocean water. In contrast, *icebergs* are bodies of land ice that have broken free from glaciers that terminate in the ocean. Aside from differences in origin, a major difference between sea ice and icebergs is thickness. Sea ice does not exceed 5 m (15 ft) in thickness, while icebergs may be hundreds of meters thick.

Pack ice is sea ice that completely covers the sea surface. Under the forces of wind and currents, pack ice breaks up into individual patches called ice floes. The narrow strips of open water between such floes are known as *leads*. Where ice floes are forcibly brought together by winds, the ice margins buckle and turn upward into pressure ridges that often resemble walls of ice. Travel on foot across the polar sea ice is extremely difficult because of such obstacles. The surface zone of

sea ice is composed of fresh water, while the deeper ice is salty.

When a valley glacier or tongue of an ice sheet terminates in sea water, blocks of ice break off to form icebergs (Figure 18.12). An iceberg may be as thick as several hundred meters. Because it is only slightly less dense than sea water, the iceberg floats very low in the water, with about five-sixths of its bulk submerged. The ice is composed of fresh water since it is formed from compacted and recrystallized snow.

THE ICE AGE

The period during which continental ice sheets grow and spread outward over vast areas is known as a **glaciation**. Glaciation is associated with a general cooling of average air temperatures over the regions where the ice sheets originate. At the same time, ample snowfall must persist over the growth areas to allow the ice masses to build in volume.

When the climate warms or snowfall decreases, ice sheets become thinner and cover less area. Eventually, the ice sheets may melt completely. This period is called a *deglaciation*. Following a deglaciation, but preceding the next glaciation, is a period in which a mild

Figure 18.11   A Landsat image of a portion of the Canadian arctic archipelago. An ice cap on a landmass is visible in the center of the photo, with exposed mountainous ridges on the uncovered portions of the landmass. A branching glacial trough, now a water-filled fiord, is seen in the lower part of the image. In the upper left are huge chunks of free-floating sea ice.

period of 1 to 10 million years or more, constitutes an *ice age.*

Throughout the past 3 million years or so, the earth has been experiencing the **Late-Cenozoic Ice Age** (or, simply, the **Ice Age**). As you may recall from Chapter 11, the Cenozoic Era has seven epochs (see Table 11.1). The Ice Age falls within the last three epochs: Pliocene, Pleistocene, and Holocene. These three epochs comprise only a small fraction—about one-twelfth—of the total duration of the Cenozoic Era.

During the first half of this century, most geologists associated the Ice Age with the Pleistocene Epoch, which began about 1.6 million years ago. However, new evidence obtained from deep-sea sediments shows that the glaciations of the Ice Age began in late Pliocene time, perhaps 2.5 to 3.0 million years ago.

At present, we are within an interglaciation of the Late-Cenozoic Ice Age, following a deglaciation that set in quite rapidly about 15,000 years ago. In the preceding glaciation, called the *Wisconsinan Glaciation*, ice sheets covered much of North America and Europe, as well as parts of northern Asia and southern South America. The maximum ice advance of the Wisconsinan Glaciation was reached about 18,000 years ago.

Glaciation During the Ice Age

Figures 18.13 and 18.14 show the maximum extent to which North America and Europe were covered during the last advance of the ice. Most of Canada was en-

Figure 18.12 An iceberg off the coast of Antarctica near Elephant Island.

climate prevails—an **interglaciation**. The last interglaciation began about 140,000 years ago and ended between 120,000 and 110,000 year ago. A succession of alternating glaciations and interglaciations, spanning a

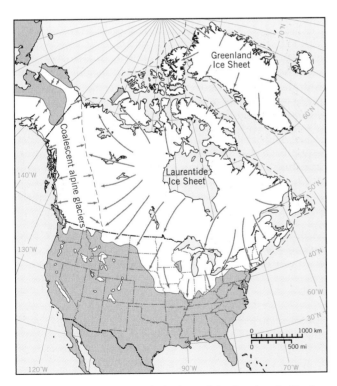

Figure 18.13 Continental glaciers of the Ice Age in North America at their maximum extent reached as far south as the present Ohio and Missouri rivers. Note that during glaciations sea level was much lower. The present coastline is shown for reference only. (Based on data of R. F. Flint, *Glacial and Pleistocene Geology*, John Wiley & Sons, New York.)

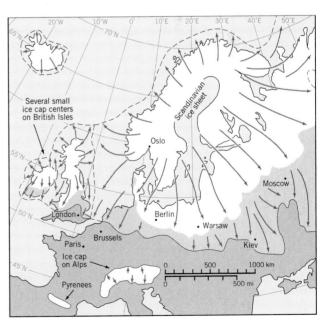

Figure 18.14 The Scandinavian Ice Sheet dominated Northern Europe during the Ice Age glaciations. As noted in Figure 18.13, the present coastline is far inland from the coastline that prevailed during glaciations. (Based on data of R. F. Flint, *Glacial and Pleistocene Geology*, John Wiley & Sons, New York.)

gulfed by the vast Laurentide Ice Sheet. It spread south into the United States, covering most of the land lying north of the Missouri and Ohio rivers, as well as northern Pennsylvania and all of New York and New England. Alpine glaciers of the western ranges coalesced into a single ice sheet that spread to the Pacific shores and met the Laurentide sheet on the east. Notice that an area in southwestern Wisconsin escaped inundation. Known as the Driftless Area, it was apparently bypassed by glacial lobes moving on either side.

In Europe, the Scandinavian Ice Sheet centered on the Baltic Sea, covering the Scandinavian countries. It spread south into central Germany and far eastward to cover much of Russia. In north-central Siberia, large ice caps formed over the northern Ural Mountains and highland areas farther east. Ice from these centers grew into a large sheet covering much of central Siberia. The European Alps were capped by enlarged alpine glaciers. The British Isles were mostly covered by a small ice sheet that had several centers on highland areas and spread outward to coalesce with the Scandinavian Ice Sheet.

At the maximum spread of these ice sheets, sea level was as much as 125 m (410 ft) lower than today, exposing large areas of the continental shelf on both sides of the Atlantic basin. The shelf supported a vegetated landscape populated with animal life, including Pleistocene elephants (mastodons and mammoths). The drawdown of sea level explains why the ice sheets shown on our maps extend far out into what is now the open ocean.

South America, too, had an ice sheet. It grew from ice caps on the southern Andes Range south of about latitude 40° S and spread westward to the Pacific shore, as well as eastward, to cover a broad belt of Patagonia. It covered all of Tierra del Fuego, the southern tip of the continent. The South Island of New Zealand, which today has a high spine of alpine mountains with small relict glaciers, developed a massive ice cap in late Pleistocene time. All high mountain areas of the world underwent greatly intensified alpine glaciation at the time of maximum ice sheet advance. Today, most remaining alpine glaciers are small ones. In less favorable locations, the Ice Age alpine glaciers are entirely gone.

The weights of the continental ice sheets, covering vast areas with ice masses several kilometers thick, exerted downward forces on the crust, causing depressions of the crust of hundreds of meters at some locations. *Working It Out 18.2 • Isostatic Rebound* provides a more detailed look at this phenomenon in two locations from Canada and one from Norway.

LANDFORMS MADE BY ICE SHEETS

Landforms made by the last ice advance and recession are very fresh in appearance and show little modification by erosion processes. It is to these landforms that we now turn our attention.

Erosion by Ice Sheets

Like alpine glaciers, ice sheets are highly effective eroding agents. The slowly moving ice scraped and ground away much solid bedrock, leaving behind smoothly rounded rock masses. These bear countless grooves and scratches trending in the general direction of ice movement (see Figure 18.2). Sometimes the ice polishes the rock to a smooth, shining surface. The evidence of ice abrasion is common throughout glaciated regions of North America and may be seen on almost any hard rock surface that is freshly exposed. Conspicuous knobs of solid bedrock shaped by the moving ice are also common features (Figure 18.15). The side from which the ice approached is usually smoothly rounded. The lee side, where the ice plucked out angular joint blocks, is irregular and blocky.

The ice sheets also excavated enormous amounts of rock at locations where the bedrock was weak and the flow of ice was channeled by the presence of a valley trending in the direction of ice flow. Under these conditions, the ice sheet behaved like a valley glacier, scooping out a deep, U-shaped trough. The Finger Lakes of western New York State are fine examples (Figure 18.16). Here, a set of former stream valleys lay largely parallel to the southward spread of the ice, and a set of long, deep basins was eroded. Blocked at their

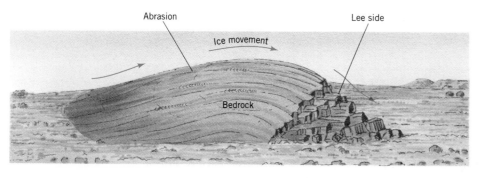

Figure 18.15 A glacially abraded rock knob. Glacial action abrades the rock into a smooth form as it rides over the rock summit, then plucks bedrock blocks from the lee side, producing a steep, rocky slope. (Copyright © A. N. Strahler.)

Figure 18.16 The Finger Lakes region of New York is shown in this photo taken by astronauts aboard the Space Shuttle. The lakes occupy valleys that were eroded and deepened by glacial ice.

north ends by glacial debris, the basins now hold lakes. Many hundreds of lake basins were created by glacial erosion and deposition over the glaciated portion of North America.

Deposits Left by Ice Sheets

The term **glacial drift** includes all varieties of rock debris deposited in close association with glaciers. Drift is of two major types. *Stratified drift* consists of layers of sorted and stratified clays, silts, sands, or gravels. These materials were deposited by meltwater streams or in bodies of water adjacent to the ice. **Till** is an unstratified mixture of rock fragments, ranging in size from clay to boulders, that is deposited directly from the ice without water transport. Where till forms a thin, more or less even cover, it is referred to as *ground moraine.*

Over those parts of North America formerly covered by late-Cenozoic ice sheets, glacial drift thickness averages from 6 m (about 20 ft) over mountainous terrain, such as New England, to 15 m (about 50 ft) and more over the lowlands of the north-central United States. Over Iowa, drift thickness is from 45 to 60 m (about 150 to 200 ft), and over Illinois it averages more than 30 m (about 100 ft). In some places where deep stream valleys existed prior to glacial advance, as in parts of Ohio, drift is much thicker.

To understand the form and composition of deposits left by ice sheets, it will help to examine the conditions prevailing at the time of the ice sheet's existence. Figure 18.17*a* shows a region partly covered by an ice sheet with a stationary front edge. This condition occurs when the rate of ice ablation balances the amount of ice brought forward by spreading of the ice sheet. Although the ice fronts of the Ice Age advanced and receded in many minor and major fluctuations, there were long periods when the front was essentially stable and thick deposits of drift accumulated.

Moraines

The transportational work of an ice sheet resembles that of a huge conveyor belt. Anything carried on the belt is dumped off at the end and, if not constantly removed, will pile up in increasing quantity. Rock fragments brought within the ice are deposited at its outer edge as the ice evaporates or melts. There is no possibility of return transportation.

Glacial till that accumulates at the immediate ice edge forms an irregular, rubbly heap—the terminal moraine. After the ice has disappeared (Figure 18.17*b*), the moraine appears as a belt of knobby hills interspersed with basinlike hollows, or kettles, some of which hold small lakes. The name *knob-and-kettle* is often applied to morainal belts.

Terminal moraines form great curving patterns. The outward curvature is southward and indicates that the ice advanced as a series of great *ice lobes*, each with a curved front (Figure 18.18). Where two lobes come together, the moraines curve back and fuse together into a single *interlobate moraine* pointed northward. In its general recession accompanying disappearance, the ice front paused for some time along a number of positions, causing morainal belts similar to the terminal moraine belt to be formed. These belts are known as recessional moraines (Figure 18.17*b*). They run roughly parallel with the terminal moraine but are often thin and discontinuous.

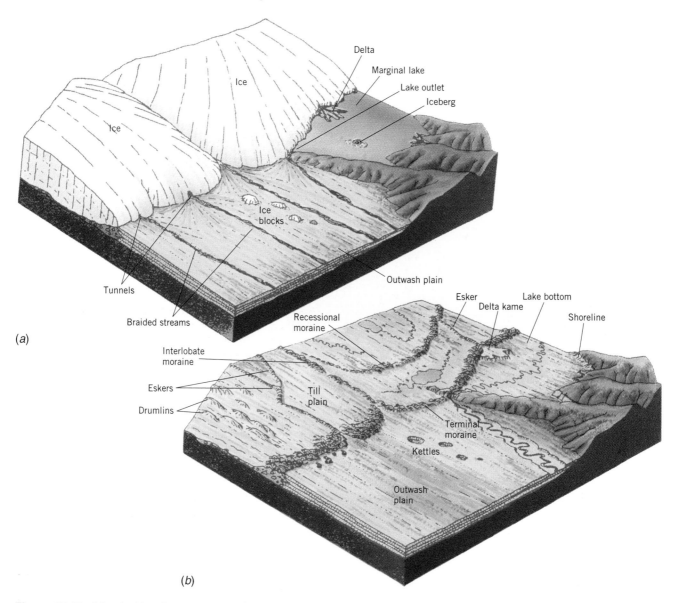

Figure 18.17 Marginal landforms of continental glaciers. (*a*) With the ice front stabilized and the ice in a wasting, stagnant condition, various depositional features are built by meltwater. (*b*) The ice has wasted completely away, exposing a variety of new landforms made under the ice. (Drawn by A. N. Strahler.)

Outwash and Eskers

Figure 18.17 shows a smooth, sloping plain lying in front of the ice margin. This is the *outwash plain*, formed of stratified drift left by braided streams issuing from the ice. The plain is built of layer upon layer of sands and gravels.

Large streams carrying meltwater issue from tunnels in the ice. These form when the ice front stops moving for many kilometers back from the front. After the ice has gone, the position of a former ice tunnel is marked by a long, sinuous ridge of sediment known as an esker (Figure 18.19). The esker is the deposit of sand and gravel laid on the floor of the former ice tunnel. After the ice has melted away, only the streambed deposit re-

mains, forming a ridge. Many eskers are several kilometers long.

Drumlins and Till Plains

Another common glacial form is the *drumlin*, a smoothly rounded, oval hill resembling the bowl of an inverted teaspoon. It consists in most cases of glacial till (Figure 18.20). Drumlins invariably lie in a zone behind the terminal moraine. They commonly occur in groups or swarms and may number in the hundreds. The long axis of each drumlin parallels the direction of ice movement. The origin of drumlins is not well understood. They seem to have been formed under mov-

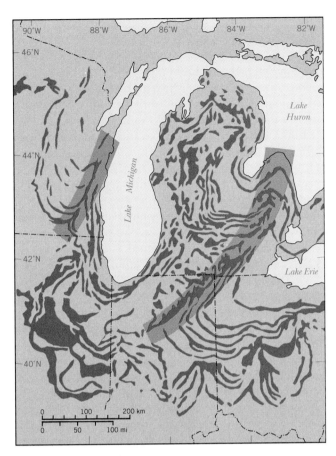

Figure 18.18 Moraine belts of the north-central United States have a curving pattern left by ice lobes. Some regions of interlobate moraines are shown by the color overlay. (Based on data of R. F. Flint and others, *Glacial Map of North America*, Geological Society of America.)

Figure 18.19 The curving ridge of sand and gravel in this photo is an esker, marking the bed of a river of meltwater flowing underneath a continental ice sheet near its margin. Kettle-Moraine State Park, Wisconsin.

When the ice withered away, the lakes drained, leaving a flat floor exposed. Here, layers of fine clay and silt had accumulated. Glacial lake plains often contain extensive areas of marshland. The deltas are now curiously isolated, flat-topped landforms known as *delta kames*, composed of well-washed and well-sorted sands and gravels (Figure 18.17*b*).

ing ice by a plastering action in which layer upon layer of bouldery clay was spread on the drumlin.

Between moraines, the surface overridden by the ice is covered by glacial till. This cover is often inconspicuous since it forms no prominent landscape feature. The till layer may be thick and may obscure, or entirely bury, the hills and valleys that existed before glaciation. Where thick and smoothly spread, the layer forms a level *till plain*. Plains of this origin are widespread throughout the central lowlands of the United States and southern Canada.

Marginal Lakes and Their Deposits

When the ice advanced toward higher ground, valleys that may have opened out northward were blocked by ice. Under such conditions, marginal glacial lakes formed along the ice front (see Figure 18.17*a*). These lakes overflowed along the lowest available channel between the ice and the rising ground slope, or over some low pass along a divide. Streams of meltwater from the ice built *glacial deltas* into these marginal lakes.

Figure 18.20 This small drumlin, located south of Sodus, New York, shows a tapered form from upper right to lower left, indicating that the ice moved in that direction (north to south).

Working It Out 18.2 • Isostatic Rebound

In *Focus on Systems 16.2 • A Model Denudation System,* we described the principle of *isostasy,* which states that the crust floats on a plastic asthenosphere. According to this principle, a crustal block rises as erosion strips away its rock mass through time, a phenomenon we termed *isostatic compensation.*

During the Ice Age, large continental ice sheets of several kilometers in thickness covered large areas of North America, Europe, and Asia. The weight of these ice sheets depressed the crust, causing sinking. Figure 1 diagrams this process. With the onset of the most recent glaciation at about 80,000 years before present, ice sheets grew rapidly, depressing the crust underneath. At the close of the last glacial period, about 12,000 years ago, deglaciation was rapid, requiring only a few thousand years. Relieved of the weight of the ice, the crust began to move upward, quickly at first, then more slowly. This crustal movement is known as *isostatic re-*

bound. Upward movement is still occurring today in some locations.

An important line of evidence for isostatic rebound is the position of ancient shorelines, now raised above sea level by the crustal motion. How did these shorelines form? As the ice sheets melted, sea level rose rapidly to near its present position. The crust, still depressed, was flooded with ocean waters, providing wave action at the shoreline that created beaches. As the crust rebounded upward, the beaches were elevated, leaving them hundreds of meters above present levels. By dating these ancient shorelines, geologists have established the amount and rate of uplift occurring in response to the unloading of ice. Many elevated shorelines can be found today in the coastal belts of Hudson Bay, located in north-central Canada, and in the Bay of Bothnia, which separates Sweden and Finland.

Figure 2 provides plots of shoreline elevations with age for three lo-

cations. The points are taken from published graphs. North Bay is in Ontario on Lake Huron, located at the south end of Georgian Bay. James Bay is the southernmost projection of Hudson Bay and lies near the center of crustal depression for the Laurentide Ice Sheet (see Figure 18.13). Oslofjord is the narrow glacial trench that provides ocean access to Oslo, Norway (see Figure 18.14). Note that both axes on the graphs are numerically reversed. For the *x*-axis, time is in years before present (B.P.), so time increases from left to right. For the *y*-axis, shoreline height is taken as the crustal depression at the time of formation of the shoreline. Upward crustal movement is shown by a higher position on this graph.

The upward motion of the crust in isostatic rebound is most rapid immediately following the melting of the ice. With time, rebound continues, but the rate of uplift decreases as the present level is approached. North Bay, located significantly south of the center of the ice sheet, shows the earliest rebound, followed by James Bay and then Oslofjord. Note the steeper rebound for James Bay. Located near the center of the ice sheet, this region was depressed most by the ice and so experienced the strongest rebound immediately after deglaciation.

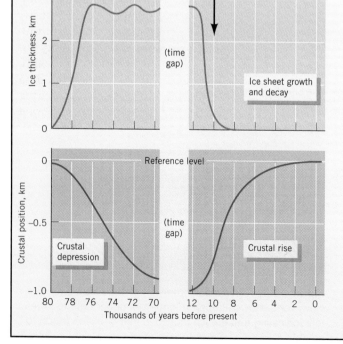

Figure 1 Crustal depression and rise in response to the growth and decay of continental ice sheets. (After A. N. Strahler.)

The middle part of the figure shows the rebound plotted on a semilog graph. The points show a good fit to a straight line. For this to occur, the points follow a line with the equation

$$y = ba^x \qquad (1)$$

where b and a are constants. (Note that in evaluating this expression, a is raised to the power x before the result is multiplied by b.) By taking the log transform, we can see how this equation becomes a straight line:

$$\log y = \log b + x \log a$$

If we substitute $y' = \log y$, $b' = \log b$, and $a' = \log a$, the equation becomes

$$y' = b' + a'x \qquad (2)$$

which is a straight line. Since $y' = \log y$, we use a logarithmic scale on the y-axis. But since x is untransformed, we just use a normal arithmetic scale for the x-axis.

It is not difficult to determine the equation for a line that best fits points in a semilog graph. A simple way is to use two points that lie on or close to the line, and from their values determine the slope, which will give a', and the intercept, which will give b'. The process is diagrammed in the bottom graph for the Oslofjord data. Two points, 1 and 2, are compared: 50 m at 3900 years B.P., and 220 m at 8000 years B.P.

The slope of the line is simply the vertical change divided by the horizontal change for the two points. Normally, we would determine the change by subtracting the coordi-nates of point 2 from those of point 1, but in this case the axes are re-versed, changing the sign of the val-ues. So here we subtract the coordi-nates of point 1 from point 2. Note also that we will need to use log val-ues for y, the shoreline height. In addition, it will be simpler to work in units of thousands of years, as shown on the x-axis. The slope is then

$$a' = \frac{\log 220 - \log 50}{8.0 - 3.9} = \frac{0.64}{4.1} = 0.156$$

For the intercept, b', we can sim-ply solve (2) for b', giving $b' = y' - a'x$, and substitute the values of y' and x from one of the points, for ex-ample, point 1, given the value of a':

$$b' = y' - a'x = \log 50 - 0.156 \times 3.9 = 1.09$$

In semilog form, the equation is then

$$\log y = 1.09 + 0.156x$$

To obtain the exponential form of the equation (1), we first find a. Since $a' = \log a$, $a = 10^{a'} = 10^{0.156} = 1.43$. Similarly, $b' = \log b$, so $b = 10^{b'} = 10^{1.09} = 12.3$. The equation is then

$$y = 12.3 \times (1.43)^x$$

Plotting data and fitting func-tions to the points is one of the im-portant tools of science, including physical geography. Geophysicists have derived more complex formu-las for crustal rebound using the theory of physics of elastic solids, but for many purposes, including exploring data and comparing trends, simple methods like the one above are often all that is needed.

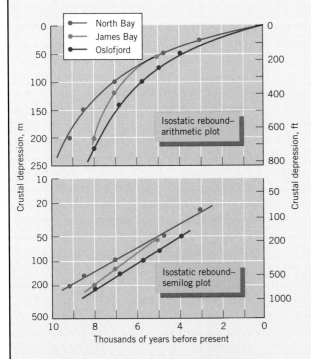

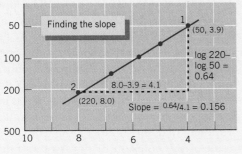

Figure 2 Plots of isostatic rebound. Upper plots: arith-metic and semilogarithmic. The lowest plot shows how the slope of the semilog line is determined. (Data from R. F. Flint, W. Ferrand, and W. A. Heiskanen.)

Pluvial Lakes

During the Ice Age, some regions experienced a cooler, moister climate. In the western United States, closed basins filled with water, forming pluvial lakes. The largest of these, glacial Lake Bonneville, was about the size of Lake Michigan and occupied a vast area of western Utah. With the warmer and drier climate of the present interglacial period, these lakes shrank greatly in volume. Lake Bonneville became the present-day Great Salt Lake. Many other lakes dried up completely, forming desert playas. The history of these pluvial lakes is known from their ancient shorelines, some as high as 300 m (about 1000 ft) above present levels.

EYE ON THE ENVIRONMENT:
Environmental Aspects of Glacial Deposits

Because much of Europe and North America was glaciated by the Pleistocene ice sheets, landforms associated with the ice are of major environmental importance. Agricultural influences of glaciation are both favorable and unfavorable, depending on preglacial topography and the degree and nature of ice erosion and deposition.

In hilly or mountainous regions, such as New England, the glacial till is thinly distributed and extremely stony. Till cultivation is difficult because of countless boulders and cobbles in the clay soil. Till accumulations on steep mountain or roadside slopes are subject to mass movement as earthflows when clay in the till becomes weakened after absorbing water from melting snows and spring rains. Along moraine belts, the steep slopes, the irregularity of knob-and-kettle topography, and the abundance of boulders conspired to prevent crop cultivation but invited use as pasture.

Flat till plains, outwash plains, and lake plains, on the other hand, can sometimes provide very productive agricultural land. Bordering the Great Lakes, fertile soils have formed on till plains and on exposed lakebeds. This fertility is enhanced by a blanket of wind-deposited silt (loess) that covers these plains (Chapter 17).

Stratified drift deposits are of great commercial value. The sands and gravels of outwash plains, deltas, and eskers provide necessary materials for both concrete manufacture and highway construction. Where it is thick, stratified drift forms an excellent aquifer and is a major source of groundwater supplies. 🐾

INVESTIGATING THE ICE AGE

A great scientific breakthrough in the study of Ice Age glacial history came in the 1960s. First, scientists learned how to measure the absolute age of certain types of water-laid sediments by means of ancient magnetism. The earth's magnetic field experienced many sudden reversals of polarity in Cenozoic time, and the absolute ages of these reversals have been firmly established. Second, techniques were developed to take long sample cores of undisturbed fine-textured sediments of the deep ocean floor. Within each core, scientists could determine the age of sediment layers at various control points by identifying magnetic polarity reversals. By further studying the composition and chemistry of the layers within the core, a record of ancient temperature cycles in the air and ocean could be established.

Deep-sea cores reveal a long history of alternating glaciations and interglaciations going back at least as far as 2 million years and possibly 3 million years before present. The cores show that in late-Cenozoic time more than 30 glaciations occurred, spaced at time intervals of about 90,000 years. How much longer this sequence will continue into the future is not known, but perhaps for 1 or 2 million years, or even longer.

Possible Causes of the Late-Cenozoic Ice Age

What caused the earth to enter into an Ice Age with its numerous cycles of glaciation and interglaciation? Three causes seem possible. First is a change in the placement of continents on the earth's surface through plate tectonic activity. Second is an increase in the number and severity of volcanic eruptions. Third is a reduction in the sun's energy output.

Perhaps the answer lies in plate tectonics, through the motions of lithospheric plates following the breakup of Pangea. Recall from Chapter 11 that in Permian time, only the northern tip of the Eurasian continent projected into the polar zone. But as the Atlantic basin opened up, North America moved westward and poleward to a position opposite Eurasia, while Greenland took up a position between North America and Europe.

The effect of these plate motions was to bring an enormous landmass area to a high latitude and to surround a polar ocean with land. Because the flow of warm ocean currents into the polar ocean was greatly reduced, or at times totally cut off, this arrangement was favorable to the growth of ice sheets. The polar ocean was ice-covered much of the time, and average air temperatures in high latitudes were at times lowered enough to allow ice sheets to grow on the encircling continents. Furthermore, Antarctica moved southward during the breakup of Pangea and took up a position over the south pole. In that location, it was ideally situated to develop a large ice sheet. Some scientists have also proposed that the uplift of the Himalayan Plateau, a result of the collision of the Austral-Indian and Eurasian plates, could have modified weather patterns sufficiently to trigger the Ice Age.

The second geological mechanism suggested as a basic cause of the Ice Age is increased volcanic activity on a global scope in late-Cenozoic time. Volcanic eruptions produce dust veils that linger in the stratosphere and reduce the intensity of solar radiation reaching the ground (Chapter 2). Temporary cooling of near-surface air temperatures follows such eruptions. Although the geologic record shows periods of high levels of volcanic activity in the Miocene and Pliocene epochs, their role in initiating the Ice Age has not been convincingly demonstrated on the basis of present evidence.

Another possible cause of the Ice Age is a slow decrease in the sun's energy output over the last several million years, perhaps as part of a cycle of slow increase and decrease over many million years' duration. As yet, data are insufficient to identify this mechanism as a possible basic cause. However, research on this topic is being stepped up as new knowledge of the sun is acquired from satellites that probe the sun's atmosphere and map its surface features.

Possible Causes of Glaciation Cycles

What timing and triggering mechanisms are responsible for the many cycles of glaciation and interglaciation that the earth is experiencing during the present Ice Age? Although many causes for glacial cycles have been proposed, we will limit our discussion here to one major contender called the **astronomical hypothesis**. It has been under consideration for about 40 years and is based on well-established motions of the earth in its orbit around the sun.

Two cyclic factors are involved—the changing distance between earth and sun, and the changing angle of tilt of the earth's axis of rotation. As to the first factor, recall from Chapter 1 that the earth's orbit is not perfectly circular, but is slightly elliptical. The shape of the ellipse varies slowly through time. As a result, the distance that separates the earth and sun at summer solstice, June 21, undergoes a cyclic variation. A single cycle lasts 21,000 years. During the cycle, the average distance between earth and sun may vary from 1 to 5 percent greater to 1 to 5 percent less than the normal long-term average. When the distance is greater, less solar energy will reach earth and our earth will grow colder.

Another cycle slowly changes the tilt of the earth's axis. Just as a spinning top sometimes develops a wobble, so the earth's axis also has a slight wobble that changes the earth's tilt from 24° to 22° and back again. This cycle takes about 40,000 years. When the tilt is smaller (22°), it tends to make the polar regions slightly colder and the equatorial regions slightly warmer. When the tilt is greater (24°), the effect is opposite—warmer polar regions and a cooler equatorial belt.

Since both cycles are going on simultaneously, they can either cancel or reinforce one another. When both cycles combine to make the polar regions colder and reduce summer snowmelt, ice cap formation is enhanced. When one cycle warms the poles while the other cools the planet, snowmelt keeps the ice caps from growing quickly, and continental glaciers form more slowly or begin to waste away. If both cycles combine to warm the polar regions, the ice cap melts and shrinks away.

If we pick a point on the earth at, say, lat. 65° N, we can calculate the cycle of total change in the intensity of incoming solar radiation, or insolation, on a day in summer (June 21), resulting from changing earth–sun distance and the cycle of axial tilt. The changing insolation, plotted against time on a graph, yields a series of peaks and valleys (Figure 18.21). The graph is known as the Milankovitch curve, named for Milutin Milankovitch, the astronomer who first calculated it in

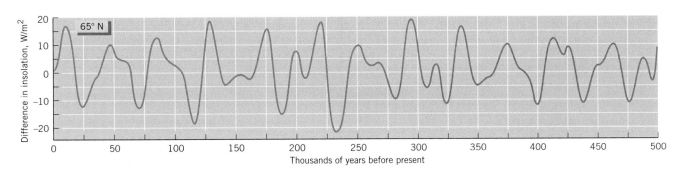

Figure 18.21   The Milankovitch curve. The vertical axis shows fluctuations in incoming solar energy at lat. 65° N on June 21, the summer solstice. These are calculated from mathematical models of the change in earth–sun distance and change in axial tilt with time. The zero value represents the present value. (Based on calculations by A. D. Vernekar, 1968. Copyright © A. N. Strahler.)

1938. The major peaks on the graph are repeated at intervals of about 80,000 to 90,000 years. Each peak seems to have triggered a deglaciation, ending a glaciation and bringing on an interglaciation. During the longer periods of reduced insolation, ice sheets formed and grew in volume. However, the glacial events may have lagged behind the insolation peaks and valleys by some thousands of years.

The actual mechanisms by which insolation changes cause ice sheets to grow or to disappear are unclear. The entire subject is so complex that it is difficult for even a research scientist to grasp fully. Interactions between the atmosphere, the oceans, and the continental surfaces (including the ice sheets) are numerous and closely interrelated with many threads of cause and effect. Changes that occur in one earth realm are fed back to the other realms in a most complex manner.

Holocene Environments

The elapsed time span of about 10,000 years since the Wisconsinan Glaciation ended is called the *Holocene Epoch*. It began with a rapid warming of ocean surface temperatures. Continental climate zones then quickly shifted poleward, and plants soon became reestablished in glaciated areas.

Three major climatic periods occurred during the Holocene Epoch leading up to the last 2000 years. These periods are inferred from studies of changes in vegetation cover types as observed in fossil pollen and spores preserved in glacial bogs. The earliest of the three is known as the Boreal stage, and was characterized by boreal forest vegetation in midlatitude regions. There followed a general warming until the Atlantic stage, with temperatures somewhat warmer than today, was reached about 8000 years ago (-8000 years). Next came a period of temperatures that were below average, the Subboreal stage. This stage spanned the age range -5000 to -2000 years.

Through the availability of historical records and of more detailed evidence, the climate of the past 2000 years can be described on a finer scale. A secondary warm period occurred in the period A.D. 1000 to 1200 (-100 to -800 years). This warm episode was followed by the Little Ice Age, A.D. 1450-1850 (-550 to -50 years). During the latter, valley glaciers made new advances and extended to lower elevations. More recent temperature cycles are discussed in Chapter 3.

Cycles of glaciation and interglaciation, as well as the lesser climatic cycles of the Holocene epoch, have proceeded without human influence for millions of years. They demonstrate the power of natural forces to make drastic swings from cold to warm climates, and they confound our efforts to understand human impact on climate.

EYE ON THE ENVIRONMENT: ICE SHEETS AND GLOBAL WARMING

A subject of much scientific study and speculation is the effect of climatic warming on the ice sheets of Greenland and Antarctica. The Antarctic Ice Sheet holds 91 percent of the earth's ice. If it were to melt entirely, the earth's mean sea level would rise by about 40 m (about 200 ft). Additional water could be released by the melting of the Greenland Ice Sheet, which holds most of the remaining volume of land ice. It seems unlikely that global warming will melt these ice sheets completely, unless the warming is much greater than anticipated. Still, the ice sheets might be greatly affected by only a small warming.

In the 1970s climatologist John Mercer pointed out that a portion of the Antarctic Ice Sheet lying west of the Transantarctic Mountains was in a potentially unstable condition and that it might surge rapidly into the ocean, causing a sea-level rise of about 5 m (16 ft). Mercer's theory depends heavily on the role of the Ross ice shelf of western Antarctica, which merges on its landward sides with the ice sheet. Because the ice shelf is already afloat, its detachment or melting would not change sea level. It does, however, hold back the downslope flow of ice from the western Antarctic Ice Sheet.

The shelf ice is in contact with the bottom (grounded) at a number of points. If sea level were to rise slightly, owing to thermal expansion, it could lift the ice shelf and allow sea water to circulate more freely underneath it, melting the underside of the floating ice shelf. The ice shelf would then move out to sea, allowing the unsupported land ice to quickly surge forward and enter the ocean. Thus far, there are no indications that this activity has begun. On the other hand, if Mercer's prediction should materialize, it could result in a 5 m sea-level rise (16 ft)—a disaster of monumental proportions.

In 1989 other scientists active in glaciological research—among them geophysicist Charles R. Bentley—presented a different view of the effects of global warming on ice sheets. They reasoned that, although global warming will accelerate glacier melting, increased precipitation over the ice sheets will produce a net growth in their thickness. Measurements made by radar on satellites have shown that the Greenland Ice Sheet has increased in thickness by 23 cm (9 in.) in the past decade. Growth in bulk of the Antarctic Ice Sheet, by itself, could cause a lowering of global sea level. In balance with total melting, ice sheet growth would result in only a modest rate of rise of sea level—perhaps no faster than the rate that has been documented over the past decade.

Of course, other factors may be at work as well. A team of scientists led by Donald D. Blankenship has re-

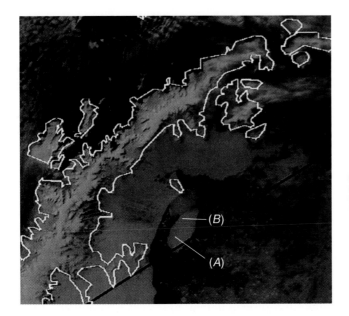

Figure 18.22 Satellite image of the Palmer Peninsula acquired on February 22, 1995, showing the Larson Ice Shelf. Land areas are outlined in white. A large iceberg (A, center of image) has broken away from the shelf. Above (north) of the berg is a region of disintegrated ice (B) that has produced a swarm of many smaller bergs. (The black line crossing the image diagonally is a data gap.) (National Snow and Ice Data Center.)

cently discovered evidence that several volcanoes may underlie the western Antarctic Ice Sheet at a position near the ice shelves. The heat from the volcanoes melts just enough ice to create a layer of slippery mud under the ice sheet, which lubricates the flow of the ice above. The result is the formation of streams of rapidly moving ice within the ice sheet that feed the ice shelves. If these streams slowed, the ice shelf might retreat, which could release the unstable western ice sheet to flow into the ocean, producing catastrophic flooding.

Although it is not certain what the fate of climatic warming may be on the Antarctic Ice Sheet as a whole, satellite images acquired in 1995 showed that a large region of the Larsen Ice Shelf (Figure 18.10), about 2000 km² (about 770 sq mi), had disintegrated into small icebergs during a storm (Figure 18.22). This unusual event was accompanied by the breaking off of a very large iceberg — nearly the size of the state of Rhode Island — that was strikingly visible on satellite images. Although more spectacular, formation of an iceberg of this size is not unusual, and is part of the normal process of preserving the mass balance of the ice sheet.

Scientists hypothesized that the unusual disintegration of the Larson Ice Shelf was caused by extreme surface melting during several warm seasons of the 1990s

and by a regional climatic warming of about 2.5°C (4.5°F) observed over the past 50 years at a nearby station. However, other measurements show that the temperature of the Antarctic continent as a whole has not increased significantly since the mid-1960s. The Larson Ice Shelf is somewhat different from the larger Ross and Filcher ice shelves, in that it is not fed primarily by glacial ice, but rather by snowfall. As a result, it is thinner and likely to be more sensitive to small changes in climate.

Thus, possible effects of climate change on global ice sheets are far from certain. We can only hope that human-induced global warming will result in slow, progressive changes to which human civilization can readily adapt. 🦫

The group of chapters we have now completed has reviewed landform-making processes that operate on the surface of the continents. These have ranged from mass wasting to the erosion and deposition brought about by glacial ice. The great variety and complexity of landforms we have described are not difficult to understand when each agent of denudation is examined in turn.

Human influence on landforms is felt most strongly on surfaces of fluvial denudation because of the severity of surface changes caused by agriculture and urbanization. Landforms shaped by wind and by waves and currents are also highly sensitive to changes induced by human activity. Only glaciers maintain their integrity and are thus far largely undisturbed by human activity. Perhaps even this last realm of nature's superiority will eventually fall prey to human interference through climate changes induced by industrial activity.

CHAPTER SUMMARY

Glaciers form when snow accumulates to a great depth, creating a mass of ice that is plastic in lower layers and flows outward or downhill from a center in response to gravity. As they move, glaciers can deeply erode bedrock by abrasion and plucking. The eroded fragments, incorporated into the flowing ice, leave depositional landforms when the ice melts.

Alpine glaciers develop in cirques in high mountain locations. Alpine glaciers flow downvalley on steep slopes, picking up rock debris and depositing it in lateral and terminal moraines. Through erosion, glaciers carve U-shaped glacial troughs that are distinctive features of glaciated mountain regions. They become fiords if later submerged by rising sea level.

Ice sheets are huge plates of ice that cover vast areas. They are present today in Greenland and Antarctica. The Antarctic Ice Sheet includes ice shelves—great plates of floating glacial ice. Icebergs form when glacial ice flowing into an ocean breaks into great chunks and floats free. Sea ice, which is much thinner and more continuous, is formed by direct freezing of ocean water and accumulation of snow.

An ice age includes alternating periods of glaciation, deglaciation, and interglaciation. During the past 2 to 3 million years, the earth has experienced the Late-Cenozoic Ice Age. During this ice age, continental ice sheets have grown and melted as many as 30 times. The most recent glaciation is the Wisconsinan Glaciation, in which ice sheets covered much of North America and Europe, as well as parts of northern Asia and southern South America.

Moving ice sheets create many types of landforms. Bedrock is grooved and scratched. Where rocks are weak, long valleys can be excavated to depths of hundreds of meters. The melting of glacial ice deposits glacial drift, which may be stratified by water flow or deposited directly as till. Moraines accumulate at ice edges. Outwash plains are built up by meltwater streams. Tunnels within the ice leave streambed deposits as eskers. Till may be spread smooth and thick under an ice sheet, leaving a till plain. This may be studded with elongated till mounds, termed drumlins. Meltwater streams build deltas into lakes formed at the ice margin and line lake bottoms with clay and silt. When the lakes drain, these features remain.

Several factors have been proposed to explain the cause of present ice age glaciations and interglaciations. These include ongoing change in the global position of continents, an increase in volcanism, and a reduction in the sun's energy output. Individual cycles of glaciation seem strongly related to small, cyclic changes in earth–sun distance and axial tilt.

The effects of global warming on the earth's ice sheets are not well understood at present. Some scientists fear that warming could release vast volumes of Antarctic ice that would quickly slide into the ocean and increase sea level by 5 m (16 ft) or more. Others have noted that global warming enhances precipitation, which causes snow to accumulate atop ice sheets and may cancel some of the potential rise in sea level due to slow melting.

KEY TERMS

ablation	moraine	interglaciation	astronomical
alpine glacier	glacial trough	Late-Cenozoic Ice Age	hypothesis
ice sheet	fiord	glacial drift	
cirque	glaciation	till	

REVIEW QUESTIONS

1. How does a glacier form? What factors are important? Why does a glacier move?

2. Distinguish between alpine glaciers and ice sheets.

3. What are some typical features of an alpine glacier? Sketch a cross section along the length of an alpine glacier and label it.

4. What is a glacial trough and how is it formed? What is its basic shape? In what ways can a glacial trough appear after glaciation is over?

5. Where are ice sheets present today? How thick are they?

6. Contrast sea ice and icebergs, including the processes by which they form.

7. Identify the Late-Cenozoic Ice Age. When did it begin? What was the last glaciation in this cycle? When did it end?

8. What areas were covered with ice sheets by the last glaciation? How was sea level affected?

9. What are moraines? How are they formed? What types of moraines are there?

10. Identify the landforms and deposits associated with stream action at or near the front of an ice sheet.

11. Identify the landforms and deposits associated with deposition underneath a moving ice sheet.

12. Identify the landforms and deposits associated with lakes that form at ice sheet margins.

13. What two cycles are known to affect the amount of solar radiation received by polar regions of the earth?

14 What is the Milankovitch curve? What does it show about warm and cold periods during the last 500,000 years?

15. How have environments changed during the Holocene Epoch? What periods are recognized, and what are their characteristics?

16. Under what conditions would global warming create a sudden rise in sea level of as much as 5 m (16 ft)?

Focus on Systems 18.1 • A Glacier as a Flow System of Matter and Energy

1. Describe the matter flow system of an alpine glacier. How can a glacier achieve a steady state?

2. What two energy flow subsystems can be recognized in the flow of an alpine glacier? Describe each.

ESSAY QUESTIONS

1. Imagine that you are planning a car trip to the Canadian Rockies. What glacial landforms would you expect to find there? Where would you look for them?

2 At some time during the latter part of the Pliocene Epoch, the earth entered an ice age. Describe the nature of this ice age and the cycles that occur within it. What explanations are proposed for causing an ice age and its cycles? What cycles have been observed since the last ice sheets retreated?

PROBLEMS

Working It Out 18.2 • Isostatic Rebound

1. The two extreme points for the James Bay data are 200 m at 8000 yr B.P., and 55 m at 5000 yr B.P. Find the equation fitting these points, and express it in both semilogarithmic and exponential form. How does the slope compare with that of Oslofjord? What does the comparison show?

2. For the Oslofjord equation, $b = 12.3$. What is the physical meaning of this value? (*Hint:* For the equation $y = ba^x$, when $x = 0$, $a^x = a^0 = 1$, so $y = b$. Thus, b is the value of y when $x = 0$, or in this case, the position on the uplift scale for the present time.)

3. A geomorphologist observes that isostatic rebound at a location fits the equation

$$y = 9.2 \times (1.7)^x$$

where y is the elevation of the shoreline in meters and x is the age, in thousands of years B.P. What is the slope of this function on a semilog plot? Is the rebound rate he observes greater or lesser than Oslofjord? than James Bay?

Part IV:

Systems and Cycles of Soils and the Biosphere

Part IV, the final part of our examination of physical geography, focuses on two important components of the life layer—soils and ecosystems. Both are complex components that are influenced by many of the processes we have already described in Parts I, II, and III.

Climate, which we examined in Part I, is a very important factor in soil development. When precipitation is abundant, water moves downward in soils, carrying with it nutrients and clay particles that influence both the physical characteristics and fertility of the soil. When precipitation is scarce, water moves downward into the soil after rainfall events, then upward as the soil dries out. This sequence also directly influences soil characteristics. Climate is a key factor for ecosystems as well. Since ecosystems are ultimately powered by photosynthesis, they are conditioned by such climatic factors as sunlight, temperature, and water availability. The tolerances of individual species for climatic extremes of temperature and drought influence the species composition of ecosystems. Furthermore, certain types of vegetation covers are associated with specific climate types, as grasslands with semiarid climates and forests with moist climates.

The processes and systems we examined in Parts II and III are also important for soils and ecosystems. The rock material that is cycled to the surface by plate tectonic mechanisms provides the raw material for weathering and soil-forming processes. The lithosphere is also the original source of the supplies of nutrient elements within the life layer that are required for life processes, including the essential atmospheric gases of oxygen and carbon dioxide. Landforms and soils are intimately related. By eroding, transporting, and depositing weathered rock materials, the geomorphic processes we described in Part III add or remove soil materials from a particular location, thus influencing the nature of the soil. Landforms and ecosystems are also strongly related on the local level. Individual landforms provide habitats for ecosystems that are often uniquely adjusted to the conditions of their locations.

Soils and ecosystems are strongly interrelated. Soil characteristics obviously influence the types of plants and animals that are rooted in, or live within, the soil. But ecosystems also influence the soil by contributing organic matter as plant roots and by aerating the soil through the burrowing activities of insects, earthworms, and higher animals. Soils and ecosystems are linked in other important ways as well.

From these observations, you can see why we have chosen to conclude our study of physical geography with soils systems, biospheric systems and cycles, and global ecosystems. Because these topics refer to many of the systems, cycles and processes that we have introduced in earlier chapters, we can finally do them justice. Chapter 19 provides an overview of the key characteristics of soils, the important processes of soil formation, the classification of soils, and the global distribution of soil types. Chapter 20 describes ecosystems, focusing on how ecosystems are powered and on the cycling of major nutrient elements by ecosystems and their environments. Chapter 21 examines ecosystems in a global context, including the nature of biomes and formation classes of vegetation and their global distribution patterns, especially as they are related to climate types. In the epilogue that follows these chapters, we examine global change in the context of the environmental issues that arise when human activity intersects the natural processes that we have examined in the study of physical geography.

Chapter 19

Soil Systems

This chapter is devoted to soil systems, including the processes that form soils and give them their distinctive characteristics, soil classification, and global distribution of soil types. **Soil** is the uppermost layer of the land surface that plants use and depend on for nutrients, water, and physical support. Soils can vary greatly from continent to continent, region to region, and even from field to field. This is because they are influenced by factors and processes that can vary widely from place to place. For example, a field near a large river that floods regularly may acquire a layer of nutrient-rich silt, making its soil very productive. A nearby field at a higher elevation, without the benefit of silt enrichment, may be sandy or stony and require the addition of fertilizers to grow crops productively.

Vegetation is an important factor in determining soil qualities. For example, some of America's richest soils developed in the Middle West under a cover of thick grass sod. The deep roots of the grass, in a cycle of growth and decay, deposited nutrients and organic matter throughout the thick soil layer. In the Northeast, conifer forests provided a surface layer of decaying needles that kept the soil quite acid. This acidity allowed nutrients to be washed below root depth, out of the reach of plants. When farmed today, these soils need applications of lime to reduce their acidity and enhance their fertility.

Climate, measured by precipitation and temperature, is also an important determiner of soil properties. Precipitation controls the downward movement of nutrients and other chemical compounds in soils. If precipitation is abundant, water tends to wash soluble compounds, including nutrients, deeper into the soil and out of reach of plant roots.

Temperature acts to control the rate of decay of organic matter that falls to the soil from the plant cover or that is provided to the soil by the death of roots.

Gully erosion near Phenix City, Alabama.

When conditions are warm and moist, decay organisms work efficiently, consuming organic matter readily. Thus, organic matter and nutrients in soils of the tropical and equatorial zones are generally low. Where conditions are cooler, decay proceeds more slowly, and organic matter is more abundant in the soil. Of course, if the climate is very dry or desertlike, then vegetation growth is slow or absent. No matter what desert temperatures are like, organic matter will be low.

Time is also an important factor. The characteristics and properties of soils require time for development. For example, a fresh deposit of mineral matter, like the clean, sorted sand of a dune, may require hundreds to thousands of years to acquire the structure and properties of a sandy soil.

Geographers are keenly interested in the differences in soils from place to place over the globe. The ability of the soils and climate of a region to produce food largely determines the size of the population it will support. In spite of the growth of cities, most of the world's inhabitants still live close to the soil that furnishes their food. And many of those same inhabitants die prematurely when the soil does not furnish enough food for all.

THE NATURE OF THE SOIL

Soil, as the term is used in soil science, is a natural surface layer that contains living matter and can support plants. The soil consists of matter in all three states—solid, liquid, and gas. It includes both *mineral matter* and *organic matter*. Mineral matter is largely derived from rock material, whereas organic matter is of biological origin and may be living or dead. Living matter in the soil consists not only of plant roots, but also of many kinds of organisms, including microorganisms.

Soil scientists use the term *humus* to describe finely divided, partially decomposed organic matter in soils. Some humus rests on the soil surface, and some is mixed through the soil. Humus particles of the finest size are gradually carried downward to lower soil layers by rainfall that sinks in and moves through the soil. When abundant, humus particles can give the soil a brown or black coloration.

Both air and water are found in soil. Water may tend to contain high levels of dissolved substances, such as nutrients. Air in soils may have high levels of such gases as carbon dioxide or methane, and low levels of oxygen.

The solid, liquid, and gaseous matter of the soil are constantly changing and interacting through chemical and physical processes. This makes the soil a very dynamic layer. Because of these processes, soil science, often called *pedology*, is a highly complex body of knowledge.

Although we may think of soil as occurring everywhere, large expanses of continents possess a surface layer that cannot be called soil. For example, dunes of moving sand, bare rock surfaces of deserts and high mountains, and surfaces of fresh lava near active volcanoes do not have a soil layer.

The characteristics of soils are developed over a long period of time through a combination of many processes acting together. Physical processes act to break down rock fragments of regolith into smaller and smaller pieces. Chemical processes alter the mineral composition of the original rock, producing new minerals. Taken together, these physical and chemical processes are referred to as *weathering*, which we described in Chapter 13. Weathering occurs in soils and is part of the process by which soils develop their properties and characteristics.

In most soils, the inorganic material of the soil consists of fine particles of mineral matter. The term **parent material** describes all forms of mineral matter that are suitable for transformation into soil. Parent material may be derived from the underlying *bedrock*, which is solid rock below the soil layer (Figure 19.1). Over time, weathering processes soften, disintegrate, and break bedrock apart, forming a layer of *regolith*, or residual mineral matter. Regolith is one of the common forms of parent material. Other kinds of regolith consist of mineral particles transported to a place of rest by the action of streams, glaciers, waves and water currents, or winds. For example, dunes formed of sand transported by wind are a type of regolith on which soil may be formed.

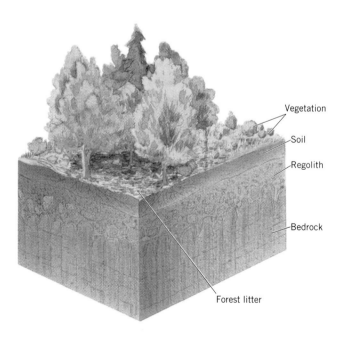

Vegetation

Soil

Regolith

Bedrock

Forest litter

Figure 19.1 In this cross section through the land surface, vegetation and forest litter lie atop the soil. Below is regolith, produced by the breakup of the underlying bedrock.

Soil Color and Texture

The most obvious feature of a soil or soil layer is probably its color. Some color relationships are quite simple. For example, the soils of the Midwest prairies have a black or dark brown color because they contain abundant particles of humus. Red or yellow colors often mark the soils of the Southeast. These colors are created by the presence of iron-containing oxides.

In some areas, soil color may be inherited from the mineral parent material, but, more generally, soil color is generated by soil-forming processes. For example, a white surface layer in soils of dry climates often indicates the presence of mineral salts, brought upward by evaporation. A pale, ash-gray layer near the top of soils of the boreal forest climate results when organic matter and various colored minerals are washed downward, leaving only pure, light-colored mineral matter behind. As we explain soil-forming processes and describe the various classes of soils, soil color will take on more meaning.

The mineral matter of the soil consists of individual mineral particles that vary widely in size. The term **soil texture** refers to the proportion of particles that fall into each of three size grades—*sand, silt, and clay*. The diameter range of each of these grades is shown in Figure 19.2. Millimeters are the standard units. Each unit on the scale represents a power of ten, so that clay particles of 0.000,001 millimeter diameter are one-millionth the size of sand grains 1 mm in diameter. The finest of all soil particles are termed *colloids*. In measuring soil texture, gravel and larger particles are eliminated, since these play no important role in soil processes.

Soil texture is described by a series of names that emphasize the dominant particle size, whether sand, silt, or clay (including colloids). Figure 19.3 gives examples

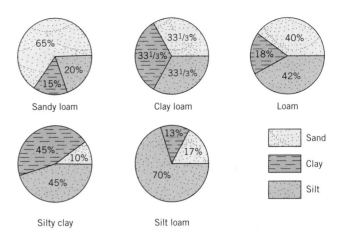

Figure 19.3 These diagrams show the proportions of sand, silt, and clay in five different soil texture classes.

of five soil textures with typical percentage compositions. A *loam* is a mixture containing a substantial proportion of each of the three grades. Loams are classified as sandy, silty, or clay-rich when one of these grades is dominant.

Why is soil texture important? Texture largely determines the ability of the soil to retain water. Coarse-textured (sandy) soils have many small passages between touching mineral grains that quickly conduct water through to deeper layers. If the soil consists of fine particles, passages and spaces are much smaller. Thus, water will penetrate more slowly and also tend to be retained. We will return to the important topic of the water-holding ability of soils in a later section.

Soil Colloids

Soil colloids consist of particles smaller than one hundred-thousandth of a millimeter (0.000,01 mm, 0.000,000,4 in.). Like other soil particles, some colloids are mineral, while others are organic. Mineral colloids are usually very fine particles of clay minerals. If you examine mineral colloids under a microscope, you will find that they consist of thin, platelike bodies (Figure 19.4). When well mixed in water, particles this small remain suspended indefinitely, giving the water a murky appearance. Organic colloids are tiny bits of organic matter that are resistant to decay.

Soil colloids are important because their surfaces attract soil nutrients, which are in the form of ions dissolved in soil water. Figure 19.5 diagrams a colloidal particle, showing this effect. Colloid surfaces tend to be negatively charged because of their molecular structure, and thus attract and hold positively charged ions. Among the many ions in soil water, one important group consists of **bases**, which are ions of four elements: calcium (Ca^{++}), magnesium (Mg^{++}), potassium (K^+), and sodium (Na^+). Because plants require these

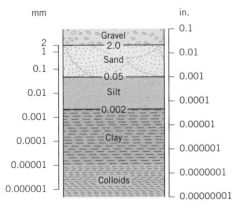

Figure 19.2 Size grades, which are names like sand, silt, and clay, refer to mineral particles within a specific size range. They are defined using the metric system. English equivalents are also shown.

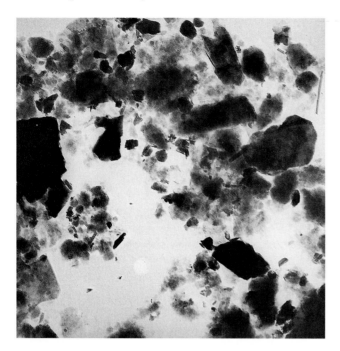

Figure 19.4 Seen here enlarged about 20,000 times are tiny flakes of clay minerals of colloidal dimensions. These particles have settled from suspension in San Francisco Bay.

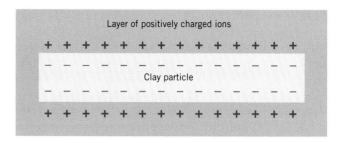

Figure 19.5 Schematic diagram of a thin, flat colloidal particle with negative surface charges and a layer of positively charged ions held to the surface.

Soil Acidity and Alkalinity

The soil solution also contains hydrogen (H^+) and aluminum (Al^{+++}) ions. But unlike the bases, they are not considered to be plant nutrients. The presence of these acid ions in the soil solution tends to make the solution acid in chemical balance.

An important principle of soil chemistry is that the acid ions have the power to replace the nutrient bases clinging to the surfaces of the soil colloids. As acid ions accumulate, the bases are released to the soil solution. They are gradually washed downward below rooting level, reducing soil fertility. When this happens, the soil acidity is increased.

The degree of acidity or alkalinity of a solution is designated by the pH value. The lower the pH value, the greater the degree of acidity. A pH value of 7 represents a neutral state—for example, pure water has a pH of 7. Lower values are in the acid range, while higher values are in the alkaline range.

Table 19.1 shows the natural range of acidity and alkalinity found in soils. High soil acidity is typical of cold, humid climates. In arid climates, soils are typically alkaline. Acidity can be corrected by the application of lime, a compound of calcium, carbon, and oxygen ($CaCO_3$), which removes acid ions and replaces them with the base calcium.

elements, they are among the *plant nutrients*—ions or chemical compounds that are needed for plant growth. Colloids hold these ions, but also give them up to plants when in close contact with root membranes. Without this ion-holding ability of soil colloids, most of the vital nutrients would be carried out of the soil by percolating water and would be taken out of the region in streams, eventually reaching the sea. This leaching process goes on continually in moist climates, but loss is greatly retarded by the ion-holding capacity of soil colloids.

Table 19.1   Soil Acidity and Alkalinity

pH	4.0 4.5	5.0	5.5	6.0 6.5 6.7	7.0	8.0	9.0	10.0	11.0
Acidity	Very strongly acid	Strongly acid	Moderately acid	Slightly acid	Neutral	Weakly alkaline	Alkaline	Strongly alkaline	Excessively alkaline
Lime requirements	Lime needed except for crops requiring acid soil	Lime needed for all but acid-tolerant crops		Lime generally not required	No lime needed				
Occurrence	Rare	Frequent	Very common in cultivated soils of humid climates			Common in sub-humid and arid climates		Limited areas in deserts	

Source: Based on data of C. E. Millar, L. M. Turk, and H. D. Foth, *Fundamentals of Soil Science,* John Wiley and Sons, New York.

Figure 19.6 This soil shows a granular texture. The grains are referred to as peds.

Soil Structure

Soil structure refers to the way in which soil grains are grouped together into larger masses, called *peds*. Peds range in size from small grains to large blocks. They are bound together by soil colloids. Small peds, roughly shaped like spheres, give the soil a granular structure or crumb structure (see Figure 19.6). Larger peds provide an angular, blocky structure. Peds form when colloid-rich clays shrink in volume as they dry out. Shrinkage results in formation of soil cracks, which define the surfaces of the peds.

Soils with a well-developed granular or blocky structure are easy to cultivate. This is an important agricultural factor in lands where primitive plows, drawn by animals, are still widely used. Soils with a high clay content can lack peds. These soils are sticky and heavy when wet and are difficult to cultivate. When dry, they become too hard to be worked.

Minerals of the Soil

Soil scientists recognize two classes of minerals abundant in soils: primary minerals and secondary minerals. The *primary minerals* are compounds present in unaltered rock. These are mostly silicate minerals—compounds of silicon and oxygen, with varying proportions of aluminum, calcium, sodium, iron, and magnesium. (The silicate minerals are described more fully in Chapter 10.) Primary minerals form a large fraction of the solid matter of many kinds of soils, but they play no important role in sustaining plant or animal life.

When primary minerals are exposed to air and water at or near the earth's surface, they are slowly altered in chemical composition. This process is part of *mineral alteration*, a chemical weathering process that is ex-

plained in more detail in Chapter 10. The primary minerals are altered into **secondary minerals**, which are essential to soil development and to soil fertility.

In terms of the properties of soils, the most important secondary minerals are the *clay minerals*. They form the majority of fine mineral particles in soils. From the viewpoint of soil fertility, the ability of a clay mineral to hold base ions is its most important property. This ability varies with the particular type of clay mineral; some hold bases tightly, and others loosely.

The nature of the clay minerals in a soil determines its *base status*. If the clay minerals can hold abundant base ions, the soil is of *high base status* and generally will be highly fertile. If the clay minerals hold a smaller supply of bases, the soil is of *low base status* and is generally less fertile. Humus colloids have a high capacity to hold bases so that the presence of humus is usually associated with potentially high soil fertility.

Mineral oxides are secondary minerals of importance in soils. They occur in many kinds of soils, particularly those that remain in place in areas of warm, moist climates over very long periods of time (hundreds of thousands of years). Under these conditions, minerals are ultimately broken down chemically into simple oxides, compounds in which a single element is combined with oxygen.

Oxides of aluminum and iron are the most important oxides in soils. Two atoms of aluminum are combined with three atoms of oxygen to form the *sesquioxide* of aluminum (Al_2O_3). (The prefix *sesqui-* means "one and a half" and refers to the chemical composition of one and one-half atoms of oxygen for every atom of aluminum.) In soils, aluminum oxide forms the mineral *bauxite*, which is a combination of aluminum sesquioxide and water molecules bound together. It occurs as hard, rocklike lumps and layers below the soil surface. Where bauxite layers are thick and uniform, they are sometimes strip-mined as aluminum ore.

Sesquioxide of iron (Fe_2O_3), again held in combination with water molecules, is *limonite*, a yellowish to reddish mineral that supplies the typical reddish to chocolate-brown colors of soils and rocks. Some shallow accumulations of limonite were formerly mined as a source of iron. Limonite and bauxite occur in close association in soils of warm, moist climates in low latitudes.

Soil Moisture

Besides providing nutrients for plant growth, the soil layer serves as a reservoir for the moisture that plants require. Soil moisture is a key factor in determining how the soils of a region support vegetation and crops.

The soil receives water from rain and from melting snow. Where does this water go? First, some of the water can run off the soil surface and not sink in. Instead, it flows into brooks, streams, and rivers, eventually reaching the sea. What of the water that sinks into the soil? Some of this water is returned to the atmosphere as water vapor. This happens when soil water evaporates and when transpiration by plants lifts soil water from roots to leaves, where it evaporates. Taken together, we can term these last two losses *evapotranspiration*. Some water can also flow completely through the soil layer to recharge supplies of groundwater at depths below the reach of plant roots.

When precipitation infiltrates the soil, the water wets the soil layer. This process is called *soil water recharge*.

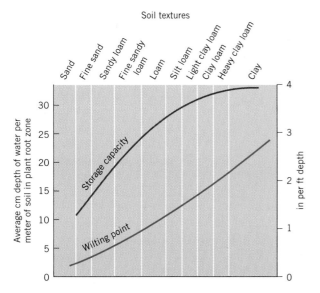

Figure 19.7 Storage capacity and wilting point vary according to soil texture. Finer textured soils hold more water. They also hold water more tightly, so that plants wilt more quickly.

Eventually, the soil layer holds the maximum possible quantity of water, even though the larger pores may remain filled with air. Water movement then continues downward.

Suppose now that no further water enters the soil for a time. Excess soil water continues to drain downward, but some water clings to the soil particles. This water resists the pull of gravity because of the force of *capillary tension*. To understand this force, think about a droplet of condensation that has formed on the cold surface of a glass of ice water. The water droplet seems to be enclosed in a "skin" of surface molecules, drawing the droplet together into a rounded shape. The "skin" is produced by capillary tension. This force keeps the drop clinging to the side of the glass indefinitely, defying the force of gravity. Similarly, tiny films of water adhere to soil grains, particularly at the points of grain contacts. They remain until they evaporate or are absorbed by plant rootlets.

When a soil has first been saturated by water and then allowed to drain under gravity until no more water moves downward, the soil is said to be holding its *storage capacity* of water. For most soils drainage takes no more than two or three days. Most excess water is drained out within one day.

Storage capacity is measured in units of depth, usually centimeters or inches, as with precipitation. It depends largely on the texture of the soil, as shown in Figure 19.7. Finer textures hold more water than coarser textures. This occurs because fine particles have a much larger surface area in a unit of volume than coarse particles. Thus, a sandy soil has a small storage capacity, while a clay soil has a large storage capacity.

The figure also shows the *wilting point*, which is the water storage level below which plants will wilt. The wilting point depends on soil texture. Because fine particles hold water more tightly, it is more difficult for plants to extract moisture from fine soils. Thus, plants can wilt in fine-textured soils, even though more soil water is present than in coarse-textured soils. The difference between the storage capacity of a soil and its wilting point is the available water capacity—that is, the maximum amount of water available to plants when the soil is at storage capacity. The available water capacity is greatest in loamy soils.

THE SOIL WATER BALANCE

Water in the soil is a critical resource needed for plant growth. The amount of water available at any given time is determined by the *soil water balance*, which includes the gain, loss, and storage of soil water. Figure 19.8 is a pictorial flow diagram that illustrates the components of the balance. Water held in storage in the soil water zone is increased by recharge during precipitation, but decreased by use through evapotranspira-

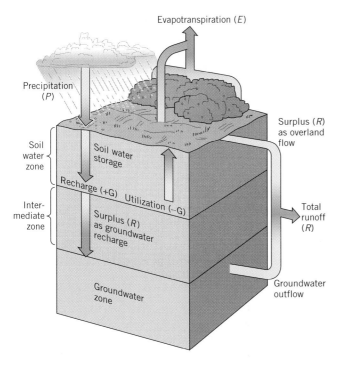

Figure 19.8 Schematic diagram of the soil water balance in a soil column. (Copyright © A. N. Strahler.)

tion. Surplus water is disposed of by downward percolation to the groundwater zone or by overland flow.

To proceed, we must recognize two ways to define evapotranspiration. First is *actual evapotranspiration (Ea)*, which is the true or real rate of water vapor return to the atmosphere from the ground and its plant cover. Second is *potential evapotranspiration (Ep)*, representing the water vapor loss under an ideal set of conditions. One condition is that there is present a complete (or closed) cover of uniform vegetation consisting of fresh green leaves and no bare ground exposed through that cover. A second condition is that there is an adequate water supply, so that the storage capacity of the soil is maintained at all times. This condition can be fulfilled naturally by abundant and frequent precipitation, or artificially by irrigation.

To simplify the ponderous terms we have just defined, they may be transformed as follows:

actual evapotranspiration (*Ea*) is **water use**
potential evapotranspiration (*Ep*) is **water need**

The word "need" signifies the quantity of soil water needed if plant growth is to be maximized for the given conditions of solar radiation and air temperature and the available supply of nutrients. The most important factor in determining water need is temperature. In warmer months, water need will be greater, while in cooler months, it will be less.

The difference between water use and water need is the *soil water shortage*, or *deficit*. This is the quantity of

water that must be furnished by irrigation to achieve maximum crop growth within an agricultural system.

A Simple Soil Water Budget

We now turn to a simple accounting of the monthly and annual quantities of the components of the soil water balance. The numerical accounting is called a *soil water budget*, and it involves only simple addition and subtraction of monthly mean values for a given observing station. All terms of the soil water budget are stated in centimeters of water depth, the same as for precipitation.

A simplified soil water budget is shown in Figure 19.9. The seven terms we need for a complete budget are listed as follows, along with abbreviations used on the graph:

Precipitation, *P*
Water need, *Ep*
Water use, *Ea*
Storage withdrawal, *−G*
Storage recharge, *+G*
Soil water shortage, *D*
Water surplus, *R*

Points on the graph represent average monthly values. They are connected by smooth curves to enhance annual cycles of change. In our example, precipitation (*P*) is much the same in all months, with no strong annual cycle. In contrast, water need (*Ep*) shows a strong seasonal cycle, with low values in winter and a high summer peak. This example would fit a typical moist midlatitude climate with mild winters.

At the start of the year, precipitation greatly exceeds water use and a large water surplus (*R*) exists. This surplus is disposed of by runoff. By May, water use exceeds

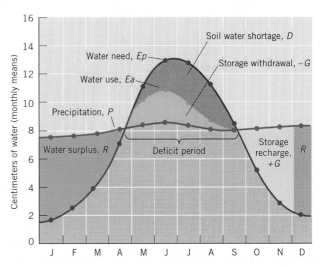

Figure 19.9 A simplified soil water budget typical of a moist climate in middle latitudes.

precipitation, and a water deficit occurs. In this month plants begin to withdraw soil water from storage.

Storage withdrawal (–G) is represented by the difference between the water-use curve and the precipitation curve. As storage withdrawal continues, however, plants draw soil water only with increasing difficulty. Thus, water use (*Ea*) is less than water need (*Ep*) during this period. Storage withdrawal continues throughout the summer. The deficit period lasts through September. The area labeled soil water shortage (*D*) is the difference between water need and water use. It represents the total quantity of water needed by irrigation to ensure maximum growth throughout the deficit period.

In October, precipitation (*P*) again begins to exceed water need (*Ep*), but the soil must first absorb an amount equal to the summer storage withdrawal. So there follows a period of *storage recharge (+G)* that lasts through November. In December the soil reaches its full storage capacity, arbitrarily fixed at 30 cm (11.8 in.). Now, a water surplus (*R*) again sets in, lasting through the winter.

The soil water budget was developed by C. Warren Thornthwaite, a distinguished climatologist and geographer who was concerned with practical problems of crop irrigation. He developed the calculation of the soil water budget in order to place irrigation on a precise, accurate basis. Only in the equatorial zone and in a few parts of the tropical and midlatitude zones is precipitation ample to fulfill the water need during the growing season. In a world beset by severe and prolonged food shortages, the Thornthwaite concepts and calculations are of great value in assessing the benefits to be gained by increased irrigation.

Calculating the annual soil water budget for a station is not difficult, given monthly values of precipitation, water use, and water need. *Working It Out 19.1 • Calculating a Simple Soil Water Budget* shows you how.

SOIL DEVELOPMENT

How do soils develop their distinctive characteristics? Let's turn to the processes that act to form soils and soil layers. We begin with soil horizons.

Soil Horizons

Most soils possess **soil horizons**—distinctive horizontal layers that differ in physical composition, chemical composition, or organic content or structure (Figure 19.10). Soil horizons are developed by the interactions through time of climate, living organisms, and the configuration of the land surface. Horizons usually develop by either selective removal or accumulation of certain ions, colloids, and chemical compounds. The removal or accumulation is normally produced by water seeping down through the soil profile from the surface to deeper layers. Horizons are often distinguished by their color. The display of horizons on a cross section through the soil is termed a **soil profile**.

Let's review briefly the main types of horizons and their characteristics. For now, this discussion will apply to the types of horizons and processes that are found in moist forest climates. These are shown in Figure 19.11. Soil horizons are of two types: organic and mineral. Organic horizons, designated by the capital letter *O*, overlie the mineral horizons and are formed from accumulations of organic matter derived from plants and animals. Soil scientists recognize two possible layers. The upper O_i horizon contains decomposing organic matter that is recognizable as leaves or twigs. The lower O_a horizon contains material that is broken down beyond recognition by eye. This material is humus, which we mentioned earlier.

Mineral horizons lie below the organic horizons. Four main horizons are important—*A*, *E*, *B*, and *C*. Plant roots readily penetrate *A*, *E*, and *B* horizons and influence soil development within them. Soil scientists limit the term *soil* to refer to the *A*, *E*, and *B* horizons. The *A* horizon is the uppermost mineral horizon. It is rich in organic matter, consisting of numerous plant roots and downwashed humus from the organic horizons above. Next is the *E* horizon. Clay particles and oxides of aluminum and iron are removed from the *E*

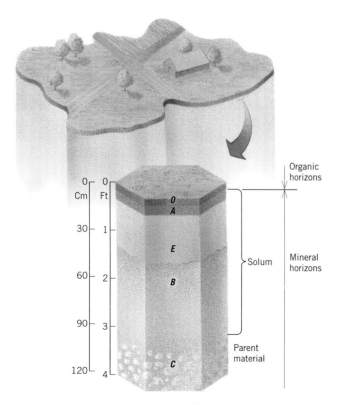

Figure 19.10 A column of soil will normally show a series of horizons, which are horizontal layers with different properties.

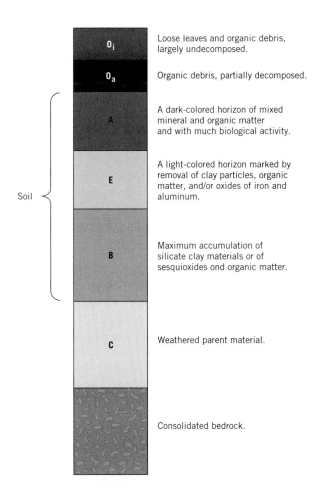

O$_i$	Loose leaves and organic debris, largely undecomposed.
O$_a$	Organic debris, partially decomposed.
A	A dark-colored horizon of mixed mineral and organic matter and with much biological activity.
E	A light-colored horizon marked by removal of clay particles, organic matter, and/or oxides of iron and aluminum.
B	Maximum accumulation of silicate clay materials or of sesquioxides ond organic matter.
C	Weathered parent material.
	Consolidated bedrock.

Soil { (bracket encompassing A, E, B horizons)

Figure 19.11 A sequence of horizons that might appear in a forest soil developed under a cool, moist climate. (Soil Conservation Service, U.S. Department of Agriculture.)

mulate on the soil surface. Organic enrichment occurs when humus accumulating in *O* horizons is carried downward to enrich the *A* horizon below.

The second class includes processes that remove material from the soil body. This *removal* occurs when surface erosion carries sediment away from the uppermost layer of soil. Another important process of loss is *leaching*, in which seeping water dissolves soil materials and moves them to deep levels or to groundwater.

The third class of soil-forming processes involves *translocation*, in which materials are moved within the soil body, usually from one horizon to another. Two processes of translocation that operate simultaneously are eluviation and illuviation. **Eluviation** consists of the downward transport of fine particles, particularly the clays and colloids, from the uppermost part of the soil. Eluviation leaves behind grains of sand or coarse silt, forming the *E* horizon. **Illuviation** is the accumulation of materials that are brought down downward, normally from the *E* horizon to the *B* horizon. The materials that accumulate may be clay particles, humus, or sesquioxides of iron and aluminum.

Figure 19.12 is a soil profile developed under a cool, humid forest climate. It shows the effects of both soil

horizon by downward-seeping water, leaving behind pure grains of sand or coarse silt.

The *B horizon* receives the clay particles, aluminum, and iron oxides, as well as organic matter washed down from the *A* and *E* horizons. It is made dense and tough by the filling of natural spaces with clays and oxides. Transitional horizons are present at some locations.

Beneath the *B* horizon is the *C horizon*, which is not considered part of the soil. It consists of the parent mineral matter of the soil and has been described earlier in this chapter as regolith (see Figure 19.1). Below the regolith lies bedrock or sediments of much older age than the soil.

Soil-Forming Processes

There are four classes of soil-forming processes. The first includes processes of *soil enrichment*, which add material to the soil body. For example, inorganic enrichment occurs when sediment is brought from higher to lower areas by overland flow. Stream flooding also deposits fine mineral particles on low-lying soil surfaces. Wind is another source of fine material that can accu-

Figure 19.12 A forest soil profile on outer Cape Cod. The pale grayish *E* horizon overlies a reddish *B* horizon. A thin layer of wind-deposited silt and dune sand (pale brown layer) has been deposited on top.

Working It Out 19.1 • Calculating a Simple Soil Water Budget

The soil water balance is actually a matter flow system in which an input of precipitation flows through soil pathways to be output as evapotranspiration or runoff (see Figure 19.8). The amount of water returned to the atmosphere through evapotranspiration (*Ea*) is determined largely by temperature and by the amount of vegetation cover.

With warm temperatures and a thick layer of transpiring leaves, water use will be larger. The amount of runoff will depend in the long term on how water use (*Ea*) compares with precipitation (*P*). If precipitation is greater, then runoff will occur, both as overland flow and as groundwater outflow.

An important characteristic of the soil water matter flow system is the storage of water (*S*) in the soil layer. This storage provides a reserve of water for plants to draw upon, so that water use can exceed precipitation for some period of the year. Later, when precipitation exceeds water use, the reserve can be recharged, and only after the reserve is replenished can runoff occur again in the annual cycle.

The calculation of a monthly soil water budget is a simple exercise, once given the values of precipitation, water use, and water need for a station. Precipitation is easily observed by measurement with a rain gauge. Water need (*Ep*) is calculated from a formula that uses temperature and atmospheric moisture content. Water use is either measured directly or estimated from similar formulas calibrated by actual measurements.

The table and figure demonstrate the calculation of the monthly soil water budget for a station in the marine west-coast climate ⑧, which exhibits wet, cool winters and warm, drier summers. For each month, three items of basic data are first provided—precipitation (*P*), water use (*Ea*), and water need (*Ep*). From these, the other necessary quantities can be found. In any

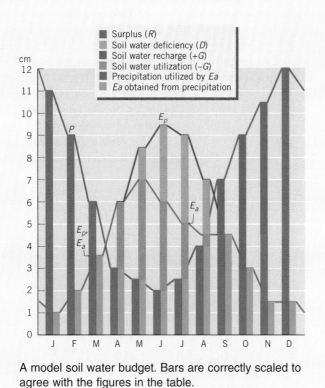

A model soil water budget. Bars are correctly scaled to agree with the figures in the table.

enrichment and translocation processes. The topmost layer of the soil is a thin deposit of wind-blown silt and dune sand, which has enriched the soil profile. In translocation processes, eluviation has removed colloids and sesquioxides from the whitened *E* horizon. Illuviation has added them to the *B* horizon, which displays the orange-red colors of iron sesquioxide.

The translocation of calcium carbonate is another important process. In pure form, this secondary mineral is calcite ($CaCO_3$). In many areas, the parent material of the soil contains a substantial proportion of calcium carbonate derived from the disintegration of limestone, a common variety of bedrock. Carbonic acid, which forms when carbon dioxide gas dissolves in rainwater or soil water, readily reacts with calcium carbonate. The products of this reaction remain dissolved in solution as ions.

In moist climates, a large amount of surplus soil water moves downward to the groundwater zone. This water movement leaches calcium carbonate from the entire soil in a process called *decalcification*. Soils that have lost most of their calcium are also usually acid in chemical balance and so are low in bases. Addition of lime or pulverized limestone not only corrects the acid condition, but also restores the calcium, which is used as a plant nutrient.

In dry climates, calcium carbonate is dissolved in upper layers of the soil during periods of rain or snowmelt when soil water recharge is taking place. The dissolved carbonate matter is carried down to the *B*

month, precipitation can only follow three pathways—to the atmosphere as water use (Ea), to streams and rivers as runoff (R), or to storage ($+G$) within the soil layer. In January, precipitation is 11.0 cm, and water use is 1.0 cm, leaving 10.0 cm to go to storage or runoff. At this time of the year, storage is full, however, so this entire amount goes to runoff.

The flow of runoff persists until April, when water use (6.0 cm) exceeds precipitation (3.0 cm) by 3.0 cm. The needed 3.0 cm is withdrawn from storage, giving the value of –3.0 for –G. (Separate columns are used for –G and +G.) Storage withdrawals continue through August, yielding a total withdrawal of 14.5 cm.

In September, precipitation (7.0 cm) again exceeds water use (4.5 cm), leaving a surplus (2.5 cm) for +G. This surplus begins to recharge the soil water storage. In October, soil water storage is also increased. The increase continues into November, when it is 9.0 cm. However, only 6.0 cm is required to balance the total summer withdrawal of 14.5 cm, leaving 9.0 – 6.0 = 3.0 cm for runoff (R). In December, the surplus goes entirely to runoff.

The soil water shortage (D) is simply the difference between water need (Ep) and water use (Ea). The two quantities are equal until May, when water use drops below water need as significant water storage withdrawals begin to occur. This condition persists until September, when abundant precipitation allows water use to equal water need.

These principles can be used to calculate the soil water budget for any station. It is important to note, however, that the depth of storage of water available to plants in the soil is limited. In calculating soil water budgets, a limit of 30 cm is normally used for field crops. Thus, total withdrawal and recharge may not exceed this amount.

The soil water balance not only influences human agricultural activities, but is also strongly linked to natural vegetation and is a key factor in climate classification. The soil water budget is an example of how systems principles of identifying flow pathways and budgeting flow rates can be used to describe and quantify phenomena of such great importance.

Simplified Example of a Soil-Water Budget

Equation	P	=	Ea	+	G	+	R	Ep	(Ep – Ea) = D
January	11.0	=	1.0				+10.0	1.0	0.0
February	9.0	=	2.0				+ 7.0	2.0	0.0
March	6.0	=	3.5				+ 2.5	3.5	0.0
April	3.0	=	6.0	–3.0				6.0	0.0
May	2.5	=	7.0	–4.5				8.5	1.5
June	2.0	=	6.0	–4.0				9.5	3.5
July	2.5	=	5.0	–2.5				9.0	4.0
August	4.0	=	4.5	–0.5				7.0	2.5
September	7.0	=	4.5			+2.5		4.5	0.0
October	9.0	=	3.0			+6.0		3.0	0.0
November	10.5	=	1.5			+6.0	+ 3.0	1.5	0.0
December	12.0	=	1.5				+10.5	1.5	0.0
Totals	78.5	=	45.5	–14.5		+14.5	+33.0	57.0	11.5
	78.5	=	78.5						

horizon, where water penetration reaches its limits. Here, the carbonate matter is precipitated (deposited in crystalline form) in the *B* horizon, a process called *calcification*. Calcium carbonate deposition takes the form of white or pale-colored grains, nodules, or plates in the *B* or *C* horizons.

A last process of translocation occurs in desert climates. In some low areas, a layer of groundwater lies close to the surface, producing a flat, poorly drained area. Evaporation of water at or near the soil surface draws up a continual flow of groundwater by capillary tension, much like a cotton wick draws oil upward in an oil lamp. Moreover, the groundwater is often rich in dissolved salts. When evaporation occurs, the salts precipitate and accumulate as a distinctive *salic horizon*.

This process is called *salinization*. Most of the salts are compounds of sodium, of which ordinary table salt (sodium chloride, or halite, $NaCl$) is a familiar example. Sodium in large amounts is associated with highly alkaline conditions and is toxic to many kinds of plants. When salinization occurs in irrigated lands in a desert climate, the soil can be ruined for further agricultural use (see Chapter 14).

The last class of soil-forming processes involves the *transformation* of material within the soil body. An example is the conversion of minerals from primary to secondary types, which we have already described. Another example is decomposition of organic matter to produce humus, a process termed *humification*. In warm, moist climates, humification can decompose organic matter completely to yield carbon dioxide and water, leaving virtually no organic matter in the soil.

Soil Temperature

Soil temperature is another important factor in determining the chemical development of soils and the formation of horizons. Temperature acts as a control over biologic activity and also influences the intensity of chemical processes affecting soil minerals. Below 10°C (50°F), biological activity is slowed, and at or below the freezing point (0°C, 32°F), biological activity stops. Chemical processes affecting minerals are inactive. The root growth of most plants and germination of their seeds require soil temperatures above 5°C (41°F). For plants of the warm, wet low-latitude climates, germination of seeds may require a soil temperature of at least 24°C (75°F).

The temperature of the uppermost soil layer and the soil surface strongly affects the rate at which organic matter is decomposed by microorganisms. Thus, in cold climates, where decomposition is slow, organic matter in the form of fallen leaves and stems tends to accumulate to form a thick *O* horizon. As we have already described, this material becomes humus, which is carried downward to enrich the *A* horizon.

In warm, moist climates of low latitudes, the rate of decomposition of plant material is rapid, so that nearly all the fallen leaves and stems are disposed of by bacterial activity. Under these conditions, the *O* horizon may be missing and the entire soil profile will contain very little organic matter.

Soil and Surface Configuration

The configuration, or shape, of the ground surface is an important factor in soil formation. Configuration includes the steepness of the ground surface, or *slope*, as well as its compass orientation, or *aspect*. Generally speaking, soil horizons are thick on gentle slopes but thin on steep slopes. This is because the soil is more rapidly removed by erosion processes on the steeper slopes.

Aspect acts to influence soil temperatures and the soil water regime. Slopes facing away from the sun are sheltered from direct insolation and tend to have cooler, moister soils. Slopes facing toward the sun are exposed to direct solar rays, raising soil temperatures and increasing evapotranspiration.

Biological Processes in Soil Formation

The presence and activities of living plants and animals, as well as their nonliving organic products, have an important influence on soil. We have already noted the role that organic matter as humus plays in soil fertility. The colloidal structure of humus holds bases,

which are needed for plant growth. In this way, humus helps keep nutrients cycling through plants and soils, and ensures that nutrients will not be lost by leaching to groundwater. Humus also helps bind the soil into crumbs and clumps. This structure allows water and air to penetrate the soil freely, providing a healthy environment for plant roots.

Animals living in the soil include many species and come in many individual sizes—from bacteria to burrowing mammals. The total role of animals in soil formation is extremely important in soils that are warm and moist enough to support large animal populations. For example, earthworms continually rework the soil not only by burrowing, but also by passing soil through their intestinal tracts. They ingest large amounts of decaying leaf matter, carry it down from the surface, and incorporate it into the mineral soil horizons. Many forms of insect larvae perform a similar function. Tube-like openings are made by larger animals—moles, gophers, rabbits, badgers, prairie dogs, and other species.

Human activity also influences the physical and chemical nature of the soil. Large areas of agricultural soils have been cultivated for centuries. As a result, both the structure and composition of these agricultural soils have undergone great changes. These altered soils are often recognized as distinct soil classes that are just as important as natural soils.

THE GLOBAL SCOPE OF SOILS

An important aspect of soil science for physical geography is the classification of soils into major types and subtypes that are recognized in terms of their distribution over the earth's land surfaces. Geographers are particularly interested in the linkage of climate, parent material, time, biologic process, and landform with the distribution of types of soils. Geographers are also interested in the kinds of natural vegetation associated with each of the major soil classes. The geography of soils is thus essential in determining the quality of environments of the globe. It is important because soil fertility, along with availability of fresh water, is a basic measure of the ability of an environmental region to produce food for human consumption.

The soils of the world have been classified according to a system developed by scientists of the U.S. Soil Conservation Service, in cooperation with soil scientists of many other nations. Here we are concerned only with the two highest levels of this classification system. The top level contains 11 **soil orders** summarized in Table 19.2. The second level consists of *suborders*, of which we need to mention only a few.

Soil orders and suborders are often distinguished by the presence of a diagnostic horizon. Each diagnostic horizon has some unique combination of physical properties (color, structure, texture) or chemical properties (minerals present or absent). The two basic kinds of diagnostic horizons are (1) a horizon formed at the surface and called an *epipedon* (from Greek *epi*, meaning "over" or "upon") and (2) a subsurface horizon formed by the removal or accumulation of matter. In our descriptions of soil orders, we will refer to a number of diagnostic horizons.

We can recognize three groups of soil orders. The largest group includes seven orders with well-developed horizons or fully weathered minerals. A second group includes a single soil order that is very rich in organic matter. The last group includes three soil orders with poorly developed horizons or no horizons.

The world soils map presented in Figure 19.13 shows the major areas of occurrence of the soil orders. The Alfisols (Table 19.2) have been subdivided into four important suborders that correspond well to four basic climate zones. The map is quite general, indicating those areas where a given soil order is likely to be found. The map does not show many important areas of Entisols, Inceptisols, Histosols, and Andisols (Table 19.2). These orders are largely of local occurrence, since they are found on recent deposits such as floodplains, glacial landforms, sand dunes, marshlands, bogs, or volcanic ash deposits. The map also shows areas of highlands. In these regions, the soil patterns are too complex to show at a global scale.

Soil Orders

Table 19.3 explains the names of the soil orders. The formative element is a syllable used in the names of suborders and lower groups. Although each order has several suborders, we will refer to only a few.

Three soil orders dominate the vast land areas of low latitudes: Oxisols, Ultisols, and Vertisols. Soils of these orders have developed over long time spans in an envi-

Table 19.2 Soil Orders

Group I	
Soils with well-developed horizons or with fully weathered minerals, resulting from long-continued adjustment to prevailing soil temperature and soil water conditions.	
Oxisols	Very old, highly weathered soils of low latitudes, with a subsurface horizon of accumulation of mineral oxides and very low base status.
Ultisols	Soils of equatorial, tropical, and subtropical latitude zones, with a subsurface horizon of clay accumulation and low base status.
Vertisols	Soils of subtropical and tropical zones with high clay content and high base status. Vertisols develop deep, wide cracks when dry, and the soil blocks formed by cracking move with respect to each other.
Alfisols	Soils of humid and subhumid climates with a subsurface horizon of clay accumulation and high base status. Alfisols range from equatorial to subarctic latitude zones.
Spodosols	Soils of cold, moist climates, with a well-developed *B* horizon of illuviation and low base status.
Mollisols	Soils of semiarid and subhumid midlatitude grasslands, with a dark, humus-rich epipedon and very high base status.
Aridisols	Soils of dry climates, low in organic matter, and often having subsurface horizons of accumulation of carbonate minerals or soluble salts.
Group II	
Soils with a large proportion of organic matter.	
Histosols	Soils with a thick upper layer very rich in organic matter.
Group III	
Soils with poorly developed horizons or no horizons, and capable of further mineral alteration.	
Entisols	Soils lacking horizons, usually because their parent material has accumulated only recently.
Inceptisols	Soils with weakly developed horizons, having minerals capable of further alteration by weathering processes.
Andisols	Soils with weakly developed horizons, having a high proportion of glassy volcanic parent material produced by erupting volcanoes.

ronment of warm soil temperatures and soil water that is abundant in a wet season or lasts throughout the year. We will discuss these orders first.

Oxisols

Oxisols have developed in equatorial, tropical, and subtropical zones on land surfaces that have been stable over long periods of time. During soil development, the climate has been moist, with a large water surplus. Oxisols have developed over vast areas of South America and Africa in the wet equatorial climate ①. Here the native vegetation is rainforest. The wet-dry tropical climate ③ with its large seasonal water surplus is also associated with Oxisols in South America and Africa.

Oxisols usually lack distinct horizons, except for darkened surface layers. Soil minerals are weathered to an extreme degree and are dominated by stable sesquioxides of aluminum and iron. Red, yellow, and yellowish-brown colors are normal (Figures 19.14*a* and 19.15). The base status of the Oxisols is very low, since nearly all the bases required by plants have been removed from the soil profile. A small store of nutrient bases occurs very close to the soil surface. The soil is quite easily broken apart and allows easy penetration by rainwater and plant roots.

Ultisols

Ultisols are quite closely related to the Oxisols in outward appearance and environment of origin. Ultisols are reddish to yellowish in color (Figures 19.14*b* and

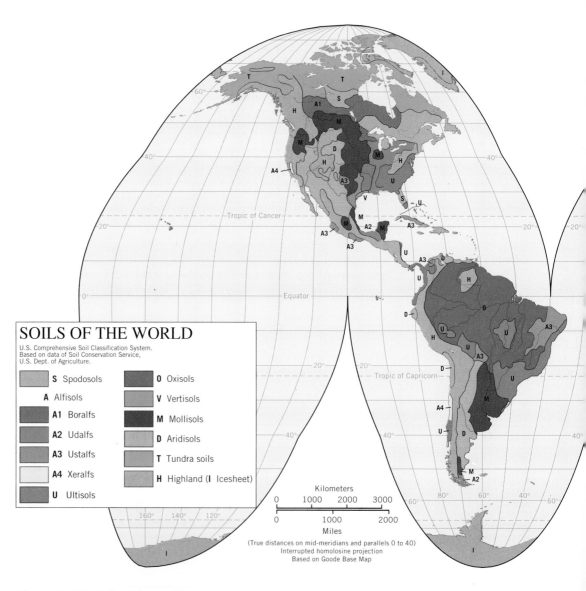

Figure 19.13   Soils of the world.

19.16). They have a subsurface horizon of clay accumulation, called an *argillic horizon*, which is not found in the Oxisols. It is a *B* horizon and has developed through accumulation of clay in the process of illuviation. Although forest is the characteristic native vegetation, the base status of the Ultisols is low. As in the Oxisols, most of the bases are found in a shallow surface layer where they are released by the decay of plant matter. They are quickly taken up and recycled by the shallow roots of trees and shrubs.

In a few areas, the Ultisol profile contains a subsurface horizon of sesquioxides. This horizon is capable of hardening to a rocklike material if it becomes exposed at the surface and is subjected to repeated wetting and drying. This material is referred to as *plinthite* (from the Greek word *plinthos*, meaning "brick"). In the hard-

ened state, plinthite is referred to as *laterite* (from the Latin *later*, or brick). In Southeast Asia, plinthite is quarried and cut into building blocks (Figure 19.17). These blocks harden into laterite blocks when they are exposed to the air.

Ultisols are widespread throughout Southeast Asia and the East Indies. Other important areas are in eastern Australia, Central America, South America, and the southeastern United States. Ultisols extend into the lower midlatitude zone in the United States, where they correspond quite closely in extent with the area of moist subtropical climate ⑥. In lower latitudes, Ultisols are identified with the wet-dry tropical climate ③ and the monsoon and trade-wind coastal climate ②. Note that all these climates have a dry season, even though it may be short.

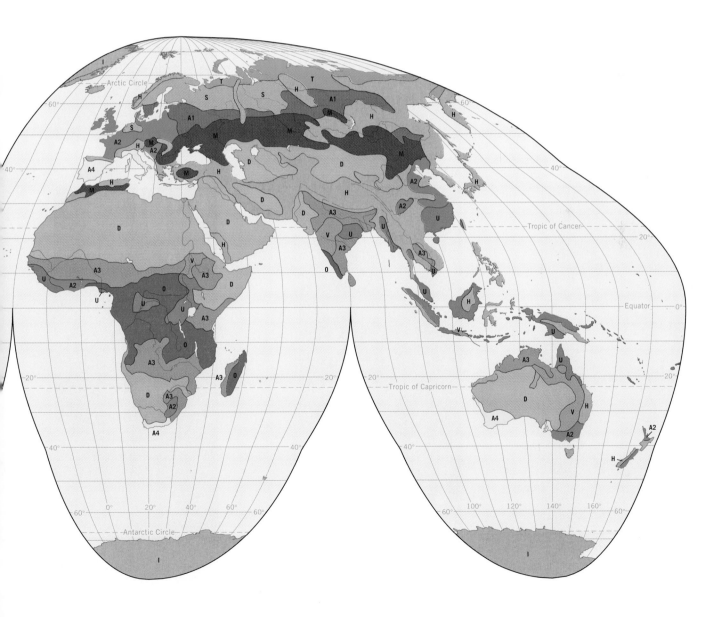

OXISOLS

A   Torrox, Hawaii

ULTISOLS

B   Udult, Virginia

VERTISOLS

C   Ustert, India

ALFISOLS

D   Udalf, Michigan

ALFISOLS

E   Ustalf, Texas

SPODOSOLS

F   Orthod, France

Figure 9.14   Soil profiles of several soil orders.

MOLLISOLS

G   Boroll, USSR

MOLLISOLS

H   Udoll, Argentina

MOLLISOLS

I   Ustoll, Colorado

MOLLISOLS

J   Rendoll, Argentina

ARDISOLS

K   Argid, Colorado

HISTOSOLS

L   Fibrist, Minnesota

Figure 19.14 *(continued)*

Table 19.3 Formative Elements in Names of Soil Orders

Name of Order	Formative Element	Derivation of Formative Element	Pronunciation of Formative Element
Entisol	ent	Meaningless syllable	recent
Inceptisol	ept	L. *inceptum*, beginning	inept
Histosol	ist	Gr. *histos*, tissue	histology
Oxisol	ox	F. *oxide*, oxide	ox
Ultisol	ult	L. *ultimus*, last	ultimate
Vertisol	ert	L. *verto*, turn	invert
Alfisol	alf	Meaningless syllable	alfalfa
Spodosol	od	Gr. *spodos*, wood ash	odd
Mollisol	oll	L. *mollis*, soft	mollify
Aridisol	id	L. *aridus*, dry	arid
Andisol	and	Eng. *andesite*, a volcanic rock type	and

Both Oxisols and Ultisols of low latitudes were used for centuries under shifting agriculture prior to the advent of modern agricultural technology. This primitive agricultural method, known as slash-and-burn, is still widely practiced. Without fertilizers, these soils can sustain crops on freshly cleared areas for only two or three years, at most, before the nutrient bases are exhausted and the garden plot must be abandoned. Substantial use of lime, fertilizers, and other industrial inputs is necessary for high, sustained crop yields. Furthermore, the exposed soil surface of the Ultisols is vulnerable to devastating soil erosion, particularly on steep hill slopes.

Vertisols

Vertisols have a unique set of properties that stand in sharp contrast to the Oxisols and Ultisols. Vertisols are black in color and have a high clay content (Figures 19.14*c* and 19.18). Much of the clay consists of a particular mineral that shrinks and swells greatly with seasonal changes in soil water content. Wide, deep vertical cracks develop in the soil during the dry season. As the dry soil blocks are wetted and softened by rain, some fragments of surface soil drop into the cracks before they close, so that the soil "swallows itself" and is constantly being mixed.

Vertisols typically form under grass and savanna vegetation in subtropical and tropical climates with a pronounced dry season. These climates include the semiarid subtype of the dry tropical steppe climate ④ and the wet-dry tropical climate ③. Because Vertisols require a particular clay mineral as a parent material, the major areas of occurrence are scattered and show no distinctive pattern on the world map. An important region of Vertisols is the Deccan Plateau of western India, where basalt, a dark variety of igneous rock, supplies the silicate minerals that are altered into the necessary clay minerals.

Vertisols are high in base status and are particularly rich in such nutrient bases as calcium and magnesium. The soil solution is nearly neutral in pH, and a moderate content of organic matter is distributed through the soil. The soil retains large amounts of water because of its fine texture, but much of this water is held tightly by the clay particles and is not available to plants. Where soil cultivation depends on human or animal power, as it does in most of the developing nations where the soil occurs, agricultural yields are low. This is because the moist soil becomes highly plastic and is difficult to till with primitive tools. For this reason, many areas of Vertisols have been left in grass or shrub cover, providing grazing for cattle. Soil scientists think that the use of modern technology, including heavy farm machinery, could result in substantial production of food and fiber from Vertisols that are not now in production.

Alfisols

The **Alfisols** are soils characterized by an argillic horizon, produced by illuviation. Unlike the argillic hori-

Figure 19.15 An Oxisol in Hawaii. Sugarcane is being cultivated here.

Figure 19.16 Ultisol profile in North Carolina. The thin, pale layer at the top is an *E* horizon, showing the effects of removal of materials by eluviation. Below the base of the thick, reddish *B* horizon is a blotchy-colored zone of plinthite.

zon of the Ultisols, this *B* horizon is enriched by silicate clay minerals that have an adequate capacity to hold bases such as calcium and magnesium. The base status of the Alfisols is therefore generally quite high.

Above the *B* horizon of clay accumulation is a horizon of pale color, the *E* horizon, that has lost some of the original bases, clay minerals, and sesquioxides by the process of eluviation. These materials have become concentrated by illuviation in the *B* horizon. Alfisols also have a gray, brownish, or reddish surface horizon.

The world distribution of Alfisols is extremely wide in latitude (see Figure 19.13). Alfisols range from latitudes as high as 60° N in North America and Eurasia to the equatorial zone in South America and Africa. Obviously, the Alfisols span an enormous range in climate types. For this reason, we need to recognize four of the important suborders of Alfisols, each with its own climate affiliation.

Boralfs are Alfisols of cold (boreal) forest lands of North America and Eurasia. They have a gray surface horizon and a brownish subsoil. *Udalfs* are brownish Alfisols of the midlatitude zone. They are closely associated with the moist continental climate ⑩ in North America, Europe, and eastern Asia (Figure 19.14*d*).

Ustalfs are brownish to reddish Alfisols of the warmer climates (Figure 19.14*e*). They range from the subtropical zone to the equator and are associated with the wet-

dry tropical climate ③ in Southeast Asia, Africa, Australia, and South America. *Xeralfs* are Alfisols of the Mediterranean climate ⑦, with its cool moist winter and dry summer. The Xeralfs are typically brownish or reddish in color.

Spodosols

Poleward of the Alfisols in North America and Eurasia lies a great belt of soils of the order **Spodosols**, formed in the cold boreal forest climate ⑪ beneath a needle-leaf forest. Spodosols have a unique property—a *B* horizon of accumulation of reddish mineral matter with a low capacity to hold bases (Figure 19.14*f*). This horizon is called the spodic horizon (Figure 19.19). It is made up of a dense mixture of organic matter and compounds of aluminum and iron, all brought downward by eluviation from an overlying *E* horizon. Be-

Figure 19.17   Laterite, formed by hardening of plinthite, is being quarried for building stone in this scene from India.

Figure 19.18   A Vertisol in Texas. The clay minerals that are abundant in Vertisols shrink when they dry out, producing deep cracks in the soil surface.

cause of the intensive removal of matter from the *E* horizon, it has a bleached, pale gray to white appearance (Figure 19.12). This conspicuous feature led to the naming of the soil as *podzol* (ash-soil) by Russian peasants. In modern terminology, this pale layer is an *albic horizon*. The *O* horizon, a thin, very dark layer of organic matter, overlies the *A* horizon.

Spodosols are strongly acid and are low in plant nutrients such as calcium and magnesium. They are also low in humus. Although the base status of the Spodosols is low, forests of pine and spruce are supported through the process of recycling of the bases.

Spodosols are closely associated with regions recently covered by the great ice sheets of the Pleistocene Epoch. These soils are therefore very young. Typically, the parent material is coarse sand consisting largely of the mineral quartz. This mineral cannot weather to form clay minerals.

Spodosols are naturally poor soils in terms of agricultural productivity. Because they are acid, application of lime is essential. Heavy applications of fertilizers are also required. With proper management and the input of the required industrial products, Spodosols can be highly productive, if the soil texture is favorable. High yields of potatoes from Spodosols in Maine and New

Brunswick are examples. Another factor unfavorable to agriculture is the shortness of the growing season in the more northerly parts of the Spodosol belt.

Histosols

Throughout the northern regions of Spodosols are countless patches of **Histosols**. This unique soil order has a very high content of organic matter in a thick, dark upper layer (Figures 19.14*l*). Most Histosols go by such common names as peats or mucks. They have formed in shallow lakes and ponds by accumulation of partially decayed plant matter. In time, the water is replaced by a layer of organic matter, or *peat*, and becomes a *bog* (Figure 19.20). Peat bogs are used extensively for cultivation of cranberries (cranberry bogs). Sphagnum peat from bogs is dried and baled for sale as a mulch for use on suburban lawns and shrubbery beds. For centuries, Europe has used dried peat from bogs of glacial origin as a low-grade fuel.

Some Histosols are *mucks*—organic soils composed of fine black materials of sticky consistency. These are agriculturally valuable in midlatitudes, where they occur as beds of former lakes in glaciated regions. After appropriate drainage and application of lime and fer-

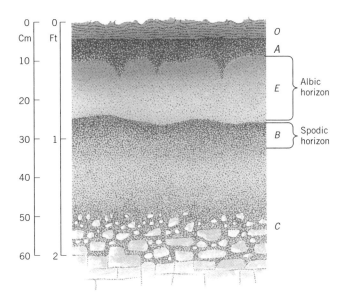

Figure 19.19 Diagram of a Spodosol profile.

Entisols

Entisols have in common the combination of a mineral soil and the absence of distinct horizons. Entisols are soils in the sense that they support plants, but they may be found in any climate and under any vegetation. Entisols lack distinct horizons for two reasons. It may be the result of a parent material, such as quartz sand, in which horizons do not readily form. Or it may be the result of lack of time for horizons to form in recent deposits of alluvium or on actively eroding slopes.

Entisols occur from equatorial to arctic latitude zones. From the standpoint of agricultural productivity, Entisols of the subarctic zone and tropical deserts (along with arctic areas of Inceptisols) are the poorest of all soils. In contrast, Entisols and Inceptisols of floodplains and delta plains in warm and moist climates are among the most highly productive agricultural soils in the world because of their favorable texture, ample nutrient content, and large soil water storage.

Inceptisols

Inceptisols are soils with horizons that are weakly developed, usually because the soil is quite young. These areas occur within some of the regions shown on the world map as Ultisols and Oxisols. Especially important are the Inceptisols of river floodplains and delta plains

tilizers, these mucks are remarkably productive for truck garden vegetables (Figure 19.21). Histosols are also found in low latitudes, where conditions of poor drainage have favored thick accumulations of plant matter.

Figure 19.20 This peat bog in Galway, Ireland, has been trenched to reveal a Histosol profile. Peat blocks, seen to the sides of the trench, are drying for use as fuel.

Figure 19.21   Garden crops cultivated on a Histosol of a former glacial lake bed, Southern Ontario, Canada.

in Southeast Asia that support dense populations of rice farmers.

In these regions, annual river floods cover low-lying plains and deposit layers of fine silt. This sediment is rich in primary minerals that yield bases as they weather chemically over time. The constant enrichment of the soil explains the high soil fertility in a region where uplands develop only Ultisols of low fertility. Inceptisols of these floodplain and delta lands are of a suborder called *Aquepts*—Inceptisols of wet places. Much closer to home is another prime example of Aquepts within the domain of the Ultisols—the lower Mississippi River floodplain and delta plain.

Andisols

Andisols are soils in which more than half of the parent mineral matter is volcanic ash, spewed high into the air from the craters of active volcanoes and coming to rest in layers over the surrounding landscape. The fine ash particles are glasslike shards. A high proportion of carbon, formed by the decay of plant matter, is also typical, so that the soil usually appears very dark in color. Andisols form over a wide range of latitudes and climates. They are for the most part fertile soils, and in moist climates they support a dense natural vegetation cover.

Andisols do not appear on our world map because they are found in small patches associated with individual volcanoes that are located mostly in the "Ring of Fire"—the chain of volcanic mountains and islands that surrounds the great Pacific Ocean. Andisols are also found on the island of Hawaii, where volcanoes are presently active.

Mollisols

Mollisols are soils of grasslands that occupy vast areas of semiarid and subhumid climates in midlatitudes. Mollisols are unique in having a very thick, dark brown to black surface horizon called the *mollic epipedon* (Figures 19.14*g* to *j* and 19.22). This layer lies within the *A* horizon and is always more than 25 cm (9.8 in.) thick. The

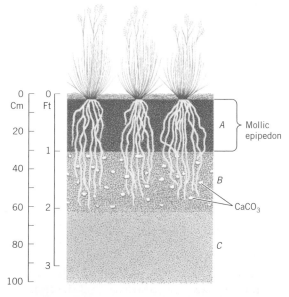

Figure 19.22   Schematic diagram of a Mollisol profile.

soil has a loose, granular structure (Figure 19.6) or a soft consistency when dry. Other important qualities of the Mollisols are the dominance of calcium among the bases of the *A* and *B* horizons and the very high base status of the soil.

Most areas of Mollisols are closely associated with the semiarid subtype of the dry midlatitude climate ⑨ and the adjacent portion of the moist continental climate ⑩. In North America, Mollisols dominate the Great Plains region, the Columbia Plateau, and the northern Great basin. In South America, a large area of Mollisols covers the Pampa region of Argentina and Uruguay. In Eurasia, a great belt of Mollisols stretches from Rumania eastward across the steppes of Russia, Siberia, and Mongolia. Russians refer to the Mollisols as *chernozems*, a term that has gained widespread use throughout the Western world as well.

Because of their loose texture and very high base status, Mollisols are among the most naturally fertile soils in the world. They now produce most of the world's commercial grain crop. Most of these soils have been used for crop production only in the last century. Prior to that time, they were used mainly for grazing by nomadic herds. The Mollisols have favorable properties for growing cereals in large-scale mechanized farming and are relatively easy to manage. Production of grain varies considerably from one year to the next because seasonal rainfall is highly variable.

A brief mention of four suborders of the Mollisols as they occur in the United States and Canada will help you to understand important regional soil differences related to climate. *Borolls*, the cold-climate suborder of the Mollisols, are found in a large area extending on both sides of the U.S.-Canadian border east of the Rocky Mountains (Figure 19.14*g*). *Udolls* are Mollisols of a relatively moist climate, as compared with the other suborders. Formerly, the Udolls supported tall-grass prairie, but today they are closely identified with the corn belt in the American Midwest (Figure 19.14*h*).

Ustolls are Mollisols of the semiarid subtype of the dry midlatitude climate ⑨, with a substantial soil water shortage in the summer months (Figure 19.14*i*). The Ustolls underlie much of the short-grass prairie region east of the Rockies (Figure 19.23). *Xerolls* are Mollisols of the Mediterranean climate ⑦, with its tendency to cool, moist winters and rainless summers.

Desert and Tundra Soils

Desert and tundra soils are soils of extreme environments. Aridisols characterize the desert climate. As might be expected, they are low in organic matter and high in salts. Tundra soils are poorly developed because they are formed on very recent parent material, left in place by glacial activity during the Ice Age. Cold temperatures have also restricted soil development in tundra regions.

Figure 19.23   A Mollisol developed on wind-blown silt. Dry prairie grasses are seen on the surface.

Aridisols

Aridisols, soils of the desert climate, are dry for long periods of time. Because the climate supports only a very sparse vegetation, humus is lacking and the soil color ranges from pale gray to pale red (Figure 19.14*k*). Soil horizons are weakly developed, but there may be important subsurface horizons of accumulated calcium carbonate (petrocalcic horizon) or soluble salts (salic horizon) (Figure 19.24). The salts, of which sodium is a dominant constituent, give the soil a very high degree of alkalinity. The Aridisols are closely correlated with the arid subtype of the dry tropical ④, dry subtropical ⑤, and dry midlatitude ⑨ climates.

Most Aridisols are used for nomadic grazing, as they have been through the ages. This use is dictated by the limited rainfall, which is inadequate for crops without irrigation. Locally, where water supplies from mountain streams or groundwater permit, Aridisols can be highly productive for a wide variety of crops under irri-

Figure 19.24   A salic horizon, appearing as a white layer, lies close to the surface in this Aridisol profile in the Nevada desert. The scale is marked in feet (1 ft = 30.5 cm).

gation (Figure 19.25). Great irrigation systems, such as those of the Imperial Valley of the United States, the Nile Valley of Egypt, and the Indus Valley of Pakistan, have made Aridisols highly productive, but not without problems of salt buildup and waterlogging (see Chapter 14).

A Midcontinental Transect from Aridisols to Alfisols

Figure 19.26 provides a diagram showing generalized changes in soil profile and soil type as they might occur on a transect from a cool, dry desert to a cool moist climate, starting in the northern desert of the Great Basin and crossing the midwest. Starting with the Aridisols in the west, horizons are thin, but show an accumulation of clay minerals in the *B* horizon. As precipitation increases, grasses dominate and Mollisols are formed. The *B* horizon thickens and includes calcium accumulation. In the east, where alfisols occur, enhanced precipitation generates an *E* horizon that takes its place between the *A* and *B* horizons.

Tundra Soils

Soils of the arctic tundra fall largely into the order of Inceptisols, soils with weakly developed horizons that are usually associated with a moist climate. Inceptisols of the tundra climate belong to the suborder of Aquepts, which as we noted earlier are Inceptisols of wet places. More specifically, the tundra soils can be assigned to the *Cryaquepts*, a subdivision within the Aquepts. The prefix *cry* is derived from the Greek word *kryos*, meaning "icy cold." We may refer to these soils simply as *tundra soils*.

Tundra soils are formed largely of primary minerals ranging in size from silt to clay that are broken down by frost action and glacial grinding. Layers of peat are often present between mineral layers. Beneath the tundra soil lies perennially frozen ground (permafrost), described in Chapter 18. Because the annual summer thaw affects only a shallow surface layer, soil water cannot easily drain away. Thus, the soil is saturated with water over large areas. Repeated freezing and thawing of this shallow surface layer disrupts plant roots, so that only small, shallow-rooted plants can maintain a hold.

Figure 19.25   This gray desert soil, an Aridisol, has proved highly productive when cultivated and irrigated. The locality is near Palm Springs, California, in the Coachella Valley.

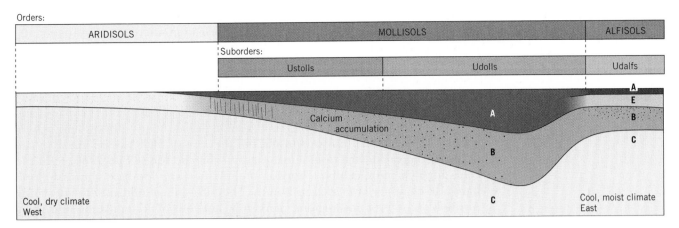

Figure 19.26 A schematic diagram of the changing soil profile from a cool dry desert on the west to a cool moist climate on the east. (After C. E. Millar, L. M. Turk, and H. D. Foth, *Fundamentals of Soil Science*, John Wiley and Sons, New York.)

The soil layer is a complex body subject to many influences, including the parent materials from which it is derived, the vegetation that it harbors, and the water regime of precipitation and evapotranspiration that it experiences. In many environments, soil-forming processes operate very slowly. Thus, some soils can be the products of complex histories involving climatic changes. By inducing soil erosion (Chapter 15), human activities can rapidly strip uppermost soil horizons, leaving less productive horizons at the surface. Soil horizons and properties can also vary strongly over short distances, thereby exhibiting important local variation. Soil science is an interesting and complex topic, and it is difficult to do justice to it in a single textbook chapter.

From the topic of soils, we turn next to ecosystems in Chapters 20 and 21. Like soil science, ecology—the science of ecosystems—is both broad and deep. But it is not difficult to highlight the ecological systems and processes of importance to geographers. This is the objective of the following chapters.

CHAPTER SUMMARY

The soil layer is a complex mixture of solid, liquid, and gaseous components. It is derived from parent material, or regolith, that is produced from rock by weathering. The major factors influencing soil and soil development are parent material, climate, vegetation, and time. Soil texture refers to the proportions of sand, silt, and clay that are present. Colloids are the finest particles in soils and are important because they help retain nutrients, or bases, that are used by plants. Soils show a wide range of pH values, from acid to alkaline. Soils with granular or blocky structures are most easily cultivated.

In soils, primary minerals are chemically altered to secondary minerals, which include oxides and clay minerals. The nature of the clay minerals determines the soil's base status. If base status is high, the soil retains nutrients. If low, the soil can lack fertility. When a soil is fully wetted by heavy rainfall or snowmelt and allowed to drain, it reaches its storage capacity. Evaporation from the surface and transpiration from plants draws down the soil water store until precipitation occurs again.

The soil water balance describes the gain, loss, and storage of soil water. It depends on water need (potential evapotranspiration), water use (actual evapotranspiration), and precipitation. Monitoring these values on a monthly basis provides a soil water budget for the year.

Soils possess distinctive horizontal layers called horizons. These layers are developed by processes of enrichment, removal, translocation, and transformation. In downward translocation, materials such as humus, clay particles, and mineral oxides are removed by eluviation from an upper horizon and accumulate by illuviation in a lower one. In salinization, salts are translocated upward by evaporating water to form a salic horizon. In humification, a transformation process, organic matter is broken down by bacterial decay. Where soil temperatures are warm, this process can be highly effective, leaving a soil low in organic content. Animals, such as earthworms, can be very important in soil formation where they are abundant.

Global soils are classified into 11 soil orders, often by the presence of one of more diagnostic horizons. Oxisols are old, highly weathered soils of low latitudes. They have a horizon of mineral oxide accumulation and a low base status. Ultisols are also found in low latitudes. They have a horizon of clay accumulation and are also of low base status. Vertisols are rich in a type of clay mineral that expands and contracts with wetting and drying, and has a high base status. Alfisols have a horizon of clay accumulation like Ultisols, but they are of high base status. They are found in moist climates from equatorial to subarctic zones. Spodosols, found in cold, moist climates, exhibit a horizon of illuviation and low base status. Mollisols have a thick upper layer rich in humus. They are soils of midlatitude grasslands. Aridisols are soils of arid regions, marked by horizons of accumulation of carbonate minerals or salts. Histosols have a thick upper layer formed almost entirely of organic matter.

Three soil orders have poorly developed horizons or no horizons—Entisols, Inceptisols, and Andisols. Entisols are composed of fresh parent material and have no horizons. The horizons of Inceptisols are only weakly developed. Andisols are weakly developed soils occurring on young volcanic deposits.

KEY TERMS

soil	soil horizons	Alfisols
parent material	soil profile	Spodosols
soil texture	eluviation	Histosols
soil colloids	illuviation	Entisols
bases	soil orders	Inceptisols
secondary minerals	Oxisols	Andisols
water use	Ultisols	Mollisols
water need	Vertisols	Aridisols

REVIEW QUESTIONS

1. Which important factors condition the nature and development of the soil?

2. Soil color, soil texture, and soil structure are used to describe soils and soil horizons. Identify each of these three terms, showing how they are applied.

3. Explain the concepts of acidity and alkalinity as they apply to soils.

4. Identify two important classes of secondary minerals in soils and provide examples of each class.

5. How does the ability of soils to hold water vary, and how does this ability relate to soil texture?

6. Define water need (potential evapotranspiration) and water use (actual evapotranspiration). How are they used in the soil water balance?

7. Identify the following terms as used in the soil water budget: storage, withdrawal, storage recharge, soil water shortage, water surplus.

8. What is a soil horizon? How are soil horizons named? Provide two examples.

9. Identify four classes of soil-forming processes and describe each.

10. What are translocation processes? Identify and describe four translocation processes.

11. How many soil orders are there? Try to name them all.

12. Name three soil orders that are especially associated with low latitudes. For each order, provide at least one distinguishing characteristic and explain it.

13. Compare Alfisols and Spodosols. What features do they share? What features differentiate them? Where are they found?

14. Where are Mollisols found? How are the properties of Mollisols related to climate and vegetation cover? Name four suborders within the Mollisols.

15. Desert and tundra are extreme environments. Which soil order is characteristic of each environment? Briefly describe desert and tundra soils.

ESSAY QUESTIONS

1. Document the important role of clay particles and clay mineral colloids in soils. What is meant by the term *clay*? What are colloids? What are their properties? How does the type of clay mineral influence soil fertility? How does the amount of clay influence the water-holding capacity of the soil? What is the role of clay minerals in horizon development?

2. Using the world maps of global soils and global climate, compare the pattern of soils on a transect along the 20° E longitude meridian with the patterns of climate encountered along the same meridian. What conclusions can you draw about the relationship between soils and climate? Be specific.

PROBLEMS

Working It Out 19.1 • Calculating a Simple Soil Water Budget

1. Below are monthly values and annual totals for precipitation, water need (*Ep*), and water use (*Ea*) at Urbana, Illinois, in the moist continental climate ⑩. Prepare and plot a soil water budget similar to the figure in *Working It Out 19.1* for this station. This will involve determining soil water shortage (*D*), soil water utilization (−*G*), soil water recharge (+*G*), and water surplus (*R*) for each month following the method described in the text of the feature.

Soil Water Budget for Urbana, Illinois

Month	P	=	Ea	+	G	+	R	Ep	D
Jan	5.7		0.0					0.0	
Feb	4.5		0.0					0.0	
Mar	8.2		1.4					1.4	
Apr	10.0		4.4					4.4	
May	9.9		8.8					8.8	
Jun	8.4		12.4					12.6	
Jul	8.0		13.4					14.9	
Aug	9.0		11.6					13.0	
Sep	8.3		7.8					7.8	
Oct	6.6		4.8					4.8	
Nov	5.7		1.4					1.4	
Dec	5.5		0.0					0.0	
Total	89.8		66.0		−12.0		+12.1	23.7	69.1	3.1

Chapter 20

Systems and Cycles of the Biosphere

This chapter focuses on the systems and cycles of the *biosphere*—the domain of living organisms, in which they interact with the hydrosphere, atmosphere, and lithosphere. The study of the interactions between life forms and their environment is the science of **ecology**. The total assemblage of components entering into the interactions of a group of organisms is known as an ecological system, or more simply, an **ecosystem.**

Ecosystems have inputs of matter and energy that are used by plants and animals to grow, reproduce, and maintain life. Matter and energy are also exported from ecosystems. An ecosystem tends to achieve a set of balances of the various processes and activities that occur within it. Although many of these balances are robust and self-regulating, some are quite sensitive and can be easily upset or destroyed. Biogeographers are concerned with studying the nature and functions of ecosystems and how they vary from place to place.

Ecosystems are part of the physical composition of the life layer. A forest, for example, is not only a living community of organisms, but also a characteristic assemblage of plants and animals that occurs on the land surface. A forest of tall trees is physically very different from a prairie of short grasses and other low plants. Biogeographers focus on this physical aspect of ecosystems in mapping the global distributions of ecosystem types.

Geographers also view ecosystems as natural resource systems. Food, fiber, fuel, and structural material are products of ecosystems that are built up by organisms through the expenditure of energy derived ultimately from the sun. Geographers are also interested in the influence of climate on ecosystem productivity, since human habitation of the lands depends directly on it. Your basic understanding of global climates will prove most useful in explaining the global pattern of ecosystems on the lands, a topic that we cover in Chapter 21.

ENERGY FLOW IN ECOSYSTEMS

We begin our examination of systems and cycles of the biosphere by focusing on the flows of energy that take place within ecosystems. These flows form the basis for

The Alaskan brown bear—a predator at the top of the food chain. Katmai National Park, Alaska.

studying the global productivity of ecosystems, a topic of concern to biogeographers and global ecologists alike.

The Food Web

A salt marsh provides a good example of an ecosystem (Figure 20.1). A variety of organisms are present—algae and aquatic plants, microorganisms, insects, snails, and crayfish, as well as such larger organisms as fishes, birds, shrews, mice, and rats. Inorganic components will be found as well—water, air, clay particles and organic sediment, inorganic nutrients, trace elements, and light energy. Energy transformations in the ecosystem occur by means of a series of steps or levels, referred to as a **food chain** or **food web**.

The plants and algae in the food web are the **primary producers**. They use light energy to convert carbon dioxide and water into carbohydrates (long chains of sugar molecules) and eventually into other biochemical molecules needed for the support of life. This process of energy conversion is called *photosynthesis*, and we will return to it in more detail shortly. Organisms engaged in photosynthesis form the base of the food web.

The primary producers support the **consumers**—organisms that ingest other organisms as their food source. At the lowest level of consumers are the *primary*

consumers (the snails, insects, and fishes). At the next level are the *secondary consumers* (the mammals, birds, and larger fishes), which feed on the primary consumers. Still higher levels of feeding occur in the salt marsh ecosystem as marsh hawks and owls consume the smaller animals below them in the food web. The **decomposers** feed on *detritus*, or decaying organic matter, derived from all levels. They are largely microscopic organisms (microorganisms) and bacteria.

The food web is really an energy flow system, tracing the path of solar energy through the ecosystem. Solar energy is absorbed by the primary producers and stored in the chemical products of photosynthesis. As these organisms are eaten and digested by consumers, chemical energy is released. This chemical energy is used to power new biochemical reactions, which again produce stored chemical energy in the bodies of the consumers.

At each level of energy flow in the food web, energy is lost to *respiration*. Respiration can be thought of as the burning of fuel to keep the organism operating. It will be discussed in more detail in the next section. Energy expended in respiration is ultimately lost as waste heat and cannot be stored for use by other organisms higher up in the food chain. This means that, generally, both the numbers of organisms and their total amount of living tissue must decrease greatly up the food chain. In general, only 10 to 50 percent of the en-

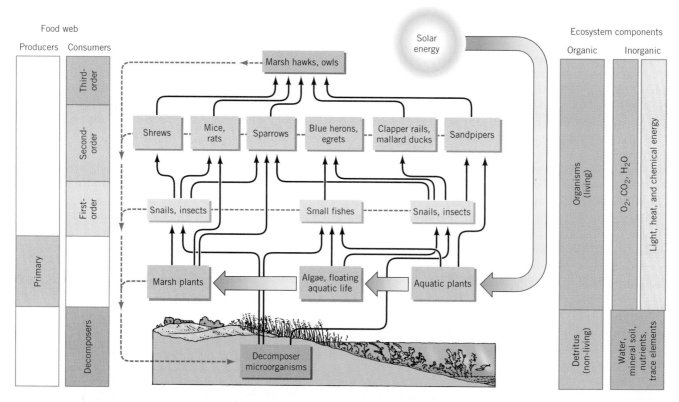

Figure 20.1 Flow diagram of a salt-marsh ecosystem in winter. The arrows show how energy flows from the sun to producers, consumers, and decomposers. (Food chain after R. L. Smith, *Ecology and Field Biology*, Harper and Row, New York.)

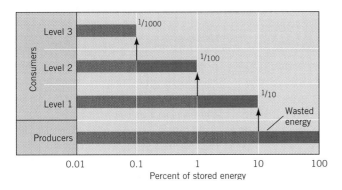

Figure 20.2 Percentage of energy passed up the steps of the food chain, assuming 90 percent is lost energy at each step.

ergy stored in organic matter at one level can be passed up the chain to the next level. Normally, there are about four levels of consumers.

Figure 20.2 is a bar graph showing the percentage of energy passed up the chain when only 10 percent moves from one level to the next. The horizontal scale is in powers of 10. In ecosystems of the lands, the mass of organic matter and the number of individuals of the consuming animals decrease with each upward step. In the food chain shown in Figure 20.1, there are only a few marsh hawks and owls in the third level of consumers, while countless individuals are found in the primary level.

For individual species, the number of individuals of a species present in an ecosystem ultimately depends on the level of resources available to support the species population. If these resources provide a steady supply of energy, the population size will normally attain a steady level. In some cases, however, resources vary with time, for example, in an annual cycle. The population size of a species depending on these resources may then also fluctuate in a corresponding cycle. Population sizes and the change in size of species populations with time have been studied extensively by ecologists. *Working It Out 20.1 • Logistic Population Growth* provides more information on a type of population growth that approaches and slowly reaches a stable carrying capacity.

Photosynthesis and Respiration

Stated in the simplest possible terms, **photosynthesis** is the production of carbohydrate. *Carbohydrate* is a general term for a class of organic compounds consisting of the elements carbon, hydrogen, and oxygen. Carbohydrate molecules are composed of short chains of carbon bonded to one another. Also bonded to each carbon are hydrogen (H) atoms and hydroxyl (OH) molecules. We can symbolize a single carbon atom with

its attached hydrogen atom and hydroxyl molecule as –CHOH–. The leading and trailing dashes indicate that the unit is just one portion of a longer chain of connected carbon atoms.

Photosynthesis of carbohydrate requires a series of complex biochemical reactions using water (H_2O) and carbon dioxide (CO_2) as well as light energy. A simplified chemical reaction for photosynthesis can be written as follows:

$$H_2O + CO_2 + \text{light energy} \rightarrow \text{–CHOH–} + O_2$$

Oxygen in the form of gas molecules (O_2) is a byproduct of photosynthesis. Photosynthesis is also referred to as *carbon fixation*, since in the process gaseous carbon as CO_2 is "fixed" to a solid form in carbohydrate.

Respiration is the process opposite to photosynthesis, in which carbohydrate is broken down and combined with oxygen to yield carbon dioxide and water. The overall reaction is as follows:

$$\text{–CHOH–} + O_2 \rightarrow CO_2 + H_2O + \text{chemical energy}$$

As in the case of photosynthesis, the actual reactions are far from simple. The chemical energy released is stored in several types of energy-carrying molecules in living cells and used later to synthesize all the biological molecules necessary to sustain life.

At this point, it is helpful to link photosynthesis and respiration in a continuous cycle involving both the primary producer and the decomposer. (For now, we omit consumers from the cycle.) Figure 20.3 shows one closed loop for hydrogen (H), one for carbon (C), and two loops for oxygen (O). We are not taking into account that there are two atoms of hydrogen in each molecule of water and carbohydrate, or that there are two atoms of oxygen in each molecule of carbon dioxide and oxygen gas. Only the flow pattern counts in this representation.

A good place to start is the soil, from which water is drawn up into the body of a living plant. In the green leaves of the plant, photosynthesis takes place while light energy is absorbed by the leaf cells. Carbon dioxide is brought in from the atmosphere at this point. Oxygen is also liberated here and begins its atmospheric cycle. The plant tissue then dies and falls to the ground, where it is acted on by the decomposer. Through respiration, oxygen is taken out of the atmosphere or soil air and combined with the decomposing carbohydrate. Energy is now liberated. Here both carbon dioxide and water enter the atmosphere as gases.

An important concept emerges from this flow diagram. Energy passes through the system. It comes from the sun and returns eventually to outer space. On the other hand, the material components—hydrogen, oxygen, and carbon—are recycled within the total system. Of course, many other material components are recycled in the same way. These are plant nutrients, essen-

Working It Out 20.1 • Logistic Population Growth

The size of a population of plants or animals in an ecosystem is determined by the *growth rate* of the population. If the growth rate is positive, the population is increasing. If it is negative, the population is decreasing. If the growth rate is zero, the population size is stable, neither increasing nor decreasing. Growth rates are usually expressed as the percent or proportion of increase in the population in a given unit of time. As we saw in Chapter 3, this type of growth is *exponential growth,* which can be positive or negative. For example, a population of mosquitoes in the late spring might be increasing at a growth rate of 10 percent per week. Or, a population of alligators in the Everglades might be decreasing at 5 percent per year.

In most natural ecosystems, populations tend to be of a stable size, and therefore exhibit a growth rate near zero, when taken over a long time period. Why does this stability occur? One important reason is that individuals of the same species are in competition with each other for the same resources. These resources could be light and nutrients in the case of plants, or individuals of a prey species for predators higher up on the food chain. Normally, the population will expand exponentially until the resources become increasingly scarce. As this happens, an increasing proportion of the population is unable to sustain itself, and so population growth slows and eventually stops. This type of growth, in which the population growth rate slows and eventually goes to zero is called *logistic growth*. At this point, the *carrying capacity* has been reached.

As an example of a population experiencing logistic growth, consider a microecosystem — a population of yeast bacteria growing in a culture medium. The table shows the results of an experiment in which yeast bacteria were grown over a period of 18 hours. Each hour a sample of yeast was removed, allowing calculation of the biomass of the whole population, shown in the second column. (Since it is too laborious to count the yeast bacteria in a sample to obtain the number of individuals in the culture medium, the biomass of the bacteria was used as a measure of the population size.) A scan down the column shows how the population increased rapidly at first and then stabilized at a value approaching about 665 grams. This value corresponds to the carrying capacity.

Column 3 shows the increase in biomass during the period. A scan of this column shows that biomass increments grew and then became smaller as the population biomass reached the carrying capacity. The largest increment to the population occurred in hour 8.

Column 4 shows the increase expressed as a percentage of the population at the start of each period. These values are quite high at first, showing that the population grew at rates of about 50 to 90 percent per hour during the first five hours. After that time the growth rate slowed, eventually decreasing to less than 1 percent per hour for hours 16–18.

The left-hand graph shows the yeast biomass plotted with time. The graph reveals an "S" shape that is characteristic of logistic growth. The population grew exponentially at first, then leveled off as it approached the carrying capacity. Plotted as a semilogarithmic func-

Growth of Yeast in a Culture Medium

(1) Time (hours)	(2) Biomass (grams)	(3) Increase (grams)	(4) Increase (percent)	(5) Logistic Model	(6) Difference
0	9.6			9.6	0.0
1	18.3	8.7	90.6	16.3	2.0
2	29.0	10.7	58.5	27.5	1.5
3	47.2	18.2	62.8	45.8	1.4
4	71.1	23.9	50.6	74.9	−3.8
5	119.1	48.0	67.5	119.0	0.1
6	174.6	55.5	46.6	181.0	−6.4
7	257.3	82.7	47.4	260.0	−2.7
8	350.7	93.4	36.3	348.5	2.2
9	441.0	90.3	25.7	434.9	6.1
10	513.3	72.3	16.4	508.3	5.0
11	559.7	46.4	9.0	563.7	−4.0
12	594.8	35.1	6.3	602.0	−7.2
13	629.4	34.6	5.8	626.8	2.6
14	640.8	11.4	1.8	642.2	−1.4
15	651.1	10.3	1.6	651.5	−0.4
16	655.9	4.8	0.7	657.1	−1.2
17	659.6	3.7	0.6	660.4	−0.8
18	661.8	2.2	0.3	662.3	−0.5

Source: Data of T. Carlson, *Biochemische Zeitschrift*, vol. 57.

tion (right-hand graph), the growth follows a straight line at first, indicating initial exponential growth. However, the line curves smoothly toward the horizontal and levels out at the carrying capacity.

Logistic growth follows this equation:

$$P(t) = \frac{C}{1+Ae^{-kt}}$$

$P(t)$ is the population size at time t, which depends on the value of t and three constants: C, A, and k. C is the carrying capacity, or the stable level of the population. For the yeast example, this would be about 665 grams.

The constant A is found by the relation

$$A = \frac{C-P_0}{P_0}$$

where P_0 is the initial population, which for the yeast example is 9.6 grams. This constant scales the factor e^{-kt} according to the range between the starting value, P_0, and the final value, C, the carrying capacity. For the yeast example,

$$A = \frac{C-P_0}{P_0} = \frac{665-9.6}{9.6} = 68.3$$

The constant k in the exponential expression of the denominator is the exponential growth rate of the population before the decrease sets in. It is related to the growth observed in a single time period by the equation

$$k = \ln(1+R)$$

where R is the growth rate experienced at the beginning of growth, when the population size is far from the carrying capacity. Note that this formula uses the growth rate as a proportion rather than as a percent. This formula actually applies only to a very small time interval at the start of growth of a very small population, but it provides a useful approximation for other cases. Scanning column 4 of the table shows that the population increased at rates ranging from 51 to 91 percent in the first five hours. As it turns out, a value of 72 percent for R, or 0.72 expressed as a proportion, fits the data best. The value of k is then

$$k = \ln(1+R) = \ln(1+0.72)$$
$$= \ln(1.72) = 0.542$$

With the values of C, A, and k

known, we can now write the equation for the logistic growth of the yeast culture as

$$P(t) = \frac{C}{1+Ae^{-kt}} = \frac{665}{1+(68.3)e^{-0.542t}}$$

Column 5 of the table shows the values of biomass predicted by this logistic equation, and column 6 shows the difference between these predictions and the observations in column 2. The differences are very small, indicating that the model fits the data very well.

Logistic growth, in which a population increases exponentially and then slows its growth to reach a sustained carrying capacity, is only a very simple model of population growth. Other growth patterns characterize many natural populations. For example, a common pattern called *J-shaped growth* provides rapid growth followed by a destructive population crash to low levels. Populations then increase only to crash again and again. At present, the human population is increasing exponentially at an ever-increasing rate. Let's hope that human population growth will be logistic, not J-shaped!

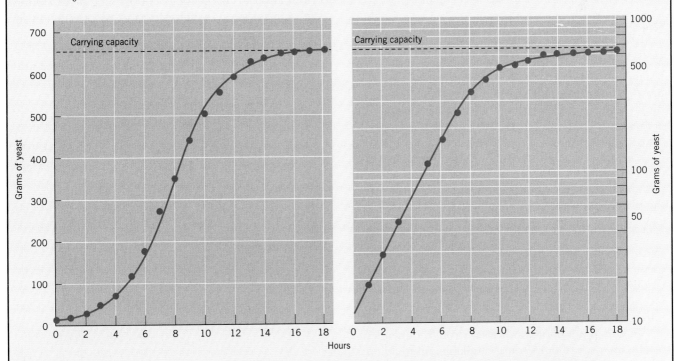

Growth of yeast in a culture medium. (*left*) Growth plotted linearly, fitting the logistic model. (*right*) Growth plotted semilogarithmically. (Data of T. Carlson, *Biochemische Zeitschrift*, vol. 57.)

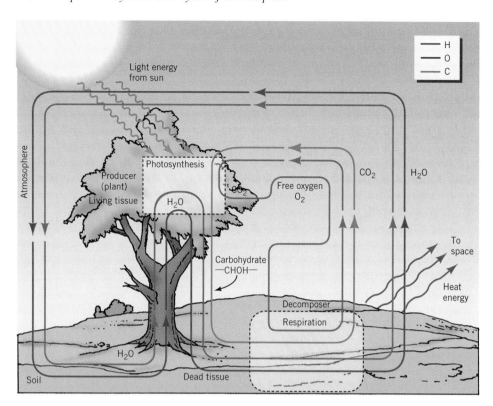

Figure 20.3 A simplified flow diagram of the essential components of photosynthesis and respiration through the biosphere.

tial in the growth of plants. Nutrients are constantly re-cycled. Because the earth as a planet is a closed system, the material components never leave the total system. However, they can be stored in other ways and forms where they are unavailable for use by organisms for prolonged periods of geologic time. We will develop this concept more fully later in this chapter.

Net Photosynthesis

Because both photosynthesis and respiration occur simultaneously in a plant, the amount of new carbohydrate placed in storage is less than the total carbohydrate being synthesized. We must thus distinguish between gross photosynthesis and net photosynthesis. **Gross photosynthesis** is the total amount of carbohydrate produced by photosynthesis. **Net photosynthesis** is the amount of carbohydrate remaining after respiration has broken down sufficient carbohydrate to power the plant. Stated as an equation,

Net photosynthesis = Gross photosynthesis – Respiration

Because both photosynthesis and respiration occur in the same cell, gross photosynthesis cannot be measured readily. Instead, we will deal with net photosynthesis. In most cases, respiration will be held constant, so use of the net instead of the gross will show the same trends.

The rate of net photosynthesis is strongly dependent on the intensity of light energy available, up to a limit. Figure 20.4 shows this principle. The rate of net photosynthesis is indicated on the vertical axis by the rate at which a plant takes up carbon dioxide. On the horizontal axis, light intensity increases from left to right. At first, net photosynthesis rises rapidly as light intensity increases. The rate then slows and reaches a maximum value, shown by the plateau in the curve. Above this maximum, the rate falls off because the incoming light is also causing heating. This heating increases the rate of respiration, which offsets gross production by photosynthesis and decreases the net.

Light intensity sufficient to allow maximum net photosynthesis is only 10 to 30 percent of full summer sun-

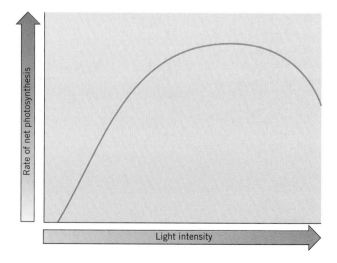

Figure 20.4 The curve of net photosynthesis shows a steep initial rise, then levels off as light intensity rises.

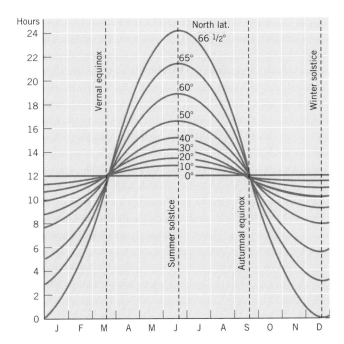

Figure 20.5 Duration of the daylight period at various latitudes throughout the year. The vertical scale gives the number of hours the sun is above the horizon.

light for most green plants. Additional light energy is simply ineffective. Duration of daylight then becomes the important factor in the rate at which products of photosynthesis accumulate as plant tissues. On this subject, you can draw on your knowledge of the seasons and the changing angle of the sun's rays with latitude. Figure 20.5 shows the duration of the daylight period with changing seasons for a wide range of latitudes in the northern hemisphere. At low latitudes, days are not far from the average 12-hour length throughout the year. At high latitudes, days are short in winter but long in summer. The seasonal contrast in day length increases with latitude. In subarctic latitudes, photosynthesis can go on in summer during most of the 24-hour day, a factor that can compensate partly for the shortness of the growing season.

The rate of photosynthesis also increases as air temperature increases, up to a limit. Figure 20.6 shows the results of a laboratory experiment in which sphagnum moss was grown under constant illumination. Gross photosynthesis increased rapidly to a maximum at about 20°C (68°F), then leveled off. Respiration increased quite steadily to the limit of the experiment. Net photosynthesis, which is the difference between the values in the two curves, peaked at about 18°C (64°F), then fell off rapidly.

Net Primary Production

Plant ecologists measure the accumulated net production by photosynthesis in terms of the **biomass**, which is

the dry weight of organic matter. This quantity could, of course, be stated for a single plant or animal, but a more useful measurement is the biomass per unit of surface area within the ecosystem—that is, kilograms of biomass per square meter or (metric) tons of biomass per hectare (1 hectare = 10^4 m^2). Of all ecosystems, forests have the greatest biomass because of the large amount of wood that the trees accumulate through time. The biomass of grasslands and croplands is much smaller in comparison. For freshwater bodies and the oceans, the biomass is even smaller—on the order of one-hundredth that of the grasslands and croplands.

Although the amount of biomass present per unit area is an important indicator of the amount of photosynthetic activity, it can be misleading. In some ecosystems, biomass is broken down very quickly by consumers and decomposers, so the amount maintained is less. From the viewpoint of ecosystem productivity, what is important is the annual yield of useful energy produced by the ecosystem, or the *net primary production.*

Table 20.1 provides the net primary production of various ecosystems in units of kilograms of dry organic matter produced annually from one square meter of surface. The figures are rough estimates, but they are nevertheless highly meaningful. Note that the highest values are in two quite unlike environments: forests and wetlands (estuaries). Agricultural land compares favorably with grassland, but the range is very large in agricultural land, reflecting many factors such as availability of soil water, soil fertility, and use of fertilizers and machinery.

Productivity of the oceans is generally low. The deep-water oceanic zone, which comprises about 90 percent of the world ocean area, is the least productive of the

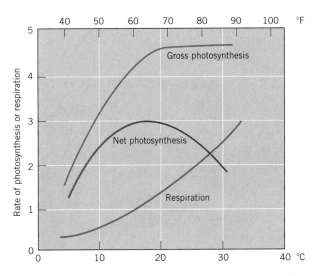

Figure 20.6 Respiration and gross and net photosynthesis vary with temperature. (Data of Stofelt, in A. C. Leopold, *Plant Growth and Development*, McGraw-Hill, New York.)

Table 20.1 Net Primary Production for Various Ecosystems

	Grams per Square Meter per Year	
	Average	Typical Range
Lands		
Rainforest of the equatorial zone	2000	1000–5000
Fresh water swamps and marshes	2500	800–4000
Midlatitude forest	1300	600–2500
Midlatitude grassland	500	150–1500
Agricultural land	650	100–4000
Lakes and streams	500	100–1500
Extreme desert	3	0–10
Oceans		
Algal beds and reefs	2000	1000–3000
Estuaries (tidal)	1800	500–4000
Continental shelf	360	300–600
Open ocean	125	1–400

marine ecosystems. Continental shelf areas are a good deal more productive and support much of the world's fishing industry (Figure 20.7).

Upwelling zones are also highly productive. Upwelling of cold water from ocean depths brings nutrients to the surface and greatly increases the growth of microscopic floating plants known as *phytoplankton*.

These, in turn, serve as food sources for marine animals in the food chain. Consequently, zones of upwelling near habitable coastlines are highly productive fisheries. An example is the Peru Current off the west coast of South America. Here, countless individuals of a single species of small fish, the anchoveta, provide food for larger fish and for birds. The birds, in turn, excrete their wastes on the mainland coast of Peru. The accumulated deposit, called guano, is a rich source of nitrate fertilizer that is now severely depleted.

Within the past decade or so, remote sensing has come into use as a tool for measuring and mapping primary productivity on a global scale. Figure 20.8 shows a polar view of the polar hemisphere, displaying in image format satellite measurements of phytoplankton pigment concentration, which is an index of photosynthetic activity near the ocean surface. The red and yellow colors, marking high pigment concentrations, emphasize the importance of the polar seas in ocean productivity. The data were collected over a period of years, and show averages. It is important to realize that the productivity is highly seasonal, because it is limited by light. Thus polar productivity falls to near zero during the polar winter.

Net Production and Climate

What climatic factors control net primary productivity? We have already identified light intensity and duration,

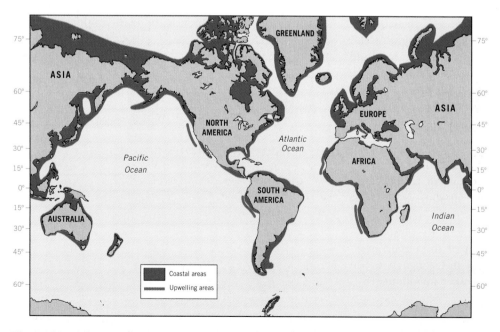

Figure 20.7 Distribution of world fisheries. Coastal areas and upwelling areas together supply over 99 percent of world production. (Compiled by the National Science Board, National Science Foundation.)

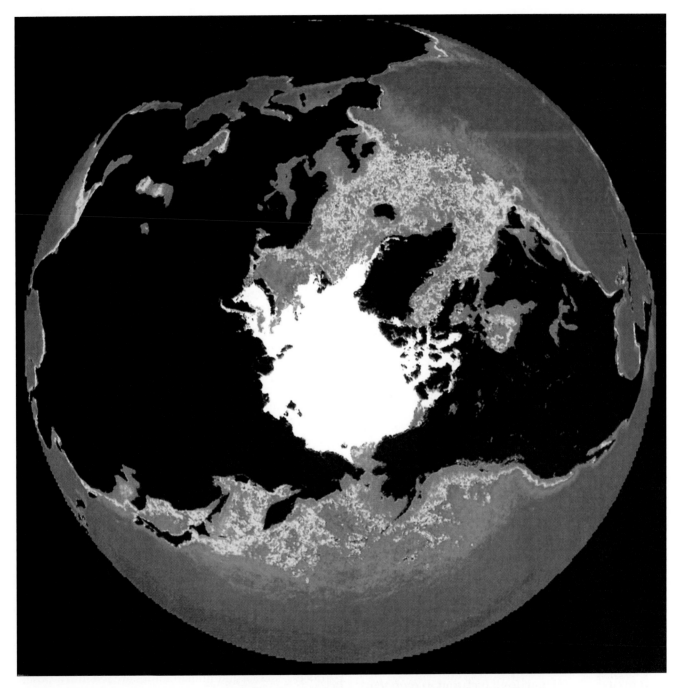

Figure 20.8 Ocean productivity as inferred using remote sensing. This polar image shows northern hemisphere phytoplankton pigment concentrations averaged over the period 1978 to 1986. They were derived by processing data obtained from the Coastal Zone Color Scanner instrument on the *Nimbus-7* satellite. Pigment concentrations are indicated by colors: red (highest) through yellow, green, and blue, to purple (lowest).

as well as temperature, as influencing net photosynthesis. Another important factor is the availability of water. A shortage or surplus of soil water might be the best climatic factor to examine, but data are not available. Ecologists have related net annual primary production to mean annual precipitation, as shown in Figure 20.9. The production values are for plant structures above the ground surface. Although the productivity increases rapidly with precipitation in the lower range from desert through semiarid to subhumid climates, it seems to level off in the humid range. Apparently, a large soil water surplus carries with it some counteractive influence, such as removal of plant nutrients by leaching.

Combining the effects of light intensity, temperature, and precipitation, we can assign rough values of productivity to each of the climates as follows (units are kilograms of carbon per square meter per year):

Highest (over 800)	Wet equatorial ①
Very high (600–800)	Monsoon and trade-wind coastal ②
	Wet-dry tropical ③
High (400–600)	Wet-dry tropical ③ (Southeast Asia)
	Moist subtropical ⑥
	Marine west coast ⑧
Moderate (200–400)	Mediterranean ⑦
	Moist continental ⑩
Low (100–200)	Dry tropical, semiarid ④s
	Dry midlatitude, semiarid ⑨s
	Boreal forest ⑩
Very low (0–100)	Dry tropical, desert ④d
	Dry midlatitude, desert ⑨d
	Boreal forest ⑩
	Tundra ⑫

Remote sensing can also be used to monitor net primary productivity of the earth's land surface. Figure 20.10 shows oceanic phytoplankton pigment concentration (as in Figure 20.8), in addition to which a land vegetation productivity index has been overlain on the continents. The dark-green colors are assigned to the most productive areas, while the yellow-brown tones indicate the least productive. The world's deserts show clearly and present an obvious contrast to equatorial rainforest regions. Tropical savanna and midlatitude grassland are intermediate in productivity, whereas boreal forest and tundra are low.

For natural ecosystems, productivity is largely dependent on climate and soils. For agricultural ecosystems, however, productivity is strongly influenced by the flow of energy, in the form of fertilizer, agricultural chemicals, and irrigation water, that is provided to the crop by human agents. Much if not all of this energy is derived from the burning of fossil fuels and so represents a conversion of fossil fuel energy to human foodstuffs that is not always very efficient. *Focus on Systems 20.2 • Agricultural Ecosystems* provides more information on this topic.

EYE ON THE ENVIRONMENT: BIOMASS ENERGY

Net primary production represents a source of renewable energy derived from the sun that can be exploited to fill human energy needs. The use of biomass as an energy source involves releasing solar energy that has been fixed in plant tissues through photosynthesis.

This process can take place in a number of ways—the simplest is direct burning of plant matter as fuel, as in a campfire or a wood-burning stove. Other approaches involve the generation of intermediate fuels from plant matter—methane gas, charcoal, and alcohol, for example. Biomass energy conversion is not highly energy efficient. Typical values of net annual primary production of plant communities range from 1 to 3 percent of available solar energy. However, the abundance of terrestrial biomass is so great that biomass utilization could provide the energy equivalent to 3 million barrels of oil per day for the United States with proper development.

One important use of biomass energy is the burning of firewood for cooking (and some space heating) in developing nations. The annual growth of wood in the forest of developing countries totals about half the world's energy production—plenty of firewood is thus available. However, fuelwood use exceeds production in many areas, creating local shortages and severe strains on some forest ecosystems. The forest-desert transition areas of thorntree, savanna, and desert scrub in central Africa south of the Sahara Desert are examples.

Even in closed stoves, wood burning is not very efficient, ranging from 10 to 15 percent for cooking. However, the conversion of wood to charcoal or gas can boost efficiencies to values as high as 70 to 80 percent with appropriate technology. In this process, termed *pyrolysis*, controlled partial burning in an oxygen-deficient environment reduces carbohydrate to free carbon (charcoal), and yields flammable gases such as carbon monoxide and hydrogen. Charcoal is more energy efficient than wood, burns more cleanly, and is easier to transport. As an added advantage, charcoal can be made from waste fibers and agricultural residues that would normally be discarded. Thus, charcoal is an efficient fuel that can help extend the firewood supply in areas where wood is in high demand.

A second method of extracting energy from biomass uses anaerobic digestion to produce *biogas*. In this process, animal and human wastes are fed into a closed digesting chamber, where anaerobic bacteria break down the waste to produce a gas that is a mixture of methane and carbon dioxide. The biogas can be easily burned for cooking or heating, or may be used to generate electric power. The digested residue is a sweet-smelling fertilizer. China now maintains a vigorous program of construction of biogas digesters for the use of small family units. The benefits include better sanitation and reduced air pollution, as well as more efficient fuel usage.

Another use of biomass of increasing importance is the conversion of agricultural wastes to alcohol. In this process, yeast microorganisms are used to convert the carbohydrate to alcohol through fermentation. An advantage of alcohol is that it can serve as a substitute and

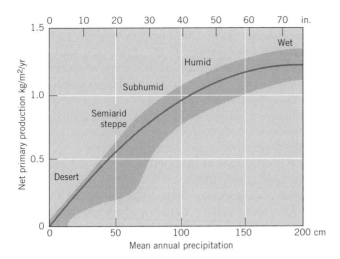

Figure 20.9 New primary production increases rapidly with increasing precipitation, but levels off in the higher values. Observed values fall mostly within the shaded zone. (Data of Whittaker, 1970.)

extender for gasoline. Gasohol, a mixture of up to 10 percent alcohol in gasoline, can be burned in conventional engines without adjustment.

Brazil, a country without adequate petroleum production, has relied heavily on alcohol fuel derived from sugarcane. In 1988, alcohol provided 63 percent of Brazil's automotive fuel needs. Distillation of alcohol, however, requires heating, thus greatly reducing the net energy yield. Alcohol, charcoal, and firewood are all alternatives to fossil fuels that will become increasingly important as petroleum becomes scarcer and more costly in the coming decades.

Relying on biomass energy can also yield important benefits in reducing carbon dioxide emissions. However, burning biomass does not reduce the CO_2 flow to the atmosphere directly. The burning of biomass quickly releases CO_2 that would normally be released more slowly, as the biomass decays. But the energy obtained in the biomass burning will in all likelihood substitute for some fossil fuel burning. Because this fossil fuel is not burned, its CO_2 is not released to the atmosphere, thus reducing overall carbon dioxide emissions.

BIOCHEMICAL CYCLES IN THE BIOSPHERE

We have seen how energy of solar origin flows through ecosystems, passing from one part of the food chain to the next, until it is ultimately lost from the biosphere as energy radiated to space. Matter also moves through

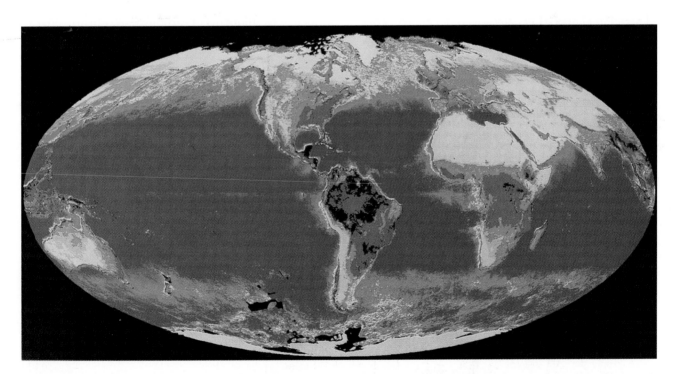

Figure 20.10 Global productivity as inferred using remote sensing. *Oceans:* From NOAA's Coastal Zone Color Scanner instrument as described for Figure 20.8. *Lands:* Displayed is a vegetation greenness index for the period April 1982 to March 1985, obtained from the Advanced Very High Resolution Radiometer (AVHRR) instrument aboard the NOAA-7 spacecraft. The most productive regions are shown in dark green tones, ranging through medium and light green tones to yellow-brown tones (least productive).

ecosystems, but because gravity keeps surface material earthbound, matter cannot be lost in the global ecosystem. As molecules are formed and reformed by chemical and biochemical reactions within an ecosystem, the atoms that compose them are not changed or lost. Thus matter is conserved within an ecosystem, and atoms and molecules can be used and reused, or cycled, within ecosystems.

Atoms and molecules move through ecosystems under the influence of both physical and biological processes. The pathways of a particular type of matter through the earth's ecosystem comprise a **biogeochemical cycle** (sometimes referred to as a *material cycle*, or *nutrient cycle*).

Ecologists recognize two types of biogeochemical cycles—gaseous and sedimentary. In the *sedimentary cycle*, the compound or element is released from rock by weathering, then follows the movement of running water either in solution or as sediment to the sea. Eventually, by precipitation and sedimentation, these materials are converted into rock. When the rock is uplifted and exposed to weathering, the cycle is completed.

In a *gaseous cycle*, a shortcut is provided—the element or compound can be converted into a gaseous form. The gas diffuses throughout the atmosphere and thus arrives over land or sea, to be reused by the biosphere, in a much shorter time. The primary constituents of living matter—carbon, hydrogen, oxygen, and nitrogen—all move through gaseous cycles.

The major features of a biogeochemical cycle are diagrammed in Figure 20.11. Any area or location of concentration of a material is a **pool**. There are two types of pools: *active pools,* where materials are in forms and places easily accessible to life processes, and *storage pools,* where materials are more or less inaccessible to life. A system of pathways of material flows connects the various active and storage pools within the cycle. Pathways between active pools are usually controlled by life processes, whereas pathways between storage pools are usually controlled by physical processes.

The magnitudes of the total storage and total active pools can be very different. In many cases, the active pools are much smaller than storage pools, and materials move more rapidly between active pools than between storage pools or in and out of storage. Taking an example from the carbon cycle, photosynthesis and respiration will cycle all the carbon dioxide in the atmosphere (active pool) through plants in about 10 years. But it may be many millions of years before the carbonate sediments (storage pool) now forming as rock will be uplifted and decomposed to release carbon dioxide.

Nutrient Elements in the Biosphere

Since all elements are more or less available at the earth's surface, we could actually identify cycles for them all. However, only a limited number of elements are important to life forms. Table 20.2 lists the 15 elements that are most abundant in global living matter and show their percentage representation. The three principal components of carbohydrate—hydrogen, car-

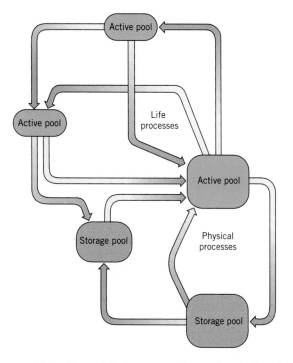

Figure 20.11 General features of a biogeochemical cycle.

Table 20.2 Elements Comprising Global Living Matter, Taking 100 Percent as the Total of the 15 Most Abundant Elements

Basic Carbohydrate

Hydrogen (M)	49.74
Carbon (M)	24.90
Oxygen (M)	24.83
	Subtotal 99.47

Other Nutrients

Nitrogen (M)	0.272
Calcium (M)	0.072
Potassium (M)	0.044
Silicon	0.033
Magnesium (M)	0.031
Sulfur (M)	0.017
Aluminum	0.016
Phosphorus (M)	0.013
Chlorine	0.011
Sodium	0.006
Iron	0.005
Manganese	0.003

M—macronutrient

Source: E. S. Deevey, Jr., *Scientific American*, Vol. 223.

bon, and oxygen—account for 99.5 percent of all living matter and are called **macronutrients**. Macronutrients are elements required in substantial quantities for organic life to thrive. The remaining one-half percent is divided among 12 elements. Six of these are also macronutrients: nitrogen, calcium, potassium, magnesium, sulfur, and phosphorus. The first three macronutrients—hydrogen, carbon, oxygen—are materials whose pathways we have already followed in the photosynthesis-respiration circuits (Figure 20.3). We will undertake a more detailed analysis of the gaseous cycles of carbon and oxygen because they are influenced by human combustion of hydrocarbon compounds. We will also give special attention to nitrogen, the fourth most abundant element in the composition of living matter. Of the remaining macronutrients, three—calcium, potassium, and magnesium—are elements derived from silicate rocks through mineral weathering. Two other macronutrients derived from rock weathering are sulfur and phosphorus.

The Carbon Cycle

The movements of carbon through the life layer are of great importance because all life is composed of carbon compounds of one form or another. Of the total carbon available, most lies in storage pools as carbonate sediments below the earth's surface. Only about two-tenths of 1 percent are readily available to orga-

nisms as CO_2 or as decaying biomass in active pools.

Some details of the *carbon cycle* are shown in a schematic diagram, Figure 20.12. In the gaseous portion of the cycle, carbon moves largely as carbon dioxide (CO_2), which is a free gas in the atmosphere and a dissolved gas in fresh and saltwater. In the sedimentary portion of its cycle, carbon resides in carbohydrate molecules in organic matter, as hydrocarbon compounds in rock (petroleum, coal), and as mineral carbonate compounds such as calcium carbonate ($CaCO_3$). The world supply of atmospheric carbon dioxide is represented in Figure 20.12 by a box. It is a small portion of the carbon in active pools, constituting less than 2 percent. This atmospheric pool is supplied by plant and animal respiration in the oceans and on the lands. Under natural conditions, some new carbon enters the atmosphere each year from volcanoes by outgasing in the form of CO_2 and carbon monoxide (CO). Industry injects substantial amounts of carbon into the atmosphere through combustion of fossil fuels. This increment from fuel combustion and its probable effects on global air temperatures were discussed in Chapter 3.

Carbon dioxide leaves the atmospheric pool to enter the oceans, where it is used in photosynthesis by phytoplankton. These organisms are primary producers in the ocean ecosystem and are consumed by marine animals in the food chain. Phytoplankton also build skeletal structures of calcium carbonate. This mineral mat-

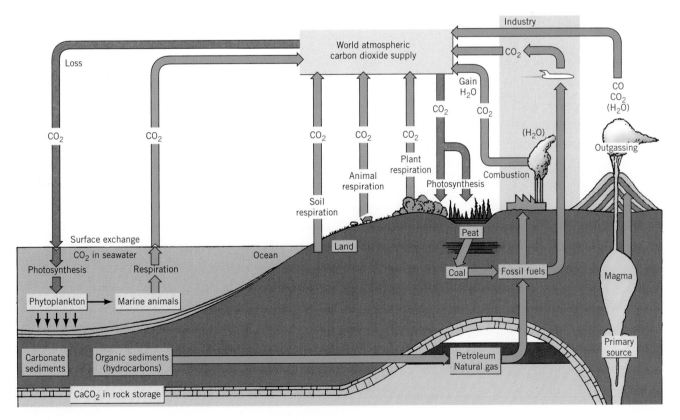

Figure 20.12 The carbon cycle. (Copyright © A. N. Strahler.)

Focus on Systems 20.2 • Agricultural Ecosystems

The principles of energy use and flow in natural ecosystems also apply to agricultural ecosystems. Important differences exist, however, between natural ecosystems and those that are highly managed for agriculture. The first major difference is the reliance of agricultural ecosystems on inputs of energy that are ultimately derived from fossil fuels. The most obvious of these inputs is the fuel that runs the machinery used to plant, cultivate, and harvest crops. Another is the application of fertilizers and pesticides, which require large expenditures of fuel to extract, synthesize, and transport. Yet another is the use of electricity, powered by fossil fuel burning, for irrigation pumps to bring groundwater or surface water to agricultural fields. Fossil fuels also indirectly power such activities as the breeding of plants that have higher yields and are resistant to disease, as well as the development of new chemicals to combat insect pests. Fuel is expended in transporting crops to distant sources of consumption, thus

enabling large areas of similar climate and soils to be used for the same crop. In these and many other ways, the high yields obtained today are brought about only at the cost of a large energy input derived from fossil fuels.

A schematic flow diagram (to the right) shows how fuel-energy input enters into both the purchased inputs and the operations performed on the farm to raise and harvest crops. Solar energy of photosynthesis is, of course, a "free" input. But to be delivered to the humans or animals that consume it, the raw food or feed product of this flow system requires the expenditure of fossil fuel energy.

Agricultural ecosystems, unlike natural ecosystems, are simple in structure and function. They often consist of one genetic strain of one species. Such ecosystems are overly sensitive to attacks by one or two well-adapted insects that can multiply rapidly to take advantage of an abundant food source. Thus, pesticides are constantly needed to reduce insect populations. Weeds,

too, are a problem, adapted as they are to rapid growth on disturbed soil in sunny environments. Weeds can divert much of the productivity to undesirable forms. Herbicides are often the immediate solution to this problem.

In natural ecosystems, nutrient elements are returned to the soil following the death of the plants that concentrate them. In agricultural ecosystems, this recycling is usually interrupted by harvesting the crop for consumption at a distant location. Continuous removal of nutrients in crop biomass introduces a new pathway in the nutrient cycle. To preserve soil fertility, still another pathway must be introduced—from the fertilizer plant to the farm. Keeping this new nutrient subcycle functioning requires considerable fossil fuel energy input.

Excluding the solar energy of photosynthesis, energy expended on the production of a raw food or feed crop can be referred to as *cultural energy*. Let us examine the cultural energy input needed to produce various feed and food crops.

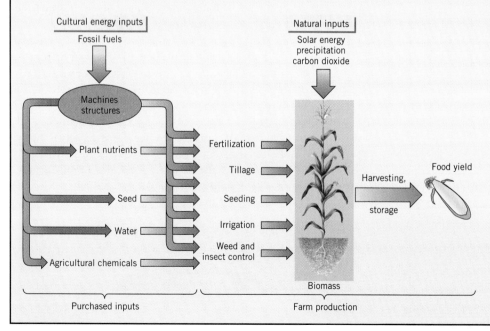

A schematic diagram of the inputs of cultural energy into various stages in the agriculture production system. (After G. H. Heichel, *American Scientist*, Vol. 64.)

The upper graph below shows the relative efficiency of food production of 24 crops. The horizontal scale shows cultural energy expended in terms of thousands of joules per square meter per year (kJ/m²/yr). The vertical scale shows the ratio of food energy produced to input of cultural energy—the higher the ratio, the greater the efficiency (and the lower the cultural input required). Notice that field crops, such as sorghum and corn, which are used largely as animal feeds, have the highest levels of efficiency, whereas foods consumed directly by humans have relatively low efficiencies. Garden crops require a very high cultural energy input per unit area of land and also have low efficiencies.

The lower graph shows the protein derived from each unit of cultural energy on the vertical scale. Alfalfa and soybeans rate very high on this scale. (Soybean protein is presently being processed in various forms for human diet as a meat substitute.) Notice that oats, wheat, and corn have intermediate values of protein yield, but that rice ranks very low. Protein deficiency is a serious health problem for peoples subsisting largely on rice. A small area of the graph at the lower left is labeled "chicken, beef, pork." This insertion serves to show that protein obtained from meat has an energy efficiency only about one-tenth as great as that from soybeans. Indeed, the energy available from edible meat represents only about 10 percent of the energy expended in animal feed, a fact previously pointed out in our discussion of energy flow in the food chain.

The data thus indicate clearly that our food production system is not efficient in terms of cultural energy expended to furnish food to humans. The most highly coveted part of the Western diet, meat and garden vegetables, is extremely wasteful of cultural energy as compared with, say, diets based largely on grain foods (bread, cereals) and soybean products. In a future world where energy will be more costly and the human population will be many times larger, the human diet may well be forced to rely heavily on these more fuel-efficient foods.

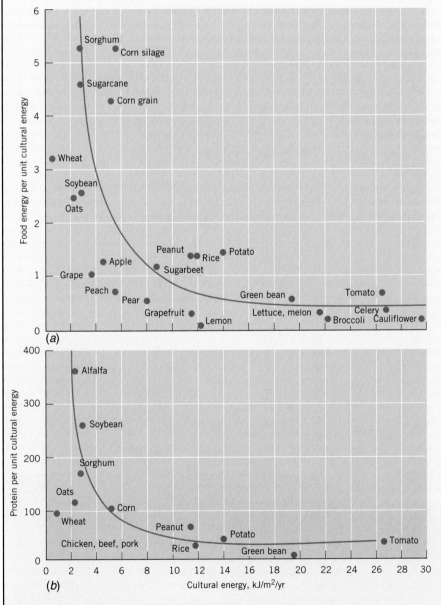

(a) Cultural energy used to produce certain food crops in relation to yield of food energy. (b) Protein yield in relation to cultural energy for several kinds of crops. (After G. H. Heichel, *American Scientist,* Vol. 64.)

ter settles to the ocean floor to accumulate as sedimentary strata, an enormous storage pool not available to organisms until released later by rock weathering. Organic compounds synthesized by phytoplankton also settle to the ocean floor and eventually are transformed into the hydrocarbon compounds making up petroleum and natural gas. On the lands, plant matter accumulating over geologic time forms layers of peat that are ultimately transformed into coal. Petroleum, natural gas, and coal comprise the fossil fuels, and these represent huge storage pools of carbon.

The Oxygen Cycle

Details of the *oxygen cycle* are shown in schematic form in Figure 20.13. The complete picture of the cycling of oxygen also includes its movements and storages when combined with carbon as carbon dioxide and as organic and inorganic compounds. These we have covered in the carbon cycle.

The world supply of atmospheric free oxygen is shown in Figure 20.13 by a box at the top of the diagram. Oxygen enters this active pool through release in photosynthesis, both in the oceans and on the lands. Each year a small amount of new oxygen comes from volcanoes through outgasing, principally as CO_2 and H_2O (shown in Figure 20.12). Balancing the input to

the atmospheric pool is loss through organic respiration and mineral oxidation. Adding to the withdrawal from the atmospheric oxygen pool is industrial activity through the combustion (oxidation) of wood and fossil fuels. Forest fires and grass fires (not shown) are another means of oxygen consumption. The oceans also contain a small, active pool of dissolved gaseous oxygen. Some oxygen is continuously placed in storage in mineral carbonate form in ocean-floor sediments.

Human activity reduces the amount of oxygen in the air by (1) burning fossil fuels; (2) clearing and draining land, which speeds the oxidation of soils and soil organic matter; and (3) reducing photosynthesis by clearing forests for agriculture and by paving and covering previously productive surfaces. The importance of urbanization can be appreciated by the fact that every six months a land area about the size of Rhode Island is covered by new construction in the United States alone. Fortunately, the oxygen pool is so large that the human impact or potential impact is very small, at least at this time.

The Nitrogen Cycle

Nitrogen moves through the biosphere in a gaseous *nitrogen cycle* in which the atmosphere, containing 78 percent nitrogen as N_2 by volume, is a vast storage pool

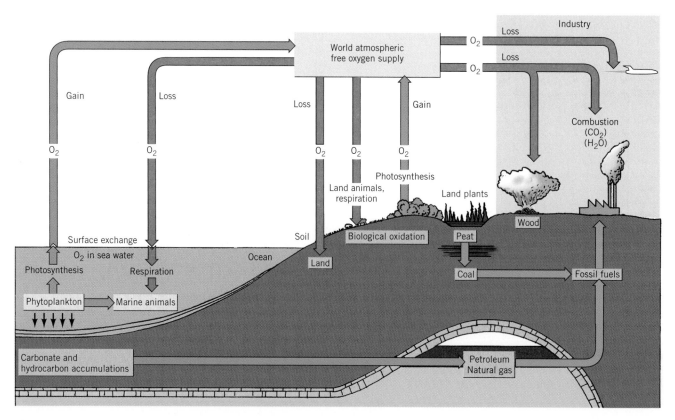

Figure 20.13   The oxygen cycle. (Copyright © A. N. Strahler.)

(Figure 20.14). Nitrogen in the atmosphere in the form of N_2 cannot be assimilated directly by plants or animals. Only certain microorganisms possess the ability to utilize N_2 directly, a process termed *nitrogen fixation*. One class of such microorganisms consists of certain species of free-living soil bacteria. Some blue-green algae can also fix nitrogen.

Another class consists of the symbiotic nitrogen fixers. In a symbiotic relationship, two species of organisms live in close physical contact, each contributing to the life processes or structures of the other. Symbiotic nitrogen fixers are bacteria of the genus *Rhizobium*. These bacteria are associated with some 190 species of trees and shrubs as well as almost all members of the legume family. Legumes important as agricultural crops are clover, alfalfa, soybeans, peas, beans, and peanuts. *Rhizobium* bacteria infect the root cells of these plants in root nodules produced jointly by action of the plant and the bacteria. The bacteria supply the nitrogen to the plant through nitrogen fixation, while the plant supplies nutrients and organic compounds needed by the bacteria. Crops of legumes are often planted in seasonal rotation with other food crops to ensure an adequate nitrogen supply in the soil. Both the action of nitrogen-fixing crops and of soil bacteria are shown in the nitrogen cycle diagram (Figure 20.14).

Nitrogen is lost to the biosphere by *denitrification*, a process in which certain soil bacteria convert nitrogen from usable forms back to N_2. This process is also shown in the diagram. Denitrification completes the organic portion of the nitrogen cycle, as nitrogen returns to the atmosphere.

At the present time, nitrogen fixation is far exceeding denitrification, and usable nitrogen is accumulating in the life layer. This excess of fixation is produced almost entirely by human activities. Human activity fixes nitrogen in the manufacture of nitrogen fertilizers and by oxidizing nitrogen in the combustion of fossil fuels. Widespread cultivation of legumes has also greatly increased worldwide nitrogen fixation. At present rates, nitrogen fixation attributable to human activity nearly equals all natural biological fixation.

Much of the nitrogen fixed by human activities is carried from the soil into rivers and lakes and ultimately reaches the ocean. Major water pollution problems can arise when nitrogen stimulates the growth of algae and phytoplankton. The respiration of these organisms can then reduce quantities of dissolved oxygen to levels that are detrimental to desirable forms of aquatic life. These problems will be accentuated in years to come because industrial fixation of nitrogen in fertilizer manufacture is doubling about every six years at present. The global impact of such large amounts of nitrogen reaching rivers, lakes, and oceans on the earth's global ecosystem remains uncertain.

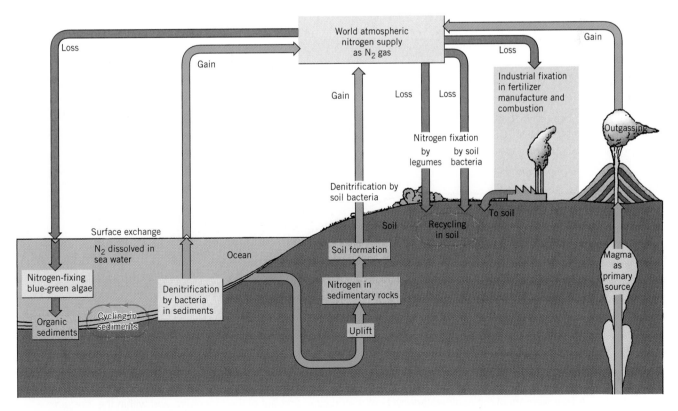

Figure 20.14 The nitrogen cycle.

Sedimentary Cycles

The oxygen, carbon, and nitrogen cycles are all referred to as gaseous cycles because they possess a gaseous phase in which the element involved is present in significant quantities in the atmosphere. Many other elements move in sedimentary cycles, that is, from the land to ocean in running water, returning after millions of years as uplifted terrestrial rock. These elements are not present in the atmosphere except in small quantities as blowing dust or condensation nuclei in precipitation.

Figure 20.15 shows how some important macronutrients move in sedimentary cycles. Within the large box representing the lithosphere are smaller compartments representing the parent matter of the soil and the soil itself. In the soil, nutrients are held as ions on the surfaces of soil colloids and are readily available to plants. (This was explained in Chapter 19.)

The nutrient elements are also held in enormous storage pools, where they are unavailable to organisms. These storage pools include sea water (unavailable to land organisms), sediments on the sea floor, and enormous accumulations of sedimentary rock beneath both lands and oceans. Eventually, elements held in the geologic storage pools are released into the soil by weathering. Soil particles are lifted into the atmosphere by winds and fall back to earth or are washed down by precipitation. Chlorine and sulfur are shown as passing from the ocean into the atmosphere and entering the soil by the same mechanisms of fallout and washout.

The organic realm, or biosphere, is shown in three compartments: producers, consumers, and decomposers. Considerable element recycling occurs between organisms of these three classes and the soil. The elements used in the biosphere, however, are continually escaping to the sea as ions dissolved in stream runoff and groundwater flow.

The organizing principles of energy flow and matter cycling can greatly aid our understanding of the processes of the biosphere and their implications for human activity. Just as solar energy is the driving force for the circulation of global atmospheric and ocean fluids, so it also provides the power source for photosynthesis, on which all the world's organisms ultimately depend for sustenance. Humans are no different from other organisms. Without this input of solar energy and its conversion to consumable biomass, we would soon perish.

Of course, we are also dependent on the smooth functioning of material cycles. Without the tiny fraction of the earth's atmosphere that is carbon dioxide, no terrestrial photosynthesis would occur. If that component is enhanced by human activity through CO_2 release, it is sure to impact the productivity of the biosphere directly, as well as indirectly through climate change. Similarly, we influence the nitrogen and oxy-

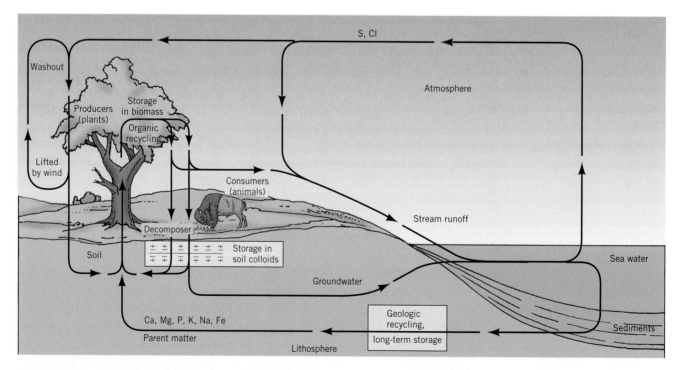

Figure 20.15 A flow diagram of the sedimentary cycle of materials in and out of the biosphere and within the inorganic realm of the lithosphere, hydrosphere, and atmosphere.

gen cycles without knowing or understanding the consequences.

Since the processes of the biosphere are driven by solar energy and conditioned by water availability, it is no surprise that world ecosystems are much influenced by climate. However, soil, topography, fire, and human influences are also important in shaping the global patterns of ecosystem types. Our next chapter examines the characteristics and global distributions of the major ecosystem types—the biomes of the world.

CHAPTER SUMMARY

Ecology is the science of interactions among organisms and their environment. Its focus is the ecosystem, which by interaction among components provides pathways for flows of energy and cycles of matter. The food web of an ecosystem details how food energy flows from primary producers through consumers and on to decomposers. Because energy is lost at each level, only a relatively few top-level consumers are normally present.

Photosynthesis is the production of carbohydrate from water, carbon dioxide, and light energy by primary producers. Respiration is the opposite process, in which carbohydrate is broken down into carbon dioxide and water to yield chemical energy and thus power organisms. Net photosynthesis is the amount of carbohydrate remaining after respiration has reduced gross photosynthesis. Net photosynthesis increases with increasing light and temperature, up to a point. Forests and estuaries are ecosystems with high rates of net primary production, while grasslands and agricultural lands are generally lower. Oceans are most productive in coastal and upwelling zones near continents. Among climate types, those with abundant rainfall and warm temperatures are most productive.

Biomass is an attractive form of solar-powered energy. Charcoal, biogas, and alcohol are biomass products that can be used as fuels.

Biogeochemical cycles are of two types—gaseous, in which the element has an important gaseous phase and moves within the atmosphere, and sedimentary, when no important gaseous phase is involved. Biogeochemical cycles consist of active pools and storage pools linked by flow paths. Of most concern are biogeochemical cycles of the macronutrients, which include carbon, hydrogen, oxygen, nitrogen, calcium, potassium, magnesium, potassium, and magnesium.

The carbon cycle includes an active pool of biospheric carbon and atmospheric atmospheric CO_2, with a large storage pool of carbonate in sediments. Human activities have provided a pathway from storage to active pools by the burning of fossil fuel. The oxygen cycle features an active pool of atmospheric O_2, which is increased by photosynthesis and reduced by respiration, combustion, and mineral oxidation.

The nitrogen cycle also has an important gas phase, but the nitrogen is largely held in the form of N_2, which cannot be used directly by most organisms. Nitrogen fixation occurs when N_2 is converted to more useful forms by bacteria or blue-green algae, often in symbiosis with higher plants. Human activity has doubled the rate of nitrogen fixation, largely through fertilizer manufacture. Sedimentary cycles involve macronutrients that do not have an important gas phase. These elements are held in active pools in living and decaying organisms and in soils. Storage pools include sea water, sediments, and sedimentary rocks.

KEY TERMS

ecology
ecosystem
food chain
food web
primary producers

consumers
decomposers
photosynthesis
respiration
biomass

biogeochemical cycle
pool
macronutrients

REVIEW QUESTIONS

1. Define the terms *biosphere, ecology,* and *ecosystem.*

2. What is a food web or food chain? What are its essential components? How does energy flow through the food web of an ecosystem?

3. Compare and contrast the processes of photosynthesis and respiration. What classes of organisms are associated with each?

4. Distinguish between gross and net photosynthesis. How is net photosynthesis influenced by light and temperature?

5. How is net primary production related to biomass? Identify some types of terrestrial ecosystems that have a high rate of net primary production and some with a low rate.

6. Which areas of oceans and land are associated with high net primary productivity? How is net primary production on land related to climate?

7. Why is biomass energy a desirable energy source? Identify and describe the most useful forms of biomass energy as fuel.

8. What is a *biogeochemical cycle*? What are its essential features? Identify and compare two types of biogeochemical cycles.

9. List nine macronutrients and identify those associated with gaseous and with sedimentary cycles.

10. What are the essential features and flow pathways of the carbon cycle? How have human activities impacted the carbon cycle?

11. What are the essential features and flow pathways of the oxygen cycle? What are the effects of human activity on the oxygen cycle?

12. What are the essential features and flow pathways of the nitrogen cycle? What role do bacteria play? How has human activity modified the nitrogen cycle?

13. What are the essential features and flow pathways of macronutrients in sedimentary cycles?

Focus on Systems 20.2 • Agricultural Ecosystems

1. In what ways is agricultural production dependent on fossil fuels?

2. How do agricultural ecosystems differ from natural ecosystems?

3. Use the concept of *cultural energy* to discuss the efficiency of production of different food crops and the implications for human diets.

ESSAY QUESTIONS

1. Select one of the cycles described in the text (carbon, oxygen, nitrogen, sedimentary). Identify and describe the power sources for each of the major pathways in the cycle.

2. Suppose atmospheric carbon dioxide concentration doubles. What will be the effect on the carbon cycle? How will flows change? Which pools will increase? Decrease?

PROBLEMS

Working It Out 20.1 • Logistic Population Growth

1. An ecologist repeats the experiment of yeast growing in a culture medium, but this time alters the medium to see what effect the change may have on population growth. She plots the data, shown in columns 1 and 2 below, and observes that the carrying capacity (*C*) is about 440 grams. She then calculates the increase in each time period in columns 3 and 4, and estimates an initial growth rate (*R*) of about 60 percent. She uses these values to model the growth using the logistic model, calculating the values in columns 5 and 6. The following table omits the results of her calculations for times of 3, 7, 9, 13, and 17 hours. Find the missing values. For increases, calculate these by comparing the observed biomass with the observed biomass of the previous period. For the logistic model, find the constants *A* and *k*, then evaluate the logistic equation to provide the modeled result. Subtract the modeled biomass from the observed biomass to find the difference.

Yeast Growth, Alternative Medium

(1) Time (hours)	(2) Biomass (grams)	(3) Increase (grams)	(4) Increase (percent)	(5) Logistic Model	(6) Difference
0	10.0			10.0	0.0
1	16.2	6.2	62.0	15.8	0.4
2	26.2	10.0	61.7	24.7	1.5
3	37.4				
4	61.0	23.6	63.1	58.2	2.8
5	85.7	24.7	40.5	86.3	−0.6
6	127.2	41.5	48.4	123.5	3.7
7	165.8				
8	222.0	56.2	33.9	219.9	2.1
9	271.7				
10	316.8	45.1	16.6	316.3	0.5
11	350.4	33.6	10.6	353.6	−3.2
12	383.8	33.4	9.5	381.7	2.1
13	399.9				
14	419.1	19.2	4.8	415.2	3.9
15	428.0	8.9	2.1	424.2	3.8
16	430.5	2.5	0.6	430.0	0.5
17	433.6				
18	443.5	9.9	2.3	436.0	7.5

2. (*a*) Compare the pattern of increase in grams with the pattern of increase in percent. (*b*) Do the data seem to fit the model well? (*c*) Compare the outcome of this experiment with that in the *Working It Out* box. Do the two different culture media seem to be different? What other explanations might there be for this result?

Chapter 21

Global Ecosystems

Most of the earth's land surface has some sort of plant cover. Except for ice sheet surfaces and the most barren of deserts, where a plant cover is absent, the vegetation is a visible and obvious part of the landscape. Thus, vegetation is an important concern of geographers. Plants are also important because humans depend on them for food, medicines, fuel, clothing, shelter, and many other life essentials. Plants are the most obvious part of terrestrial ecosystems, and since they are the primary producers on which consumers depend, they also determine many of the characteristics of terrestrial ecosystems. Ecologists recognize major classes of ecosystems—the biomes—largely by the types of vegetation cover that are associated with them.

A Mediterranean landscape of fields, orchards, vineyards, and grazing lands surrounds the village of Olvera, Andalusia, Spain.

This chapter is devoted to a global study of vegetation. As we will see, vegetation is often strongly related to climate, so the knowledge of climate you obtained in Chapters 7–9 will be very useful as you study the world's vegetation.

NATURAL VEGETATION

Over the last few thousand years, human societies have come to dominate much of the land area of our planet. In many regions, humans have changed the natural vegetation—sometimes drastically, other times subtly. What do we mean by natural vegetation? **Natural vegetation** is a plant cover that develops with little or no human interference. It is subject to natural forces of modification and destruction, such as storms or fires. Natural vegetation can still be seen over vast areas of the wet equatorial climate, although the rainforests there are being rapidly cleared. Much of the arctic tundra and the boreal forest of the subarctic zones is in a natural state.

In contrast to natural vegetation is *human-influenced vegetation*, which is modified by human activities. Much of the land surface in midlatitudes is totally under human control, through intensive agriculture, grazing, or urbanization. Some areas of natural vegetation appear to be untouched but are actually dominated by human activity in a subtle manner. For example, most national parks and national forests have been protected from fire for many decades. When lightning starts a forest fire, the firefighters put out the flames as fast as possible. However, periodic burning is part of the natural cycle in many regions. One vital function of fire is to release nutrients that are stored in plant tissues. When the vegetation burns, the ashes containing the nutrients remain. These enrich the soil for the next cycle of vegetation cover. In recent years, managers of some parks and forests have stopped suppressing wildfires, allowing the return of more natural periodic burning.

Our species has influenced vegetation in yet another way—by moving plant species from their original habitats to foreign lands and foreign environments. The eucalyptus tree is a striking example. From Australia, various species of eucalyptus have been transplanted to such far-off lands as California, North Africa, and India. Sometimes exported plants thrive like weeds, forcing out natural species and becoming a major nuisance. Few of the grasses that clothe the coastal ranges of California are native species, yet a casual observer might think that these represent native vegetation.

Even so, all plants have limited tolerance to the environmental conditions of soil water, heat and cold, and soil nutrients. Consequently, the structure and outward appearance of the plant cover conforms to basic environmental controls, and each vegetation type is associated with a characteristic geographical region—whether forest, grassland, or desert.

STRUCTURE AND LIFE-FORM OF PLANTS

Plants come in many types, shapes, and sizes. Botanists recognize and classify plants by species. However, the plant geographer is less concerned with individual species and more concerned with the plant cover as a whole. In describing the plant cover, plant geographers refer to the **life-form** of the plant—its physical structure, size, and shape. Although the life-forms go by common names and are well understood by almost everyone, we will review them to establish a uniform set of meanings.

Both trees and shrubs are erect, woody plants (Figure 21.1). They are *perennial*, meaning that their woody tissues endure from year to year. Most have life spans of many years. *Trees* are large, woody perennial plants having a single upright main trunk, often with few branches in the lower part, but branching in the upper part to form a crown. *Shrubs* are woody perennial plants that have several stems branching from a base near the soil surface, so as to place the mass of foliage close to ground level.

Lianas are also woody plants, but they take the form of vines supported on trees and shrubs. Lianas include not only the tall, heavy vines of the wet equatorial and tropical rainforests, but also some woody vines of midlatitude forests. Poison ivy and the tree-climbing form of poison oak are familiar North American examples of lianas.

Herbs comprise a major class of plant life-forms. They lack woody stems and so are usually small, tender plants. They occur in a wide range of shapes and leaf types. Some are *annuals*, living only for a single sea-

Figure 21.1 This schematic diagram shows the layers of a beech-maple-hemlock forest. The vertical dimensions of the lower layers are greatly exaggerated. (After P. Dansereau.)

Figure 21.2 Reindeer moss, a white variety of lichen, seen here on rocky tundra of Alaska.

son—while others are perennials—living for multiple seasons. Some herbs are broad-leaved, and others are narrow-leaved, such as grasses. Herbs as a class share few characteristics in common except that they usually form a low layer as compared with shrubs and trees.

Forest is a vegetation structure in which trees grow close together. Crowns are in contact, so that the foliage largely shades the ground. Many forests in moist climates show at least three layers of life-forms (Figure 21.1). Tree crowns form the uppermost layer, shrubs an intermediate layer, and herbs a lower layer. There is sometimes a fourth, lowermost layer that consists of mosses and related very small plants. In **woodland**, crowns of trees are mostly separated by open areas, usually having a low herb or shrub layer.

Lichens are another life-form seen in a layer close to the ground. They are plant forms in which algae and fungi live together to form a single plant structure. In some alpine and arctic environments, lichens grow in profusion and dominate the vegetation (Figure 21.2).

PLANTS AND ENVIRONMENT

A primary objective of this chapter is to describe the earth's vegetation cover as part of the earth's physical geography. But before we begin this task, we will examine some concepts from *plant ecology*—the study of the interrelationships among plants and their environment. These will be helpful in understanding how vegetation is related to climate and soils.

Plant Habitats

As we travel through a hilly, wooded area, it is easy to see that the vegetation is strongly influenced by landform and soil. (As we will see in Chapter 10, landform refers to the configuration of the land surface, including features such as hills, valleys, ridges, or cliffs.) Vegetation on an upland—relatively high ground with thick soil through which water drains easily—is quite different from that on an adjacent valley floor, where soils are wet much of the time. Vegetation is also strikingly different in form on rocky ridges and on steep cliffs, where water drains away rapidly and soil is thin or largely absent.

The total vegetation cover is actually a collection of small patches with different conditions of slope, water drainage, and soil type. Such subdivisions of the plant environment are described as *habitats*. Figure 21.3 presents an example taken from a Canadian forest. Here, there are six distinctive habitats: upland, bog, bottomland, ridge, cliff, and active sand dune. Each habitat supports a different type of vegetation cover.

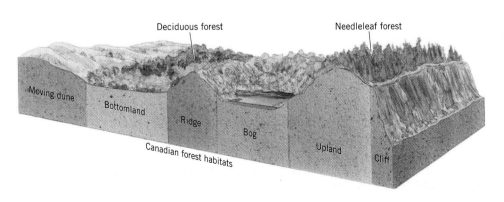

Deciduous forest   Needleleaf forest

Moving dune   Bottomland   Ridge   Bog   Upland   Cliff

Canadian forest habitats

Figure 21.3 Habitats within the Canadian forest. (After P. Dansereau.)

Plants and Water Need

The two most important environmental factors influencing plant growth are water availability and temperature. As we have seen in several earlier chapters, green plants give off large quantities of water to the atmosphere through **transpiration**. In this process, water from the soil is absorbed by root tissues and moves through the plant's stems to the leaves, where it evaporates. This water flow carries nutrients to the leaves and also helps keep the leaves cool. Evaporation at the leaf is controlled by specialized leaf pores, which provide openings in the outer layer of cells. When soil water is depleted, the pores are closed and evaporation is greatly reduced.

As you know from the study of climate in Chapters 7–9, many regions of the world have climates that are dry or have a significant dry season. Many species of plants have developed adaptations to help survive dry conditions. Plants that are adapted to drought conditions are termed **xerophytes**. The word "xerophyte" comes from the Greek roots *xero-*, meaning "dry," and *phyton*, meaning "plant." Because they are highly tolerant of drought, xerophytes can survive in habitats that dry quickly following rapid drainage of precipitation (for example, sand dunes, beaches, and bare rock surfaces). The adaptations of desert plants to dry conditions make them xerophytes as well.

In some xerophytes, water loss is reduced by a thick layer of wax or waxlike material on leaves and stems. The wax helps to seal water vapor inside the leaf or stem. Still other xerophytes adapt to a desert environment by greatly reducing their leaf area or by bearing no leaves at all. Needlelike leaves, or spines in place of leaves, are also adaptations of plants to conserve water. In cactus plants, the foliage leaf is not present, and transpiration is limited to thickened, water-filled stems that store water for use during long, dry periods.

Adaptations of plants to water-scarce environments also include improved abilities to obtain and store water. Roots may extend deeply to reach soil moisture far from the surface. In cases where the roots reach to the groundwater zone, a steady supply of water is assured. Plants drawing from groundwater may be found along dry stream channels and valley floors in desert regions. In these environments, groundwater is usually near the surface. Other desert plants produce a widespread, but shallow, root system. This enables them to absorb water from short desert downpours that saturate only the uppermost soil layer.

Another adaptation to extreme aridity is a very short life cycle. Many small desert plants will germinate from seed, then leaf out, bear flowers, and produce seed in the few weeks immediately following a heavy rain shower. In this way, they complete their life cycle when soil moisture is available, and they survive the dry period as seeds that require no moisture.

Certain climates, such as the wet-dry tropical climate ③ and the moist continental climate ⑩, have a yearly cycle with one season in which water is unavailable to plants because of lack of precipitation or because the soil water is frozen. This season alternates with one in which there is abundant water. Many plants respond to this pattern by dropping their leaves at the close of the moist season and becoming dormant during the dry season. When water is again available, they leaf out and grow at a rapid rate. Trees and shrubs that shed their leaves seasonally are termed *deciduous*. In contrast are *evergreen* plants, which retain most of their leaves in a green state through one or more years.

The Mediterranean climate ⑦ also has a strong seasonal wet-dry alternation, with dry summers and wet winters. Plants in this climate are often xerophytic and characteristically have hard, thick, leathery leaves. An example is the live oak, which holds most of its leaves through the dry season (Figure 21.4). Such hard-leaved evergreen trees and woody shrubs are called *sclerophylls*. (The prefix *scler-* is from the Greek root for "hard" and is combined with the Greek word for leaf, *phyllon*.) Plants that hold their leaves through a dry or cold season have the advantage of being able to resume photosynthesis immediately when growing conditions become favorable, whereas the deciduous plants must grow a new set of leaves.

Plants and Temperature

Temperature is the second important climatic factor that influences global vegetation. It acts directly on plants by influencing the rates at which physiological processes take place in plant tissues. In general, each plant species has an optimum temperature associated with each of its functions, such as photosynthesis (see Chapter 20), flowering, fruiting, or seed germination. There are also limiting lower and upper temperatures for these individual functions and for the total survival of the plant itself.

Temperature can also act indirectly on plants. For example, higher air temperatures increase the water-vapor holding capacity of the air. This means that liquid water will evaporate more readily if air temperature is increased. Therefore, plants lose more water through increased transpiration at higher temperatures.

In general, the colder the climate, the fewer the species that are capable of surviving. A large number of tropical plant species cannot survive below-freezing temperatures for more than a few hours. In the severely cold arctic and alpine environments of high latitudes and high altitudes, only a few plant species are found. This principle explains why a forest in the equatorial zone has many species of trees, whereas a forest of the subarctic zone will be dominated by only a few.

Freezing damages plant tissues largely by causing ice crystals to form inside the cells. These crystals physically disrupt the internal structure of the cell, damaging the fine cell structure. Cold-tolerant species can expel excess water from cells to spaces between cells, where freezing does no damage.

Plant geographers recognize that there is a critical level of climatic stress beyond which a plant species cannot survive. Where this critical level occurs, a geographical boundary will exist that marks the limit of the distribution of the species. Such a boundary is sometimes referred to as a *climatic frontier*. Although the frontier is determined by a complex of climatic elements, it is sometimes possible to single out one climate element related to soil water or temperature that coincides with the plant's frontier.

Ecological Succession and Human Impact on Vegetation

Although climatic influences are important in determining the general type of vegetation cover within a region, local vegetation patterns often depend on the developmental stage of the vegetation cover. For example, a country drive in the Southeast reveals patches of vegetation in many stages of development—from open fallow fields, to grassy shrublands, to forests. As we will see shortly, these are stages in a process by which abandoned fields become mature forests. On a longer time scale, clear lakes gradually fill in with sediment from the rivers that drain into them and become marshes. The marshes may eventually dry out as drainage improves and be replaced by a tall, mature forest. The development process, in which plant communities succeed one another on the way to a stable endpoint, make up an *ecological succession*.

In general, succession leads to formation of the most complex community of organisms possible in an area, given its physical controlling factors of climate, soil, and water. The stable community that is the endpoint of succession is the *climax*. Succession may begin on a newly constructed mineral deposit, such as a sand dune or river bar. It may also occur on a previously vegetated area that has been recently disturbed by agents such as fire, flood, and windstorms, or by human activity.

An important case of succession occurs on farmlands that are intensively cultivated for decades and then abandoned to the natural sequence of vegetation development. An example of this "old-field" succession comes from the southeastern United States, where the moist subtropical climate ⑥ provides ample soil water much of the year and the growing season is long. Figure 21.5 shows the succession, starting at the left with the bare field.

The first plants to take hold in the hostile environment are called *pioneers*. They are annual herbs that most persons would call weeds. When they die, these plants supply organic matter to the soil. They also begin to shade the soil and reduce the extremes of soil temperature. Next, grasses and shrubs move in, and these occupy the old field during the first two decades or so. Pine seedlings now enter the habitat, and, as these grow, a shade cover develops. This greatly changes the climate near the ground.

In the next half-century, as the pine forest matures, broad-leaved deciduous trees begin to displace the pines. By the middle of the second century, a mature forest of oak and hickory has dominated the habitat.

Figure 21.4 This California live oak is an example of a sclerophyll—a plant with thick, leathery leaves that is adapted to an environment with a very dry season.

Focus on Systems 21.1 • Forests and Global Warming

In Chapter 3, and again in Chapter 6, we documented the influence of carbon dioxide on the earth's climate. This greenhouse gas is steadily increasing in the atmosphere, thus enhancing the absorption of longwave radiation from the earth's surface by the atmosphere and increasing counterradiation back to the surface. The result is an increase in global surface temperatures that climatologists have now begun to detect.

Although atmospheric concentrations of CO_2 have been steadily increasing for the better part of a century, scientists have long been puzzled by the fact that only about half of the CO_2 produced by fossil fuel burning is retained by the atmosphere. When CO_2 is added to the atmosphere by fossil fuel or biomass burning, it can have three fates. First, it may remain in the atmosphere. Second, it may be absorbed by the ocean. Third, it may be taken up by photosynthesis and fixed in organic matter.

The figure to the right diagrams the annual flows of carbon in this simplified portion of the carbon cycle. The magnitudes of the annual flows are labeled in gigatons (Gt) per year (1 gigaton = 1 billion metric tons = 10^{12} kg). However, these flows are estimates, each of which is subject to error. The values known with most certainty are the increase of atmospheric carbon, 3.6 Gt, and the release of carbon by fossil fuel burning, 6.0 Gt. The value for CO_2 uptake by the oceans, 2.0 Gt, is also thought to be reasonably accurate. What is known far less accurately are the two flows of carbon to and from land biomass—the release of biomass to the atmosphere by burning (1.6 Gt) and the uptake of carbon by net primary production of terrestrial ecosystems (2.0 Gt). Of these two values, the former

(biomass burning) is better known. The value for uptake is then simply taken as the value needed to balance the budget—6.0 + 1.6 – 3.6 – 2.0 = 2.0 Gt.

If these values are correct, somehow the amount of terrestrial biomass is increasing by 2.0 Gt/yr. Considering that equatorial and tropical forests are diminishing in area as they are logged or converted to farmland or grazing land, the midlatitude forests must be increasing in area or biomass at a faster rate. Independent evidence suggests that this may be the case. In Europe, for example, forest statistics show an increase of growing stock—the volume of living trees— of about 25 percent from 1970 to 1990. This increase has been sustained in spite of damage to forests by air pollution, especially in Eastern Europe. Some of the increase in biomass may also be the effect of enhancement of photosynthesis by warmer temperatures and increased CO_2 concentrations (discussed in more detail shortly). Another factor proposed to account for increased productivity is nitrogen fertilization of soils by the washout of nitrogen pollutant gases in the atmosphere. In the United States, forest areas and ages are increasing in many regions as agricultural production, by shifting to smaller amounts of more intensively managed land, has abandoned marginal areas to natural forest growth. New England is a good example of this trend.

Some foresters have observed that harvesting mature forests and replacing them with young, fast-growing timber should increase the rate of withdrawal of CO_2 from the atmosphere. Since the lumber of the mature forests goes into semipermanent storage in dwellings and structures where it is protected from decay and oxidation to CO_2, it

represents a withdrawal of CO_2 from the atmosphere. The young forests that replace the mature ones grow quickly, fixing carbon at a much faster rate than the older, mature forest, in which annual growth has slowed.

A 1990 report of research scientists at the College of Forestry of Oregon State University showed, however, that the conversion of old-growth forests to young, fast-growing forests will not significantly decrease atmospheric CO_2. They calculated that although 42 percent of the harvested timber goes into comparatively long-term storage (greater than five years) in building structures, much of the remainder is directly discarded on the logging site where it is burned or rapidly decomposes. In addition, some biomass becomes waste in factory processing of the lumber, where sawdust and scrap are burned as fuel. Similarly, the manufacture of paper also results in short-term conversion of a large proportion of the harvested trees to CO_2. In sum, harvesting of old-growth forests as now practiced actually contributes substantially to atmospheric CO_2.

Some environmentalists have advocated increased tree planting as a way of enhancing CO_2 fixation. To take up the quantity of carbon now being released by fossil fuel burning would require some 7 million square kilometers of new closed-crown broadleaved deciduous forest—an area about the size of Australia. To absorb the net increase in atmospheric carbon would require about half that area. This would be a daunting task at best.

Another factor is that increased CO_2 concentration in the atmosphere might enhance photosynthesis and thus increase the rate of carbon fixation. This would serve as a negative feedback loop in the carbon flow system, because it reduces the increase of CO_2 concentration (see figure at right). The enhancement of photosynthesis by increased CO_2 concentrations has been observed for many plants and has been demonstrated as a way of increasing yields of some crops. However, CO_2 is only one factor in photosynthesis—light, temperature, nutrients, and water are also needed, and restrictions in any of these will reduce photosynthesis. In one research study, forest photosynthesis under enriched CO_2 conditions was stimulated, but no net increase in carbon storage was observed. On the other hand, the CO_2 fertilization seemed to en-

hance root growth and nitrogen fixation by root nodules, thus making the trees better able to withstand stress.

Whether produced by CO_2 enrichment, maturing of young forests, or increases in forested land, the increases in biomass by forest growth and areal expansion are only temporary. Eventually, the forests will mature, the trees will die, and the carbon they have fixed will be released to the atmosphere by decay. A new equilibrium will be reached in which there is more overall terrestrial biomass, but production and decay rates are balanced. The period of net increase in carbon fixation and storage will

be over, and CO_2 concentrations will grow at a more rapid rate as fossil fuel burning continues.

In the long term, the dependence of human society on fossil fuels will likely continue for the foreseeable future, unless and until the world's inhabitants as a whole agree that global warming must be countered and are willing to pay the price of energy conservation and conversion to clean energy sources, such as solar power. Although steps have been made in this direction, an effective global commitment to reduction of CO_2 releases still awaits us.

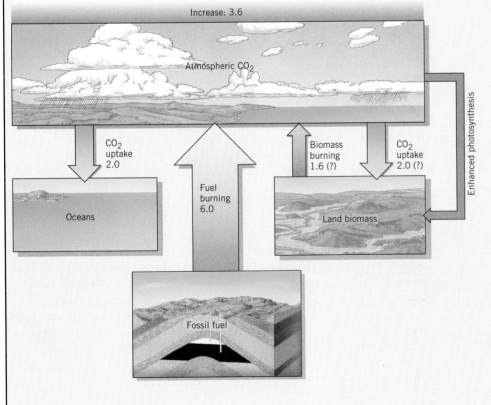

Flows of carbon to and from the atmosphere. Values are gigatons/yr.

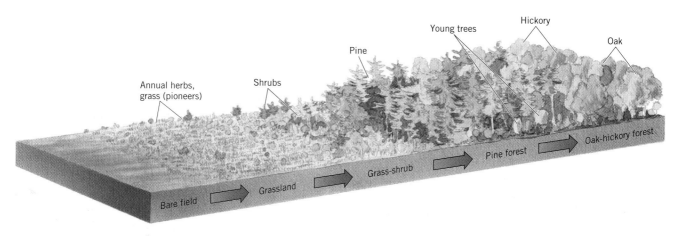

Figure 21.5 Old-field succession in the southeastern United States, following abandonment of cultivated fields. This is a pictorial graph of continuously changing plant composition spanning about 150 years.

Figure 21.6 Natural vegetation of the world.

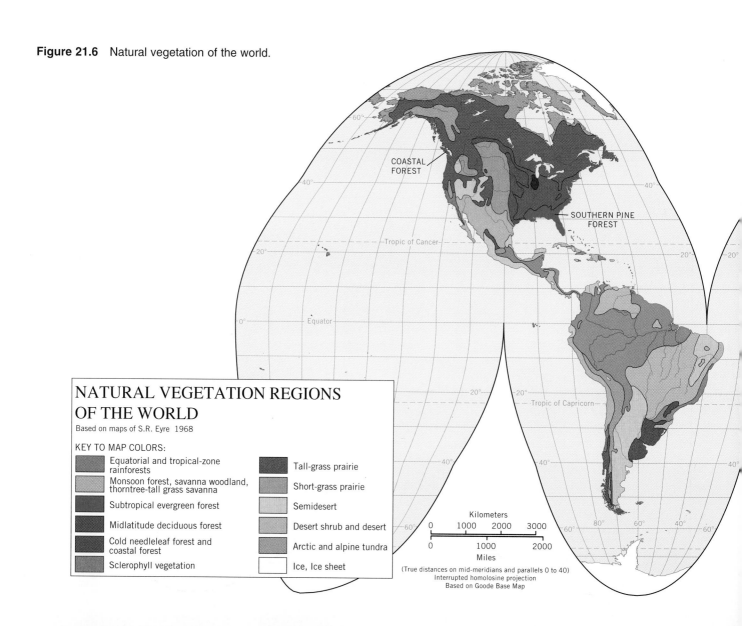

NATURAL VEGETATION REGIONS OF THE WORLD

Based on maps of S.R. Eyre 1968

KEY TO MAP COLORS:

Equatorial and tropical-zone rainforests

Monsoon forest, savanna woodland, thorntree-tall grass savanna

Subtropical evergreen forest

Midlatitude deciduous forest

Cold needleleaf forest and coastal forest

Sclerophyll vegetation

Tall-grass prairie

Short-grass prairie

Semidesert

Desert shrub and desert

Arctic and alpine tundra

Ice, Ice sheet

Kilometers
0 1000 2000 3000

Miles
0 1000 2000

(True distances on mid-meridians and parallels 0 to 40)
Interrupted homolosine projection
Based on Goode Base Map

This climax forest has a lower layer of shrubs and small trees, a basal herb layer, and a substantial accumulation of organic matter on the ground in various stages of decomposition. In this environment, soil water tends to be conserved, and the thermal environment is protected from extremes of heat or cold.

Although the oak–hickory climax forest is generally stable in composition and structure, serious natural upsets to the successional sequence may occur through fires and severe storms. Insect hordes and disease epidemics may also radically alter the climax forest. Such disturbances will be followed by an appropriate succession, with stages leading to the return of the oak–hickory forest.

Disturbances may also result from human activity. Cutting and clearing of forest trees to reclaim the land for agriculture removes the climax forest. Or a disease introduced from a foreign continent may cause the extinction of a particular plant species, changing the climax. An example is the chestnut blight, introduced from Europe, which eliminated the American chestnut from the forests of the northeastern United States. Imported insects may also wipe out most of the mature individuals of a plant species if no native predator is available to combat the invasion. These are just a few of the ways in which humans influence natural vegetation and the succession process.

Human impact on the natural landscape now looms large as the human population continues to expand. By clearing and burning forests, carbon dioxide is released to the atmosphere, adding to the large-volume carbon dioxide generated by fossil fuel burning. At the same time, succession of forests on abandoned land removes carbon from the atmosphere, thus reducing the increase of CO_2 generated by combustion of coal, oil, and natural gas. *Focus on Systems 21.1 • Forests and Global Warming* develops the atmospheric carbon balance in more detail.

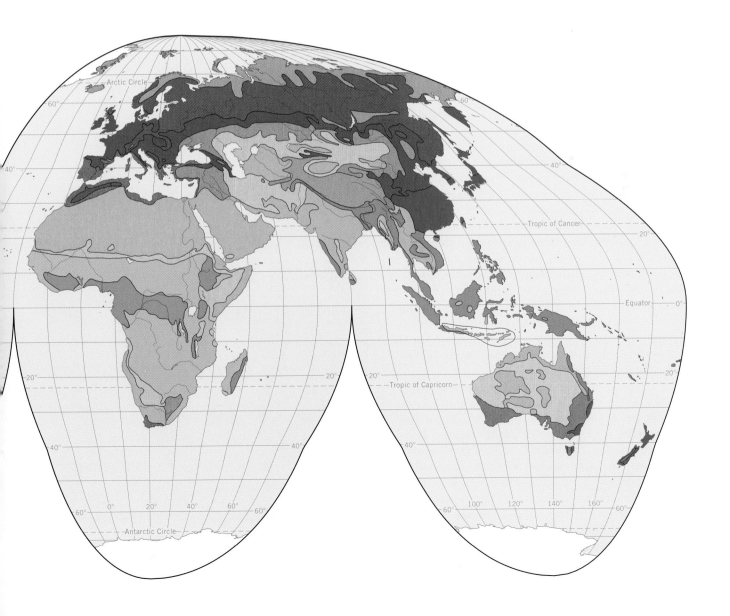

TERRESTRIAL ECOSYSTEMS—THE BIOMES

From the viewpoint of human use, ecosystems are natural resource systems. Food, fiber, fuel, and structural material are products of ecosystems and are manufactured by organisms using energy derived from the sun. Humans harvest that energy by using ecosystem products. The products and productivity of ecosystems depend to a large degree on climate. Where temperature and rainfall cycles permit, ecosystems provide a rich bounty for human use. Where temperature or rainfall cycles restrict ecosystems, human activities can also be limited. Of course, humans, too, are part of the ecosystem. A persistent theme of human geography is the study of how human societies function within ecosystems, utilizing ecosystem resources and modifying ecosystems for human benefit.

Ecosystems fall into two major groups—aquatic and terrestrial. *Aquatic ecosystems* include marine environments and the freshwater environments of the lands. Marine ecosystems include the open ocean, coastal estuaries, and coral reefs. Freshwater ecosystems include lakes, ponds, streams, marshes, and bogs. Our survey of physical geography will not include these aquatic environments. Instead, we will focus on the **terrestrial ecosystems**, which are dominated by land plants spread widely over the upland surfaces of the continents. The terrestrial ecosystems are directly influenced by climate and interact with the soil. In this way, they are closely woven into the fabric of physical geography.

Within terrestrial ecosystems, the largest recognizable subdivision is the **biome**. Although the biome includes the total assemblage of plant and animal life interacting within the life layer, the green plants dominate the biome physically because of their enormous biomass, as compared with that of other organisms. Plant geographers concentrate on the characteristic life-form of the green plants within the biome. These life-forms are principally trees, shrubs, lianas, and herbs, but other life-forms are important in certain biomes. Biomes are recognized and mapped using their climax vegetation type. Areas intensively modified by agriculture and urbanization are assigned to a biome according to the climax vegetation assumed to have once been present.

There are five principal biomes. The **forest biome** is dominated by trees, which form a closed or nearly closed canopy. Forest requires an abundance of soil water, so forests are found in the moist climates. Temperatures must also be suitable, requiring at least a warm season, if not warm temperatures the year round. The **savanna biome** is transitional between forest and grassland. It exhibits an open cover of trees with grasses and herbs underneath. The **grassland biome** develops in regions with moderate shortages of soil water. The semiarid regions of the dry tropical, dry subtropical, and dry midlatitude climates are the home of the grass-land biome. Temperatures must also provide adequate warmth during the growing season. The **desert biome** includes organisms that can survive a moderate to severe water shortage for most, if not all, of the year. Temperatures can range from very hot to cool. Plants are often xerophytes, showing adaptations to the dry environment. The **tundra biome** is limited by cold temperatures. Only small plants that can grow quickly when temperatures warm above freezing in the warmest month or two can survive.

Biogeographers break the biomes down further into smaller vegetation units, called *formation classes*, using the life-form of the plants. For example, at least four and perhaps as many as six kinds of forests are easily distinguished within the forest biome. At least three kinds of grasslands are easily recognizable. Deserts, too, span a wide range in terms of the abundance and life-form of plants. The formation classes described in the remaining portion of this chapter are major, widespread types that are clearly associated with specific climate types. Figure 21.6 (preceding page) is a generalized world map of the formation classes. It simplifies the very complex patterns of natural vegetation to show large uniform regions in which a given formation class might be expected to occur.

Forest Biome

Within the forest biome, we can recognize six major formations: low-latitude rainforest, monsoon forest,

Figure 21.7 An aerial view of low-latitude rainforest. Middle Mazaruni River near Bartica, Guyana.

Figure 21.8 This diagram shows the typical structure of equatorial rainforest. (After J. S. Beard, *The Natural Vegetation of Trinidad,* Clarendon Press, Oxford.)

subtropical evergreen forest, midlatitude deciduous forest, needleleaf forest, and sclerophyll forest.

Low-Latitude Rainforest

Low-latitude rainforest, found in the equatorial and tropical latitude zones, consists of tall, closely set trees. Crowns form a continuous canopy of foliage and provide dense shade for the ground and lower layers (Figure 21.7). The trees are characteristically smooth-barked and unbranched in the lower two-thirds. Tree leaves are large and evergreen—thus, the equatorial rainforest is often described as "broadleaf evergreen forest."

Crowns of the trees of the low-latitude rainforest tend to form two or three layers (Figure 21.8). The highest layer consists of scattered "emergent" crowns that protrude from the closed canopy below, often rising to 40 m (130 ft). Some emergent species develop wide buttress roots, which aid in their physical support (Figure 21.9). Below the layer of emergents is a second, continuous layer, which is 15 to 30 m (about 50 to 100 ft) high. A third, lower layer consists of small, slender trees 5 to 15 m (about 15 to 50 ft) high with narrow crowns.

Typical of the low-latitude rainforest are thick, woody lianas supported by the trunks and branches of trees. Some are slender, like ropes, while others reach thicknesses of 20 cm (8 in.). They climb high into the trees to the upper canopy, where light is available, and develop numerous branches of their own. *Epiphytes* ("air plants") are also common in low-latitude rainforest. These plants attach themselves to the trunk, branches, or foliage of trees and lianas. Their "host" is used solely

Figure 21.9 Buttress roots at the base of a large rainforest tree. From Daintree National Park, Queensland, Australia.

Figure 21.10 In this photo from the El Yunque rainforest, Caribbean National Forest, Puerto Rico, red-flowering epiphytes adorn the trunks of sierra palms.

as a means of physical support. Epiphytes include plants of many different types—ferns, orchids, mosses, and lichens (Figure 21.10).

A particularly important characteristic of the low-latitude rainforest is the large number of species of trees that coexist. In equatorial regions of rainforest, as many as 3000 species may be found in a few square kilometers. Individuals of a given species are often widely separated. Many species of plants and animals in this very diverse ecosystem still have not been identified or named by biologists.

Equatorial and tropical rainforests are not "jungles" of impenetrable plant thickets. Rather, the floor of the low-latitude rainforest is usually so densely shaded that plant foliage is sparse close to the ground. This gives the forest an open aspect, making it easy to travel within its interior. The ground surface is covered only by a thin litter of leaves. Dead plant matter rapidly decomposes because the warm temperatures and abundant moisture promote its breakdown by bacteria. Nutrients released by decay are quickly absorbed by roots. As a result, the soil is low in organic matter.

Low-latitude rainforest develops in a climate that is continuously warm, frost-free, and has abundant precipitation in all months of the year (or, at most, has only one or two dry months). These conditions occur in the wet equatorial climate ① and the monsoon and trade-wind coastal climate ②. In the absence of a cold or dry season, plant growth goes on continuously throughout the year. In this uniform environment, some plant species grow new leaves continuously, shedding old ones continuously as well. Still other plant species shed their leaves according to their own sea-

Figure 21.11 World map of low-latitude rainforest, showing equatorial and tropical rainforest types. (Data source same as Figure 21.6. Based on Goode Base Map.)

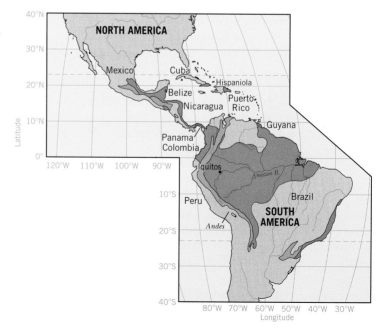

sons, responding to the slight changes in day length that occur with the seasons.

World distribution of the low-latitude rainforest is shown in Figure 21.11. A large area of rainforest lies astride the equator. In South America, this equatorial rainforest includes the Amazon lowland. In Africa, the *equatorial rainforest* is found in the Congo lowland and in a coastal zone extending westward from Nigeria to Guinea. In Indonesia, the island of Sumatra, on the west, bounds a region of equatorial rainforest that stretches eastward to the islands of the western Pacific.

The low-latitude rainforest extends poleward through the tropical zone (lat. 10° to 25° N and S) along monsoon and trade-wind coasts. The monsoon and trade-wind coastal climate in which this *tropical-zone rainforest* thrives has a short dry season. However, the dry season is not intense enough to deplete soil water.

In the northern hemisphere, trade-wind coasts that have rainforests are found in the Philippine Islands and also along the eastern coasts of Central America and the West Indies. These highlands receive abundant orographic rainfall in the belt of the trade winds. A good example of tropical-zone rainforest is the rainforest of the eastern mountains of Puerto Rico. In Southeast Asia, tropical-zone rainforest is extensive in monsoon coastal zones and highlands that have heavy rainfall and a very short dry season. These occur in Vietnam and Laos, southeastern China, and on the western coasts of India and Myanmar. In the southern hemisphere, belts of tropical-zone rainforest extend down the eastern Brazilian coast, the Madagascar coast, and the coast of northeastern Australia.

Within the regions of low-latitude rainforest are many islandlike highland regions where climate is cooler and rainfall is increased by the orographic effect. Here, rainforest extends upward on the rising mountain slopes. Between 1000 and 2000 m (about 3300 and 6500 ft), the rainforest gradually changes in structure and becomes *montane forest* (Figure 21.11). The canopy of montane forest is more open, and tree heights are lower, than in the rainforest. With increasing elevation, the forest canopy height becomes even lower. Tree ferns and bamboos are numerous, and epiphytes are particularly abundant. As elevation increases, mist and fog become persistent, giving high-elevation montane forest the name *cloud forest*.

Monsoon Forest

Monsoon forest of the tropical latitude zone differs from tropical rainforest because it is deciduous, with most of the trees of the monsoon forest shedding their leaves during the dry season. Shedding of leaves results from the stress of a long dry season that occurs at the time of low sun and cooler temperatures. In the dry season, the forest resembles the deciduous forests of the midlatitudes during their leafless winter season.

Monsoon forest is typically open. It grades into woodland, with open areas occupied by shrubs and grasses (Figure 21.12). Because of its open nature, light easily reaches the lower layers of the monsoon forest. As a result, these lower layers are better developed than in the rainforest. Tree heights are also lower. Typically, many tree species are present—as many as 30 to 40 species in a small tract—although the rainforest has many more. Tree trunks are massive, often with thick, rough bark.

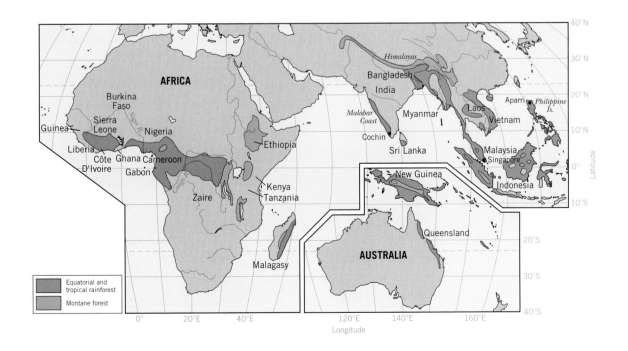

Figure 21.12 Monsoon woodland in the Bandipur Wild Animal Sanctuary in the Nilgiri Hills of southern India. The scene is taken in the rainy season, with trees in full leaf.

Branching starts at a comparatively low level and produces large, round crowns.

Figure 21.13 is a world map of the monsoon forest and closely related formation classes. Monsoon forest develops in the wet-dry tropical climate ③ in which a long rainy season alternates with a dry, rather cool season. These conditions, though most strongly developed in the Asiatic monsoon climate, are not limited to that area. The typical regions of monsoon forest are in Myanmar, Thailand, and Cambodia. In the monsoon forest of southern Asia, the teakwood tree was once abundant and was widely exported to the Western world to make furniture, paneling, and decking. Now this great tree is logged out, and the Indian elephant, once trained to carry out this logging work, is unemployed. Large areas of monsoon forest also occur in south central Africa and in Central and South America, bordering the equatorial and tropical rainforests.

Subtropical Evergreen Forest

Subtropical evergreen forest is generally found in regions of moist subtropical climate ⑥, where winters are mild and there is ample rainfall throughout the year. This

Figure 21.13 World map of monsoon forest and related types—savanna woodland and thorntree-tall grass savanna. (Data source same as Figure 21.6. Based on Goode Base Map.)

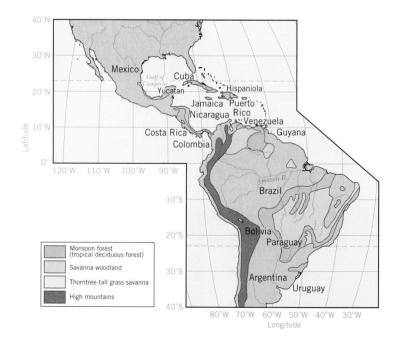

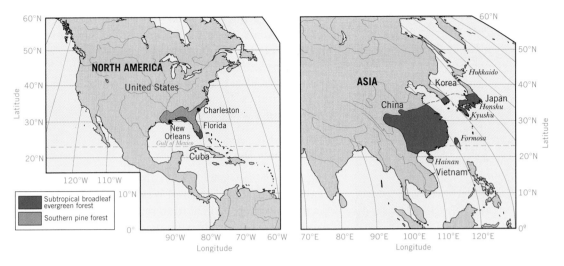

Figure 21.14 Northern hemisphere map of subtropical evergreen forests, including the southern pine forest. (Data source same as Figure 21.6. Based on Goode Base Map.)

forest occurs in two forms: broadleaf and needleleaf. The *subtropical broadleaf evergreen forest* differs from the low-latitude rainforests, which are also broadleaf evergreen types, in having relatively few species of trees. Trees are not as tall as in the low-latitude rainforests. Their leaves tend to be smaller and more leathery, and the leaf canopy less dense. The subtropical broadleaf evergreen forest often has a well-developed lower layer of vegetation. Depending on the location, this layer may include tree ferns, small palms, bamboos, shrubs, and herbaceous plants. Lianas and epiphytes are abundant.

Figure 21.14 is a map of the subtropical evergreen forests of the northern hemisphere. Here subtropical evergreen forest consists of broad-leaved trees such as evergreen oaks and trees of the laurel and magnolia families. The name "laurel forest" is applied to these forests, which are associated with the moist subtropical climate ⑥ in the southeastern United States, southern China, and southern Japan. However, these regions are under intense crop cultivation because of their favorable climate. In these areas, the land has been cleared of natural vegetation for centuries, and little natural laurel forest remains.

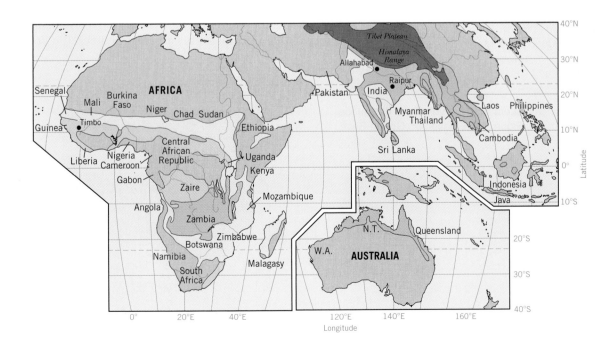

Figure 21.15 This plantation of longleaf pine grows on the sandy soil of the southeastern coastal plain. Near Waycross, Georgia.

The *subtropical needleleaf evergreen forest* occurs only in the southeastern United States (Figure 21.14). Here it is referred to as the southern pine forest, since it is dominated by species of pine. It is found on the wide belt of sandy soils that fringes the Atlantic and Gulf coasts. Because the soils are sandy, water drains away

quickly, leaving the soils quite dry. In infrequent drought years, these forests may burn. Since pines are well adapted to droughts and fires, they form a stable vegetation cover for large areas of the region. Timber companies have taken advantage of this natural preference for pines, creating many plantations yielding valuable lumber and pulp (Figure 21.15).

Midlatitude Deciduous Forest

Midlatitude deciduous forest is the native forest type of eastern North America and Western Europe. It is dominated by tall, broadleaf trees that provide a continuous and dense canopy in summer but shed their leaves completely in the winter. Lower layers of small trees and shrubs are weakly developed. In the spring, a lush layer of lowermost herbs quickly develops but soon fades after the trees have reached full foliage and shaded the ground.

Figure 21.16 is a map of midlatitude deciduous forests, which are found almost entirely in the northern hemisphere. Throughout much of its range, this forest type is associated with the moist continental climate ⑩. Recall from Chapter 7 that this climate receives adequate precipitation in all months, normally with a summer maximum. There is a strong annual temperature cycle with a cold winter season and a warm summer.

Common trees of the deciduous forest of eastern North America, southeastern Europe, and eastern Asia are oak, beech, birch, hickory, walnut, maple, elm, and ash. A few needleleaf trees are often present as well— hemlock, for example (Figure 21.17). Where the deciduous forests have been cleared by lumbering, pines readily develop as second-growth forest.

In Western Europe, the midlatitude deciduous forest is associated with the marine west-coast climate ⑧. Here the dominant trees are mostly oak and ash, with beech in cooler and moister areas. In Asia, the midlatitude deciduous forest occurs as a belt between the bo-

Figure 21.16 Northern hemisphere map of midlatitude deciduous forests. (Data source same as Figure 21.6. Based on Goode Base Map.)

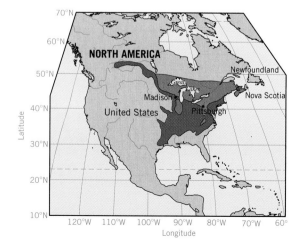

real forest to the north and steppelands to the south. A small area of deciduous forest is found in Patagonia, near the southern tip of South America.

Needleleaf Forest

Needleleaf forest refers to a forest composed largely of straight-trunked, cone-shaped trees with relatively short branches and small, narrow, needlelike leaves. These trees are conifers. Most are evergreen, retaining their needles for several years before shedding them. When the needleleaf forest is dense, it provides continuous and deep shade to the ground. Lower layers of vegetation are sparse or absent, except for a thick carpet of mosses that may occur. Species are few—in fact, large tracts of needleleaf forest consist almost entirely of only one or two species.

Boreal forest is the cold-climate needleleaf forest of high latitudes. It occurs in two great continental belts, one in North America and one in Eurasia (Figure 21.18). These belts span their landmasses from west to east in latitudes 45° N to 75° N, and they closely correspond to the region of boreal forest climate ⑪. The boreal forest of North America, Europe, and western Siberia is composed of such evergreen conifers as spruce and fir (Figure 21.19). The boreal forest of north-central and eastern Siberia is dominated by larch. The larch tree sheds its needles in winter and is thus a deciduous needleleaf tree. Broadleaf deciduous trees, such as aspen, balsam poplar, willow, and birch, tend to take over rapidly in areas of needleleaf forest that have been burned over. These species can also be found bordering streams and in open places. Between the boreal forest and the midlatitude deciduous forest lies a broad transition zone of mixed boreal and deciduous forest.

Needleleaf evergreen forest extends into lower latitudes wherever mountain ranges and high plateaus exist. For example, in western North America this for-

Figure 21.17 Deciduous forest of the Catskill Mountains, New York. Some needleleaf trees—pine and hemlock—are also present.

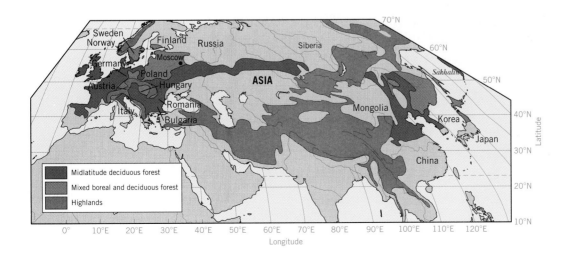

Figure 21.18 Northern hemisphere map of cold-climate needleleaf forests, including coastal forest. (Data source same as Figure 21.6. Based on Goode Base Map.)

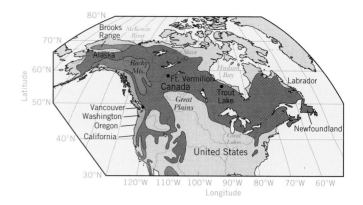

mation class extends southward into the United States on the Sierra Nevada and Rocky Mountain ranges and over parts of the higher plateaus of the southwestern states (Figure 21.20, *left*). In Europe, needleleaf evergreen forests flourish on all the higher mountain ranges.

In its northernmost range, the boreal needleleaf forest grades into cold woodland. This form of vegetation is limited to the northern portions of the boreal forest climate ⑪ and the southern portions of the tundra climate ⑫. Trees are low in height and spaced well apart. A shrub layer may be well developed. The ground cover of lichens and mosses is distinctive. Cold woodland is often referred to as *taiga*. It is transitional into the treeless tundra at its northern fringe.

Coastal forest is a distinctive needleleaf evergreen forest of the Pacific Northwest coastal belt, ranging in latitude from northern California to southern Alaska (Figure 21.18). Here, in a band of heavy orographic precipitation, mild temperatures, and high humidity, are perhaps the densest of all conifer forests, with the world's largest trees (Figure 21.20, *right*). Individual redwood trees attain heights of over 100 m (about 330 ft) and girths of over 20 m (about 65 ft).

Figure 21.19 A view of the boreal forest, Denali National Park, Alaska, pictured here just after the first snowfall of the season. At this location near the northern limits of the boreal forest, the tree cover is sparse. The golden leaves of aspen mark the presence of this deciduous species.

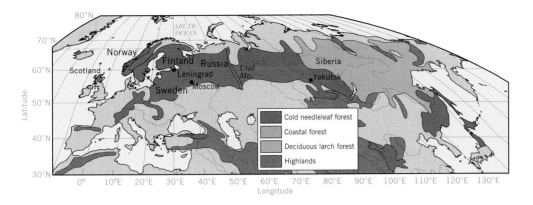

Figure 21.20 Two kinds of needleleaf forest of the western United States. (*left*) Open forest of western yellow pine (ponderosa pine), in the Kaibab National Forest, Arizona. (*right*) A grove of great redwood trees in Humboldt State Park, California.

Sclerophyll Forest

The native vegetation of the Mediterranean climate ⑦ is adapted to survival through the long summer drought. Shrubs and trees that can survive such drought are characteristically equipped with small, hard, or thick leaves that resist water loss through transpiration. As we noted earlier in the chapter, these plants are called sclerophylls.

Sclerophyll forest consists of trees with small, hard, leathery leaves. The trees are often low-branched and gnarled, with thick bark. The formation class includes *sclerophyll woodland*, an open forest in which only 25 to 60 percent of the ground is covered by trees. Also included are extensive areas of *scrub*, a plant formation type consisting of shrubs covering somewhat less than half of the ground area. The trees and shrubs are evergreen, retaining their thickened leaves despite a severe annual drought.

Our map of sclerophyll vegetation, Figure 21.21, includes forest, woodland, and scrub types. Sclerophyll forest is closely associated with the Mediterranean climate ⑦ and is narrowly limited to west coasts between 30° and 40° or 45° N and S latitude. In the Mediterranean lands, the sclerophyll forest forms a narrow, coastal belt ringing the Mediterranean Sea. Here, the Mediterranean forest consists of such trees as cork oak, live oak, Aleppo pine, stone pine, and olive. Over the centuries, human activity has reduced the sclerophyll forest to woodland or destroyed it entirely. Today, large areas of this former forest consist of dense scrub.

The other northern hemisphere region of sclerophyll vegetation is the California coast ranges. Here, the sclerophyll forest or woodland is typically dominated by live oak and white oak. Grassland occupies the open ground between the scattered oaks (see Figure 9.10). Much of the remaining vegetation is sclerophyll scrub or "dwarf forest," known as *chaparral*. It varies in composition with elevation and exposure. Chaparral may contain wild lilac, manzanita, mountain mahogany, poison "oak," and live oak.

In central Chile and in the Cape region of South Africa, sclerophyll vegetation has a similar appearance, but the dominant species are quite different. Important areas of sclerophyll forest, woodland, and scrub are also found in southeast, south central, and southwest Australia, including many species of eucalyptus and acacia.

EYE ON THE ENVIRONMENT:

Exploitation of the Low-Latitude Rainforest Ecosystem

Many of the world's equatorial and tropical regions are home to the rainforest ecosystem. This ecosystem is perhaps the most diverse on earth. That is, it possesses more species of plants and animals than any other. Very large tracts of rainforest still exist in South America, south Asia, and some parts of Africa. Ecologists regard this ecosystem as a genetic reservoir of many species of plants and animals. But as human populations expand and the quest for agricultural land continues, low-latitude rainforests are being threatened with clearing, logging, and cultivation for cash crops and animal grazing.

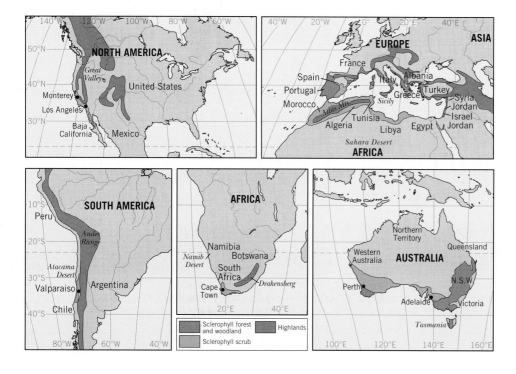

Figure 21.21 World map of sclerophyll vegetation. (Data source same as Figure 21.6. Based on Goode Base Map.)

Figure 21.22 This rainforest in Maranhao, Brazil, has been felled and burned in preparation for cultivation.

In the past, low-latitude rainforests were farmed by native peoples using the *slash-and-burn* method—cutting down all the vegetation in a small area, then burning it (Figure 21.22). In a rainforest ecosystem, most of the nutrients are held within living plants rather than in the soil. Burning the vegetation on the site releases the trapped nutrients, returning a portion of them to the soil. Here, the nutrients are available to growing crops. The supply of nutrients derived from the original vegetation cover is small, however, and the harvesting of crops rapidly depletes the nutrients. After a few seasons of cultivation, a new field is cleared, and the old field is abandoned. Rainforest plants reestablish their hold on the abandoned area. Eventually, the rainforest returns to its original state. This cycle shows that primitive, slash-and-burn agriculture is fully compatible with the maintenance of the rainforest ecosystem.

On the other hand, modern intensive agriculture uses large areas of land and is not compatible with the rainforest ecosystem. When large areas are abandoned, seed sources are so far away that the original forest species cannot take hold. Instead, secondary species dominate, often accompanied by species from other vegetation types. These species are good invaders, and once they enter an area, they tend to stay. The dominance of these secondary species is permanent, at least

on the human time scale. Thus, we can regard the rainforest ecosystem as a resource that, once cleared, will never return. The loss of low-latitude rainforest will result in the disappearance of thousands of species of organisms from the rainforest environment—a loss of millions of years of evolution, together with the destruction of the most complex ecosystem on earth.

In Amazonia, transformation of large areas of rainforest into agricultural land uses heavy machinery to carve out major highways, such as the Trans-Amazon Highway in Brazil, and innumerable secondary roads and trails. Large fields for cattle pasture or commercial crops are created by cutting, bulldozing, clearing, and burning the vegetation. In some regions, the great broadleaved rainforest trees are removed for commercial lumber.

What are the effects of large-scale clearing? A study using a computer model, completed in 1990 by scientists of the University of Maryland and the Brazilian Space Research Institute, indicated that when the Amazon rainforest is entirely removed and replaced with pasture, surface and soil temperatures will be increased 1 to 3°C (2 to 5°F). Precipitation in the region will decline by 26 percent and evaporation by 30 percent. The deforestation will change weather and wind patterns so that less water vapor enters the Amazon basin from out-

Figure 21.23 World map of the grassland biome in subtropical and midlatitude zones. (Data source same as Figure 21.6. Based on Goode Base Map.)

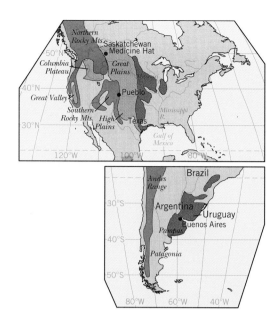

side sources, making the basin ever drier. In areas where a marked dry season occurs, that season will be lengthened. Although such models contain simplifications and are subject to error, the results confirm the pessimistic conclusion that once large-scale deforestation has occurred, artificial restoration of a rainforest comparable to the original one may be impossible to achieve.

According to a 1993 report by the United Nations Food and Agricultural Organization, about 0.6 percent of the world's rainforest is lost annually by conversion to other uses. More rainforest land, 2.2 million hectares (about 8500 sq mi.), is lost annually in Asia than in Latin America and the Caribbean, where 1.9 million hectares (about 7300 sq mi) are converted. Africa's loss of rainforest was estimated at about 470,000 hectares per year (1800 sq mi). Among individual countries, Brazil and Indonesia are the loss leaders, accounting for nearly half of the rainforest area converted to other uses. Note that these values do not include even larger losses of moist deciduous forests in these regions. Deforestation in low-latitude dry deciduous forests and hill and montane forests is also very serious.

Although deforestation rates are very rapid in some regions, many nations are now working to reduce the rate of loss of rainforest environment. However, because the rainforest can provide a source of agricultural land, minerals, and timber, the pressure to allow deforestation continues. 🦥

Savanna Biome

The savanna biome is usually associated with the tropical wet-dry climate ③ of Africa and South America. It includes vegetation formation classes ranging from woodland to grassland. In *savanna woodland*, the trees are spaced rather widely apart, because soil moisture during the dry season is not sufficient to support a full tree cover. The open spacing permits development of a dense lower layer, which usually consists of grasses. The woodland has an open, parklike appearance. Savanna woodland usually lies in a broad belt adjacent to equatorial rainforest.

In the tropical savanna woodland of Africa, the trees are of medium height. Tree crowns are flattened or umbrella-shaped, and the trunks have thick, rough bark (see Figure 8.10). Some species of trees are xerophytic forms with small leaves and thorns. Others are broad-leaved deciduous species that shed their leaves in the dry season. In this respect, savanna woodland resembles monsoon forest.

Fire is a frequent occurrence in the savanna woodland during the dry season, but the tree species of the savanna are particularly resistant to fire. Many geographers hold the view that periodic burning of the savanna grasses maintains the grassland against the invasion of forest. Fire does not kill the underground parts of grass plants, but it limits tree growth to individuals of fire-resistant species. Many rainforest tree species that might otherwise grow in the wet-dry climate regime are prevented by fires from invading. The browsing of animals, which kills many young trees, is also a factor in maintaining grassland at the expense of forest.

The regions of savanna woodland are shown in Figure 21.13, along with monsoon forest (discussed earlier). In Africa, the savanna woodland grades into a belt of *thorntree-tall grass savanna*, a formation class transitional to the desert biome. The trees are largely of thorny species. Trees are more widely scattered, and the open grassland is more extensive than in the savanna woodland. One characteristic tree is the flat-

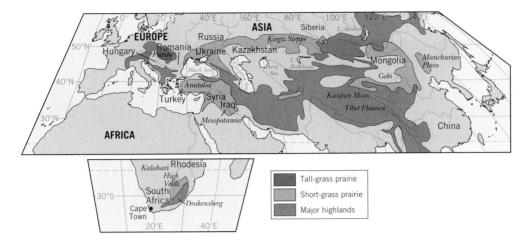

topped acacia, seen in Figure 8.10. Elephant grass is a common species. It can grow to a height of 5 m (16 ft) to form an impenetrable thicket.

The thorntree-tall-grass savanna is closely identified with the semiarid subtype of the dry tropical and subtropical climates (④s, ⑤s). In the semiarid climate, soil water storage is adequate for the needs of plants only during the brief rainy season. The onset of the rains is quickly followed by the greening of the trees and grasses. For this reason, vegetation of the savanna biome is described as rain-green, an adjective that also applies to the monsoon forest.

Grassland Biome

The grassland biome includes two major formation classes that we will discuss here—tall-grass prairie and steppe (Figure 21.23). *Tall-grass prairie* consists largely of tall grasses. *Forbs*, which are broad-leaved herbs, are also present. Trees and shrubs are absent from the prairie but may occur in the same region as narrow patches of forest in stream valleys. The grasses are deeply rooted and form a thick and continuous turf (Figure 21.24).

Prairie grasslands are best developed in regions of the midlatitude and subtropical zones with well-developed winter and summer seasons. The grasses flower in spring and early summer, and the forbs flower in late summer. Tall-grass prairies are closely associated with the drier areas of moist continental climate ⑩. Here, soil water is in short supply during the summer months.

When European settlers first arrived in North America, the tall-grass prairies were found in a belt extending from the Texas Gulf coast northward to southern Saskatchewan (Figure 21.23). A broad peninsula of tall-grass prairie extended eastward into Illinois, where

Figure 21.24 Tall-grass prairie, Iowa. In addition to grasses, tall-grass prairie vegetation includes many forbs, such as the flowering species shown in this photo.

Figure 21.25 World map of the desert biome, including desert and semidesert formation classes. (Data source same as Figure 21.6. Based on Goode Base Map.)

conditions are somewhat more moist. Since the time of settlement, these prairies have been converted almost entirely to agricultural land. Another major area of tallgrass prairie is the Pampa region of South America, which occupies parts of Uruguay and eastern Argentina. The Pampa region falls into the moist subtropical climate ⑥, with mild winters and abundant precipitation.

Recall from Chapter 7 that **steppe**, also called *shortgrass prairie*, is a vegetation type consisting of short grasses occurring in sparse clumps or bunches. Scattered shrubs and low trees may also be found in the steppe. The plant cover is poor, and much bare soil is exposed. Many species of grasses and forbs occur. A typical grass of the American steppe is buffalo grass. Other typical plants are the sunflower and loco weed. Steppe grades into semidesert in dry environments and into prairie where rainfall is higher.

Our map of the grassland biome (Figure 21.23) shows that steppe grassland is concentrated largely in the midlatitude areas of North America and Eurasia. The only southern hemisphere occurrence shown on the map is the "veldt" region of South Africa, a highland steppe surface in Orange Free State and Transvaal.

Steppe grasslands correspond well with the semiarid subtype of the dry continental climate ⑨. Spring rains nourish the grasses, which grow rapidly until early summer. By midsummer, the grasses are usually dormant. Occasional summer rainstorms cause periods of revived growth.

Desert Biome

The desert biome includes several formation classes that are transitional from grassland and savanna biomes into vegetation of the arid desert. Here we recognize two basic formation classes: semidesert and dry desert. Figure 21.25 is a world map of the desert biome.

Semidesert is a transitional formation class found in a wide latitude range—from the tropical zone to the midlatitude zone. It is identified primarily with the arid subtypes of all three dry climates. Semidesert consists of sparse xerophytic shrubs. One example is the sagebrush vegetation of the middle and southern Rocky Mountain region and Colorado Plateau (Figure 21.26). Recently, as a result of overgrazing and trampling by livestock, semidesert shrub vegetation seems to have expanded widely into areas of the western United States that were formerly steppe grasslands.

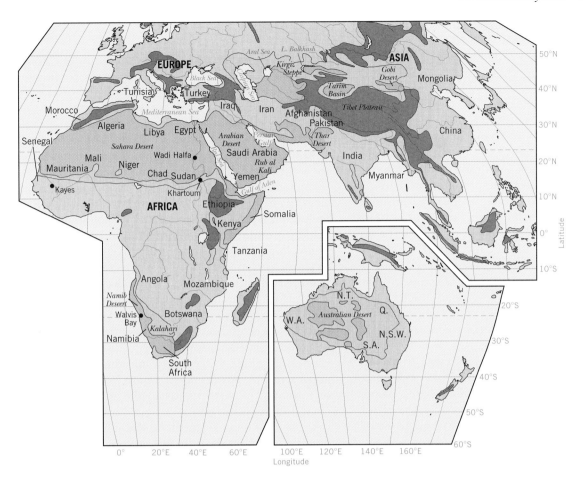

Figure 21.26   Sagebrush semidesert in Monument Valley, Utah.

Thorntree semidesert of the tropical zone consists of xerophytic trees and shrubs that are adapted to a climate with a very long, hot dry season and only a very brief, but intense, rainy season. These conditions are found in the semiarid and arid subtypes of the dry tropical ④ and dry subtropical ⑤ climates. The thorny trees and shrubs are known locally as thorn forest, thornbush, or thornwoods (see Figure 8.19). Many of these are deciduous plants that shed their leaves in the dry season. The shrubs may be closely intergrown to form dense thickets. Cactus plants are present in some localities.

Dry desert is a formation class of xerophytic plants that are widely dispersed over only a very small proportion of the ground. The visible vegetation of dry desert consists of small, hard-leaved, or spiny shrubs, succulent plants (such as cactus), or hard grasses. Many species of small annual plants may be present but appear only after a rare, but heavy, desert downpour. Much of the world map area assigned to desert vegetation has no plant cover at all, because the surface consists of shifting dune sands or sterile salt flats.

Desert plants differ greatly in appearance from one part of the world to another. In the Mojave and Sonoran deserts of the southwestern United States, plants are often large, giving the appearance of a woodland (Figure 21.27). Examples are the treelike saguaro cactus, the prickly pear cactus, the ocotillo, creosote bush, and smoke tree.

Tundra Biome

Arctic tundra is a formation class of the tundra climate ⑫. (See Figure 9.27 for a polar map of the arctic tundra.) In this climate, plants grow during the brief summer of long days and short (or absent) nights. At this time, air temperatures rise above freezing, and a shallow surface layer of ground ice thaws. The permafrost beneath, however, remains frozen, keeping the meltwater at the surface. These conditions create a marshy environment for at least a short time over wide areas. Because plant remains decay very slowly within the cold meltwater, layers of organic matter can build up in the marshy ground. Frost action in the soil fractures and breaks large roots, keeping tundra plants small. In winter, wind-driven snow and extreme cold also injure plant parts that project above the snow.

Plants of the arctic tundra are mostly low herbs, although dwarf willow, a small woody plant, occurs in places. Sedges, grasses, mosses, and lichens dominate the tundra in a low layer (Figure 9.29). Typical plant species are ridge sedge, arctic meadow grass, cotton grasses, and snow lichen. There are also many species of forbs that flower brightly in the summer. Tundra composition varies greatly as soils range from wet to well drained. One form of tundra consists of sturdy hummocks of plants with low, water-covered ground between. Some areas of arctic scrub vegetation com-

Figure 21.27 A desert scene near Phoenix, Arizona. The tall, columnar plant is saguaro cactus; the delicate wandlike plant is ocotillo. Small clumps of prickly pear cactus are seen between groups of hard-leaved shrubs.

posed of willows and birches are also found in tundra.

Tundra vegetation is also found at high elevations. This *alpine tundra* develops above the limit of tree growth and below the vegetation-free zone of bare rock and perpetual snow (Figure 21.28). Alpine tundra resembles arctic tundra in many physical respects.

Altitude Zones of Vegetation

In earlier chapters, we described the effects of increasing elevation on climatic factors, particularly air temperature and precipitation. We noted that, with elevation, temperatures decrease and precipitation generally increases. These changes produce systematic changes in the vegetation cover as well, yielding a sequence of vegetation zones related to altitude.

The vegetation zones of the Colorado Plateau region in northern Arizona and adjacent states provide a striking example of this altitude zonation. Figure 21.29 is a diagram showing a cross section of the land surface in this region. Elevations range from about 700 m (about 2300 ft) at the bottom of the Grand Canyon to 3844 m (12,608 ft) at the top of San Francisco Peak. The vegetation cover and rainfall range are shown on the left. The vegetation zonation includes desert shrub, grassland, woodland, pine forest, Douglas fir forest, Engelmann spruce forest, and alpine meadow. Annual rainfall ranges from 12 to 25 cm (about 5 to 10 in.) in the desert shrub vegetation type to 80–90 cm (about 30–35 in.) in the Engelmann spruce forest.

Ecologists have used the vegetation zonation to set up a series of *life zones*, which are also shown on the figure. These range from the lower Sonoran life zone, which includes the desert shrubs typical of the Sonoran

Figure 21.28 Alpine tundra near the summit of the Snowy Range, Wyoming. Flag-shaped spruce trees (on left side of photo), shaped by prevailing winds, mark the upper limit of tree growth.

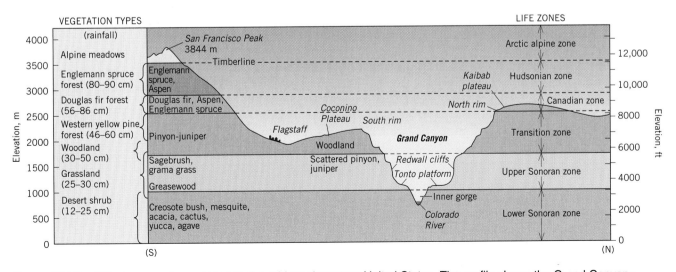

Figure 21.29 Altitude zone of vegetation in the arid southwestern United States. The profile shows the Grand Canyon–San Francisco Mountain district of northern Arizona. (Based on data of G. A. Pearson, C. H. Merriam, and A. N. Strahler.)

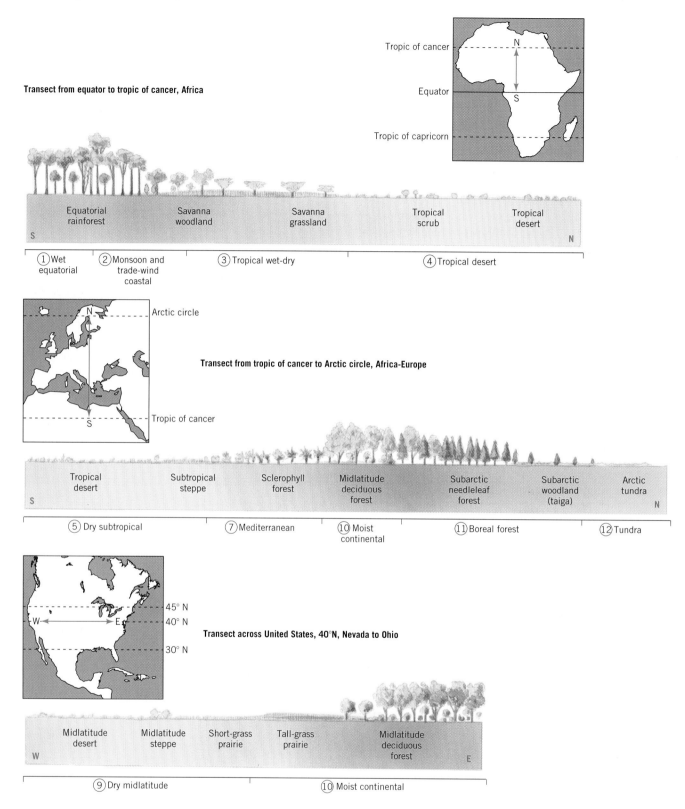

Figure 21.30   Three continental transects showing the succession of plant formation classes across climatic gradients.

Desert, to the arctic-alpine life zone, which includes the alpine meadows at the top of San Francisco Peak. Other life zones take their names from typical regions of vegetation cover. For example, the Hudsonian zone, 2900 to 3500 m (about 9500 to 11,500 ft), bears a needleleaf forest quite similar to needleleaf boreal forest of the subarctic zone near Hudson's Bay, Canada.

Climatic Gradients and Vegetation Types

In discussing the major formation classes of vegetation, we have emphasized the importance of climate. As climate changes with latitude or longitude, vegetation will also change. Figure 21.30 shows three transects across portions of continents that illustrate this principle. (For these transects, we will ignore the effects of mountains or highland regions on climate and vegetation.)

The upper transect stretches from the equator to the tropic of cancer in Africa. Across this region, climate ranges through all four low-latitude climates: wet equatorial ①, monsoon and trade-wind coastal ②, wet-dry tropical ③, and dry tropical ④. Vegetation grades from equatorial rainforest, savanna woodland, and savanna grassland to tropical scrub and tropical desert.

The middle transect is a composite from the tropic of cancer to the arctic circle in Africa and Eurasia. Climates include many of the mid- and high-latitude types: dry subtropical ⑤, Mediterranean ⑦, moist continental ⑩, boreal forest ⑪, and tundra ⑫. The vegetation cover grades from tropical desert through subtropical steppe to sclerophyll forest in the Mediterranean. Further north is the midlatitude deciduous forest in the region of moist continental climate ⑩, which grades into boreal needleleaf forest, subarctic woodland, and finally tundra.

The lower transect ranges across the United States, from Nevada to Ohio. On this transect, the climate begins as dry midlatitude ⑨. Precipitation gradually increases eastward, reaching moist continental ⑩ near the Mississippi River. The vegetation changes from midlatitude desert and steppe to short-grass prairie, tall-grass prairie, and midlatitude deciduous forest.

The changes on these transects are largely gradational rather than abrupt. Yet, the global maps of both vegetation and climate show distinct boundaries from one region to the next. Which is correct? The true situation is gradational rather than abrupt. Maps must necessarily have boundaries to communicate information. But climate and vegetation know no specific boundaries. Instead, they are classified into specific types for convenience in studying their spatial patterns. When studying any map of natural features, keep in mind that boundaries are always approximate and gradational.

This chapter concludes our overview of physical geography as the science of global environments. Our study began with an examination of weather and climate systems, powered largely by solar energy. We then turned to systems of the solid earth, powered by earth forces. Next we presented the systems of landforms and landscape evolution, driven both by earth forces and solar power. Lastly, we covered systems of soils and ecosystems, grouped together by their dependence on climate and substrate, and powered largely by solar energy. Throughout, we have maintained a focus on environmental problems, especially those associated with global change, and stressed the advantages of viewing many of the phenomena we have described as the products of systems of matter and energy flow.

Physical geography is, of course, concerned with the earth—our home planet, at least for the foreseeable future. As a discipline, physical geography stresses the interrelationships among its many component sciences and focuses on our environment in an integrated way. Yet it emphasizes processes that have, thus far, not felt the major effects of human activity. As the human population continues to grow, our influence on our home planet will grow more and more profound. The Epilogue, our final chapter, provides an overview of some of the most important issues at the intersection of human activities and natural systems.

CHAPTER SUMMARY

Natural vegetation is a plant cover that develops with little or no human interference. Although much vegetation appears to be in a natural state, humans influence the vegetation cover by fire suppression and introduction of new species. The life-form of a plant refers to its physical structure, size, and shape. Life-forms include trees, shrubs, lianas, herbs, and lichens.

Plants require water, which they take up and transpire. Xerophytes are plants adapted to dry environments, typically by small, wax-coated leaves or thickened, spongy stems. Other adaptations to drought include deep roots, or extensive and shallow roots, or a short life cycle. Deciduous plants drop their leaves in the dry or cold season. Temperature also limits plants. Only a relatively few plant species can endure freezing. In the process of ecological succession on cleared land or new ground, one vegetation type succeeds another until the stable climax type is reached.

The largest unit of terrestrial ecosystems is the biome: forest, grassland, savanna, desert, and tundra. The forest biome includes a number of important forest formation classes. The low-latitude rainforest exhibits a dense canopy and open floor with a very large number of species. Subtropical evergreen forest occurs in broadleaf and needleleaf forms in the moist subtropical climate ⑥. Monsoon forest is largely deciduous, with most species shedding their leaves after the wet season. Midlatitude deciduous forest is associated with the moist continental climate ⑩. Its species shed their leaves before the cold season. Needleleaf forest consists largely of evergreen conifers. It includes the coastal forest of the Pacific Northwest, the boreal forest of high latitudes, and needle-leaved mountain forests. Sclerophyll forest is comprised of trees with small, hard, leathery leaves, and is found in the Mediterranean climate ⑦ region.

The savanna biome consists of widely spaced trees with an understory, often of grasses. Dry-season fire is frequent in the savanna biome, limiting the number of trees and encouraging the growth of grasses. The grassland biome of midlatitude regions includes tall-grass prairie, in moister environments, and short-grass prairie, or steppe, in semiarid areas. Vegetation of the desert biome ranges from thorny shrubs and small trees to dry desert vegetation comprised of drought-adapted species. Tundra biome vegetation is limited largely to low herbs that are adapted to the severe drying cold experienced on the fringes of the Arctic Ocean. Since climate changes with altitude, vegetation typically occurs in altitudinal zones. Climate also changes gradually with latitude, and so biome changes are typically gradual, without abrupt boundaries.

KEY TERMS

natural vegetation	xerophyte	grassland biome
life-form	terrestrial ecosystem	desert biome
forest	biome	tundra biome
woodland	forest biome	steppe
transpiration	savanna biome	

REVIEW QUESTIONS

1. What is natural vegetation? How do humans influence vegetation?

2. Plant geographers describe vegetation by its overall structure and by the life-forms of individual plants. Define and differentiate the following terms: forest, woodland, tree, shrub, herb, liana, perennial, deciduous, evergreen, broadleaf, needleleaf.

3. The climates of the globe provide great variations in annual moisture and temperature cycles. Discuss the various adaptations that plants have developed for life in stressful climates.

4. What is plant succession? Where and why does it occur? Describe the succession on old fields in the southeastern United States using a sketch of the vegetation sequence.

5. What are the five main biome types that ecologists and biogeographers recognize? Describe each briefly.

6. Low-latitude rainforests occupy a large region of the earth's land surface. What are the characteristics of these forests? Include forest structure, types of plants, diversity, and climate in your answer.

7. Monsoon forest and midlatitude deciduous forest are both deciduous, but for different reasons. Compare the characteristics of these two formation classes and their climates.

8. Subtropical broadleaf evergreen forest and tall-grass prairie are two vegetation formation classes that have been greatly altered by human activities. How was this done and why?

9. Distinguish among the types of needleleaf forest. What characteristics do they share? How are they different? How do their climates compare?

10. Which type of forest, with related woodland and scrub types, is associated with the Mediterranean climate? What are the features of these vegetation types? How are they adapted to the Mediterranean climate?

11. How do traditional agricultural practices in the low-latitude rainforest compare to present-day practices? What are the implications for the rainforest environment?

12. What are the effects of large-scale clearing on the rainforest environment?

13. Describe the formation classes of the savanna biome. Where is this biome found and in what climate types? What role does fire play in the savanna biome?

14. Compare the two formation classes of the grassland biome. How do their climates differ?

15. Describe the vegetation types of the desert biome.

16. What are the features of arctic and alpine tundra? How does the cold tundra climate influence the vegetation cover?

17. How does elevation influence vegetation? Provide an example of how vegetation zonation is related to elevation.

Focus on Systems 21.1 • Forests and Global Warming

1. Identify the major flow paths of carbon into and out of the atmospheric pool. Explain how and why incoming flows balance outgoing flows.

2. What factors could account for an increase in terrestrial biomass that is implied by studies of the atmospheric carbon balance?

ESSAY QUESTIONS

1. Figure 21.30 presents a vegetation transect from Nevada to Ohio. Expand the transect on the west so that it begins in Los Angeles. On the east, extend it northeast from Ohio through Pennsylvania, New York, western Massachusetts, and New Hampshire, to end in Maine. Sketch the vegetation types in your additions and label them, as in the diagram. Below your vegetation transect, draw a long bar subdivided to show the climate types.

2. Construct a similar transect of climate and vegetation from Miami to St. Louis, Minneapolis, and Winnipeg.

EPILOGUE

Physical Geography, Environment, and Global Change

The theme of global change now appears in everyday life. It's unusual to watch a television news report or open a newspaper without encountering some aspect of global change. Perhaps it's a report that the present year is the hottest on record, or a confirmation that the Antarctic ozone hole is deeper this year than in previous years. Or it might be a prediction that another El Niño year is on the way, or even the observation that an Antarctic ice shelf has broken off and is floating out to sea. What is happening? Is our world really changing so fast? What are the reasons?

Change is an inherent part of the natural environment. Over geologic time, continents have converged and collided, only to split and separate. Huge meteorites have impacted the earth's surface, devastating vast areas and darkening skies with great volumes of dust to blot out the sun for years. Within the last few hundreds of thousands of years, natural cycles in the earth's rotation and revolution around the sun have induced cycles in solar radiation received at high latitudes, causing continental ice sheets to wax and wane. Within the span of human history, volcanic eruptions, floods, and severe storms have leveled entire cities and produced change that was felt for decades.

Now, however, global change has become inextricably linked to the influence of human populations on the earth. Thousands of years of overgrazing and cutting of firewood in the Mediterranean lands of Europe have permanently reduced many open woodlands of oaks, grasses, and shrubs to rocky, eroded barren landscapes suited only for goats to graze. Hundreds of years of irrigation in semiarid regions fed by such rivers as the Nile and Indus induced waterlogging and salinization of soils, reducing agricultural productivity to low levels. Within the last century, strip mines yielding coal and metal ores have pockmarked the planet's land surface with scars oozing toxic effluents. Over the past few decades, the Aral Sea, the world's fourth-largest lake, has lost two-thirds of its volume by diversion of its water for irrigation.

Global Climate Change

Perhaps the greatest human impact on the earth has yet to be strongly felt—global climate change. It has the potential to change the lives of many, if not most, of the planet's human inhabitants. In Chapter 3, we discussed a 1995 report by the United Nations Intergovernmental Panel on Climate Change that identified the climatic impacts of increases in greenhouse gases on

This solar-powered home in Massachusetts symbolizes human habitation of our planet in harmony with nature.

global climate. This report concluded that although many uncertainties remained to be resolved, "the balance of evidence suggests that there is a discernible human influence on global climate." This simple pronouncement marked the fact that the preponderance of scientific opinion on the question of whether human influence on global climate has yet been felt had shifted from a verdict of "not proven" to one of "probably guilty."

The culprit in global climate change is, of course, the release of greenhouse gases to the atmosphere by industrial activity and fossil fuel burning. As we showed in Figure 3.22, carbon dioxide the more important greenhouse gas, has been increasing steadily in the atmosphere since 1940 and will double its concentration sometime around the middle of the twenty-first century. Although some scientists still contend that the global warming signal from increased greenhouse gases has not yet been felt, nearly all agree that by the time concentrations are doubled, the global climate system will be impacted.

The latest predictions are that global temperature will increase by the year 2100 by 1°C to 3.5°C (1.8°F to 6.3°F) with a best estimate of 2°C (3.6°F). Furthermore, sea level will rise by 0.15 to 0.9 m (0.5 to 3.0 ft) with a best estimate of 0.48 m (1.6 ft). What are the possible effects of these changes? As we saw in Chapter 3, they include the following:

- Climatic extremes of temperature, dryness, and precipitation would become larger in some regions. These extremes—from record snowfalls to record droughts—would have marked effects on all phases of human activity.
- Beneficial agricultural effects include milder winters in northern regions, as well as increased precipitation and faster crop growth in some areas. North American and Russian grain belts would expand.
- Deserts would expand and continental interiors would in general become drier. In sub-Saharan Africa, tropical Latin America, Southeast Asia, and Australia, crop yields would be reduced.
- Midlatitude climate zones could shift poleward by as much as 550 km (about 350 mi), causing conversion of forest to grassland and scrubland. Some species of forest trees would not be able to migrate quickly enough and would die out.
- One-third of all ecosystems would shift to another type. The rapid changes would greatly reduce biodiversity in many environments.
- Expansion of tropical and subtropical zones containing disease-bearing mosquitoes could lead to 50 to 80 million cases of malaria per year—an increase of 10 to 15 percent. Other diseases, such as dengue fever, yellow fever, and viral encephalitis would also increase.

- Rising sea level would increase coastal flooding and place 92 to 118 million people at risk each year by 2100. Many heavily populated river deltas and their cities would be inundated. Broad ocean beaches in regions such as the east coast of the United States could disappear.
- Mountain glaciers would shrink, and one-third to one-half of mountain glaciers would be lost. Snow cover in the northern hemisphere would also be reduced.

Because the effects of climatic warming could be so far-reaching, scientists have managed to alert the public about the danger of continued increases of CO_2 and other gases that contribute to global warming. At the 1992 Earth Summit in Rio de Janiero, most of the world's nations agreed on a treaty to reduce human impact on climate. The industrialized countries will reduce the levels of greenhouse gases they release, thus slowing the rapid increases in atmospheric CO_2, methane, chlorofluorocarbons, and nitrous oxide (N_2O) that have been measured. The United States and other developed countries agreed to reduce greenhouse gas emissions voluntarily in the year 2000 to 1990 levels. By late 1995, it appeared that Britain, Denmark, The Netherlands, Switzerland, and Germany would meet this goal. But nine other industrialized countries, including the United States, expected to fall short of the goal.

Ozone Layer

The thinning of the ozone layer presents another case of broad-scale change brought about by human activity. As noted in Chapter 2, ozone protects life at the earth's surface by absorbing damaging solar ultraviolet radiation. The thinning is the result of the release of common industrial chemicals—chlorofluorocarbons in particular—that diffuse into the stratosphere where ozone is located. Although the thinning is primarily felt in polar regions during the early spring, reductions in annual ozone concentration of as much as 7 percent have been observed over broad regions of the northern hemisphere.

Recognizing the threat to the ozone layer and its implications, an international coalition of scientists, politicians, and diplomats worked to ban the manufacture of ozone-destroying compounds by industrial nations through a series of treaties signed and amended in Montreal in 1987, London in 1990, and Copenhagen in 1992. As a result of these efforts, ozone concentrations are expected to continue to decline into the first or second decade of the next century, but then begin a 50-year recovery period.

High biodiversity is a feature of equatorial rainforests, such as this example in the Danum River gorge, Sabah, Borneo.

Biodiversity

Another dimension of global change is biodiversity—the assemblage of species occurring within a region or ecosystem. By altering natural habitats, human activity can endanger species or cause their extinction. This reduces the global gene pool in a way that cannot be easily reversed. Recent estimates of extinction rates in well-known groups of organisms suggest that current extinction rates are from 100 to 1000 times greater than in prehuman times.

An action of the Rio Earth Summit, mentioned above, was the adoption of a treaty on biodiversity. In the treaty, nations pledged to protect the genetic resources in their plant and animal populations by shielding them from extinction. Forest preservation was addressed as a need as well. The low-latitude rainforests cover only about 7 percent of the earth's surface, but they contain over half of the world's species. They are especially rich in insect species and flowering plants.

Plants of these regions are a particularly valuable resource. They have provided most of the species on which humans rely for food. Many highly effective medical drugs are produced from these plants.

Not only are low-latitude rainforests valuable resources for human exploitation, but they are also very complex ecosystems that have evolved over millions of years. Many environmentalists believe that humans have a responsibility to preserve our planet's natural systems and maintain their diversity. It is encouraging that the world's nations have recognized the importance of ecological resources such as the low-latitude rainforests and are working toward their preservation.

Extreme Events

Although human activity is responsible for many dimensions of global change, extreme natural processes continue to shape the face of the earth. A few of these cited in our text are especially relevant in human terms—earthquakes at Northridge, California, and Kobe, Japan, that destroyed many structures and claimed many lives; the Mississippi floods of 1993, which drove tens of thousands of people from their homes and caused huge losses of agricultural crops; and Hurricane Andrew, which devastated a wide swath of the south Florida peninsula and claimed property damage in the tens of billions of dollars.

Earthquakes, floods, and tropical storms are, of course, not new phenomena. What is new is that human civilization has come to dominate most areas of the planet, so that these natural occurrences now have major impacts on human societies in the regions in which they occur. Taken in the aggregate, they become a factor of global change.

Note that some of these events—earthquakes, for example—are not influenced by human activities. However, others are. Although hard data are lacking, it is easy to believe that the frequency and severity of tropical storms will be affected by global climate change. Presently, it appears that El Niño years are associated with fewer hurricanes and that El Niño years have become more common as our climate has warmed in the past two decades. Thus, global warming may have the effect of reducing the threat of hurricanes to east coast populations. However, it may also mean that groundwater tables in Caribbean island aquifers will go unreplenished or that coral reefs will lose a source of rejuvenation from wave action. As for floods, it seems certain that changing patterns of rainfall induced by global warming will enhance the frequency and magnitude of floods in some regions, while reducing them in others. So even these rare and extreme events are likely to be influenced by changes on an earth that is increasingly being dominated by our species.

Population

A key force behind the increasing impact of humans on the earth is their sheer numbers. Recent estimates are that the world's population will nearly double from the present value of about 5.75 billion to around 10 billion in 2050. Humans are the most powerful force transforming the surface of the earth today. About 15 to 20 percent of the world's forest area has been cleared since human civilization began to exploit forest resources. In the last 300 years, humans have increased the area of cropland by 450 percent. Populations of the largest urban areas have increased 25 times in that same time span. These changes have directly influenced the global energy balance, the hydrologic cycle, and the dynamics of the atmosphere. Furthermore, as economic development proceeds, the technological demands of each person increase. This translates to an increasing consumption of natural resources and energy that, in turn, places increasing demands on the environment. Thus, as population expands, the impact on natural environments increases.

Recognizing these trends, the nations of the world have come together in an effort to promote sustainable development, a policy in which the needs and aspirations of present and future generations are met without impacting the ability of future generations to meet their needs. The idea is to focus on social and economic progress that is compatible with the environment. The concept draws on the premise that human knowledge, sophisticated technology, and access to resources are now so well developed that they can be used to allow developing countries to increase the well-being of their populations in a way that does not dam-

Climatic change may alter the frequency with which extreme events occur. Here, Hurricane Alicia lashes the Texas coast at Galveston.

The crowded humanity of Bangkok, Thailand, spills over into this floating market.

age the environment. In this effort, much help and co-operation is required from developed nations, to provide not only the funds to stimulate sustainable development, but also the technological base that will be required.

There are some hopeful signs that the long-continued increase of human population will slow. Reports from early 1996 revealed for the first time that fertility rates had declined in all regions of the world. A survey of population experts predicted that the earth's population will peak at around 11 billion by 2075 and remain level or decrease slightly from then to 2100. Other studies of the economic impact of population growth have noted that technological innovation seems to naturally lead to increasing use of technologies that have a smaller environmental impact. Although the citizens of underdeveloped countries will want and need to advance their use of resources, their consumption will perhaps be less damaging to the environment.

The social and economic aspects of population growth and policies such as sustainable development are far beyond the scope of physical geography, but they are concerns of which every responsible citizen on our planet must be aware. This book has provided a background and context on the workings of the earth and the environment that we, as authors, hope will be valuable to you, its readers, as you face the challenges of world citizenship in the twenty-first century.

Appendix 1

Maps and Mapping

Maps play an essential role in the study of physical geography, because much of the information content of geography is stored and displayed on maps. Map literacy—the ability to read and understand what a map shows—is a basic requirement for day-to-day functioning in our society. Maps appear in almost every issue of a newspaper and in nearly every TV newscast. Most people routinely use highway maps and street maps. The purpose of this appendix is to supplement the map-reading skills you have already acquired with some additional information on the subject of cartography, the science of maps and their construction. In understanding the following material, it will be helpful to review the section "Map Projections" in Chapter 1 of the text.

MAP PROJECTIONS

Recall from Chapter 1 that a *map projection* is an orderly system of parallels and meridians used as a base on which to draw a map on a flat surface. A projection is needed because the earth's surface is not flat but rather is curved in a shape that is very close to the surface of a sphere. All map projections misstate the shape of the earth is some way. It's simply impossible to transform a spherical surface to a flat (planar) surface without some violation of the true surface as a result of cutting, stretching, or otherwise distorting the information that lies on the sphere.

Perhaps the simplest of all map projections is a grid of perfect squares. In this simple map, horizontal lines are parallels, and vertical lines are meridians. This grid is often used in modern computer-generated world maps displaying data that consist of a single number for each square, representing, for example, 1, 5, or 10 de-

grees of latitude and longitude. A grid of this kind can show the true spacing (approximately) of the parallels, but it fails to show how the meridians converge toward the two poles. This grid fails dismally in high latitudes, and the map usually has to be terminated at about 70° to 80° north and south.

Early attempts to find satisfactory map projections made use of a simple concept. Imagine the spherical earth grid as a cage of wires (a kind of bird cage). A tiny light source is placed at the center of the sphere, and the image of the wire grid is cast on a surface outside the sphere. This situation is like a reading lamp with a lampshade. Basically, three kinds of "lampshades" can be used, as shown in Figure A1.1.

First is a flat paper disk balanced on the north pole. The shadow of the wire grid on this plane surface will appear as a combination of concentric circles (parallels) and radial straight lines (meridians). Here we have a polar-centered projection. Second is a cone of paper resting point-up on the wire grid. The cone can be slit down the side, unrolled, and laid flat to produce a map that is some part of a full circle. This is called a conic projection. Parallels are arcs of circles, and meridians are radiating straight lines. Third, a cylinder of paper can be wrapped around the wire sphere so as to be touching all around the equator. When slit down the side along a meridian, the cylinder can be unrolled to produce a cylindrical projection, which is a true rectangular grid.

Take note that none of these three projection methods can show the entire earth grid, no matter how large a sheet of paper is used to receive the image. Obviously, if the entire earth grid is to be shown, some quite different system must be devised. Many such alternative solutions have been proposed. In Chapter 1, we described three types of projections used throughout the book—the *polar projection*; the *Mercator projection*, which

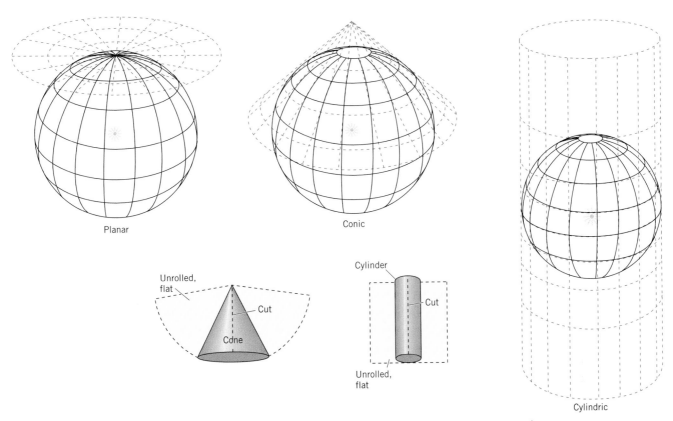

Figure A1.1 Simple ways to generate map projections. Rays from a central light source cast shadows of the spherical geographic grid on target screens. The conical and cylindrical screens can be unrolled to become flat maps. (A. N. Strahler.)

is a cylindrical projection; and the *Goode projection*, which uses a special mathematical principle.

SCALES OF GLOBES AND MAPS

All globes and maps depict the earth's features in much smaller size than the true features they represent. Globes are intended in principle to be perfect scale models of the earth itself, differing from the earth only in size. The *scale* of a globe is therefore the ratio between the size of the globe and the size of the earth, where "size" is some measure of length or distance (but not of area or volume).

Take, for example, a globe 20 cm (about 8 in.) in diameter, representing the earth, with a diameter of about 13,000 km. The scale of the globe is the ratio between 20 cm and 13,000 km. Dividing 13,000 by 20, we see that this ratio reduces to a scale stated as follows: One centimeter on the globe represents 650 kilometers on the earth. This relationship holds true for distances between any two points on a globe.

Scale is more usefully stated as a simple fraction, termed the *fractional scale*, or *representative fraction*. It can be obtained by reducing both earth and globe distances to the same unit of measure, which in this case is

centimeters. (There are 100,000 centimeters in one kilometer.) The advantage of the representative fraction is that it is entirely free of any specified units of measure, such as the foot, mile, meter, or kilometer. Persons of any nationality can understand the fraction, regardless of their language or units of measure.

Being a true-scale model of the earth, a globe has a constant scale everywhere on its surface, but this is not true of a map projection drawn on a flat surface. In flattening the curved surface of the sphere to conform to a plane surface, all map projections stretch the earth's surface in a nonuniform manner, so that the map scale changes from place to place. So we can't say about any world map: "Everywhere on this map the scale is 1:65,000,000." It is, however, possible to select a meridian or parallel—the equator, for example—for which a fractional scale can be given, relating the map to the globe it represents. For example, in Figure 1.8, the scale of the Mercator projection along its equator is about 1:325,000,000.

Small-Scale and Large-Scale Maps

From maps on the global scale, showing the geographic grid of the entire earth or a full hemisphere, we turn to maps that show only small sections of the earth's sur-

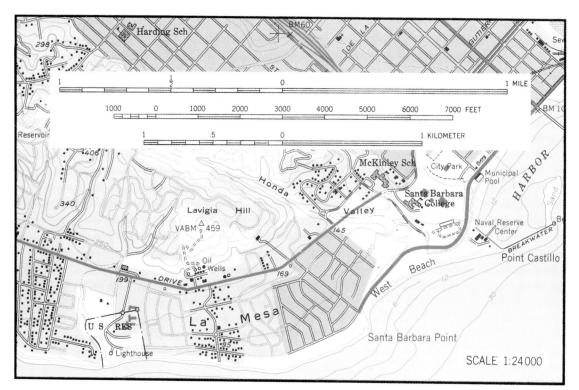

Figure A1.2 A portion of a modern, large-scale map for which three graphic scales have been provided. (U.S. Geological Survey.)

face. These maps of large scale are capable of carrying the enormous amount of geographic information that is available and must be shown in a convenient and effective manner. For practical reasons, maps are printed on sheets of paper usually less than a meter (3 ft) wide, as in the case of the ordinary highway map or navigation chart. Bound books of maps—atlases, that is—consist of pages that are usually no larger than 30 by 40 cm (about 12 by 16 in.), whereas maps found in textbooks and scientific journals are even smaller.

Actually, it is not simply the size of map sheet or page that determines how much information a map can carry. Instead, it is the scale on which the surface is depicted. Keep in mind that the relative magnitude of two different map scales is determined according to which representative fraction is the larger and which the smaller. For example, a scale of 1:10,000 is twice as large as a scale of 1:20,000. Don't make the mistake of supposing that the fraction with the larger denominator represents the larger scale. If in doubt, ask yourself, "Which fraction is the larger: 1/4 or 1/2?"

In the case of globes and maps of the entire earth, the fractional scale may range from 1:100,000,000 to 1:10,000,000; these are *small-scale maps. Large-scale maps* have fractional scales generally greater than 1:100,000. Stated in words, this fraction means: "One centimeter

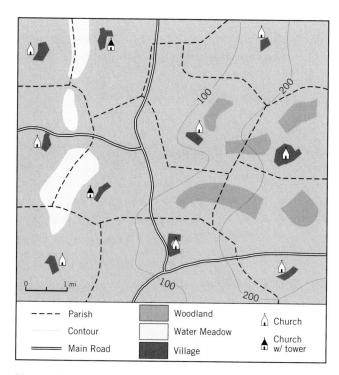

Figure A1.3 Multipurpose map of an imaginary area with 10 villages. (After J. P. Cole and C.A.M. King, *Quantitative Geography*, copyright © John Wiley & Sons, London. Used by permission.)

on the map represents one kilometer on the ground." Most large-scale maps carry a *graphic scale*, which is a line marked off into units representing kilometers or miles. Figure A1.2 shows a portion of a large-scale map on which sample graphic scales in miles, feet, and kilometers are superimposed. Graphic scales make it easy to measure ground distances.

INFORMATIONAL CONTENT OF MAPS

Up to this point, our investigation of cartography has yielded a geographic grid with only the barest outlines of the continents. Information conveyed by a map projection grid system is limited to one category only: absolute location of points on the earth's surface. What other categories of information can be shown on the flat map sheet? The number of categories is almost infinite, since the map is capable of carrying information about anything that occupies a particular location at a specified time. For the needs of physical geography, which is focused primarily on the physical environment of life in the surface layer, the list is greatly reduced, but still enormously long.

Multipurpose maps show a number of different types of information. A simple example is shown in Figure A1.3. Map sheets published by national governments are usually multipurpose maps. Using a great variety of symbols, patterns, and colors, these maps carry a high information content available on demand to the general user, who may select and use only one information category, or *theme*, at any one time. A real example of a multipurpose map is shown on the endpapers of this book, located on the inside of the back cover. It is a portion of a U.S. Geological Survey topographic quadrangle map for San Rafael, California.

In contrast to the multipurpose map is the *thematic map*, which shows only one type of information. Some examples are distributions of the human population, boundaries of election districts, geologic age of exposed rock, elevation of the land surface above sea level, number of days of frost in the year, and speed of winds in the upper atmosphere.

Map Symbols

Symbols on maps associate information with points, lines, and areas. To express points as symbols, they can be renamed "dots." Broadly defined, a dot can be any small device to show point location. It might be a closed circle, open circle, letter, numeral, or a little picture of the object it represents (see "church with tower"

in Figure A1.3). A line can vary in width and can be single or double. The line can also consist of a string of dots or dashes. A specific area of surface can be referred to as a "patch." The patch can be shown simply by a line marking its edge, or it can be depicted by a distinctive pattern or a solid color. Patterns are highly varied. Some consist of tiny dots and others of parallel, intersecting, or wavy lines.

A map consisting of dots, lines, and patches can carry a great deal of information, as Figure A1.3 shows. In this case, the map uses two kinds of dot symbols (both symbolic of churches), three kinds of line symbols, and three kinds of patch symbols. Altogether, eight types of information are offered. Line symbols freely cross patches, and dots can appear within patches. Two different kinds of patches can overlap.

The relation of map symbols to map scale is of prime importance in cartography. Maps of very large scale, along with architectural and engineering plans, can show objects to their true outline form. As map scale is decreased, representation becomes more and more generalized. In physical geography an excellent example is the depiction of a river, such as the lower Mississippi. Figure A1.4 shows the river channel at three scales, starting with a detailed plan, progressing to a double-line analog that generalizes the channel form, and ending with a single-line symbol. As generalization develops, the details of the river banks and channel bends are simplified as well. The level of depiction of fine details is described by the term *resolution*. Maps of large scale have much greater resolving power than maps of small scale.

Presenting Numerical Data on Thematic Maps

How are numerical data shown on maps? In physical geography, much of the information collected about particular areas is in the form of numbers. The numbers might represent readings taken from a scientific instrument at various places throughout the study area. A simple example is the collection of weather data, such as air temperature, air pressure, wind speed, and amount of rainfall. Another example is the set of measurements of the elevation of a land surface, given in meters above sea level. Another category of information consists merely of the presence or absence of a quantity or attribute. In such cases, we can simply place a dot to mean "present," so that when entries are completed, we have before us a field of scattered dots (Figure A1.5).

In some scientific programs, measurements are taken uniformly, for example, at the centers of grid squares laid over a map. For many classes of data, however, the locations of the observation points are prede-

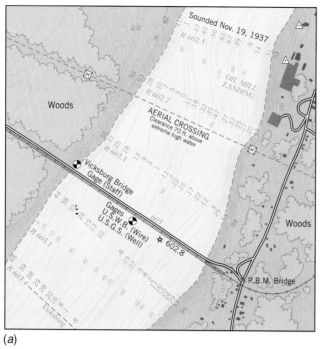

(a)

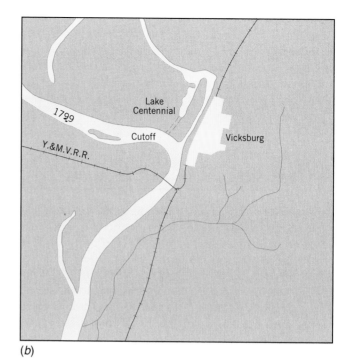

(b)

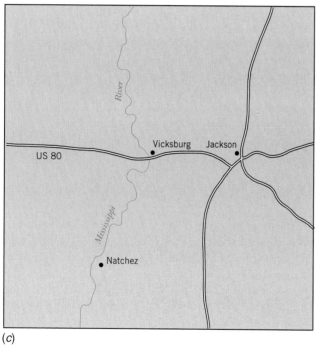

(c)

Figure A1.4 Maps of the Mississippi River on three scales. (*a*) 1:20,000. Channel contours give depth below mean water level. (*b*) 1:250,000. Waterline is shown only to depict channel. (*c*) 1:3,000,000. Channel is shown as a solid line symbol. (Maps slightly enlarged for reproduction.) (Modified from U.S. Army Corps of Engineers.)

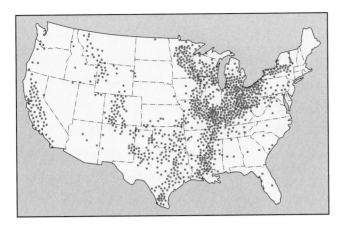

Figure A1.5 A dot map showing the distribution of soils of the order Alfisols in the United States. (From P. Gersmehl, *Annals of the Assoc. of Amer. Geographers*, vol. 67. Copyright © Association of American Geographers. Used by permission.)

termined by a fixed and nonuniform set of observing stations. For example, data of weather and climate are collected at stations typically located at airports. Whatever the sampling method used, we end up with an array of numbers and dots indicating their location on the field of the base map.

Although the numbers and locations may be accurate, it may be difficult to see the spatial pattern present in the data being displayed. For this reason, cartographers often simplify arrays of point values into isopleth maps. An *isopleth* is a line of equal value (from the Greek *isos*, "equal," and *plethos*, "fullness" or "quantity"). Figure 3.17 shows how an isopleth map is constructed for temperature data. In this case, the isopleth is an *isotherm*, or line of constant temperature. In drawing an isopleth, the line is routed among the points in a way that best indicates a uniform value, given the observations at hand.

Isopleth maps are important in various branches of physical geography. Table A1.1 gives a partial list of isopleths of various kinds used in the earth sciences, together with their special names and the kinds of information they display. Examples are cited from our text.

A special kind of isopleth, the *topographic contour* (or isohypse), is shown on the maps in Figures A1.2, A1.4(*a*) and in the rear endpapers. Topographic contours show the configuration of land surface features, such as hills, valleys, and basins.

Table A1.1 Examples of Isopleths

Name of Isopleth	Greek Root	Property Described	Examples in Figures
Isobar	*baros*, weight	Barometric pressure	5.13, 5.14
Isotherm	*therme*, heat	Temperature of air, water, or soil	3.19
Isotach	*tachos*, swift	Fluid velocity	5.22
Isohyet	*hyetos*, rain	Precipitation	4.17
Isohypse	*hypso*, height	Elevation	18.9, 18.10

Appendix 2

Remote Sensing and Geographic Information Systems

A primary activity of geographers is the acquisition, processing, and display of *spatial data*—pieces of information that are in some way associated with a specific location or area of the earth's surface. Two important areas of scientific and technical development have greatly extended the ability of geographers to acquire, examine, and manipulate spatial data in recent years—**remote sensing** and **geographic information systems**. Utilizing the power of computers to acquire and process large volumes of spatial data, these technologies have allowed geographers, geologists, geophysicists, ecologists, planners, landscape architects, and others to develop applications of spatial data processing ranging from planning land subdivisions on the fringes of suburbia to monitoring the deforestation of the Amazon Basin. This appendix provides a brief introduction to some of the key concepts in these two dynamic areas of the geographic sciences.

REMOTE SENSING

Remote sensing refers to gathering information from great distances and over broad areas, usually through instruments mounted on aircraft or orbiting space vehicles that are known collectively as *remote sensors*. These instruments measure electromagnetic radiation coming from the earth's surface and atmosphere as probed from the aircraft or spacecraft platform. As we saw in Chapter 2, all substances, whether naturally occurring or synthetic, are capable of reflecting, absorbing, and emitting electromagnetic energy. The spatial data acquired by remote sensors are typically displayed as images—photographs or similar depictions on a computer screen or color printer—but are often processed further to provide many other types of outputs, as we will see in examples below.

Radar—An Active Remote Sensing System

There are two classes of remote sensor systems: active systems and passive systems. *Active systems* use a beam of wave energy as a source, sending the beam toward an object or a surface. Part of the energy is reflected back to the source, where it is recorded by a detector. A simple analogy would be the use of a spotlight on a dark night to illuminate a target, which reflects light back to the eye.

Radar is an example of an active sensing system. Radar uses radiation from the *microwave* portion of the electromagnetic spectrum, so named because the waves have a short wavelength. Radar systems emit short pulses of microwave radiation and then "listen" for a returning microwave echo. Many radar systems can penetrate clouds to provide images of the earth's surface in any weather. At short wavelengths, however, microwaves can be scattered by water droplets and produce a return signal sensed by the radar apparatus. This effect is the basis for weather radars, which can detect rain and hail and are used in local weather forecasting.

Figure A2.1 shows a radar image of the folded Appalachian Mountains in south-central Pennsylvania. It is produced by an airborne radar instrument that sends

Figure A2.1 Radar image of a portion of the folded Appalachians in south-central Pennsylvania. The area shown is about 40 km (25 mi) wide. Compare this with Figure 16.15, which is a Landsat image of this region.

pulses of radio waves downward and sideward as the airplane flies forward. Surfaces oriented most nearly at right angles to the slanting radar beam will return the strongest echo and therefore appear lightest in tone. In contrast, those surfaces facing away from the beam will appear darkest. The effect is to produce an image resembling a shaded relief map. Various types of surfaces, such as forest, rangeland, and agricultural fields, can also be identified by variations in image tone and pattern.

Passive Remote Sensing Systems

Passive systems measure radiant energy that is reflected or emitted by an object. Reflected energy falls mostly in the visible light and near-infrared regions, while emit-

ted energy lies in the longer thermal infrared region. The most familiar instrument of the passive type is the camera, which uses film that is sensitive to the light energy reflected from the scene. Aerial photographs, taken by downward-pointing cameras from aircraft, have been in wide use since before World War II. Commonly, the field of one photograph overlaps the next along the plane's flight path, so that the photographs can be viewed stereoscopically for a three-dimensional effect. Because of its high resolution (degree of sharpness), aerial photography remains one of the most valuable of the older remote-sensing techniques.

Photography can also extend into infrared wavelengths. Within the infrared spectrum, there is a region in the near-infrared, immediately adjacent to the visible red region, in which reflected rays can be recorded by cameras with suitable film and filter combinations. One example is color infrared film (Figure A2.2). In this type of film, the red color is produced as a response to infrared light, the green color is produced by red light, and the blue color by green light. Because healthy, growing vegetation reflects much more strongly in the near-infrared than in the red or green regions of the spectrum, vegetation has a characteristic red appearance. Figure 15.23, which shows a portion of the Mississippi River and its floodplain, is an example of an air photo taken with color-infrared film.

Satellite images are often processed to give the appearance of color-infrared film. Examples are Figures 16.15 and 17.15. On these images, agricultural crops appear as color shades ranging from pink to orange-red to deep red. Mature crops and dried vegetation (dormant grasses) appear yellow or brown. Urbanized areas typically appear in tones of blue and gray. Shallow water areas appear blue; deep water appears dark blue to blue-black.

Photography has been extended to greater distances through the use of cameras on orbiting space vehicles. A recent Space Shuttle flight included a specially constructed large-format camera, designed to produce very large, very detailed transparencies of the earth's surface suitable for precise topographic mapping. An excellent example is shown in Figure A2.3.

Scanning Systems

Passive remote sensing also includes images acquired by scanning systems, which may be mounted in aircraft or orbiting space vehicles. *Scanning* is the process of receiving information instantaneously from only a very small portion of the area being imaged (Figure A2.4). The scanning instrument senses a very small field of view that runs rapidly across the ground scene. Light from the field of view is focused on a detector that measures its intensity at very short time intervals and

Figure A2.2 High-altitude infrared photograph of an area near Bakersfield, California, in the southern San Joaquin Valley, taken by a NASA U-2 aircraft flying at approximately 18 km (60,000 ft). Photos such as this can be used to study problems associated with agriculture. Problems affecting crop yields arise in perched groundwater areas (A), which appear dark, and areas of high soil salinity (B), which appear light. The various red tones are associated with different types of crops.

Figure A2.3 This color infrared photograph was taken with a large-format camera carried on a flight of the NASA Space Shuttle. It gives an extremely detailed picture of the terrain and is suitable for precision mapping. The area shown, about 125 km in width, includes the Sulaiman Range in West Pakistan (upper left). The Indus and Sutlej rivers, fed by snowmelt from the distant Hindu Kush and Himalaya ranges, cross the scene flowing from northeast to southwest. A mottled pattern of green fields (red) interspersed with barren patches of saline soil (white) covers much of the lower fourth of the area.

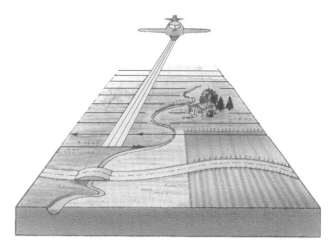

Figure A2.4 Multispectral scanning from aircraft. As the aircraft flies forward, the scanner sweeps side to side. The result is a digital image covering the overflight area. (A. H. Strahler.)

records the intensities as a series of numbers that can be processed by a computer. Later, the computer reconstructs an image of the ground scene from the measurements acquired by the scanning system.

Most scanning systems in common use are *multispectral scanners*. These devices measure brightness in several wavelength regions simultaneously. An example is the Multispectral Scanning System (MSS) used aboard the Landsat series of earth-observing satellites. This instrument simultaneously collects reflectance data in four spectral bands. Two are bands in the visible light spectrum (one green, one red), and two in the infrared region. A successor to this system, the Landsat Thematic Mapper (TM), collects data in seven spectral bands. The text contains a number of fine examples of Landsat images. (See Figures 14.22, 16.15, 17.15, 17.16, and 18.11.)

Another type of scanner provides images from the *thermal infrared* portion of the spectrum, which is sensitive to the temperature of the objects in the scene. Since warmer objects emit more infrared radiation than do cooler ones, the warmer objects will appear lighter on thermal infrared imagery. An example is shown in Figure 2.3. Here, a color spectrum has been used so that the warmest surfaces are shown in red, while cooler surfaces appear in blue, violet, or black.

Orbiting Earth Satellites

With the development of orbiting earth satellites carrying remote sensing systems, remote sensing has expanded into a major branch of geographic research, going far beyond the limitations of conventional aerial photography. One reason for this advance is that most satellite remote sensors are scanners that provide data instead of photographs. These data can be processed and enhanced by computers much more effectively than photographic images. Another reason is the ability of orbiting satellites to monitor nearly all the earth's surface. As a result, global and regional studies that could not have been carried out with air photographs have become possible.

How does a satellite sensor cover the entire earth? Nearly all satellites used for remote sensing are placed in a *sun-synchronous orbit.* In this orbit, all the satellite images of a location are acquired at about the same time of day. For the Landsat series, this is normally between 9:30 and 10:30 A.M. local time. This means that images collected on different dates can easily be compared, since they are all illuminated in a similar way by the sun. The orbit is also designed so that images are acquired from all locations on the globe before the orbit repeats. This can take from two to twenty-seven days, depending on how large an area is covered by each image.

Although in the past NASA has dominated the field of earth-observing satellites with its Landsat series, new satellites have been developed and launched by other nations. An example is the French satellite system, SPOT (an acronym for the French title *Système Probatoire d'Observation de la Terre*). The first SPOT satellite was launched in 1986, and the series continues to the present. Figure A2.5 shows a SPOT color image of the New York harbor area. The resolution, which is finer than that of the Landsat series, shows how much small detail can be seen with new-generation sensing systems.

Many technical advances will continue to be made in remote sensing. An international effort, led by NASA, is developing the *Earth Observing System,* which includes many satellites and sensing instruments to be used to study global change during the next few decades. Using these instruments, research in physical geography will be significantly enhanced. Remote sensing will be extended not only into remote areas of the earth that have never been mapped in detail, but also to global studies of the earth as a whole, utilizing the vantage point of earth orbit as never before in the history of science.

GEOGRAPHIC INFORMATION SYSTEMS

Another new area of technical development in the processing of spatial data is that of *geographic information systems.* What is a geographic information system, or *GIS?* In essence, it is a system for acquiring, processing, storing, querying, creating, and displaying spatial data.

A simple example of a GIS is a map overlay system. Imagine that a planner is deciding how to divide a tract

Figure A2.5 Digital image of New York Harbor, acquired by the SPOT satellite in June 1986. The data are presented as a color infrared picture. The tip of Manhattan is at the top center; the blocky, dark structures at the tip are Wall Street skyscrapers and their shadows. Just below the tip is Governor's Island, a Coast Guard military reservation, and immediately to the west (left) is small Liberty Island, home of the Statue of Liberty. The long container-loading piers of Port Newark are visible on the left, as are piers on the Brooklyn shoreline to the right. At the center bottom, the Verrazano Narrows bridge connects Brooklyn with Staten Island. The fine detail of the street grid pattern, and even of the wakes of ships in the harbor, shows the fine resolution of the SPOT's imaging instrument.

of land into building lots. Appropriate inputs would include a topographic map, giving contour lines of equal elevation; a vegetation map, showing the type of existing vegetation cover; a map of existing roads and trails; a map of streams and watercourses; a map of wetlands areas; a map of utility corridors crossing the area (power transmission lines and gas pipelines); and so forth. From these, the planner would locate the positions of roads and bridges; identify steep slopes, utility corridors, wetlands, and regions near streams as conservation areas; and then lay out individual lots having a minimum area of land suitable for building, access to roads and utilities, and other necessary attributes.

To do this, the planner needs to be able to overlay the various maps. Normally, this would be done by tracing them onto clear layers, then stacking the layers as needed while drawing the subdivision plans on the top layer. However, some problems may be encountered in this process. One common one is that the maps may not be drawn to the same scale or compiled on the same map projection. Another is that sometimes some maps need further work, as, for example, to identify areas of steep slopes given a map of topographic contours. Modern, computer-based geographic information systems are designed to solve these problems and allow ready manipulation of spatial data for diverse applications.

SPATIAL OBJECTS IN GEOGRAPHIC INFORMATION SYSTEMS

Geographic information systems are designed to manipulate spatial objects. A *spatial object* is a geographic area to which some information is attached. This information may be as simple as a place name or as complicated as a large data table with many types of information. A spatial object will normally have a boundary, which is described by a *polygon* that outlines the object. Some spatial objects are illustrated in Figure A2.6.

Spatial objects also include points and lines. A *point* can be thought of as a special type of spatial object with no area. A *line* is also a spatial object with no area, but it has two points associated with it, one for each end of the line. These special points are often referred to as *nodes*. Normally a line is straight, but it can also be defined as a smooth curve having a certain shape. If the two nodes marking the ends of the line are differentiated as starting and ending, then the line has a direction. If the line has a direction, then its two sides can be distinguished. This allows information to be attached to each side—for example, labels differentiating land from water. Lines connect to other lines when they share a common node. A series of connected lines that form a closed chain is a polygon.

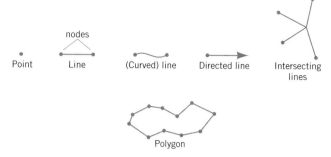

Figure A2.6 Some different types of spatial objects.

By defining spatial objects in this way, computer-based geographic information systems allow easy manipulation of the objects and permit many different types of operations to compare objects and generate new objects. As an example, suppose we have a GIS data layer composed of conservation land in a region represented as polygons and another layer containing the location of preexisting water wells as points within the region (Figure A2.7). It is very simple to use the GIS to identify the wells that are on conservation land. Or the conservation polygons containing wells may also be identified, and even output as a new data layer. By comparing the conservation layer with a road network layer portrayed as a series of lines, we could identify the conservation polygons containing roads. We could also compare the conservation layer to a layer of polygons showing vegetation type, and tabulate the amount of conservation land in forest, grassland, brush, and so forth. We could even calculate distance zones around a spatial object, for example, to create a map of buffer zones that are located within, say, 100 meters of conservation land. Many other possible manipulations exist.

Key Elements of a GIS

There are five elements of a geographic information system: data acquisition, preprocessing, data management, data manipulation and analysis, and product generation. Each is a component or process needed to ensure the functioning of the system as a whole. In the *data acquisition* process, data are gathered together for the particular application. These may include maps, air photos, tabular data, and other forms as well. In *preprocessing*, the assembled spatial data are converted to forms that can be ingested by the GIS to produce data layers of spatial objects and their associated information. The *data management* component creates, stores, retrieves, and modifies data layers and spatial objects. It is essential to proper functioning of all parts of the GIS. The *manipulation and analysis* component is the real

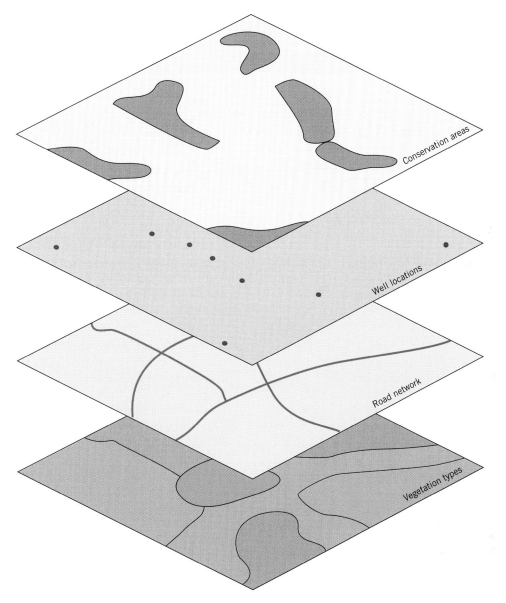

Figure A2.7 Data layers in a simple geographic information system.

workhorse of the GIS. Utilizing this component, the user asks and answers questions about spatial data and creates new data layers of derived information. The last component of the GIS, *product generation*, produces output products in the form of maps, graphics, tabulations, statistical reports, and the like that are the end products desired by the users. Taken together, these components provide a system that can serve many geographic applications at many scales.

Many new and exciting areas of geographic research are associated with geographic information systems, ranging from algorithms for manipulation of spatial data to the modeling of spatial processes using a GIS.

An especially interesting area is understanding how outputs are affected by errors and uncertainty in spatial data inputs, and how to communicate this information effectively to users.

Remote sensing and geographic information systems are two new, emerging fields of geographic research and application. Both rely heavily on computers and computer technology. Given the rate at which computers become ever more powerful as the technology improves, we can expect great strides in these fields in future years.

Appendix 3

The Canadian System of Soil Classification*

The formation of the National Soil Survey Committee of Canada in 1940 was a milestone in the development of soil classification and of pedology generally in Canada. Prior to that time, Canada used the 1938 USDA system. Although the Canadian experience showed that the concept of zonal soils was useful in the western plains, it proved less applicable in eastern Canada where parent materials and relief factors had a dominant influence on soil properties and development in many areas.

Canadian pedologists observed closely the evolution of the U.S. Comprehensive Soil Classification System (CSCS) during the 1950s and 1960s and ultimately adopted several important features of that system. Nevertheless, the special needs of a workable Canadian national system required that a completely independent classification system be established. Because Canada lies entirely in a latitude zone poleward of the 40th parallel, there was no need to incorporate those soil orders found only in lower latitudes. Furthermore, the vast expanse of Canadian territory lying within the boreal forest and tundra climates necessitated the recognition at the highest taxonomic level (the order) of soils of cold regions that appear only as suborders and even as great groups in the CSCS—for example, the Cryaquents and Cryorthents that occupy much of the soils map of northern Canada above the 50th parallel (Figure 19.13).

The overall philosophy of the Canadian system is pragmatic: the aim is to organize knowledge of soils in a reasonable and usable way. The system is a natural or taxonomic one in which the classes (taxa) are based on properties of the soils themselves and not on interpretations of the soils for various uses. Thus, the taxa are concepts based on generalization of properties of real bodies of soils rather than idealized concepts of the kinds of soils that would result from the action of presumed genetic processes. In this respect, the philosophy agrees with that used in the CSCS. Although taxa in the Canadian system are defined on the basis of actual soil properties that can be observed and measured, the system has a genetic bias in that properties or combinations of properties that reflect genesis are favored in distinguishing among the higher taxa. Thus, the soils brought together under a single soil order are seen as the product of a similar set of dominant soil-forming processes resulting from broadly similar climatic conditions.

The Canadian system recognizes the *pedon* as the basic unit of soils; it is defined as in the CSCS. Major mineral horizons of the soil (A, B, C) are defined in much the same way as in the U.S. system. Thus, the Canadian system of soil taxonomy is more closely related to the U.S. system than to any other. Both are hierarchical, and the taxa are defined on the basis of measurable soil properties. However, they differ in several respects. The Canadian system is designed to classify only the soils that occur in Canada and is not a comprehensive system. The U.S. system includes the suborder, a taxon not recognized in the Canadian system. Because 90 percent of the area of Canada is not likely to be cultivated, the Canadian system does not recognize as diagnostic those horizons strongly affected by plowing and application of soil conditioners and fertilizers.

*Thoughout this section numerous sentences and phrases are taken verbatim or paraphrased from the following work: Canada Soil Survey Committee, *The Canadian System of Soil Classification*, Research Branch, Canada Department of Agriculture, Publication 1646, 1978. Table A3.1 and Figure A3.2 are also compiled from this source.

Table A3.1   Subhorizons and Organic Horizons of the Canadian System of Soil Classification

Subhorizons; Lowercase Suffixes

b   Buried soil horizon.

c   Cemented (irreversible) pedogenic horizon.

ca   Horizon of secondary carbonate enrichment in which the concentration of lime exceeds that in the unenriched parent material.

e   Horizon characterized by the eluviation of clay, Fe, Al, or organic matter alone or in combination.

f   Horizon enriched with amorphous material, principally Al and Fe combined with organic matter; reddish near upper boundary, becoming yellower at depth.

g   Horizon characterized by gray colors, or prominent mottling, or both, indicative of permanent or intense reduction.

h   Horizon enriched with organic matter.

j   Used as a modifier of suffixes e, f, g, n, and t to denote an expression of, but failure to meet, the specified limits of a suffix it modifies.

k   Denotes presence of carbonate as indicated by visible effervescence when dilute HCl is added.

m   Horizon slightly altered by hydrolysis, oxidation, or solution, or all three to give a change in color or structure, or both.

n   Horizon in which the ratio of exchangeable Ca to exchangeable Na is 10 or less. It must also have following distinctive morphological characteristics: prismatic or columnar structure, dark coatings on ped surfaces, and hard to very hard consistence when dry.

p   Horizon disturbed by human activities such as cultivation, logging, and habitation.

s   Horizon of salts, including gypsum, which may be detected as crystals or veins or as surface crusts of salt crystals.

sa   Horizon with secondary enrichment of salts more soluble than Ca and Mg carbonates; the concentration of salts exceeds that in the unenriched parent material.

t   Illuvial horizon enriched with silicate clay.

u   Horizon that is markedly disrupted by physical or faunal processes other than cryoturbation.

x   Horizon of fragipan character. (A fragipan is a loamy subsurface horizon of high bulk density and very low organic matter content. When dry it is hard and seems to be cemented.)

y   Horizon affected by cryoturbation as manifested by disrupted or broken horizons, incorporation of materials from other horizons, and mechanical sorting.

z   A frozen layer.

Organic Horizons

O   Organic horizon developed mainly from mosses, rushes, and woody materials.

L   Organic horizon characterized by an accumulation of organic matter derived mainly from leaves, twigs, and woody materials in which the organic structures are easily discernible.

F   Same as L, above, except that original structures are difficult to recognize.

H   Organic horizon characterized by decomposed organic matter in which the original structures are indiscernible.

SOIL HORIZONS AND OTHER LAYERS

The definitions of classes in the Canadian system are based mainly on kinds, degrees of development, and sequence of soil horizons and other layers in pedons. The major mineral horizons are A, B, and C. The major organic horizons are L, F, and H, which are mainly forest litter at various stages of decomposition, and O, which is derived mainly from bog, marsh, or swamp vegetation. Subdivisions of horizons are labeled by adding lowercase suffixes to the major horizon symbols—for example, Ah or Ae.

Besides the horizons, nonsoil layers are recognized. Two such layers are R, rock, and W, water. Lower mineral layers not affected by pedogenic processes are also identified. In organic soils, layers are described as tiers.

The principal mineral horizons, A, B, and C, are defined as follows:

A   Mineral horizon found at or near the surface in the zone of leaching or eluviation of materials in solution or suspension, or of maximum *in situ* accumulation of organic matter or both.

B   Mineral horizon characterized by enrichment in organic matter, sesquioxides, or clay; or by the development of soil structure; or by change of color denoting hydrolysis, reduction, or oxidation.

C   Mineral horizon comparatively unaffected by the pedogenic processes operative in A and B horizons. The processes of gleying and the accumulation of calcium and magnesium and more soluble salts can occur in this horizon.

Lowercase suffixes, used to designate subdivisions of horizons, are shown in Table A3.1.

SOIL ORDERS OF THE CANADIAN SYSTEM

Nine soil orders make up the highest taxon of the Canadian System of Soil Classification. Listed in alphabetical order, they are as follows.

Brunisolic	Gleysolic	Podzolic
Chernozemic	Luvisolic	Regosolic
Cryosolic	Organic	Solonetzic

Table A3.2 lists the great groups within each order.

Brunisolic Order

The central concept of the *Brunisolic order* is that of soils under forest having brownish-colored Bm horizons. Most Brunisolic soils are well to imperfectly drained. They occur in a wide range of climatic and vegetative environments, including boreal forest; mixed forest, shrubs, and grass; and heath and tundra. As compared

Table A3.2 Great Groups of the Canadian Soil Classification System

Order	Great Group
Brunisolic	Melanic Brunisol Eutric Brunisol Sombric Brunisol Dystric Brunisol
Chernozemic	Brown Dark Brown Black Dark Gray
Cryosolic	Turbic Cryosol Static Cryosol Organic Cryosol
Gleysolic	Humic Gleysol Gleysol Luvic Gleysol
Luvisolic	Gray Brown Luvisol Gray Luvisol
Organic	Fibrisol Mesisol Humisol Folisol
Podzolic	Humic Podzol Ferro-Humic Podzol Humo-Ferric Podzol
Regosolic	Regosol Humic Regosol
Solonetzic	Solentz Solodized Solonetz Solod

with the Chernozemic soils, the Brunisolic soils show a weak B horizon of accumulation attributable to their moister environment. Brunisolic soils lack the diagnostic podzolic B horizon of the Podzolic soils, in which accumulation in the B horizon is strongly developed. The Melanic Brunisol shown in profile Figure A3.1 can be found in the St. Lawrence Lowlands, surrounded by Podzolic soils (Figure A3.2).

Chernozemic Order

The general concept of the *Chernozemic order* is that of well to imperfectly drained soils having surface horizons darkened by the accumulation of organic matter from the decomposition of xerophytic or mesophytic grasses and forms representative of grassland communities or of grassland-forest communities with associated shrubs and forbs. The major area of Chernozemic soils is the cool, subarid to subhumid interior plains of western Canada. Most Chernozemic soils are frozen during some period each winter, and the soil is dry at some period each summer. The mean annual temperature is higher than O°C and usually less than 5.5°C. The associated climate is typically the semiarid (steppe) variety of the dry midlatitude climate ⑨ (Figure A3.2).

Essential to the definition of soils of the Chernozemic order is that they must have an A horizon (typically, Ah) in which organic matter has accumulated, and they must meet several other requirements. The A horizon is at least 10 cm thick; its color is dark brown to black. It usually has sufficiently good structure that it is neither massive and hard nor single-grained when dry. The profile shown in Figure A3.1 is that of the Orthic subgroup of the Brown great group; it shows a Bm horizon that is typically of prismatic structure. The C horizon is one of lime accumulation (Cca). Clearly, the Chernozemic soils can be closely correlated with the Mollisols of the CSCS.

Cryosolic Order

Soils of the *Cryosolic order* occupy much of the northern third of Canada where permafrost remains close to the surface of both mineral and organic deposits. Cryosolic soils predominate north of the tree line, are common in the subarctic forest area in fine-textured soils, and extend into the boreal forest in some organic materials and into some alpine areas of mountainous regions. Cryoturbation (intense disturbance by freeze–thaw activity) of these soils is common, and it may be indicated by patterned ground features such as sorted and nonsorted nets, circles, polygons, stripes, and earth hummocks.

Cryosolic soils are found in either mineral or organic materials that have permafrost either within 1 m of the

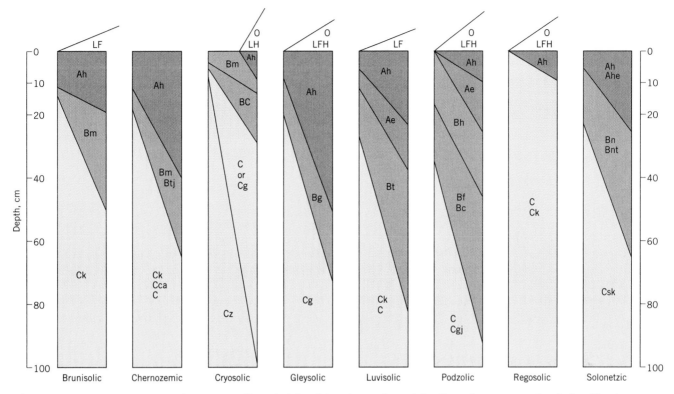

Figure A3.1 Representative schematic profiles of eight of the nine orders of the Canadian system of soil classification. Slanting lines show the range in depth and thickness of each horizon. (The horizon planes are actually approximately horizontal within the pedon.) See Table A3.1 for an explanation of symbols. (From Canada Soil Survey Committee, Research Branch, Canada Department of Agriculture, 1978.)

surface or within 2 m if more than one-third of the pedon has been strongly cryoturbated, as indicated by disrupted, mixed, or broken horizons. The profile shown in Figure A3.1 is that of the Orthic subgroup of the Static Cryosol great group. Note the presence of organic L, H, and O surface horizons and the thin Ah horizon. The Cryosolic soils are closely correlated with the Cryaquepts of the CSCS.

Gleysolic Order

Soils of the *Gleysolic order* have features indicating periodic or prolonged saturation with water and reducing conditions. They commonly occur in patchy association with other soils in the landscape. Gleysolic soils are usually associated with either a high groundwater table at some period of the year or temporary saturation above a relatively impermeable layer. Some Gleysolic soils may be submerged under shallow water throughout the year. The profile shown in Figure A3.1 is that of the Gleysol great group. It has a thick Ah horizon. The underlying Bg horizon is grayish and shows mottling typical of reducing conditions.

Luvisolic Order

Soils of the *Luvisolic order* generally have light-colored, eluvial horizons (Ae), and they have illuvial B horizons (Bt) in which silicate clay has accumulated. These soils develop characteristically in well to imperfectly drained sites, in sandy loam to clay base-saturated parent materials under frost vegetation in subhumid to humid, mild to very cold climates. The genesis of Luvisolic soils is thought to involve the suspension of clay in the soil solution near the soil surface, downward movement of the suspended clay with the soil solution, and deposition of the translocated clay at a depth where downward motion of the soil solution ceases or becomes very slow. The representative profile shown in Figure A3.1 is that of the Orthic subgroup of the Gray Brown Luvisol great group.

Luvisolic soils occur from the southern extremity of Ontario to the zone of permafrost and from Newfoundland to British Columbia. The largest area of these soils are Gray Luvisols occurring in the central to northern interior plains under deciduous, mixed, and coniferous forest. In this location they appear to corre-

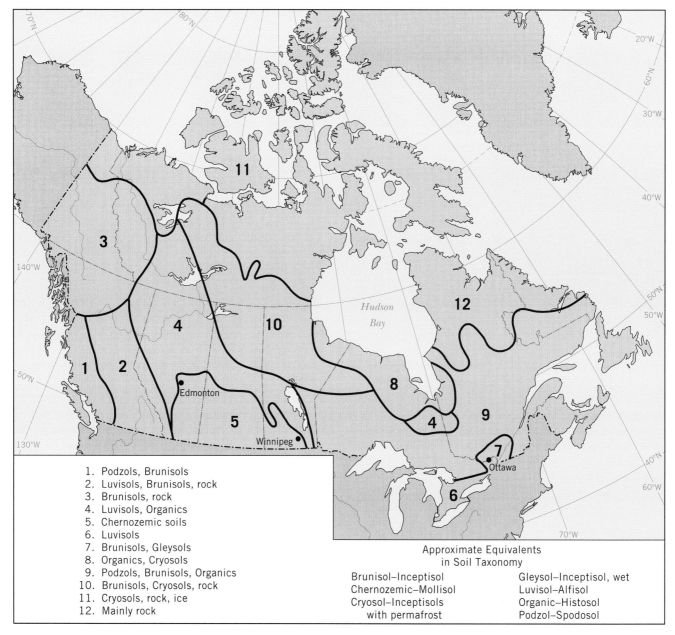

1. Podzols, Brunisols
2. Luvisols, Brunisols, rock
3. Brunisols, rock
4. Luvisols, Organics
5. Chernozemic soils
6. Luvisols
7. Brunisols, Gleysols
8. Organics, Cryosols
9. Podzols, Brunisols, Organics
10. Brunisols, Cryosols, rock
11. Cryosols, rock, ice
12. Mainly rock

Approximate Equivalents
in Soil Taxonomy

Brunisol–Inceptisol	Gleysol–Inceptisol, wet
Chernozemic–Mollisol	Luvisol–Alfisol
Cryosol–Inceptisols	Organic–Histosol
with permafrost	Podzol–Spodosol

Figure A3.2   Generalized map of soil regions of Canada. (Courtesy of Land Resources Research Institute, Agriculture Canada.) (Illustration is taken from *Fundamentals of Soil Science*, 7th ed., by Henry D. Foth, John Wiley & Sons.)

late with the Boralfs of the Alfisol order in the CSCS. Gray-Brown Luvisolic soils of southern Ontario would correlate with the suborder of Udalfs.

Organic Order

Soils of the *Organic order* are composed largely of organic materials. They include most of the soils commonly known as peat, muck, or bog soils. Organic soils contain 17 percent or more organic carbon (30 percent organic matter) by weight. Most Organic soils are saturated with water for prolonged periods. They occur widely in poorly and very poorly drained depressions and level areas in regions of subhumid to humid climate and are derived from vegetation that grows in such sites. However, one group of Organic soils consists of leaf litter overlying rock or fragmental material; soils of this group may occur on steep slopes and may rarely be saturated with water. (No profile of the Organic soils is shown in Figure A3.1.) Organic soils can be correlated with the Histosols of the CSCS.

Podzolic Order

Soils of the *Podzolic order* have B horizons in which the dominant accumulation product is amorphous material composed mainly of humified organic matter in varying degrees with Al and Fe. Typically, Podzolic soils occur in coarse- to medium-textured, acid parent materials, under forest and heath vegetation in cool to very cold humid to very humid climates. Podzolic soils can usually be readily recognized in the field. Generally, they have organic surface horizons that are commonly L, F, and H. Most Podzolic soils have a reddish brown to black B horizon (Bh) with an abrupt upper boundary. The profile shown in Figure A3.1 is that of the Orthic subgroup of the Humic Podzol great group.

The Podzolic soils correspond closely to the Spodosols (Orthods) of the CSCS.

Regosolic Order

Regosolic soils have weakly developed horizons. The lack of development of genetic horizons may be due to any number of factors: youthfulness of the parent material, for example, recent alluvium; instability of the material, for example, colluvium on slopes subject to mass wasting; nature of the material, for example, nearly pure quartz sand; climate, for example, dry cold conditions. Regosolic soils are generally rapidly to imperfectly drained. They occur in a wide range of vegetation and climates. The profile shown in Figure A3.1 is that of the Orthic subgroup of the Regosol great group. It has only a thin humic A horizon (Ah) and a surface horizon of organic materials.

Regosolic soils correspond with the Entisols of the CSCS.

Solonetzic Order

Soils of the *Solonetzic order* have B horizons that are very hard when dry and swell to a sticky mass of very low permeability when wet. Typically, the Solonetzic B horizon has prismatic or columnar macrostructure that breaks into hard to extremely hard, blocky peds with dark coatings. Solonetzic soils occur on saline parent materials in some areas of the semiarid to subhumid interior plains in association with Chernozemic soils and to a lesser extent with Luvisolic and Gleysolic soils. Most Solonetzic soils are associated with a vegetative cover of grasses and forbs. The profile shown in Figure A3.1 is that of the Brown subgroup of the Solonetz great group.

Solonetzic soils are thought to have developed from parent materials that were more or less uniformly salinized with salts high in sodium. Leaching of salts by descending rainwater presumably mobilizes the sodium-saturated colloids. The colloids are apparently carried downward and deposited in the B horizon. Further leaching results in depletion of alkali cations in the A horizon, which becomes acidic, and a platy Ahe horizon usually develops. The underlying Solonetzic B horizon (Bn, Bnt) usually consists of darkly stained, fused, intact columnar peds. This stage is followed by the structural breakdown of the upper part of the B horizon and eventually its complete destruction in the most advanced stage, known as solodization. Solonetzic soils are correlated with the suborder of Argids in the order of Aridisols under the CSCS.

Appendix 4

Climate Definitions and Boundaries

The following table summarizes the definitions and boundaries of climates and climate subtypes based on the soil-water balance, as described in Chapter 14 and shown on the world climate map, Figure 7.9. All definitions and boundaries are provisional.

Group I: Low-Latitude Climates

1. Wet equatorial climate ①
Ep ≥ 10 cm in every month, and
S ≥ 20 cm in 10 or more months.

2. Monsoon and trade-wind coastal climate ②
Ep ≥ 4 cm in every month, or
Ep > 130 cm annual total, or both, and
S ≥ 20 cm in 6, 7, 8, or 9 consecutive
    months, or, if
S > 20 cm in 10 or more months, then Ep ≤
    10 cm in 5 or more consecutive months.

3. Wet-dry tropical climate ③
D ≥ 20cm, and
R ≥ 10 cm, and
Ep ≥ 130 cm annual total, or Ep ≥ 4 cm in
    every month, or both,
    and
S ≥ 20 cm in 5 months or fewer, or mini-
    mum monthly S < 3 cm.

4. Dry tropical climate ④
D ≥ 15 cm, and
R = 0, and
Ep ≥ 130 cm annual total, or Ep ≥ 4 cm in
    every month, or both.
Subtypes of dry climates (④, ⑤, ⑦, and ⑨)

s Semiarid subtype (Steppe subtype)
    At least 1 month with S > 2 cm.
a Desert subtype
    No month with S > 2 cm.

Group II: Midlatitude Climates

5. Dry subtropical climate ⑤
D ≥ 15 cm, and
R = 0, and Ep < 130 cm annual total, and
Ep ≥ 0.8 cm in every month, and
Ep < 4 cm in 1 month.
(Subtypes ⑤a and ⑤s as defined under ④.)

6. Moist subtropical climate ⑥
D < 15 cm when R = 0, and
Ep < 4 cm in at least 1 month, and
Ep ≥ 0.8 cm in every month.

7. Mediterranean climate ⑦
D ≥ 15 cm, and
R ≥ 0, and
Ep ≥ 0.8 cm in every month, and storage index >
    75%, or P/Ea × 100 < 40%.
(Subtypes ⑦a and ⑦s as defined under ④.)

8. Marine west-coast climate ⑧
D < 15 cm, and
Ep < 80 cm annual total, and
Ep ≥ 0.8 cm in every month.

9. Dry midlatitude climate ⑨
D ≥ 15 cm, and
R = 0, and
Ep ≤ 0.7 cm in at least 1 month, and
Ep > 52.5 cm annual total.
(Subtypes ⑨a and ⑨s as defined under ④.)

10. Moist continental climate ⑩

D < 15 cm when R = 0, and

Ep ≤ 0.7 cm in at least 1 month, and

Ep > 52.5 cm annual total.

Group III: High-Latitude Climates

11. Boreal forest climate ⑪

52.5 cm > Ep > 35 cm annual total, and

Ep = 0 in fewer than 8 consecutive months.

12. Tundra climate ⑫

Ep < 35 cm annual total, and

Ep = 0 in 8 or more consecutive months.

13. Ice-sheet climate ⑬

Ep = 0 in all months.

GLOSSARY

This glossary contains definitions of terms shown in the text in italics or boldface. Terms that are *italicized* within the definitions will be found as individual entries elsewhere in the glossary.

A horizon   *mineral* horizon of the *soil*, overlying the *E* and *B horizons*.

ablation   a wastage of glacial ice by both *melting* and *evaporation*.

abrasion   erosion of *bedrock* of a *stream channel* by impact of particles carried in a stream and by rolling of larger *rock* fragments over the stream bed; abrasion is also an activity of glacial ice, waves, and wind.

abrasion platform   sloping, nearly flat *bedrock* surface extending out from the foot of a *marine cliff* under the shallow water of breaker zone.

absorption of radiation   transfer of *electromagnetic energy* into heat *energy* within a *gas* or *liquid* through which the radiation is passing or at the surface of a *solid* struck by the radiation.

abyssal plain   large expanse of very smooth, flat ocean floor found at depths of 4600 to 5500 m (15,000 to 18,000 ft).

accelerated erosion   *soil erosion* occurring at a rate much faster than *soil horizons* can be formed from the parent *regolith*.

accretion of lithosphere   production of new *oceanic lithosphere* at an active *spreading plate boundary* by the rise and solidification of *magma* of basaltic composition.

accretionary prism   mass of deformed trench *sediments* and ocean floor *sediments* accumulated in wedgelike slices on the underside of the overlying plate above a plate undergoing *subduction*.

acid deposition   the *deposition* of acid raindrops or dry acidic dust particles on vegetation and ground surfaces.

acid mine drainage   sulfuric acid effluent from *coal* mines, mine tailings, or spoil ridges made by *strip mining*.

acid rain   rainwater having an abnormally low *pH*, between 2 and 5, as a result of air pollution by sulfur oxides and nitrogen oxides.

active continental margins   continental margins that coincide with tectonically active plate boundaries. (See also *continental margins, passive continental margins*.)

active layer   shallow surface layer subject to seasonal thawing in *permafrost* regions.

active pool   type of pool in the *biogeochemical cycle* in which the materials are in forms and places easily accessible to life processes. (See also *storage pool*.)

active systems   *remote sensing* systems that emit a beam of wave *energy* at a source and measure the intensity of that *energy* reflected back to the source.

actual evapotranspiration (water use)   actual rate of *evapotranspiration* at a given time and place.

adiabatic lapse rate   (See *dry adiabatic lapse rate, wet adiabatic lapse rate*.)

adiabatic process   change of temperature within a *gas* because of compression or expansion, without gain or loss of heat from the outside.

advection fog   *fog* produced by *condensation* within a moist basal air layer moving over a cold land or water surface.

aerosols   tiny dust particles present in the *atmosphere*, so small and light that the slightest movements of air keep them aloft.

aggradation   raising of *stream channel* altitude by continued *deposition* of *bed load*.

air   a mixture of gases that surrounds the earth.

air mass   extensive body of air within which upward gradients of temperature and moisture are fairly uniform over a large area.

air pollutant   an unwanted substance injected into the *atmosphere* from the earth's surface by either natural or human activities; includes *aerosols, gases*, and *particulates*.

air temperature   temperature of air, normally observed by a *thermometer* under standard conditions of shelter and height above the ground.

albedo   percentage of *electromagnetic energy* reflected from a surface.

albic horizon   pale, often sandy *soil horizon* from which *clay* and free iron oxides have been removed. Found in the profile of the *Spodosols*.

Alfisols   *soil order* consisting of *soils* of humid and subhumid climates, with high *base status* and an *argillic horizon*.

alluvial fan   gently sloping, conical accumulation of coarse *alluvium* deposited by a *braided stream* undergoing *aggradation* below the point of emergence of the channel from a narrow *gorge* or *canyon*.

alluvial meanders   sinuous bends of a *graded stream* flowing in the alluvial deposit of a *floodplain*.

alluvial river   *stream* of low *gradient* flowing upon thick deposits of *alluvium* and experiencing approximately annual overbank flooding of the adjacent *floodplain*.

alluvial terrace   benchlike landform carved in *alluvium* by a *stream* during *degradation*.

alluvium   any stream-laid *sediment* deposit found in a *stream channel* and in low parts of a stream valley subject to flooding.

alpine chains   high mountain ranges that are narrow belts of *tectonic activity* severely deformed by *folding* and thrusting in comparatively recent geologic time.

alpine glacier   long, narrow, mountain *glacier* on a steep downgrade, occupying the floor of a troughlike valley.

alpine permafrost   *permafrost* occurring at high altitudes equatorward of the normal limit of *permafrost*.

alpine tundra   a plant *formation class* within the *tundra biome*, found at high altitudes above the limit of *tree* growth.

amphibole group   *silicate minerals* rich in calcium, magnesium, and iron, dark in color, high in *density*, and classed as *mafic minerals*.

andesite   *extrusive igneous rock* of diorite composition, dominated by *plagioclase feldspar*; the extrusive equivalent of *diorite*.

Andisols a *soil order* that includes *soils* formed on volcanic ash; often enriched by organic matter, yielding a dark soil color.

anemometer weather instrument used to indicate *wind* speed.

aneroid barometer *barometer* using a mechanism consisting of a partially evacuated air chamber and a flexible diaphragm.

angle of repose natural surface inclination (*dip*) of a *slope* consisting of loose, coarse, well-sorted *rock* or *mineral* fragments; for example, the *slip face* of a *sand dune*, a *talus slope*, or the sides of a *cinder cone*.

annuals plants that live only a single growing season, passing the unfavorable season as a seed or spore.

annular drainage pattern a stream network dominated by concentric (ringlike) major *subsequent streams*.

antarctic circle *parallel of latitude* at 66 1/2°S.

antarctic front zone frontal zone of interaction between antarctic *air masses* and polar air masses.

antarctic zone *latitude* zone in the latitude range 60° to 75°S (more or less), centered on the *antarctic circle*, and lying between the *subantarctic zone* and the *polar zone*.

anticlinal valley valley eroded in weak *strata* along the central line or axis of an eroded *anticline*.

anticline upfold of *strata* or other layered *rock* in an archlike structure; a class of *folds*. (See also *syncline*.)

anticyclone center of high *atmospheric pressure*.

aphelion point on the earth's elliptical orbit at which the earth is farthest from the sun.

aquatic ecosystem *ecosystem* of a *lake*, bog, pond, river, *estuary*, or other body of water.

Aquepts *suborder* of the *soil order Inceptisols*; includes Inceptisols of wet places, seasonally saturated with water.

aquiclude *rock* mass or layer that impedes or prevents the movement of *groundwater*.

aquifer *rock* mass or layer that readily transmits and holds *groundwater*.

arc curved line that forms a portion of a circle.

arc-continent collision collision of a volcanic arc with *continental lithosphere* along a *subduction* boundary.

arctic circle *parallel of latitude* at 66 1/2°N.

arctic front zone frontal zone of interaction between arctic *air masses* and polar air masses.

arctic tundra a plant *formation class* within the *tundra biome*, consisting of low, mostly herbaceous plants, but with some very small stunted *trees*, associated with the *tundra climate* ⑫.

arctic zone *latitude* zone in the latitude range 60° to 75° N (more or less), centered about on the *arctic circle*, and lying between the *subarctic zone* and the *polar zone*.

arête sharp, knifelike divide or crest formed between two *cirques* by alpine glaciation.

argillic horizon *soil horizon*, usually the *B horizon*, in which *clay minerals* have accumulated by *illuviation*.

arid (dry climate subtype) subtype of the dry climates that is extremely dry and supports little or no vegetation cover.

Aridisols *soil order* consisting of soils of dry climates, with or without *argillic horizons*, and with accumulations of *carbonates* or soluble salts.

artesian well drilled well in which water rises under hydraulic pressure above the level of the surrounding *water table* and may reach the surface.

aspect compass orientation of a *slope* as an inclined element of the ground surface.

asthenosphere soft layer of the upper *mantle*, beneath the rigid *lithosphere*.

astronomical hypothesis explanation for glaciations and interglaciations making use of cyclic variations in the form of solar *energy* received at the earth's surface.

atmosphere envelope of gases surrounding the earth, held by *gravity*.

atmospheric pressure pressure exerted by the atmosphere because of the force of *gravity* acting on the overlying column of air.

atoll circular or closed-loop *coral reef* enclosing an open *lagoon* with no island inside.

atomic mass number total number of protons and *neutrons* within the nucleus of an atom.

atomic number number of protons within the nucleus of an atom; determines element name and chemical properties of the atom.

autumnal equinox *equinox* occurring on September 22 or 23.

axial rift narrow, trenchlike depression situated along the center line of the *mid-oceanic ridge* and identified with active seafloor spreading.

axis of rotation center line around which a body revolves, as the earth's axis of rotation.

B horizon mineral *soil horizon* located beneath the A *horizon*, and usually characterized by a gain of *mineral matter* (such as *clay minerals* and oxides of aluminum and iron) and organic matter (*humus*).

backswamp area of low, swampy ground on the *floodplain* of an *alluvial river* between the *natural levee* and the *bluffs*.

backwash return flow of *swash* water under the influence of *gravity*.

badlands rugged land surface of steep *slopes*, resembling miniature mountains, developed on weak *clay* formations or clay-rich *regolith* by fluvial erosion too rapid to permit plant growth and soil formation.

bar low ridge of *sand* built above water level across the mouth of a *bay* or in shallow water paralleling the shoreline. May also refer to embankment of sand or gravel on floor of a *stream channel*.

bar (pressure) unit of pressure or *capillary tension* equal to 1 million dynes per sq cm; it is approximately equal to the pressure of the earth's *atmosphere* at sea level.

barchan dune *sand dune* of crescentic base outline with a sharp crest and a steep lee *slip face*, with crescent points (horns) pointing downwind.

barometer instrument for measurement of *atmospheric pressure*.

barrier island long, narrow island, built largely of beach *sand* and dune sand, parallel with the mainland and separated from it by a *lagoon*.

barrier reef *coral reef* separated from mainland *shoreline* by a *lagoon*.

barrier-island coast *coastline* with broad zone of shallow water offshore (a *lagoon*) shut off from the ocean by a *barrier island*.

basalt *extrusive igneous rock* of *gabbro* composition; occurs as *lava*.

base flow that portion of the *discharge* of a *stream* contributed by *groundwater* seepage.

base level lower limiting surface or level that can ultimately be attained by a *stream* under conditions of stability of the earth's crust and sea level; an imaginary surface equivalent to sea level projected inland.

base status of soils quality of a *soil* as measured by the presence or absence of *clay minerals* capable of holding large numbers of *bases*. Soils of high *base status* are rich in base-holding *clay minerals*; soils of low *base status* are deficient in such minerals.

bases certain positively charged *ions* in the *soil* that are also plant nutrients; the most important are calcium, magnesium, potassium, and sodium.

batholith large, deep-seated body of *intrusive igneous rock*, usually with an area of surface exposure greater than 100 km² (40 mi²).

bauxite mixture of several *clay minerals*,

consisting largely of aluminum oxide and water with impurities; a principal ore of aluminum.

bay   a body of water sheltered from strong wave action by the configuration of the *coast*.

beach   thick, wedge-shaped accumulation of *sand*, *gravel*, or cobbles in the zone of breaking waves.

beach drift   transport of *sand* on a beach parallel with a *shoreline* by a succession of landward and seaward water movements at times when *swash* approaches obliquely.

bearing   direction angle between a line of interest and a reference line, which is usually a line pointing north.

bed load   that portion of the *stream load* moving close to the stream bed by rolling and sliding.

bedrock   solid *rock* in place with respect to the surrounding and underlying rock and relatively unchanged by *weathering* processes.

biogas   mixture of methane and *carbon dioxide* generated by action of anaerobic bacteria in animal and human wastes enclosed in a digesting chamber.

biogeochemical cycle   total system of *pathways* by which a particular type of matter (a given element, compound, or ion, for example) moves through the earth's *ecosystem* or *biosphere*; also called a *material cycle* or *nutrient cycle*.

biomass   dry weight of living organic matter in an *ecosystem* within a designated surface area; units are kilograms of organic matter per square meter.

biome   largest recognizable subdivision of *terrestrial ecosystems*, including the total assemblage of plant and animal life interacting within the *life layer*.

biosphere   all living organisms of the earth and the environments with which they interact.

bitumen   combustible mixture of hydrocarbons that is highly viscous and will flow only when heated; considered a form of petroleum.

bituminous sand   (See *bitumen*.)

blackbody   ideal object or surface that is a perfect radiator and absorber of *energy*; absorbs all radiation it intercepts and emits radiation perfectly according to physical theory.

block mountains   class of mountains produced by block faulting and usually bounded by *normal faults*.

block separation   separation of individual joint blocks during the process of *physical weathering*.

blowout   shallow depression produced by continued *deflation*.

bluffs   steeply rising ground slopes marking the outer limits of a *floodplain*.

bog   a shallow depression filled with organic matter, for example, a glacial lake or pond basin filled with *peat*.

Boralfs   *suborder* of the *soil order Alfisols*; includes Alfisols of *boreal forests* or high mountains.

boreal forest   variety of *needleleaf forest* found in the *boreal forest climate* ⑪ regions of North America and Eurasia.

boreal forest climate ⑪   cold climate of the *subarctic zone* in the northern *hemisphere* with long, extremely severe winters and several consecutive months of zero *potential evapotranspiration (water need)*.

Borolls   *suborder* of the *soil order Mollisols*; includes Mollisols of cold-winter semiarid plants (*steppes*) or high mountains.

braided stream   *stream* with shallow channel in coarse *alluvium* carrying multiple threads of fast flow that subdivide and rejoin repeatedly and continually shift in position.

breaker   sudden collapse of a steepened water wave as it approaches the *shoreline*.

broadleaf deciduous forest   *forest* type consisting of broadleaf *deciduous trees* and found in the *moist subtropical climate* ⑥ and in parts of the *marine west-coast climate* ⑧. (See also *midlatitude deciduous forest*.)

broadleaf evergreen forest   *forest* type consisting of broadleaf *evergreen trees* and found in the wet equatorial and tropical climates. (See also *low-latitude rainforest*.)

Brunisolic order   a class of *forest soils* in the Canadian soil classification system with brownish *B horizon*.

budget   in flow systems, an accounting of *energy* and matter flows that enter, move within, and leave a system.

bush-fallow farming   agricultural system practiced in the African *savanna woodland* in which *trees* are cut and burned to provide cultivation plots.

butte   prominent, steep-sided hill or peak, often representing the final remnant of a resistant layer in a region of flat-lying *strata*.

C horizon   *soil horizon* lying beneath the *soil solum (A, E, and B horizons)*; it is a layer of *sediment* or *regolith* that is the *parent material* of the solum.

calcification   accumulation of *calcium carbonate* in a soil, usually occurring in the *B horizon*, or in the *C horizon* below the *soil solum*.

calcite   mineral having the composition *calcium carbonate*.

calcium carbonate   compound consisting of calcium (Ca) and carbonate (CO_3)

ions, formula $CaCO_3$, occurring naturally as the mineral *calcite*.

caldera   large, steep-sided circular depression resulting from the explosion and subsidence of a *stratovolcano*.

canyon   (See *gorge*.)

capillary tension   a cohesive force among surface molecules of a *liquid* that gives a droplet its rounded shape.

carbohydrate   class of organic compounds consisting of the elements carbon, hydrogen and oxygen.

carbon cycle   *biogeochemical cycle* in which carbon moves through the *biosphere*; includes both *gaseous cycles* and *sedimentary cycles*.

carbon dioxide   the chemical compound CO_2, formed by the union of two atoms of oxygen and one atom of carbon; normally a gas present in low concentration in the *atmosphere*.

carbon fixation   (See *photosynthesis*.)

carbonates (carbonate minerals, carbonate rocks)   *minerals* that are carbonate compounds of calcium or magnesium or both, that is, *calcium carbonate* or magnesium carbonate. (See also *calcite*.)

carbonic acid   a weak acid created when CO_2 gas dissolves in water.

carbonic acid action   chemical reaction of *carbonic acid* in rainwater, *soil water*, and *groundwater* with *minerals*; most strongly affects carbonate minerals and *rocks*, such as limestone and marble; an activity of *chemical weathering*.

cartography   the science and art of making maps.

Celsius scale   temperature scale in which the *freezing* point of water is 0°, the boiling point 100°.

Cenozoic Era   last (youngest) of the *eras* of geologic time.

channel   (See *stream channel*.)

chaparral   sclerophyll scrub and dwarf *forest* plant *formation class* found throughout the coastal mountain ranges and hills of central and southern California.

chemical energy   *energy* stored within an organic molecule and capable of being transformed into *heat* during metabolism.

chemical weathering   chemical change in *rock*-forming minerals through exposure to atmospheric conditions in the presence of water; mainly involving *oxidation*, *hydrolysis*, *carbonic acid action*, or direct solution.

chemically precipitated sediment   *sediment* consisting of *mineral matter* precipitated from a water solution in which the matter has been transported in the dissolved state as *ions*.

chernozem   type of *soil order* closely equivalent to *Mollisol*; an order of the Canadian Soil Classification System.

Chernozemic order   a class of grassland *soils* in the Canadian soil classification system with a thick *A horizon* rich in organic matter.

chert   *sedimentary rock* composed largely of silicon dioxide and various impurities, in form of nodules and layers, often occurring with *limestone* layers.

chinook wind   a *local wind* occurring at certain times to the lee of the Rocky Mountains; a very dry wind with a high capacity to evaporate *snow*.

chlorofluorocarbons (CFCs)   synthetic chemical compounds containing chlorine, fluorine, and carbon atoms that are widely used as coolant fluids in refrigeration systems.

cinder cone   conical hill built of coarse *tephra* ejected from a narrow volcanic vent; a type of *volcano*.

circle of illumination   great circle that divides the globe at all times into a sunlit *hemisphere* and a shadowed hemisphere.

circum-Pacific belt   chains of andesite *volcanoes* making up mountain belts and *island arcs* surrounding the Pacific Ocean basin.

cirque   bowl-shaped depression carved in *rock* by glacial processes and holding the *firn* of the upper end of an *alpine glacier*.

clastic sediment   *sediment* consisting of particles broken away physically from a parent *rock* source.

clay   *sediment* particles smaller than 0.004 mm in diameter.

clay minerals   class of *minerals* produced by alteration of *silicate minerals*, having plastic properties when moist.

claystone   *sedimentary rock* formed by lithification of *clay* and lacking *fissile* structure.

cliff   sheer, near-vertical *rock* wall formed from flat-lying resistant layered *rocks*, usually *sandstone*, *limestone*, or *lava* flows; may refer to any near-vertical rock wall. (See also *marine cliff*.)

climate   generalized statement of the prevailing weather conditions at a given place, based on statistics of a long period of record and including mean values, departures from those means, and the probabilities associated with those departures.

climatic frontier   a geographical boundary that marks the limit of survival of a plant species subjected to climatic stress.

climax   stable community of plants and animals reached at the end point of *ecological succession*.

climograph   a graph on which two or more climatic variables, such as monthly mean temperature and monthly mean precipitation, are plotted for each month of the year.

closed flow system   flow system that is completely self-contained within a boundary through which no matter or *energy* is exchanged with the external environment. (See also *open flow system*.)

cloud forest   a type of low evergreen rainforest that occurs high on mountain slopes, where *clouds* and *fog* are frequent.

clouds   dense concentrations of suspended water or ice particles in the diameter range 20 to 50 µm. (See *cumuliform clouds*, *stratiform clouds*.)

coal   *rock* consisting of hydrocarbon compounds, formed of compacted, lithified, and altered accumulations of plant remains (*peat*).

coarse textured (rock)   having *mineral* crystals sufficiently large that they are at least visible to the naked eye or with low magnification.

coast   (See *coastline*.)

coastal blowout dune   high *sand dune* of the *parabolic dunes* class formed adjacent to a beach, usually with a deep *deflation* hollow (*blowout*) enclosed within the dune ridge.

coastal forest   subtype of *needleleaf evergreen forest* found in the humid coastal zone of the northwestern United States and western Canada.

coastal plain   coastal belt, emerged from beneath the sea as a former *continental shelf*, underlain by *strata* with gentle *dip* seaward.

coastline (coast)   zone in which coastal processes operate or have a strong influence.

cold front   moving weather *front* along which a cold *air mass* moves underneath a warm air mass, causing the latter to be lifted.

cold-core ring   circular eddy of cold water, surrounded by warm water and lying adjacent to a warm, poleward-moving *ocean current*, such as the Gulf Stream. (See also *warm-core ring*.)

colloids   particles of extremely small size, capable of remaining indefinitely in suspension in water. May be mineral or organic in nature.

colluvium   deposit of *sediment* or *rock* particles accumulating from overland flow at the base of a *slope* and originating from higher slopes where *sheet erosion* is in progress. (See also *alluvium*.)

component   in flow systems, a part of the system, such as a *pathway*, connection, or flow of matter or *energy*.

compression (tectonic)   squeezing together, as horizontal compression of crustal layers by *tectonic* processes.

condensation   process of change of matter in the gaseous state (*water vapor*) to the *liquid* state (liquid water) or *solid* state (ice).

condensation nucleus   a tiny bit of solid matter (*aerosol*) in the *atmosphere* on which *water vapor* condenses to form a tiny water droplet.

conduction of heat   transmission of *sensible heat* through matter by transfer of *energy* from one atom or molecule to the next in the direction of decreasing temperature.

cone of depression   conical configuration of the lowered *water table* around a well from which water is being rapidly withdrawn.

conformal projection   *map projection* that preserves without shearing the true shape or outline of any small surface feature of the earth.

conglomerate   a *sedimentary rock* composed of pebbles in a matrix of finer *rock* particles.

conic projections   a group of *map projections* in which the *geographic grid* is transformed to lie on the surface of a developed cone.

consequent stream   *stream* that takes its course down the slope of an *initial landform*, such as a newly emerged *coastal plain* or a *volcano*.

consumers   animals in the *food chain* that live on organic matter formed by *primary producers* or by other *consumers*. (See also *primary consumers*, *secondary consumers*.)

consumption (of a lithospheric plate)   destruction or disappearance of a subducting *lithospheric plate* in the *asthenosphere*, in part by *melting* of the upper surface, but largely by softening because of heating to the temperature of the surrounding *mantle rock*.

continental collision   event in *plate tectonics* in which subduction brings two segments of the *continental lithosphere* into contact, leading to formation of a *continental suture*.

continental crust   crust of the continents, of felsic composition in the upper part; thicker and less dense than *oceanic crust*.

continental drift   hypothesis, introduced by Alfred Wegener and others early in the 1900s, of the breakup of a parent continent, *Pangea*, starting near the close of the *Mesozoic Era*, and resulting in the present arrangement of *continental shields* and intervening *ocean-basin floors*.

continental lithosphere   *lithosphere* bearing *continental crust* of *felsic igneous rock*.

continental margins   (1) topographic: one of three major divisions of the ocean basins, being the zones directly adjacent to the continent and including the *continental shelf*, *continental slope*, and *continental rise*; (2) tectonic: marginal belt of continental crust and lithosphere that is in contact with *oceanic crust* and *lithosphere*, with or without an active plate boundary being present at the contact. (See also *active continental margins, passive continental margins.*)

continental rise   gently sloping seafloor lying at the foot of the *continental slope* and leading gradually into the *abyssal plain.*

continental rupture   crustal spreading apart affecting the *continental lithosphere,* so as to cause a *rift valley* to appear and to widen, eventually creating a new belt of *oceanic lithosphere.*

continental scale   scale of observation at which we recognize continents and other large earth surface features, such as ocean currents.

continental shelf   shallow, gently sloping belt of seafloor adjacent to the continental shoreline and terminating at its outer edge in the *continental slope.*

continental shields   ancient crustal *rock* masses of the continents, largely *igneous rock* and *metamorphic rock,* and mostly of *Precambrian age.*

continental slope   steeply descending belt of seafloor between the *continental shelf* and the *continental rise.*

continental suture   long, narrow zone of crustal deformation, including underthrusting and intense *folding,* produced by a *continental collision.* Examples: Himalayan Range, European Alps.

continuous permafrost   *permafrost* that extends without gaps or interruptions under all surface features.

convection (atmospheric)   air motion consisting of strong updrafts taking place within a *convection cell.*

convection cell   individual column of strong updrafts produced by atmospheric *convection.*

convection loop   circuit of moving *fluid,* such as *air* or water, created by unequal heating of the *fluid.*

convectional precipitation   a form of *precipitation* induced when warm, moist air is heated at the ground surface, rises, cools, and condenses to form water droplets, raindrops, and eventually, rainfall.

converging boundary   boundary between two crustal plates along which *subduction* is occurring and *lithosphere* is being consumed.

coral reef   rocklike accumulation of *carbonates* secreted by corals and algae in shallow water along a marine shoreline.

coral-reef coast   *coast* built out by accumulations of *limestone* in *coral reefs.*

core of earth   spherical central mass of the earth composed largely of iron and consisting of an outer *liquid* zone and an interior *solid* zone.

Coriolis effect   effect of the earth's rotation tending to turn the direction of motion of any object or *fluid* toward the right in the northern *hemisphere* and to the left in the southern hemisphere.

corrosion   erosion of *bedrock* of a *stream channel* (or other *rock* surface) by chemical reactions between solutions in stream water and *mineral* surfaces.

counterradiation   *longwave radiation* of atmosphere directed downward to the earth's surface.

covered shields   areas of *continental shields* in which the ancient *rocks* are covered beneath a thin layer of sedimentary *strata.*

crater   central summit depression associated with the principal vent of a *volcano.*

crescentic dune   (See *barchan dune.*)

crevasse   gaping crack in the brittle surface ice of a *glacier.*

crude oil   liquid fraction of *petroleum.*

crust of earth   outermost solid shell or layer of the earth, composed largely of *silicate minerals.*

Cryaquepts   great group within the soil *suborder* of *Aquepts;* includes Aquepts of cold climate regions and particularly the *tundra climate* ⑫.

Cryosolic order   a class of *soils* in the Canadian soil classification system associated with strong frost action and underlying *permafrost.*

cuesta   *erosional landform* developed on resistant *strata* having low to moderate *dip* and taking the form of an asymmetrical low ridge or hill belt with one side a steep slope and the other a gentle slope; usually associated with a *coastal plain.*

cultural energy   *energy* in forms exclusive of solar *energy* of *photosynthesis* that is expended on the production of raw food or feed crops in agricultural *ecosystems.*

cumuliform clouds   *clouds* of globular shape, often with extended vertical development.

cumulonimbus cloud   large, dense *cumuliform cloud* yielding *precipitation.*

cumulus   cloud type consisting of low-lying, white cloud masses of globular shape well separated from one another.

cutoff   cutting-through of a narrow neck of land, so as to bypass the stream flow in an *alluvial meander* and cause it to be abandoned.

cycle   in flow systems, a closed flow system of matter. Example: *biogeochemical cycle.* (See *closed flow system.*)

cycle of rock change   total cycle of changes in which *rock* of any one of the three major rock classes—*igneous rock, sedimentary rock, metamorphic rock*—is transformed into rock of one of the other classes.

cyclone   center of low *atmospheric pressure.* (See *tropical cyclone, wave cyclone.*)

cyclonic precipitation   a form of *precipitation* that occurs as warm moist air is lifted by air motion occurring in a *cyclone.*

cyclonic storm   intense weather disturbance within a moving *cyclone* generating strong winds, cloudiness, and *precipitation.*

cylindric projections   group of *map projections* in which the *geographic grid* is transformed to lie on the surface of a developed cylinder.

data acquisition component   component of a *geographic information system* in which data are gathered together for input to the system.

data management component   component of a *geographic information system* that creates, stores, retrieves, and modifies data layers and *spatial objects.*

daughter product   new *isotope* created by decay of an *unstable isotope.*

daylight saving time   time system under which time is advanced by one hour with respect to the *standard time* of the prevailing *standard meridian.*

debris flood (debris flow)   streamlike flow of muddy water heavily charged with *sediment* of a wide range of size grades, including boulders, generated by sporadic torrential rains upon steep mountain watersheds.

decalcification   removal of *calcium carbonate* from a *soil horizon* or *soil solum* as *carbonic acid* reacts with *carbonate mineral matter.*

December solstice   (See *winter solstice.*)

deciduous plant   *tree* or *shrub* that sheds its leaves seasonally.

declination of sun   latitude at which the sun is directly overhead; varies from $-23^1/_2°$ ($23^1/_2°$ S lat.) to $+23^1/_2°$ ($23^1/_2°$ N lat.)

décollement   detachment and extensive sliding of a *rock* layer, usually *sedimentary,* over a near-horizontal basal *rock* surface; a special form of low-angle thrust *faulting.*

decomposers   organisms that feed on dead organisms from all levels of the *food chain;* most are microorganisms and bacteria that feed on decaying organic matter.

deep sea cone   a fan-shaped accumulation of undersea *sediment* on the *continental rise* produced by sediment-rich currents flowing down the *continental slope.*

deficit (soil-water shortage) in the soil-water budget, the difference between *water use* and *water need*; the quantity of irrigation water required to achieve maximum growth of agricultural crops.

deflation lifting and transport in *turbulent suspension* by wind of loose particles of *soil* or *regolith* from dry ground surfaces.

deglaciation widespread recession of *ice sheets* during a period of warming global climate, leading to an interglaciation. (See also *glaciation, interglaciation.*)

degradation lowering or downcutting of a *stream channel* by *stream erosion* in *alluvium* or *bedrock.*

degree of arc measurement of the angle associated with an *arc*, in degrees.

delta *sediment* deposit built by a stream entering a body of standing water and formed of the *stream load.*

delta coast *coast* bordered by a *delta.*

delta kame flat-topped hill of *stratified drift* representing a glacial *delta* constructed adjacent to an *ice sheet* in a marginal glacial lake.

dendritic drainage pattern *drainage pattern* of treelike branched form, in which the smaller streams take a wide variety of directions and show no parallelism or dominant trend.

denitrification biochemical process in which nitrogen in forms usable to plants is converted into molecular nitrogen in the gaseous form and returned to the atmosphere—a process that is part of the *nitrogen cycle.*

density of matter quantity of mass per unit of volume, stated in gm/cc.

denudation total action of all processes whereby the exposed *rocks* of the continents are worn down and the resulting *sediments* are transported to the sea by the *fluid agents*; includes also *weathering* and *mass wasting.*

deposition (atmosphere) the change of state of a substance from a *gas* (*water vapor*) to a *solid* (ice); in the science of *meteorology*, the term sublimation is used to describe both this process and the change of state from solid to vapor. (See *sublimation.*)

deposition (of sediment) (See *stream deposition.*)

depositional landform *landform* made by *deposition* of *sediment.*

desert biome *biome* of the dry climates consisting of thinly dispersed plants that may be *shrubs*, grasses, or perennial *herbs*, but lacking in *trees.*

desert pavement surface layer of closely fitted pebbles or coarse *sand* from which finer particles have been removed.

desertification (See *land degradation.*)

detritus decaying organic matter on which *decomposers* feed.

dewpoint lapse rate rate at which the dewpoint of an air mass decreases with elevation; typical value is 1.8°C/1000 m (1.0°F/1000 ft).

dewpoint temperature temperature of an *air mass* at which the air holds its full capacity of water vapor.

diagnostic horizons *soil horizons*, rigorously defined, that are used as diagnostic criteria in classifying *soils.*

diffuse radiation solar radiation that has been *scattered* (deflected or reflected) by minute dust particles or cloud particles in the *atmosphere.*

diffuse reflection solar *radiation* scattered back to space by the earth's *atmosphere.*

digital image numeric representation of a picture consisting of a collection of numeric brightness values (pixels) arrayed in a fine grid pattern.

dike thin layer of *intrusive igneous rock*, often near-vertical or with steep *dip*, occupying a widened fracture in the surrounding *rock* and typically cutting across older rock planes.

diorite *intrusive igneous rock* consisting dominantly of plagioclase feldspar and pyroxene; a *felsic igneous rock.*

dip acute angle between an inclined natural *rock* plane or surface and an imaginary horizontal plane of reference; always measured perpendicular to the *strike.* Also a verb, meaning to incline toward.

discharge volume of flow moving through a given cross section of a stream in a given unit of time; commonly given in cubic meters (feet) per second.

discontinuous permafrost *permafrost* that occurs in patches separated by frost-free zones under lakes and rivers.

distributary branching *stream channel* that crosses a *delta* to discharge into open water.

diurnal adjective meaning "daily."

doldrums belt of calms and variable winds occurring at times along the *equatorial trough.*

dolomite carbonate mineral or *sedimentary rock* having the composition calcium magnesium carbonate.

dome (See *sedimentary dome.*)

drainage basin total land surface occupied by a *drainage system*, bounded by a *drainage divide* or watershed.

drainage divide imaginary line following a crest of high land such that overland flow on opposite sides of the line enters different *streams.*

drainage pattern the plan of a network of interconnected *stream channels.*

drainage system a branched network of *stream channels* and adjacent land *slopes*, bounded by a *drainage divide* and converging to a single channel at the outlet.

drainage winds *winds*, usually cold, that flow from higher to lower regions under the direct influence of *gravity.*

drawdown (of a well) difference in height between base of cone of depression and original water table surface.

drought occurrence of substantially lower-than-average *precipitation* in a season that normally has ample precipitation for the support of food-producing plants.

drumlin hill of glacial *till*, oval or elliptical in basal outline and with smoothly rounded summit, formed by plastering of till beneath moving, debris-laden glacial ice.

dry adiabatic lapse rate rate at which rising air is cooled by expansion when no *condensation* is occurring; 10°C per 1000 m (5.5°F per 1000 ft).

dry desert plant *formation class* in the *desert biome* consisting of widely dispersed xerophytic plants that may be small, hardleaved or spiny *shrubs*, succulent plants (cacti), or hard grasses.

dry midlatitude climate ⑨ dry climate of the *midlatitude zone* with a strong annual cycle of *potential evapotranspiration* (*water need*) and cold winters.

dry subtropical climate ⑤ dry climate of the *subtropical zone*, transitional between the *dry tropical climate* ⑨ and the *dry midlatitude climate* ⑨.

dry tropical climate ④ dry climate of the *tropical zone* with large total annual *potential evapotranspiration* (*water need*).

dune (See *sand dune.*)

dust storm heavy concentration of dust in a turbulent *air mass*, often associated with a *cold front.*

E horizon mineral horizon of the *soil solum* lying below the *A horizon* and characterized by the loss of *clay minerals* and oxides of iron and aluminum; it may show a concentration of *quartz* grains and is often pale in color.

earth's crust (See *crust of earth.*)

earthflow moderately rapid downhill flowage of masses of water-saturated *soil*, *regolith*, or weak *shale*, typically forming a steplike terrace at the top and a bulging toe at the base.

earthquake a trembling or shaking of the ground produced by the passage of *seismic waves.*

earthquake focus point within the earth at which the *energy* of an *earthquake* is first released by rupture and from which *seismic waves* emanate.

easterly wave weak, slowly moving trough of low pressure within the belt of *tropical easterlies*; causes a weather disturbance with rain showers.

ebb current oceanward flow of *tidal current* in a *bay* or tidal stream.

ecological succession time-succession (sequence) of distinctive plant and animal communities occurring within a given area of newly formed land or land cleared of plant cover by burning, clear cutting, or other agents.

ecology science of interactions between life forms and their environment; the science of *ecosystems*.

ecosystem group of organisms and the environment with which the organisms interact.

El Niño episodic cessation of the typical *upwelling* of cold deep water off the coast of Peru; literally, "The Christ Child," for its occurrence in the Christmas season once every few years.

electromagnetic radiation (electromagnetic energy) wavelike form of *energy* radiated by any substance possessing heat; it travels through space at the speed of light.

electromagnetic spectrum the total *wavelength* range of *electromagnetic energy*.

eluviation soil-forming process consisting of the downward transport of fine particles, particularly the *soil colloids* (both mineral and organic), carrying them out of an upper *soil horizon*.

emergence exposure of submarine landforms by a lowering of sea level or a rise of the crust, or both.

energy the capacity to do work, that is, to bring about a change in the state or motion of matter.

energy balance (global) balance between *shortwave* solar *radiation* received by the earth–atmosphere system and radiation lost to space by shortwave reflection and *longwave radiation* from the earth–atmosphere system.

energy balance (of a surface) balance between the flows of *energy* reaching a surface and the flows of energy leaving it.

energy flow system *open system* that receives an input of *energy*, undergoes internal energy flow, energy transformation, and energy storage, and has an energy output.

Entisols *soil order* consisting of mineral soils lacking *soil horizons* that would persist after normal plowing.

entrenched meanders winding, sinuous valley produced by *degradation* of a *stream*

with trenching into the *bedrock* by downcutting.

environmental temperature lapse rate rate of temperature decrease upward through the *troposphere*; standard value is 6.4 C°/km (3 F°/1000 ft).

epipedon *soil horizon* that forms at the surface.

epiphytes plants that live above ground level out of contact with the soil, usually growing on the limbs of *trees* or *shrubs*; also called air plants.

epoch a subdivision of geologic time.

equal-area projections class of *map projections* on which any given area of the earth's surface is shown to correct relative areal extent, regardless of position on the globe.

equator *parallel of latitude* occupying a position midway between the earth's poles of *rotation*; the largest of the parallels, designated as *latitude* 0°.

equatorial current westward-flowing *ocean current* in the belt of the *trade winds*.

equatorial easterlies upper-level easterly air flow over the *equatorial zone*.

equatorial rainforest plant *formation class* within the *forest biome*, consisting of tall, closely set broadleaf *trees* of evergreen or semideciduous habit.

equatorial trough atmospheric low-pressure trough centered more or less over the *equator* and situated between the two belts of *trade winds*.

equatorial zone *latitude* zone lying between lat. 10° S and 10° N (more or less) and centered on the *equator*.

equilibrium in flow systems, a state of balance in which flow rates remain unchanged.

equinox instant in time when the *subsolar point* falls on the earth's *equator* and the *circle of illumination* passes through both poles. *Vernal equinox* occurs on March 20 or 21; *autumnal equinox* on September 22 or 23.

era major subdivision of geologic time consisting of a number of geologic periods. The three *eras* following *Precambrian time* are *Paleozoic*, *Mesozoic*, and *Cenozoic*.

erg large expanse of active *sand dunes* in the Sahara Desert of North Africa.

erosional landforms class of the *sequential landforms* shaped by the removal of *regolith* or *bedrock* by agents of erosion. Examples: *gorge*, glacial *cirque*, marine *cliff*.

esker narrow, often sinuous embankment of coarse gravel and boulders deposited in the bed of a meltwater *stream* enclosed in a tunnel within stagnant ice of an *ice sheet*.

estuary *bay* that receives fresh water from a river mouth and salt water from the ocean.

Eurasian-Indonesian belt mountain arc system extending from southern Europe across southern Asia and Indonesia.

eustatic referring to a true change in sea level, as opposed to a local change created by upward or downward *tectonic* motion of land.

eutrophication excessive growth of algae and other related organisms in a *stream* or *lake* as a result of the input of large amounts of nutrient *ions*, especially phosphate and nitrate.

evaporation process in which water in liquid state or solid state passes into the vapor state.

evaporites class of *chemically precipitated sediment* and *sedimentary rock* composed of soluble salts deposited from saltwater bodies.

evapotranspiration combined water loss to the atmosphere by *evaporation* from the soil and *transpiration* from plants.

evergreen plant *tree* or *shrub* that holds most of its green leaves throughout the year.

exfoliation dome smoothly rounded *rock* knob or hilltop bearing rock sheets or shells produced by spontaneous expansion accompanying *unloading*.

exotic river *stream* that flows across a region of dry climate and derives its *discharge* from adjacent uplands where a *water surplus* exists.

exponential growth increase in number or value over time in which the increase is a constant proportion or percentage within each time unit.

exposed shields areas of *continental shields* in which the ancient basement *rock*, usually of *Precambrian* age, is exposed to the surface.

extension (tectonic) drawing apart of crustal layers by *tectonic activity* resulting in *faulting*.

extrusion release of molten *rock magma* at the surface, as in a flow of *lava* or shower of volcanic ash.

extrusive igneous rock *rock* produced by the solidification of *lava* or ejected fragments of *igneous rock* (*tephra*).

Fahrenheit scale temperature scale in which the *freezing* point of water is 32°, and the boiling point 212°.

fallout *gravity* fall of atmospheric particles of *particulates* reaching the ground.

fault sharp break in *rock* with a displacement (slippage) of the block on one side with respect to an adjacent block. (See *normal fault, overthrust fault, strike-slip fault, transform fault.*)

fault coast *coast* formed when a *shoreline* comes to rest against a *fault scarp.*

fault creep more or less continuous slippage on a *fault plane*, relieving some of the accumulated strain.

fault plane surface of slippage between two earth blocks moving relative to each other during faulting.

fault scarp clifflike surface feature produced by faulting and exposing the *fault plane*; commonly associated with a *normal fault.*

fault-line scarp erosion scarp developed on an inactive *fault* line.

feedback in flow systems, a linkage between flow paths such that the flow in one *pathway* acts either to reduce or increase the flow in another pathway.

feldspar group of *silicate minerals* consisting of silicate of aluminum and one or more of the metals potassium, sodium, or calcium. (See *plagioclase feldspar, potash feldspar.*)

felsenmeer expanse of large blocks of *rock* produced by *joint* block separation and shattering by *frost action* at high altitudes or in high latitudes; from the German for "rock sea."

felsic igneous rock *igneous rock* dominantly composed of *felsic minerals.*

felsic minerals (felsic mineral group) *quartz* and *feldspars* treated as a mineral group of light color and relatively low *density.* (See also *mafic minerals.*)

fine textured (rock) having *mineral* crystals too small to be seen by eye or with low magnification.

fiord narrow, deep ocean embayment partially filling a *glacial trough.*

fiord coast deeply embayed, rugged coast formed by partial *submergence* of *glacial troughs.*

firn granular old *snow* forming a surface layer in the zone of accumulation of a *glacier.*

fissile adjective describing a *rock*, usually *shale*, that readily splits up into small flakes or scales.

flood stream flow at a stream *stage* so high that it cannot be accommodated within the *stream channel* and must spread over the banks to inundate the adjacent *floodplain.*

flood basalts large-scale outpourings of basalt *lava* to produce thick accumulations of *basalt* over large areas.

flood current landward flow of a *tidal current.*

flood stage designated stream-surface level

for a particular point on a *stream*, higher than which overbank flooding may be expected.

floodplain belt of low, flat ground, present on one or both sides of a *stream channel*, subject to inundation by a *flood* about once annually and underlain by alluvium.

fluid substance that flows readily when subjected to unbalanced stresses; may exist as a *gas* or a *liquid.*

fluid agents *fluids* that erode, transport, and deposit *mineral matter* and organic matter; they are running water, waves and currents, glacial ice, and *wind.*

fluvial landforms *landforms* shaped by running water.

fluvial processes geomorphic processes in which running water is the dominant *fluid* agent, acting as *overland flow* and *stream flow.*

focus (See *earthquake focus.*)

fog cloud layer in contact with land or sea surface, or very close to that surface. (See *advection fog, radiation fog.*)

folding process by which *folds* are produced; a form of *tectonic activity.*

folds wavelike corrugations of *strata* (or other layered *rock* masses) as a result of crustal *compression.*

food chain (food web) organization of an *ecosystem* into steps or levels through which *energy* flows as the organisms at each level consume *energy* stored in the bodies of organisms of the next lower level.

forb broad-leaved *herb*, as distinguished from the grasses.

forearc trough in plate tectonics, a shallow trough between a *tectonic arc* and a continent; accumulates *sediment* in a basinlike structure.

foredunes ridge of irregular *sand dunes* typically found adjacent to *beaches* on low-lying *coasts* and bearing a partial cover of plants.

foreland folds *folds* produced by *continental collision* in *strata* of a *passive continental margin.*

forest assemblage of *trees* growing close together, their crowns forming a layer of foliage that largely shades the ground.

forest biome *biome* that includes all regions of *forest* over the lands of the earth.

formation classes subdivisions within a *biome* based on the size, shape, and structure of the plants that dominate the vegetation.

fossil fuels naturally occurring hydrocarbon compounds that represent the altered remains of organic materials enclosed in *rock*; examples are *coal, petroleum (crude oil),* and *natural gas.*

fractional scale (See *scale fraction.*)

freezing change from *liquid* state to *solid* state accompanied by release of *latent heat*, becoming *sensible heat.*

fringing reef *coral reef* directly attached to land with no intervening *lagoon* of open water.

front surface of contact between two unlike *air masses.* (See *cold front, occluded front, polar front, warm front.*)

frost action *rock* breakup by forces accompanying the *freezing* of water.

gabbro *intrusive igneous rock* consisting largely of pyroxene and *plagioclase feldspar*, with variable amounts of *olivine*; a *mafic igneous rock.*

gas (gaseous state) *fluid* of very low density (as compared with a *liquid* of the same chemical composition) that expands to fill uniformly any small container and is readily compressed.

gaseous cycle type of *biogeochemical cycle* in which an element or compound is converted into *gaseous* form, diffuses through the *atmosphere*, and passes rapidly over land or sea where it is reused in the *biosphere.*

geographic grid complete network of parallels and meridians on the surface of the globe, used to fix the locations of surface points.

geographic information system (GIS) a system for acquiring, processing, storing, querying, creating, and displaying *spatial data*; normally computer-based.

geologic norm stable natural condition in a moist climate in which slow *soil erosion* is paced by maintenance of *soil horizons* bearing a plant community in an equilibrium state.

geology science of the solid earth, including the earth's origin and history, materials comprising the earth, and the processes acting within the earth and upon its surface.

geomorphology science of *landforms*, including their history and processes of origin.

geostrophic wind *wind* at high levels above the earth's surface blowing parallel with a system of straight, parallel *isobars.*

geyser periodic jetlike emission of hot water and steam from a narrow vent at a geothermal locality.

glacial abrasion *abrasion* by a moving *glacier* of the *bedrock* floor beneath it.

glacial delta *delta* built by meltwater streams of a *glacier* into standing water of a marginal glacial lake.

glacial drift general term for all varieties and forms of *rock* debris deposited in close association with *ice sheets* of the *Pleistocene Epoch.*

glacial plucking   removal of masses of *bedrock* from beneath an *alpine glacier* or *ice sheets* as ice moves forward suddenly.

glacial trough   deep, steep-sided *rock* trench of U-shaped cross section formed by *alpine glacier* erosion.

glaciation   (1) general term for the total process of glacier growth and *landform* modification by *glaciers;* (2) single episode or time period in which *ice sheets* formed, spread, and disappeared.

glacier   large natural accumulation of land ice affected by present or past flowage. (See *alpine glacier.*)

Gleysolic order   a class of *soils* in the Canadian soil classification system characterized by indicators of periodic or prolonged water saturation.

global scale   scale at which we are concerned with the earth as a whole, for example, in considering earth–sun relationships.

gneiss   variety of *metamorphic rock* showing banding and commonly rich in *quartz* and *feldspar.*

Gondwana   a *supercontinent* of the Permian *Period* including much of the regions that are now South America, Africa, Antarctica, Australia, New Zealand, Madagascar, and peninsular India.

Goode projection   an equal-area *map projection,* often used to display areal thematic information, such as *climate* or *soil* type.

gorge (canyon)   steep-sided *bedrock* valley with a narrow floor limited to the width of a *stream channel.*

graben   trenchlike depression representing the surface of a crustal block dropped down between two opposed, infacing *normal faults.* (See *rift valley.*)

graded profile   smoothly descending profile displayed by a *graded stream.*

graded stream   *stream* (or *stream channel*) with *stream gradient* so adjusted as to achieve a balanced state in which average *bed load* transport is matched to average bed load input; an average condition over periods of many years' duration.

gradient   degree of *slope,* as the gradient of a river or a flowing glacier.

granite   *intrusive igneous rock* consisting largely of *quartz, potash feldspar,* and *plagioclase feldspar,* with minor amounts of biotite and hornblende; a *felsic igneous rock.*

granitic rock   general term for *rock* of the upper layer of the *continental crust,* composed largely of *felsic igneous* and *metamorphic rock; rock* of composition similar to that of *granite.*

granular disintegration   grain-by-grain breakup of the outer surface of coarse-grained *rock,* yielding *sand* and gravel and leaving behind rounded boulders.

graphic scale   map scale as shown by a line divided into equal parts.

grassland biome   *biome* consisting largely or entirely of *herbs,* which may include grasses, grasslike plants, and *forbs.*

gravitation   mutual attraction between any two masses.

gravity   gravitational attraction of the earth upon any small mass near the earth's surface. (See *gravitation.*)

gravity gliding   the sliding of a *thrust sheet* away from the center of an *orogen* under the force of *gravity.*

great circle   circle formed by passing a plane through the exact center of a perfect sphere; the largest circle that can be drawn on the surface of a sphere.

greenhouse effect   accumulation of heat in the lower *atmosphere* through the absorption of *longwave radiation* from the earth's surface.

greenhouse gases   atmospheric gases such as CO_2 and *cholorofluorocarbons (CFCs)* that absorb outgoing *longwave radiation,* contributing to the *greenhouse effect.*

groin   wall or embankment built out into the water at right angles to the *shoreline.*

gross photosynthesis   total amount of *carbohydrate* produced by *photosynthesis* by a given organism or group of organisms in a given unit of time.

ground ice   frozen water within the pore spaces of *soils* and *regolith.*

ground moraine   *moraine* formed of *till* distributed beneath a large expanse of land surface covered at one time by an *ice sheet.*

groundwater   *subsurface water* occupying the *saturated zone* and moving under the force of *gravity.*

growth rate (of a population)   rate at which a population grows or shrinks with time; usually expressed as a percentage or proportion of increase or decrease in a given unit of time.

gullies   deep, V-shaped trenches carved by newly formed *streams* in rapid headward growth during advanced stages of *accelerated soil erosion.*

guyot   sunken remnant of a volcanic island.

gyres   large, circular *ocean current* systems centered on the oceanic subtropical *high-pressure cells.*

habitat   subdivision of the plant environment having a certain combination of *slope,* drainage, *soil* type, and other controlling physical factors.

Hadley cell   atmospheric circulation cell in low latitudes involving rising air over the *equatorial trough* and sinking air over the *subtropical high-pressure belts.*

hail   form of *precipitation* consisting of pellets or spheres of ice with a concentric layered structure.

half-life   time required for an initial quantity at time-zero to be reduced by one-half in an exponential decay system.

hanging valley   stream valley that has been truncated by marine erosion so as to appear in cross section in a *marine cliff,* or truncated by glacial erosion so as to appear in cross section in the upper wall of a *glacial trough.*

haze   minor concentration of *pollutants* or natural forms of *aerosols* in the atmosphere causing a reduction in visibility.

heat   (See *latent heat, sensible heat.*)

heat island   persistent region of higher air temperatures centered over a city.

hemisphere   half of a sphere; that portion of the earth's surface found between the *equator* and a pole.

herbs   tender plants, lacking woody stems, usually small or low; may be annual or perennial.

heterosphere   region of the *atmosphere* above about 100 km in which *gas* molecules tend to become increasingly sorted into layers by molecular weight and electric charge.

high base status   (See *base status of soils.*)

high-latitude climates   group of climates in the *subarctic zone, arctic zone,* and *polar zone,* dominated by arctic *air masses* and polar air masses.

high-level temperature inversion   condition in which a high-level layer of warm air overlies a layer of cooler air, reversing the normal trend of cooling with altitude.

high-pressure cell   center of high barometric pressure; an *anticyclone.*

Histosols   *soil order* consisting of *soils* with a thick upper layer of organic matter.

hogbacks   sharp-crested, often sawtooth ridges formed of the upturned edge of a resistant *rock* layer of *sandstone, limestone,* or *lava.*

Holocene Epoch   last *epoch* of geologic time, commencing about 10,000 years ago; it followed the *Pleistocene Epoch* and includes the present.

homosphere   the lower portion of the *atmosphere,* below about 100 km altitude, in which atmospheric *gases* are uniformly mixed.

horse latitudes   *subtropical high-pressure belt* of the North Atlantic Ocean, coincident with the central region of the Azores high; a belt of weak, variable winds and frequent calms.

horst crustal block uplifted between two *normal faults.*

hot springs springs discharging heated *groundwater* at a temperature close to the boiling point; found in geothermal areas and thought to be related to a magma body at depth.

hotspot center of intrusive *igneous* and *volcanic* activity thought to be located over a rising *mantle plume.*

human habitat the lands of the earth that support human life.

human-influenced vegetation vegetation that has been influenced in some way by human activity, for example, through cultivation, grazing, timber cutting, or urbanization.

humidity general term for the amount of *water vapor* present in the air. (See *relative humidity, specific humidity.*)

humification *pedogenic process* of transformation of plant tissues into *humus.*

humus dark brown to black organic matter on or in the *soil,* consisting of fragmented plant tissues partly digested by organisms.

hurricane *tropical cyclone* of the western North Atlantic and Caribbean Sea.

hydraulic action *stream erosion* by impact force of the flowing water on the bed and banks of the *stream channel.*

hydrograph graphic presentation of the variation in *stream discharge* with elapsed time, based on data of stream gauging at a given station on a stream.

hydrologic cycle total plan of movement, exchange, and storage of the earth's free water in *gaseous* state, *liquid* state, and *solid* state.

hydrology science of the earth's water and its motions through the *hydrologic cycle.*

hydrolysis chemical union of water molecules with *minerals* to form different, more stable mineral compounds.

hydrosphere total water realm of the earth's surface zone, including the oceans, surface waters of the lands, *groundwater,* and water held in the *atmosphere.*

hygrometer instrument that measures the *water vapor* content of the *atmosphere;* some types measure *relative humidity* directly.

ice age span of geologic time, usually on the order of 1 to 3 million years, or longer, in which glaciations alternate with interglaciations repeatedly in rhythm with cyclic global climate changes. (See also *glaciation, interglaciation.*)

Ice Age (Late-Cenozoic Ice Age) the present ice age, which began in late Pliocene time, perhaps 2.5 to 3 million years ago.

ice lobes (glacial lobes) broad tonguelike extensions of an *ice sheet* resulting from more rapid ice motion where terrain was more favorable.

ice sheet large thick plate of glacial ice moving outward in all directions from a central region of accumulation.

ice shelf thick plate of floating glacial ice attached to an *ice sheet* and fed by the ice sheet and by *snow* accumulation.

ice storm occurrence of heavy glaze of ice on solid surfaces.

ice wedge vertical, wall-like body of ground ice, often tapering downward, occupying a shrinkage crack in *silt* of *permafrost* areas.

ice-sheet climate ⑬ severely cold climate, found on the Greenland and Antarctic *ice sheets,* with *potential evapotranspiration (water need)* effectively zero throughout the year.

ice-wedge polygons polygonal networks of *ice wedges.*

iceberg mass of glacial ice floating in the ocean, derived from a *glacier* that extends into tidal water.

igneous rock *rock* solidified from a high-temperature molten state; *rock* formed by cooling of *magma.* (See *extrusive igneous rock, felsic igneous rock, intrusive igneous rock, mafic igneous rock, ultramafic igneous rock.*)

illuviation accumulation in a lower *soil horizon* (typically, the *B horizon*) of materials brought down from a higher horizon; a soil-forming process.

image processing mathematical manipulation of digital images, for example, to enhance contrast or edges.

Inceptisols *soil order* consisting of soils having weakly developed *soil horizons* and containing weatherable *minerals.*

induced deflation loss of *soil* by wind erosion that is triggered by human activity such as cultivation or overgrazing.

induced mass wasting *mass wasting* that is induced by human activity, such as creation of waste *soil* and *rock* piles or undercutting of *slopes* in construction.

infiltration absorption and downward movement of *precipitation* into the *soil* and *regolith.*

infrared imagery images formed by *infrared radiation* emanating from the ground surface as recorded by a remote sensor.

infrared radiation *electromagnetic energy* in the *wavelength* range of 0.7 to about 200 μm.

initial landforms *landforms* produced directly by internal earth processes of *volcanism* and *tectonic activity.* Examples: *volcano, fault scarp.*

inner lowland on a *coastal plain,* a shallow valley lying between the first *cuesta* and the area of older *rock* (oldland).

insolation interception of solar *energy (shortwave radiation)* by an exposed surface.

inspiral horizontal inward spiral or motion, such as that found in a *cyclone.*

interglaciation within an *ice age,* a time interval of mild global climate in which continental *ice sheets* were largely absent or were limited to the Greenland and Antarctic ice sheets; the interval between two glaciations. (See also *deglaciation, glaciation.*)

interlobate moraine *moraine* formed between two adjacent lobes of an *ice sheet.*

International Date Line the 180° *meridian of longitude,* together with deviations east and west of that meridian, forming the time boundary between adjacent *standard time zones* that are 12 hours fast and 12 hours slow with respect to Greenwich standard time.

interrupted projection projection subdivided into a number of sectors (gores), each of which is centered on a different central meridian.

intertropical convergence zone (ITC) zone of convergence of *air masses* of *tropical easterlies (trade winds)* along the axis of the *equatorial trough.*

intrusion body of *igneous rock* injected as *magma* into preexisting crustal *rock.* Examples: *dike* or *sill.*

intrusive igneous rock *igneous rock* body produced by solidification of *magma* beneath the surface, surrounded by preexisting *rock.*

inversion (See *temperature inversion.*)

ion atom or group of atoms bearing an electrical charge as the result of a gain or loss of one or more electrons.

island arcs curved lines of volcanic islands associated with active *subduction* zones along the boundaries of *lithospheric plates.*

isobars lines on map passing through all points having the same *atmospheric pressure.*

isohyet line on a map drawn through all points having the same numerical value of *precipitation.*

isopleth line on a map or globe drawn through all points having the same value of a selected property or entity.

isostasy principle describing the flotation of the *lithosphere,* which is less dense, on the plastic *asthenosphere,* which is more dense.

isostatic compensation crustal rise or sinking in response to unloading by *denuda-*

tion or loading by sediment deposition, following the principle of *isostasy*.

isostatic rebound   local crustal rise after the melting of ice sheets, following the principle of *isostasy*.

isotherm   line on a map drawn through all points having the same air temperature.

isotope   form of an element with a unique *atomic mass number*.

jet stream   high-speed air flow in narrow bands within the *upper-air westerlies* and along certain other global *latitude* zones at high levels.

joints   fractures within *bedrock*, usually occurring in parallel and intersecting sets of planes.

June solstice   (See *summer solstice*.)

karst   landscape or topography dominated by surface features of *limestone* solution and underlain by a *limestone cavern* system.

Kelvin scale (K)   temperature scale on which the starting point is absolute zero, equivalent to –273°C.

kinetic energy   form of *energy* represented by matter (mass) in motion.

knob and kettle   terrain of numerous small knobs of *glacial drift* and deep depressions usually situated along the *moraine* belt of a former *ice sheet*.

lag time   interval of time between occurrence of precipitation and peak discharge of a *stream*.

lagoon   shallow body of open water lying between a *barrier island* or a *barrier reef* and the mainland.

lahar   rapid downslope or downvalley movement of a tonguelike mass of water-saturated *tephra* (volcanic ash) originating high up on a steep-sided volcanic cone; a variety of mudflow.

lake   terrestrial body of standing water surrounded by land or glacial ice.

laminar flow   smooth, even flow of a *fluid* shearing in thin layers without *turbulence*.

land breeze   local *wind* blowing from land to water during the night.

land degradation   *degradation* of the quality of plant cover and *soil* as a result of overuse by humans and their domesticated animals, especially during periods of *drought*.

landforms   configurations of the land surface taking distinctive forms and produced by natural processes. Examples: hill, valley, plateau. (See *depositional landforms, erosional landforms, initial landforms, sequential landforms*.)

landmass   large area of *continental crust*

lying above sea level (base level) and thus available for removal by *denudation*.

landmass rejuvenation   episode of rapid fluvial *denudation* set off by a rapid crustal rise, increasing the available *landmass*.

landslide   rapid sliding of large masses of *bedrock* on steep mountain slopes or from high *cliffs*.

lapse rate   rate at which temperature decreases with increasing altitude (See *dry adiabatic lapse rate, environmental temperature lapse rate, wet adiabatic lapse rate*.)

large-scale map   map with *fractional scale* greater than 1:100,000; usually shows a small area.

Late-Cenozoic Ice Age   the series of *glaciations, deglaciations,* and *interglaciations* experienced during the late *Cenozoic Era*.

latent heat   heat absorbed and held in storage in a *gas* or *liquid* during the processes of *evaporation*, or *melting*, or *sublimation;* distinguished from *sensible heat*.

latent heat transfer   flow of *latent heat* that results when water absorbs heat to change from a *liquid* or solid to a *gas* and then later releases that heat to new surroundings by *condensation* or *deposition*.

lateral moraine   *moraine* forming an embankment between the ice of an *alpine glacier* and the adjacent valley wall.

laterite   rocklike layer rich in *sequioxides* and iron, including the minerals *bauxite* and *limonite*, found in low latitudes in association with *Ultisols* and *Oxisols*.

latitude   *arc* of a *meridian* between the *equator* and a given point on the globe.

Laurasia   a *supercontinent* of the Permian *Period* including much of the regions that are now North America and western Eurasia.

lava   *magma* emerging on the earth's solid surface, exposed to air or water.

leaching   *pedogenic process* in which material is lost from the *soil* by downward washing out and removal by percolating surplus soil water.

leads   narrow strips of open ocean water between ice floes.

level of condensation   elevation at which an upward-moving parcel of moist air cools to the *dewpoint* and *condensation* begins to occur.

liana   woody vine supported on the trunk or branches of a *tree*.

lichens   plant forms in which algae and fungi live together (in a symbiotic relationship) to create a single structure; they typically form tough, leathery coatings or crusts attached to *rocks* and tree trunks.

life cycle   continuous progression of stages in a growth or development process, such as that of a living organism.

life layer   shallow surface zone containing the *biosphere*, a zone of interaction between *atmosphere* and land surface, and between atmosphere and ocean surface.

life zones   series of vegetation zones describing vegetation types that are encountered with increasing elevation, especially in the southwestern United States.

life-form   characteristic physical structure, size, and shape of a plant or of an assemblage of plants.

limestone   nonclastic *sedimentary rock* in which *calcite* is the predominant *mineral,* and with varying minor amounts of other *minerals* and *clay*.

limestone caverns   interconnected subterranean cavities formed in *limestone* by *carbonic acid action* occurring in slowly moving *groundwater*.

limonite   mineral or group of *minerals* consisting largely of iron oxide and water, produced by *chemical weathering* of other iron-bearing minerals.

line   type of *spatial object* in a *geographic information system* that has starting and ending *nodes*; may be directional.

liquid   *fluid* that maintains a free upper surface and is only very slightly compressible, as compared with a *gas*.

lithosphere   (1) general term for the entire solid earth realm; (2) in *plate tectonics*, the strong, brittle outermost *rock* layer lying above the *asthenosphere*.

lithospheric plate   segment of *lithosphere* moving as a unit, in contact with adjacent lithospheric plates along plate boundaries.

littoral drift   transport of *sediment* parallel with the *shoreline* by the combined action of *beach drift* and *longshore current* transport.

loam   soil-texture class in which no one of the three size grades (*sand, silt, clay*) dominates over the other two.

local scale   scale of observation of the earth in which local processes and phenomena are observed.

local winds   general term for *winds* generated as direct or immediate effects of the local terrain.

loess   accumulation of yellowish to buff-colored, fine-grained *sediment*, largely of *silt* grade, upon upland surfaces after transport in the air in *turbulent suspension* (i.e., carried in a *dust storm*).

logistic growth   growth according to a mathematical model in which the *growth rate* smoothly decreases to near zero.

longitude   *arc* of a *parallel* between the *prime meridian* and a given point on the globe.

longitudinal dunes   class of *sand dunes* in

which the dune ridges are oriented parallel with the prevailing wind.

longshore current current in the breaker zone, running parallel with the *shoreline* and set up by the oblique approach of waves.

longshore drift *littoral drift* caused by action of a *longshore current*.

longwave radiation *electromagnetic energy* emitted by the earth, largely in the range from 3 to 50 μm.

low base status (See *base status of soils*.)

low-angle overthrust fault *overthrust fault* in which the *fault plane* or fault surface has a low angle of *dip* or may be horizontal.

low-latitude climates group of climates of the *equatorial zone* and *tropical zone* dominated by the subtropical high-pressure belt and the *equatorial trough*.

low-latitude rainforest evergreen broadleaf forest of the wet equatorial and tropical climate zones.

low-latitude rainforest environment low-latitude environment of warm temperatures and abundant *precipitation* that characterizes rainforest in the *wet equatorial* ① and *monsoon and trade-wind coastal* ② climates.

low-level temperature inversion atmospheric condition in which temperature near the ground increases, rather than decreases, with elevation.

low-pressure trough zone of low pressure between two *anticyclones*.

lowlands broad, open valleys between two *cuestas* of a *coastal plain*. (The term may refer to any low areas of land surface.)

Luvisolic order a class of *forest soils* in the Canadian soil classification system in which the *B horizon* accumulates *clay*.

mafic igneous rock *igneous rock* dominantly composed of *mafic minerals*.

mafic minerals (mafic mineral group) *minerals*, largely *silicate minerals*, rich in magnesium and iron, dark in color, and of relatively great density.

magma mobile, high-temperature molten state of *rock*, usually of *silicate mineral* composition and with dissolved *gases*.

manipulation and analysis component component of a *geographic information system* that responds to spatial queries and creates new data layers.

mantle *rock* layer or shell of the earth beneath the *crust* and surrounding the *core*, composed of *ultramafic igneous rock* of silicate mineral composition.

mantle plume a columnlike rising of heated *mantle rock*, thought to be the cause of a *hot spot* in the overlying *lithospheric plate*.

map projection any orderly system of parallels and meridians drawn on a flat surface to represent the earth's curved surface.

marble variety of *metamorphic rock* derived from *limestone* or dolomite by recrystallization under pressure.

marine cliff *rock* cliff shaped and maintained by the undermining action of breaking waves.

marine scarp steep seaward *slope* in poorly consolidated *alluvium*, *glacial drift*, or other forms of *regolith*, produced along a *coastline* by the undermining action of waves.

marine terrace former *abrasion platform* elevated to become a steplike coastal *landform*.

marine west-coast climate ⑧ cool moist climate of west coasts in the *midlatitude zone*, usually with a substantial annual *water surplus* and a distinct winter *precipitation* maximum.

mass number (See *atomic mass number*.)

mass wasting spontaneous downhill movement of *soil*, *regolith*, and *bedrock* under the influence of *gravity*, rather than by the action of *fluid* agents.

material cycle (See *biogeochemical cycle*.)

matter flow system total system of *pathways* by which a particular type of matter (a given element, compound, or ion, for example) moves through the earth's *ecosystem* or *biosphere*.

mean annual temperature mean of daily air temperature means for a given year or succession of years.

mean daily temperature sum of daily maximum and minimum air temperature readings divided by two.

mean monthly temperature mean of daily air temperature means for a given calendar month.

mean velocity mean, or average, speed of flow of water through an entire stream cross section.

meanders (See *alluvial meanders*.)

mechanical energy *energy* of motion or position; includes *kinetic energy* and *potential energy*.

medial moraine long, narrow deposit of fragments on the surface of a *glacier*, created by the merging of *lateral moraines* when two glaciers join into a single stream of ice flow.

Mediterranean climate ⑦ climate type of the *subtropical zone*, characterized by the alternation of a very dry summer and a mild, rainy winter.

melting change from *solid* state to *liquid* state, accompanied by absorption of *sensible heat* to become *latent heat*.

Mercator projection conformal map projection with horizontal parallels and vertical meridians and with map scale rapidly increasing with increase in *latitude*.

mercury barometer *barometer* using the Torricelli principle, in which *atmospheric pressure* counterbalances a column of mercury in a tube.

meridian of longitude north-south line on the surface of the global *oblate ellipsoid*, connecting the *north pole* and *south pole*.

meridional transport flow of *energy* (heat) or matter (water) across the *parallels of latitude*, either poleward or equatorward.

mesa table-topped *plateau* of comparatively small extent bounded by *cliffs* and occurring in a region of flat-lying *strata*.

mesopause upper limit of the *mesosphere*.

mesosphere atmospheric layer of upwardly diminishing temperature, situated above the stratopause and below the mesopause.

Mesozoic Era second of three geologic *eras* following *Precambrian time*.

metamorphic rock *rock* altered in physical structure or chemical (*mineral*) composition by action of heat, pressure, *shearing stress*, or infusion of elements, all taking place at substantial depth beneath the surface.

meteorology science of the *atmosphere*, particularly the physics of the lower or inner atmosphere.

mica group aluminum-silicate *mineral* group of complex chemical formula having perfect cleavage into thin sheets.

microburst brief onset of intense *winds* close to the ground beneath the downdraft zone of a *thunderstorm* cell.

microcontinent fragment of *continental crust* and its *lithosphere* of subcontinental dimensions that is embedded in an expanse of *oceanic lithosphere*.

micrometer metric unit of length equal to one-millionth of a meter (0.000001 m); abbreviated μm.

microwaves waves of the *electromagnetic radiation* spectrum in the *wavelength* band from about 0.03 cm to about 1 cm.

mid-oceanic ridge one of three major divisions of the ocean basins, being the central belt of submarine mountain topography with a characteristic *axial rift*.

midlatitude climates group of climates of the *midlatitude zone* and *subtropical zone*, located in the *polar front zone* and dominated by both tropical *air masses* and polar air masses.

midlatitude deciduous forest plant *formation class* within the *forest biome* dominated by tall, broadleaf deciduous *trees*, found mostly in the *moist continental climate* ⑩ and *marine west-coast climate* ⑧.

midlatitude zones latitude zones occupying the *latitude* range 35° to 55° N and S (more or less) and lying between the *subtropical zones* and the *subarctic (subantarctic) zones*.

millibar unit of *atmospheric pressure;* one-thousandth of a bar. *Bar* is a force of 1 million dynes per square centimeter.

mineral naturally occurring inorganic substance, usually having a definite chemical composition and a characteristic atomic structure. (See *felsic minerals, mafic minerals, silicate minerals.*)

mineral alteration chemical change of *minerals* to more stable compounds upon exposure to atmospheric conditions; same as *chemical weathering.*

mineral matter (soils) component of *soil* consisting of weathered or unweathered mineral grains.

mineral oxides (soils) secondary *minerals* found in *soils* in which original minerals have been altered by chemical combination with oxygen.

minute (of arc) 1/60 of a degree.

mistral local drainage wind of cold air affecting the Rhone Valley of southern France.

Moho contact surface between the earth's *crust* and *mantle;* a contraction of Mohorovic, the name of the seismologist who discovered this feature.

moist continental climate ⑩ moist climate of the *midlatitude zone* with strongly defined winter and summer seasons, adequate *precipitation* throughout the year, and a substantial annual *water surplus.*

moist subtropical climate ⑥ moist climate of the *subtropical zone,* characterized by a moderate to large annual *water surplus* and a strongly seasonal cycle of *potential evapotranspiration (water need).*

mollic epipedon relatively thick, dark-colored surface *soil horizon,* containing substantial amounts of organic matter (*humus*) and usually rich in *bases.*

Mollisols *soil order* consisting of *soils* with a *mollic horizon* and high *base status.*

monadnock prominent, isolated mountain or large hill rising conspicuously above a surrounding peneplain and composed of a *rock* more resistant than that underlying the peneplain; a *landform* of *denudation* in moist climates.

monsoon and trade-wind coastal climate ② moist climate of low latitudes showing a strong rainfall peak in the season of high sun and a short period of reduced rainfall.

monsoon forest *formation class* within the *forest biome* consisting in part of deciduous

trees adapted to a long dry season in the *wet-dry tropical climate* ③.

monsoon system system of low-level *winds* blowing into a continent in summer and out of it in winter, controlled by *atmospheric pressure* systems developed seasonally over the continent.

montane forest plant *formation class* of the *forest biome* found in cool upland environments of the *tropical zone* and *equatorial zone.*

moraine accumulation of *rock* debris carried by an *alpine glacier* or an *ice sheet* and deposited by the ice to become a *depositional landform.* (See *lateral moraine, terminal moraine.*)

mountain arc curving section of an *alpine chain* occurring on a *converging boundary* between two crustal plates.

mountain roots erosional remnants of deep portions of ancient *continental sutures* that were once *alpine chains.*

mountain winds daytime movements of air up the *gradient* of valleys and mountain slopes; alternating with nocturnal *valley winds.*

mucks organic *soils* largely composed of fine, black, sticky organic matter.

mud *sediment* consisting of a mixture of *clay* and *silt* with water, often with minor amounts of *sand* and sometimes with organic matter.

mudflow a form of *mass wasting* consisting of the downslope flowage of a mixture of water and *mineral* fragments (*soil, regolith,* disintegrated *bedrock*), usually following a natural drainage line or *stream channel.*

mudstone *sedimentary rock* formed by the lithification of *mud.*

multipurpose map map containing several different types of information.

multispectral image image consisting of two or more images, each of which is taken from a different portion of the spectrum (e.g., blue, green, red, infrared).

multispectral scanner *remote sensing* instrument, flown on an aircraft or spacecraft, that simultaneously collects multiple *digital images* (*multispectral images*) of the ground. Typically, images are collected in four or more spectral bands.

nappe overturned recumbent *fold* of *strata,* usually associated with *thrust sheets* in a collision *orogen.*

natural bridge natural *rock* arch spanning a *stream channel,* formed by cutoff of an *entrenched meander* bend.

natural flow systems flow systems of *energy* or naturally occurring substances that are powered largely or completely by natural power sources.

natural gas naturally occurring mixture of hydrocarbon compounds (principally methane) in the gaseous state held within certain porous *rocks.*

natural levee belt of higher ground paralleling a meandering *alluvial river* on both sides of the *stream channel* and built up by *deposition* of fine *sediment* during periods of overbank flooding.

natural vegetation stable, mature plant cover characteristic of a given area of land surface largely free from the influences and impacts of human activities.

needleleaf evergreen forest needleleaf *forest* composed of evergreen tree species, such as spruce, fir, and pine.

needleleaf forest plant *formation class* within the *forest biome,* consisting largely of needleleaf *trees.* (See also *boreal forest.*)

negative exponential mathematical form of a curve that smoothly decreases to approach a steady value, usually zero.

negative feedback in flow systems, a linkage between flow paths such that the flow in one *pathway* acts to reduce the flow in another *pathway.* (See also *feedback, positive feedback.*)

net photosynthesis *carbohydrate* production remaining in an organism after *respiration* has broken down sufficient *carbohydrate* to power the metabolism of the organism.

net primary production Rate at which *carbohydrate* is accumulated in the tissues of plants within a given *ecosystem;* units are kilograms of dry organic matter per year per square meter of surface area.

net radiation difference in intensity between all incoming *energy* (positive quantity) and all outgoing *energy* (negative quantity) carried by both *shortwave radiation* and *longwave radiation.*

neutron atomic particle contained within the nucleus of an atom; similar in mass to a proton but without a magnetic charge.

nitrogen cycle *biogeochemical cycle* in which nitrogen moves through the *biosphere* by the processes of *nitrogen fixation* and *denitrification.*

nitrogen fixation chemical process of conversion of *gaseous* molecular nitrogen of the *atmosphere* into compounds or ions that can be directly utilized by plants; a process carried out within the *nitrogen cycle* by certain microorganisms.

node point marking the end of a *line* or the intersection of *lines* as *spatial objects* in a *geographic information system.*

noon (See *solar noon.*)

noon angle (of the sun) angle of the sun above the horizon at its highest point during the day.

normal fault variety of *fault* in which the

fault plane inclines (*dips*) toward the downthrown block and a major component of the motion is vertical.

north pole   point at which the northern end of the earth's *axis of rotation* intersects the earth's surface.

northeast trade winds   surface *winds* of low latitudes that blow steadily from the northeast. (See also *trade winds*.)

nuclei (atmospheric)   minute particles of solid matter suspended in the *atmosphere* and serving as cores for *condensation* of water or ice.

nutrient cycle   (See *biogeochemical cycle*.)

O₁ horizon   surface *soil horizon* containing decaying organic matter that is recognizable as leaves, twigs, or other organic structures.

Oₐ horizon   *soil horizon* below the O₁ *horizon* containing decaying organic matter that is too decomposed to recognize as specific plant parts, such as leaves or twigs.

oasis   desert area where *groundwater* is tapped for crop irrigation and human needs.

oblate ellipsoid   geometric solid resembling a flattened sphere, with polar axis shorter than the equatorial diameter.

occluded front   weather *front* along which a moving *cold front* has overtaken a *warm front*, forcing the warm *air mass* aloft.

ocean basin floors   one of the major divisions of the ocean basins, comprising the deep portions consisting of *abyssal plains* and low hills.

ocean current   persistent, dominantly horizontal flow of ocean water.

ocean tide   periodic rise and fall of the ocean level induced by gravitational attraction between the earth and moon in combination with earth *rotation*.

oceanic crust   crust of basaltic composition beneath the ocean floors, capping *oceanic lithosphere*. (See also *continental crust*.)

oceanic lithosphere   *lithosphere* bearing *oceanic crust*.

oceanic trench   narrow, deep depression in the seafloor representing the line of *subduction* of an oceanic *lithospheric plate* beneath the margin of a continental lithospheric plate; often associated with an *island arc*.

olivine   *silicate mineral* with magnesium and iron but no aluminum, usually olive-green or grayish-green; a *mafic mineral*.

open flow system   system of interconnected flow paths of *energy* or matter with a boundary through which energy or matter can enter and leave the system.

organic matter (soils)   material in *soil* that

was originally produced by plants or animals and has been subjected to decay.

Organic order   a class of *soils* in the Canadian soil classification system that is composed largely of organic materials.

organic sediment   *sediment* consisting of the organic remains of plants or animals.

orogen   the mass of tectonically deformed *rocks* and related *igneous rocks* produced during an *orogeny*.

orogeny   major episode of *tectonic activity* resulting in *strata* being deformed by folding and faulting.

orographic   pertaining to mountains.

orographic precipitation   *precipitation* induced by the forced rise of moist air over a mountain barrier.

outcrop   surface exposure of *bedrock*.

outspiral   horizontal outward spiral or motion, such as that found in an *anticyclone*.

outwash   glacial deposit of stratified drift left by *braided streams* issuing from the front of a *glacier*.

outwash plain   flat, gently sloping plain built up of *sand* and gravel by the *aggradation* of meltwater *streams* in front of the margin of an *ice sheet*.

overburden   *strata* overlying a layer or *stratum* of interest, as overburden above a *coal* seam.

overland flow   motion of a surface layer of water over a sloping ground surface at times when the *infiltration* rate is exceeded by the *precipitation* rate; a form of *runoff*.

overthrust fault   *fault* characterized by the overriding of one crustal block (or *thrust sheet*) over another along a gently inclined *fault plane*; associated with crustal *compression*.

oxbow lake   crescent-shaped lake representing the abandoned channel left by the *cutoff* of an *alluvial meander*.

oxidation   chemical union of free oxygen with metallic elements in *minerals*.

oxide   chemical compound containing oxygen; in *soils*, iron oxides and aluminum oxides are examples.

Oxisols   *soil order* consisting of very old, highly weathered *soils* of low latitudes, with an oxic horizon and *low base status*.

oxygen cycle   *biogeochemical cycle* in which oxygen moves through the *biosphere* in both *gaseous* and sedimentary forms.

ozone   a form of oxygen with a molecule consisting of three atoms of oxygen, O_3.

ozone layer   layer in the *stratosphere*, mostly in the altitude range 20 to 35 km (12 to 31 mi), in which a concentration of *ozone* is produced by the action of solar *ultraviolet radiation*.

pack ice   floating *sea ice* that completely covers the sea surface.

Paleozoic Era   first of three geologic *eras* comprising all geologic time younger than *Precambrian time*.

Pangea   hypothetical parent continent, enduring until near the close of the *Mesozoic Era*, consisting of the *continental shields* of *Laurasia* and *Gondwana* joined into a single unit.

parabolic dunes   isolated low *sand dunes* of parabolic outline, with points directed into the prevailing *wind*.

parallel of latitude   east-west circle on the earth's surface, lying in a plane parallel with the *equator* and at right angles to the *axis of rotation*.

parent material   inorganic, *mineral* base from which the *soil* is formed; usually consists of *regolith*.

particulates   solid and *liquid* particles capable of being suspended for long periods in the *atmosphere*.

passive continental margins   continental margins lacking active plate boundaries at the contact of *continental crust* with *oceanic crust*. A passive margin thus lies within a single *lithospheric plate*. Example: Atlantic continental margin of North America. (See also *active continental margins, continental margins*.)

passive systems   electromagnetic *remote sensing* systems that measure radiant *energy* reflected or emitted by an object or surface.

pathway   in an *energy flow system*, a mechanism by which matter or *energy* flows from one part of the system to another.

patterned ground   general term for a ground surface that bears polygonal or ringlike features, including stone polygons, and ice wedge polygons; typically produced by *frost action* in severe climates.

peat   partially decomposed, compacted accumulation of plant remains occurring in a bog environment.

ped   individual natural *soil* aggregate.

pediment   gently sloping, rock-floored land surface found at the base of a mountain mass or *cliff* in an arid region.

pedogenic processes   group of recognized basic soil-forming processes, mostly involving the gain, loss, *translocation*, or transformation of materials within the *soil* body.

pedology   science of the *soil* as a natural surface layer capable of supporting living plants; synonymous with *soil science*.

peneplain   land surface of low elevation and slight relief produced in the late stages of *denudation* of a *landmass*.

percolation   slow, downward flow of water

by *gravity* through *soil* and subsurface layers toward the *water table.*

perennials plants that live for more than one growing season.

peridotite *igneous rock* consisting largely of *olivine* and *pyroxene,* an *ultramafic igneous rock* occurring as a *pluton,* also thought to compose much of the upper *mantle.*

periglacial located near the margin of *alpine glaciers* or large *ice sheets,* in an environment of intense *frost action.*

periglacial system a distinctive set of landforms and land-forming processes that are created by intense frost action.

perihelion point on the earth's elliptical orbit at which the earth is nearest to the sun.

period of geologic time time subdivision of the *era,* each ranging in duration between about 35 and 70 million years.

permafrost condition of permanently frozen water in the *soil, regolith,* and *bedrock* in cold climates of subarctic and arctic regions.

permafrost table in *permafrost,* the upper surface of perennially frozen ground; lower surface of the *active layer.*

petroleum (crude oil) natural *liquid* mixture of many complex hydrocarbon compounds of organic origin, found in accumulations (oil pools) within certain *sedimentary rocks.*

pH measure of the concentration of hydrogen ions in a solution. (The number represents the logarithm to the base 10 of the reciprocal of the weight in grams of hydrogen ions per liter of water.) Acid solutions have pH values less than 6, and basic solutions have pH values greater than 6.

photosynthesis production of carbohydrate by the union of water with *carbon dioxide* while absorbing light *energy.*

physical geography the study and synthesis of selected subject areas from the natural sciences—especially atmospheric science, *hydrology,* physical oceanography, *geology, geomorphology, soil science,* and *plant ecology*—in order to gain a complete picture of the physical environment of humans and to examine the interactions of humans with that environment.

physical weathering breakup of massive rock (*bedrock*) into small particles through the action of physical forces acting at or near the earth's surface. (See *weathering.*)

phytoplankton microscopic floating plants found largely in the uppermost layer of ocean water.

pingo conspicuous conical mound or circular hill, having a core of ice, found on

plains of the arctic tundra where *permafrost* is present.

pioneer plants plants that first invade an environment of new land or a *soil* that has been cleared of vegetation cover; often these are annual *herbs.*

plagioclase feldspar aluminum-silicate *mineral* with sodium or calcium, or both.

plane of the ecliptic imaginary plane in which the earth's orbit lies.

plant ecology the study of the relationships between plants and their environment.

plant nutrients *ions* or chemical compounds that are needed for plant growth.

plate tectonics theory of *tectonic activity* dealing with *lithospheric plates* and their activity.

plateau upland surface, more or less flat and horizontal, upheld by resistant beds of *sedimentary rock* or *lava* flows and bounded by a steep *cliff.*

playa flat land surface underlain by fine *sediment* or evaporite minerals deposited from shallow lake waters in a dry climate in the floor of a closed topographic depression.

Pleistocene Epoch *epoch* of the *Cenozoic Era,* often identified as the Ice Age; it preceded the *Holocene Epoch.*

plinthite iron-rich concentrations present in some kinds of *soils* in deeper *soil horizons* and capable of hardening into rocklike material with repeated wetting and drying.

plucking (See *glacial plucking.*)

pluton any body of *intrusive igneous rock* that has solidified below the surface, enclosed in preexisting *rock.*

pocket beach *beach* of crescentic outline located at a *bay* head.

podzol type of *soil order* closely equivalent to *Spodosol;* an order of the Canadian soil classification system.

Podzolic order a class of *forest* and heath *soils* in the Canadian soil classification system in which an amorphous material of humified organic matter with Al and Fe accumulates.

point *spatial object* in a *geographic information system* with no area.

point bar deposit of coarse bed-load *alluvium* accumulated on the inside of a growing *alluvial meander.*

polar easterlies system of easterly surface winds at high latitude, best developed in the southern *hemisphere,* over Antarctica.

polar front *front* lying between cold polar *air masses* and warm tropical air masses, often situated along a *jet stream* within the *upper-air westerlies.*

polar front jet stream *jet stream* found along the *polar front,* where cold polar air and warm tropical air are in contact.

polar front zone broad zone in midlatitudes and higher latitudes, occupied by the shifting *polar front.*

polar high persistent low-level center of high *atmospheric pressure* located over the *polar zone* of Antarctica.

polar outbreak tongue of cold polar air, preceded by a *cold front,* penetrating far into the *tropical zone* and often reaching the *equatorial zone;* it brings rain squalls and unusual cold.

polar projection map projection centered on earth's *north pole* or *south pole.*

polar zones *latitude* zones lying between 75° and 90° N and S.

poleward heat transport movement of heat from equatorial and tropical regions toward the poles, occurring as *latent* and *sensible heat transfer.*

pollutants in air pollution studies, foreign matter injected into the lower *atmosphere* as *particulates* or as chemical pollutant *gases.*

pollution dome broad, low dome-shaped layer of polluted air, formed over an urban area at times when winds are weak or calm prevails.

pollution plume (1) the trace or path of pollutant substances, moving along the flow paths of *groundwater;* (2) trail of polluted air carried downwind from a pollution source by strong *winds.*

polygon type of *spatial object* in a *geographic information system* with a closed chain of connected *lines* surrounding an area.

polypedon smallest distinctive geographic unit of the *soil* of a given area.

pool in flow systems, an area or location of concentration of matter. (See also *active pool, storage pool.*)

positive feedback in flow systems, a linkage between flow paths such that the flow in one *pathway* acts to increase the flow in another pathway. (See also *feedback, negative feedback.*)

potash feldspar aluminum-silicate *mineral* with potassium the dominant metal.

potential energy *energy* of position; produced by *gravitational* attraction of the earth's mass for a smaller mass on or near the earth's surface.

potential evapotranspiration (water need) ideal or hypothetical rate of *evapotranspiration* estimated to occur from a complete canopy of green foliage of growing plants continuously supplied with all the *soil water* they can use; a real condition reached in those situations where *precipi-*

tation is sufficiently great or irrigation water is supplied in sufficient amounts.

pothole cylindrical cavity in hard *bedrock* of a *stream channel* produced by *abrasion* of a rounded *rock* fragment rotating within the cavity.

prairie plant f*ormation class* of the *grassland biome*, consisting of dominant tall grasses and subdominant *forbs*, widespread in sub-humid continental climate regions of the *subtropical zone* and *midlatitude zone*. (See *short-grass prairie, tall-grass prairie*.)

Precambrian time all of geologic time older than the beginning of the Cambrian Period, that is, older than 600 million years.

precipitation particles of *liquid* water or ice that fall from the atmosphere and may reach the ground. (See *convectional precipitation, cyclonic precipitation, orographic precipitation*.)

preprocessing component component of a *geographic information system* that prepares data for entry to the system.

pressure gradient change of *atmospheric pressure* measured along a line at right angles to the *isobars*.

pressure gradient force force acting horizontally, tending to move air in the direction of lower *atmospheric pressure*.

prevailing westerly winds (westerlies) surface *winds* blowing from a generally westerly direction in the *midlatitude zone*, but varying greatly in direction and intensity.

primary consumers organisms at the lowest level of the *food chain* that ingest *primary producers* or *decomposers* as their *energy* source.

primary minerals in *pedology* (*soil science*), the original, unaltered *silicate minerals* of *igneous rocks* and *metamorphic rocks*.

primary producers organisms that use light *energy* to convert *carbon dioxide* and water to *carbohydrates* through the process of *photosynthesis*.

prime meridian reference meridian of zero *longitude*; universally accepted as the Greenwich meridian.

product generation component component of a *geographic information system* that provides output products such as maps, images, or tabular reports.

progradation shoreward building of a *beach, bar,* or *sandspit* by addition of coarse *sediment* carried by *littoral drift* or brought from deeper water offshore.

pyroxene group complex aluminum-silicate *minerals* rich in calcium, magnesium, and iron, dark in color, high in density, classed as *mafic minerals*.

quartz mineral of silicon dioxide composition.

quartzite *metamorphic rock* consisting largely of the mineral *quartz*.

quick clays *clay* layers that spontaneously change from a solid condition to a near-liquid condition when disturbed.

radar an active *remote sensing* system in which a pulse of radiation is emitted by an instrument, and the strength of the echo of the pulse is recorded.

radial drainage pattern stream pattern consisting of *streams* radiating outward from a central peak or highland, such as a *sedimentary dome* or a *volcano*.

radiant energy transfer net flow of radiant *energy* between an object and its surroundings.

radiation (See *electromagnetic radiation*.)

radiation balance condition of balance between incoming *energy* of solar *shortwave radiation* and outgoing *longwave radiation* emitted by the earth into space.

radiation fog *fog* produced by radiational cooling of the basal air layer.

radioactive decay spontaneous change in the nucleus of an atom that leads to the emission of matter and *energy*.

radiogenic heat heat from the earth's interior that is slowly released by the *radioactive decay* of *unstable isotopes*.

radiometric dating a method of determining the geologic age of a *rock* or *mineral* by measuring the proportions of certain of its elements in their different isotopic forms.

rain form of *precipitation* consisting of falling water drops, usually 0.5 mm or larger in diameter.

rain gauge instrument used to measure the amount of *rain* that has fallen.

rain-green vegetation vegetation that puts out green foliage in the wet season, but becomes largely dormant in the dry season; found in the *tropical zone*, it includes the *savanna biome* and *monsoon forest*.

rainshadow belt of arid climate to lee of a mountain barrier, produced as a result of adiabatic warming of descending air.

raised shoreline former *shoreline* lifted above the limit of wave action; also called an elevated shoreline.

rapids steep-*gradient* reaches of a *stream channel* in which *stream* velocity is high.

recumbent overturned, as a folded sequence of *rock* layers in which the folds are doubled back upon themselves.

reflection outward scattering of *radiation* toward space by the *atmosphere* or earth's surface.

reg desert surface armored with a pebble layer, resulting from long-continued *defla-*

tion; found in the Sahara Desert of North Africa.

regional scale the scale of observation at which subcontinental regions are discernible.

regolith layer of *mineral* particles overlying the *bedrock*; may be derived by *weathering* of underlying bedrock or be transported from other locations by *fluid* agents. (See *residual regolith, transported regolith*.)

Regosolic order a class of *soils* in the Canadian soil classification system that exhibits weakly developed *horizons*.

relative humidity ratio of *water vapor* present in the air to the maximum quantity possible for *saturated air* at the same temperature.

remote sensing measurement of some property of an object or surface by means other than direct contact; usually refers to the gathering of scientific information about the earth's surface from great heights and over broad areas, using instruments mounted on aircraft or orbiting space vehicles.

remote sensor instrument or device measuring *electromagnetic radiation* reflected or emitted from a target body.

removal in soil science, the set of processes that result in the removal of material from a *soil horizon*, such as surface erosion or *leaching*.

representative fraction (R.F.) (See *scale fraction*.)

residual regolith *regolith* formed in place by alteration of the *bedrock* directly beneath it.

resolution on a map, power to resolve small objects present on the ground.

respiration the oxidation of organic compounds by organisms that powers bodily functions.

retrogradation cutting back (retreat) of a *shoreline, beach, marine cliff,* or *marine scarp* by wave action.

reverse fault type of *fault* in which one fault block rides up over the other on a steep *fault plane*.

revolution motion of a planet in its orbit around the sun, or of a planetary satellite around a planet.

rhyolite *extrusive igneous rock* of *granite* composition; it occurs as *lava* or *tephra*.

ria coastal embayment or *estuary*.

ria coast deeply embayed *coast* formed by partial *submergence* of a *landmass* previously shaped by fluvial *denudation*.

Richter scale scale of magnitude numbers describing the quantity of *energy* released by an *earthquake*.

ridge-and-valley landscape assemblage of *landforms* developed by *denudation* of a system of open *folds* of *strata* and consisting of long, narrow ridges and valleys arranged in parallel or zigzag patterns.

rift valley trenchlike valley with steep, parallel sides; essentially a *graben* between two *normal faults*; associated with crustal spreading.

rill erosion form of *accelerated erosion* in which numerous, closely spaced miniature channels (rills) are scored into the surface of exposed *soil* or *regolith*.

rock natural aggregate of *minerals* in the solid state; usually hard and consisting of one, two, or more mineral varieties.

rock terrace terrace carved in *bedrock* during the *degradation* of a *stream channel* induced by the crustal rise or a fall of the sea level. (See also *alluvial terrace, marine terrace.*)

Rossby waves horizontal undulations in the flow path of the *upper-air westerlies*; also known as upper-air waves.

rotation spinning of an object around an axis.

runoff flow of water from continents to oceans by way of *stream flow* and *groundwater* flow; a term in the water balance of the *hydrologic cycle*. In a more restricted sense, runoff refers to surface flow by *overland flow* and channel flow.

Sahel (Sahelian zones) belt of *wet-dry tropical* ③ and *semiarid dry tropical* ④ climate in Africa in which *precipitation* is highly variable from year to year.

salic horizon *soil horizon* enriched by soluble salts.

salinization precipitation of soluble salts within the *soil*.

salt flat shallow basin covered with salt deposits formed when stream input to the basin is subjected to severe *evaporation*; may also form by evaporation of a saline lake when climate changes.

salt marsh *peat*-covered expanse of *sediment* built up to the level of high tide over a previously formed tidal mud flat.

salt-crystal growth a form of *weathering* in which *rock* is disintegrated by the expansive pressure of growing salt crystals during dry weather periods when *evaporation* is rapid.

saltation leaping, impacting, and rebounding of sand grains transported over a *sand* or pebble surface by *wind*.

sand *sediment* particles between 0.06 and 2 mm in diameter.

sand dune hill or ridge of loose, well-sorted *sand* shaped by *wind* and usually capable of downwind motion.

sand sea field of *transverse dunes*.

sandspit narrow, fingerlike embankment of *sand* constructed by *littoral drift* into the open water of a *bay*.

sandstone variety of *sedimentary rock* consisting largely of mineral particles of sand grade size.

Santa Ana easterly *wind*, often hot and dry, that blows from the interior desert region of southern California and passes over the coastal mountain ranges to reach the Pacific Ocean.

saturated air air holding the maximum possible quantity of *water vapor* at a given temperature and pressure.

saturated zone zone beneath the land surface in which all pores of the *bedrock* or *regolith* are filled with *groundwater*.

savanna a vegetation cover of widely spaced *trees* with a grassland beneath.

savanna biome *biome* that consists of a combination of *trees* and grassland in various proportions.

savanna woodland plant *formation class* of the *savanna biome* consisting of a *woodland* of widely spaced *trees* and a grass layer, found throughout the *wet-dry tropical climate* ③ regions in a belt adjacent to the *monsoon forest* and *low-latitude rainforest*.

scale fraction ratio that relates distance on the earth's surface to distance on a map or surface of a globe.

scale of globe ratio of size of a globe to size of the earth, where size is expressed by a measure of length or distance.

scale of map ratio of distance between two points on a map and the same two points on the ground.

scanning systems *remote sensing* systems that make use of a scanning beam to generate images over the frame of surveillance.

scarification general term for artificial excavations and other land disturbances produced for purposes of extracting or processing mineral resources.

scattering turning aside by reflection of solar *shortwave radiation* by gas molecules of the *atmosphere*.

schist foliated *metamorphic rock* in which mica flakes are typically found oriented parallel with foliation surfaces.

sclerophyll forest plant *formation class* of the *forest biome*, consisting of low sclerophyll *trees*, and often including sclerophyll woodland or *scrub*, associated with regions of *Mediterranean climate* ⑦.

sclerophyll woodland plant *formation class* of the *forest biome* composed of widely spaced sclerophyll *trees* and *shrubs*.

sclerophylls hardleaved evergreen *trees* and *shrubs* capable of enduring a long, dry summer.

scoria *lava* or *tephra* containing numerous cavities produced by expanding gases during cooling.

scrub plant *formation class* or subclass consisting of *shrubs* and having a canopy coverage of about 50 percent.

sea arch archlike *landform* of a rocky, cliffed coast created when waves erode through a narrow headland from both sides.

sea breeze local *wind* blowing from sea to land during the day.

sea cave cave near the base of a *marine cliff*, eroded by breaking waves.

sea fog *fog* layer formed at sea when warm moist air passes over a cool ocean current and is chilled to the *condensation* point.

sea ice floating ice of the oceans formed by direct *freezing* of ocean water.

second of arc 1/60 of a minute, or 1/3600 of a degree.

secondary consumers animals that feed on *primary consumers*.

secondary minerals in *soil science, minerals* that are stable in the surface environment, derived by *mineral alteration* of the *primary minerals*.

sediment finely divided *mineral matter* and organic matter derived directly or indirectly from preexisting *rock* and from life processes. (See *chemically precipitated sediment, organic sediment.*)

sediment yield quantity of sediment removed by *overland flow* from a land surface of given unit area in a given unit of time.

sedimentary cycle type of biogeochemical cycle in which the compound or element is released from *rock* by *weathering*, follows the movement of running water either in solution or as *sediment* to reach the sea, and is eventually converted into *rock*.

sedimentary dome up-arched *strata* forming a circular structure with domed summit and flanks with moderate to steep outward *dip*.

sedimentary rock *rock* formed from accumulation of *sediment*.

seismic sea wave (tsunami) train of sea waves set off by an *earthquake* (or other seafloor disturbance) traveling over the ocean surface.

seismic waves waves sent out during an *earthquake* by faulting or other crustal disturbance from an *earthquake focus* and propagated through the solid earth.

semiarid (dry climate subtype) subtype of the dry climates exhibiting a short wet season supporting the growth of grasses and *annual* plants.

semiarid (steppe) climate subtype subtype of the dry climate in which soil-water storage equals or exceeds 6 cm in at least two months of the year.

semidesert plant *formation class* of the *desert biome*, consisting of xerophytic *shrub* vegetation with a poorly developed herbaceous lower layer; subtypes are semidesert scrub and *woodland.*

sensible heat heat measurable by a *thermometer*; an indication of the intensity of *kinetic energy* of molecular motion within a substance.

sensible heat transfer flow of heat from one substance to another by direct contact.

sequential landforms *landforms* produced by external earth processes in the total activity of *denudation.* Examples: *alluvial fan, floodplain, gorge.*

sesquioxides oxides of aluminum or iron with a ratio of two atoms of aluminum or iron to three atoms of oxygen.

shale fissile, *sedimentary rock* of *mud* or *clay* composition, showing lamination.

shearing (of rock) slipping motion between very thin *rock* layers, like a deck of cards fanned with the sweep of a palm.

sheet erosion type of *accelerated soil erosion* in which thin layers of *soil* are removed without formation of rills or *gullies.*

sheet flow overland flow taking the form of a continuous thin film of water over a smooth surface of *soil, regolith,* or *rock.*

sheeting structure thick, subparallel layers of massive *bedrock* formed by spontaneous expansion accompanying *unloading.*

shield volcano low, often large, domelike accumulation of basalt lava flows emerging from long radial fissures on flanks.

shoreline shifting line of contact between water and land.

short-grass prairie plant *formation class* in the *grassland biome* consisting of short grasses sparsely distributed in clumps and bunches and some *shrubs*, widespread in areas of semiarid climate in continental interiors of North America and Eurasia; also called *steppe.*

shortwave infrared *infrared radiation* with wavelengths shorter than 3µm.

shortwave radiation *electromagnetic energy* in the range from 0.2 to 3 µm, including most of the *energy* spectrum of solar radiation.

shrubs woody perennial plants, usually small or low, with several low-branching stems and a foliage mass close to the ground.

silica silicon dioxide in any of several mineral forms.

silicate minerals (silicates) *minerals* containing silicon and oxygen atoms, linked in the crystal space lattice in units of four oxygen atoms to each silicon atom.

sill *intrusive igneous rock* in the form of a plate where *magma* was forced into a nat-

ural parting in the *bedrock*, such as a bedding surface in a sequence of *sedimentary rocks.*

silt *sediment* particles between 0.004 and 0.06 mm in diameter.

sinkhole surface depression in *limestone*, leading down into *limestone caverns.*

slash-and-burn agricultural system, practiced in the *low-latitude rainforest*, in which small areas are cleared and the *trees* burned, forming plots that can be cultivated for brief periods.

slate compact, fine-grained variety of *metamorphic rock*, derived from *shale*, showing well-developed cleavage.

sleet form of *precipitation* consisting of ice pellets, which may be frozen raindrops.

sling psychrometer form of *hygrometer* consisting of a wet-bulb thermometer and a dry-bulb thermometer.

slip face steep face of an active *sand dune*, receiving sand by *saltation* over the dune crest and repeatedly sliding because of oversteepening.

slope (1) degree of inclination from the horizontal of an element of ground surface, analogous to *dip* in the geologic sense; (2) any portion or element of the earth's solid surface; (3) verb meaning "to incline."

small-scale map map with *fractional scale* of less than 1:100,000; usually shows a large area.

smog mixture of *aerosols* and chemical *pollutants* in the lower *atmosphere*, usually found over urban areas.

snow form of *precipitation* consisting of ice particles.

soil natural terrestrial surface layer containing living matter and supporting or capable of supporting plants.

soil colloids mineral particles of extremely small size, capable of remaining suspended indefinitely in water; typically, they have the form of thin plates or scales.

soil creep extremely slow downhill movement of *soil* and *regolith* as a result of continued agitation and disturbance of the particles by such activities as *frost action*, temperature changes, or wetting and drying of the soil.

soil enrichment additions of materials to the *soil* body; one of the *pedogenic processes.*

soil erosion erosional removal of material from the *soil* surface.

soil horizon distinctive layer of the *soil*, more or less horizontal, set apart from other soil zones or layers by differences in physical and chemical composition, organic content, structure, or a combination of those properties, produced by soil-forming processes.

soil orders those 11 *soil* classes forming the highest category in the classification of soils.

soil profile display of *soil horizons* on the face of a freshly cut vertical exposure through the *soil.*

soil science (See *pedology.*)

soil solum that part of the *soil* made up of the A, E, and B soil horizons; the soil zone in which living plant roots can influence the development of soil horizons.

soil structure presence, size, and form of aggregations (lumps or clusters) of *soil* particles.

soil texture descriptive property of the *mineral* portion of the *soil* based on varying proportions of *sand, silt,* and *clay.*

soil water water held in the *soil* and available to plants through their root systems; a form of *subsurface water.*

soil-water balance balance among the component terms of the *soil-water budget*; namely, *precipitation, evapotranspiration,* change in *soil-water storage,* and *water surplus.*

soil-water belt *soil* layer from which plants draw *soil water.*

soil-water budget accounting system evaluating the daily, monthly, or yearly amounts of *precipitation, evapotranspiration, soil-water storage,* water deficit, and water surplus.

soil-water recharge restoring of depleted *soil water* by *infiltration* of *precipitation.*

soil-water shortage (See *deficit.*)

soil-water storage actual quantity of water held in the *soil-water belt* at any given instant; usually applied to a soil layer of given depth, such as 300 cm (about 12 in.).

solar constant intensity of solar radiation falling upon a unit area of surface held at right angles to the sun's rays at a point outside the earth's *atmosphere*, equal to an *energy* flow of about 1400 W/m².

solar day average time required for the earth to complete one *rotation* with respect to the sun; time elapsed between one solar noon and the next, averaged over the period of one year.

solar noon instant at which the *subsolar point* crosses the *meridian of longitude* of a given point on the earth; instant at which the sun's shadow points exactly due north or due south at a given location.

solids substances in the solid state; they resist changes in shape and volume, are usually capable of withstanding large unbalanced forces without yielding, but will ultimately yield by sudden breakage.

solifluction tundra (arctic) variety of *earthflow* in which the saturated thawed layer over *permafrost* flows slowly downhill to produce multiple terraces and solifluction lobes.

solifluction lobe bulging mass of saturated *regolith* with steep curved front moved downhill by *solifluction*.

solifluction terrace mass of saturated *regolith* formed by *solifluction* into a flat-topped terrace.

Solonetzic order a class of *soils* in the Canadian soil classification system with a *B horizon* of sticky *clay* that dries to a very hard condition.

sorting separation of one grade size of *sediment* particles from another by the action of currents of air or water.

source region extensive land or ocean surface over which an *air mass* derives its temperature and moisture characteristics.

south pole point at which the southern end of the earth's *axis of rotation* intersects the earth's surface.

southeast trade winds surface *winds* of low latitudes that blow steadily from the southeast. (See also *trade winds*.)

Southern Oscillation episodic reversal of prevailing barometric pressure differences between two regions, one centered on Darwin, Australia, in the eastern Indian Ocean, and the other on Tahiti in the western Pacific Ocean; a precursor to the occurrence of an El Niño event. (See also *El Ñino*.)

southern pine forest subtype of *needleleaf forest* dominated by pines and occurring in the *moist subtropical climate*.

spatial data information associated with a specific location or area of the earth's surface.

spatial object a geographic area, *line*, or *point* to which information is attached.

specific heat property of a substance that governs the temperature change of the substance with a given input of heat *energy*.

specific humidity mass of *water vapor* contained in a unit mass of air.

spit (See *sandspit*.)

splash erosion *soil erosion* caused by direct impact of falling raindrops on a wet surface of *soil* or *regolith*.

spodic horizon *soil horizon* containing precipitated amorphous materials composed of organic matter and *sesquioxides* of aluminum, with or without iron.

Spodosols *soil order* consisting of *soils* with a *spodic horizon*, an *albic horizon*, with low *base status*, and lacking in *carbonate* materials.

spreading plate boundary *lithospheric plate* boundary along which two plates of *oceanic lithosphere* are undergoing separation, while at the same time new lithos-

phere is being formed by *accretion*. (See also *transform plate boundary*.)

stable air mass *air mass* in which the *environmental temperature lapse rate* is less than the *dry adiabatic lapse rate*, inhibiting *convectional* uplift and mixing.

stack (marine) isolated columnar mass of *bedrock* left standing in front of a retreating *marine cliff*.

stage height of the surface of a river above its bed or a fixed level near the bed.

standard meridians *standard time* meridians separated by 15° of *longitude* and having values that are multiples of 15°. (In some cases meridians are used that are multiples of $7^1/2°$.)

standard time system time system based on the local time of a *standard meridian* and applied to belts of *longitude* extending $7^1/2°$ (more or less) on either side of that meridian.

standard time zone zone of the earth in which all inhabitants keep the same time, which is that of a *standard meridian* within the zone.

star dune large, isolated *sand dune* with radial ridges culminating in a peaked summit; found in the deserts of North Africa and the Arabian Peninsula.

steppe (See *short-grass prairie*.)

steppe climate (See *semiarid (steppe) climate subtype*.)

stone polygons linked ringlike ridges of cobbles or boulders lying at the surface of the ground in arctic and alpine tundra regions.

storage capacity maximum capacity of *soil* to hold water against the pull of *gravity*.

storage pool type of pool in a *biogeochemical cycle* in which materials are largely inaccessible to life. (See also *active pool*.)

storage recharge restoration of stored soil water during periods when *precipitation* exceeds *potential evapotranspiration* (*water need*).

storage withdrawal depletion of stored soil water during periods when *evapotranspiration* exceeds *precipitation*, calculated as the difference between *actual evapotranspiration* (*water use*) and precipitation.

storm surge rapid rise of coastal water level accompanying the onshore arrival of a *tropical cyclone*.

strata layers of *sediment* or *sedimentary rock* in which individual beds are separated from one another along bedding planes.

stratified drift *glacial drift* made up of sorted and layered *clay*, *silt*, *sand*, or *gravel* deposited from meltwater in *stream channels*, or in marginal lakes close to the ice front.

stratiform clouds clouds of layered, blanketlike form.

stratosphere layer of *atmosphere* lying directly above the *troposphere*.

stratovolcano volcano constructed of multiple layers of *lava* and *tephra* (volcanic ash).

stratus cloud type of the low-height family formed into a dense, dark gray layer.

stream long, narrow body of flowing water occupying a *stream channel* and moving to lower levels under the force of *gravity*. (See *consequent stream*, *graded stream*, *subsequent stream*.)

stream capacity maximum *stream load* of solid matter that can be carried by a *stream* for a given *discharge*.

stream channel long, narrow, troughlike depression occupied and shaped by a *stream* moving to progressively lower levels.

stream deposition accumulation of transported particles on a *stream* bed, upon the adjacent *floodplain*, or in a body of standing water.

stream erosion progressive removal of mineral particles from the floor or sides of a *stream channel* by drag force of the moving water, or by *abrasion*, or by *corrosion*.

stream flow water flow in a *stream channel*; same as channel flow.

stream gradient rate of descent to lower elevations along the length of a *stream channel*, stated in m/km, ft/mi, degrees, or percent.

stream load solid matter carried by a *stream* in dissolved form (as *ions*), in *turbulent suspension*, and as *bed load*.

stream profile a graph of the elevation of a *stream* plotted against its distance downstream.

stream transportation downvalley movement of eroded particles in a *stream channel* in solution, in *turbulent suspension*, or as *bed load*.

strike compass direction of the line of intersection of an inclined *rock* plane and a horizontal plane of reference. (See *dip*.)

strike-slip fault variety of *fault* on which the motion is dominantly horizontal along a near-vertical *fault plane*.

strip mining mining method in which overburden is first removed from a seam of *coal*, or a sedimentary ore, allowing the coal or ore to be extracted.

subantarctic low-pressure belt persistent belt of low *atmospheric pressure* centered about at lat. 65°S over the Southern Ocean.

subantarctic zone *latitude* zone lying between lat. 55° and 60° S (more or less) and occupying a region between the *midlatitude zone* and the *antarctic zone*.

subarctic zone *latitude* zone between lat. 55° and 60° N (more or less), occupying a region between the *midlatitude zone* and the *arctic zone.*

subduction descent of the downbent edge of a *lithospheric plate* into the *asthenosphere* so as to pass beneath the edge of the adjoining plate.

sublimation process of change of ice (solid state) to *water vapor* (*gaseous* state); in *meteorology*, sublimation also refers to the change of state from water vapor (liquid) to ice (solid), which is referred to as *deposition* in this text.

submergence inundation or partial drowning of a former land surface by a rise of sea level or a sinking of the *crust* or both.

suborder a unit of *soil* classification representing a subdivision of the *soil order.*

subsea permafrost *permafrost* lying below sea level, found in a shallow offshore zone fringing the arctic seacoast.

subsequent stream *stream* that develops its course by *stream erosion* along a band or belt of weaker *rock.*

subsolar point point on the earth's surface at which solar rays are perpendicular to the surface.

subsurface water water of the lands held in *soil, regolith,* or *bedrock* below the surface.

subtropical broadleaf evergreen forest a formation class of the *forest biome* composed of broadleaf evergreen *trees*; occurs primarily in the regions of the *moist subtropical climate* ⑥.

subtropical evergreen forest a subdivision of the *forest biome* composed of both broadleaf and needleleaf evergreen *trees.*

subtropical high-pressure belts belts of persistent high *atmospheric pressure* trending east-west and centered about on lat. 30° N and S.

subtropical jet stream *jet stream* of westerly winds forming at the *tropopause*, just above the *Hadley cell.*

subtropical needleleaf evergreen forest a *formation class* of the *forest biome* composed of needleleaf evergreen *trees* occurring in the *moist subtropical climate* ⑥ of the southeastern United States; also referred to as the southern pine forest.

subtropical zones *latitude* zones occupying the region of lat. 25° to 35° N and S (more or less) and lying between the *tropical zones* and the *midlatitude zones.*

summer monsoon inflow of maritime air at low levels from the Indian Ocean toward the Asiatic low-pressure center in the season of high sun; associated with the rainy season of the *wet-dry tropical climate* ③ and the Asiatic monsoon climate.

summer solstice solstice occurring on June 21 or 22, when the *subsolar point* is located at 23 ¹/₂°N.

sun-synchronous orbit satellite orbit in which the orbital plane remains fixed in position with respect to the sun.

supercontinent single world continent, formed when *plate tectonic* motions move continents together into a single, large landmass. (See also *Pangea.*)

supercooled water water existing in the liquid state at a temperature lower than the normal *freezing* point.

surface energy balance equation equation expressing the balance among *heat* flows to and from a surface.

surface water water of the lands flowing freely (as *streams*) or impounded (as ponds, *lakes*, marshes).

surges episodes of very rapid downvalley movement within an *alpine glacier.*

suspended load that part of the *stream load* carried in *turbulent suspension.*

suspension (See *turbulent suspension.*)

suture (See *continental suture.*)

swash surge of water up the *beach* slope (landward) following collapse of a breaker.

synclinal mountain steep-sided ridge or elongate mountain developed by erosion of a syncline.

synclinal valley valley eroded on weak *strata* along the central trough or axis of a *syncline.*

syncline downfold of *strata* (or other layered *rock*) in a troughlike structure; a class of *folds.* (See also *anticline.*)

systems theory body of knowledge explaining how systems work.

taiga plant *formation class* consisting of *woodland* with low, widely spaced *trees* and a ground cover of lichens and mosses, found along the northern fringes of the region of *boreal forest climate* ⑪; also called cold woodland.

tall-grass prairie a *formation class* of the *grassland biome* that consists of tall grasses with broad-leaved *herbs.*

talus accumulation of loose *rock* fragments derived by fall of *rock* from a *cliff.*

talus slope slope formed of *talus.*

tar sand (See *bitumen.*)

tarn small *lake* occupying a *rock* basin in a *cirque* of *glacial trough.*

tectonic activity process of bending (folding) and breaking (faulting) of crustal mountains, concentrated on or near active *lithospheric plate* boundaries.

tectonic arc long, narrow chain of islands or mountains or a narrow submarine ridge adjacent to a *subduction* boundary and its trench, formed by *tectonic processes,* such as the construction and rise of an *accretionary prism.*

tectonic crest ridgelike summit line of a *tectonic arc* associated with an *accretionary prism.*

tectonics branch of *geology* relating to tectonic activity and the features it produces. (See also *plate tectonics, tectonic activity.*)

temperature gradient rate of temperature change along a selected line or direction.

temperature inversion upward reversal of the normal *environmental temperature lapse rate*, so that the air temperature increases upward. (See *high-level temperature inversion, low-level temperature inversion.*)

temperature regime distinctive type of annual temperature cycle.

tephra collective term for all size grades of solid *igneous rock* particles blown out under gas pressure from a volcanic vent.

terminal moraine *moraine* deposited as an embankment at the terminus of an *alpine glacier* or at the leading edge of an *ice sheet.*

terrane continental crustal *rock* unit having a distinctive set of lithologic properties, reflecting its geologic history, that distinguish it from adjacent or surrounding *continental crust.*

terrestrial ecosystems *ecosystems* of land plants and animals found on upland surfaces of the continents.

thematic map map showing a single type of information.

theme category or class of information displayed on a map.

thermal erosion in regions of permafrost, the physical disruption of the land surface by melting of *ground ice*, brought about by removal of a protective organic layer.

thermal infrared a portion of the *infrared radiation wavelength* band, from approximately from 3 to 20 μm, in which objects at temperatures encountered on the earth's surface (including fires) emit *electromagnetic radiation.*

thermal pollution form of water pollution in which heated water is discharged into a *stream* or *lake* from the cooling system of a power plant or other industrial heat source.

thermistor electronic device that measures (air) temperature.

thermocline water layer of a lake or the ocean in which temperature changes rapidly in the vertical direction.

thermokarst in arctic environments, a hummocky terrain produced by thawing of *permafrost* and subsequent water erosion when the natural surface cover is disturbed.

thermometer instrument measuring temperature.

thermometer shelter louvered wooden cabinet of standard construction used to hold *thermometers* and other weather-monitoring equipment.

thermosphere atmospheric layer of upwardly increasing temperature, lying above the *mesopause*.

thorntree semidesert *formation class* within the *desert biome*, transitional from *grassland biome* and *savanna biome* and consisting of xerophytic *trees* and *shrubs*.

thorntree–tall-grass savanna plant *formation class*, transitional between the *savanna biome* and the *grassland biome*, consisting of widely scattered *trees* in an open grassland.

thrust sheet sheetlike mass of *rock* moving forward over a *low-angle overthrust fault*.

thunderstorm intense, local convectional storm associated with a *cumulonimbus cloud* and yielding heavy *precipitation*, also with lightning and thunder, and sometimes the fall of *hail*.

tidal current current set in motion by the *ocean tide*.

tidal inlet narrow opening in a *barrier island* or baymouth *bar* through which *tidal currents* flow.

tide (See *ocean tide*.)

tide curve graphical presentation of the rhythmic rise and fall of ocean water because of *ocean tides*.

till heterogeneous mixture of *rock* fragments ranging in size from *clay* to boulders, deposited beneath moving glacial ice or directly from the *melting* in place of stagnant glacial ice.

till plain undulating, plainlike land surface underlain by glacial *till*.

time cycle in flow systems, a regular alternation of flow rates with time.

time zones zones or belts of given east-west (*longitudinal*) extent within which *standard time* is applied according to a uniform system.

topographic contour *isopleth* of uniform elevation appearing on a map.

tornado small, very intense wind vortex with extremely low air pressure in center, formed beneath a dense *cumulonimbus cloud* in proximity to a *cold front*.

trade winds (trades) surface winds in low latitudes, representing the low-level air flow within the *tropical easterlies*.

transcurrent fault *fault* on which the relative motion is dominantly horizontal, in the direction of the *strike* of the fault; also called a *strike-slip fault*.

transform fault special case of a *strike-slip fault* making up the boundary of two moving *lithospheric plates*; usually found along an offset of the *mid-oceanic ridge* where seafloor spreading is in progress.

transform plate boundary *lithospheric plate* boundary along which two plates are in contact on a *transform fault*; the relative motion is that of a *strike-slip fault*.

transform scar linear topographic feature of the ocean floor taking the form of an irregular scarp or ridge and originating at the offset *axial rift* of the *mid-oceanic ridge*; it represents a former *transform fault* but is no longer a plate boundary.

transformation (soils) a class of soil-forming processes that transform materials within the soil body; examples include *mineral alteration* and *humification*.

translocation a soil-forming process in which materials are moved within the soil body, usually from one horizon to another.

transpiration evaporative loss of water to the *atmosphere* from leaf pores of plants.

transportation (See *stream transportation*.)

transported regolith *regolith* formed of *mineral matter* carried by *fluid* agents from a distant source and deposited on the *bedrock* or on older regolith. Examples: floodplain silt, lake clay, beach sand.

transverse dunes field of wavelike *sand dunes* with crests running at right angles to the direction of the prevailing *wind*.

traveling anticyclone center of high pressure and outspiraling *winds* that travels over the earth's surface; often associated with clear, dry weather.

traveling cyclone center of low pressure and inspiraling winds that travels over the earth's surface; includes *wave cyclones, tropical cyclones*, and *tornadoes*.

travertine *carbonate mineral matter*, usually *calcite*, accumulating on *limestone cavern* surfaces situated in the *unsaturated zone*.

tree large erect woody perennial plant typically having a single main trunk, few branches in the lower part, and a branching crown.

trellis drainage pattern *drainage pattern* characterized by a dominant parallel set of major *subsequent streams*, joined at right angles by numerous short tributaries; typical of *coastal plains* and belts of eroded *folds*.

tropic of cancer *parallel of latitude* at 23 1/2°N.

tropic of capricorn *parallel of latitude* at 23 1/2°S.

tropical cyclone intense *traveling cyclone* of tropical and subtropical latitudes, accompanied by high *winds* and heavy rainfall.

tropical easterlies low-latitude wind system of persistent air flow from east to west between the two *subtropical high-pressure belts*.

tropical easterly jet stream upper-air *jet stream* of seasonal occurrence, running east to west at very high altitudes over Southeast Asia.

tropical high-pressure belt a high-pressure belt occurring in tropical latitudes at a high level in the *troposphere*; extends downward and poleward to form the *subtropical high-pressure belt*, located at the surface.

tropical zones *latitude* zones centered on the *tropic of cancer* and the *tropic of capricorn*, within the latitude ranges 10° to 25° N and 10° to 25° S, respectively.

tropical-zone rainforest plant *formation class* within the *forest biome* similar to *equatorial rainforest*, but occurring farther poleward in tropical regions.

tropopause boundary between *troposphere* and *stratosphere*.

troposphere lowermost layer of the *atmosphere* in which air temperature falls steadily with increasing altitude.

tsunami (See *seismic sea wave*.)

tundra biome *biome* of the cold regions of *arctic tundra* and *alpine tundra*, consisting of grasses, grasslike plants, flowering *herbs*, dwarf *shrubs*, mosses, and *lichens*.

tundra climate ⑫ cold climate of the *arctic zone* with eight or more consecutive months of zero *potential evapotranspiration* (*water need*).

tundra soils soils of the arctic *tundra climate* ⑫ regions.

turbulence in *fluid* flow, the motion of individual water particles in complex eddies, superimposed on the average downstream flow path.

turbulent flow mode of *fluid* flow in which individual *fluid* particles (molecules) move in complex eddies, superimposed on the average downstream flow path.

turbulent suspension *stream transportation* in which particles of *sediment* are held in the body of the *stream* by turbulent eddies. (Also applies to wind transportation.)

typhoon *tropical cyclone* of the western North Pacific and coastal waters of Southeast Asia.

Udalfs suborder of the *soil order Alfisols*; includes Alfisols of moist regions, usually in the *midlatitude zone*, with deciduous forest as the natural vegetation.

Udolls suborder of the *soil order Mollisols*; includes Mollisols of the moist soil-water regime in the *midlatitude zone* and with no horizon of *calcium carbonate* accumulation.

Ultisols *soil order* consisting of *soils* of warm soil temperatures with an *argillic horizon* and low *base status*.

ultramafic igneous rock *igneous rock* composed almost entirely of *mafic minerals*, usually *olivine* or *pyroxene group*.

ultraviolet radiation *electromagnetic energy* in the *wavelength* range of 0.2 to 0.4 μm.

unloading process of removal of overlying *rock* load from *bedrock* by processes of *denudation*, accompanied by expansion and often leading to the development of *sheeting structure*.

unsaturated zone *subsurface water* zone in which pores are not fully saturated, except at times when *infiltration* is very rapid; lies above the *saturated zone*.

unstable air air with substantial content of *water vapor*, capable of breaking into spontaneous convectional activity leading to the development of heavy showers and *thunderstorms*.

unstable isotope elemental isotope that spontaneously decays to produce one or more new isotopes. (See also *daughter product*.)

upper-air westerlies system of westerly *winds* in the upper *atmosphere* over middle and high latitudes.

upwelling upward motion of cold, nutrient-rich ocean waters, often associated with cool equatorward currents occurring along *continental margins*.

Ustalfs suborder of the *soil order Alfisols*; includes Alfisols of semiarid and seasonally dry climates in which the *soil* is dry for a long period in most years.

Ustolls suborder of the *soil order Mollisols*; includes Mollisols of the semiarid climate in the *midlatitude zone*, with a horizon of *calcium carbonate* accumulation.

valley winds air movement at night down the *gradient* of valleys and the enclosing mountainsides; alternating with daytime *mountain winds*.

veins small, irregular, branching network of *intrusive rock* within a preexisting *rock* mass.

vernal equinox *equinox* occurring on March 20 or 21, when the *subsolar point* is at the *equator*.

Vertisols *soil order* consisting of *soils* of the *subtropical zone* and the *tropical zone* with high *clay* content, developing deep, wide cracks when dry, and showing evidence of movement between aggregates.

visible light *electromagnetic energy* in the *wavelength* range of 0.4 to 0.7 μm.

volcanic bombs boulder-sized, semisolid masses of *lava* that are ejected from an erupting *volcano*.

volcanic neck isolated, narrow steep-sided peak formed by erosion of *igneous rock* previously solidified in the feeder pipe of an extinct *volcano*.

volcanism general term for *volcano* building and related forms of extrusive igneous activity.

volcano conical, circular structure built by accumulation of *lava* flows and *tephra*. (See *shield volcano*, *stratovolcano*.)

volcano coast *coast* formed by *volcanoes* and *lava* flows built partly below and partly above sea level.

warm front moving weather *front* along which a warm *air mass* is sliding up over a cold air mass, leading to production of *stratiform clouds* and *precipitation*.

warm-core ring circular eddy of warm water, surrounded by cold water and lying adjacent to a warm, poleward moving ocean current, such as the Gulf Stream. (See also *cold-core ring*.)

washout downsweeping of atmospheric *particulates* by *precipitation*.

water gap narrow transverse *gorge* cut across a narrow ridge by a *stream*, usually in a region of eroded *folds*.

water need (See *potential evapotranspiration*.)

water surplus water disposed of by *runoff* or percolation to the groundwater zone after the *storage capacity* of the *soil* is full.

water table upper boundary surface of the *saturated zone*; the upper limit of the *groundwater* body.

water use (See *actual evapotranspiration*.)

water vapor the gaseous state of water.

waterfall abrupt descent of a *stream* over a *bedrock* downstep in the *stream channel*.

waterlogging rise of a *water table* in *alluvium* to bring the zone of saturation into the root zone of plants.

watt unit of power equal to the quantity of work done at the rate of one joule per second, or 10^7 ergs per second.

wave cyclone traveling, vortexlike *cyclone* involving interaction of cold and warm *air masses* along sharply defined *fronts*.

wave-cut notch *rock* recess at the base of a *marine cliff* where wave impact is concentrated.

wavelength distance separating one wave crest from the next in any uniform succession of traveling waves.

weak equatorial low weak, slowly moving low-pressure center (*cyclone*) accompanied by numerous convectional showers and *thunderstorms*; it forms close to the *intertropical convergence zone* in the rainy season, or *summer monsoon*.

weather physical state of the *atmosphere* at a given time and place.

weather system organized state of the *atmosphere* associated with a characteristic weather pattern, such as a *cyclone* or *anticyclone*.

weathering total of all processes acting at or near the earth's surface to cause physi-

cal disruption and chemical decomposition of *rock*. (See *chemical weathering*, *physical weathering*.)

west-wind drift ocean drift current moving eastward in zone of *prevailing westerlies*.

westerlies (See *prevailing westerly winds*, *upper-air westerlies*.)

wet adiabatic lapse rate reduced *adiabatic lapse rate* when *condensation* is taking place in rising air; value ranges between 4 and 9°C per 1000 m (2.2 and 4.9°F per 1000 ft).

wet equatorial climate ① moist climate of the *equatorial zone* with a large annual *water surplus*, and with uniformly warm temperatures throughout the year.

wetlands land areas of poor surface drainage, such as marshes and swamps.

wet–dry tropical climate ③ climate of the *tropical zone* characterized by a very wet season alternating with a very dry season.

Wilson Cycle *plate tectonic* cycle in which continents rupture and pull apart, forming oceans and *oceanic crust*, then converge and collide with accompanying subduction of *oceanic crust*.

wilting point quantity of stored *soil water*, less than which the foliage of plants not adapted to *drought* will wilt.

wind air motion, dominantly horizontal relative to the earth's surface.

wind abrasion mechanical wearing action of wind-driven *mineral* particles striking exposed *rock* surfaces.

wind vane weather instrument used to indicate *wind* direction.

winter monsoon outflow of continental air at low levels from the Siberian high, passing over Southeast Asia as a dry, cool northerly *wind*.

winter solstice solstice occurring on December 21 or 22, when the *subsolar point* is at 23 ½°S.

Wisconsinan Glaciation last *glaciation* of the *Pleistocene Epoch*.

woodland plant *formation class*, transitional between *forest biome* and *savanna biome*, consisting of widely spaced *trees* with canopy coverage between 25 and 60 percent.

Xeralfs suborder of the *soil order Alfisols*; includes Alfisols of the *Mediterranean climate* ⑦.

Xerolls suborder of the *soil order Mollisols*; includes Mollisols of the *Mediterranean climate* ⑦.

xerophytes plants adapted to a dry environment.

ANSWERS TO PROBLEMS

Working It Out 1.2 • Distances from Latitude and Longitude

1. Since the two cities are separated by 11° of latitude, and each latitude degree is 111 km or 69 mi, we have

$$11° \text{ lat} \times \frac{111 \text{ km}}{1° \text{ lat}} = 1221 \text{ km}$$

$$11° \text{ lat} \times \frac{69 \text{ mi}}{1° \text{ lat}} = 759 \text{ mi}$$

2. The two cities are nearly on the 45° parallel, and a degree of longitude there is equivalent to $\cos(45°) \times 111$ km $= 0.707 \times 111 = 78.5$ km (0.707×69 mi $= 48.8$ mi). The longitude difference is $124° - 76° = 48°$, so

$$48° \text{ long} \times \frac{78.5 \text{ km}}{1° \text{ long}} = 3768 \text{ km}$$

$$48° \text{ long} \times \frac{48.8 \text{ mi}}{1° \text{ long}} = 2342 \text{ mi}$$

3. Close to the equator, a degree of latitude is about equal to a degree of longitude, so both are about 111 km (or 69 mi). A square area 111 km (69 mi) on a side then has the following area:

$$111 \text{ km} \times 111 \text{ km} = 12{,}321 \text{ km}^2 \text{, or}$$
$$69 \text{ mi} \times 69 \text{ mi} = 4761 \text{ mi}^2$$

At Winnipeg, a degree of longitude has length equal to $\cos(50°) \times 111$ km $= 0.643 \times 111 = 71.3$ km ($0.643 \times 69 = 44.3$ mi). Thus, at Winnipeg, a region of 1° longitude by 1° latitude will be of area

$$71.3 \text{ km} \times 111 \text{ km} = 7914 \text{ km}^2 \text{, or}$$
$$44.3 \text{ mi} \times 69 \text{ mi} = 3057 \text{ mi}^2.$$

Working It Out 1.3 • Global Timekeeping

1. Applying the formula, $D = Z_{HOME} - Z_{AWAY} = +6 - (-8) = +14$, so add 14 hours to Chicago time to get Beijing time. When you depart, at 3:30 P.M. on Saturday, June 14 in Chicago, it will be 5:30 A.M. on Sunday in Beijing. Add 16 hours and 30 minutes for the flight, and you get a scheduled arrival time of 10:00 P.M. Sunday, June 15, in Beijing.

2. From Figure 1.12, Anchorage is in zone +9, and Washington is in zone +5. Applying the formula, $D = Z_{HOME} - Z_{AWAY} = +9 - (+5) = +4$, so when you leave Anchorage at 1:59 A.M., the time in Washington will be 5:59 A.M. If the plane arrives at 3:51 P.M., that will be 9 hours, 52 minutes later. Subtracting the 1 hour, 25 minute layover, the air time is then 8 hours, 27 minutes.

Working It Out 2.2 • Radiation Laws

1. The Stefan-Boltzmann Law describes the flow rate of energy leaving a surface, M, depending on its temperature. Since the temperature of the sun's surface is 5950°K, the energy flow rate will be

$$M = \sigma T^4 = (5.67 \times 10^{-8} \text{ W/m}^2\text{K}^4) \times (5950 \text{ K})^4$$
$$= 5.67 \times 10^{-8} \text{ W/m}^2\text{K}^4 \times 1.25 \times 10^{15} \text{ K}^4$$
$$= 7.11 \times 10^7 \text{ W/m}^2$$

The wavelength of greatest radiance, λ_{max}, follows Wein's Law. Thus,

$$\lambda_{max} = \frac{b}{T} = \frac{2898 \ \mu\text{m K}}{5950 \ K} = 0.487 \ \mu\text{m}$$

2. This problem is the same as Problem 1, except that the surface temperature is different—15.4°C. In absolute temperature, this value is $15.4 + 273 = 288.4$ K. So,

$$M = \sigma T^4 = (5.67 \times 10^{-8} \text{ W/m}^2\text{K}^4) \times (288.4 \text{ K})^4$$
$$= 5.67 \times 10^{-8} \text{ W/m}^2\text{K}^4 \times 6.92 \times 10^9 \text{ K}^4$$
$$= 392 \text{ W/m}^2$$

and

$$\lambda_{max} = \frac{b}{T} = \frac{2898 \ \mu\text{m K}}{288.4 \text{ K}} = 10.1 \ \mu\text{m}$$

3. From Problem 1, we have the sun's surface energy emission rate as 7.11×10^7 W/m^2, and from Problem 2 we have the earth's surface energy emission rate as 392 W/m^2. The ratio is therefore

$$\frac{7.11 \times 10^7 \text{ W/m}^2}{392 \text{ W/m}^2} = 1.81 \times 10^5 = 181{,}000$$

Thus, the flow of energy from the sun's surface is about 181,000 times as large as that from the earth's surface.

Working It Out 2.3 • Calculating the Global Radiation Balance

1. Since the radiation flow to Venus is 1.92 times greater than the radiation flow to the earth, the solar constant for Venus will be 1.92×1.40 kW/m$^2 = 2.69$ kW/m^2. If the radius of Venus is 6050 km, the area the planet presents to the sun will be that of a disk with area $\pi r^2 = 3.14 \times (6050 \text{ km})^2 = 3.14 \times 3.65 \times 10^7 \text{ km}^2 = 1.15 \times 10^8 \text{ km}^2 = 1.15 \times 10^{14} \text{ m}^2$. The total energy flow intercepted is equal to Venus's solar constant times the area it presents to the sun, or $= 2.69 \text{ kW/m}^2 \times 1.15 \times 10^{14} \text{ m}^2 = 3.09 \times 10^{14}$ kW. For outflows, 65 percent of the incoming solar radiation is directly reflected back, providing a shortwave radiation flow rate to space of $0.65 \times 3.09 \times 10^{14}$ kW $= 2.01 \times 10^{14}$ kW. The remaining portion, 35 percent, or $0.35 \times 3.09 \times 10^{14}$ kW $= 1.08 \times 10^{14}$ kW, flows outward to space as longwave radiation emitted by the planet and its atmosphere.

Working It Out 3.1 • Temperature Conversion

1. Using the formula $F = \frac{9}{5}C + 32$, we have $F = \frac{9}{5}(38) + 32 = 68.4 + 32 = 100.4°$F for Toronto. For Buffalo, $C = \frac{5}{9}(F - 32) = \frac{5}{9}(38 - 32) = \frac{5}{9}(6) = 3.3°$C.

2. For 46°F, $C = \frac{5}{9}(46 - 32) = \frac{5}{9}(14) = 7.8°$C, and for 28°F, $C = \frac{5}{9}(28 - 32) = \frac{5}{9}(-4) = -2.2$. Thus, the range is $7.8 - (-2.2) = 10°$C. An easier way is to note that when comparing differences, only the fraction is needed for conversion. Thus, $\Delta 18°\text{F} = \Delta \frac{5}{9}(18)°\text{C} = \Delta 10°$C, where the symbol Δ denotes difference.

3. Let X be the unknown temperature in °F and °C. Since the temperature is the same, we substitute X for F and C and equate the two formulas:

$$\frac{9}{5}X + 32 = \frac{5}{9}(X - 32)$$

$$\frac{9}{5}X + 32 = \frac{5}{9}X - \frac{5}{9}(32)$$

$$\frac{9}{5}X - \frac{5}{9}X = -\frac{5}{9}(32) - 32$$

$$\frac{9}{5}X - \frac{5}{9}X = -\frac{5}{9}(32) - \frac{9}{9}(32)$$

$$\frac{81}{45}X - \frac{25}{45}X = -\frac{14}{9}(32)$$

$$\frac{56}{45}X = -\frac{448}{9}$$

$$X = -\frac{448 \times 45}{9 \times 56}$$

$$= -40$$

Thus, the thermometers will have an identical reading at −40°.

Working It Out 3.3 • *Exponential Growth*

1. For the 2 percent rate, the multiplier will be
$$M = e^{(R \times T)} = e^{(0.02 \times 50)} = 2.718^{1.00} = 2.72$$
so $2.72 \times 350 = 951$ ppm. For the 3 percent rate,
$$M = e^{(0.03 \times 50)} = 2.718^{1.50} = 4.48$$
and $4.48 \times 350 = 1570$ ppm.

2. Doubling time for Singapore is $70 \div 1.3 = 53.8$ yrs, and for Republic of Congo it is $70 \div 3.0 = 23.3$ yrs. For Singapore,
$$M = e^{(0.013 \times 25)} = 2.718^{0.325} = 1.38$$
and so the population will be $2.8 \times 1.38 = 3.88$ million. For Congo,
$$M = e^{(0.03 \times 25)} = 2.718^{0.75} = 2.12$$
and the population will be $2.12 \times 2.4 = 5.08$ million.

Working It Out 4.1 • *Energy and Latent Heat*

1. Following the example, we can easily find that $4.19 \times (100 - 15) + 2260 = 4.19 \times 85 + 2260 = 2616$ kJ/kg are required. Thus, we have
$$490 \text{ km}^3 \times \left[\frac{10^3 \text{ m}}{1 \text{ km}}\right]^3 \times \frac{10^3 \text{ kg}}{1 \text{ m}^3} \times \frac{2616 \text{ kJ}}{1 \text{ kg}} = 1.28 \times 10^{18} \text{ kJ}$$

2. The ratio of annual energy in evaporation to annual energy consumption is then
$$\frac{1.28 \times 10^{18} \text{ kJ}}{8.5 \times 10^{13} \text{ kJ}} = 1.51 \times 10^4 = 15,100$$
Thus, the annual evaporation is about 15 thousand times larger than U.S. annual energy consumption. Put another way,
$$\frac{8.5 \times 10^{13} \text{ kJ}}{1.28 \times 10^{18} \text{ kJ}} \times 100 = 0.00663\%$$
or U.S. annual energy consumption is about 7 one-thousandths of 1 percent of annual evaporation.

Working It Out 4.3 • *The Lifting Condensation Level*

1.
$$H = 1000 \times \frac{25 - 18}{8.2} = 1000 \times \frac{7}{8.2} = 854 \text{ m}$$
$$T = 25 - \frac{854}{1000} \times 10 = 25 - 8.5 = 16.5°C$$

2.
$$H = 1000 \times \frac{30 - 18}{8.2} = 1000 \times \frac{12}{8.2} = 1463 \text{ m}$$
$$T = 25 - \frac{1463}{1000} \times 10 = 25 - 14.6 = 15.5°C$$

3. Equation (1) states
$$T_0 - \frac{H}{1000}R_{DRY} = T_{DEW} - \frac{H}{1000}R_{DEW}$$
Placing terms with T on the left and terms with H on the right,
$$T_0 - T_{DEW} = \frac{H}{1000}R_{DRY} - \frac{H}{1000}R_{DEW}$$
Factoring,
$$T_0 - T_{DEW} = \frac{H}{1000}(R_{DRY} - R_{DEW})$$
Solving for H, we have
$$H = 1000 \frac{T_0 - T_{DEW}}{R_{DRY} - R_{DEW}} = 1000 \frac{T_0 - T_{DEW}}{10 - 1.8}$$
$$= 1000 \frac{T_0 - T_{DEW}}{8.2}$$
which is Equation (2).

Working It Out 5.1 • *Pressure and Density in the Oceans and Atmosphere*

1. Since P (in t/m^2) $= D$ (in m) for the ocean, the pressure at the bottom of the diving pool will be 5 t/m^2. Adding atmospheric pressure (1 t/m^2) gives 6 t/m^2. The fraction due to water is then 5/6, while the fraction due to the atmosphere is 1/6. For the deep-sea diver, the pressure will be $100 + 1 = 101$, with fractions 100/101 for ocean water and 1/101 for the atmosphere.

2. For Mount Washington, $Z = 1917$ m $= 1.917$ km, so
$$P_z = 1014 \times [1 - (0.0226 \times 1.917)]^{5.26}$$
$$= 1014 \times (1 - 0.0433)^{5.26}$$
$$= 1014 \times (0.9567)^{5.26}$$
$$= 1014 \times 0.7925 = 804 \text{ mb}$$
For Mount Whitney, $Z = 4418$ m $= 4.418$ km, and
$$P_z = 1014 \times [1 - (0.0226 \times 4.418)]^{5.26}$$
$$= 583 \text{ mb}$$
The percentages are then $804/1014 \times 100 = 79.3\%$ for Mount Washington and $583/1014 \times 100 = 57.5\%$ for Mount Whitney. Reading from the graph should give about the same result.

3. At 350 m $= 0.35$ km, the barometer would read
$$P_z = 1014 \times [1 - (0.0226 \times 0.35)]^{5.26}$$
$$= 973 \text{ mb}$$
for a change of $1014 - 973 = 41$ mb, so the answer is yes.

Working It Out 7.2 • *Averaging in Time Cycles*

1.

J	F	M	A	M	J	J	A	S	O	N	D
4.9	5.4	7.1	12.8	10.5	7.1	9.8	8.0	8.7	8.8	11.5	7.8

2. This sequence of years seems somewhat drier than the average. With the exception of March and July, all five-year monthly means are lower than the long-term monthly means.

Working It Out 8.1 • *Cycles of Rainfall in the Low Latitudes*

1. The mean is 16.4 cm, and the mean deviation is 4.3 cm.

2. The relative variability is 0.26, which makes San Juan about the same as Padang, but less variable than Abbassia and Bombay.

Working It Out 9.1 • *Standard Deviation and Coefficient of Variation*

1.

Statistic	June	September
P	18.0	11.1
s_P	8.0	4.4
CV_P	0.45	0.40

Working It Out 10.1 • *Radioactive Decay*

1. The half-life is 1.26 b.y., $k = 0.693/1.28 = 0.550$, so
$$P(t = 1) = e^{-0.550 \times 1} = e^{-0.550} = 0.577 = 57.7\%$$
$$P(t = 3) = e^{-0.550 \times 3} = e^{-1.650} = 0.192 = 19.2\%$$

2. Here, the half-life is 14 b.y., $k = 0.693/14 = 0.0495$, and
$$P(t = 5) = e^{-0.0495 \times 5} = e^{-0.248} = 0.781 = 78.2\%$$
$$P(t = 10) = e^{-0.0495 \times 10} = e^{-0.495} = 0.610 = 61.2\%$$
$$P(t = 15) = e^{-0.0495 \times 15} = e^{-0.743} = 0.476 = 47.6\%$$

3. For ^{14}C, $k = 0.693/5730 = 1.21 \times 10^{-4}$. Then
$$P(t) = e^{-1.21 \times 10^{-4}t}$$

In this case, $P(t)$ is known and equal to 0.1, while t is unknown. Solving for t, we start with
$$P(t) = e^{-kt}$$
and taking the log to the base e of both sides, we obtain
$$\ln(P(t)) = -kt$$
Then we can simply solve for t:
$$t = -\frac{\ln(P(t))}{k}$$

Substituting values for $P(t)$ and k in this formula yields
$$t = -\frac{\ln(P(t))}{k} = -\frac{\ln(0.1)}{1.21 \times 10^{-4}} = -\frac{-2.30}{1.21 \times 10^{-4}} = 1.90 \times 10^4$$
$$= 19{,}000 \text{ yrs}$$

Working It Out 11.1 • *Radiometric Dating*

1. $t = \dfrac{1}{k} \ln\left[\dfrac{D}{M} + 1\right] = \dfrac{1}{0.155} \ln[0.448 + 1] = \dfrac{0.370}{0.155} = 2.39$ b.y.

2. For this decay sequence, $H = 7.04 \times 10^8$ yr $= 0.704$ b.y., and $k = 0.693/H = 0.693/0.704 = 0.985$ for time in b.y. Thus,
$$t = \frac{1}{0.985} \ln[9.56 + 1] = \frac{2.36}{0.985} = 2.40 \text{ b.y.}$$

The results of the two analyses are therefore quite consistent.

Working It Out 12.2 • *The Richter Scale*

1. Using the formula, we have
$$\log_{10}E = 4.8 + 1.5M = 4.8 + (1.5 \times 5.2) = 12.6$$
so $E = 10^{12.6} = 3.98 \times 10^{12}$ joules.

2. Again applying the formula,
$$\log_{10}E = 4.8 + (1.5 \times 6.0) = 13.8,$$
so $E = 10^{13.8} = 6.31 \times 10^{13}$ joules. If that energy is released in a 60-second period, then the flow rate is $6.31 \times 10^{13} \div 60 = 1.05 \times 10^{12}$ J/sec $= 1.05 \times 10^{12}$ W. Since each bulb consumes an energy flow of 100 W, $1.05 \times 10^{12} \div 10^2 = 1.05 \times 10^{10}$, or about 10 billion bulbs are needed. Compared to the average U.S electric power consumption rate, the flow rate of 1.05×10^{12} is about 1/3.

3. The ratio will be
$$\frac{E(7.3)}{E(7.1)} = \frac{10^{4.8 + (1.5 \times 7.3)}}{10^{4.8 + (1.5 \times 7.1)}} = \frac{10^{15.75}}{10^{15.45}} = 10^{15.75 - 15.45}$$
$$= 10^{0.3} = 2.00$$
That is, the energy release estimate has doubled.

Working It Out 13.1 • *The Power of Gravity*

1. To apply the formula, we first need to convert the velocity of the rock mass from km/hr to m/sec:

$$\frac{150 \text{ km}}{\text{hr}} \times \frac{1 \text{ hr}}{60 \text{ min}} \times \frac{1 \text{ min}}{60 \text{ s}} \times \frac{10^3 \text{ m}}{\text{km}} = 41.7 \text{ m/s}$$
Then we have
$$E = \frac{1}{2}mv^2 = \frac{1}{2} \times 7.56 \times 10^{10} \text{ kg} \times \left(\frac{41.7 \text{ m}}{\text{s}}\right)^2$$
$$= 6.57 \times 10^{13} \text{ J}$$

The total energy released is 2.82×10^{14} J, so the ratio of kinetic to total energy is
$$\frac{6.57 \times 10^{13} \text{ J}}{2.82 \times 10^{14} \text{ J}} = 0.223 = 22.3\%$$

2. The free-fall velocity will be
$$v = \sqrt{2gd} = \sqrt{2 \times \frac{9.8 \text{ m}}{\text{s}^2} \times 500 \text{ m}} = 99.0 \text{ m/s}$$

The ratio of the velocity of the Madison Slide to the free-fall velocity is then
$$\frac{41.7 \text{ m/s}}{99.0 \text{ m/s}} = .421 = 42.1\%$$

Thus, the slide moves less than half as fast as a free-falling body. The slide moves more slowly because it encounters friction in moving.

Working It Out 14.2 • *Magnitude and Frequency of Flooding*

1. The ranked flows with recurrence intervals are shown in the table. The magnitudes of floods with recurrence intervals closest to 1, 2, 5, 10, and 35 years are 113, 234, 337, 487, and 711 m³/s, respectively.

Rank	Flow (m³/s)	Recurrence (yrs)	Rank	Flow (m³/s)	Recurrence (yrs)	Rank	Flow (m³/s)	Recurrence (yrs)
1	711	34.0	12	279	2.8	23	210	1.5
2	677	17.0	13	278	2.6	24	209	1.4
3	487	11.3	14	266	2.4	25	194	1.4
4	419	8.5	15	264	2.3	26	193	1.3
5	357	6.8	16	245	2.1	27	184	1.3
6	351	5.7	17	234	2.0	28	181	1.2
7*	337	4.5	18	234	1.9	29	172	1.2
8*	337	4.5	19	229	1.8	30	164	1.1
9	312	3.8	20	225	1.7	31	139	1.1
10	309	3.4	21	217	1.6	32	119	1.1
11	292	3.1	22	216	1.5	33	113	1.0

*Indicates tied ranking.

2. For 250 m³/s, the table shows the recurrence interval for 245 m³/s to be 2.1 years, and 264 m³/s to be 2.3 years. The percent probabilities for these flows are then $100/2.1 = 47.6$ percent and $100/2.3 = 43.5$ percent. So the percent probability will be less than 47.6 percent but not as little as 43.5 percent. A good guess would be about 46.5 percent. In the case of 500 m³/s, the values for 487 and 677 m³/s are 8.85 and 5.88 percent, so a good guess would be about 8.5 percent. If a flow is to be equaled or exceeded in 25 percent of all years, its recurrence interval must be $100/25 = 4$ yrs. From the table, we see that flows of 312 m³/s and 337 m³/s are associated with recurrence intervals of 3.8 and 4.5 years, respectively. So this flow will be more than 312 m³/s, but less than 337 m³/s. A reasonable guess would be about 320 m³/s.

Working It Out 15.1* • *River Discharge and Suspended Sediment

1. Applying the formula for this river,
$$S = 178Q^{1.75} = 178(50)^{1.75} = 178 \times 940 = 167,000 \text{ t/d}$$

   and

$$S = 178(5)^{1.75} = 178 \times 16.7 = 2980 \text{ t/d}$$

2. Let's assume that the discharge is 1 m³/s and doubles to 2 m³/s. The ratio of suspended sediment transported at 2 m³/s to that at 1 m³/s is then
$$\frac{S(2)}{S(1)} = \frac{178(2)^{1.75}}{178(1)^{1.75}} = \frac{2^{1.75}}{1^{1.75}} = \frac{2^{1.75}}{1} = 3.36$$

   Thus, we would expect that the sediment load would more than triple with a doubling of discharge.

3. To convert the equation from logarithmic form to exponential form, raise 10 to the quantity on each side of the equation:
$$\log S = 2.50 + 2.12 \log Q$$
$$10^{\log S} = 10^{(2.50 + 2.12 \log Q)}$$
$$S = 10^{2.50} \times 10^{(2.12 \log Q)}$$
$$= 316 \times Q^{2.12}$$

   Thus, $s = 316Q^{2.12}$.

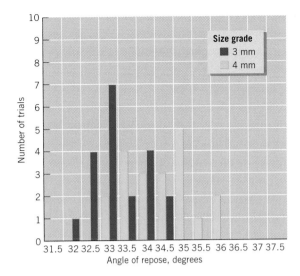

Working It Out 16.1* • *Properties of Stream Networks

1. From the problem statement, we have $R_b = 3.5$ and $k = 6$. Thus, the formula is
$$N_u = R_b^{(k-u)} = 3.5^{(6-u)}$$

   so we have
$$N_1 = 3.5^{(6-1)} = 3.5^5 = 525$$
$$N_2 = 3.5^{(6-2)} = 3.5^4 = 150$$
$$N_3 = 3.5^{(6-3)} = 3.5^3 = 42.9$$
$$N_4 = 3.5^{(6-4)} = 3.5^2 = 12.3$$
$$N_5 = 3.5^{(6-5)} = 3.5^1 = 3.5$$
$$N_6 = 3.5^{(6-6)} = 3.5^0 = 1$$

2. Substituting $L_1 = 0.12$ and $R_L = 3.2$ into the formula for expected cumulative mean length from the text gives
$$L_u^* = L_1 R_L^{(u-1)} = 0.12 \times 3.2^{(u-1)}$$

   so we have
$$L_1^* = 0.12 \times 3.2^{(1-1)} = 0.12 \times 3.2^0 = 0.12 \text{ km}$$
$$L_2^* = 0.12 \times 3.2^{(2-1)} = 0.12 \times 3.2^1 = 0.384 \text{ km}$$
$$L_3^* = 0.12 \times 3.2^{(3-1)} = 0.12 \times 3.2^2 = 1.23 \text{ km}$$
$$L_4^* = 0.12 \times 3.2^{(4-1)} = 0.12 \times 3.2^3 = 3.93 \text{ km}$$

   Using the formula $L_u = L_u^* - L_{u-1}$ to obtain the (noncumulative) mean lengths, we have
$$L_1 = L_1^* = 0.12$$
$$L_2 = L_2^* - L_1^* = 0.384 - 0.12 = 0.264 \text{ km}$$
$$L_3 = L_3^* - L_2^* = 1.23 - 0.384 = 0.846 \text{ km}$$
$$L_4 = L_4^* - L_3^* = 3.93 - 1.23 = 2.70 \text{ km}$$

Working It Out 17.2* • *Angle of Repose of Dune Sands

1. Your graph should look something like the one in the right column. The samples are not as well separated, since there is significant overlap between them.

2. First, find the standard deviation of the difference in sample means using the formula:
$$s_{\bar{X}_1 - \bar{X}_2} = \sqrt{\frac{s_1^2 + s_2^2}{n}} = \sqrt{\frac{(0.91)^2 + (0.72)^2}{20}}$$
$$= \sqrt{\frac{0.828 + 0.518}{20}} = \sqrt{\frac{1.346}{20}} = \sqrt{0.0673} = 0.26$$

   Then apply the formula for t:
$$t = \frac{|\bar{X}_1 - \bar{X}_2|}{s_{\bar{X}_1 - \bar{X}_2}} = \frac{|34.4 - 33.3|}{0.26} = \frac{1.1}{0.26} = 4.23$$

   Comparing this value with those given in the problem, we see that there is less than 1 chance in 1000 that the two means are actually the same and that the difference arose by chance. So we could conclude that grain size is definitely related to angle of repose, with the larger grain size having the smaller angle of repose.

Working It Out 18.2* • *Isostatic Rebound

1. For the slope, we have
$$a' = \frac{\log 200 - \log 55}{8.0 - 5.0} = \frac{0.56}{3.0} = 0.187$$

   For the intercept,
$$b' = y' - a'x = \log 55 - 0.187 \times 5.0 = 0.805$$

   Thus, $\log y = 0.805 + 0.187x$. For the exponential form, $a = 10^{a'} = 10^{0.187} = 1.54$, $b = 10^{b'} = 10^{0.805} = 6.38$, and so $y = 6.38(1.54)^x$. Comparing the two slopes, we see that the slope for James Bay, 0.187, is steeper than that of Oslofjord, at 0.156. This agrees with the slopes as plotted in the figure. The comparison shows that the rate of uplift due to isostatic rebound was greater for James Bay than for Oslofjord.

2. If b is the position on the uplift scale for the present time, it means that 12.3 m of uplift remains to be accomplished.

3. To find the slope, $a' = \log a = \log 1.7 = 0.230$. This rebound rate is greater than those of both Oslofjord (0.156) and James Bay (0.187).

Working It Out 19.1 • ***Calculating a Simple Soil Water Budget***

1.

Soil-water Budget for Urbana, Illinois

Month	P	Ea	-G	+G	R	Ep	D
Jan	5.7	0.0			+5.7	0.0	0.0
Feb	4.5	0.0			+4.5	0.0	0.0
Mar	8.2	1.4			+6.8	1.4	0.0
Apr	10.0	4.4			+5.6	4.4	0.0
May	9.9	8.8			+1.1	8.8	0.0
Jun	8.4	12.4	−4.0			12.6	0.2
Jul	8.0	13.4	−5.4			14.9	1.5
Aug	9.0	11.6	−2.6			13.0	1.4
Sep	8.3	7.8		+0.5		7.8	0.0
Oct	6.6	4.8		+1.8		4.8	0.0
Nov	5.7	1.4		+4.3		1.4	0.0
Dec	5.5	0.0		+5.5		0.0	0.0
Total	89.8	66.0	−12.0	+12.1	23.7	69.1	3.1

Working It Out 20.1 • ***Logistic Population Growth***

1. Taking C at 440 and reading P_0 from the value in the table at $t = 0$, we have for A,

$$A = \frac{C - P_0}{P_0} = \frac{440 - 10.0}{10.0} = 43.0$$

For k, we have

$$k = \ln(1 + R) = \ln(1 + 0.60) = \ln(1.60) = 0.470$$

The logistic equation for these data is then

$$P(t) = \frac{C}{1 + Ae^{-kt}} = \frac{440}{1 + (43)e^{-0.47t}}$$

Missing values for the table:

(1) Time (hours)	(3) Increase (grams)	(4) Increase (percent)	(5) Logistic model	(6) Difference
3	11.2	42.7	38.3	−0.9
7	38.6	30.3	169.1	−3.3
9	49.7	22.4	270.6	1.1
13	16.1	4.2	401.7	−1.8
17	3.1	0.7	433.7	−0.1

2. (*a*) For the increase in grams, values increase fairly steadily until a peak is reached at 8 hours, and then decline to small values. In contrast, the increase in percent is high at first, and then there is a decrease to very small values. (*b*) Judging by the differences in column 6, the model fits the data quite well. Most differences are only a few grams. (c) Compared to the data in the box, the yeast growth pattern here shows a lower carrying capacity and lower growth rate. This could indicate that the medium is less nutritive. Other factors that might explain this result include the use of a different strain of yeast, or incubation of the yeast at a different temperature.

PHOTO CREDITS

bert Moldvay/Eriako Associates. Figure 15.26: ©T. A. Wiewandt/DRK Photo. Figure 15.29: ©Mark A. Melton. Figure 15.31: ©John S. Shelton. Figure 15.32: ©Mark A. Melton.

Chapter 16 Opener: ©Jeff Gnass. Figure 16.5: ©Larry Ulrich/DRK Photo. Figure 16.11: ©John S. Shelton. Figure 16.15: Earth Satellite Corporation. Figure 16.19: ©J.A. Kraulis/Masterfile. Figure 16.25: ©John S. Shelton. Figure 16.26: ©Alex McLean/Landslides. Figure 16.28: ©Larry Ulrich.

Chapter 17 Opener: ©Mike Yamashita/Woodfin Camp & Associates. Figure 17.2: ©Alex McLean/Landslides. Figure17.4: A. N. Strahler. Figure 17.6: ©Steve Dunnell/The Image Bank. Figure 17.8: ©Cliff deBear/Newsday. Figure 17.11: ©David Muench Photography. Figure 17.15 & 17.16: Courtesy NASA. Figure 17.17: ©David Hiser/Photographers Aspen. Figure 17.19: ©John S. Shelton. Figure 17.20: ©Tom Bean. Figure 17.21: ©M. J. Coe/Animals/ Animals. Figure 17.24: ©John S. Shelton. Figure 17.25: ©John S. Shelton. Figure 17.27: ©William E. Ferguson. Figure 17.29: ©J. A. Kraulis/ Masterfile. Figure 17.31: Alan H. Strahler.

Chapter 18 Opener: ©Nancy Simmerman/Bruce Coleman, Inc. Figure 18.1: ©Fred Hirschmann Wilderness Photography. Figure 18.2: ©Carr Clifton/Tony Stone Images/ Seattle. Figure 18.6: ©John S. Shelton. Figure 18.8: ©Floyd L. Norgaard/Ric Ergenbright Photography. Figure 18.11: Courtesy NASA. Figure 18.12: ©Wolfgang Kaehler. Figure 18.16: Courtesy NASA. Figure 18.19: ©C. Wolinsky/Stock, Boston. Figure 18.20: A. N. Strahler. Figure 18.22: Courtesy National Snow and Ice Data Center.

Part 4 Opener: ©Ric Ergenbright Photography.

Chapter 19 Opener: ©David Schwimmer/Bruce Coleman, Inc. Figure 19.4: Harry Gold, Courtesy R. B. Krone, San Francisco District Corps. of Engineers, U. S. Army. Figure 19.6: R. Schaetzl. Figure 19.12: A. N. Strahler. Figure 19.14: Henry D. Foth. Figure 19.15: Alan H. Strahler. Figure 19.16: Soil Conservation Service. Figure 19.17: Henry D. Foth. Figure 19.18: ©William E. Ferguson. Figure 19.20: ©Ric Ergenbright/Tony Stone Images/Seattle. Figure 19.21: ©Robin White/Fotolex Associates. Figure 19.23: R. Schaetzl. Figure 19.24: Courtesy Soil Conservation Service. Figure 19.25: Ned L. Reglein

Chapter 20 Opener: ©Kevin Schafer. Figure 20.8 & 20.10: Courtesy NOAA/NESDIS/NCDC/SDSD.

Chapter 21 Opener: ©Ron Stroud/Masterfile. Figure 21.2: ©Steve McCutcheon. Figure 21.4: ©Josef Muench. Figure 21.7: ©M. Freeman/Bruce Coleman, Inc. Figure 21.9: ©Ferrero/Labat/Auscape International. Figure 21.10: ©Tom Bean. Figure 21.12: ©John S. Shelton. Figure 21.15: ©Kenneth Murray/Photo Researchers. Figure 21.17: ©Jake Rajs. Figure 21.19: ©Michael Townsend/Tony Stone Images/ Seattle. Figure 21.20a: A. N. Strahler. Figure 21.20b: Alan H. Strahler. Figure 21.22: ©Jacques Jangoux/Peter Arnold, Inc. Figure 21.24: ©Annie Griffiths Belt/DRK Photo. Figure 21.26: ©Brian A. Vikander. Figure 21.27 & 21.28: A. N. Strahler.

Epilogue Opener: ©Antonio M. Rosario/The Image Bank. Figure E-1: ©Frans Lanting/Minden Pictures, Inc. Figure E-2: ©Alicia Hurr/ Weatherstock. Figure E-3: ©Bob Krist/Black Star.

Appendixes Figure A2.1: SAR image Courtesy of Intera Technologies Corporation, Calgary, Alberta, Canada. Figure A2.2: Courtesy NASA, compiled and annotated by John E. Estes and Leslie W. Senger. Figure A2.3: Chicago Aerial Survey, Inc. Figure A2.5: ©1993 by CNES; Courtesy SPOT Image Corporation, Reston, Virginia.

INDEX

Page references followed by lowercase "f" indicate illustrations, while lowercase "t" indicate tables.

TOPOGRAPHIC MAP SYMBOLS

VARIATIONS WILL BE FOUND ON OLDER MAPS

Hard surface, heavy duty road, four or more lanes	Boundary, national ....
Hard surface, heavy duty road, two or three lanes	State ....
Hard surface, medium duty road, four or more lanes	County, parish, municipio ....
Hard surface, medium duty road, two or three lanes	Civil township, precinct, town, barrio ....
Improved light duty road	Incorporated city, village, town, hamlet ....
Unimproved dirt road and trail	Reservation, national or state ....
Dual highway, dividing strip 25 feet or less	Small park, cemetery, airport, etc. ....
Dual highway, dividing strip exceeding 25 feet	Land grant ....
Road under construction	Township or range line, United States land survey .....
	Township or range line, approximate location ......
Railroad, single track and multiple track	Section line, United States land survey ....
Railroads in juxtaposition	Section line, approximate location ....
Narrow gage, single track and multiple track	Township line, not United States land survey ....
Railroad in street and carline	Section line, not United States land survey ....
Bridge road and railroad	Section corner, found and indicated ..... + ... +
Drawbridge, road and railroad	Boundary monument: land grant and other .... □ ...... □
Footbridge	United States mineral or location monument .... ▲
Tunnel, road and railroad	
Overpass and underpass	Index contour .... ——— Intermediate contour ..
Important small masonry or earth dam	Supplementary contour ..... Depression contours ..
Dam with lock	Fill ...... Cut ....
Dam with road	Levee ...... Levee with road ..
Canal with lock	Mine dump ...... Wash ....
	Tailings ...... Tailings pond ....
Buildings (dwelling, place of employment, etc.)	Strip mine ...... Distorted surface ..
School, church, and cemetery .... Cem	Sand area ...... Gravel beach .....
Buildings (barn, warehouse, etc.)	
Power transmission line	Perennial streams .... Intermittent streams ..
Telephone line, pipeline, etc. (labeled as to type)	Elevated aqueduct .... Aqueduct tunnel ......
Wells other than water (labeled as to type) ..... o Oil ..... o Gas	Water well and spring . o .... o~ Disappearing stream .
Tanks; oil, water, etc. (labeled as to type) .... ● ● ⊘ Water	Small rapids .... Small falls ......
Located or landmark object; windmill .... o ...... ☒	Large rapids .... Large falls ......
Open pit, mine, or quarry; prospect .... ✕ x	Intermittent lake ..... Dry lake ......
Shaft and tunnel entrance .... ◪ ...... Y	Foreshore flat ...... Rock or coral reef ...
	Sounding, depth curve . 10 Piling or dolphin .. o
Horizontal and vertical control station:	Exposed wreck ...... Sunken wreck ......
Tablet, spirit level elevation .... BM△ 5653	Rock, bare or awash; dangerous to navigation ....
Other recoverable mark, spirit level elevation .... △ 5455	
Horizontal control station: tablet, vertical angle elevation VABM△ 9519	Marsh (swamp) ...... Submerged marsh ..
Any recoverable mark, vertical angle or checked elevation △3775	Wooded marsh ...... Mangrove ....
Vertical control station: tablet, spirit level elevation ..... BM✕ 957	Woods or brushwood .. Orchard ....
Other recoverable mark, spirit level elevation .... ✕ 954	Vineyard ...... Scrub ....
Checked spot elevation .... ✕4675	Inundation area ...... Urban area ....
Unchecked spot elevation and water elevation .... ✕ 5657 ... 870	

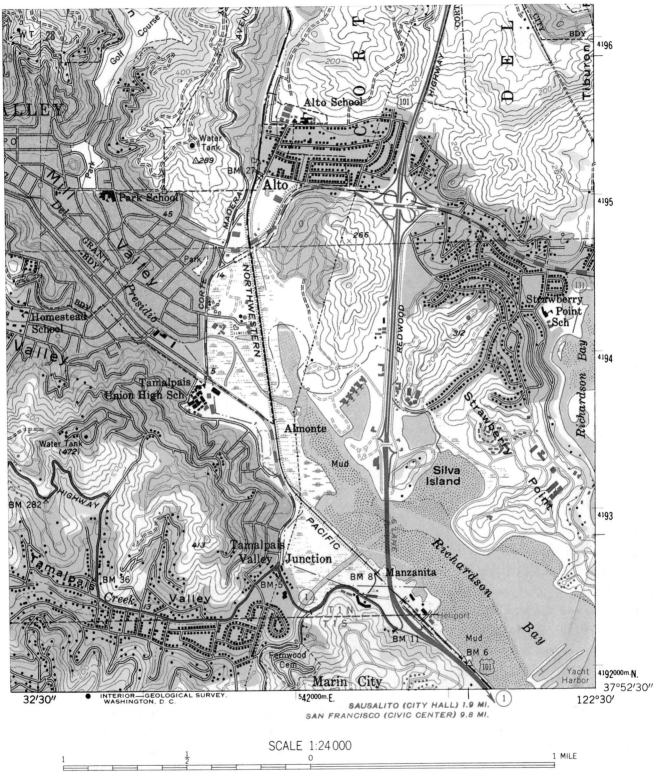

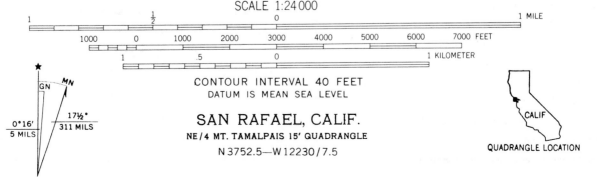

SCALE 1:24 000

CONTOUR INTERVAL 40 FEET
DATUM IS MEAN SEA LEVEL

SAN RAFAEL, CALIF.

NE/4 MT. TAMALPAIS 15' QUADRANGLE

N 3752.5—W 12230/7.5

QUADRANGLE LOCATION

CONVERSION FACTORS

Metric to English

Metric Measure	Multiply by*	English Measure
LENGTH		
Millimeters (mm)	0.0394	Inches (in.)
Centimeters (cm)	0.394	Inches (in.)
Meters (m)	3.28	Feet (ft)
Kilometers (km)	0.621	Miles (mi)
AREA		
Square centimeters (cm^2)	0.155	Square inches (in^2)
Square meters (m^2)	10.8	Square feet (ft^2)
Square meters (m^2)	1.12	Square yards (yd^2)
Square kilometers (km^2)	0.386	Square miles (mi^2)
Hectares (ha)	2.47	Acres
VOLUME		
Cubic centimeters (cm^3)	0.0610	Cubic inches (in^3)
Cubic meters (m^3)	35.3	Cubic feet (ft^3)
Cubic meters (m^3)	1.31	Cubic yards (yd^3)
Milliliters (ml)	0.0338	Fluid ounces (fl oz)
Liters (l)	1.06	Quarts (qt)
Liters (l)	0.264	Gallons (gal)
MASS		
Grams (g)	0.0353	Ounces (oz)
Kilograms (kg)	2.20	Pounds (lb)
Kilograms (kg)	0.00110	Tons (2000 lb)
Tonnes (t)	1.10	Tons (2000 lb)

English to Metric

English Measure	Multiply by*	Metric Measure
LENGTH		
Inches (in.)	2.54	Centimeters (cm)
Feet (ft)	0.305	Meters (m)
Yards (yd)	0.914	Meters (m)
Miles (mi)	1.61	Kilometers (km)
AREA		
Square inches (in^2)	6.45	Square centimeters (cm^2)
Square feet (ft^2)	0.0929	Square meters (m^2)
Square yards (yd^2)	0.836	Square meters (m^2)
Square miles (mi^2)	2.59	Square kilometers (km^2)
Acres	0.405	Hectares (ha)
VOLUME		
Cubic inches (in^3)	16.4	Cubic centimeters (cm^3)
Cubic feet (ft^3)	0.0283	Cubic meters (m^3)
Cubic yards (yd^3)	0.765	Cubic meters (m^3)
Fluid ounces (fl oz)	29.6	Milliliters (ml)
Pints (pt)	0.473	Liters (l)
Quarts (qt)	0.946	Liters (l)
Gallons (gal)	3.79	Liters (l)
MASS		
Ounces (oz)	28.4	Grams (g)
Pounds (lb)	0.454	Kilograms (kg)
Tons (2000 lb)	907	Kilograms (kg)
Tons (2000 lb)	0.907	Tonnes (t)

*Conversion factors shown to 3 decimal-digit precision.